Telecommunication System Engineering

WILEY SERIES IN TELECOMMUNICATIONS

Donald L. Schilling, Editor
City College of New York

Worldwide Telecommunications Guide for the Business Manager
Walter L. Vignault

Expert Systems Applications to Telecommunications
Jay Liebowitz

Business Earth Stations for Telecommunications
Walter L. Morgan and Denis Rouffet

Introduction to Communications Engineering, 2nd Edition
Robert M. Gagliardi

Satellite Communications: The First Quarter Century of Service
David W. E. Rees (in preparation)

Synchronization in Digital Communications, Volume 1

Synchronization in Digital Communications, Volume 2
Heinrich Meyr and G. Ascheid (in preparation)

Telecommunication System Engineering, 2nd Edition
Roger L. Freeman

Digital Signal Estimation
Robert J. Mammone (in preparation)

Microwave in Private Networks: A Guide to Corporately Owned Microwave
James B. Pruitt (in preparation)

Telecommunication System Engineering

Second Edition

Roger L. Freeman
Raytheon Company

A Wiley-Interscience Publication
JOHN WILEY & SONS
New York • Chichester • Brisbane • Toronto • Singapore

Library of Congress Cataloging in Publication Data:

Freeman, Roger L.
Telecommunication system engineering/Roger L. Freeman.—2nd ed.
p. cm. —(Wiley series in telecommunications)
"A Wiley-Interscience publication."
Includes bibliographical references and index.
ISBN 0-471-63423-9
1. Telecommunication systems—Design and construction.
2. Telephone systems—Design and construction. I. Title.
II. Series.
TK5103.F68 1989
621.382—dc20 89-31275
CIP

Printed in the United States of America

10 9 8 7 6 5 4 3 2 1

To my father,
Andrew A. Freeman

CONTENTS

PREFACE

The purpose of this book is two-fold: in basic terms, it describes how a telecommunications network operates, and it presents the general engineering considerations necessary to design a practical network. My approach carries out this dual goal by starting out the reader with conventional analog speech communications. The first seven chapters do just this. The chapter order is purposeful; I'm trying to build a foundation. Speech telephony is emphasized because still today the vast majority of traffic carried on telecommunication networks is speech. However, these same networks are being used more and more extensively to carry other types of information such as data and facsimile. I introduce these topics in Chapter 8.

Another change that is taking place at a rapid pace is the conversion of the public switched network (PSN) from analog to digital. With an analog foundation and an introduction to data transmission, I now feel comfortable introducing the reader to digital transmission, switching, and networks in Chapters 9 and 10. Chapter 11 sets the stage for the remainder of the book where the reader is shown that there is more to data communication than the development and transmission of an electrical signal. Open System Interconnection (OSI) is a cornerstone of this chapter. Chapter 12 is an introduction to local area network (LAN) design and operation. Chapter 13 follows with the integration of voice and data on a common digital network called Integrated Services Digital Networks (ISDN). Most texts include CCITT Signaling System No. 7 (SS No. 7) with their description of ISDN. I purposefully separated the subjects because SS No. 7 is the first attempt to implement a common worldwide signaling system. Further, ISDN is only one service that SS No. 7 supports. The final chapter, in a manner of speaking, ties the book together with a discussion of telecommunication planning with emphasizing the needs of a telecommunication manager.

I define telecommunication as a service that permits people and/or machines to communicate at a distance. It involves many disciplines that work together as a system. Traditionally, telecommunication has been broken down into two major engineering specialties: transmission and switching. Each major specialty in itself is broken down into loosely definable disciplines as shown below. (The numbers shown in the figure are chapter references.)

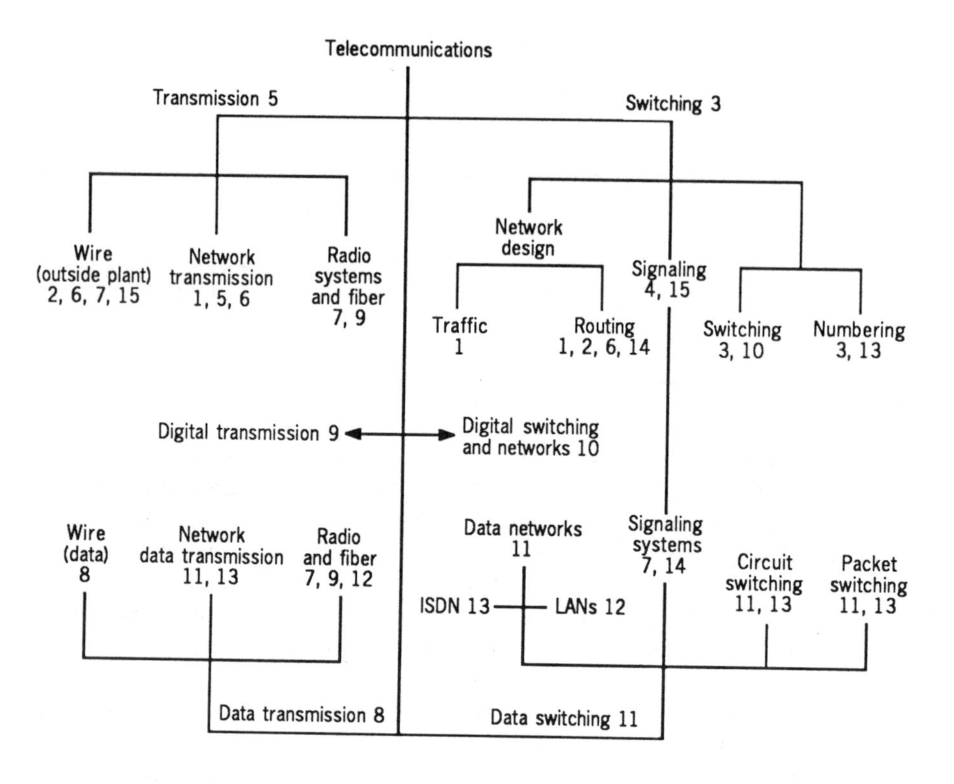

This tradition and conventionality are maintained in the first seven chapters. However, the advent of data communications on one hand, and digital telephony on the other, caused the distinct separation of disciplines to become hazy and ill-defined. In fact, with ISDN and SS No. 7 I cannot discern any dividing lines. Now enter software engineering into the picture. Many of my colleagues try to draw a hard-and-fast line between the telecommunication engineer and programmer, software engineer or (which some misname) a system engineer. I try to address the issue where the communicator's responsibility leaves off and the programmer takes over. I do stress that each should have an appreciation of the other's job and how the decision of one will impact the other; that to me is system engineering.

To obtain maximum benefit of the book, the reader should have some grounding in electrical communications and algebra including logarithms. This book has been carefully written to impart practical knowledge to a wide readership: those interested in telecommunications.

ROGER L. FREEMAN

Sudbury, Massachusetts
July 1989

ACKNOWLEDGMENTS

My colleagues and friends were very supportive during the preparation of this second edition. Support varied from a pat on the back or a word of encouragement to review and critique, as well as helping me carry out rewrites. To cite everyone would be a lengthy list indeed; however, several people really went out of their way to assist me.

Don Schilling set me straight on content and carried out a last-minute review of the manuscript before submission to the publisher. Don is one of those few that I know who uniquely bridges industry and academia. He is founder and president of SCS Telecom and, at the same time, is the Herbert Kayser Professor of Electrical Engineering at City College of New York (CCNY).

I am grateful to Preston Peek and Brian Murphy of Northern Telecomm for their contributions and review of digital switching. L. P. (Phil) Fabiano, formerly technical director for R & D at ITT World Headquarters, kindly reviewed and helped me rewrite several of the new chapters. Dr. Jim Mullen of Raytheon's Research Division was willing and able to help when I needed it.

I am also grateful to Don Marsh, vice president of CONTEL, and John Lawlor, president of John Lawlor Associates, for their overall support when it most counted; and to Jim Metzler of Prime Computer and faculty advisor, and to Janice Halpern, director of Northeastern University's state-of-the-art engineering program, for critique of my topical outline.

My wife and children remained patient and understanding allowing me priority over term papers on our home computer, as this second edition was developed and finalized. I depended on my son, Bob, to help me with the computer program mechanics in the assembly of the vital index.

Let me also offer a special note of gratitude, posthumously, to Joanne Portsch. Joanne was a very special person and supportive of my efforts in preparation of both the first and second editions of this book. She was manager of the Technical Information Centers of Raytheon's Equipment Division until her death last year.

R. L. F.

Telecommunication System Engineering

1

SOME BASICS IN CONVENTIONAL TELEPHONY

1 BACKGROUND AND CONCEPT

Telecommunications deals with the service of providing electrical communication at a distance. The service is supported by an industry that depends on a large body of increasingly specialized scientists and engineers. The service may be private or open to public correspondence (i.e., access). The most cogent example of a service open to public correspondence is the telephone embodied in the telephone company, when based on private enterprise, or telephone administration, when government owned. Consider that by the early 1990s there will be more than 750 million telephones in the international network, with intercommunication between each and every telephone in that network.

A primary concern of this book is to describe the development of a telephone network, why it is built as it is, and how it is evolving. We also intend to show how it will expand and carry other than voice communication and how special services will evolve as based originally on the existing telephone network, where certain split-offs will occur at a later date. The bulk of the telecommunication industry is dedicated to the telephone network. Telecommunication engineering traditionally has been broken down into two basic segments, transmission and switching. This division is most apparent in telephony. Transmission concerns the carrying of an electrical signal from point X to point Y. Let us say that switching connects X to Y, rather than to Z. Until several years ago transmission and switching were two very separate and distinct disciplines. Today that distinction is disappearing. As we proceed, we deal with both disciplines and show in later chapters how the two are starting to meld together.

2 THE SIMPLE TELEPHONE CONNECTION

The common telephone as we know it today is a device connected to the outside world by a pair of wires. It consists of a handset and its cradle with a signaling device, consisting of either a dial or push buttons. The handset is made up of two electroacoustic transducers, the earpiece or receiver and the mouthpiece or transmitter. There is also a sidetone circuit that allows some of the transmitted energy to be fed back to the receiver.

The transmitter or mouthpiece converts acoustic energy into electric energy by means of a carbon granule transmitter. The transmitter requires a direct-current (dc) potential, usually on the order of 3–5 V, across its electrodes. We call this the *talk battery*, and in modern telephone systems it is supplied over the line (central battery) from the switching center and has been standardized at −48 V dc. Current from the battery flows through the carbon granules or grains when the telephone is lifted from its cradle or goes "off hook." When sound impinges on the diaphragm of the transmitter, variations of air pressure are transferred to the carbon, and the resistance of the electrical path through the carbon changes in proportion to the pressure. A pulsating direct current results.

The typical receiver consists of a diaphragm of magnetic material, often soft iron alloy, placed in a steady magnetic field supplied by a permanent magnet, and a varying magnetic field caused by voice currents flowing through the voice coils. Such voice currents are alternating (ac) in nature and originate at the far-end telephone transmitter. These currents cause the magnetic field of the receiver to alternately increase and decrease, making the diaphragm move and respond to the variations. Thus an acoustic pressure wave is set up, more or less exactly reproducing the original sound wave from the distant telephone transmitter. The telephone receiver, as a converter of electrical energy to acoustic energy, has a comparatively low efficiency, on the order of 2–3%.

Sidetone is the sound of the talker's voice heard in his (or her) own receiver. Sidetone level must be controlled. When the level is high, the natural human reaction is for the talker to lower his or her voice. Thus by regulating sidetone, talker levels can be regulated. If too much sidetone is fed back to the receiver, the output level of the transmitter is reduced as a result of the talker lowering his or her voice, thereby reducing the level (voice volume) at the distant receiver and deteriorating performance.

To develop our discussion, let us connect two telephone handsets by a pair of wires, and at middistance between the handsets a battery is connected to provide that all-important talk battery. Such a connection is shown diagrammatically in Figure 1.1. Distance D is the overall separation of the two handsets and is the sum of distances d_1 and d_2; d_1 and d_2 are the distances from each handset to the central battery supply. The exercise is to extend the distance D to determine limiting factors given a fixed battery voltage, say, 48 V dc. We find that there are two limiting factors to the extension of the wire pair between the handsets. These are the IR drop, limiting the voltage across the handset transmitter, and the attenuation. For 19-gauge wire, the limiting distance is about 30 km, depending on the efficiency of the handsets. If the limiting characteristic is attenuation and we desire to extend the pair farther, amplifiers could be used in the line. If the battery voltage is limiting, then the

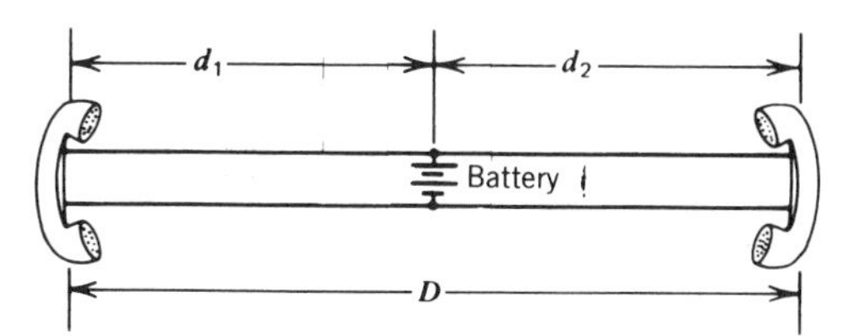

Figure 1.1 A simple telephone connection.

battery voltage could be increased. With the telephone system depicted in Figure 1.1, only two people can communicate. As soon as we add a third person, some difficulties begin to arise. The simplest approach would be to provide each person with two handsets. Thus party A would have one set to talk to B, another to talk to C, and so forth. Or the sets could be hooked up in parallel. Now suppose A wants to talk to C and doesn't wish to bother B. Then A must have some method of selectively alerting C. As stations are added to the system, the alerting problem becomes quite complex. Of course, the proper name for this selection and alerting is *signaling*. If we allow that the pair of wires through which current flows is a loop, we are dealing with loops. Let us also call the holder of a telephone station a *subscriber*. The loops connecting them are subscriber loops.

Let us now look at an eight-subscriber system, each subscriber connected directly to every other subscriber. This is shown in Figure 1.2. When we connect each and every station with every other one in the system, this is called a *mesh* connection, or sometimes full mesh. Without the use of amplifiers and with 19-gauge copper wire size, the limiting distance is 30 km. Thus any connecting segment of the octagon may be no greater than 30 km. The only way we can justify a mesh connection of subscribers economically is when each and every subscriber wishes to communicate with every other subscriber in the network for virtually all the day (full period). As we know, however, most telephone subscribers do not use their telephones on a full-time

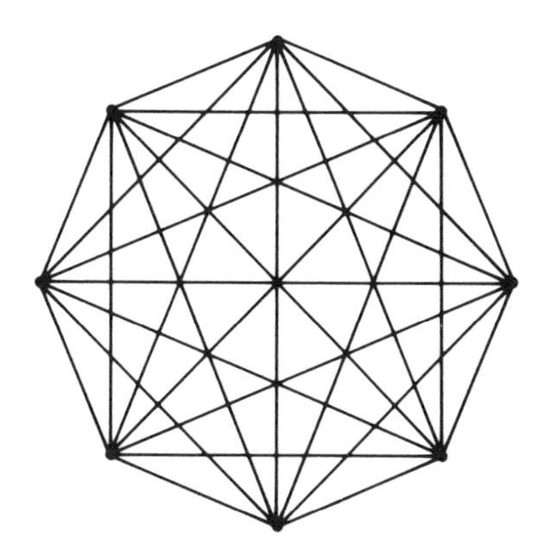

Figure 1.2 An 8-point mesh connection.

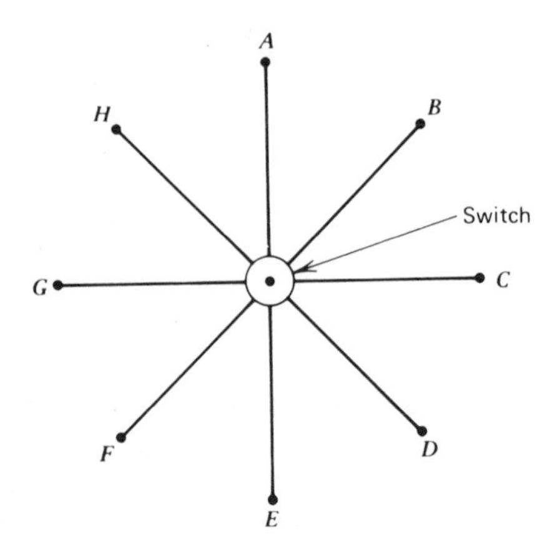

Figure 1.3 Subscribers connected in a star arrangement.

basis. The telephone is used at what appear to be random intervals throughout the day. Further, the ordinary subscriber or telephone user will normally talk to only one other subscriber at a time. He/she will not need to talk to all other subscribers simultaneously.

If more subscribers are added and the network is extended beyond about 30 km, it is obvious that transmission costs will spiral, for that is what we are dealing with exclusively here—transmission. We are connecting each and every subscriber together with wire transmission means, requiring many amplifiers and talk batteries. Thus it would seem wiser to share these facilities in some way and cut down on the transmission costs. We now discuss this when switch and switching enter the picture. Let us define a *switch* as a device that connects inlets to outlets. The inlet may be a calling subscriber line and the outlet, the line of a called subscriber. The techniques of switching and the switch as a concept are widely discussed later in this text. Switching devices and how they work are covered in Chapters 3 and 9. Consider Figure 1.3, which shows our subscribers connected in a *star* network with a switch at the center. All the switch really does in this case is to reduce the transmission cost outlay. Actually, this switch reduces the number of links between subscribers, which really is a form of concentration. Later in our discussion it becomes evident that switching is used to concentrate traffic, thus reducing the cost of transmission facilities.

3 SOURCES AND SINKS*

Traffic is a term that quantifies usage. A subscriber *uses* the telephone when he/she wishes to talk to somebody. We can make the same statement for a telex (teleprinter service) subscriber or a data-service subscriber. But let us stay with the telephone.

*The traffic engineer may wish to use the terminology "origins and destinations."

A network is a means of connecting subscribers. We have seen two simple network configurations, the mesh and star connections, in Figures 1.2 and 1.3. When talking about networks, we often talk of sources and sinks. A call is initiated at a traffic source and received at a traffic sink. Nodal points or nodes in a network are the switches.

4 TELEPHONE NETWORKS: INTRODUCTORY TERMINOLOGY

From our discussion we can say that a telephone network can be regarded as a systematic development of interconnecting transmission media arranged so that one telephone user can talk to any other within that network. The evolving layout of the network is primarily a function of economics. For example, subscribers share common transmission facilities; switches permit this sharing by concentration.

Consider a very simplified example. Two towns are separated by, say, 20 mi, and each town has 100 telephone subscribers. Logically, most of the telephone activity (the traffic) will be among the subscribers of the first town and among those of the second town. There will be some traffic, but considerably less, from one town to the other. In this example let each town have its own switch. With the fairly low traffic volume from one town to the other, perhaps only six lines would be required to interconnect the switch of the first town to that of the second. If no more than six people want to talk simultaneously between the two towns, a number as low as six can be selected. Economics has mandated that we install the minimum number of connecting telephone lines from the first town to the second to serve the calling needs between the two towns. The telephone lines connecting one telephone switch or exchange with another are called *trunks* in North America and *junctions* in Europe. The telephone lines connecting a subscriber to the switch or exchange that serves the subscriber are called *lines*, *subscriber lines*, or *loops*. Concentration is a line-to-trunk ratio. In the simple case above it was 100 lines to six trunks (or junctions), or about at 16 : 1 ratio.

A telephone subscriber looking into the network is served by a *local exchange*. This means that the subscriber's telephone line is connected to the network via the local exchange or central office, in North American parlance. A local exchange has a serving area, which is the geographical area in which the exchange is located; all subscribers in that area are served by that exchange.

The term *local area*, as opposed to *toll area*, is that geographical area containing a number of local exchanges and inside which any subscriber can call any other subscriber without incurring tolls (extra charges for a call). Toll calls and long-distance calls are synonymous. For instance, a local call in North America, where telephones have detailed billing, shows up on the bill as a time-metered call or is covered by a flat monthly rate. Toll calls in North America appear as separate detailed entries on the telephone bill. This is not so in most European countries and in those countries following European practice. In these countries there is no detailed billing on direct-distance-dialed (subscriber-trunk-dialed) calls. All such subscriber-dialed calls, even international ones, are just metered, and the subscriber pays for the meter steps used per billing period, which is often one or two months. In European practice a long-distance call, a toll call if you will, is one involving the dialing of additional digits (e.g., more than six or seven digits).

Let us call a network a *grouping of interworking telephone exchanges*. As the discussion proceeds, the differences between local networks and national networks are shown. Two other types of network are also discussed. These are specialized versions of a local network and are the rural network (rural area) and metropolitan network (metropolitan area).

5 ESSENTIALS OF TRAFFIC ENGINEERING

5.1 Introduction and Terminology

As we have already mentioned, telephone exchanges are connected by trunks or junctions. The number of trunks connecting exchange X with exchange Y are the number of voice pairs or their equivalent used in the connection. One of the most important steps in telecommunication engineering practice is to determine the number of trunks required on a route or connection between exchanges. We could say we are *dimensioning* the route. To dimension a route correctly, we must have some idea of its usage, that is, how many people will wish to talk at once over the route. The usage of a transmission route or a switch brings us into the realm of traffic engineering, and the usage may be defined by two parameters: (1) *calling rate*, or the number of times a route or traffic path is used per unit period; or, more properly defined, "the call intensity per traffic path during the busy hour";* and (2) *holding time*, or "the duration of occupancy of a traffic path by a call,"* or sometimes, "the average duration of occupancy of one or more paths by calls."* A *traffic path* is "a channel, time slot, frequency band, line, trunk, switch, or circuit over which individual communications pass in sequence."* *Carried traffic* is the volume of traffic actually carried by a switch, and *offered traffic* is the volume of traffic offered to a switch.

To dimension a traffic path or size a telephone exchange, we must know the traffic intensity representative of the normal busy season. There are weekly and daily variations in traffic within the busy season. Traffic is very random in nature. However, there is a certain consistency we can look for. For one thing, there usually is more traffic on Mondays and Fridays and a lower volume on Wednesdays. A certain consistency can also be found in the normal workday hourly variation. Across the typical day the variation is such that a 1-h period shows greater usage than any other. From the hour with least traffic to the

*Reference Data for Radio Engineers [1], p. 31-8.

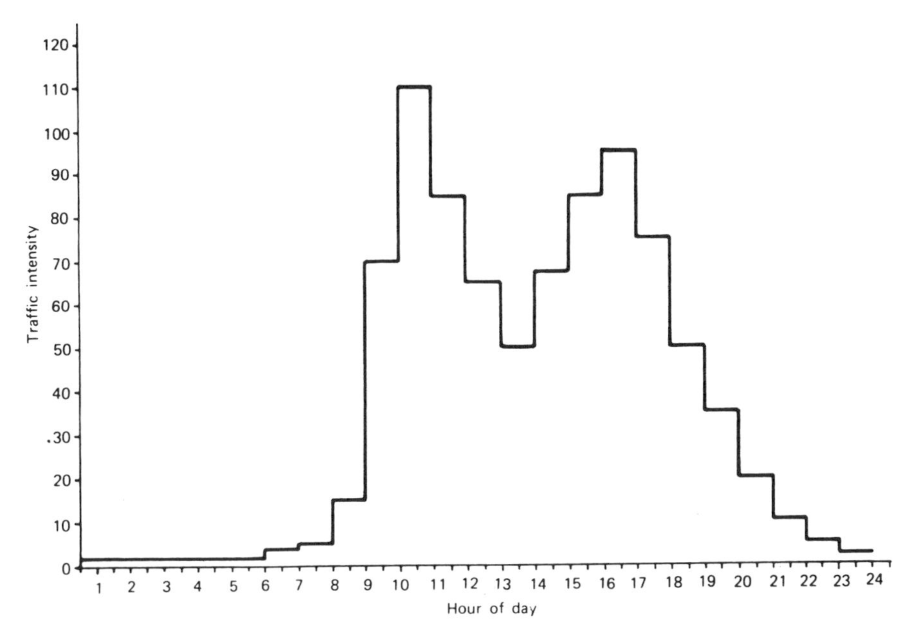

Figure 1.4 Bar chart of traffic intensity over a typical working day (US, mixed business and residential).

hour of greatest traffic, the variation can exceed 100 : 1. Figure 1.4 shows a typical hour-by-hour traffic variation for a serving switch in the United States. It can be seen that the busiest period, the *busy hour* (BH), is between 10 A.M. and 11 A.M. From one workday to the next, originating BH calls can vary as much as 25%. To these fairly "regular" variations, there are also unpredictable peaks caused by stock market or money market activity, weather, natural disaster, international events, sporting events, and so on. Normal system growth must also be taken into account. Nevertheless, suitable forecasts of BH traffic can be made. However, before proceeding, consider the five most common definitions of BH:

Busy Hour definitions (CCITT Rec. E.600).

1. *Busy Hour.* The busy hour refers to the traffic volume or number of call attempts, and is that continuous 1-h period lying wholly in the time interval concerned for which this quantity (i.e., traffic volume or call attempts) is greatest.
2. *Peak Busy Hour.* The busy hour each day; it usually is not the same over a number of days.
3. *Time Consistent Busy Hour.* The 1-h period starting at the same time each day for which the average traffic volume or call-attempt count of the exchange or resource group concerned is greatest over the days under consideration.

From *Engineering and Operations in the Bell System*, 2nd ed. (Ref. 23):

4. The engineering period (where the grade of service criteria is applied) as the Busy Season Busy Hour (BSBH). This is the busiest clock hour of the busiest weeks of the year.
5. The Average Busy Season Busy Hour (ABSBH) is used for trunk groups and always has a grade of service criterion applied. For example, for the ABSBH load, a call requiring a circuit in a trunk group should encounter "all trunks busy" (ATB) no more than 1% of the time.

Reference 23 goes on to state that peak loads are of more concern than average loads when engineering switching equipment and engineering periods other than the ABSBH are defined. Examples of these are the highest BSBH and the average of the ten highest BSBHs. Sometimes the engineering period is the weekly peak hour (which may not even be the BSBH).

When dimensioning telephone exchanges and transmission routes, we shall be working with BH traffic levels and care must be used in the definition of busy hour.

5.2 Measurement of Telephone Traffic

If we define *telephone traffic* as the aggregate of telephone calls over a group of circuits or trunks with regard to the duration of calls as well as their number [2], we can say that traffic flow (A)

$$A = \mathrm{C} \times \mathrm{T}$$

where C is the calling rate per hour and T is the average holding time per call. From this formula it would appear that the traffic unit would be call-minutes or call-hours.

Suppose that the average holding time is 2.5 min and the calling rate in the BH for a particular day is 237. The traffic flow would then be 237 × 2.5, or 592.5 call-minutes (Cm), or 592.5/60, or about 9.87 call-hours (Ch).

The preferred unit of traffic is the erlang, named after the Danish mathematician A. K. Erlang [5]. The erlang is a dimensionless unit. One erlang of traffic intensity on one traffic circuit means a continuous occupancy of that circuit. Considering a group of circuits, traffic intensity in erlangs is the number of call-seconds per second or the number of call-hours per hour. If we knew that a group of 10 circuits had a call intensity of 5 erlangs, we would expect half of the circuits to be busy at the time of measurement.

Other traffic units are not dimensionless. For instance: *call-hour* (CH)—1 Ch is the quantity represented by one or more calls having an aggregate duration of 1 h; *call-second* (Cs)—1 Cs is the quantity represented by one or more calls having an aggregate duration of 1 s; "*cent*" *call-second* (ccs)—1 ccs is the quantity represented by one 100-s call or by an aggregate of 100 Cs of traffic (the term ccs derives from "cent" call seconds; with cent representing 100 from the French); and the *equated busy hour call* (EBHC) is a European unit of traffic intensity (1 EBHC is the average intensity in one or more traffic paths occupied in the BH by one 2-min call or for an aggregate duration of 2 min). Thus we can relate our terms as follows:

$$1 \text{ erlang} = 30 \text{ EBHC} = 36 \text{ ccs} = 60 \text{ Cm}$$

assuming a 1-h time-unit interval.

5.3 Congestion, Lost Calls, and Grade of Service

Assume that an isolated telephone exchange serves 5000 subscribers and that no more than 10% of the subscribers wish service simultaneously. Therefore, the exchange is dimensioned with sufficient equipment to complete 500 simultaneous connections. Each connection would be, of course, between any two of the 5000 subscribers. Now let subscriber 501 attempt to originate a call. He cannot because all the connecting equipment is busy, even though the line he wishes to reach may be idle. This call from subscriber 501 is termed a *lost call* or *blocked call*. He has met congestion. The probability of meeting congestion is an important parameter in traffic engineering of telecommunication systems. If congestion conditions are to be met in a telephone system, we can expect that those conditions will usually be met during the BH. A switch is engineered (dimensioned) to handle the BH load. But how well? We could, indeed, far overdimension the switch such that it could handle any sort of traffic peaks. However, that is uneconomical. So with a well-designed switch, during the busiest of BHs we may expect some moments of congestion such that additional call attempts will meet blockage. *Grade of service* expresses the probability of meeting congestion during the BH and is expressed by the letter p. A typical grade of service is $p = 0.01$. This means that an average of one call in 100 will be blocked or "lost" during the BH. Grade of service, a term in the Erlang formula, is more accurately defined as the *probability of congestion*. It is important to remember that lost calls (blocked calls) refer to calls that fail at *first* trial. We discuss attempts (at dialing) later, that is, the way blocked calls are handled.

We exemplify grade of service by the following problem. If we know that there are 354 seizures (lines connected for service) and 6 blocked calls (lost calls) during the BH, what is the grade of service?

$$\text{Call congestion} = \frac{\text{Number of lost calls}}{\text{Total number of offered calls}} \tag{1.1}$$

$$= \frac{6}{354 + 6} = \frac{6}{360}$$

or

$$p = 0.017$$

The average grade of service for a network may be obtained by adding the grade of service contributed by each constituent switch, switching network, or trunk group. The *Reference Data for Radio Engineers* [Ref. 1, Section 31] states that the grade of service provided by a particular group of trunks or circuits of specified size and carrying a specified traffic intensity is the probability that a call offered to the group will find available trunks already occupied on first attempt. That probability depends on a number of factors, the most important of which are (1) the distribution in time and duration of offered traffic (e.g., random or periodic arrival and constant or exponentially distributed holding time), (2) the number of traffic sources [limited or high (infinite)], (3) the availability of trunks in a group to traffic sources (full or restricted availability), and (4) the manner in which lost calls are "handled."

Several new concepts are suggested in these four factors. These must be explained before continuing.

5.4 Availability

Switches were previously discussed as devices with lines and trunks, but better terms for describing a switch are "inlets" and "outlets." When a switch has full availability, each inlet has access to any outlet. When not all the free outlets in a switching system can be reached by inlets, the switching system is referred to as one with "limited availability." Examples of switches with limited and full availability are shown in Figures 1.5A and 1.5B.

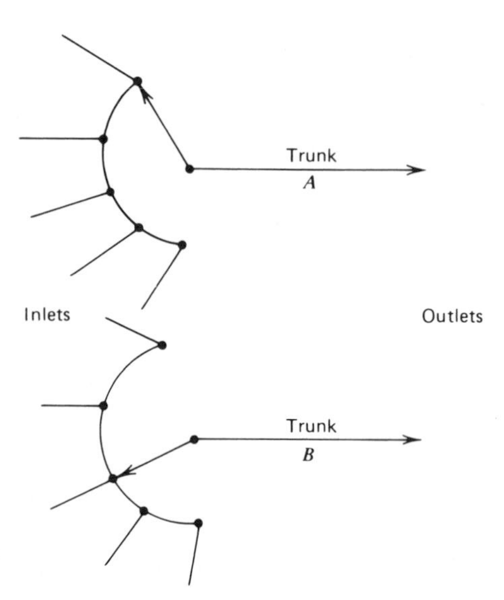

Figure 1.5A An example of a switch with limited availability.

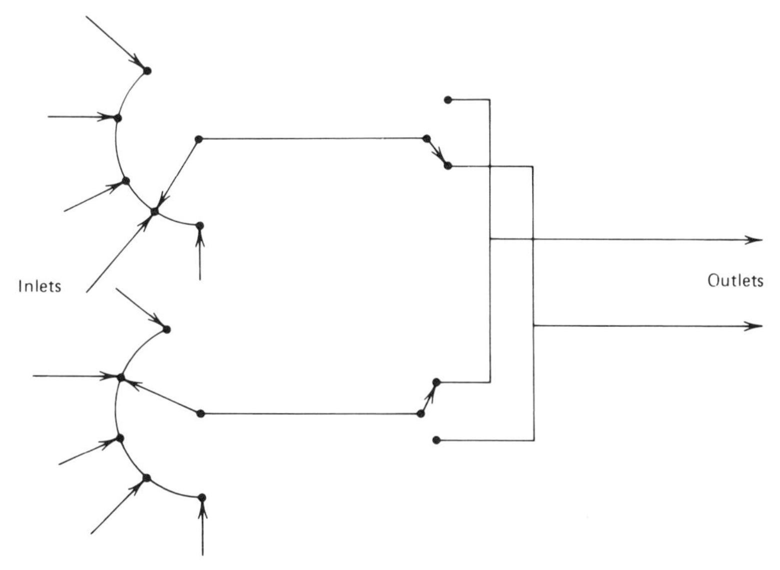

Figure 1.5B An example of a switch with full availability.

Of course, full availability switching is more desirable than limited availability but is more expensive for larger switches. Thus full availability switching is generally found only in small switching configurations and in many new digital switches (see Chapter 10). *Grading* is one method of improving the traffic-handling capacities of switching configurations with limited availability. Grading is a scheme for interconnecting switching subgroups to make the switching load more uniform.

5.5 "Handling" of Lost Calls

In conventional telephone traffic theory three methods are considered for the handling or dispensing of lost calls: (1) lost calls held (LCH), (2) lost calls cleared (LCC), and (3) lost calls delayed (LCD). The LCH concept assumes that the telephone user will immediately reattempt the call on receipt of a congestion signal and will continue to redial. The user hopes to seize connection equipment or a trunk as soon as switching equipment becomes available for the call to be handled. It is the assumption in the LCH concept that lost calls are held or waiting at the user's telephone. This concept further assumes that such lost calls extend the average holding time theoretically, and in this case the average holding time is zero, and all the time is waiting time. The principal traffic formula used in North America is based on the LCH concept.

The LCC concept, which is used primarily in Europe or those countries accepting European practice, assumes that the user will hang up and wait some time interval before reattempting if the user hears the congestion signal on the first attempt. Such calls, it is assumed, disappear from the system. A reattempt (after the delay) is considered as initiating a new call. The Erlang formula is based on this criterion.

The LCD concept assumes that the user is automatically put in queue (a waiting line or pool). For example, this is done when the operator is dialed. It is also done on some modern computer-controlled switching systems, generally referred to under the blanket term *stored program control* (SPC). The LCD category may be broken down into three subcategories, depending on how the queue or pool of waiting calls is handled. The waiting calls may be handled last in first out, first in line first served, or at random.

5.6 Infinite and Finite Traffic Sources

We can assume that traffic sources are infinite or finite. For the infinite-traffic-sources case the probability of call arrival is constant and does not depend on the state of occupancy of the system. It also implies an infinite number of call arrivals, each with an infinitely small holding time. An example of finite traffic sources is when the number of sources offering traffic to a group of trunks or circuits is comparatively small in comparison to the number of circuits. We can also say that with a finite number of sources the arrival rate is proportional to the number of sources that are not already engaged in sending a call.

5.7 Probability-Distribution Curves

Telephone-call originations in any particular area are random in nature. We find that originating calls or call arrivals at an exchange closely fit a family of probability-distribution curves following a Poisson distribution. The Poisson distribution is fundamental to traffic theory.

Most of the common probability distribution curves are two-parameter curves; that is, they may be described by two parameters, mean and variance. The mean is a point on the probability-distribution curve where an equal number of events occur to the right of the point as to the left of the point. "Mean" is synonymous with "average." We define mean as the x-coordinate of the center of the area under the probability-density curve for the population. The small Greek letter mu (μ) is the traditional indication of the mean; $\bar{x}$ is also used.

The second parameter used to describe a distribution curve is the dispersion, which tells us how the values or population are dispersed about the center or mean of the curve. There are several measures of dispersion. One is the familiar *standard deviation*, where "the standard deviation s of a sample of n observations $x_1, x_2, \ldots, x_n$ is

$$s = \sqrt{\frac{1}{n-1} \sum_{i=1}^{n} (x_i - \bar{x})^2} \tag{1.2}$$

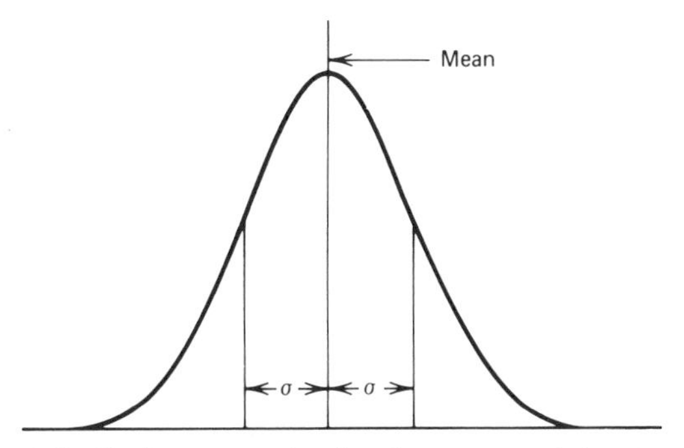

Figure 1.6 A normal distribution curve showing the mean and the standard deviation, σ.

The *variance* V of the sample values is the square of s. The parameters for dispersion s and s^2, the standard deviation and variance, respectively, are usually denoted σ and σ^2 and give us an idea of the squatness of a distribution curve. Mean and standard deviation of a normal distribution curve are shown in Figure 1.6, where we can see that σ^2 is another measure of dispersion, the variance, or essentially the average of the squares of the distances from mean aside from the factor $n/(n-1)$.

We have introduced two distribution functions describing the probability of distribution, often called the *distribution* of or just $f(x)$. Both functions are used in traffic engineering. But before proceeding, the variance-to-mean ratio (VMR) must also be introduced. Sometimes VMR is called the *coefficient of overdispersion*. The formula for VMR is

$$\alpha = \frac{\sigma^2}{\mu} \tag{1.3}$$

5.8 Smooth, Rough, and Random Traffic

Traffic probability distributions can be divided into three distinct categories: (1) smooth, (2) rough, and (3) random. Each may be defined by α, the VMR. For smooth traffic, α is less than 1. For rough traffic, α is greater than 1. When α is equal to 1, the traffic distribution is called *random*. The Poisson distribution function is an example of random traffic where the VMR = 1. Rough traffic tends to be peakier than random or smooth traffic. For a given grade of service more circuits are required for rough traffic because of the greater spread of the distribution curve (greater dispersion).

Smooth traffic behaves like random traffic that has been filtered. The filter is the local exchange. The local exchange looking out at its subscribers sees call arrivals as random traffic, assuming that the exchange has not been overdimensioned. The smooth traffic is the traffic on the local exchange outlets. The filtering or limiting of the peakiness is done by call blockage during the BH. Of course, the blocked traffic may actually overflow to alternative routes. Smooth traffic is characterized by a positive binomial distribution function, perhaps better known to traffic people as the *Bernoulli distribution*. An example of the Bernoulli distribution is as follows [6]. If we assume subscribers make calls independently of each other and that each has a probability p of being engaged in conversation, then if n subscribers are examined, the probability that x of them will be engaged is

$$B(x) = C_x^n p^x (1-p)^{n-x}; \qquad 0 < x < n$$

$$\text{Its mean} = np \tag{1.4}$$

$$\text{Its variance} = np(1-p)$$

where the symbol C_x^n means the number of ways that x entities can be taken n at a time. Smooth traffic is assumed in dealing with small groups of subscribers; the number 200 is often used as the breakpoint [6]. That is, groups of subscribers are considered small when the subscribers number less than 200. And as mentioned, smooth traffic is also used with carried traffic. In this case the rough or random traffic would be the offered traffic.

Let's consider the binomial distribution for rough traffic. This is characterized by a negative index. Therefore, if the distribution parameters are k and q, where k is a positive number representing a hypothetical number of traffic sources and q represents the occupancy per source and may vary between 0 and 1, then

$$R'(x, k, q) = \binom{x+k-1}{k-1} q^x (1-q)^k \tag{1.5}$$

where R' is the probability of finding x calls in progress for the parameters k and q [2]. Rough traffic is used in dimensioning toll trunks with alternative routing. The symbol B (Bernoulli) is used by traffic engineers for smooth traffic and R for rough traffic. Although P may designate probability, in traffic engineering it designates Poissonian, and hence we have "P" tables such as those in Table 1.2.

The Bernoulli formula is

$$B'(x, s, h) = C_s^x h^x (1-h)^{s-x} \tag{1.6}$$

where C_s^x indicates the number of combinations of s things taken x at a time, h is the probability of finding the first line busy of an exchange, $1 - h$ is the probability of finding the first line idle, and s is the number of subscribers. The probability of finding two lines busy is h^2, the probability of finding s lines busy is h^s, and so on. We are interested in finding the probability of x of the s subscribers with busy lines.

The Poisson probability function can be derived from the binomial distribution, assuming that the number of subscribers s is very large and the calling rate per line h is low* such that the product $sh = m$ remains constant and letting s increase to infinity in the limit

$$P(x) = \frac{m^x}{x!} e^{-m} \tag{1.7}$$

where

$$x = 0, 1, 2, \ldots$$

For most of our future discussion, we consider call-holding times to have a negative exponential distribution in the form

$$P = e^{-t/h} \tag{1.8}$$

where t/h is the average holding time and in this case P is the probability of a call lasting longer than t, some arbitrary time interval.

6 ERLANG AND POISSON TRAFFIC FORMULAS

When dimensioning a route, we want to find the number of circuits that serve the route. There are several formulas at our disposal to determine that number of circuits based on the BH traffic load. In Section 5.3 four factors were discussed that will help us to determine which traffic formula to use given a particular set of circumstances. These factors primarily dealt with (1) call arrivals and holding-time distribution, (2) number of traffic sources, (3) availability, and (4) handling of lost calls.

The Erlang B loss formula is probably the most common one used today outside the United States. Loss here means the probability of blockage at the switch due to congestion or to "all trunks busy" (ATB). This is expressed as grade of service E_B or the probability of finding x channels busy. The other two factors in the Erlang B formula are the mean of the *offered* traffic and the number of trunks or servicing channels available. Thus

$$E_B = \frac{A^n/n!}{1 + A - A^2/2! + \cdots + A^n/n!} \tag{1.9}$$

where n is the number of trunks or servicing channels, A is the mean of the offered traffic, and E_B is the grade of service using the Erlang B formula. This

*For example, less than 50 millierlangs (mE).

TABLE 1.1 Trunk-Loading Capacity, Based on Erlang B Formula, Full Availability

	Grade of Service 1 in 1000		Grade of Service 1 in 500		Grade of Service 1 in 200		Grade of Service 1 in 100		Grade of Service 1 in 50		Grade of Service 1 in 20	
Trunks	UC	TU	UC	TU	UC	TU	UC	TU	UC	TU	UC	TU
1	0.04	0.001	0.07	0.002	0.2	0.005	0.4	0.01	0.7	0.02	1.8	0.05
2	1.8	0.05	2.5	0.07	4	0.11	5.4	0.15	7.9	0.22	14	0.38
3	6.8	0.19	9	0.25	13	0.35	17	0.46	22	0.60	32	0.90
4	16	0.44	19	0.53	25	0.70	31	0.87	39	1.09	55	1.52
5	27	0.76	32	0.90	41	1.13	49	1.36	60	1.66	80	2.22
6	41	1.15	48	1.33	58	1.62	69	1.91	82	2.28	107	2.96
7	57	1.58	65	1.80	78	2.16	90	2.50	106	2.94	135	3.74
8	74	2.05	83	2.31	98	2.73	113	3.13	131	3.63	163	4.54
9	92	2.56	103	2.85	120	3.33	136	3.78	156	4.34	193	5.37
10	111	3.09	123	3.43	143	3.96	161	4.46	183	5.08	224	6.22
11	131	3.65	145	4.02	166	4.61	186	5.16	210	5.84	255	7.08
12	152	4.23	167	4.64	190	5.28	212	5.88	238	6.62	286	7.95
13	174	4.83	190	5.27	215	5.96	238	6.61	267	7.41	318	8.83
14	196	5.45	213	5.92	240	6.66	265	7.35	295	8.20	350	9.73
15	219	6.08	237	6.58	266	7.38	292	8.11	324	9.01	383	10.63
16	242	6.72	261	7.26	292	8.10	319	8.87	354	9.83	415	11.54
17	266	7.38	286	7.95	318	8.83	347	9.65	384	10.66	449	12.46
18	290	8.05	311	8.64	345	9.58	376	10.44	414	11.49	482	13.38
19	314	8.72	337	9.35	372	10.33	404	11.23	444	12.33	515	14.31
20	339	9.41	363	10.07	399	11.09	433	12.03	474	13.18	549	15.25
21	364	10.11	388	10.79	427	11.86	462	12.84	505	14.04	583	16.19
22	389	10.81	415	11.53	455	12.63	491	13.65	536	14.90	617	17.13
23	415	11.52	442	12.27	483	13.42	521	14.47	567	154.76	651	18.08
24	441	12.24	468	13.01	511	14.20	550	15.29	599	16.63	685	19.03
25	467	12.97	495	13.76	540	15.00	580	16.12	630	17.50	720	19.99
26	493	13.70	523	14.52	569	15.80	611	16.96	662	18.38	754	20.94
27	520	14.44	550	15.28	598	16.60	641	17.80	693	19.26	788	21.90

28	546	15.18	578	16.05	627	17.41	671	18.64	725	20.15	823	22.87
29	573	15.93	606	16.83	656	18.22	702	19.49	757	21.04	858	23.83
30	600	16.68	634	17.61	685	19.03	732	20.34	789	21.93	893	24.80
31	628	17.44	662	18.39	715	19.85	763	21.19	822	22.83	928	25.77
32	655	18.20	690	19.18	744	20.68	794	22.05	854	23.73	963	26.75
33	683	18.97	719	19.97	774	21.51	825	22.91	887	24.63	998	27.72
34	711	19.74	747	20.76	804	22.34	856	23.77	919	25.53	1033	28.70
35	739	20.52	776	21.56	834	23.17	887	24.64	951	26.43	1068	29.68
36	767	21.30	805	22.36	864	24.01	918	25.51	984	27.34	1104	30.66
37	795	22.03	834	23.17	895	24.85	950	26.38	1017	28.25	1139	31.64
38	823	22.86	863	23.97	925	25.69	981	27.25	1050	29.17	1175	32.63
39	851	23.65	892	24.78	955	26.53	1013	28.13	1083	30.08	1210	33.61
40	880	24.44	922	25.60	986	27.38	1044	29.01	1116	31.00	1246	34.60
41	909	25.24	951	26.42	1016	28.23	1076	29.89	1149	31.92	1281	35.59
42	937	26.04	981	27.24	1047	29.08	1108	30.77	1182	32.84	1317	36.58
43	966	26.84	1010	28.06	1078	29.94	1140	31.66	1215	33.76	1353	37.57
44	995	27.64	1040	28.88	1109	30.80	1171	32.54	1248	34.68	1388	38.56
45	1024	28.45	1070	29.71	1140	31.66	1203	33.43	1282	35.61	1424	39.55
46	1053	29.26	1099	30.54	1171	32.52	1236	34.32	1315	36.53	1459	40.54
47	1083	30.07	1129	31.37	1202	33.38	1268	35.21	1349	37.46	1495	41.54
48	1111	30.88	1159	32.20	1233	34.25	1300	36.11	1382	38.39	1531	42.54
49	1141	31.69	1189	33.04	1264	35.11	1332	37.00	1415	39.32	1567	43.54
50	1170	32.51	1220	33.88	1295	35.98	1364	37.90	1449	40.25	1603	44.53
51		1200	33.33	1250		34.72	1327		36.85	1397		38.80
52		1229	34.15	1280		35.56	1358		37.72	1429		39.70
53		1259	34.98	1310		36.40	1390		38.60	1462		40.60
54		1289	35.80	1341		37.25	1421		39.47	1494		41.50

TABLE 1.1 Continued

Trunks	Grade of Service 1 in 1000		Grade of Service 1 in 500		Grade of Service 1 in 200		Grade of Service 1 in 100	
	UC	TU	UC	TU	UC	TU	UC	TU
55	1319	36.63	1371	38.09	1453	40.35	1527	42.41
56	1349	37.46	1402	38.94	1484	41.23	1559	43.31
57	1378	38.29	1432	39.79	1516	42.11	1592	44.22
58	1408	39.12	1463	40.64	1548	42.99	1625	45.13
59	1439	39.96	1494	41.50	1579	43.87	1657	46.04
60	1468	40.79	1525	42.35	1611	44.76	1690	46.95
61	1499	41.63	1556	43.21	1643	45.64	1723	47.86
62	1529	42.47	1587	44.07	1675	46.53	1756	48.77
63	1559	43.31	1617	44.93	1707	47.42	1789	49.69
64	1590	44.16	1648	45.79	1739	48.31	1822	50.60
65	1620	45.00	1679	46.65	1771	49.20	1855	51.52
66	1650	45.84	1710	47.51	1803	50.09	1888	52.44
67	1681	46.69	1742	48.38	1835	50.98	1921	53.35
68	1711	47.54	1773	49.24	1867	51.87	1954	54.27
69	1742	48.39	1804	50.11	1900	52.77	1987	55.19
70	1773	49.24	1835	50.98	1932	53.66	2020	56.11
71	1803	50.09	1867	51.85	1964	54.56	2053	57.03
72	1834	50.94	1898	52.72	1996	55.45	2087	57.96
73	1865	51.80	1929	53.59	2029	56.35	2120	58.88
74	1895	52.65	1960	54.46	2061	57.25	2153	59.80
75	1926	53.51	1992	55.34	2093	58.15	2186	60.73
76	1957	54.37	2024	56.21	2126	59.05	2219	61.65
77	1988	55.23	2055	57.09	2159	59.96	2253	62.58
78	2019	56.09	2087	57.96	2191	60.86	2286	63.51
79	2050	56.95	2118	58.84	2223	61.76	2319	64.43

80	2081	57.81	2150	59.72	2256	62.67	2353	65.36
81	2112	58.67	2182	60.60	2289	63.57	2386	66.29
82	2143	59.54	2213	61.48	2321	64.48	2420	67.22
83	2174	60.40	2245	62.36	2354	65.38	2453	68.15
84	2206	61.27	2277	63.24	2386	66.29	2487	69.08
85	2237	62.14	2308	64.13	2419	67.20	2521	70.02
86	2268	63.00	2340	65.01	2452	68.11	2554	70.95
87	2299	63.87	2372	65.90	2485	69.02	2588	71.88
88	2330	64.74	2404	66.78	2517	69.93	2621	72.81
89	2362	65.61	2436	67.67	2550	70.84	2655	73.75
90	2393	66.48	2468	68.56	2583	71.76	2688	74.68
91	2425	67.36	2500	69.44	2616	72.67	2722	75.62
92	2456	68.23	2532	70.33	2650	73.58	2756	76.56
93	2488	69.10	2564	71.22	2682	74.49	2790	77.49
94	2519	69.98	2596	72.11	2715	75.41	2823	78.43
95	2551	70.85	2628	73.00	2748	76.32	2857	79.37
96	2582	71.73	2660	73.90	2781	77.24	2891	80.31
97	2614	72.61	2692	74.79	2814	78.16	2925	81.24
98	2645	73.48	2724	75.68	2847	79.07	2958	82.18
99	2677	74.36	2757	76.57	2880	79.99	2992	83.12
100	2709	75.24	2789	77.47	2913	80.91	3026	84.06
101	2740	76.12	2821	78.36	2946	81.83	3060	85.00
102	2772	77.00	2853	79.26	2979	82.75	3094	85.95
103	2804	77.88	2886	80.16	3012	83.67	3128	86.89
104	2836	78.77	2918	81.05	3045	84.59	3162	87.83
105	2867	79.65	2950	81.95	3078	85.51	3196	88.77
106	2899	80.53	2983	82.85	3111	86.43	3230	89.72
107	2931	81.42	3015	83.75	3145	87.35	3264	90.66
108	2963	82.30	3047	84.65	3178	88.27	3298	91.60
109	2995	83.19	3080	85.55	3211	89.20	3332	92.55
110	3027	84.07	3112	86.45	3244	90.12	3366	93.49
111	3059	84.96	3145	87.35	3277	91.04	3400	94.44
112	3091	85.85	3177	88.25	3311	91.97	3434	95.38

TABLE 1.1 **Continued**

Trunks	Grade of Service 1 in 1000		Grade of Service 1 in 500		Grade of Service 1 in 200		Grade of Service 1 in 100	
	UC	TU	UC	TU	UC	TU	UC	TU
113	3122	86.73	3209	89.15	3344	92.89	3468	96.33
114	3154	87.62	3242	90.06	3378	93.82	3502	97.28
115	3186	88.51	3275	90.96	3411	94.74	3536	98.22
116	3218	89.40	3307	91.86	3444	95.67	3570	99.17
117	3250	90.29	3340	92.77	3478	96.60	3604	100.12
118	3282	91.18	3372	93.67	3511	97.53	3639	101.07
119	3315	92.07	3405	94.58	3544	98.45	3673	102.02
120	3347	92.96	3437	95.48	3578	99.38	3707	102.96
121	3379	93.86	3470	96.39	3611	100.31	3741	103.91
122	3411	94.75	3503	97.30	3645	101.24	3775	104.86
123	3443	95.64	3535	98.20	3678	102.17	3809	105.81
124	3475	96.54	3568	99.11	3712	103.10	3843	106.76
125	3507	97.43	3601	100.02	3745	104.03	3878	107.71
126	3540	98.33	3633	100.93	3779	104.96	3912	108.66
127	3572	99.22	3666	101.84	3812	105.89	3946	109.62
128	3604	100.12	3699	102.75	3846	106.82	3981	110.57
129	3636	101.01	3732	103.66	3879	107.75	4015	111.52
130	3669	101.91	3765	104.57	3912	108.68	4049	112.47

131	3701	102.81	3797	105.48	3946	109.62	4083	113.42
132	3733	103.70	3830	106.39	3980	110.55	4118	114.38
133	3766	104.60	3863	107.30	4013	111.48	4152	115.33
134	3798	105.50	3896	108.22	4047	112.42	4186	116.28
135	3830	106.40	3929	109.13	4081	113.35	4221	117.24
136	3863	107.30	3961	110.04	4114	114.28	4255	118.19
137	3895	108.20	3994	110.95	4148	115.22	4289	119.14
138	3928	109.10	4027	111.87	4181	116.15	4324	120.10
139	3960	110.00	4060	112.78	4215	117.09	4358	121.05
140	3992	110.90	4093	113.70	4249	118.02	4392	122.01
141	4025	111.81	4126	114.61	4283	118.96	4427	122.96
142	4058	112.71	4159	115.53	4316	119.90	4461	123.92
143	4090	113.61	4192	116.44	4350	120.83	4496	124.88
144	4122	114.51	4225	117.36	4384	121.77	4530	125.83
145	4155	115.42	4258	118.28	4418	122.71	4564	126.79
146	4188	116.32	4291	119.19	4451	123.64	4599	127.74
147	4220	117.22	4324	120.11	4485	124.58	4633	128.70
148	4253	118.13	4357	121.03	4519	125.52	4668	129.66
149	4285	119.03	4390	121.95	4552	126.46	4702	130.62
150	4318	119.94	4423	122.86	4586	127.40	4737	131.58

Source: Courtesy of GTE Automatic Electric Company (Bulletin No. 485).

TABLE 1.2 Trunk Loading Capacity Based on Poisson Formula, Full Availability

	Grade of Service 1 in 1000		Grade of Service 1 in 100		Grade of Service 1 in 50		Grade of Service 1 in 20		Grade of Service 1 in 10	
Trunks	UC	TU	UC	TU	UC	TU	UC	TU	UC	TU
1	0.1	0.003	0.4	0.01	0.7	0.02	1.9	0.05	3.8	0.10
2	1.6	0.05	5.4	0.15	7.9	0.20	12.9	0.35	19.1	0.55
3	6.9	0.20	16	0.45	20	0.55	29.4	0.80	39.6	1.10
4	15	0.40	30	0.85	37	1.05	49	1.35	63	1.75
5	27	0.75	46	1.30	56	1.55	71	1.95	88	2.45
6	40	1.10	64	1.80	76	2.10	94	2.60	113	3.15
7	55	1.55	84	2.35	97	2.70	118	3.25	140	3.90
8	71	1.95	105	2.90	119	3.30	143	3.95	168	4.65
9	88	2.45	126	3.50	142	3.95	169	4.70	195	5.40
10	107	2.95	149	4.15	166	4.60	195	5.40	224	6.20
11	126	3.50	172	4.80	191	5.30	222	6.15	253	7.05
12	145	4.05	195	5.40	216	6.00	249	6.90	282	7.85
13	166	4.60	220	6.10	241	6.70	277	7.70	311	8.65
14	187	5.20	244	6.80	267	7.40	305	8.45	341	9.45
15	208	5.80	269	7.45	293	8.15	333	9.25	370	10.30
16	231	6.40	294	8.15	320	8.90	362	10.05	401	11.15
17	253	7.05	320	8.90	347	9.65	390	10.85	431	11.95
18	276	7.65	346	9.60	374	10.40	419	11.65	462	12.85
19	299	8.30	373	10.35	401	11.15	448	12.45	492	13.65
20	323	8.95	399	11.10	429	11.90	477	13.25	523	14.55
21	346	9.60	426	11.85	458	12.70	507	14.10	554	15.40
22	370	10.30	453	12.60	486	13.50	536	14.90	585	16.25
23	395	10.95	480	13.35	514	14.30	566	15.70	616	17.10
24	419	11.65	507	14.10	542	15.05	596	16.55	647	17.95
25	444	12.35	535	14.85	572	15.90	626	17.40	678	18.85

26	469	13.05	562	15.60	599	16.65	656	18.20	710	19.70
27	495	13.75	590	16.40	627	17.40	686	19.05	741	20.60
28	520	14.45	618	17.15	656	18.20	717	19.90	773	21.45
29	545	15.15	647	17.95	685	19.05	747	20.75	805	22.35
30	571	15.85	675	18.75	715	19.85	778	21.60	836	23.20
31	597	16.60	703	19.55	744	20.65	809	22.45	868	24.10
32	624	17.35	732	20.35	773	21.45	840	23.35	900	25.00
33	650	18.05	760	21.10	803	22.30	871	24.20	932	25.90
34	676	18.80	789	21.90	832	23.10	902	25.05	964	26.80
35	703	19.55	818	22.70	862	23.95	933	25.90	996	27.65
36	729	20.25	847	23.55	892	24.80	964	26.80	1028	28.55
37	756	21.00	876	24.35	922	25.60	995	27.65	1060	29.45
38	783	21.75	905	25.15	951	26.40	1026	28.50	1092	30.35
39	810	22.50	935	25.95	982	27.30	1057	29.35	1125	31.25
40	837	23.25	964	26.80	1012	28.10	1088	30.20	1157	32.14
41	865	24.05	993	27.60	1042	28.95	1120	31.10	1190	33.05
42	892	24.80	1023	28.40	1072	29.80	1151	31.95	1222	33.95
43	919	25.55	1052	29.20	1103	30.65	1183	32.85	1255	34.85
44	947	26.30	1082	30.05	1133	31.45	1214	33.70	1287	35.75
45	975	27.10	1112	30.90	1164	32.35	1246	34.60	1320	36.65
46	1003	27.85	1142	31.70	1194	33.15	1277	35.45	1352	37.55
47	1030	28.60	1171	32.55	1225	34.05	1309	36.35	1385	38.45
48	1058	29.40	1201	33.35	1255	34.85	1340	37.20	1417	39.35
49	1086	30.15	1231	34.20	1286	35.70	1372	38.10	1450	40.30
50	1115	30.95	1261	35.05	1317	36.60	1403	38.95	1482	41.15

TABLE 1.2 **Continued**

Trunks	Grade of Service 1 in 1000		Grade of Service 1 in 100		Grade of Service 1 in 50	
	UC	TU	UC	TU	UC	TU
51	1143	31.75	1291	35.85	1349	37.45
52	1171	32.55	1322	36.70	1380	38.35
53	1200	33.35	1352	37.55	1410	39.15
54	1228	34.10	1382	38.40	1441	40.05
55	1256	34.90	1412	39.20	1472	40.90
56	1285	35.70	1443	40.10	1503	41.75
57	1313	36.45	1473	40.90	1534	42.60
58	1342	37.30	1504	41.80	1565	43.45
59	1371	38.10	1534	42.60	1596	44.35
60	1400	38.90	1565	43.45	1627	45.20
61	1428	39.65	1595	44.30	1659	46.10
62	1457	40.45	1626	45.15	1690	46.95
63	1486	41.30	1657	46.05	1722	47.85
64	1516	42.10	1687	46.85	1752	48.65
65	1544	42.90	1718	47.70	1784	49.55
66	1574	43.70	1749	48.60	1816	50.45
67	1603	44.55	1780	49.45	1847	51.30
68	1632	45.35	1811	50.30	1878	52.15
69	1661	46.15	1842	51.15	1910	53.05
70	1691	46.95	1873	52.05	1941	53.90
71	1720	47.80	1904	52.90	1973	54.80
72	1750	48.60	1935	53.75	2004	55.65
73	1779	49.40	1966	54.60	2036	56.55
74	1809	50.25	1997	55.45	2067	57.40
75	1838	51.05	2028	56.35	2099	58.30
76	1868	51.90	2059	57.20	2130	59.15
77	1898	52.70	2091	58.10	2162	60.05
78	1927	53.55	2122	58.95	2194	60.95
79	1957	54.35	2153	59.80	2226	61.85
80	1986	55.15	2184	60.65	2258	62.70
81	2016	56.00	2215	61.55	2290	63.60
82	2046	56.85	2247	62.40	2321	64.45
83	2076	57.65	2278	63.30	2354	65.40
84	2106	58.50	2310	64.15	2386	66.30
85	2136	59.35	2341	65.05	2418	67.15
86	2166	60.15	2373	65.90	2451	68.10
87	2196	61.00	2404	66.80	2483	68.95
88	2226	61.85	2436	67.65	2515	69.85
89	2256	62.65	2467	68.55	2547	70.75
90	2286	63.50	2499	69.40	2579	71.65
91	2317	64.35	2530	70.30	2611	72.55
92	2346	65.15	2562	71.15	2643	73.40
93	2377	66.05	2594	72.05	2674	74.30
94	2407	66.85	2625	72.90	2706	75.15
95	2437	67.70	2657	73.80	2739	76.10
96	2468	68.55	2689	74.70	2771	76.95
97	2498	69.40	2721	75.60	2803	77.85
98	2528	70.20	2752	76.45	2836	78.80
99	2559	71.10	2784	77.35	2868	79.65
100	2589	71.90	2816	78.20	2900	80.55

Source: Courtesy of GTE Automatic Electric Company (Bulletin No. 485).

formula assumes that

- Traffic originates from an infinite number of sources.
- Lost calls are cleared assuming a zero holding time.
- The number of trunks or servicing channels is limited.
- Full availability exists.

At this point in our discussion of traffic we suggest that the reader learn to differentiate between time congestion and call congestion when dealing with grade of service. *Time congestion*, of course, refers to the decimal fraction of an hour during which all trunks are busy simultaneously. *Call congestion*, on the other hand, refers to the number of calls that fail at first attempt, which we term *lost calls*. Keep in mind that the Erlang B formula deals with offered traffic, which differs from carried traffic by the number of lost calls.

Table 1.1 is based on the Erlang B formula and gives trunk-dimensioning information for some specific grades of service, from 0.001 to 0.05 and from 1 to 150 trunks. Table 1.1 uses traffic-intensity units UC (unit call) and TU (traffic unit), where TU is in erlangs assuming BH and UC is in ccs (100 call-seconds); 1 erlang = 36 ccs (based on a 1-h time interval). To exemplify the use of Table 1.1, suppose that a route carried 16.68 erlangs of traffic with a desired grade of service of 0.001; then 30 trunks would be required. If the grade of service were reduced to 0.05, the 30 trunks could carry 24.80 erlangs of traffic. When sizing a route for trunks or an exchange, we often come up with a fractional number of servicing channels or trunks. In this case we would opt for the next highest integer because we cannot install a fraction of a trunk. For instance, if calculations show that a trunk route should have 31.4 trunks, it would be designed for 32 trunks.

The Erlang B formula, based on lost calls cleared, has been standardized by the CCITT (CCITT Rec. Q.87) and has been generally accepted outside the United States. In the United States the Poisson formula [2] is favored. This formula is often called the *Molina formula*. It is based on the LCH concept. Table 1.2 provides trunking sizes for various grades of service deriving from the P formula; such tables are sometimes called "P" tables, (Poisson) and assume full availability. We must remember that the Poisson equation also assumes that traffic originates from a large (infinite) number of independent subscribers or sources (random traffic input), with a limited number of trunks or servicing channels and LCH.

It is not as straightforward as it may seem when comparing grades of service between Poisson and Erlang B formulas (or tables). The grade of service $p = 0.01$ for the Erlang B formula is equivalent to a grade of service of 0.005 when applying the Poisson (Molina) formula. Given these grades of service, assuming LCC with the Erlang B formula permits up to several tenths of erlangs of less traffic when dimensioning up to 22 trunks, where the two approaches equate (e.g., where each formula allows 12.6 erlangs over the

22 trunks). Above 22 trunks the Erlang B formula permits the trunks to carry somewhat more traffic and at 100 trunks, 2.7 erlangs more than for the Poisson formula under the LCH assumption.

6.1 Alternative Traffic Formula Conventions

Some readers may be more comfortable using traffic formulas with a different convention and notation. The Erlang B and Poisson formulas were derived from reference [2]. The formulas and notation used in this subsection have been taken from Reference [20]. The following is the notation used in the formulas given below:

A = The expected traffic density, expressed in busy hour erlangs.

P = The probability that calls will be lost (or delayed) because of insufficient channels.

n = The number of channels in the group of channels.

s = The number of sources in the group of sources.

p = The probability that a single source will be busy at an instant of observation. This is equal to A/s.

x = A variable representing a number of busy sources or busy channels.

e = The Naperian logarithmic base, which is the constant 2.71828 + .

$\binom{m}{n}$ = The combination of m things taken n at a time.

$\sum_{X=m}^{n}$ = The summation of all values obtained when each integer or whole number value, from m to n inclusive, is substituted for the value x in the expression following the symbol.

∞ = The conventional symbol for infinity.

The Poisson formula has the following assumptions: (1) infinite sources, (2) equal traffic density per source, (3) lost calls held (LCH). The formula is

$$P = e^{-A} \sum_{x=n}^{\infty} \frac{A^x}{x!} \tag{1.10}$$

The Erlang B formula assumes (1) infinite sources, (2) equal traffic density per source, and (3) lost calls to cleared (LCC). The formula is

$$P = \frac{\dfrac{A^n}{n!}}{\displaystyle\sum_{x=0}^{n} \frac{A^x}{x!}} \tag{1.11}$$

The Erlang C formula assumes (1) infinite sources, (2) lost calls delayed (LCD), (3) exponential holding times, and (4) calls served in order of arrival. The formula is

$$P = \frac{\dfrac{A^n}{n!} \cdot \dfrac{n}{n-A}}{\displaystyle\sum_{x=0}^{n-1} \frac{A^x}{x!} + \frac{A^n}{n!} \frac{n}{n-A}} \tag{1.12}$$

The binomial formula assumes (1) finite sources, (2) equal traffic density per source, and (3) lost calls held (LCH). The formula is

$$P = \left(\frac{s-A}{s}\right)^{s-1} \sum_{x=n}^{s-1} \binom{s-1}{x} \left(\frac{A}{s-A}\right)^x \tag{1.13}$$

7 WAITING SYSTEMS (QUEUEING)

A short discussion follows regarding traffic in queueing systems. Queueing or waiting systems, when dealing with traffic, are based on the third assumption, namely, lost calls delayed (LCD). Of course, a queue in this case is a pool of callers waiting to be served by a switch. The term *serving time* is the time a call takes to be served from the moment of arrival in the queue to the moment of being served by the switch. For traffic calculations in most telecommunication queueing systems, the mathematics is based on the assumption that call arrivals are random and Poissonian. The traffic engineer is given the parameters of offered traffic, the size of the queue, and a specified grade of service and will determine the number of serving circuits or trunks required.

The method by which a waiting call is selected to be served from the pool of waiting calls is called *queue discipline*. The most common discipline is the first-come–first-served discipline, where the call waiting longest in the queue is served first. This can turn out to be costly because of the equipment required to keep order in the queue. Another type is random selection, where the time a call has waited is disregarded and those waiting are selected in random order. There is also the last-come–first-served discipline and bulk service discipline, where batches of waiting calls are admitted, and there are also priority service disciplines, which can be preemptive and nonpreemptive. In queueing systems the grade of service may be defined as the probability of delay. This is expressed as $P(t)$, the probability that a call is not being immediately served and has to wait a period of time greater than t. The average delay on all calls is another parameter that can be used to express grade of service, and the length of queue is another.

The probability of delay, the most common index of grade of service for waiting systems when dealing with full availability and a Poisonnian call arrival process, is calculated by using the Erlang C formula, which assumes an infinitely long queue length. Syski [3] provides a good guide to Erlang C and other, more general, waiting systems.

8 DIMENSIONING AND EFFICIENCY

By definition, if we were to dimension a route or estimate the required number of servicing channels, where the number of trunks (or servicing channels) just equaled the erlang load, we would attain 100% efficiency. All trunks would be busy with calls all the time or at least for the entire BH. This would not even allow several moments for a trunk to be idle while the switch decided the next call to service. In practice, if we engineered our trunks, trunk routes, or switches this way, there would be many unhappy subscribers.

On the other hand, we do, indeed, want to size our routes (and switches) to have a high efficiency and still keep our customers relatively happy. The goal of our previous exercises in traffic engineering was just that. The grade of service is one measure of subscriber satisfaction. As an example, let us assume that between cities X and Y there are 100 trunks on the interconnecting telephone route. The tariffs, from which the telephone company derives revenue, are a function of the erlangs of carried traffic. Suppose we allow a dollar per erlang-hour. The very upper limit of service on the route is 100 erlangs. If the route carried 100 erlangs of traffic per day, the maximum return on investment would be $2400 a day for that trunk route and the portion of the switches and local plant involved with these calls. As we well know, many of the telephone company's subscribers would be unhappy because they would have to wait excessively to get calls through from X to Y. How, then, do we optimize a trunk route (or serving circuits) and keep the customers as happy as possible?

Remember from Table 1.1, with an excellent grade of service of 0.001, that we relate grade of service to subscriber satisfaction and that the 100 circuits could carry up to 75.24 erlangs during the BH. Assuming the route did carry 75.24 erlangs for the BH, it would earn $75.24 for that hour and something far less than $2400 per day. If the grade of service were reduced to 0.01, 100 trunks would bring in $84.06 for the busy hour. Note the improvement in revenue at the cost of reducing grade of service. Another approach to saving money is to hold the erlang load constant and decrease the number of trunks and switch facilities accordingly as the grade of service is reduced. For instance, 70 erlangs of traffic at $p = 0.001$ requires 96 trunks and at $p = 0.01$, only 86 trunks.

8.1 Alternative Routing

One method of improving efficiency is to use alternative routing (called *alternate routing* in North America). Suppose that we have three serving areas, X, Y, and Z, served by three switches, X, Y, and Z, as shown in Figure 1.7.

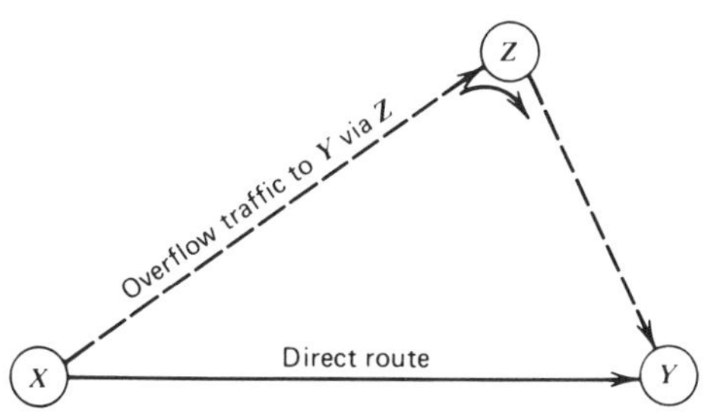

Figure 1.7 Simplified diagram of the alternative routing concept (solid line direct route, dashed line alternative route carrying the overflow from X to Y).

Let the grade of service be 0.005 (1 in 200 in Table 1.1). We found that it would require 67 trunks to carry 50 erlangs of traffic during the BH to meet that grade of service between X and Y. Suppose that we reduced the number of trunks between X and Y, still keeping the BH traffic intensity at 50 erlangs. We would thereby increase the efficiency on the $X-Y$ route at the cost of reducing the grade of service. With a modification of the switch at X, we could route the traffic bound for Y that met congestion on the $X-Y$ route via Z. Then Z would route this traffic on the $Z-Y$ link. Essentially, this is alternative routing in its simplest form. Congestion probably would only occur during very short peaky periods in the BH, and chances are that these peaks would not occur simultaneously with peaks in traffic intensity on the $Z-Y$ route. Further, the added load on the $X-Z-Y$ route would be very small. Some idea of traffic peakiness that would overflow onto the secondary route ($X + Z + Y$) is shown in Figure 1.8.

One of the most accepted methods of dimensioning switches and trunks using alternative routing is the equivalent random group (ERG) method developed by Wilkinson [11]. The Wilkinson method uses the mean M and the variance V. Here the *overflow traffic* is the "lost" traffic in the Erlang B calculations, which were discussed earlier. Let M be the mean value of that

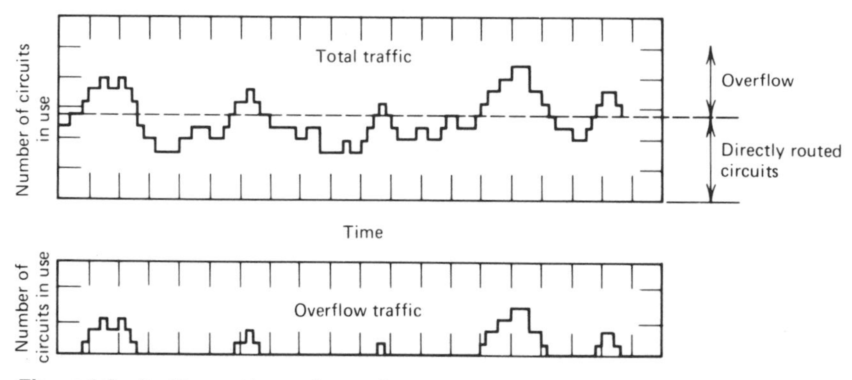

Figure 1.8 Traffic peakiness, the peaks representing overflow onto alternative routes.

overflow and A be the random traffic offered to a group of n circuits (trunks). Then

$$V = M\left(1 - M + \frac{A}{1 + n + M - A}\right) \tag{1.14}$$

When the overflow traffic from several sources is combined and offered to a single second (or third, fourth, etc.) choice of a group of circuits, both the mean and the variance of the combined traffic are the arithmetical sums of the means and variances of the contributors.

The basic problem in alternative routing is to optimize circuit group efficiency (e.g., to dimension a route with an optimum number of trunks). Thus we are to find what circuit quantities result in minimum cost for a given grade of service, or to find the optimum number of circuits (trunks) to assign to a direct route allowing the remainder to overflow on alternative choices. There are two approaches to the optimization. The first method is to solve the problem by successive approximations, and this lends itself well to the application of the computer [12]. Then there are the manual approaches, two of which are suggested in the annex to CCITT Rec. Q.88 (1976).

8.2 Efficiency versus Circuit Group Size

In the present context a *circuit group* refers to a group of circuits performing a specific function. For instance, all the trunks (circuits) routed from X to Y in Figure 1.7 make up a circuit group, irrespective of size. This circuit group should not be confused with the "group" used in transmission-engineering carrier systems.

If we assume full loading, it can be stated that efficiency improves with circuit group size. From Table 1.1, given $p = 0.01$, 5 erlangs of traffic requires a group with 11 trunks, more than a 2 : 1 ratio of trunks to erlangs, and 20 erlangs requires 30 trunks, a 3 : 2 ratio. Note how the efficiency has improved.

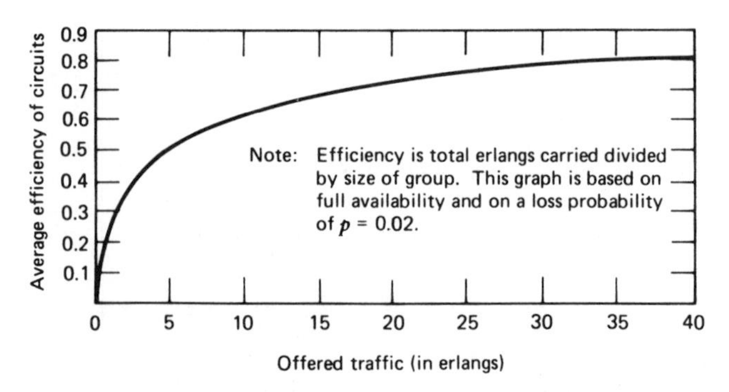

Figure 1.9 Group efficiency increases with size.

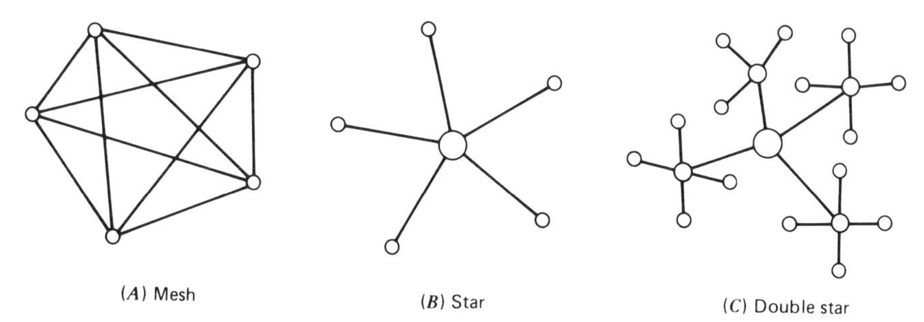

Figure 1.10 Examples of star, double star, and mesh configurations.

One hundred and twenty trunks will carry 100 erlangs, or 6 trunks for every 5 erlangs for a group of this size. Figure 1.9 shows how efficiency improves with group size.

9 BASES OF NETWORK CONFIGURATIONS

9.1 Introductory Concepts

A network in telecommunications may be defined as a method of connecting exchanges so that any one subscriber in the network can communicate with any other subscriber. For this introductory discussion, let us assume that subscribers access the network by a nearby local exchange. Thus the problem is essentially how to connect exchanges efficiently. There are three basic methods of connection in conventional telephony: (1) mesh, (2) star, and (3) double and higher-order star (see Section 2 of this chapter). The mesh connection is one in which each and every exchange is connected by trunks (or junctions) to each and every other exchange as shown in Figure 1.10A. A star connection utilizes an intervening exchange, called a *tandem exchange*, such that each and every exchange is interconnected via a *single* tandem exchange. An example of a star connection is shown in Figure 1.10B. A double star configuration is one where sets of pure star subnetworks are connected via higher-order tandem exchanges, as shown in Figure 1.10C. This trend can be carried still further, as we see later on, when hierarchical networks are discussed.

As a general rule we can say that mesh connections are used when there are comparatively high traffic levels between exchanges, such as in metropolitan networks. On the other hand, a star network may be applied when traffic levels are comparatively low.

Another factor that leads to star and multiple-star network configurations is network complexity in the trunking outlets (and inlets) of a switch in a full mesh. For instance, an area with 20 exchanges would require 380 traffic groups (or links) and 100 exchanges, 9900 traffic groups. This assumes what are called *one-way groups*. A one-way group is best defined considering the connection between two exchanges, A and B. Traffic originating at A to B is carried in one group and the traffic originating at B bound for A, in another group, as shown in the following diagram:

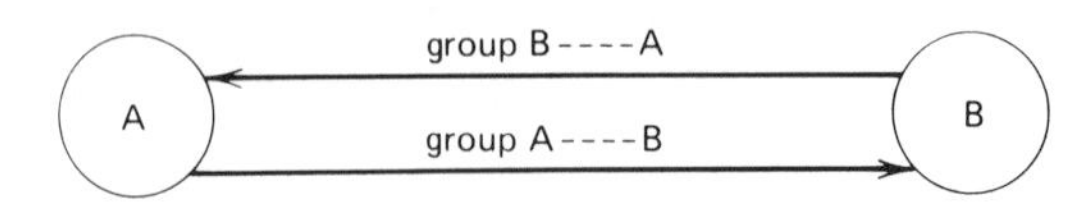

Thus, in practice, most networks are compromises between mesh and star configurations. For instance, outlying suburban exchanges may be connected to a nearby major exchange in the central metropolitan area. This exchange may serve nearby subscribers and be connected in mesh to other large exchanges in the city proper. Another example is the city's long-distance exchange, which is a tandem exchange looking into the national long-distance network, whereas the major exchanges in the city are connected to it in mesh. An example of a real-life compromise among mesh, star, and multiple-star configurations is shown in Figure 1.11.

9.2 Hierarchical Networks

To bring order out of this confusion, hierarchical networks evolved. That is, a systematic network was developed that reduces the trunk group outlets (and inlets) of a switch to some reasonable amount, permits the handling of high traffic intensities on certain routes where necessary, and allows for overflow and a means of restoral in certain circumstances. Consider Figure 1.12, which is a simplified example of a higher-order star network. The term "order" here is significant and leads to the discussion of hierarchical networks.

A hierarchical network has levels giving orders of importance of the exchanges making up the network, and certain restrictions are placed on traffic flow. For instance, in Figure 1.12 there are three levels or ranks of exchange. The smallest boxes in the diagram are the lowest-ranked exchanges, which have been marked with a "3" to indicate the third level or rank. Note the restrictions (or rules) of traffic flow. As the figure is drawn, traffic from $3A_1$ bound for $3A_2$ would have to flow through exchange $2A_1$. Likewise, traffic from exchange $2A_2$ to $2A_3$ would have to flow through exchange $1A$. Carrying the concept somewhat further, traffic from any A exchange to any B exchange would necessarily have to be routed through exchange $1A$.

The next consideration is the high-usage route. For instance, if we found that there were high traffic intensities between $2B_1$ and $2B_2$, trunks and switch gear might well be saved by establishing a high-usage route between the two (shown in dashed line). Thus we might call the high-usage route a *highly*

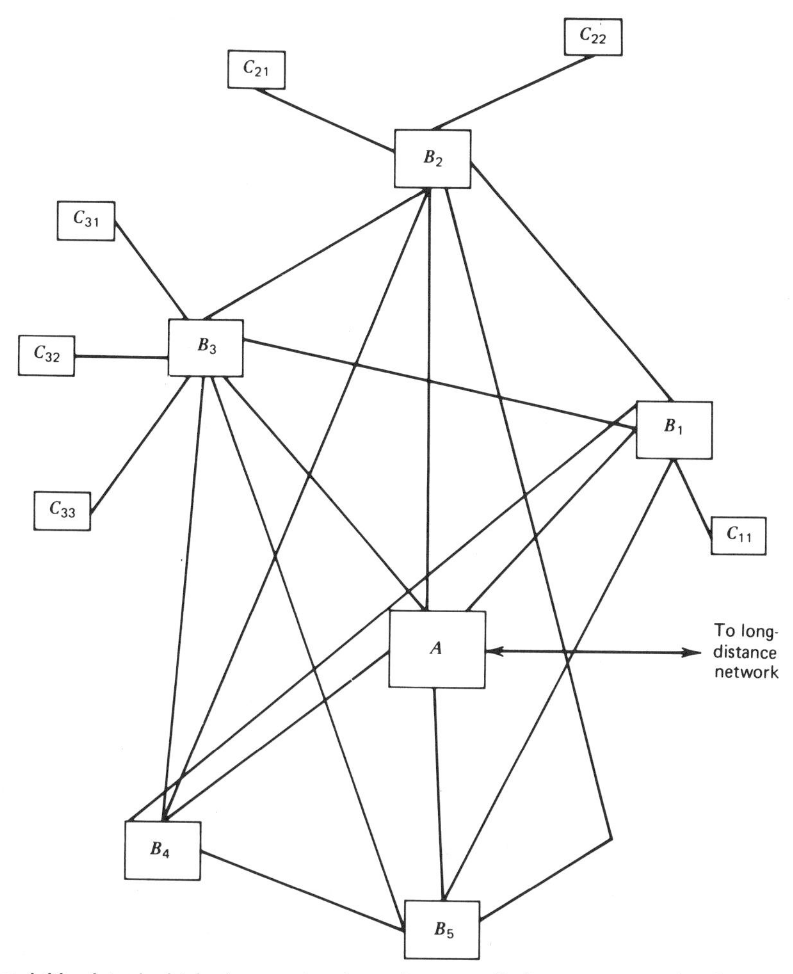

Figure 1.11 A typical telephone network serving a small city as an example of a compromise between mesh and star configuration.
A is a class 4 (ATT), primary center (CCITT).
B is a class 5 exchange, a local exchange.
C may be a satellite exchange or a concentrator.

traveled shortcut. Of course, high-usage routes could be established between any pair of exchanges in the network if traffic intensities and distances involved proved this strategy economical. When high-usage routes are established, traffic between the exchanges involved will first be offered to the high-usage route and overflow would take place through hierarchical structure. Or as shown in our Figure 1.12, up to the next level and down. If routing is through the highest level in the hierarchy, we call this route the *final route*.

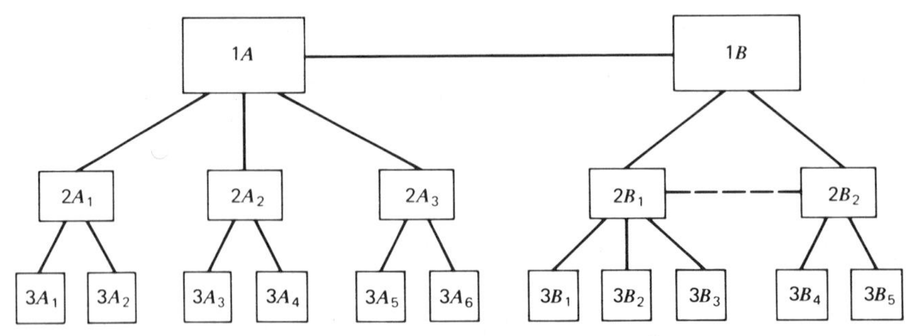

Figure 1.12 A higher-order star network.

Figure 1.12 shows traffic routed between exchanges $2B_1$ and $2B_2$ via exchange $1B$ being routed on the final route.

9.3 The ATT and CCITT Hierarchical Networks

Two types of hierarchical network exist today, each serving about 50% of the world's telephones. These are the ATT network, generally used in North America, and the CCITT network, typically used in Europe or areas of the world under European influence. Frankly, there is really little difference from the routing viewpoint. Each has five levels or ranks in the hierarchy, although CCITT allows for a sixth level. The basic difference is in the nomenclature used. Figure 1.13 illustrates the ATT hierarchy and Figure 1.14, the CCITT hierarchy.

Particularly in Europe, the terminology distinguishes tandem exchanges from transit exchanges. Although both perform the same function, the switching of trunks, a tandem exchange serves the local area, as shown at the bottom of Figure 1.14, and figures in the lowest levels of hierarchy. A transit exchange switches trunks in the toll or long-distance area. Also, in older CCITT documents we should expect to see the term "CT," meaning "central transit" in French or "central tránsito" in Spanish. In English the term is simply *transit exchange*. The CCITT usually places a number after CT, as follows: CT1, the highest-order transit exchange in CCITT routing; CT2, the next-to-highest order; and CT3, the third order from the top. You will sometimes see the term *junction*, which is a British (UK) term for a trunk serving the local area. In CCITT terminology, trunks serve as higher-order connections. Primary centers are collecting centers (exchanges) for traffic to interconnect the toll or long-distance network. The term *center* may be related to "central," meaning a switching node or exchange, usually of higher order.

Figure 1.13 presents the ATT "routing pattern." The highest order or rank in the hierarchy is the class 1 center and the lowest, a class 5 office ("office" is taken from the North American term "central office"). It should be noted that a high-usage (HU) trunk group may be established between any two switching

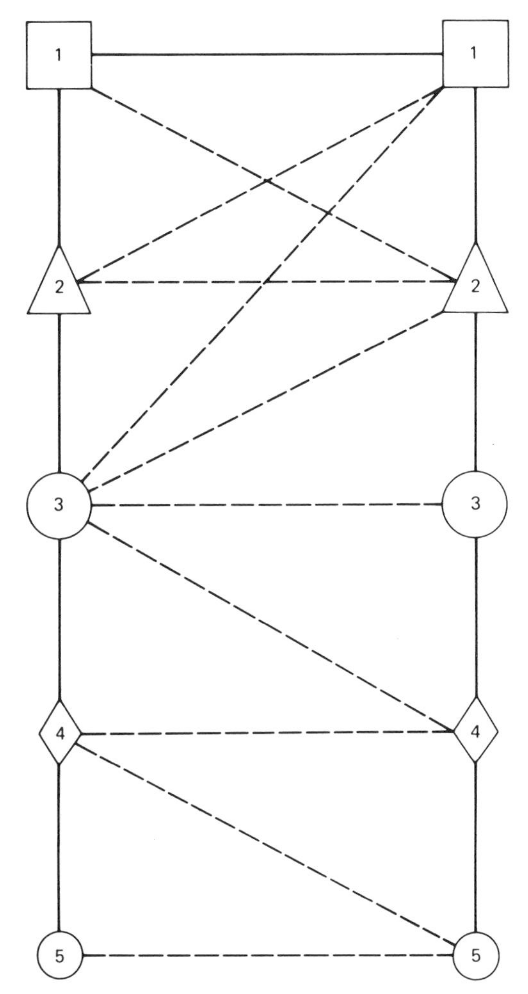

Figure 1.13 The North American (ATT) hierarchical network (dashed lines show high-usage trunks). Note how the two highest ranks are connected in mesh.

centers regardless of location or rank, whenever the traffic volume justifies it. The table that follows clarifies the comparative nomenclature of the two types of hierarchy, with the highest rank at the top.

	North American	CCITT
Class 1.	Regional center	Quaternary center
Class 2.	Sectional center	Tertiary center
Class 3.	Primary center	Secondary center
Class 4.	Toll center (toll point)	Primary center
Class 5.	End office	Local exchange

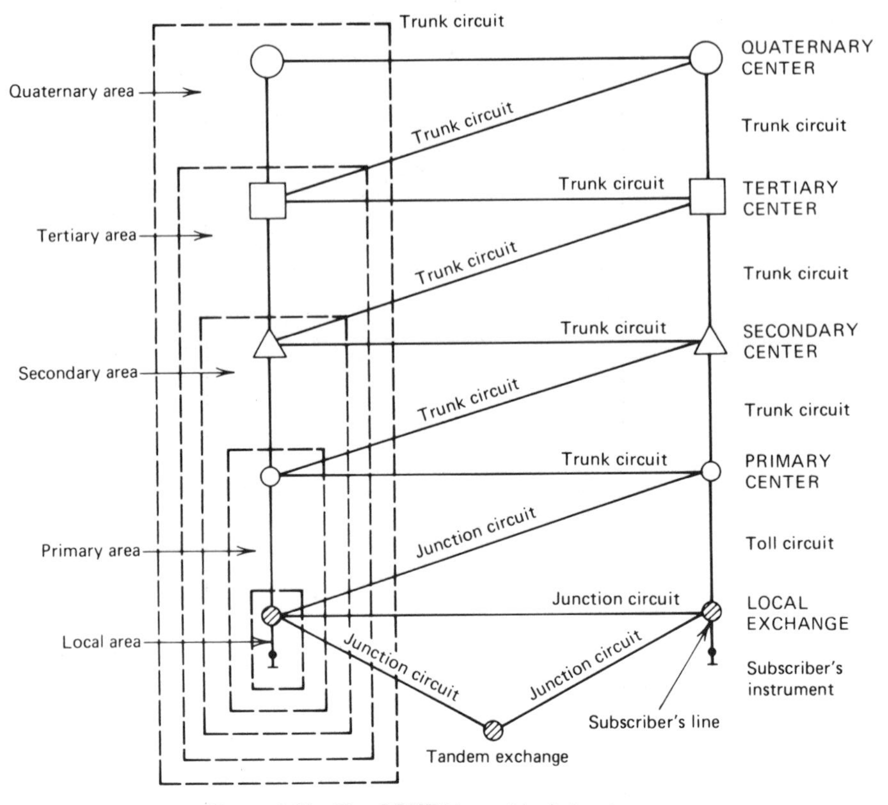

Figure 1.14 The CCITT hierarchical structure.

The first restraint on routing design derives from CCITT Rec. Q.40 (Ref. 21), the section titled "3. Number of Circuits in a Connection," which essentially states that the maximum number of circuits to be used for an international call is 12 and that the maximum number of international circuits is 4. In exceptional cases and for a low number of calls, the total number of circuits may be 14, but even in this case the maximum number of international circuits is 4 (see Chapter 6, Section 3, for further discussion).

In Figures 1.13 and 1.14 the reader will note that by proceeding up the chain, across, and down,* following final routes in every case, there are nine circuits in tandem, leaving only three for the international connection. Of course, this number becomes four because the top "across" circuit is considered an international connection.

*See also Section 9.4 for further discussion of hierarchical network rules and "up, across, and down."

9.4 Rules for Conventional Hierarchical Networks

A backbone structure to a hierarchical network is noted in Figures 1.13 and 1.14: from left to right or from right to left, the outside vertical lines connected by the top horizontal line, in either figure, which we refer to as "up, across, and down," as shown in the diagram that follows.

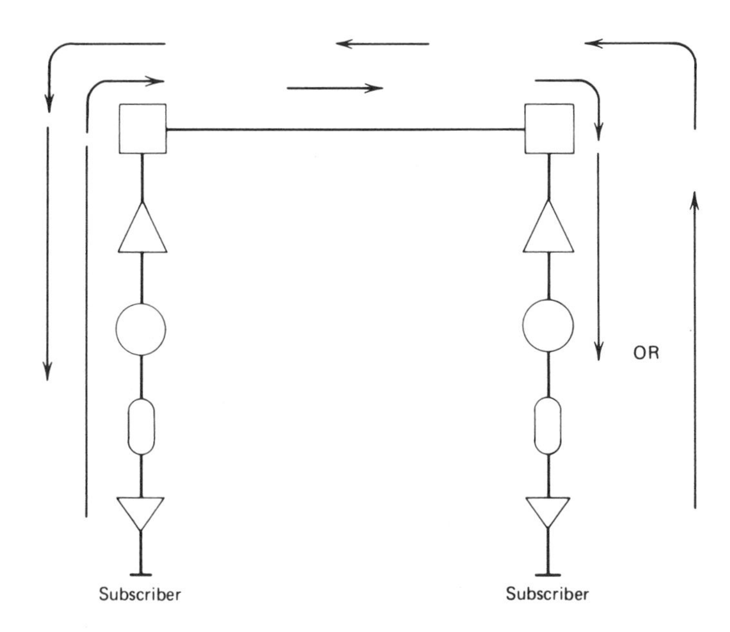

The CCITT terms these routes "theoretical final routes." For our argument, a final route is a route from which no overflow is permitted. A hierarchical network is characterized by a full set of final routes from source to sink. Any other routes are supplementary to the pure hierarchy, regardless of whether overflow is permitted on them.

A hierarchical system of routing leads to simplified switch design. A common expression used when discussing hierarchical routing and star connections is that lower-rank exchanges "home" on higher-rank exchanges. If a call is destined for an exchange of lower rank in its chain, the call proceeds down the chain. Likewise, if a call is destined for another exchange outside the chain, it proceeds up the chain. Or when such high-usage routes exist, a call may be routed on a route additional or supplementary to the pure hierarchy, proceeding to the distant transit center and then descending to the destination. Of course, at the highest level in a pure hierarchy the call crosses from one chain over to the other. In hierarchical networks only the order of each switch in the hierarchy and those additional links (high-usage routes) that provide access need be known. In such networks administration is simplified, and storage or routing information is reduced when compared to the full-mesh type of network, for example.

The CCITT (Rec. Q.13) suggests the far-to-near criterion, whereby the first choice route in advancing a call is to advance the call as far as possible from its origin using the backbone route to measure distances. The second choice is the next best and so forth. The additional routes (diagonal dashed lines in Figures 1.13 and 1.14) result in decreasing the number of links traversed, thus minimizing the total links traversed in a call connection and improving transmission and signaling characteristics and, on long international connections, keeping the total number of links required to 12 or below, as we mentioned previously.

One weakness in conventional hierarchical structured networks is circuit security. When we say "security" in this context, we mean that the loss of one (or several) links due to fire, explosion, natural disaster, cutting, or sabotage will not cause the full breakdown of communication in the network. Rather, we mean that communication can be maintained, perhaps with reduced capability and increased blockage, but nevertheless maintained. Now consider a hierarchical network. Higher in the network the nodes of each rank become fewer and fewer. For instance, take the United States and Canada. There are only 10 nodes of class 1 in the United States and 2 in Canada. Even with these 12 nodes completely interconnected in mesh, as they actually are in practice, the loss of one or more nodes or links at this level in the hierarchy might seriously jeopardize communications. Thus the current tendency is to reduce the number of levels or ranks in the hierarchy, thus increasing the number of higher-level nodal points. Improved circuit security is the result when they are fully interconnected (mesh), offering more combinations of alternative route configurations. The desirability of this trend is obvious. It becomes highly feasible as the network becomes modernized, replacing electromechanical switching with computer-controlled switching. We see future large national networks with only three levels of hierarchy.

9.5 Homing Arrangements and Interconnecting Network in North America

We spoke of "homing" of a dependent switching node to the next-higher rank nodal point in purely hierarchical networks. The use of HU routes so widely used in practice makes the rule really valid only on the backbone final routes. For example, consider the North American network where class 5, the lowest ranking switch *may* be served directly from any higher ranking location. Possible homing arrangements for each class (rank) of switching node are shown in Table 1.3.

Final trunk groups in this network are engineered for a lost-call probability of 0.01. As we have seen, direct high-usage trunk groups are provided between switching nodal points of any rank (class) when the volume of traffic and economics warrant these groups and where automatic alternative routing

TABLE 1.3 Homing Arrangements in North America

Rank and Class	May Home at Switches of Classes
End office, 5	Class 4, 3, 2, or 1
Toll center, 4	Class 3, 2, or 1
Primary center, 3	Class 2 or 1
Sectional center, 2	Class 1
Regional center, 1	All regional centers mutually interconnected

features are available. The rule is that high-usage trunk groups carry most but not all of the offered traffic in the BH. The proportion of the offered traffic carried on a direct high-usage trunk group ordinarily is determined, in part, by the relative costs of the direct route and the alternative route. The economic factors to be considered are the additional switching costs on the alternative routes. In some instances high-usage trunk groups may be designed on a "no overflow" basis with the 0.01 lost-call objective. Homing arrangements are not changed in this case, and these trunks are called "full groups." Such full groups effectively truncate or limit the hierarchical chain of final routes for the traffic offered them.

In the North American network the number of trunks connected in the final route chains from a class 4 switch to another class 4 switch is not to exceed 7. By adding these to the trunk at each end connecting to the end office (class 5), the maximum number of trunks that can be connected in tandem is 9. It is estimated that the probability of a call traversing all the final route links in two complete routing chains be only several calls out of millions. Of course, calls between high-intensity traffic locations (nodes) rarely encounter multiple switches. However, calls between low-intensity traffic points use multiple switching, or what we call "tandem switching."

10 ROUTING METHODS

There are generally three methods of routing calls from source to sink through one, several, or many intermediate switching nodes. As we have seen, there may be many possible patterns through which a given call can traverse. The problem is to *decide* how the call should proceed through the many possible path combinations in the network. The three methods are (1) right-through routing, (2) own-exchange routing, and (3) computer-controlled routing (with common-channel signaling). In right-through routing the originating exchange determines the route from source to sink. Alternative routing is not allowed at intermediate switching points. However, the initial outgoing circuit group may be arranged so that one or more alternative routes are presented. Because of its inherent limitations in alternative routing and the requirement that a change in network configuration or the addition of new exchanges entail alteration in each existing complex switch (i.e., switches with translators), right-through routing is limited almost exclusively to the local area.

Own-exchange routing allows for changes in routing as the call proceeds to its destination. This routing system is particularly suited to networks with alternative routing and changes in routing patterns in response to changes in load configuration. Another advantage in own-exchange routing is that when new exchanges are added or the network is modified, minimal switch modifications are required in the network. One disadvantage is the possibility of establishing a closed routing loop where a call may be routed such that it is eventually routed back to its originating exchange or other exchange through which it has already been routed in attempting to reach its destination. However, a hierarchical routing system ensures that such loops cannot be generated. If routing loops are established in an operating network, there can be disastrous consequences, as the reader can appreciate.

Conventional telephone networks have signaling information for a particular call carried on the same path (pair of wires or their equivalent) that carries the speech, often called the conversation path. Signaling, as we discuss later in considerable detail, is the generation and transmission of information that sets up a desired call and routes it through the network to its destination. New and more modern computer-controlled networks often use a separate path to carry the required signaling information. In this case the computer in the originating exchange or originating long-distance exchange can "optimally" route the call through the network on a separate signaling path. The originating computer would have a "map in memory" of the network with updated details of network conditions such as traffic load at the various nodes and trunks and outages. The necessary adaptive information is broadcast on the separate path that connects the various computers in the network. This is computer-controlled routing. Such routing is termed "routing with common-channel signaling" and with adaptive network management signals.

11 VARIATIONS IN TRAFFIC FLOW

In networks covering large geographic expanses and even in cases of certain local networks, there may be a variation in time of day of the BH or in the direction of traffic flow. In the United States business traffic peaks during several hours before and after the noon lunch period on weekdays, and social calls peak in early evening. Traffic flow tends to be from suburban living areas to urban centers in the morning, and the reverse in the evening.

In national networks covering several time zones where the differences in local time may be appreciable, long-distance traffic tends to be concentrated in a few hours common to BH peaks at both ends. In such cases it is possible to direct traffic so that peaks of traffic in one area fall into valleys of traffic in another (noncoincident busy hour). The network design can be made more economical if configured to take advantage of these phenomena, particularly in the design and configuration of direct routes versus overflow.

12 BOTH-WAY CIRCUITS

We defined one-way circuits in Section 9.1. Here traffic from A to B is assigned to one group of circuits, and traffic from B to A on another separate group. In both-way (or two-way) operation a circuit group may be engineered to carry traffic in both directions. The individual circuits in the group may be used in either direction, depending on which exchange seizes the circuit first.

In engineering networks it is most economical to have a combination of one-way and both-way circuits on longer routes. Signaling and control arrangements on both-way circuits are substantially more expensive. However, when dimensioning a system for a given traffic intensity, fewer circuits are needed in both-way operation, with notable savings on low-intensity routes (i.e., below about 10 erlangs in each direction). For long circuits both-way operation has obvious advantages when dealing with a noncoincident BH. During overload conditions both-way operation is also advantageous because the direction of traffic flow in these conditions is usually unequal.

The major detriment to two-way operation, besides its increased signaling cost, is the possibility of double seizure. This occurs when both ends seize a circuit at the same time. There is a period of time when double seizure can occur in a two-way circuit; this extends from the moment the circuit is seized to send a call and the moment when it becomes blocked at the other end. Signaling arrangements can help to circumvent this problem. Likewise, switching arrangements can be made such that double seizure can occur only on the last free circuit of a group. This can be done by arranging in turn the sequence of scanning circuits so that the sequence on one end of a two-way circuit is reversed from that of the other end. Of course, great care must be taken on circuits having long propagation times, such as satellite and long undersea cable circuits. By extending the time between initial seizure and blockage at the other end, these circuits are the most susceptible, just because a blocking signal takes that much longer to reach the other end.

13 QUALITY OF SERVICE

Quality of service appears at the outset to be an intangible concept. However, it is very tangible for a telephone subscriber unhappy with his or her service. The concept of service quality must be mentioned early in any all-encompassing text on telecommunications systems. System engineers should never once lose sight of the concept, no matter what segment of the system they may be responsible for. Quality of service also means *how happy* the telephone company (or other common carrier) is keeping the customer. For instance, we might find that about half the time a customer dials, the call goes awry or the caller cannot get a dial tone or cannot hear what is being said by the party at the other end. All these have an impact on quality of service. So we begin to find that quality of service is an important factor in many areas of the telecommunications business and means different things to different people. In the old days of telegraphy, a rough measure of how well the system was working was the number of service messages received at a switching center. In modern telephony we now talk about service observing (see Chapter 3, Section 18).

The transmission engineer calls quality of service "customer satisfaction," which is commonly measured by how well the customer can hear the calling party. It is called *reference equivalent*,* which is measured in decibels (dB). In our discussion of traffic, lost calls certainly constitute one measure of service quality, and if measured in decimal quantity, one target figure would be $p = 0.01$. Other items to be listed under service quality are:

- Delay before receiving dial tone ("dial-tone delay").
- Post dial(ing) delay (time from completion of dialing a number to first ring of telephone called).
- Availability of service tones (busy tone, telephone out of order, ATB, etc.).
- Correctness of billing.
- Reasonable cost to customer of service.
- Responsiveness to servicing requests.
- Responsiveness and courtesy of operators.
- Time to installation of new telephone, and by some, the additional services offered by the telephone company [22].

One way or another each item, depending on service quality goal, will have an impact on the design of the system.

Furthermore, each item on the list can be quantified—usually statistically, such as reference equivalent, or in time, such as time taken to install a telephone. In some countries this can be measured in years. Good reading can be found in CCITT Recs. Q.60, Q.60 bis, and Q.61.

REVIEW QUESTIONS

1. List at least two functions of telephone battery. Give the standard telephone battery voltage with respect to ground.

2. What is *on-hook* and *off-hook*? When a subscriber subset (the telephone) goes "off-hook," what occurs at the serving switch? List two items.

3. Suppose that the sidetone level of a telephone is increased. What is the natural reaction of the subscriber?

*Or "corrected reference equivalent."

4. A subscriber pair, with a fixed battery voltage, is extended. As we extend the loop further, two limiting performance factors come into play. Name them.

5. Define a mesh connection. Draw a star arrangement.

6. In the context of the argument presented in the chapter, what is the principal purpose of a local switch?

7. What are the two basic parameters that define "traffic"?

8. Distinguish offered traffic from carried traffic.

9. Give one valid definition of the *busy hour*.

10. On a particular traffic relation the calling rate is 461 and the average call duration is 1.5 minutes during the BH. What is the traffic intensity in ccs, in erlangs?

11. Define *grade of service*.

12. A particular exchange has been dimensioned to handle 1000 calls during the busy hour. On a certain day during the BH 1100 calls are offered. What is the resulting grade of service?

13. Distinguish a full availability switch from a limited availability switch.

14. In traffic theory there are three ways lost calls are handled. What are they?

15. Call arrivals at a large switch can be characterized by what type of mathematical distribution? Such arrivals are ______ in nature.

16. Based on the Erlang B formula and given a BH requirement for a grade of service of 0.005 and a BH traffic intensity of 25 erlangs on a certain traffic relation, how many trunks are required?

17. Carry out the same exercise as in question 16 but use the Poisson tables to determine the number of trunks required.

18. Give at least two queueing disciplines.

19. As the grade of service is improved, what is the effect on trunk efficiency?

20. What is the basic purpose of alternative routing? What does it improve?

21. How does circuit group size (number of trunks) affect efficiency for a fixed grade of service?

22. Give the three basic methods of connecting exchanges. (These are the three basic network types.)

23. At what erlang value on a certain traffic relation does it pay to use tandem routing? Is this a maximum or a minimum value?

24. Differentiate between one-way and both-way circuits.

25. What is the drawback of one-way circuits? of both-way circuits?

26. Hierarchical networks are used universally in national and international telephone networks. Differentiate between high-usage (HU) connectivities and final route.

27. Distinguish a tandem exchange from a transit exchange.

28. Define the term *homing* on hierarchical networks.

29. What advantage is there in reducing the number of hierarchical levels?

30. What is the recommended lost call probability for final trunk groups?

31. Name the three basic routing methods.

32. How can we take advantage of the noncoincident busy hour in a large national network?

33. Name at least five items that can be listed under *quality of service*.

REFERENCES

1. International Telephone and Telegraph Corporation, *Reference Data for Radio Engineers*, 5th ed., Howard W. Sams, Indianapolis, 1968.
2. Ramses R. Mina, "The Theory and Reality of Teletraffic Engineering," *Telephony*, a series of articles (April 1971).
3. R. Syski, *Introduction to Congestion Theory in Telephone Systems*, Oliver and Boyd, Edinburgh, 1960.
4. G. Dietrich et al., *Teletraffic Engineering Manual*, Standard Electric Lòrenz, Stuttgart, Germany (1971).
5. E. Brockmeyer et al., "The Life and Works of A. K. Erlang," *Acta Polytechnica Scandinavia*, The Danish Academy of Technical Sciences, Copenhagen, 1960.
6. *A Course in Telephone Traffic Engineering*, Australian Post Office, Planning Branch, 1967.
7. Arne Jensen, *Moe's Principle*, The Copenhagen Telephone Company, Copenhagen, Denmark, 1950.
8. *Networks*, Laboratorios ITT de Standard Eléctrica SA, Madrid, 1973 (limited circulation).
9. *Local Telephone Networks*, The International Telecommunications Union, Geneva, 1968.
10. *Electrical Communication System Engineering Traffic*, U.S. Department of the Army, TM-11-486-2, August 1956.
11. R. I. Wilkinson, "Theories for Toll Traffic Engineering in the USA," *BSTJ*, **35** (March 1956).

12. *Optimization of Telephone Trunking Networks with Alternate Routing*, ITT Laboratories of Standard Eléctrica (Spain), Madrid, 1974 (limited circulation).
13. John Riordan, *Stochastic Service Systems*, Wiley, New York, 1962.
14. Leonard Kleinrock, *Queueing Systems*, Vols. 1 and 2, Wiley, New York, 1975.
15. Thomas L. Saaty, *Elements of Queueing Theory with Applications*, McGraw-Hill, New York, 1961.
16. J. E. Flood, *Telecommunications Networks*, IEE Telecommunications Series 1, Peter Peregrinus, London, 1975.
17. *Notes on The Network–1980*, American Telephone and Telegraph Company, New York, 1980.
18. *National Telephone Networks for the Automatic Service*, International Telecommunications Union–CCITT, Geneva, 1964.
19. D. Bear, *Principles of Telecommunication Traffic Engineering*, IEE Telecommunications Series 2, Peter Peregrinus, London, 1976.
20. Roger L. Freeman, *Reference Manual for Telecommunications Engineering*, Wiley, New York, 1985.
21. *CCITT Red Book*, VIII Plenary Assembly, Malaga-Torremolinos, October 1984.
22. "Telecommunications Quality," *IEEE Communications Magazine*, October 1988 (entire issue).
23. *Engineering and Operations in the Bell System*, 2nd ed., AT & T Bell Laboratories, Murray Hill, N.J. 1983
24. "Notes on BOC Intra-LATA Networks—1986," Bellcore TR-NPL-000275, Livingston, NJ April 1986.

2

LOCAL NETWORKS

1 INTRODUCTION

The importance of local network design, whether standing on its own merit or part of an overall national network, cannot be overstressed. In comparison to the long-distance sector, the local sector is not the big income producer per capita invested, but there would be no national network without it. Telephone companies or administrations invest, on the average, more than 50% in their local areas. In the larger, more developed countries the investment in local plant may reach 70% of total plant investment.

The local area, as distinguished from the long distance or national network, was discussed in Section 4 of Chapter 1. In this chapter we are more precise in defining the local area itself. Let us concede that the local area includes the subscriber plant, local exchanges, the trunk plant interconnecting these exchanges as well as those trunks connecting a local area to the next level of network hierarchy, or the class 4 exchange (USA) or primary center (CCITT).

To further emphasize the importance of the local area, consider Table 2.1, which was taken from Ref. 1 (CCITT). Figure 2.1 is a simplified diagram of a local network with five local exchanges and illustrates the makeup of a typical small local area.

The design of such a network (Figure 2.1) involves a number of limiting factors, the most important of which is economic. Investment and its return

TABLE 2.1 Average Percentage of Investments in Public Telephone Equipment

Item	Average for 16 Countries
Subscriber plant	13%
Outside plant for local networks	27%
Exchanges	27%
Long-distance trunks	23%
Buildings and land	10%

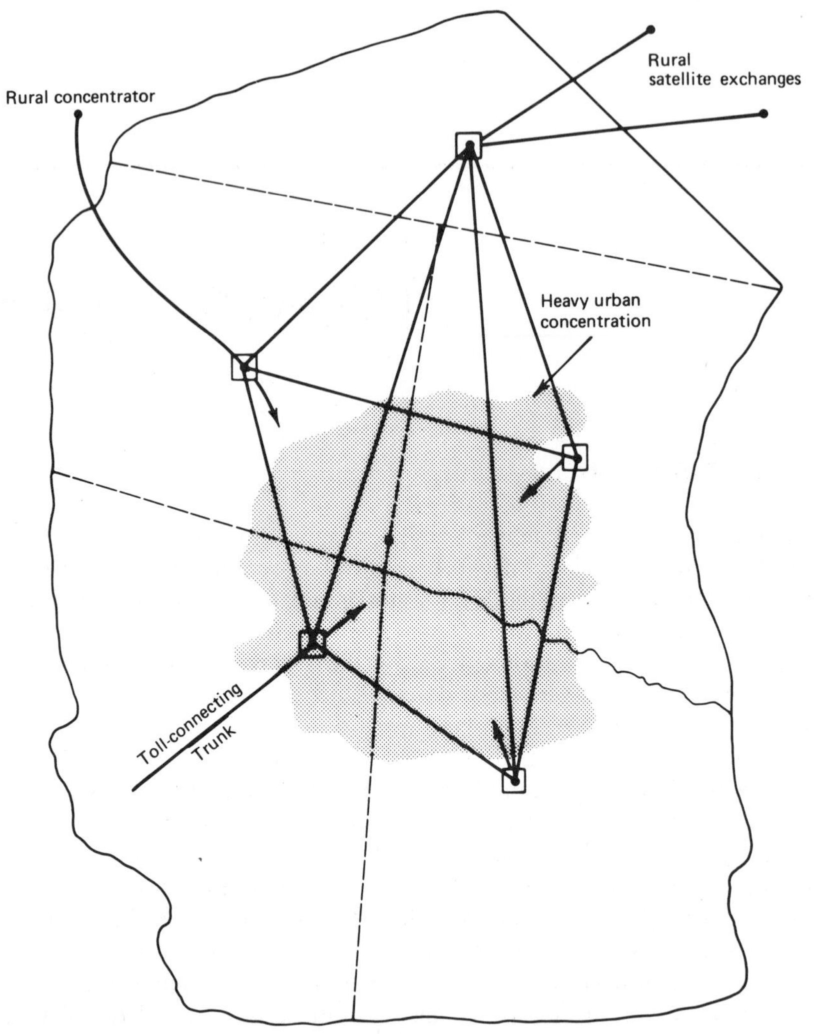

Figure 2.1 A sample local area (arrows represent trunk pull; dashed lines delineate serving areas).

are not treated in this text. However, our goal is to build the most economical network assuming an established quality of service. Considering both quality of service and economy, certain restraints will have to be placed on the design. For example, we will want to know:

- Geographic extension of the local area of interest.
- Number of inhabitants and existing telephone density.
- Calling habits.

- Percentage of business telephones.
- Location of existing telephone exchanges and extension of their serving areas.
- Trunking scheme.
- Present signaling and transmission characteristics.

Each of these criteria or limiting factors are treated separately, and interexchange signaling and switching per se are dealt with later in separate chapters. Let us also assume that each exchange in the sample will be capable of serving up to 10,000 subscribers. Also assume that all telephones in the area have seven-digit numbers, the last four of which are the subscriber number of the respective serving area of each exchange. The reasoning behind these assumptions becomes apparent in later chapters.

A further assumption is that all subscribers are connected to their respective serving exchanges by wire pairs, resulting in some limiting subscriber loop length. This leads to the first constraining factor dealing with transmission and signaling characteristics. In general terms, the subscriber should be able to hear the distant calling party reasonably well (transmission) and to "signal" that party's serving switch. These items are treated at length in the following section.

2 SUBSCRIBER LOOP DESIGN

2.1 General

The pair of wires connecting the subscriber to the local serving switch has been defined as the *subscriber loop*. It is a dc loop in that it is a wire pair supplying a metallic path for the following:

- Talk battery for the telephone transmitter (Chapter 1, Section 2).
- An ac ringing voltage for the bell on the telephone instrument supplied from a special ringing source voltage.
- Current to flow through the loop when the telephone instrument is taken out of its cradle ("off hook"), telling the serving switch that it requires "access," thus causing a line seizure at that switch.
- The telephone dial that, when operated, makes and breaks the direct current on the closed loop, which indicates to the switching equipment the number of the distant telephone with which communication is desired.

The typical subscriber loop is powered by means of a battery feed circuit at the switch. Such a circuit is shown in Figure 2.2. Telephone battery source voltage has been fairly well standardized at −48 V dc.

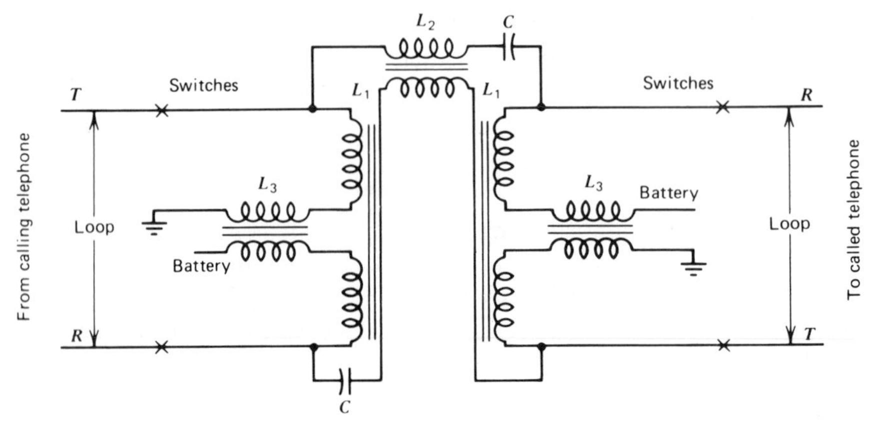

Figure 2.2 Battery feed circuit [22]. Note: Battery and ground are fed through inductors L_2 and L_1 through switch to loops. Copyright © 1961 by Bell Telephone Laboratories, Inc.

2.2 Subscriber Loop Length Limits

The two basic criteria that are considered when designing subscriber loops and that limit their length are attenuation limits and signaling limits. Attenuation in this case refers to loop loss at reference frequency measured in decibels (or nepers). Reference frequency is 1000 Hz in North America and 800 Hz in Europe and many other parts of the world. As a telephone loop is extended in length, its loss at reference frequency increases. It follows that at some point as the loop is extended, level will be attenuated such that the subscriber cannot hear *sufficiently well.*

Likewise, as a loop is extended in length while the battery (supply) voltage is kept constant, the effectiveness of signaling is ultimately lost. This limit is a function of the *IR* drop of the line. We know that *R*, the resistance, increases as length increases. With today's modern telephone sets, the first feature to suffer is usually "signaling," particularly that area of loop signaling called "supervision." In this case it is a signal sent to the switching equipment requesting "seizure" of a switch circuit and at the same time indicating to callers that the line is busy. "Off hook" is the term more commonly used to describe this signal condition. When a telephone is taken "off hook" (i.e., out of its cradle), the telephone loop is closed and current flows, closing a relay at the switch. If current flow is insufficient, the relay will not close or will close and open intermittently ("chatter") such that the line seizure cannot be effected.

Signaling limits are a function for the conductivity of the loop conductor and its diameter or gauge. For this introductory discussion, we can consider that the transmission limits are controlled by the same parameters. Consider a copper conductor. The larger the conductor, the greater the ability to conduct current and thus the longer the loop conductors may be for signaling purposes. Because copper is expensive, we cannot make the conductor as large as we would wish and extend subscriber cable loops over long distances. This is an economic constraint. Before we go further, we describe a method of measuring what a subscriber considers as "hearing sufficiently well."

2.3 The Reference Equivalent

2.3.1 Definition. "Hearing sufficiently well" on a telephone connection is a subjective matter under the blanket heading of "customer satisfaction." Various methods have been derived over the years to rate telephone connections regarding customer (subscriber) satisfaction. Regarding the received telephone signal, subscriber satisfaction is affected by the level (signal power), the signal-to-noise ratio, and the response or attenuation frequency characteristic. A common rating system internationally in use today for grading of customer satisfaction is the *reference equivalent* system. This system considers only the first criterion (viz., level). It must be emphasized that subscriber satisfaction is subjective. To measure satisfaction, the world regulative body for telecommunications, the International Telecommunications Union (ITU), devised a system of rating the level sufficient to "satisfy" using the familiar decibel as the unit of measurement. The reference equivalent is broken down into two basic parts. The first is a subjective value in decibel rating of a particular type of subset. The second part is simply the losses (measured at 800 Hz) end-to-end of the intervening network. To determine the reference equivalent of a particular circuit, we add algebraically the decibel value assigned to the subset to the losses of the connecting circuit. Let us look at how the reference equivalent system was developed, keeping in mind again that it is a subjective measurement dealing with the likes and dislikes of the "average" human being. A standard for reference equivalent was determined in Europe by a team of qualified personnel in a laboratory. A telephone connection, intended to be the most efficient telephone system known, was established in the laboratory. The original reference system or unique master reference consisted of the following:

- A solid-black telephone transmitter.
- Bell telephone receiver.
- Interconnecting these, a "zero-decibel-loss" subscriber loop.
- Connecting the loop, a manual central battery, 22-V dc telephone exchange (switch).

To avoid ambiguity of language, the test team used a test language that consisted of logatoms. A "logatom" is a one-syllable word consisting of a consonant, a vowel, and another consonant.

More accurate measurement methods have been developed since. A more modern reference system is now available at the ITU laboratory in Geneva, Switzerland, called the NOSFER. From this master reference, field test standards are available to telephone companies, administrations, and industry to establish the reference equivalents of telephone subsets in use. These field test sets are calibrated for equivalence with the NOSFER. The NOSFER is made up of a standard telephone transmitter, a receiver, and a network. The reference equivalent of a subscriber's subset, together with the associated subscriber line and feeding bridge, is a quantity obtained by balancing the loadness of receiver speech signals and is expressed relative to the whole or to a corresponding part of the NOSFER (or field) reference system.

2.3.2 Application. Most telephone companies or administrations consider that a standard telephone subset is used. The objective is to measure the capabilities of these subsets regarding loudness. Thus type tests are run on the subsets against calibrated field standards. As mentioned earlier, these may be done on the set alone or on the set plus a fixed length of subscriber loop and feed bridge of known characteristics. The tests are subjective and are carried out in a laboratory. The microphone or transmitter and the earpiece or receiver are each rated separately and are called the *transmit reference equivalent* (TRE) and the *receive reference equivalent* (RRE), respectively. The unit of measurement is the decibel, and negative values indicate that the reference equivalent is better than the laboratory standard (see CCITT Rec. P.72).

In telephone systems the *overall reference equivalent* (ORE) is the more common measurement. Simply, this is the sum of the TRE, the RRE, and the losses of the intervening network. Now consider the simplified telephone network shown in the following diagram:

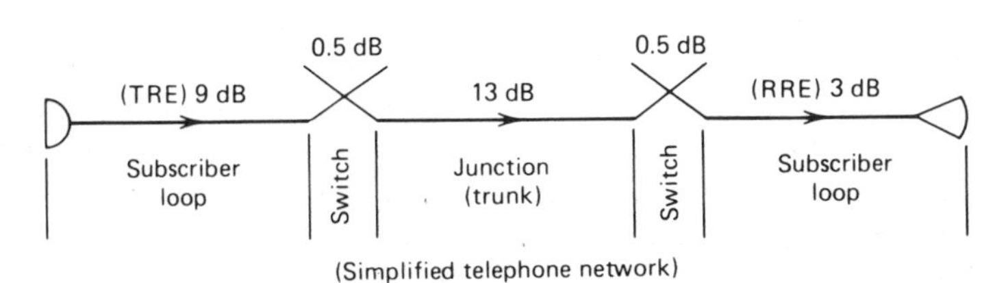

(Simplified telephone network)

The reference equivalent for this circuit is 26 dB, including a 0.5-dB loss for each switch. As defined previously, a junction or trunk is a circuit connecting two switches (exchanges). These may or may not be adjacent. The circuit shown in the preceding diagram may be called a *small transmission plan*. For this discussion, we can define a transmission plan as a method of assigning losses end to end on a telephone circuit. Further on in this text we discuss why all telephone circuits, at least all conventional analog circuits, must be lossy. The reference equivalent is a handy device for rating such a plan regarding subscriber satisfaction (see CCITT Recs. P.42 and P.72).

For more in-depth studies of transmission quality (subscriber satisfaction), the reader should review the AEN (articulation reference equivalent) method described in CCITT Recs. P.12, P.12A, and P.13.

TABLE 2.2 Reference Equivalents for Subscriber Sets in Various Countries

Country	Sending (dB)	Receiving (dB)
With limiting subscriber lines and exchange feeding bridges		
Australia	14.0[a]	6.0[a]
Austria	11.0	2.6
France	11.0	7.0
Norway	12.0	7.0
Germany	11.0	2.0
Hungary	12.0	3.0
Netherlands	17.0	4.0
United Kingdom	12.0	1.0
South Africa	9.0	1.0
Sweden	13.0	5.0
Japan	7.0	1.0
New Zealand	11.0	0.0
Spain	12.0	2.0
Finland	9.5	0.9
With no subscriber lines		
Italy	2.0	−5
Norway	3.0	−3
Sweden	3.0	−3
Japan	2.0	−1
United States (loop length 1000 ft, 83 Ω)	5.0	−1[b]

[a]Minimum acceptable performance.
[b]Freeman [8].

Source: CCITT, Local Telephone Networks, ITU, Geneva, July 1968; and National Telephone Networks for the Automatic Service.

When studying transmission plans or developing them, we usually consider that all sections of a circuit are symmetrical. Let us examine the one shown in the preceding diagram. On each end of a circuit we have a subscriber loop. Thus the same loss is assigned to each loop in the plan, which may not be the case at all in real life. From the local exchange to the first long-distance exchange, variously termed *junctions* or *toll-connecting trunks*, a loss is assigned that is identical at each end, and so forth.

To maintain the symmetry regarding reference equivalent of telephone subsets, we use the term $(T + R)/2$. As we see from the preceding diagram, the TRE and the RRE of the subset have different values. We get the $(T + R)/2$ by summing the TRE and RRE and dividing by 2. This is done to arrive at the desired symmetry. Table 2.2 gives reference equivalent data on a number of standard subscriber subsets used in various parts of the world.

It is stated in CCITT Rec. G.121 that the reference equivalent from the subscriber set to an international connection should not exceed 20.8 dB (TRE) and to the subscriber set at the other end from the same point of reference (RRE), 12.2 dB (the intervening losses are already included in these figures).

TABLE 2.3 British Post Office Survey of Subscribers for Percentage of Unsatisfactory Calls

Overall Reference Equivalent (dB)	Unsatisfactory Calls (%)
40	33.6
36	18.9
32	9.7
28	4.2
24	1.7
20	0.67
16	0.228

By adding 12.2 dB and 20.8 dB, we find 33 dB to be the ORE recommended as a maximum* for an international connection. Table 2.3 should be of interest of this regard. Supplementing the table, it should also be noted that when the overall reference equivalent drops to about 6 dB, subscribers begin to complain that calls are too loud.

The reader may ask why international connections are being discussed in a chapter on local networks. The answer is simple and the concept very important. All calls originate and terminate in a local area. If we are to follow CCITT Rec. G.121, no calls (or very few) should exceed an ORE of 33 dB. In fact, a national transmission plan should reflect the obvious results of Table 2.3 in that we could improve subscriber satisfaction by reducing the ORE from the 33-dB level. And more than half of the 33 dB can be attributed to the local area(s), source, and sink. The limitation, the constraint on local area design, is obvious; the total transmission loss should be under 9 dB (approximately half of 18 dB), with no more than 6 or 7 dB (depending on the transmission plan) for subscriber loop loss. A transmission plan assigns losses to the various segments of a telephone network to meet an ORE goal among other factors.

2.3.3 Corrected Reference Equivalent. Reference equivalent was introduced in Section 2.3.1 to familiarize the reader with a concept of transmission quality. Reference equivalents express the loudness loss of telephone connections and have been typically stated in terms of the planning value of the ORE of a complete connection. Because difficulties have been encountered in the use of reference equivalents, the planning value of the ORE has been replaced by the *corrected reference equivalent* (CRE) (CCITT Rec. G.111). This has required some adjustment in the planning values of loudness loss of complete or partial connections.

*This is the recommendation for 97% of the connections made in a country of average size.

TABLE 2.4 Subscriber Opinion Survey: Corrected Reference Equivalent

Planning Value of the Overall Corrected Reference Equivalent (dB)	Representative Opinion Results[a]	
	Percent "Good plus Excellent"	Percent "Poor plus Bad"
5 to 15	> 90	< 1
20	80	2
25	65	5
30	40	15

[a]Based on a composite opinion model.

Source: CCITT Rec. P.11 Table 1 / P.11, Page 14, Vol. V, Redbook. Copyright ITU4. (Ref. 6).

CCITT Recs. G.121 and P.11 recommended the following objectives for maximum values for sending and receiving CREs:

- For the sending system between a subscriber and the first international circuit, the CRE should not exceed 25 dB.
- For the receiving system between the same two points, the CRE should not exceed 14 dB.

For large countries with 5 national circuits forming part of the 4-wire chain, the CRE values are then 26 dB and 15 dB for sending CRE and receiving CRE, respectively. Table 2.4 gives customer opinion results for various CRE values in decibels (from CCITT Rec. P.11). Table 2.5 gives approximate equivalents between the older reference equivalent and the present corrected reference equivalent.

2.4 Subscriber Loop Design Techniques

2.4.1 Introduction. Consider the following drawing of a simplified subscriber loop:

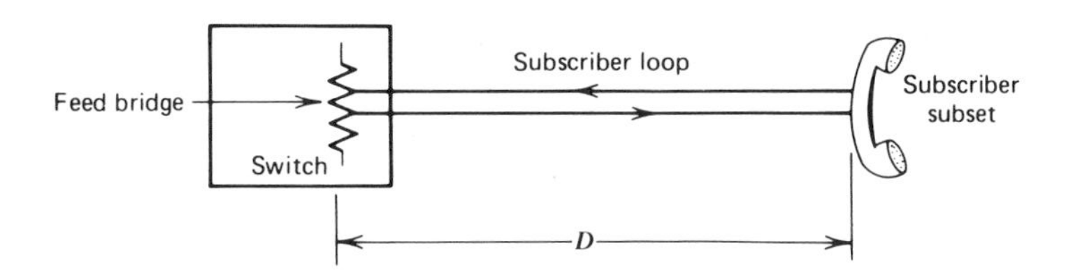

TABLE 2.5 Corresponding Values of Reference Equivalent and Corrected Reference Equivalent for Typical Connections

Connection		Previously Recommended RE(q)	Presently Recommended CRE(y)
	Min	6	5[a]
Optimum range for a connection	Optimum	9	7[a] to 11
(Rec. G.111 §3.2)	Max	18	16
Traffic weighted mean values			
Long term objectives			
Connection (Rec. G.111 § 3.2)	min	13	13
	max	18	16
National system send (Rec. G.121 § 1)	min	10	11.5
	max	13	13
National system receive (Rec. G.121 § 1)	min	2.5	2.5
	max	4.5	4
Short term objectives			
connection (Rec. G.111 § 3.2)	max	23	25.5
National system send (Rec. G.121 § 1)	max	16	19
National system receive (Rec. G.121 § 1)	max	6.5	7.5
Maximum values for national system	send	21	25
(Rec. G.121 § 2.1) of an average-sized country	receive	12	14
Minimum for the national sending system (Rec. G.121 § 3)		6	7

[a]These values apply for conditions free from echo; customers may prefer slightly larger values if some echo is present.

Source: CCITT Rec. P.11 Table 2 / P.11, Page 15, Vol. 5, redbook. copyright ITU. (Ref. 6).

Distance D, the loop length, is most important. We know from this diagram that D must be limited because of attenuation of the voice signal. Likewise, there is a limit to D due to dc resistance, so signaling the local switch can be effected.

The attenuation limit would be taken from the national transmission plan, and for our discussion 6 dB is assigned as the limit (referenced to 800 Hz). For the loop resistance limit we must look to the switch. For instance, many conventional crossbar switches will accept up to 1300 Ω.* From this figure we subtract 300 Ω, the nominal resistance for the telephone subset in series with the loop, leaving a 1000-Ω limit for the wire pair if we disregard the feed bridge resistance. Therefore, in the paragraphs that follow, the figures 6 dB (attenuation limit for loop)† and 1000 Ω (resistance limit) are used.

*Many semielectronic switches will accept 1800-Ω loops and, with special line equipment, 2400 Ω.

†In the United States this value may be as high as 9 dB.

2.4.2 Calculating the Resistance Limit. To calculate the dc loop resistance for copper conductors, the following formula is applicable:

$$R_{dc} = \frac{0.1095}{d^2} \tag{2.1}$$

where R_{dc} is the loop resistance in ohms-per mile (statute) and d is the diameter of the conductor (in inches).

If we wish a 10-mi loop and allow 100 Ω per mile of loop (for the stated 1000-Ω limit), what diameter of copper wire would be needed?

$$100 = \frac{0.1095}{d^2}$$

$$d^2 = \frac{0.1095}{100}$$

$$d = 0.033 \text{ in. or } 0.76 \text{ mm (round off to } 0.80 \text{ mm)}$$

Using Table 2.6, we can compute maximum loop lengths for 1000-Ω signaling resistance. As an example, for a 26-gauge loop, we have

$$\frac{1000}{83.5} = 11.97 \text{ or } 11{,}970 \text{ feet}$$

This, then, is the signaling limit, and not the loss (attenuation) limit, or what some call the "transmission limit," referred to in Section 2.4.3. As the reader has certainly inferred by now, resistance design is a method of designing subscriber loops using resistance limits as a basis or limiting parameter.

2.4.3 Calculating the Loss Limit. Attenuation or loop loss is the basis of transmission design of subscriber loops. The attenuation of a wire pair varies with frequency, resistance, inductance, capacitance, and leakage conductance. Also, resistance of the line will depend on temperature. For open-wire lines attenuation may vary by ±12% between winter and summer conditions. For

TABLE 2.6 Loss and Resistance per 1000 ft of Subscriber Cable[a]

Cable Gauge	Loss / 1000 ft (dB)	Ω/1000 ft of Loop
26	0.51	83.5
24	0.41	51.9
22	0.32	32.4
19	0.21	16.1

[a]Cable is low-capacitance type (i.e., ≤ 0.075 nF / mi).

TABLE 2.7 Code for Load-Coil Spacing

Code Letter	Spacing (ft)	Spacing (m)
A	700	213.5
B	3000	915.0
C	929	283.3
D	4500	1372.5
E	5575	1700.4
F	2787	850.0
H	6000	1830.0
X	680	207.4
Y	2130	649.6

buried cable, which we are more concerned with in this context, variations due to temperature are much less.

Table 2.6 gives losses of some common subscriber cable per 1000 ft. If we are limited to 6-dB (loss) on a subscriber loop, then by simple division we can derive the maximum loop length permissible for transmission design considerations for the wire gauges shown.

$$26 \qquad \frac{6}{0.51} = 11.7 \text{ kft}$$

$$24 \qquad \frac{6}{0.41} = 14.6 \text{ kft}$$

$$22 \qquad \frac{6}{0.32} = 19.0 \text{ kft}$$

$$19 \qquad \frac{6}{0.21} = 28.5 \text{ kft}$$

2.4.4 Loading. In many situations it is desirable to extend subscriber loop lengths beyond the limits described in Sections 2.2 and 2.4. Common methods to attain longer loops without exceeding loss limits are to increase conductor diameter, use amplifiers and/or range extenders,* and use the inductive loading.

Inductive loading tends to reduce transmission loss on subscriber loops and other types of voice pair at the expense of good attenuation-frequency response beyond 3000 Hz. Loading a particular voice-pair loop consists of inserting series inductances (loading coils) into the loop at fixed intervals. Adding load coils tends to decrease the velocity of propagation and increase impedance. Loaded cables are coded according to the spacing of the load coils. The standard code for load coils regarding spacing is shown in Table 2.7.

*A range extender is a device that increases battery voltage on a loop that extends its signaling range. it may also contain an amplifier, thereby extending transmission loss limits as well.

TABLE 2.8 Some Properties of Cable Conductors

Diameter (mm)	AWG No.	Mutual Capacitance (nF / km)	Type of Loading	Loop Resistance (Ω / km)	Attenuation at 1000 Hz (dB / km)
0.32	28	40	None	433	2.03
		50	None		2.27
0.40		40	None	277	1.62
		50	H-66		1.42
		50	H-88		1.24
0.405	26	40	None	270	1.61
		50	None		1.79
		40	H-66	273	1.25
		50	H-66		1.39
		40	H-88	274	1.09
		50	H-88		1.21
0.50		40	None	177	1.30
		50	H-66	180	0.92
		50	H-88	181	0.80
0.511	24	40	None	170	1.27
		50	None		1.42
		40	H-66	173	0.79
		50	H-66		0.88
		40	H-88	174	0.69
		50	H-88		0.77
0.60		40	None	123	1.08
		50	None		1.21
		40	H-66	126	0.58
		50	H-88	127	0.56
0.644	22	40	None	107	1.01
		50	None		1.12
		40	H-66	110	0.50
		50	H-66		0.56
		40	H-88	111	0.44
0.70		40	None	90	0.92
		50	H-66		0.48
		40	H-88	94	0.37
0.80		40	None	69	0.81
		50	H-66	72	0.38
		40	H-88	73	0.29
0.90		40	None	55	0.72
0.91	19	40	None	53	0.71
		50	None		0.79
		40	H-44	55	0.31
		50	H-66	56	0.29
		50	H-88	57	0.26

Source: ITT, Telecommunication Planning Documents—Outside Plant [7].

Loaded cables typically are designated 19-H-44, 24-B-88, and so forth. The first number indicates the wire gauge, with the letter taken from Table 2.7 and indicative of the spacing, and the third item is the inductance of the coil in millihenries (mH). For instance, 19-H-66 is a cable commonly used for long-distance operation in Europe. Thus the cable has 19-gauge voice pairs loaded at 1830-m intervals with coils of 66-mH inductance. The most commonly used spacings are B, D, and H.

Table 2.8 will be useful for calculation of attenuation of loaded loops for a given length. For example, for 19-H-88 cable (last entry in the table), the attenuation per kilometer is 0.26 dB (0.42 dB/statute mile). Thus for our 6-dB loop loss limit, we have 6/0.26, limiting the loop to 23 km in length (14.3 statute miles). When determining signaling limits in loop design, about 9 Ω per load coil should be added as if the coils were series resistors.

2.4.5 Summary of Limiting Conditions: Transmission and Signaling.

We have been made aware that the size of an exchange serving area is limited by factors of economy involving signaling and transmission. Signaling limitations are a function of the type of exchange and the diameter of the subscriber pairs and their conductivity, whereas transmission is influenced by pair characteristics. Both limiting factors can be extended, but that extension costs money, particularly when there may be many thousands of pairs involved. The decision boils down to:

1. If the pairs to be extended are few, they should be extended.
2. If the pairs to be extended are many, it probably is worthwhile to set up a new exchange area, a satellite exchange or use an outside plant module in the area.

These economies are linked to the cost of copper. The current tendency is to reduce the wire gauge wherever possible or even resort to the use of aluminum as the pair conductor.

3 OTHER LIMITING CONDITIONS

The size of an exchange area obviously will depend greatly on subscriber (or potential subscriber) density and distribution. The subscriber traffic is another factor to be considered. The CCITT offers the values in Table 2.9 for subscriber line traffic intensity.

Exchange sizes are often in units of 10,000 lines. Although the number of subscribers initially connected should be considerably smaller than when an exchange is installed, 10,000 is the number of subscribers that may be connected when an exchange reaches "exhaust," where it is filled and no more subscribers can be connected.

TABLE 2.9 Average Occupation Time During the Busy Hour per Subscriber Line

Subscriber Type	BH Traffic Intensity (erlangs)
Residence	0.01–0.04
Business	0.03–0.06
PABX	0.1–0.6
Coin Box	0.07

Ten thousand is not a magic number, but it is a convenient one. It lends itself to crossbar unit size and is a mean unit for subscriber densities in suburban areas and midsized towns in fairly well developed countries. More important, though, is its significance in telephone numbering (the assignment of telephone numbers). Consider a seven-digit number. Now break that down into a three-number group and a four-number group. The first three digits, that is, the first three dialed, identify the local exchange. The last four identify the individual subscriber and is called the subscriber number. Note the breakdown in the following sample:

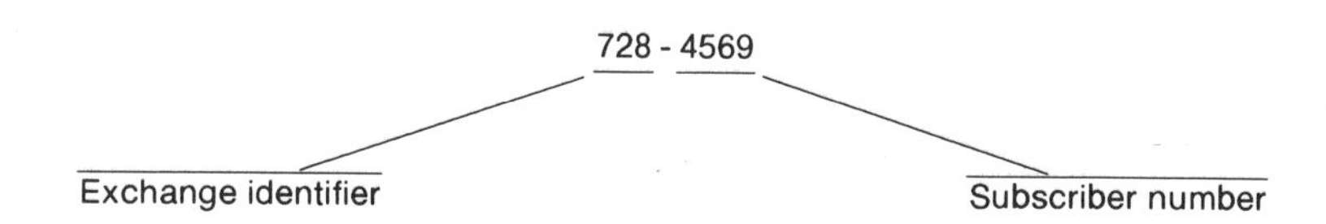

For the subscriber number there are 10,000 number combination possibilities, from 0000 to 9999. Of course there are up to 1000 possibilities for the change identifier. Numbering is discussed at length in Chapter 3 as a consideration under switching.

The foregoing discussion does not preclude exchanges larger than 10,000 lines. But we still deal in units of 10,000 lines, at least in conventional telephony. The term "wire center" is often used to denote a single location housing one or more 10,000 line exchanges. Some wire centers house up to 100,000 lines with a specific local serving area. Wire centers with an ultimate capacity of up to 140,000 lines can be economically justified under certain circumstances. Subscriber density, of course, is the key. Nevertheless, many exchanges will have extended loops requiring some sort of special conditioning, such as larger-gauge wire pairs, loading, range extenders, amplifiers, and the application of carrier techniques (Chapter 5). Leaving aside rural areas, 5% to 25% of an exchange's subscriber loops may well require such conditioning or may be called "long loops."

4 SHAPE OF A SERVING AREA

The shape of a serving area has considerable effect on optimum exchange size. If a serving area has sharply angular contours, the exchange size may have to be reduced to avoid excessively long loops (e.g., revert to the use of more exchanges in a given local geographical area of coverage).

There is an optimum trade-off between exchange size, and we mean here the economies of large exchanges (centralization) and the high cost of long subscriber loops. An equation that can assist in determining the trade-off is as follows, which is based on uniform subscriber density, a circular serving area A of radius r such that

$$C = \frac{A}{\pi r^3}(a + b d \pi r^3) + Ad(f + gL) \tag{2.2}$$

where C is the total cost of exchanges, which decreases when r increases and to which is added the cost of subscriber loops, which increases with r, and L is the average loop length, which may be related as $L = (2r/3)$, the straight-line distance. To determine the minimum cost of C with respect to r, the equation is differentiated and the result is set equal to zero. Thus

$$\frac{2Aa}{\pi r^3} = \frac{2A\,dg}{3} \tag{2.3}$$

and this may be fairly well approximated by

$$r = \left(\frac{a}{dg}\right)^{1/3} \tag{2.4}$$

The cost of exchange equipment is $a + bn$, where n is the number of lines, and the cost of subscriber loops is equal to $(f + gL)$ where, as we stated, L is the average loop length given a uniform density of subscribers, d; and a, b, f, and g are constants.

Since r varies as the cube root, its value does not change greatly for wide ranges of values of d. One flaw is that loops are seldom straight-line distances, and this can be compensated for by increasing g in the ratio (average loop length). This theory is simplified by making exchange areas into circles.

If an entire local area is to be covered, fully circular exchange serving areas are impractical. Either the circles will overlap or uncovered spaces will result, neither of which is desirable. There are then two possibilities: square serving areas or hexagonal serving areas. Of the two, hexagon more nearly approaches a circle. The size of the hexagon can vary with density with a goal of 10,000 lines per exchange as the ultimate capacity. Again, a serving area could have a wire center of 100,000 lines or more, particularly in heavily populated metropolitan areas.

Besides the hexagon, full coverage of local areas may only be accomplished using serving areas of equal triangles or squares. This assumes, of course, that the local area was *ideally* divided into identical geometric figures and would apply only under the hypothetical situation of nearly equal telephone density throughout. A typical hexagon subdivision is shown in the diagram that follows.

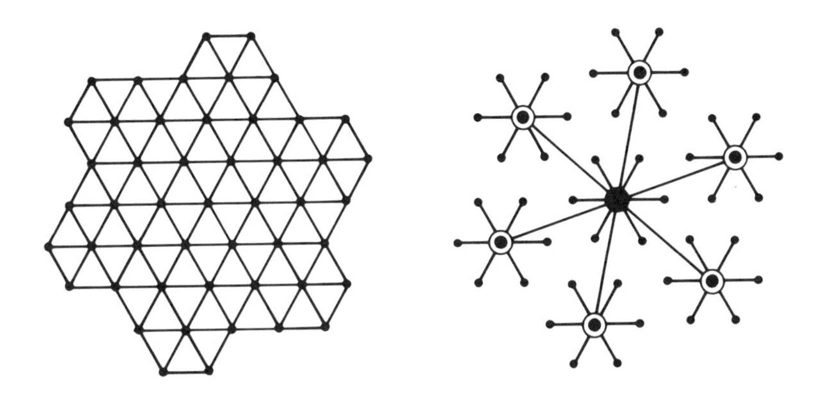

The routing problem then arises. How should the serving areas with their respective local exchanges be interconnected? From our previous discussion we know that two extremes are offered, mesh and star. We are also probably aware that as the number of exchanges involved increases, full mesh becomes very complicated and is not cost-effective. Certainly it is not as cost-effective as a simple hiearchical network of two or three levels permitting high-usage (HU) connections between selected nodes. For instance, given the hexagon formation in the preceding diagram, a full mesh or two-level star network can be derived, as shown in the diagram that follows.

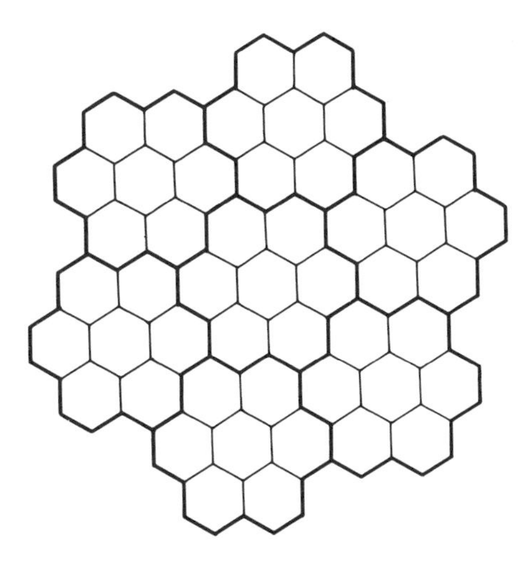

From the routing pattern in the preceding diagram it should be noted that fan outs of 6 and 8 allow symmetry, whereas fan outs of 5 and 7 lead to inequalities. A fan out of 4 is usually too small for economic routing.

It will be appreciated that some of the foregoing assumptions are rarely found in any real telecommunications environment. Uniform telephone density was one real assumption, and the implication of uniform traffic flow was another. Serving areas are not uniform geometric figures, exchanges seldom may be placed at serving-area centers, and routing will end up as a mix of star and mesh. Small local areas serving 5 to 15 exchanges or even more may well be fully mesh connected. Some cities are connected in full mesh with over 50 exchanges. But as telephone growth continues, tandem routing will become the economic alternative.

5 EXCHANGE LOCATION

A fairly simple, straightforward method for determination of the theoretical optimum exchange location is described in Ref. 1, Chapter 6. Basically, the method determines the center of subscriber density much in the same way the center of gravity would be calculated. In fact, other publications call it the center-of-gravity method.

Using a map to scale, a defined area is divided into small squares of 100 m to 500 m on a side. One guide for determination of side length would be to use a standard length of the side of a standard city block in the serving area of interest. The next step is to write in the total number of subscribers in each of the blocks. This total is the sum of three figures: (1) existing subscribers, (2) waiting list, and (3) forecast of subscribers for 15 or 20 years into the future. It follows that the squares used for this calculation should coincide with the squares used in the local forecast. The third step is to trace two lines over the subscriber area. One is a horizontal line that has approximately the same number of total subscribers above the line as below. The second is a vertical line where the number of subscribers to the left of the line is the same as that to the right. The point of intersection of these two lines is the theoretical optimum center or exchange location. A sample of this method is shown in Figure 2.3, where S_1 is the sum of subscribers across a single line and C_s is the cumulative sum.

Now that the ideal location is known, where will the real optimum location be? This will depend considerably on secondary parameters such as availability of buildings and land; existing and potential cable or feeder runs; the so-called trunk pull; and layout of streets, roads, and highways. "Trunk pull" refers to the tendency to place a new exchange near the one or several other exchanges with which it will be interconnected by trunks (junctions). Of course, this situation occurs on the fringes of urban areas where a new exchange location will tend to be placed nearer to the more populated area,

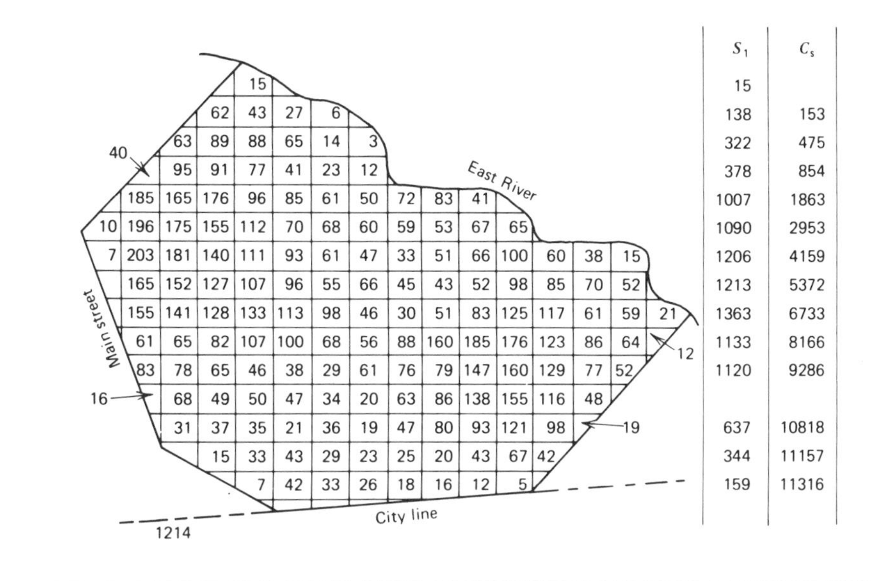

S_2	17	1104	1264	1211	1060	885	615	489	556	722	927	1072	770	399	242	33
C_s		1121	3195	3546	4686	5491	6106	6595	7151	7873	8800	9872	10642	11041	11283	11316

Figure 2.3 Sample wire centering exercise using "center-of-gravity" method.

thereby tending to shorten trunk routes. This is illustrated in Figure 2.1 and discussed in more detail below.

The preceding discussion assumed a bounded exchange area; in other words, the exchange area boundaries were known. Assume now that exchange locations are known and that the boundaries are to be determined. What follows is also valid for redistributing an entire area and cutting it up into serving areas. A great deal of this chapter has dealt with subscriber loop length limits. Thus an outer boundary will be the signaling limits of loops as described previously. The optimum cost-benefit trade-off is found when all or nearly all loops in a serving area remain nonconditioned and of small diameter, say, 26 gauge. We note that with H-66 loading the outer boundary will be just under 5 km in this case. It also would be desirable to have a hexagonal area if possible. In practice, however, natural boundaries may well be the most likely real boundaries of a serving area. "Main street," "East River," and "City line" in Figure 2.3 illustrate this point. In fact, these boundaries may set the limits such that they may be considerably greater or less than the maximum signaling (supervisory) limits suggested earlier. Of course, there are two types of serving areas where the argument does not hold. These are rural areas and densely populated urban areas. For the rural areas we can imagine very large serving areas and, for urban areas of dense

TABLE 2.10 Resistance Limits for Several Types of Exchange

Exchange Type	Resistance limit
No. 1. Step-by-step (USA)	1300 Ω
No. 1 Crossbar (USA)	1300 Ω
No. 5 Crossbar (USA)	1520 Ω
ESS (USA)	2000 Ω
Pentaconta (ITT) (crossbar)	1250 Ω
Rotary (ITT) (Europe)	1200 Ω
Metaconta (local) (ITT)	2000 Ω
Pentaconta 2000 (ITT)	1250 Ω
DMS-100	1900 Ω (4000 Ω)
Other digital local exchanges	2000–4000 Ω (Ref. 21)

population, considerably smaller serving areas than those set out with maximum supervisory signaling limits.

To determine boundaries of serving areas when dealing with an exchange that is already installed and a new exchange, we could use the so-called ratio technique. Again, we use signaling (supervisory) limits as the basis. As we are aware, these limits are basically determined by the type of exchange and copper wire gauge utilized for subscriber loops. Table 2.10 gives resistance limits for several of the more common telephone exchanges found in practice.

The ratio method is as follows. Given an existing exchange A and a new exchange B that will be established on a cable route from A, assume that A is a step-by-step exchange and that B is a Pentaconta exchange. The distance from A to B along a cable route can be computed by equating distance to the sum of the resistances of exchanges A and B. Use 26-gauge wire in this case, and take the resistance from Table 2.10. From Table 2.8 using H-66 loading, resistance can be equated to 273 Ω/km. Sum the resistance limits of exchanges $(A + B)$:

$$A + B = 1250 + 1300 = 2550\ \Omega$$

Then divide by 273:

$$2550/273 \simeq 9.34 \text{ km (distance } A \text{ to } B)$$

With no other factors influencing the decision, the boundary would be established along the feeder (cable) route (distance from A):

$$D_A = 1300 \times 9.34/(1250 + 1300)$$

or $\simeq$ 4.74 km from A. This exercise is shown diagrammatically in Figure 2.4.

Continue the exercise and examine the exchange serving area at A. Assume that the area is a square with A at its center. Allowing for *non*-crow-fly feeder routes, we can assume that the square has 8 km on a side, as shown in the following diagram:

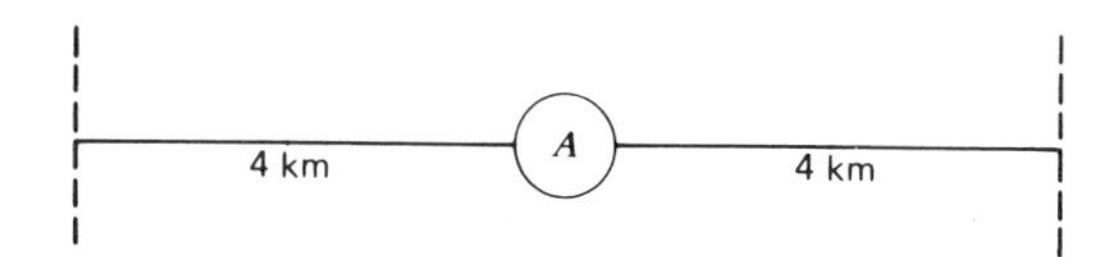

or 64 km^2. If the largest exchange wanted had 100,000 subscribers and the smallest has 10,000 at the end of the forecast period (15 to 20 years), then subscriber density would be 10,000/64 to 100,000/64 or 156 subscribers per km^2 to 1560 subscribers per km^2.

When serving areas exceed 100,000 lines at the end of a forecast period in densely populated urban areas, breaking up of the areas into smaller ones with exchanges of smaller capacity should be considered. One problem with large conventional exchanges is size; another is cable entry and mainframe size. The latter is more evident with new SPC [electronic switching system (ESS)] exchanges in that cable entry and mainframe is much larger than the exchange itself. However, the really major problem is reliability or survivability. If a very large exchange is knocked out as a result of fire, explosion, sabotage, or other disaster, it is much more catastrophic than the loss of a smaller single exchange.

Another consideration in exchange location is what is called "trunk pull" discussed briefly above. This is a secondary factor in exchange placement and refers to the tendency, in certain circumstances, to shift a proposed exchange location to shorten trunks (junctions). Trunk pull (see Fig. 2.1) becomes a significant factor only where population fringe areas around urban centers and trunks extend toward the city center and the exchanges in question may well be connected in full mesh in their respective local area. Some possible saving may be accrued by shifting the exchange more toward the center of population, thereby shortening trunks at the expense of lengthening subscriber loops in the direction of more sparse population, even with the implication of conditioning on well over 10% of the loops. The best way to determine if a

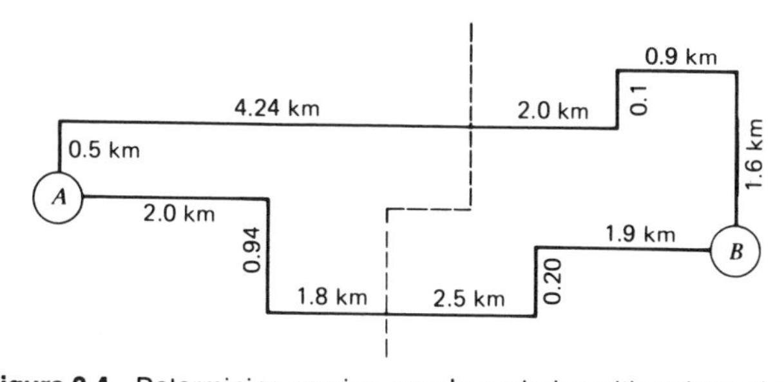

Figure 2.4 Determining serving area boundaries with ratio method.

shift is worthwhile is to carry out an economic study comparing costs in PWAC (present worth of annual charges) [18, 19], comparing the proposed exchange as located by the center-of-gravity method to the cost of the shifted exchange. Basic factors in the comparison are in the cost of subscriber plant, trunk plant, plant construction, and cost of land (or buildings). The various methods of handling the sparsely populated sections of the fringe serving area in question should also be considered. For instance, we might find that in those areas, much like rural areas, there may be a tendency for people to bunch up in little villages or other small centers of population. This situation is particularly true in Europe. In this case the system designer may find considerable savings in telephone-plant costs by resorting to the use of satellite exchanges or concentrators. Trunk connections between satellite or concentrator and main exchange may be made by using carrier techniques. Concentrator

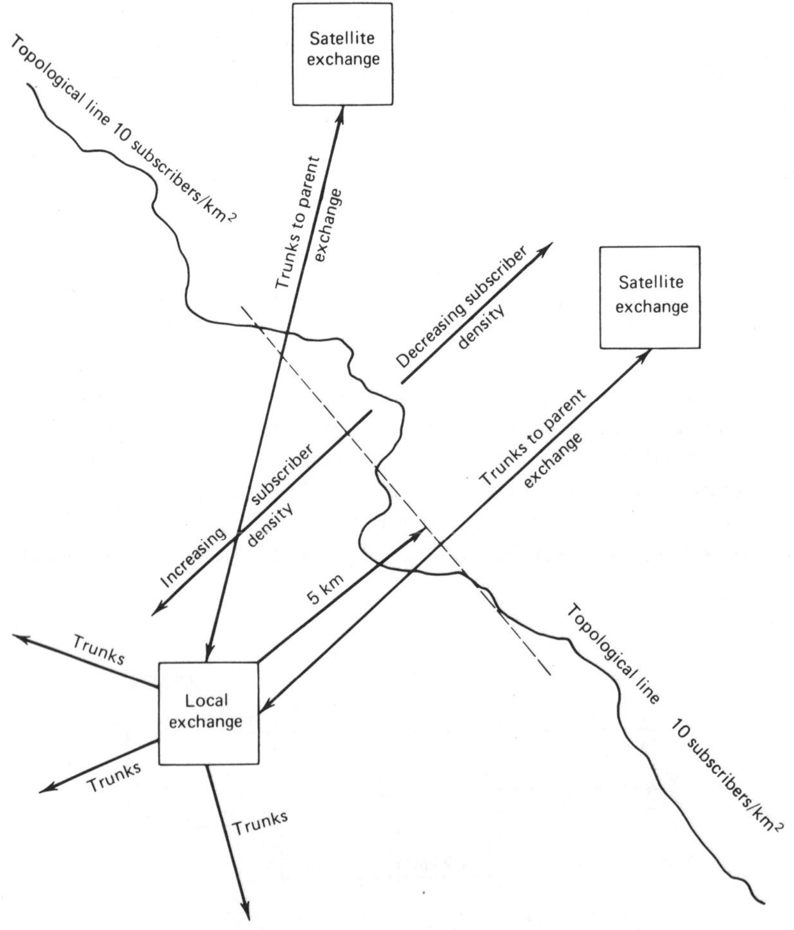

Figure 2.5 Fringe-area considerations.

and satellite exchanges are discussed in Chapter 3 and carrier techniques in Chapter 5.

A typical fringe-area situation is shown in Figure 2.5. The "bunching" and the other possibility, "thinning out," are shown in the figure. "Thinning out" is just the population density per unit area decreases as we proceed from an urban center to the countryside. A topological line of population density of 10 inhabitants per square kilometer (26 inhabitants per square mile) is a fair guideline for separation of the rural part of the fringe area from the urban–suburban part. Of course, in the latter part we would have to resort to widespread conditioning, to the use of concentrator–satellites, or to subscriber carrier or subscriber pulse-code modulation (PCM) (Chapter 9).

6 UNIGAUGE DESIGN

The administration of a serving area is often quite costly. For one thing administration costs can be reduced by ensuring that all subscribers have the same type of telephone set. Cable-pair assignment is another factor. Following resistance design or transmission design approaches to subscriber plant engineering will bring about a "mix" of various gauges of wire in loops. We also want to keep wire gauge as small as possible to minimize the investment in copper. If loading and other conditioning are to be used, "systemization" of their use will also help administration. "Unigauge design" is one approach to this systematization, and it can be applied in the urban, suburban, and fringe areas discussed earlier, as well as in many rural applications. Loops up to 52 kft (17.3 km) in length can be handled.

A typical layout of a subscriber plant based on unigauge design is shown in Figure 2.6. The example is taken from the Bell System (USA) [3]. It can be seen in Figure 2.6 that subscribers within 15,000 ft (5000 m) of the switch are connected over loops made up of 26-gauge nonloaded cable fed within the standard −48-V battery. Loop connection at the switch is conventional. Unigauge was developed in the United States, where 80% of all subscribers are within this distance of their serving exchange. Loops 15,000 ft to 30,000 ft long (5000 m to 10,000 m) are called *unigauge loops*. Subscribers in the range of 15,000 ft to 24,000 ft (5000 m to 8000 m) from the switch are connected by 26 gauge nonloaded cable as well but require a range extender to provide sufficient voltage for supervision and signaling. Figure 2.6 shows a 72-V range extender equipped with an amplifier that gives a midband gain of 5 dB. The output of the amplifier "emphasizes" the higher frequencies. This offsets that additional loss suffered at the higher frequencies of the voice channel on the long nonloaded loops. To extend the loops to 30,000 ft (10,000 m), 88-mH loading coils are added at the 15,000- and 21,000-ft points (5000- and 7000-m points).

For long loops, more than 15,000 ft (5000 m), the range extender–amplifier combinations are not connected on a line-for-line basis. It is standard practice

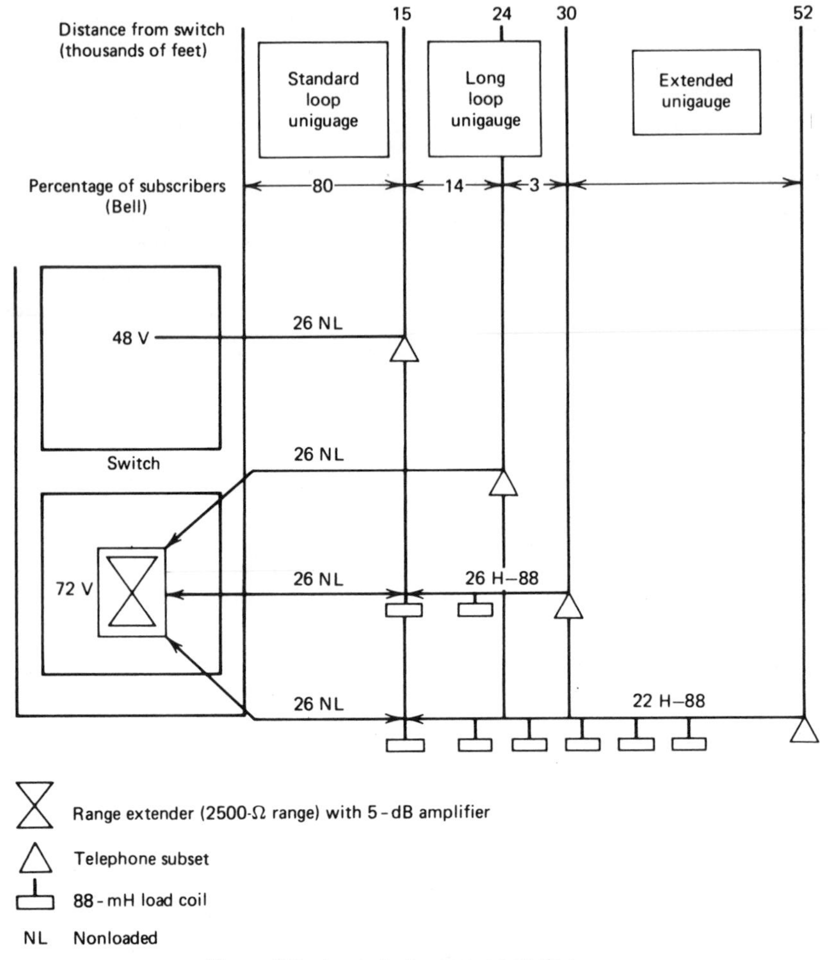

Figure 2.6 Layout of unigauge design.

to equip four or five subscriber loops with only one range extender–amplifier, which is used on a shared basis. When the subscriber goes "off hook" on a long unigauge loop, a line is seized and the range extender is switched in. This concentration is another point in favor of unigauge because of the economics involved. It should also be noted that the switch is facing into long (15,000-ft) nonloaded sections providing a fairly uniform impedance for all conditions when an amplifier is switched in. This is an important factor with regard to stability, as we discuss later.

Loops measuring more than 30,000 ft (10,000 m) may also use unigauge principle and are often referred to as "extended unigauge." Such a loop is equipped with 26-gauge nonloaded cable from the switch out to the 15,000-ft point. Beyond 15,000 ft (5000 m), 22-gauge cable is used with H-88 loading.

As with all loops measuring more than 15,000 ft in length, following the unigauge principle, a range extender–amplifier is switched in when the loop is in use. The loop length for this combination is up to 52,000 ft. Loops longer than 52,000 ft (17.3 km) may also be installed by using a gauge with a diameter larger than 22.

Another possibility is to replace switch line-relays with ones that are sensitive to up to 2500 Ω of loop resistance. Such a modification is done on long loops only. The 72 V supplied by the range extender is for pulsing. Ringing voltage (to ring the distant telephone) is superimposed on the line only when the subscriber's subset is in the "on-hook" condition. Besides the notable savings in the expenditure on copper, unigauge displays some small improvements in transmission characteristics over older design methods of subscriber loops:

- Unigauge has a slightly lower average loss, when we look at a statistical distribution of subscribers.
- There is 15-dB average return loss* on the switch side of an amplifier, compared with an average of 11 dB for older design methods.

7 OTHER LOOP DESIGN TECHNIQUES

7.1 Minigauge Techniques

Fine-gauge and "minigauge" techniques essentially are refinements of the unigauge concept. In each case the principle object is to reduce the amount of copper in the subscriber plant. Obviously, one method is to use still smaller gauge pairs on shorter loops. For instance, the use of gauges as small as 32 is being considered. Another approach is to use aluminum as the conductor. When aluminum is used, a handy rule of thumb to follow is that the ohmic and attenuation losses of aluminum may be equated to copper in that aluminum wire should always be the next "standard gauge" larger than its copper counterpart if copper were used. Some of the more common gauges are compared in the table that follows.

Copper	Aluminum
19	17
22	20
24	22
26	24

*Return loss is discussed in Chapter 5.

Aluminum has some drawbacks as well. The major ones are summed up as follows:

- It should not be used on the first 500 yd (m) of cable where the cable has a larger diameter (and here we mean more loops before branching).
- It is more difficult to splice than copper.
- It is more brittle.
- Because the equivalent conductor is larger than its copper counterpart, an equivalent aluminum cable with the same conductivity–loss characteristics will have a smaller pair count in the same sheath.

7.2 New Loop Design Plans for US Regional Bell Operating Companies*

Before 1980 US Bell operating companies designed subscriber loops in accordance with one of three loop design plans: (1) resistance design (RD), 96%; (2) unigauge design (UG), 1%; (3) long route design (LRD), 3%. (Percentages show approximate percentage of total plant for each plan.) The objective, of course, is to design subscriber loops on a global basis rather than on an individual basis, which is extremely costly. If the system engineers fully comply with these design rules, then no loop will exceed switch signaling limitations (resistance) and there will be an adequate distribution of transmission losses.

A 1980 survey (ref. 20) studied loss design in a population of loops having an average length greater than the mean length of the general population. It was found that approximately 4% of the measured losses were greater than 8.5 dB, and 2% exceeded 10 dB. With the design rules in effect at the time, losses exceeding 8.5 dB were expected. Under the new, revised resistance design (RRD) rules, no properly designed loop should display a loss greater than 8.5 dB at 1000 Hz reference frequency.

7.2.1 Previous Rules

7.2.1.1 Resistance Design (RD)

Resistance: 0–1300 Ω (includes only resistance of cable and load coils).

Load coils: H-88 beyond 18 kft [not including bridged tap (BT)].

End sections (ES) and bridged tap (BT): Nonloaded BT = 6 kft (max); loaded ES + BT = 15 kft (max), 3 kft (min), 12 kft recommended.

Transmission limitations: None.

Cable gauges: Any combination of 19–26 gauge.

*The source for this section is Ref. 20, Section 7.

7.2.1.2 Long Route Design (LRD) (Rural Areas)

Loop resistance: 1301–3000 Ω (including only resistance of cable and load coils).

Load coils: Full H-88.

End sections (ES) and bridged tap (BT): ES + BT = 12 kft.

Transmission limitations: For loop resistances > 1600 Ω, loop gain required.

Cable gauges: Any combination of 19–26 gauge.

7.2.1.3 Unigauge Design. See Section 6 of this chapter.

7.2.2 Present Loop Design Rules

7.2.2.1 Revised Resistance Design (RRD)

Loop resistance: 0–18 kft, 1300 Ω (max) (includes resistance of cable and load coils only); 18–24 kft, 1500 Ω (max) (includes resistance of cable and load coils only; > 24 kft, use digital loop carrier (DLC).

Load coils: Full H-88 for loops longer than 18 kft.

End sections (ES) and bridged tap (BT): Nonloaded total cable + BT = 18 kft, maximum BT = 6 kft. Loaded ES + BT = 3–12 kft.

Transmission limitations: None.

Cable gauges: Two gauge combinations preferred (22-, 24-, 26- gauge).

7.2.2.2 Concentrated Range Extension with Gain (CREG)

Loop resistance: 0–2800 Ω (includes only resistance of cable and load coils).

Load coils: Full H-88 > 15 kft.

End sections (ES) and bridged tap (BT): Nonloaded cable + BT = 15 kft, maximum BT = 6 kft. Loaded ES + BT = 3–12 kft.

Transmission limitations: Gain range extension required for loop resistance > 1500 Ω.

Cable gauges: Two gauge combinations preferred (22-, 24-, 26-gauge).

7.2.2.3 Modified Long Route Design (MLRD)

Loop resistance: 1501–2800 Ω.

Load coils: Full H-88.

End sections (ES) and bridged tap (BT) (max): ES + BT = 3–12 kft.

Transmission limitations: Range extension with gain required for loop resistances greater than 1500 Ω.

Cable gauges: Two gauge combinations preferred (22-, 24-, 26-gauge).

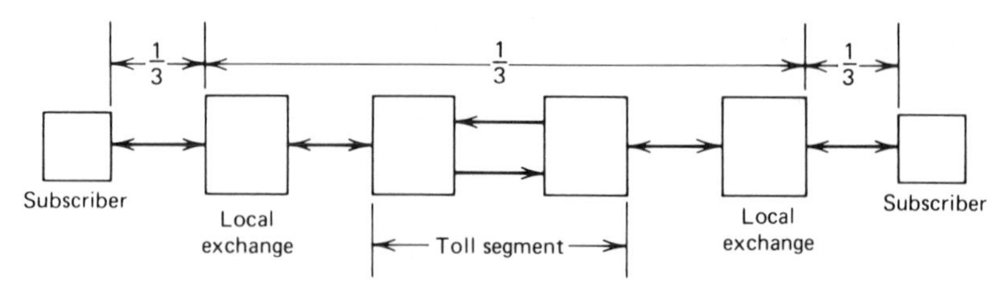

Figure 2.7 One conceptual thumb rule for network loss assignment.

8 DESIGN OF LOCAL AREA TRUNKS (JUNCTIONS)

Exchanges in a common local area are interconnected by trunks, called junctions in the United Kingdom. Depending on length and certain other economic factors, these trunks use voice-frequency (VF) transmission over wire pairs formed up in cables. In view of the relatively small number of such trunk circuits in comparison to the total number of subscriber lines* in the area, it is generally economical to minimize attenuation in this portion of the network.

One approach used by some telephone companies or administrations is to allot $\frac{1}{2}$ of the total end-to-end reference equivalent to each subscriber's loop and $\frac{1}{3}$ to the trunk network. Figure 2.7 illustrates this concept. For instance, if the transmission plan called for a 24-dB ORE, then $\frac{1}{3}$ of 24 dB, or 8 dB, would be assigned to the trunk plant. Of this we may assign 4 dB to the four-wire portion of the long distance (toll) network, leaving 4 dB for local VF trunks or 2 dB at each end. The example has been highly simplified, of course. For the toll-connecting trunks (e.g., those trunks connecting the local network to the toll network), if a good return loss cannot be maintained on all or nearly all connections, the losses on two-wire toll connecting trunks may have to be increased to reduce possibilities of echo and singing. Sometimes the range of loss for these two-wire circuits must be extended to 5 or 6 dB. It is just these circuits into which the four-wire toll network looks directly. Two-wire and four-wire circuits, echo and singing, are discussed in Chapter 5. Thus it can be seen that the approach to the design of VF trunks varies considerably from that used for subscriber loop design. Although we must ensure that signaling limits are not exceeded, the transmission limits will almost always be exceeded well before the signal limit. The tendency to use larger-diameter cable on long routes is also evident. If loading is to be used, the first load coil is installed at distance $D/2$, where D is the normal separation distance between load points. Take the case of H loading, for instance. The distance between load points is 1830 m (Table 2.7), but the first load coil from the exchange is placed at $D/2$,

*Because of the inherent concentration in local switches, approximately 1 trunk is allotted from 8 to 25 subscribers, depending on design.

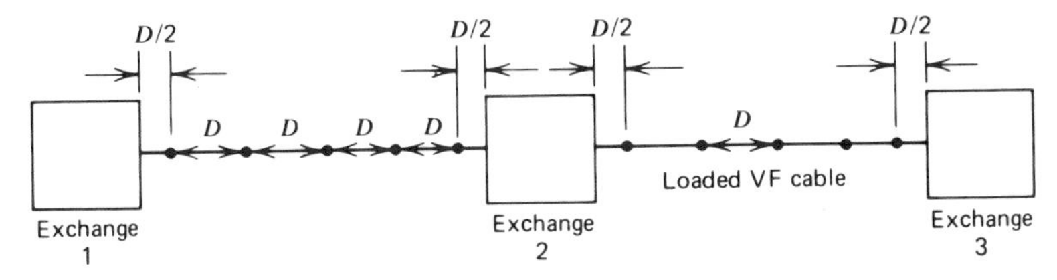

Figure 2.8 Loading of VF trunks (junctions).

or 915 m from the exchange. Then, if an exchange is bypassed, a fully-loaded section exists. This concept is illustrated in Figure 2.8.

Now consider this example. A loaded 500-pair VF trunk cable extends across town. A new switching center is to be installed along the route where 50 pairs are to be dropped and 50 pairs inserted. It would be desirable to establish the new switch midway between load points. At the switch 450 circuits will bypass the new switching center. Using this $D/2$ technique, these circuits will need no conditioning; they will be fully loaded sections (i.e., $D/2 + D/2 = 1D$, a fully loaded section). Meanwhile, the 50 circuits entering from each direction are terminated for switching and need conditioning, so each electrically resembles a fully loaded section. However, the physical distance from the switch out to the first load point is $D/2$ or, in the case of H loading, 915 m. To make the load distance electrically equivalent to 1830 m, line build-out (LBO) is used. This is done simply by adding capacitance to the line. Suppose the location of the new switching center was such that it was not midway between load points but some other fractional distance. For the section consisting of the shorter distance, LBO is used. For the other longer run, often a half load coil is installed at the switching center, and the LBO is added to trim up the remaining electrical distance.

By this time the reader has gathered what conditioning is in this context. For loaded trunks and loops, line build-out (LBO) is the action taken to change the electrical characteristics of the line by adding capacitance or inductance. It may be done at line midpoint where there is access, such as at a load-coil location. However, it is more commonly done at the switching center because of accessibility of cable pairs at the mainframe associated with the center.

9 VOICE-FREQUENCY REPEATERS

In telephone terminology *voice-frequency repeaters* imply the use of *uni*directional amplifiers at voice frequency on VF trunks. On a two-wire trunk we must resort to four-wire transmission techniques at the repeater. Two-wire and four-wire transmission are discussed in Chapter 5. Thus on a two-wire trunk two amplifiers must be used on each pair with a hybrid in and a hybrid out.

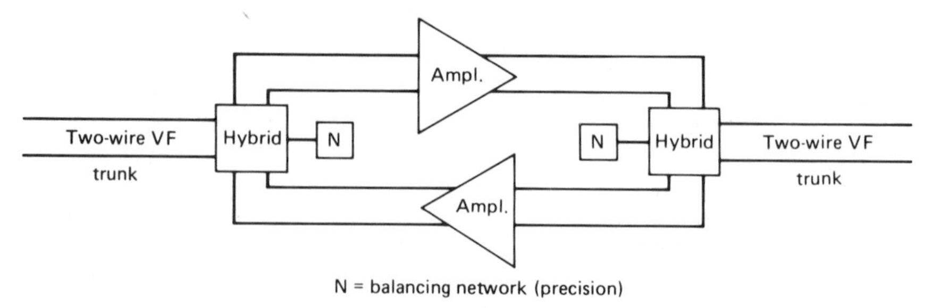

Figure 2.9 Simplified block diagram of VF repeater.

The hybrid converts the two-wire circuit to a four-wire circuit and vice versa. A simplified block diagram of a VF repeater is shown in Figure 2.9. The gain of a VF repeater can be run up as high as 20 or 25 dB, and originally they were used on 50-mi 19-gauge loaded cable in the long-distance (toll) plant. Today they are seldom found on long-distance circuits but do have application on local trunk circuits, where the gain requirements are considerably less. Trunks using VF repeaters have the repeater's gain adjusted to the equivalent loss of the circuit minus the 4-dB loss to provide the necessary singing margin (see Chapter 5). In practice, a repeater is installed at each end of the trunk circuit to simplify maintenance and power feeding. Gains may be as high as 6 to 8 dB. It can be appreciated that the application of VF repeaters would be on trunks where the losses were excessive, beyond those called for in the appropriate transmission plan, such as on long loops or long VF trunks.

Another repeater more commonly used on two-wire trunks is the negative-impedance repeater. This repeater can provide a gain as high as 12 dB, but 7 or 8 dB is more common in actual practice. The negative impedance repeater requires an LBO at each port and is a true two-way, two-wire repeater. The repeater action is based on regenerative feedback of two amplifiers. The advantage of negative impedance repeaters is that they are transparent to dc signaling. On the other hand, VF repeaters require a composite arrangement to pass the dc signaling. This consists of a transformer bypass.

10 TANDEM ROUTING

The local-area trunking scheme evolutionally has been mesh connection of exchanges, and in many areas of the world it remains full mesh. We said initially that mesh connection is desirable and viable for heavy traffic flows. As traffic flows reduce, going from one situation to another, the use of tandem routing in the local area becomes an interesting, economical alternative.

Further, it can be shown that a local trunk network can be optimized, under certain circumstances, with a mix of tandem, high-usage (overflow to alternative routes), and direct connection (mesh). We often refer to these three possibilities as THD. The system designer wishes to determine, on a particular trunk circuit, if it should be "tandem," "high-usage," or "direct" (T, H, or D). This determination is based on incremental cost of the trunk (junction), making the total network costs as low as possible. Such incremental cost can be stated:

$$B = c + (bl) \tag{2.5}$$

where c is the cost of switching equipment per circuit, b is the incremental costs for trunks per mile or kilometer, and l is the length of the trunk (or junction) circuit.

To carry out a THD decision for a particular trunk route, the input data required are the *offered* traffic between the local exchanges in question and the grade of service. We can now say that the THD decision is to a greater extent determined by the offered traffic A between the exchanges and the cost ratio ε, where

$$\varepsilon = \frac{B}{B_1 + B_3} \tag{2.6}$$

and where B is the cost for the direct route, with B_1 from exchange 1 to the proposed tandem and B_2 from the proposed tandem exchange to exchange 2, for incremental costs between direct and tandem routing. Of course, before starting such an exercise, the provisional tandem points must be known. Figure 2.10, which was taken from Ref. 4, may be helpful as a decision guide.

To approximate the number of high-usage (HU) trunk circuits required, the following formula may be used:

$$F(n, A) = AE(n, A) - E(n + 1, A) = \varepsilon\psi \tag{2.7}$$

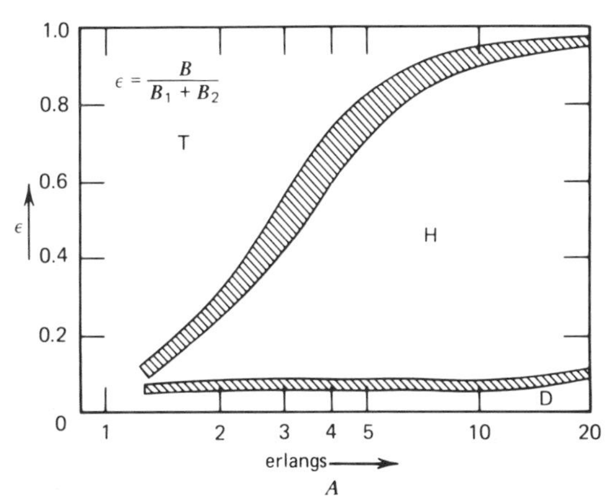

Figure 2.10 THD diagram (courtesy of the International Telecommunication Union–CCITT).

TABLE 2.11 Traffic Table: Full-Mesh Connection (120,000 Subscribers)

Number of Exchanges	Average Size of Exchange	Originated Traffic per Exchange (erlangs)	Average Traffic to Each Distant Exchange (erlangs)
10	12,000	600	60
12	10,000	500	42
15	8000	400	27
20	6000	300	15
30	4000	200	6.7
40	3000	150	3.8

where E is the grade of service, n is the number of high-usage circuits, A is the offered traffic between exchanges, $F(n, A)$ is the "improvement" function [i.e., increase in traffic carried on high-usage trunk group on increase of number of these trunk circuits from n to $(n + 1)$], ε is the cost ratio (as previously), and ψ is the efficiency of incremental trunks [marginal utilization (of magnitude 0.6 to 0.8)]. However, a still better approximation will result if the following formula is used:

$$F(n, A) = \varepsilon\left[1 - 0.3(1 - \varepsilon^2)\right] \tag{2.8}$$

The following exercise emphasizes several practical points on the application of tandem routing in the local area. Assume that there are 120,000 subscribers in a certain local area. If we allow a BH calling rate of 0.5 erlangs per subscriber (see Table 2.9) and assume full mesh connection, we can assemble Table 2.11. On the basis of Table 2.11 it would not be reasonable to expect any appreciable local area trunk economy through tandem working with less than 30 to 40 exchanges in the area. Of course, Table 2.11 assumes unity of community of interest with the more distant exchanges so that a few attractive tandem routings might exist in the 20-to-30-exchange range. Areas smaller than those in our sample will naturally reduce the number of exchanges at which tandem working is viable, and a larger area and higher calling rate will increase this number. The results of any study to determine feasibility of using tandem exchange is very sensitive to the number of exchanges in the local area and considerably less sensitive to the size of the area and calling rate. Hence we can see that the economy of tandem routing is least in areas of dense subscriber population, where exchanges can be placed without exceeding subscriber loop length limits and where the trunks will be short. Relatively sparsely populated areas are more favorable and are likely to show relatively low community-of-interest factors between distant points in the area.

TABLE 2.12 Sample Traffic Matrix[a]

From/to	Traffic (erlangs)					Toll	Total Orig.	Lines Working	Traffic per Line
	A	*B*	*C*	*D*	*E*				
A	21	20	65	2.5	2.5	1.5	54	9200	0.059
B	22	80	13	6.0	5.0	4.0	130	26000	0.050
C	5	11	7	2.5	2.5	1.0	29	7500	0.039
D	2	7	1	0.3	0.2	0.5	11	3000	0.035
E	2	5	2	0.2	0.3	0.5	10	2800	0.035
Toll	2	3	1	0.5	0.5	–	7	–	–
Total	54	126	30.5	12	11	7.5	241	48500	0.050
Total-line	0.059	0.048	0.041	0.040	0.037	0.0015	–	–	–

[a]For 8-year forecast period.

11 DIMENSIONING OF TRUNKS

A primary effort for the system engineer in the design of the local trunk network is the dimensioning of the trunks of that network. Here we simply wish to establish the economic optimum number of trunk circuits between exchanges X and Y. If we are given the traffic (in erlangs) between the two exchanges and the grade of service, we can assign the number of trunk circuits between X and Y. As discussed in Chapter 1, it is assumed that all traffic values are BH values. Once given this input information, the erlang value, the number of trunk circuits, can be derived from Table 1.1 (no overflow assumed).

It can be appreciated by the reader that the trunk traffic intensities used in the design of trunk routes have allowed for growth, that is, the increase of traffic from the present with the passage of time. This increase is attributable to several factors: (1) the increase in the number of telephones in the area that generate traffic, (2) the probable increase in telephone usage, and (3) the possibility of a change in character of the area in question, such as from rural to suburban or from residential to commercial. Thus the designer must use properly forecast future traffic values. These values in practice are for the forecast period 5 to 8 years in the future. The art of arriving at these figures is called *forecasting*. Present traffic values should always be available as a base or point of departure.

Suppose we use a sample local area with five exchanges: A, B, C, D, and E. For an 8-year forecast period we could possibly come up with the traffic matrix like that shown in Table 2.12. Thus from the traffic matrix in the direction exchange C to exchange B there is a traffic intensity of 11 erlangs. Applying the 11 erlangs to Table 1.1, we see that 23 circuits would be required. For the distance B to C, the traffic intensity is 13 erlangs, and again from

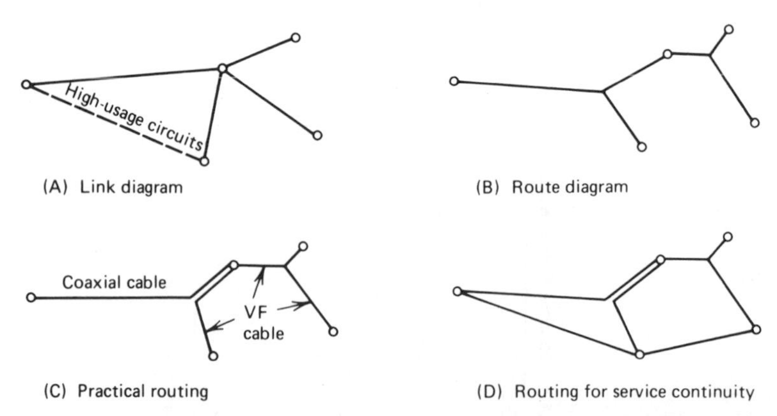

Figure 2.11 Routing diagrams and practical routing (VF = voice frequency): (A) link diagram; (B) route diagram; (C) practical routing; (D) routing for service continuity.

Table 1.1, 26 circuits would be required. These circuit figures suppose a grade of service of $p = 0.001$. For a grade of service of $p = 0.01$, 19 and 22 circuits, respectively, would be required.

The preceding discussion of routing implied an on-paper routing that would probably vary in the practice of actual cable lays and facility drawings to something quite different. In simplified terms, these differences are shown in Fig. 2.11.

12 COMMUNITY OF INTEREST

We have referred to the community-of-interest concept in passing. This is a method used as an aid to estimate calling rate and traffic distribution for a new exchange and its connecting trunks. The community-of-interest factor K can be defined as follows:

$$\text{Traffic } A - B = K \times \frac{\text{Traffic originating at } A \times \text{Traffic originating at } B}{\text{Total originating traffic in area}} \tag{2.9}$$

If all subscribers are equally likely to call all others, the proportion results:

$$\frac{\text{Traffic originating at } B}{\text{Traffic originating in area}} = \frac{\text{Traffic terminating at } B}{\text{Traffic terminating in area}} \tag{2.10}$$

This is the expected proportion of all A traffic that is directed to B. In this case $K = 1$, corresponding to equal community of interest between all subscribers. The condition of $K > 1$ or $K < 1$ indicates a greater or lesser interest than average between exchanges A and B. The factor K is affected by the type of area, whether residential or business, as well as the distances between exchanges. For instance, in metropolitan areas K may range from 4 for calls originating and terminating on the same exchange to 0.25 for calls on opposite sides of the city or metropolitan area.

The K factor is a useful reference for the installation of a new exchange where the serving area of the new exchange will be cut out from serving areas of other exchanges. The community-of-interest factors may then be taken from traffic data from the old exchanges, and the values averaged and then applied to the new exchange. The same principle may be followed for exchange extensions in a multiexchange area.

REVIEW QUESTIONS

1. Local networks can be defined in a number of ways. Give the definition of a local network that is provided in the text.
2. Considering both quality of service and economy, give at least five constraints or planning factors that go into the design of a local telephone network.
3. What are the two limiting performance factors in the design of a subscriber loop that constrain length?
4. Reference equivalent is a measure of transmission quality. What singular parameter does it measure?
5. What are the three elements (measured in decibels) that, when summed, constitute the value of ORE?
6. What important parameter contained in a national transmission plan is vital in calculating ORE?
7. If we assume that a modern local exchange is designed for a maximum of 2000 Ω of loop resistance, not including the resistance of the end instrument (i.e., the telephone subset), what is the maximum loop length of 26-gauge copper wire that can be used? Assume 83 Ω per 1000 ft of loop.
8. The National Transmission Plan allows a 7-dB maximum subscriber loop loss. Assume a 0.51-dB loss per 1000 feet of loop for a 26-gauge wire pair. What is the maximum subscriber loop length considering only loss?
9. What is the maximum subscriber loop length permissible using inputs from questions 7 and 8?
10. The exchange involved in question 9 must serve some rural subscribers who are over the maximum loop length. Name at least four expedients that can be used to serve these subscribers. Differentiate whether these expedients have an impact on resistance limit or transmission limit.
11. A subscriber connected to a local exchange commonly (not always) is identified by a seven-digit number made up of a three-digit prefix followed by a four-digit number. What does the three-digit prefix identify? What is the four-digit sequence called?
12. How many different individual subscribers can be served by a four-digit number assuming no blocked numbers.
13. What advantages accrue from implementing unigauge design?
14. Using only subscriber density, describe a method of idealized exchange placement.

15. Describe LBO and how it is used on loaded wire-pair trunks?

16. If D is the separation of trunk load coils, why is the first load coil outward from an exchange at $D/2$?

17. Name two types of VF repeaters.

18. In network design, what is the THD decision? What are the three input parameters necessary for this decision?

19. Define *community of interest.*

20. Describe a traffic matrix and what it is used for.

REFERENCES

1. *Local Telephone Networks*, International Telecommunications Union, Geneva, 1968.
2. International Telephone and Telegraph Corporation, *Reference Data for Radio Engineers*, 6th ed., Howard W. Sams, Indianapolis, 1976.
3. P. A. Gresh, L. Howson, A. F. Lowe, and A. Zarouni, "A Unigauge Design Concept for Telephone Customer Loop Plant," *IEEE Com. Tech. J.*, vol. com. 16, No. 2 (April 1968).
4. *National Networks for the Automatic Service*, International Telecommunications Union, Geneva, 1968.
5. *Networks*, Telecommunications Planning Documents, ITT Laboratories (Spain), Madrid, 1973.
6. *CCITT Red Books*, *Malaga-Torremolinos*, 1984, in particular Vols. III and V.
7. *Outside Plant*, Telecommunication Planning Documents, ITT Laboratories (Spain), Madrid, 1973.
8. Roger L. Freeman, *Telecommunication Transmission Handbook*, 2nd ed. Wiley, New York, 1981.
9. Y. Rapp, "Algunos Puntos de Vista Económicos para el Planeamiento a Largo Plazo de la Red Telefónica," L. M. Ericsson Stockholm, 1964.
10. *Placement of Exchanges in Urban Areas—Computer Program*, ITT Laboratories (Spain), Madrid, 1974.
11. J. C. Emerson, *Local Area Planning*, Telecommunications Planning Symposium [ITT Laboratories (Spain)], Boksburg, South Africa, 1972.
12. L. Alvarez Mazo and P. H. Williams, *Influence of Different Factors on the Optimum Size of Local Exchanges* [ITT Laboratories (Spain)], Boksburg, South Africa, 1972.
13. J. C. Emerson, *Factors Affecting the Use of Tandem Exchanges in the Local Area* [ITT Laboratories (Spain)], Boksburg, South Africa, 1972.
14. IEEE ComSoc, *The International Symposium on Subscriber Loops and Services*, Atlanta, Ga., 1977.
15. IEEE ComSoc, *Second International Symposium of Subscriber Loops and Services*, London, 1976.

16. J. E. Flood, *Telecommunications Networks*, IEE Telecommunications Series 1, Peter Peregrinus, London, 1975.
17. Y. Rapp, "Planning of Exchange Locations and Boundaries in Multi-Exchange Networks," Ericsson Tech., 1962, Vol. 18, p. 94.
18. "Telecommunications Planning," ITT Laboratories (Spain), Madrid, 1974.
19. O. Smidt, "Engineering Economics," Telephony Publishing Co., Chicago, 1970.
20. "Notes on The BOC Intra-Lata Networks—1986," Bell Communications Research Tech. Rep. TR-NPL-000275, issue 1, Bellcore, Morristown, N.J., April 1986.
21. *Electronic Switching: Digital Central Office Systems of the World*, Amos Joel, ed., IEEE Press, New York, 1982.
22. "Transmission Systems for Communications," Fifth ed., Bell Telephone Laboratories, 1982.

3

CONVENTIONAL SWITCHING TECHNIQUES IN TELEPHONY

1 SWITCHING IN THE TELEPHONE NETWORK

A network of telephones consists of pathways connecting switching nodes so that each telephone in the network can connect with any other telephone for which the network provides service. Today there are hundreds of millions of telephones in the world, and nearly each and every one can communicate with any other one. Chapter 1 discussed the two basic technologies in the engineering of a telephone network, transmission and switching. Transmission allows any two subscribers in the network to be heard satisfactorily. Switching permits the network to be built economically by concentration of transmission facilities. These facilities are the pathways (trunks) connecting the switching nodes.

Switching is a complex subject. To do justice to it, several volumes could be written. The intention of this chapter is to give an overview and appreciation of telephone switching. Such important factors as functional description, desirable features, trends in technology, and operational requirements are discussed.

Switching establishes a path between two specified terminals, which we call *subscribers* in *telephony*. The term *subscriber* implies a public telephone network. There is, however, no reason why these same system criteria cannot be used on private or quasipublic networks. Likewise, there is no reason why that network cannot be used to carry information other than speech telephony. In fact, in later chapters these "other" applications are discussed, as well as modifications in design and features specific for special needs.

A switch sets up a communication path on demand and takes it down when the path is no longer needed. It performs logical operations to establish the path and automatically charges the subscriber for usage. A commercial switching system satisfies, in broad terms, the following user requirements:

1. Each user has need for the capability of communicating with any other user.
2. The speed of connection is not critical, but the connection time should be relatively small compared to holding time or conversation time.
3. The grade of service, or the probability of completion of a call, is also not critical but should be high. Minimum acceptable percentage of completed calls during the BH may average as low as 95%, although the general grade of service goal for the system should be 99%* (equivalent to $p = 0.01$).
4. The user expects and assumes conversation privacy but usually does not specifically request it, nor, except in special cases, can it be guaranteed.
5. The primary mode of communication for most users will be voice (or the voice channel).
6. The system must be available to the user at any time the user may wish to use it.

1.1 Some Historical Background

The earliest telephones required no switching because they were installed on a dedicated-line basis. It quickly became evident that such a one-on-one arrangement was inefficient at best to meet growing economic and social needs. A network began to evolve in which telephones were connected to a central location where an operator could connect and disconnect telephone lines from any pair of subscribers to the service whose lines terminated in that location. In North America the location was called an *office* or *central office*. We can also see where the word *exchange* came from, for the operator changed or exchanged wire pairs.

Reference 22 describes the earliest "switching" as brass strips (mid-1870s) that looked like door hinges. These "switches" could interconnect eight or ten subscribers and were adapted from those used in the telegraph business.

The plug and jack technique with a flexible wire cord was invented by a Western Electric engineer in Chicago in 1878. An operator could interconnect 50 or more subscribers from one position. Such a technique continued to be used for some 50 years and still is used today on some older manual PBXs.

The next major development occurred in a curious way in Kansas City. An undertaker, Almon B. Strowger, had his business being compromised by the local telephone operator. Some say that she had a boyfriend who was a competing undertaker. Apparently she listened in on telephone calls from

*See CCITT Rec. Q.95; $p = 0.01$ per link on an international connection.

prospective customers and tipped off her boyfriend, who then scooped up the corpse business that ordinarily would have been Strowger's. Instead of complaining, Strowger set about inventing an automatic switchboard that would eliminate the intervention of the operator, thus ensuring his fair share of the undertaking business. The first Strowger switch went into commercial operation in 1892. This became the step-by-step switch, called the Strowger switch in the United Kingdom. Many step-by-step switches are still in operation today.

There followed the fully mechanical panel switch in the 1920s, the last of which was finally taken out of service in 1982. The No. 1 crossbar design was proposed in 1913, but the first crossbar switch did not go into operation until 1938. Crossbar switch installation peaked about 1983.

The first transistorized electronic logic switch was introduced for field trials in 1960 (No. 1 ESS) and the first digital stored-program control (SPC) switch in 1976 (4ESS). The second generation TDM-SPC switch, 5ESS, went into operation in 1982 [22].

2 NUMBERING: ONE BASIS OF SWITCHING

A telephone subscriber looking into a telephone network sees a repeatedly branching tree of links. At each branch point there are multiple choices. Assume that our calling subscriber wishes to contact one particular distant subscriber. To reach that subscriber, a connection is built up utilizing one choice at each branch point. Of course, some choices lead to the desired end point, and others lead away from it. Alternative paths are also presented. A call is directed through this maze, which we call a *telephone network*, by a telephone *number*. It is this number that activates the switch or switches at the "maze" branch points.

Actually, a telephone number performs two important operations: (1) it routes the call, and (2) it activates the necessary apparatus for proper call charging. Each telephone subscriber is assigned a distinct number, which is cross-referenced in the telephone directory with the subscriber's name and address and in the local serving exchange with a distinct subscriber line.

If a subscriber wishes to make a telephone call, she lifts her receiver "off hook" and awaits a dial tone that indicates readiness of her serving switch to receive instructions. These "instructions" are the number that the subscriber dials (or the buttons that she punches) giving the switch certain information necessary to route and charge this subscriber's call to the distant subscriber with whom she wishes to communicate.

A subscriber number is the number to be dialed or called to reach a subscriber in the same local (serving) area. Remember that our definition of a local "serving area" is the area served by a single switch (exchange).

If we had a switch with a capacity of 100 lines,

it could serve up to 100 subscribers and we could assign telephone numbers 00 through 99.

If we had a switch with a capacity of 1000 lines,

it could serve up to 1000 subscribers and we could assign telephone numbers 000 through 999.

If we had a switch with a capacity of 10,000 lines,

it could serve up to 10,000 subscribers and we could assign telephone numbers 0000 through 9999.

Thus the critical points occur where the number of subscribers reaches numbers such as 100, 1000, and 10,000.

In most present switching systems there is a top limit to the number of subscribers that can be served by one switching unit. Increase beyond this number is either impossible or uneconomical. A given switch unit is usually most economical when operating with the number of subscribers near the maximum of its design. However, it is necessary for practical purposes to hold some spare capacity in reserve. As we proceed in the discussion of switching, we consider exchanges with seven-digit subscriber numbers, such as

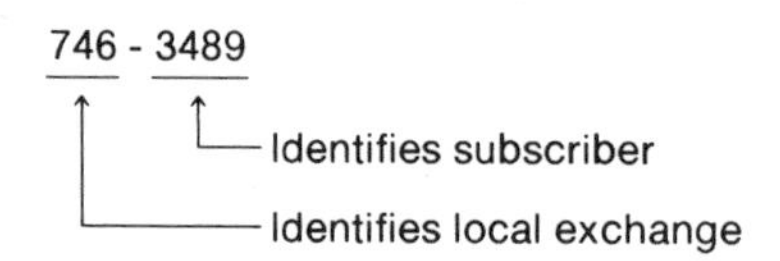

The subscriber is identified by the last four digits, permitting up to 10,000 subscribers, 0000 through 9999, allowing for no blocked numbers, such as

746-0000

The calling area has a capacity of 999 exchanges, again allowing for no blocked numbers, such as

000
911 (emergency number—USA)

Section 16 of this chapter presents a more detailed discussion of numbering.

3 CONCENTRATION

One key to switching and network design is concentration. A local switching exchange concentrates traffic. This concept is often depicted as shown in the following diagram:

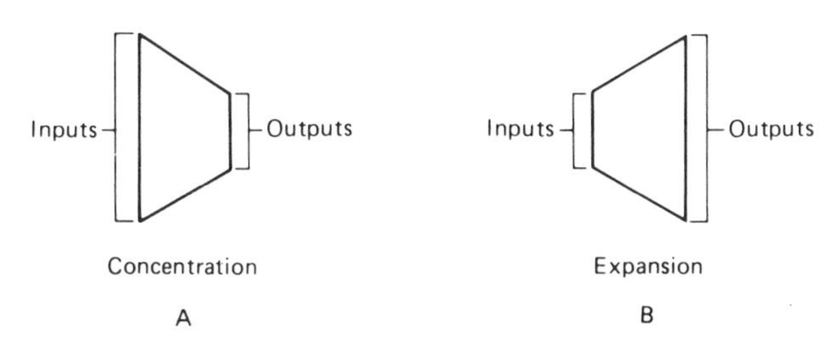

Let us dwell on the term *concentration* a bit more. Concentration reduces the number of switching paths or links *within* the exchange and the number of trunks connecting the local exchange to other exchanges. A switch also performs the function of expansion to provide all subscribers served by the exchange with access to incoming trunks and local switching paths.

Consider trunks in a long-distance network. A small number of trunks is inefficient not only in terms of loading (see Chapter 1, Section 8.2), but also in terms of economy. Cost is amortized on a per circuit basis; 100 trunks in a link or traffic relation are much more economical than 10 on the same link.* Tandem exchanges concentrate trunks in the local area for traffic relations (links) from sources of low traffic intensity, particularly below 20 erlangs, improving trunk efficiency.

4 BASIC SWITCHING FUNCTIONS

In a local exchange means are provided to connect each subscriber line to any other in the same exchange. In addition, any incoming trunk can connect to any subscriber line and any subscriber, to any outgoing trunk. The switching functions are remotely controlled by the calling subscriber, whether he is a local or long-distance subscriber. These remote instructions are transmitted to the exchange by "off-hook," "on-hook," and dial information. There are eight basic functions of a conventional switch or exchange:

1. Interconnection.
2. Control.
3. Alerting.
4. Attending.
5. Information receiving.
6. Information transmitting.
7. Busy testing.
8. Supervisory.

*Assuming an efficient traffic loading in each case.

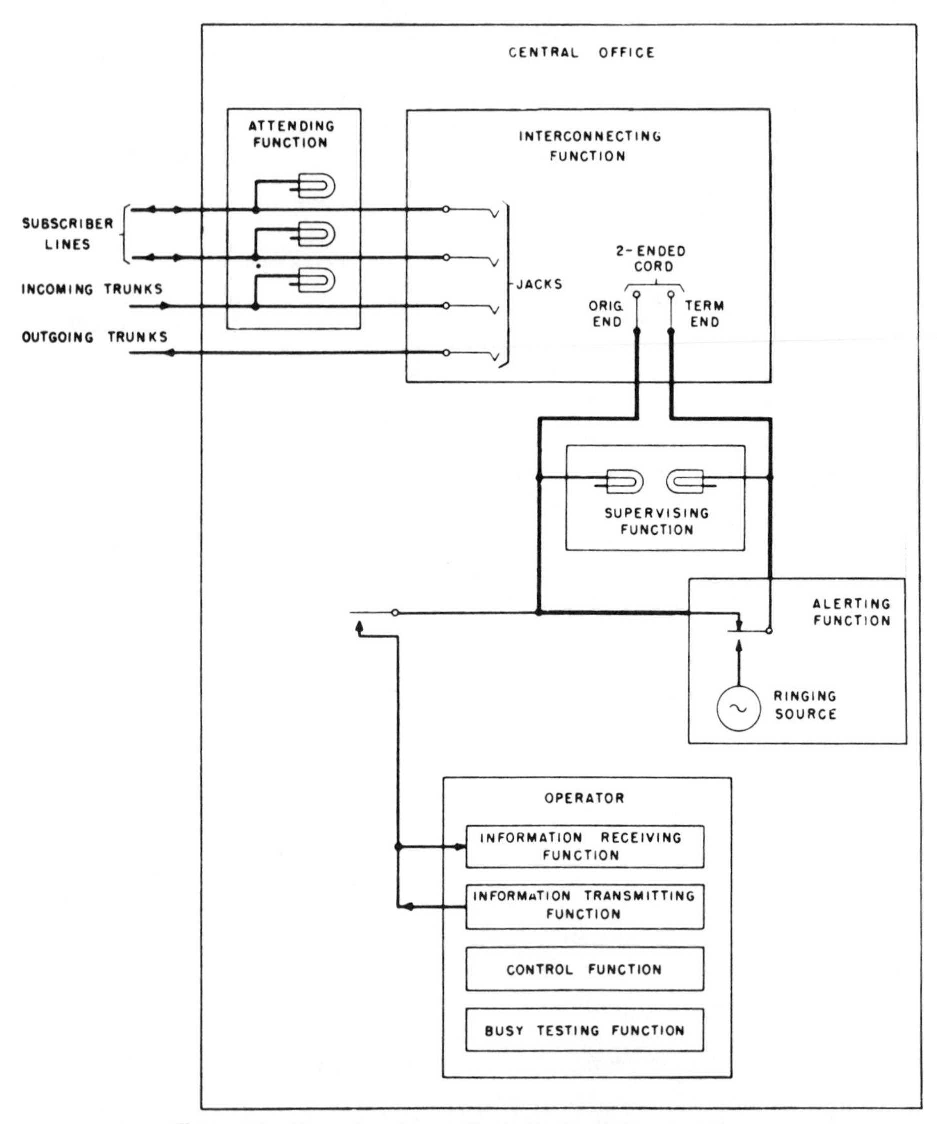

Figure 3.1 Manual exchange illustrating switching functions.

Consider a typical manual switching center (Figure 3.1) where the eight basic functions are carried on for each call. The important interconnecting function is illustrated by the jacks appearing in front of the operator, subscriber-line jacks, and jacks for incoming and outgoing trunks. The interconnection is made by double-ended connecting cords, connecting subscriber to subscriber or subscriber to trunk. The cords available are always less than half the number of jacks appearing on the board, because one interconnecting cord occupies two jacks (by definition). Concentration takes place at this point on a manual exchange. Distribution is also carried out because any cord may be used to complete a connection to any of the terminating jacks. The operator is alerted by a lamp when there is an incoming call requiring connection. This is the attending–alerting function. The operator then assumes the control function, determining an idle connecting cord and plugging into the incoming jacks. She then determines call destination, continuing her control function by plugging the cord into the terminating jack of the called subscriber or proper trunk to terminate her portion of control of the incoming call. Of course, before plugging into the terminating jack, she carries out a busy-test function to determine that the called line–trunk is not busy. To alert the called subscriber that there is a call, she uses the manual ring-down by connecting the called line to a ringing current source, as shown in Figure 3.1. Other signaling means are usually used for trunk signaling if the incoming call is destined for another exchange. On such a call the operator performs the information function orally or by dialing the call information to the next exchange.

The supervision function is performed by lamps to show when a call is completed and the cord taken down. The operator performs numerous control functions to set up a call, such as selecting a cord, plugging it into the originating jack of the calling line, connecting her headset to determine calling information, selecting (and busy testing) the called subscriber jack, and then plugging the other end of the cord into the proper terminating jack and alerting the called subscriber by ring-down. Concentration is the ratio of the field of incoming jacks to cord positions. Expansion is the number of cord positions to outgoing (terminating) jacks. The terminating jacks and originating jacks can be interchangeable. The called subscriber at another moment in time may become a calling subscriber. On the other hand, incoming and outgoing trunks may be separated. In this case they would be one-way circuits. If not separated, they would be both-way circuits, accepting both incoming and outgoing traffic.

5 SOME INTRODUCTORY SWITCHING CONCEPTS

All telephone switches have, as a minimum, three functional elements: concentration, distribution, and expansion. Concentration (and expansion) was briefly introduced in Chapter 1, Section 2 to explain the basic rationale of switching. Viewing a switch another way, we can say that it has originating-line appearances and terminating-line appearances. These are shown in the simplified conceptual drawing in Figure 3.2. Figure 3.2 shows the three different call possibilities of a typical local exchange (switch):

1. A call originated by a subscriber who is served by the exchange and bound for a subscriber who is served by the same exchange (route A–B–C–D–E).

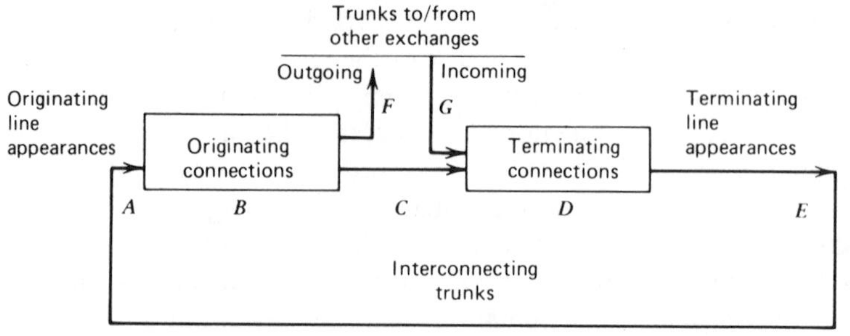

Figure 3.2 Originating and terminating line appearances.

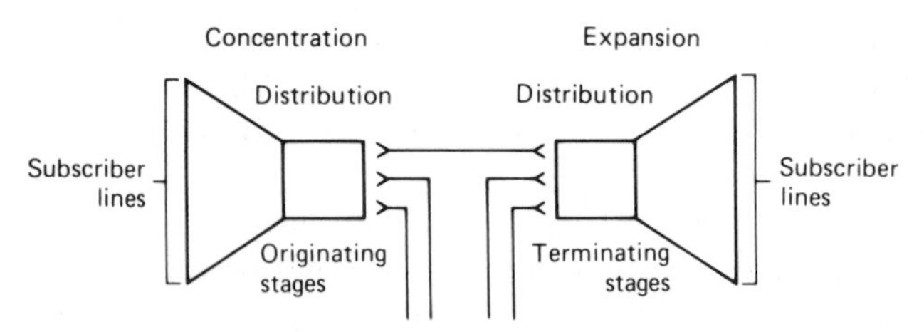

Figure 3.3 The concept of distribution.

2. A call originated by a subscriber who is served by the exchange and bound for a subscriber who is served by another exchange (route A–B–F).
3. A call originated by a subscriber who is served by another exchange and bound for a subscriber served by the exchange in question (route G–D–E).

Call concentration takes place in B and call expansion at D. Figure 3.3 is simply a redrawing of Figure 3.2 to show the concept of distribution. The distribution stage in switching serves to connect by switching the concentration stage to the expansion stage.

The symbols used in switching diagrams are as those in the following diagram, where concentration is shown on the left and expansion, on the right.

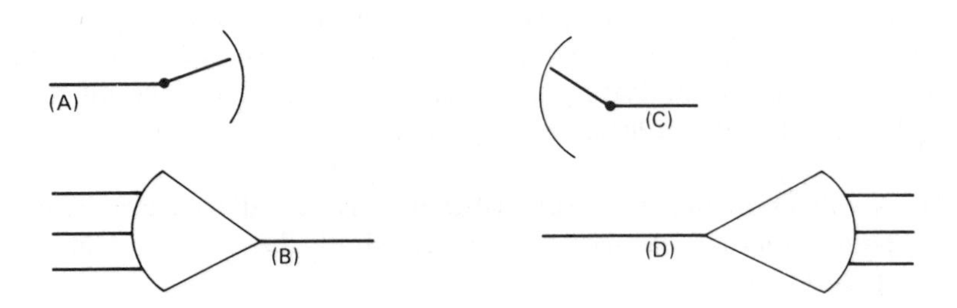

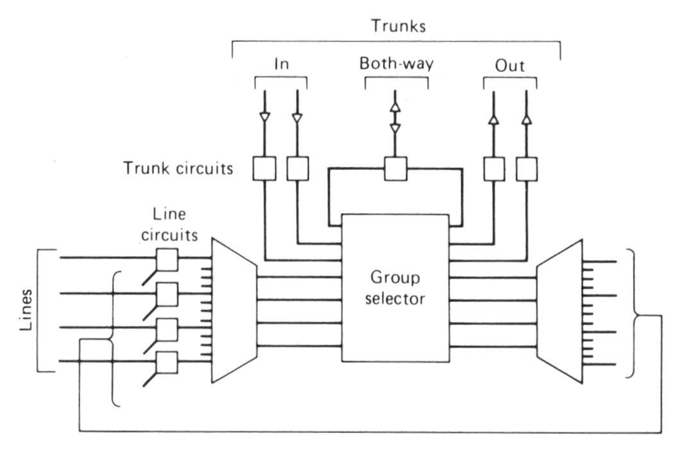

Figure 3.4 The group selector concept where both lines and trunks can be switched by the same matrix.

The number of inputs to a concentration stage is determined by the number of subscribers connected to the exchange. Likewise, the number of outputs of the expansion stage is equal to the number of connected subscribers whom the exchange serves. The outputs of the concentration stage are less than the inputs. These outputs are called *trunks* and are formed in groups; thus we refer to *trunk groups*. The sizing or dimensioning of the number of trunks per group is a major task of the systems engineer. The number is determined by the erlangs of traffic originated by the subscribers and the calling rate (see Chapter 1, Sections 6 and 8).

A group selector is used in distribution switching to switch one trunk (between concentration and expansion stages) to another and is often found not only in switches that switch subscriber lines but also where trunk switching is required. Such a requirement may be found in small cities where a switch may carry out a dual function, both subscriber switching and trunk switching from concentrators or other local switches. A group selector alone is a tandem switch. Figure 3.4 illustrates the group-selector principle.

6 TYPES OF ELECTROMECHANICAL SWITCHES

There are basically three types of electromechanical switches. Sometimes these types are broken down into gross-motion switches and fine-motion switches. All are relay (or equivalent) operated. In the gross-motion category the oldest (patented in 1891) is the Strowger switch or step-by-step switch (S × S). The second type of gross-motion switch is the rotary switch, such as the ITT 7D. Both switches are still used. The switches are operated electromechanically by ratchets such as a stepping relay. Conveniently, the relays (S × S) are in steps

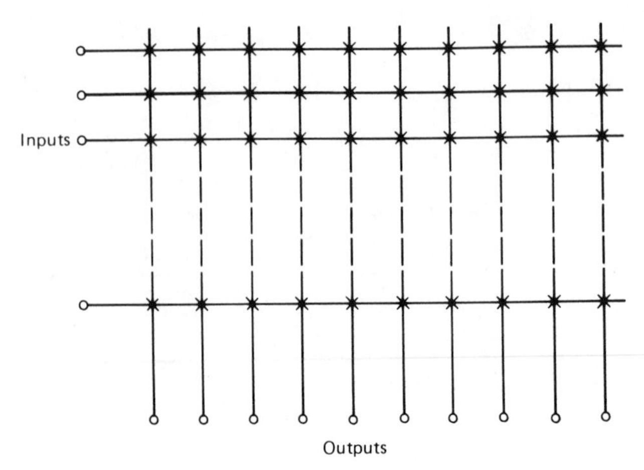

Figure 3.5 Typical diagram of a cross-point matrix.

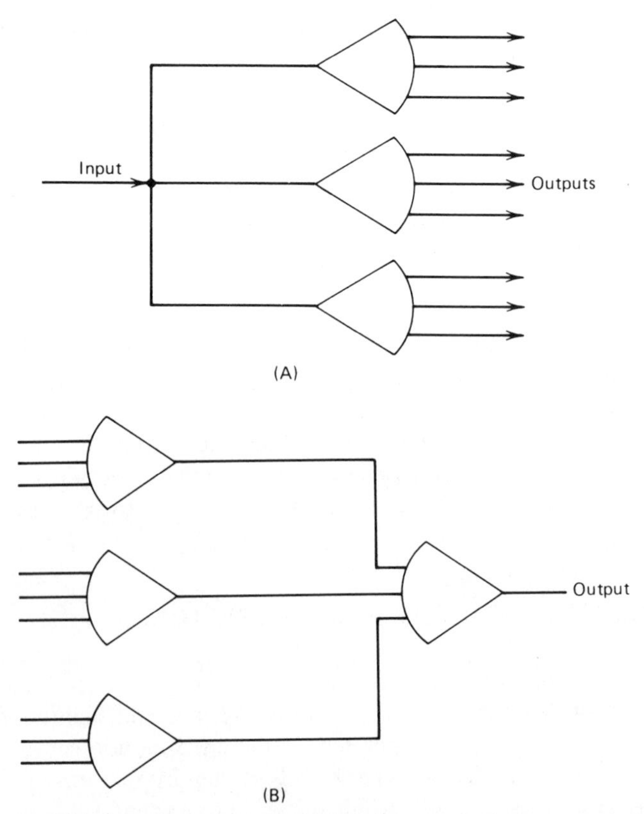

Figure 3.6 Examples of "multiple": (A) single inputs to multiple outputs; (B) multiple inputs to a single output.

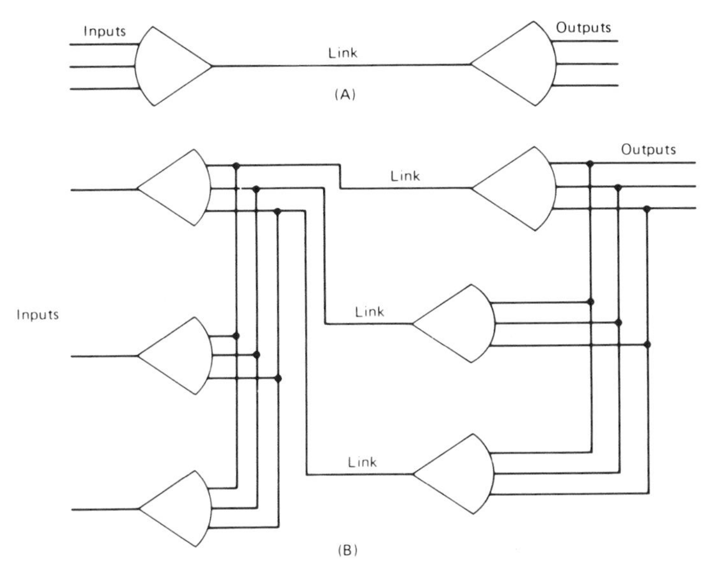

Figure 3.7 Examples of links.

of 10 to fit our decimal-number system. They were termed *gross-motion switches* because of the space traversed between terminals.

Fine-motion switches are typified by the crossbar switch. This is a coordinate switch or matrix. A speech-path connection proceeding through a switch is made by cross-points. Similar matrices can be constructed of reed relays or solid-state cross-points. Such matrices may be represented by a block diagram, such as that shown in Figure 3.5.

7 MULTIPLES AND LINKS

A multiple multiplies. It is a method of obtaining several outputs from one input. Thus access is extended (see Figure 3.6A). Or the reverse may be true, where multiple inputs gain access to one output (see Figure 3.6B). Links provide connection for a multiple of switch inputs from one stage to a multiple of switch outputs in another stage (see Figure 3.7).

8 DEFINITIONS: DEGENERATION, AVAILABILITY, AND GRADING

8.1 Degeneration

Degeneration can be expressed by the following ratio:

$$\text{Degeneration on a link} = \frac{\text{Variance of offered traffic}}{\text{Mean of offered traffic}} \tag{3.1}$$

Degeneration is a measure of the extent to which the traffic on a given link varies from pure random traffic. For pure random traffic, degeneration (the preceding ratio) equals 1. For overflow traffic,the variance is equal to or, in the majority of cases, greater than the mean. The more degenerate the traffic, the heavier the demand during peak periods and the greater the number of transmission facilities required.

8.2 Availability

At a switching array, availability describes the number of outlets that a free inlet is able to reach and test for free or busy condition:

1. *Full availability*. Every free inlet is at all times able to test every outlet.
2. *Limited availability*. The absence of full availability. The availability at a switching array can be assigned a value, namely, the number of outlets available to each inlet (see Section 8.3).

8.3 Grading

At a switching array with limited availability the inlets are arranged into groups, called *grading groups*. All the inlets in a grading group always have access to the same outlets. Grading is a method of assigning outlets to grading groups in such a way that they assist each other in handling the traffic (see Figure 3.8).

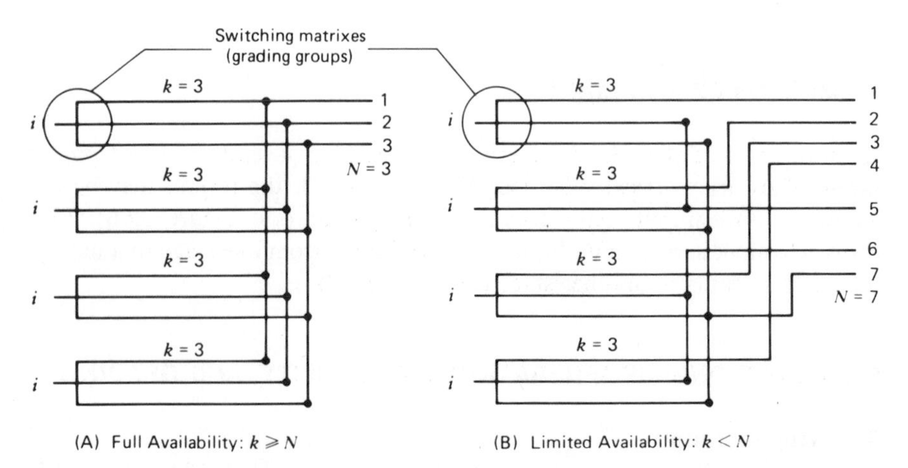

Figure 3.8 Full and limited availability. i = number of inlets per switching matrix (grading group), m = number of switching matrices (grading groups) (4, in A and B), N = number of outlets (three in A; seven in B), k = availability (number of outlets per switching matrix (grading group)). (*Note*: This is a simplified illustration. In a typical switching array, k would equal 10 or even 20.)

9 CONVENTIONAL ELECTROMECHANICAL SWITCHES

9.1 Step-by-Step or Strowger Switch

A step-by-step switch is conveniently based on a stepping relay with 10 levels. In its simplest form, which uses direct progressive control, dial pulses from the subscribers telephone activate the switch, with each pulse stepping the switch one level. If the subscriber dials a 3, three pulses are generated by the subscriber subset and transmitted to the switch. The switch then steps to level 3. We can then imagine the second digit dialed passing to the second stepping relay bank, and the third, to the third bank. A dialed number, say 375, may be stepped through three sets of banks of 10, as shown in the following diagram:

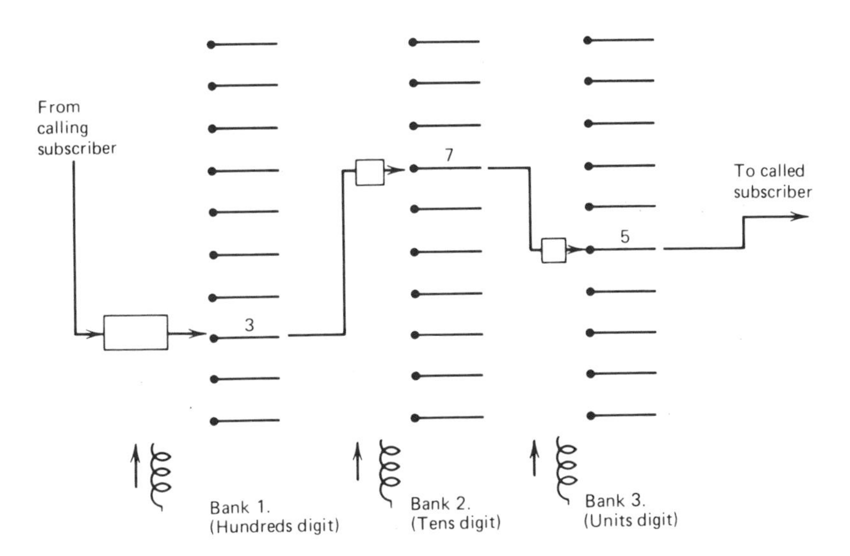

To save space and reduce postdial delay, the step-by-step switch evolved into a two-motion switch, and two banks became one. For the first digit dialed, the switch steps vertically, as shown in the preceding diagram. For the second digit, the same bank steps horizontally. Thus a bank covers 10 × 10 or 100 digits, and two banks in series can cover 10,000 (0 through 9999).

The line-finder technique is used in more modern S × S (post-1928) switches. Line finders are more simple switches, with several available per group of incoming subscriber lines. On an incoming call, when a subscriber goes "off hook," a line finder automatically seeks the line desiring service and extends the connection to a line selector. The line finder provides the first stage of concentration. The line finder, once connected, supplies dial tone to the calling subscriber. A 1000-line S × S switch is illustrated in Figure 3.9. The connector in Figure 3.9 is called the *final selector* in Europe. To extend the switch in Figure 3.9 to 10,000 lines, a second selector stage is added between the selector

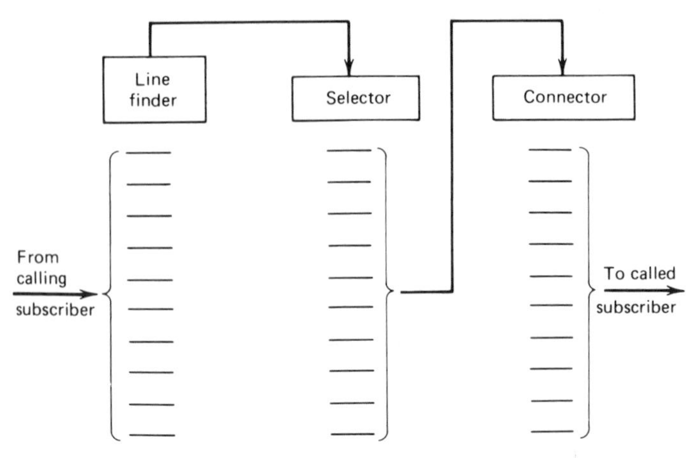

Figure 3.9 A step-by-step switch showing line finders, selectors, and connectors serving 1000 lines.

and the connector stages. The connector stage shown in Figure 3.9 may also be called the *final connecting stage*. Normally, the last two digits dialed by the calling subscriber control the connector, which typically has access to 100 lines. The connector is more complex than the preceding stages. It must busy-test the called line and if busy, return a "busy-back" (the audible "line busy" signal) to the calling subscriber. If the called line is idle, it must apply ringing current to the called line. It will supply talk battery to both calling and called subscriber once the called subscriber goes "off hook." It provides supervision, holding the talk path in operation until one or both conversing parties go "on hook."

9.2 Crossbar Switch

Crossbar switching dates from 1938 and reached a peak of installed lines in 1983. Its life has been extended by using stored-program control (SPC) rather than hard-wired control in the more conventional crossbar configuration. The crossbar is actually a matrix switch used to establish the speech path. An electrical contact is made by actuating a horizontal and a vertical relay. Consider the switch in Figure 3.10. To make contact at point B_4 on the matrix, horizontal relay B and vertical relay 4 must close to establish connection. Such a closing is usually momentary but sufficient to cause "latching." Two forms of latching are found in conventional crossbar practice, mechanical and electrical. The latch keeps the speech-path connection until an "on-hook" condition results, freeing the horizontal and vertical relays to establish other connections, whereas connection B_4 in Figure 3.10 has been "busied out."

Private industry has fielded numerous types of crossbar configuration. Early crossbar switches were made up of basic 10×10 switching matrices or blocks.

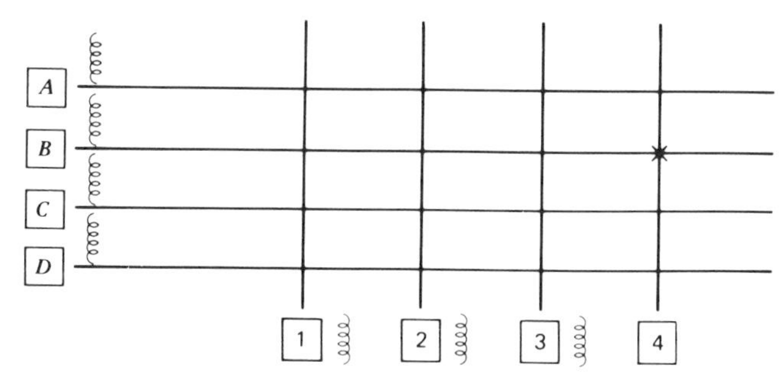

Figure 3.10 The crossbar concept.

Northern Electric (Canada) has a 10 × 20 (20 vertical) matrix in a miniswitch configuration. The ITT Pentaconta 1000 has 22 vertical and 14 horizontal bars, one of which is used as a changeover bar to provide for switching 52 outlets. The output of the basic switch block matrix for Pentaconta has 50 lines. Modern local crossbar switches, such as Pentaconta 1000 or the ATT No. 5 crossbar, handle up to 10,000 subscribers in a basic switch. In the case of Pentaconta 1000, this can be extended to 20,000. A basic 10,000-line Pentaconta has 22 primary selection switch blocks.

10 SYSTEM CONTROL

10.1 Introduction

The basic function of the control system in a switch is to establish a path through the switch matrix. Thus the control system must know the calling and called ports on the matrix and be able to find a free path between them. There are two methods of establishing a path, progressive control and common control.

10.2 Progressive Control

As the term indicates, "progressive control" implies that a speech path is set up progressively through the switch stage by stage. At each stage the selecting action chooses a group of identical paths that lead toward the ultimate call destination in the switch. Hunting action selects one of a group of competing circuits as the call progresses. The control system has no foreknowledge of conditions ahead in the next stage or step in the call setup. The call could run into blocking or a busy line and must wait through this extensive setup process before returning a "busy-back" or congestion signal.

If the control system functions directly as a result of the subscriber dial pulses, it is called *direct progressive control.* This was the case in our previous discussion of step-by-step switching. In more modern progressive control systems a register is interposed between (buffers) the subscriber dial digits. The register accepts dial digits, interprets them, and then controls switch functions. This is called *register progressive control*. With direct progressive control the called telephone number is directly associated with a specific path through the selection tree. Because the number is decimal based, progression through the switching tree is carried out in branches of 10, as shown in the diagram that follows. In direct progressive control the switching timing is in direct

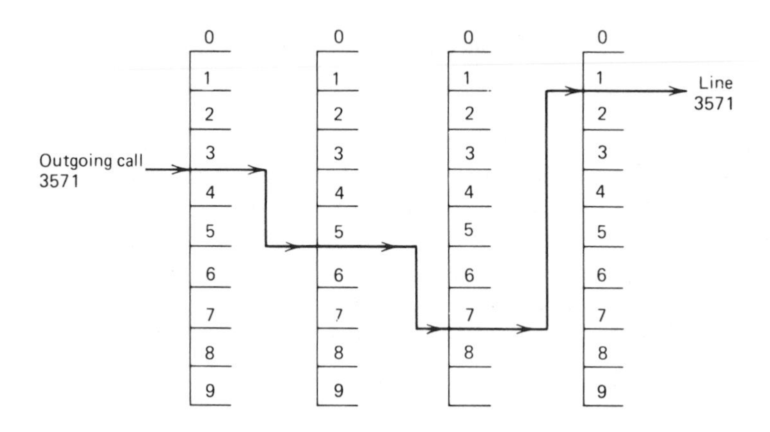

sequence and in synchronism with the subscriber-dialed digits. The principal advantage of direct progressive control is economy. Control circuitry is minimized, and such switches can still be found in small-community dial offices (exchanges). A functional block diagram of a direct-control switch is shown in Figure 3.11.

Disadvantages are that direct progressive control switches require a certain amount of overbuild to provide a satisfactory grade of service because of the stage-by-stage hunt and choose requirement without foreknowledge that a certain path is busy. The result is a lower efficiency than that in other systems on interconnecting paths. It can, however, be more tolerant of overloads concomitant with a poorer grade of service. Of course, direct progressive control requires subscriber lines to terminate directly on switch connector stages in strict correspondence with the subscribers' numbers. There is also an inherent inflexibility of trunk assignments that must conform directly with the exchange identifying digits.

Register progressive control eliminates many of the disadvantages of direct progressive control. It buffers incoming dial information, thus providing number translation. On step-by-step installations register progressive control consists of an A-unit hunter and a director. The A-unit hunter connects the incoming line to a free A-digit selector. This unit steps the selector to the group level corresponding to the first dialed digit and then rotates to a free director. The remaining two digits of the exchange code (see Section 16 of this

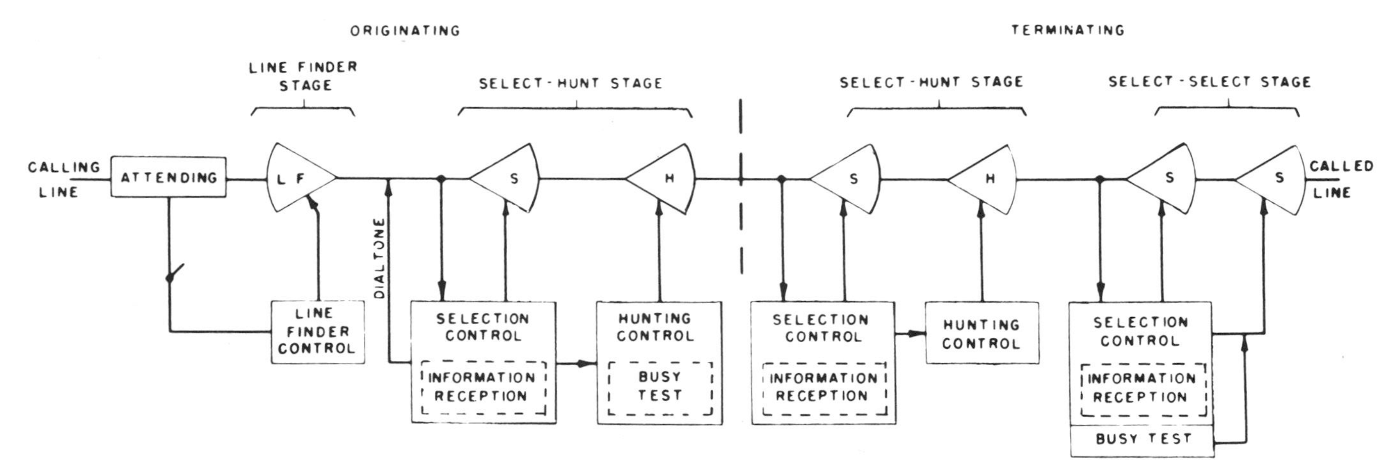

Figure 3.11 Simplified diagram of control circuits for direct progressive control. Copyright © American Telephone and Telegraph Company, 1961.

chapter) and the four digits of the subscriber number are registered in the director. The first two remaining digits route the call to (own) exchange or proper distant exchange. The director translates digits two and three to a code that is more versatile for switch operation and stores the last four digits of the number dialed. The translated code is then transmitted (second and third digits of called numbers) to the first code selector, and the call is extended to the proper exchange. The remainder of the digits (i.e., the subscriber number) are then transmitted as is for the indicated exchange to act on. Once these operations are completed, the A-digit hunter, the selector, and the director are freed to act on another incoming call. The setup of the call proceeds as in direct progressive control.

It will be noted that a seven-digit telephone number was used in the preceding example. The call setup proceeds as follows:

1. Seven incoming dialed digits.
2. A unit selector hunts/finds free director.
3. The director then:
   a. Acts on the first digit.
   b. Translates second and third digits.
   c. Transmits second and third digits.
   d. Registers fourth through seventh digits.
   e. Transmits digits four through seven (untranslated).
4. Second and third translated digits are acted on by the first selector for own exchange or indicated distant exchange.
5. Fourth through seventh digits are fed to the second selector at own exchange or the call is routed to a distant exchange.
6. A-digit switch and director is released for the next incoming call.

Register progressive control was introduced in 1923 in the United Kingdom with director system. This first attempt at number translation lacked the capability of alternative routing when congestion occurred. Call routing remained inflexible.

As switching evolved, some of these deficiencies were eliminated or, at least, alleviated by adding further translation or by associating register translation with the later stages of switches as well as the first stages.

To reduce postdial delay, a switch may be made to start a call setup once the first three dialed digits are received. Initial translations and trunk selection can take place during the period in which the calling subscriber is dialing the last four digits.

As automatic telephone service grew, switching engineers were confronted with the problem of numerous types of switches in a common area. Some would be older progressive control and others, register progressive control and

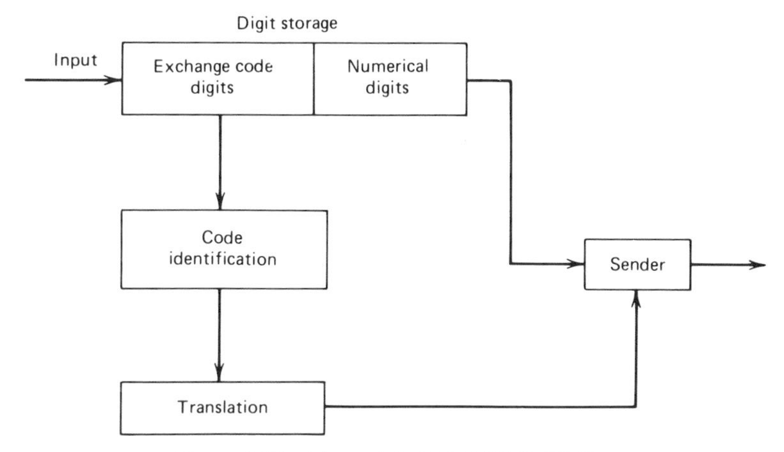

Figure 3.12 Interexchange control register.

still others, common control. Even greater variance occurred in internal switch codes, the digital codes used in a switch internally to set up and route a call.

Switching design engineers have since resorted to the technique of the control register with information transmitting function added. The register translator (control register) carries out three consecutive steps in this case: (1) information reception, (2) internal control signaling, and (3) information transmission. The translator in this case determines from the exchange code (the first three dialed digits) the outgoing trunk group *and* the type of signals required by the group and the amount of information to be transmitted. This flexibility in design and operation of multiexchange areas was vastly improved. The interexchange control register concept is shown in Figure 3.12. Outgoing trunk-control registers facilitate the use of tandem exchange operation. Also, incoming trunk calls to an exchange so equipped can be handled directly from a distant exchange if they are similarly equipped.

Interexchange control registers enabled alternative routing. When first-attempt routing is blocked, this same technique can be used for second-attempt routing through an exchange by setting up a speech path with switching components different from the first setup that failed. The sender in Figure 3.12 generates and transmits signaling information to distant exchanges.

10.3 Common Control (Hard Wired)

10.3.1 General. "Common control" is an ambiguous term. Any control circuitry in a switch that is used for more than one switching device may be termed "common control." We try to distinguish common control from the progressive control referred to in the previous section, yet common control is used in the register translators of progressive control.

For purposes of this discussion, common control is defined as providing a means of control of the interconnecting switch network, first identifying the input and output terminals of the network that are free and then establishing a path between them. This implies a busy-test of a speech path before establishment of the path. Common control may cover the entire switch or separate control for the originating and terminating halves. Common-control systems employ markers, which are discussed subsequently. Such marker systems are most applicable to grid switching networks (e.g., networks internal to the switch).

10.3.2 Grid Networks. A grid in switching networks may be defined as a combination of two or more stages of switch blocks connected together to accommodate numbers of input and output circuits as needed to meet the speech-path switching requirements of a given exchange design. The grid network differs from others previously discussed because it can be considered bidirectional.

This means that the input could also be considered an output and vice versa. Thus parts of the network may be used for both originating and terminating purposes. Concentration and expansion can be combined, and thus a subscriber line has only one appearance on the network. A basic

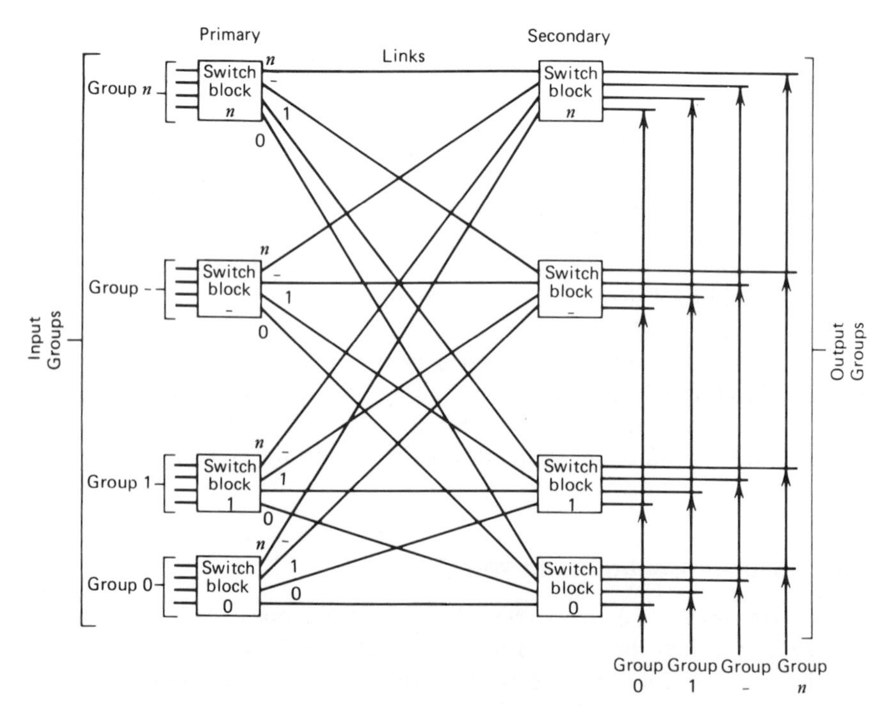

Figure 3.13 Basic two-stage grid network. Copyright © American Telephone and Telegraph Company, 1961.

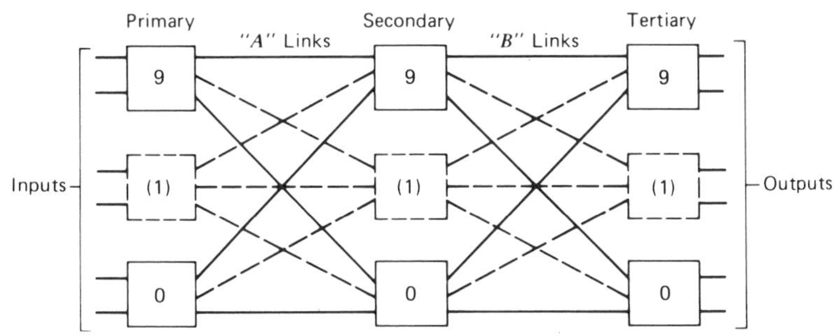

Figure 3.14 An elemental three-stage grid. Copyright © American Telephone and Telegraph Company, 1961.

two-stage primary secondary grid network is shown in Figure 3.13 and an elemental three-stage grid in Figure 3.14. Within each block in a grid network any input can connect to any output. If the blocks in Figure 3.13 had 10 inputs each, with 10 blocks in each stage, and fulfilling the basic requirement for at least 1 access from each primary group block to each secondary group, then at least 100 links would be required.

The fundamental difference between progressive and grid networks is in the number of paths a specific call looks ahead to inside the network on call setup. The progressive network is a hunt–select network, where the "tree" principle holds (see Section 10.2). As a call progresses through the network, the path is built up stage by stage. At each stage it looks at a choice of 10 branches ahead of it. Only when looking at the grid network as a whole is there a choice of 10 paths. On a stage-by-stage basis in the grid network, however, there is only a choice of one path for a given call at any intermediate stage. As we see later on, this path has been selected and busy-tested before actually being set up.

Conventional crossbar switches for large switching centers are usually four-stage grid networks. Such a network is shown in Figure 3.15. It is actually made up of two stages of primary and secondary grids. The two stages are connected by what are called *junctors*. Note that only one junctor as a minimum is required per secondary switch of the input grid to connect to each input grid of the primary switch of the output grid. One junctor matches any pair of originating and terminating links. As can be seen in Figure 3.15, it is particularly important that traffic balance be carefully maintained on the grid inputs and outputs since the junctors from each input grid are divided equally among all grids. However, the busy-test and trunk-hunting functions of common-control systems provide for many different path setups to service a particular call.

10.3.3 Common-Control Principles. The "marker" sets aside common-control systems as we define them from other more generalized common-control systems discussed previously. On grid circuits the characteristics are such

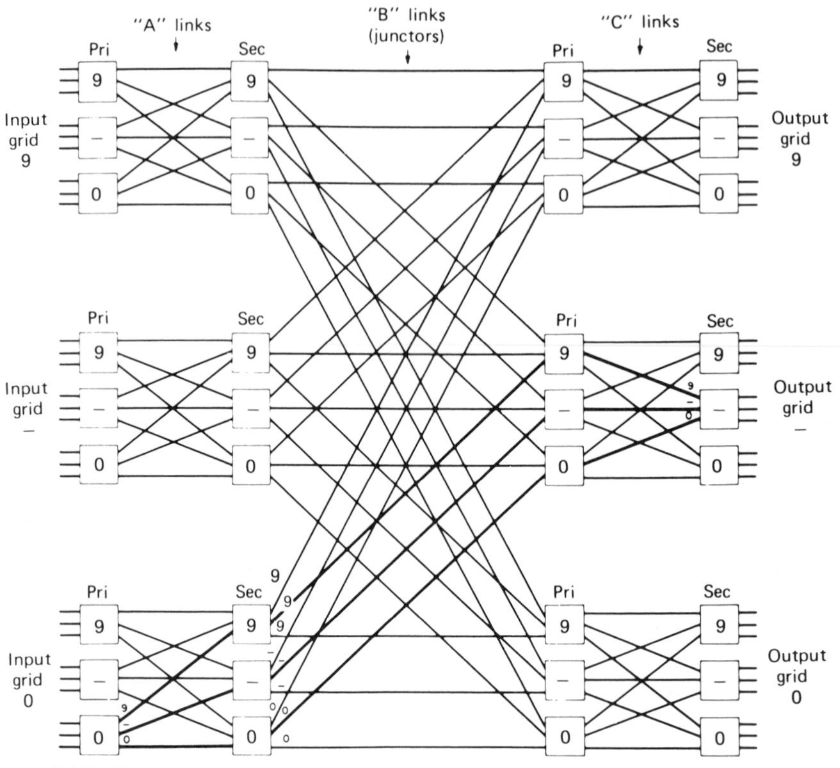

Figure 3.15 Typical four-stage grid network. Copyright © American Telephone and Telegraph Company, 1961.

that with specified input and output terminals, various sets of linkage paths exist that can provide connection between the terminals. It is the marker, with terminal points identified, that locates a path, busy-tests it, and finally sets up a particular channel through the switch grid network.

A marker always works with one or more registers. A marker is a rapidly operating device serving many calls per minute. It cannot wait on the comparatively slow input information supplied by an incoming line or trunk. Such information, whether dial pulses or interregister signaling information, is stored in the register and released to the marker on demand. The register may receive the entire dialed number and store it before dumping to the marker or may take only the exchange code or area code plus exchange code (see Section 16, this chapter) which is sufficient to identify a trunk group. The register will also identify the location of the call input or set up a control path to the input for the marker. Figure 3.16 illustrates the basic functions of a marker.

Whereas the register provides the dialed exchange number to the marker, the translator provides the marker information on access to the proper trunk group. Of course, the translator may also provide other information, such as the type of signaling required on that trunk group. The reader must bear in

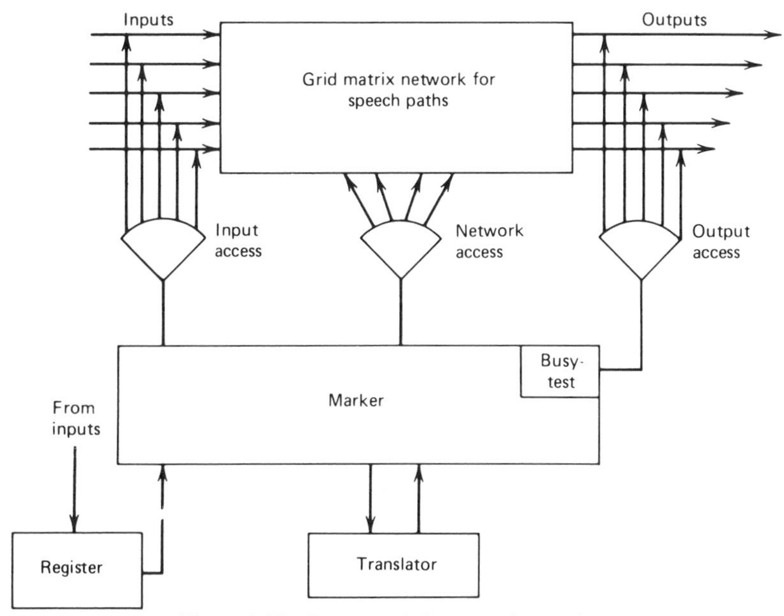

Figure 3.16 Functional diagram of a marker.

mind that modern common-control switches, particularly grid-type switches, use a control code that differs from the numerical dialed code. The dialed code is the directory number (DN) of the called subscriber. The equipment number (EN) is the number the equipment uses and is often a series of five one- or two-digit numbers to indicate location on the grid matrix of the speech-path network. The equipment codes are arbitrary and require changes from time to time to accommodate changes in number assignments. The changes are done on a patch field associated with the translator. One equipment number code is the "two out of five" code, which has ten possibilities to correspond to our decimal base number system, that used on the subscriber dial (or push buttons). As the number implies, combinations of five elements are taken two at a time. This code is shown in Table 3.1.

The number combinations are additive, corresponding to the decimal equivalent except the 4–7 combination that adds to 11 (not to 0). For

TABLE 3.1 Two Out of Five Code

Digit	Two out of Five 0–1–2–4–7	Digit	Two out of Five 0–1–2–4–7
1	0–1	6	2–4
2	0–2	7	0–7
3	1–2	8	1–7
4	0–4	9	2–7
5	1–4	0	4–7

transmission through the exchange, the numbers are represented by two out of five possible audio frequencies.

On conventional exchanges one marker cannot serve all incoming calls, particularly during the busy hour (BH). Good marker holding time per call is of the order of half a second. Call attempts may be much greater than two per second, so several markers are required. Each marker must have access from any input grid to any output grid, and thus we can see where markers might compete. This possibility of "double seizure" can cause blockage. Therefore only one control circuit is allowed into a specific grid at any one time.

To reduce marker holding time, fast action switches are required. This is one reason why common (marker) control is more applicable to fine-motion switches such as the crossbar switch. The control circuits themselves must also be fast acting. Older switches thus used all-relay devices and some vacuum tubes. Newer switches use solid-state control circuitry, which is more reliable and even faster acting.

Figure 3.17 is a simplified functional diagram of the ATT (North American) No. 5 crossbar system. The originating and terminating networks are combined. Two different types of marker are used, the dial-tone marker and the completing marker. A typical No. 5 crossbar has a maximum of four dial-tone markers and eight call-completing markers. The dial-tone marker sets up connections between the calling subscriber line and an originating register. Call-completing markers carry out the remainder of the control functions, such as trunk selection, identification of calling and called line terminals, channel busy-test and selection, junctor group and pattern control, route advance, sender link control, trunk charge control when applicable, overall timing, automatic message accounting (AMA) information, and trouble recorder control.

The marker is also used for exchange troubleshooting and fault location. It is the most convenient check of the switching network because it samples the

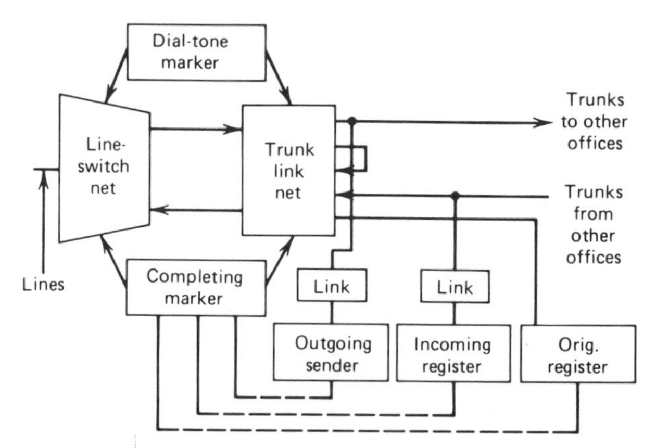

Figure 3.17 American Telephone and Telegraph No. 5 crossbar; a typical marker application. Copyright © American Telephone and Telegraph Co., 1961.

entire network at frequent intervals. Many markers provide trouble-recording circuits. Trouble on one call setup attempt may have no effect on another. The marker provides a trouble output, returns to service, and reattempts the call. Reattempts by a marker are usually limited to two so that the marker can return to service other calls.

If a marker finds an "all trunks busy" (ATB) condition on a tandem or transit call, it can ask for instructions to reroute the call. It then handles the call as a first attempt. If again an ATB condition is found and a second alternative route is available, it can make a call attempt on the second rerouting. When all possible routes are busy, the marker returns an ATB signal to the calling subscriber.

11 ROTARY SWITCHING SYSTEMS

Rotary switches are very similar in operation to step-by-step systems. One difference is that they do not have both vertical and rotary motion, a feature found in the Strowger S × S switch. Rotary systems are progressive switches with line finders and selectors. The layout is similar to S × S, but the switch is power driven with electric motors, generally continuous running. All rotary switches have register storage of dialed digits and incorporate features of common control similar to register progressive control described in Section 10.2. Some of the later versions are nearly as versatile as the marker common control described in Section 10.3.3.

Today rotary switching is used essentially in Europe or in countries using European switches and is principally manufactured by ITT*, with its 7 A-E; L. M. Ericcson, with its AGF; and Siemens, with its EMD system.

12 STORED-PROGRAM CONTROL

12.1 Introduction

Stored-program control (SPC) is a broad term designating switches where common control is carried out to a greater extent or entirely by computerware. Computerware can be a full-scale computer, minimicrocomputer, microprocessors, or other electronic logic circuits. Control functions may be entirely carried out by a central computer in one extreme for centralized processing or partially or wholly by distributed processing utilizing microprocessors. Software may be hard wired or programmable. Telephone switches are logical candidates for digital computers. A switch is digital in nature,† as it works with discrete values. Most of the control circuitry, such as the marker, work in a binary mode.

*Now ALCATEL.

†This important concept is the basis for the argument set forth in Chapter 9 regarding the rationale for digital switching.

The conventional crossbar marker requires about half a second to service a call. Up to 40 expensive markers are required on a large exchange. Strapping points on the marker are available to laboriously reconfigure the exchange for subscriber change, new subscribers, changes in traffic patterns, reconfiguration of existing trunks or their interface, and so on.

Replacing register markers with programmable logic—a computer, if you will—permits one device to carry out the work of 40. A simple input sequence on the keyboard of the computer workstation replaces strapping procedures. System faults are printed out as they occur, and circuit status may be printed out periodically. Due to the speed of the computer, postdial delay is reduced and so on. Computer-controlled exchanges permit numerous new service offerings, such as conference calls, abbreviated dialing, "camp on busy," call forwarding, and incoming-call signal to a busy line.

12.2 Basic Functions of Stored-Program Control

There are four basic functional elements of an SPC switching system: (1) switching matrix, (2) call store (memory), (3) program store (memory), and (4) central processor.

The switching matrix may be made up of electromechanical cross-points, such as in the crossbar switch, reed, correed, or ferreed cross-points or switching semiconductor diodes, often SCR (silicon-controlled rectifier). An SCR matrix is shown in Figure 3.18.

The call store is often referred to as the "scratch pad" memory. This is a temporary storage of incoming call information ready for use, on command from the central processor. It also contains availability and status information of lines, trunks, and service circuits, and internal switch-circuit conditions.

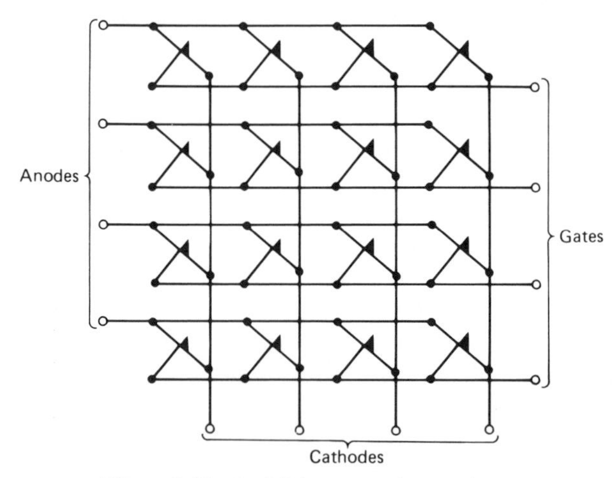

Figure 3.18 An SCR cross-point matrix.

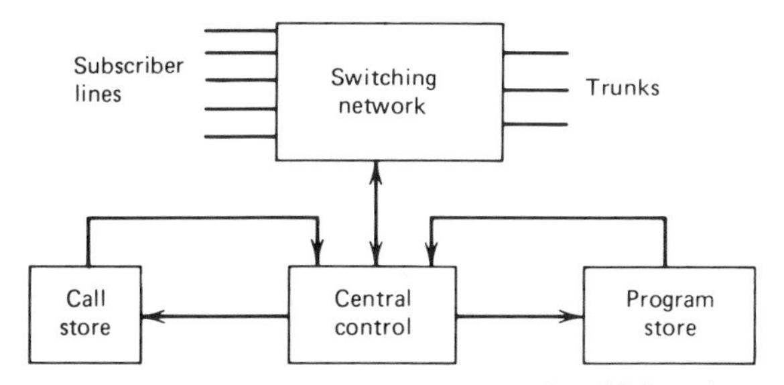

Figure 3.19 Simplified functional diagram of an SPC exchange.

Circuit status information is brought to the memory by a method of scanning. All speech circuits are scanned for a busy/idle condition.

The program store provides the basic instructions to the controller (central processor). In many installations translation information is held in this store such as DN to EN translation and trunk signaling information.

A simplified functional diagram of a basic (full) SPC system is shown in Figure 3.19.

12.3 Additional Functions in a Stored-Program-Control Exchange

Figure 3.20 is a conceptual block diagram of a typical North American SPC exchange. As we can see, an SPC exchange can be broken down into three functional levels: (1) line and trunk switching network, (2) input–output equipment, and (3) common-control equipment. Figure 3.20 is an expansion of Figure 3.19, showing, in addition, scanner circuitry and signal distribution.

The control network executes the orders given by the central control processor (computer). These orders usually consist of instructions such as "connect" or "release," along with location information on where to carry out the action in the switching network. In the ATT ESS system the network control circuits are classified in three major functional categories:

1. Selectors, which set up and release a connection on receipt of location information.
2. Identifiers, which determine the location (called an "address") of a network terminal on one side that is to be connected to a known terminal (address) on the other side of the network.
3. Enablers, which are circuits sequentially enabling junctors to be connected together.

The input–output equipment consists of a line scanner and a signal distributor. Both circuits operate under control of the central control processor. The scanner and distributor carry out the "time-sharing" concept of SPC. The term "time sharing" is the basis where one (or several) computers can control

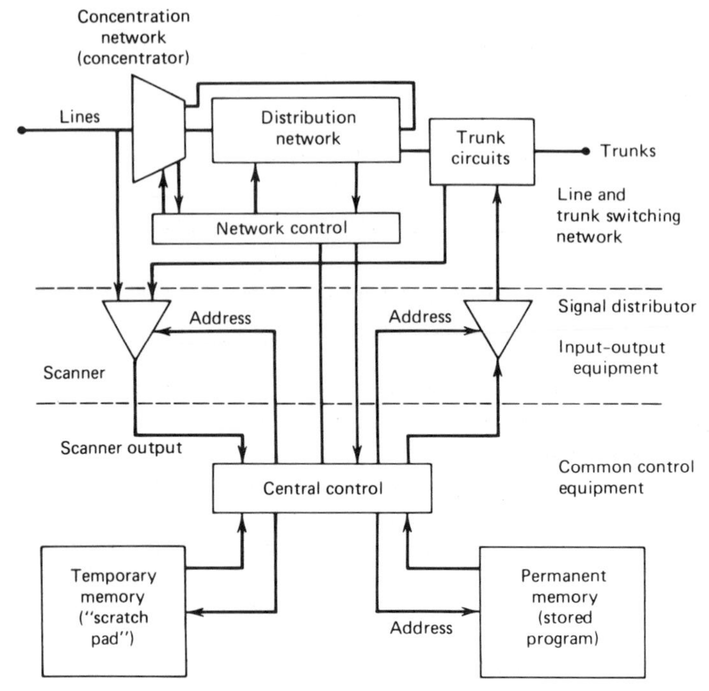

Figure 3.20 Conceptual block diagram of a typical North American semielectronic SPC exchange.

literally thousands of circuits, with each circuit being served serially. The concept of "holding time," important in SPC, is the time taken to serve a circuit, and "delay" is the time each circuit must wait to be served. A computer read–write cycle in a typical SPC is 2–5 μs, with a scanning rate of 2–5 μs per terminal, line scanning during digit reception every 10 μs, and 100-ms supervisory scan. The scanner is an input circuit used for sampling the states (idle or busy) of subscriber lines, trunks, and switch test points to permit monitoring the operation of the system. The signal distributor, on the other hand, is an output circuit directing output signals to various points in the system. In the ESS the signal distributor is primarily used on trunk circuits for supervisory and signaling actions.

Common-control equipment, as mentioned previously, is made up of the central processor, call store, and program store. The three units can be considered as making up the control computer, which is capable of transmitting orders to the system as well as detecting signals from the system. The SPC systems with centralized control that we have discussed above have a human interface (I/O = input–output) with the central controller. This, in many instances, is a teleprinter (i.e., keyboard send, printer receive) or keyboard with a visual display unit. The installation adds many advantages and conveniences not found in more conventional switching installations. Several of these are:

- Rerouting and reallocation of trunks.
- Traffic statistics.
- Renumbering of lines, subscriber move.
- Changes in subscriber class.
- Exchange status.
- Fault finding.
- Charge records.

All these functions can be carried out via the I/O equipment connected to the central processor. In the basic ATT ESS exchanges there are two teleprinter accesses.

12.4 A Typical Stored-Program-Control Exchange: European Design

The ITT Metaconta is a joint French–Spanish development installed worldwide. Its switching network can handle up to 32,000 lines. The control section can handle up to about 60,000 BH call attempts, depending on the facilities offered and the type and number of different signaling systems to be handled.

The central processor is made up of dual ITT 1600 16-bit binary parallel minicomputers under a load-sharing principle. The memory is of the random access type with a word length of 17 bits (16 bits + 1 parity bit) and a maximum capacity of $2^{16} = 65,536$ words, consisting of 8 blocks of 8192 words. The memory is a ferrite core type with an access time of 0.35 μs and a cycle time of 0.85 μs.

A Metaconta L exchange uses two classes of operational programs, consisting of a resident package and an on-demand package. The *resident package* has four types of computer programs.

1. Call-processing programs, which control all the call treatment as well as charging and statistics recording.
2. Man–machine communication programs, which handle communications between the system and operating personnel and also permit the loading of on-demand programs.
3. On-line test and defensive programs, which observe abnormal behavior of the system and either restore normal conditions or make appropriate decisions to disconnect faulty devices.
4. Start-up and recovery programs, which handle the system status evolution. In this group are programs that permit taking over call processing when one machine fails.

The on-demand package contains programs that have to be called up by the

TABLE 3.2 Service Features and Facilities: Subscriber Related

Line Classes	Subscriber Facilities

One-party lines
  With dial or push-button-type subscriber sets
  With or without special charging categories
Two-party lines
  With or without privacy, separate ringing, separate charging, revertive call facility
Multiparty lines (up to 10 main stations)
  Without privacy
  With selective or semiselective ringing
PABXs
  Unlimited number of trunks
  Uni- or bidirectional trunks
  With or without in-dialing
Coin-box lines
  Local traffic
  Toll traffic
  Special applications
Restricted lines
  To own exchange
  To urban, regional, national, or toll areas
  To some specified routes
Priority lines
  Toll essential
  Essential
  Priority during emergency, overload situations, etc.
	Abbreviated numbering
	  One or two digits
	  For individual lines, or groups of lines, or for all lines
	Transfer of terminating calls
	Conversation hold and transfer
	Calling party's ring back
	Toll-call offering
	Hot line
	Automatic wake-up
	Doctor-on-duty service
	Do-not-disturb service
	Absentee service
	Immediate time and charge information
	Conference calls
	Centrex facilities (optional)

operating personnel through teleprinter messages. The on-demand programs carry out such functions as traffic observation, charging information dump, call tracing, and routine tests. Service features and facilities offered by Metaconta L, listed in Tables 3.2 and 3.3, are broken down into two areas, subscriber related and administration related. Figure 3.21 is a functional block diagram of the Metaconta L system.

12.5 Evolutionary Stored-Program Control and Distributed Processing

12.5.1 Introduction. The presently installed switching plant represents an extremely large investment yet to be fully amortized. This plant is essentially made up of step-by-step, rotary, and crossbar equipment is quickly being replaced by digital switches. The peak in crossbar installations measured in lines served will not be reached until well into the 1980s. In 1973 SPC served

TABLE 3.3 Service Features and Facilities: Administration Related

Administration Facilities
Interoffice signaling
Direct-current signaling codes (step-by-step, rotary 7A and 7D, R6 with register or direct control, North American dc codes)
Alternating-current pulse signaling codes
MF signaling codes for register-controlled exchanges (MFCR 2 code, MF Socotel, North American MF codes)
Direct data transmission over common signaling link between processor-controlled exchanges of the time or space division types
Charging
Control of charge indicators at subscriber premises
Metering on a single fee or a multifee basis
Free number service
Numbering
Full flexibility for equipment number — directory number translation
For local calls, national toll calls, international toll calls
Private automatic branch exchanges with direct inward dialing
Routing
Prefix translation for outgoing or transit calls
Alternative or overflow routing on route busy or congestion condition
Resignaling on route busy or congestion condition
Called side release control
Maintenance
Plug-in boards
Automatic fault detection and identification means by diagnostic programs
Operation
Generalized use of teleprinters
Possibility of a remote centralized maintenance and operation center

less than 10% of ATT subscribers. One current approach is to modify existing common-control exchanges with some limited SPC techniques, and another is to install conventional crossbar with SPC control, either partial or full. The ITT Pentaconta 2000 is an example. Still another approach is to resort to distributed processing rather than use a comparatively large computer as described in Sections 12.3 and 12.4.

12.5.2 Modification of Existing Plants. Consider conventional electromechanical switching with common control (in its widest meaning). Processing by relays is distributed in registers, markers, translators, directors, and finders. Whereas implementation of new services and features available in full SPC systems is costly and difficult or even impossible because of the performance limitations of conventional electromechanical design, it is feasible or even desirable to substitute, on a limited basis, microprocessor control to replace hard-wired relay-based control elements such as registers and translators. In essence, software will replace much of the hardware in the control systems in

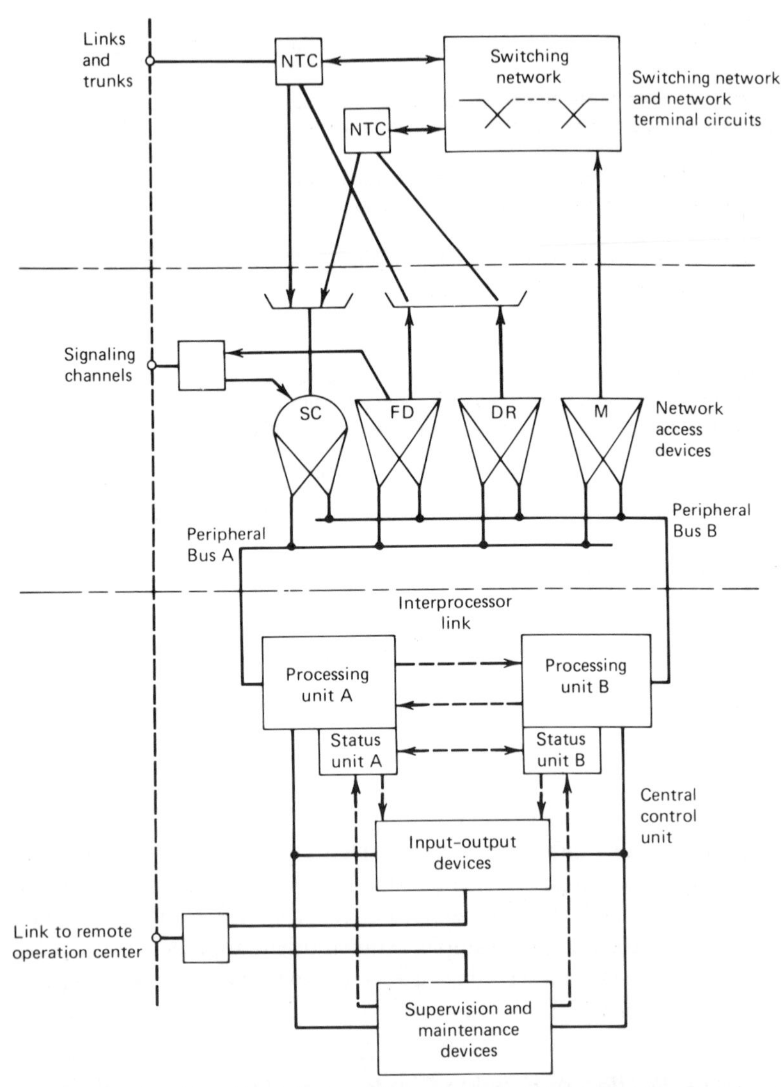

Figure 3.21 Block schematic of Metaconta L showing switching network and network terminal circuits, network access devices, and the central control unit (SC = scanner, FD = fast driver, DR = slow driver, M = marker, NTC = network terminal circuit). Courtesy of the International Telephone and Telegraph Corporation.

electromechanical switches. Many administrations and telephone companies have found this to be a cost-effective alternative to full SPC implementation. The resulting modified equipment will provide many of the new services of SPC as well as cost savings in maintenance, administration, and operation of existing installations. The basic design of crossbar and S × S features distributed control and goes against the main concept of full SPC, which features completely centralized control.

There are other incompatibilities as well, one of which is environmental. Relay circuitry can tolerate wide ranges of temperature and humidity; solid state cannot. Relay circuits use comparatively high voltage (48 V) and current (mA) ranges. Solid state, particularly integrated circuits (ICs), use low voltage and low current. "Low" in this case means < 5 V and several microamperes per circuit. Electromechanical systems are essentially immune to electromagnetic interference (EMI) and in many cases are good generators of EMI. Solid-state circuitry, on the other hand, is particularly susceptible to EMI. Replacement of electromechanical control circuits with solid-state programmable units just cannot be done at random. Interfacing costs are a major consideration.

Survivability is another argument against full SPC with centralized control. There is a certain amount of redundancy in the control circuitry of a conventional crossbar switch. Calls can be processed several at a time, each with its own marker, registers, and translator. There is "graceful degradation" in service when one control unit fails. However, let the singular central control fail in a full SPC switch, and catastrophic failure results. This is one important reason why the Metaconta L has two central processors.

To render the system cost-effective, the design engineer must select the point of interface between the conventional electromechanical switch and the SPC modules to be installed. One consideration for the interface point would be the concentration links of the register–receiver–sender–translator group.

A major area that is a prime candidate for replacement of hardware by a software-oriented device [processor(s)] is in the routing and control sector of an electromechanical switch. It is here where most changes are required in a switch during its lifetime. These are the administrative and functional changes of subscriber data, routing, and signaling. By implementing SPC, many of the advantages are achieved by carrying out changes in the stored programs rather than the laborious strapping required on the electromechanical units. There is also the advantage of substitution of one for many. One control processor can replace 64 or more conventional registers, although it is desirable to use 2 processors for the additional reliability provided. One processor in this case is used for backup with automatic switchover on failure of the operational processor. As a switch grows (i.e., as more subscribers are added) during its lifetime, additional processors may be added, as needed, to share the load.

One approach to centralization is to have the several control processors be served by a common memory for the switch as a whole. This simplifies the problem of programming new information or modifying existing programs. It also simplifies the I/O such that one keyboard teleprinter can serve the entire switch.

Administration and maintenance procedures are often carried out by a separate processor in a modified electromechanical switch. Connection is made to the routing and control processor bus as well as to electromechanical elements. A separate I/O is also provided.

12.5.3 Distributed Processing: Another Approach. The concept of distributed processing has evolved with the availability of low-cost microprocessors. The approach is economical because it overcomes some of the limitations of the central processor–common-control technique described in Section 12.4. One limitation is that of speed in time sharing. As more calls enter an exchange, the faster the central controller must process them. Processing time is a function of cost, considering even that a computer can operate up to several thousand times faster than electromechanical control. Distributed processing removes much of the burden of ultrarapid serial processing. Of course, reliability is increased notably and chances of catastrophic failure are decreased with distributed microprocessor control.

There are two principles basic to distributed control: (1) free communication between control processors via data buses and (2) one or several processor(s), termed *central control*, acting as system coordinator (not master). Control is divided on a functional basis between different types of control units, and a number of similar control units of each type are provided to share the workload. Such a modular approach permits the addition of common modules as an exchange expands service. The TXE4 [17] is an example of a local telephone exchange using distributed control. This exchange is a British development and is manufactured by Standard Telephones and Cables Ltd. Figure 3.22 is a simplified block diagram of the TXE4 showing the switching network, which is composed of reed relay matrices, and the control network, which is microprocessor based.

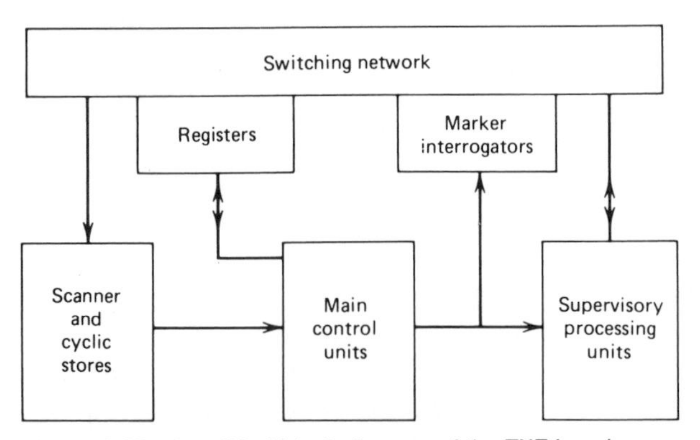

Figure 3.22 Simplified block diagram of the TXE4 exchange.

The basic control system is made up of cyclic stores, main control units, markers, and supervisory processing units. The cyclic stores are "read only" to the system. They contain the data dealing with subscriber and trunk terminations, routing codes, and numbering information relating to a particular exchange. A cyclic store uses a semiconductor memory and large shift registers up to 13k bits in length, and up to 60k bits are accommodated on one printed-circuit board. Data from the cyclic stores are broadcast to all main control units on a cyclic basis, providing the necessary multiple access. Up to 20 main control units are provided, depending on exchange size. These units supervise all call-connection processes that receive data from the cyclic stores, provide communications with subscriber lines and trunks, and instruct markers and supervising control units. Main control units have some self-checking capability and can be rendered out of service on failure. Note in Figure 3.22 that registers are independent and can handle complete digits and perform their own time-out functions, relieving the main control unit of this burden.

The markers are also microprocessor based, and up to 80 are provided on the TXE4. The main controller identifies the two ends of a required connection and the type of marking sequence to be used. The marker then finds the possible free paths in associated interrogation equipment and sets up a busy test of the best path in accordance with its internal program. The marker also has self-checking capability and will remove itself from service when a fault condition is discovered.

Supervisory processing units provide call supervision. A unit monitors the state of each call under its jurisdiction as well as its service requirements, reserving a specific area in its storage for each junctor under supervision. Junctors are scanned in rotation, as is the memory area, to determine change in call state and then process this information and issue an instruction sequence to the junctor when necessary.

Data "highways" provide communication between processors and can become complex and expensive, particularly with the call-charging functions and the additional service facilities usually offered by SPC-based exchanges. The TXE4 uses overlay processors connected to the necessary data-highway ports that can communicate with the main control units. The overlay processors carry out dedicated functions, relieving the main control units of these responsibilities.

13 CONCENTRATORS AND SATELLITES

Concentrators or "line concentrators" consolidate subscriber loops, are remotely operated, and are the concentration (and expansion) portion of a switch placed at a remote location. This is the conventional meaning of a concentrator. The concentrator may use S × S or crossbar facilities for the concentration matrix. Control may be by conventional relay or by a hard-wired

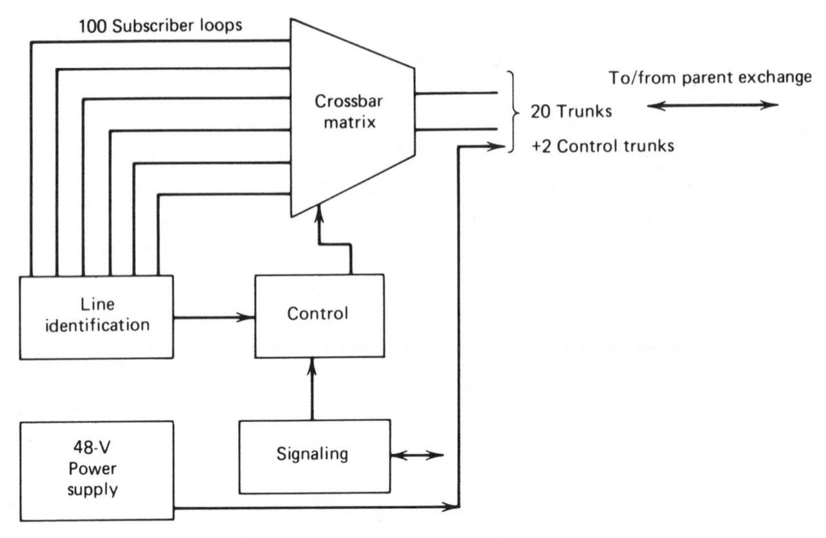

Figure 3.23 A crossbar concentrator.

processor. A typical crossbar type is shown in Figure 3.23, where 100 subscriber loops are consolidated to 20 trunks plus 2 trunks for control from a nearby exchange. Concentrators are used in sparsely populated areas that require a long trunk connection to the nearest exchange. In effect, the concentrators extend the serving area of an exchange. However, all calls originating in the concentrator serving area must be serviced by the parent exchange. Conventional concentrators can serve up to several hundred subscribers.

Concentration can also be effectively carried out using carrier techniques, either pulse-code modulation (PCM) (Chapter 9) or frequency-division multiplex (FDM) (Chapter 5).

A satellite switch originates and terminates calls from a parent exchange. It differs from a concentrator in that local calls (i.e., calls originating and terminating inside the same satellite serving area) are served by the satellite and do not have to traverse the parent exchange. A block of numbers is assigned to the satellite serving area and is usually part of the basic number block assigned to the parent exchange. Because of its numbering arrangement, the satellite can discriminate between local calls and calls to be handled by the parent exchange. The satellite can be regarded as a component of the parent exchange that has been dislocated and moved to a distant site. This is still another method of extending the serving area of a main exchange to reduce the cost of serving small groups of subscribers, which, under ordinary circumstances, could only be served by excessively long loops to the parent exchange. Satellites range in size from 300 to 2000 lines. Concentrators are usually more cost-effective than satellites when serving 300 lines or fewer.

14 CALL CHARGING: EUROPEAN VERSUS NORTH AMERICAN APPROACHES

In Europe and in countries following European practice, telephone-call charging is simple and straightforward. Each subscriber line is equipped with a meter with a stepping motor at the local exchange. Calls are metered on a time basis. The number of pulses per second actuating the meter are derived from the exchange code or the area code of the dialed number. A local call to a neighbor (same exchange) may be 1 pulse (one step) per minute, a call to a nearby city 3 pulses, or a call to a distant city 10 pulses per minute. International direct-dial calls require checking the digits of the country code that will then set the meter to pulse at an even higher rate. All completed calls are charged; thus the metering circuitry must also sense call supervision to respond to call completion, that is, when the called subscriber goes "off hook" (which starts the meter pulses) and also to respond to call termination (i.e., when either subscriber goes "on hook") to stop meter pulses. Many administrations and telephone companies refer to this as the *Karlson method* of pulse metering. Some form of number translation is required, in any event, to convert dialed number information to key the metering circuit for the proper number of pulses per unit period.

Pulse metering or "flat rate" billing requires no further record keeping other than periodic meter readings. Charging equipment cost is minimal, and administrative expenses are nearly inconsequential. Detailed billing on toll calls is used in North America (and often flat rate on local calls). Detailed billing is comparatively expensive to install and requires considerable administrative upkeep. Such billing information is recorded by determining first the calling subscriber number and then the called number. Also, of course, there is the requirement of timing call duration.

Automatic message accounting (AMA) is a term used in North American practice. When the accounting is done locally, it is called *local automatic message accounting* (LAMA), and when carried out at a central location, *centralized automatic message accounting* (CAMA). Sometimes operator intervention is required, especially when automatic number identification (ANI) is not available or for special lines such as hotels. With automatic identified outward dialing (AIOD), independent data links are sometimes used, or trunk outpulsing may be another alternative.

Only some of the complexities have been described in North American billing practice. One overriding reason why detailed billing is provided on long-distance service is customer satisfaction. In European practice it is difficult to verify a telephone bill. A customer can complain that it is too high but can do little to prove it. Ordinarily subscriber-dialed long-distance service is lumped with local service, and a bulk amount for the number of meter steps incurred for the intervening billing period is shown on the bill. With detailed billing, individual subscriber-dialed calls can be verified and the bulk local charge is separate. In this case a subscriber has verification for a complaint if it is a valid one.

15 TRANSMISSION FACTORS IN ANALOG SWITCHING

15.1 Introduction

Transmission in telephony deals with the delivery of a "quality" signal to the far-end user. If the signal is speech, it should suffer minimum distortion as well as meet certain requirements regarding level and signal-to-noise ratio. If the signal is data, it should meet a certain minimal error rate. A switch placed in a transmission path is one more element that will affect signal quality as it traverses the network. On many calls, whether speech, data, or facsimile, there could be as many as 13 switches in tandem. Thus the extent to which a switch distorts transmission must be specified.

15.2 Basic Considerations

Normally a switch sets up a speech path. Space division switching, the type of switching discussed in this chapter, provides a metallic path made up of wire and relay or semiconductor cross-points. Thus we would expect an analog space-division switch to act on the signal passing through it in the following ways:

1. It will attenuate the signal; this is called *insertion loss*.
2. It will add noise to the signal, thus deteriorating signal-to-noise ratio.
3. As longitudinal balance is improved, it will tend to decrease added noise.
4. It will be a source of impulse noise, especially with gross-motion switches.
5. It will tend to distort the attenuation response.
6. It will tend to deteriorate envelope delay characteristics.
7. It will have a return loss characteristic when looking into the switch from either direction. This may affect echo characteristics.
8. It will be a source of crosstalk.
9. It will be a source of absolute delay.
10. It will be a source of harmonic distortion.

15.3 Zero Test-Level Point

Signal level in telephony (see Chapter 5, Section 2.3) is a most important consideration. In a telephone network we often deal with relative levels measured in dBr, which can be related to the familiar dBm as follows:

$$\text{dBm} = \text{dBm0} + \text{dBr} \tag{3.2}$$

where dBm0 is an absolute unit of power in dBm referred to as the 0 TLP.

The 0 TLP can be located anywhere in the network. North American practice places it at the outgoing switch of the local switching network, whereas CCITT Rec. G.122 sets levels outgoing from the first international switch in a country (the CT; see Chapter 6) at −3.5 dBr and entering the same switch from the international network at −4.0 dBr. The CCITT then leaves it to the national telephone administration to set the 0 TLP, provided that it meets the relative levels just specified.

TABLE 3.4 Transmission Characteristics for Four-Wire Switches

Item	1 [(CCITT) Q.45][c]	2 (Ref. 18)[c]
Loss	0.5 dB	0.5 dB
Loss, dispersion[a]	< 0.8 dB	< 0.2 dB
Attenuation / frequency response	300–400 Hz: −0.2 / +0.5 dB 400–2400 Hz: −0.2 / +0.3 dB 2400–3400 Hz: −0.2 / +0.5 dB	−0.1 / +0.2 dB −0.1 / +0.2 dB −0.1 / +0.3 dB
Impulse noise	5 in 5 min above −35 dBm0	5 in 5 min, 12 dB above floor of random noise[d]
Noise		
Weighted	200 pWp	25 pWp
Unweighted	100,000 pW0	3000 pW
Unbalance against ground	300–600 Hz: 40 dB 600–3400 Hz: 46 dB	300–3000 Hz: 55 dB 3000–3400 Hz: 53 dB
Crosstalk		
Between go and return paths	60 dB	65 dB (in the band 200–3200 Hz)
Between any two paths	70 dB	80 dB (in the band 200–3200 Hz)
Harmonic distortion		50 dB down with a −10 dBm signal for second harmonic, 60 dB down for third harmonic
Impedance variation with frequency[b]	300–600 Hz: 15 dB 600–3400 Hz: 20 dB	200 Hz: 15 dB 300 Hz: 18 dB 500–2500 Hz: 20 dB 3000 Hz: 18 dB 3400 Hz: 15 dB
Group delay (600–3000 Hz)	100 μsec	

[a] Dispersion loss is the variation in loss from calls with the highest loss to those with the lowest loss. This important parameter affects circuit stability.
[b] Expressed as return loss.
[c] Reference frequency, where required, is 800 Hz for column 1 and 1000 Hz for column 2.
[d] Taken from standard measurement techniques for impulse noise.

System gain is usually achieved and levels adjusted at the inputs and outputs of carrier equipment (FDM, see Chapter 5; PCM, Chapters 9 and 10) to meet the requirements of a national transmission plan.

15.4 Transmission Specifications

Table 3.4 lists transmission characteristics for four-wire switches. Loss dispersion deals with the variation of loss through a switch. The loss of a four-wire switch (Table 3.4) is 0.5 dB. On some paths through the switch the loss may only be 0.4 dB and on others, 0.6 dB. Therefore the loss dispersion is 0.2 dB. This factor is of particular importance when dealing with network stability. If loss dispersion is too great, singing may occur, particularly when circuits operate near their singing margin. Return loss at local exchanges (two-wire) should be at least 11 dB (North American practice), with a standard deviation of 3 dB in the band 500–2500 Hz and from 250 Hz to 3200 Hz, 6 dB with a standard deviation of 2 dB. Return loss measurements on a local exchange are made against a 600-Ω or 900-Ω resistor in series with a 2.14-microfarad (μF) capacitor [12].

Longitudinal balance refers to balance to ground for both sides of the circuit. Open telephone circuit measurements from one of the two wire leads to ground can display a 100-V peak, and it is not unusual to find several milliamperes of current through a 1000-Ω terminating resistor to ground. The objective is to balance the two sides to ground, producing no net voltage between the two sides. When there is imbalance, the noise produced is often referred to as *metallic noise*. The method of measurement of longitudinal balance must be stated.

16 NUMBERING CONCEPTS FOR TELEPHONY

16.1 Introduction

Numbering was introduced in Section 2 as a basic element of switching in telephony. This section discusses, in greater detail, numbering as a factor in the design of a telephone network.

16.2 Definitions

There are four elements to an international telephone number. CCITT Rec. E.163 recommends that not more than 12 digits make up an international number. These 12 digits exclude the international prefix, which is that combination of digits used by a calling subscriber to a subscriber in another country to obtain access to the automatic outgoing international equipment; thus we have 12 digits maximum made up of 4 elements. For example, dialing from Madrid to a specific subscriber in Brussels requires only 10 digits (inside the

12-digit maximum).

07	32	2	4561234
International prefix	Country code	National significant number	Subscriber number
		Trunk code (area code)	

Thus the international number is

32 2 456 1234

According to CCITT international usage (Recs. E.160, E.161, and E.162), we define the following terms.

Numbering Area (Local Numbering Area). This is the area in which any two subscribers use the same dialing procedure to reach another subscriber in the telephone network. Subscribers belonging to the same numbering area may call one another simply by dialing the subscriber number. If they belong to different numbering areas, they must dial the trunk prefix plus the trunk code in front of the subscriber number.

Subscriber Number. This is the number to be dialed or called to reach a subscriber in the same local network or numbering area.

Trunk Prefix (Toll-Access Code). This is a digit or combination of digits to be dialed by a calling subscriber making a call to a subscriber in his own country but outside his own numbering area. The trunk prefix provides access to the automatic outgoing trunk equipment.

Trunk Code (Area Code). This is a digit or combination of digits (not including the trunk prefix) characterizing the called numbering area within a country.

Country Code. This is the combination of one, two, or three digits characterizing the called country.

Local Code. This is a digit or combination of digits for obtaining access to an adjacent numbering area or to an individual exchange (or exchanges) in that area. The national significant number is not used in this situation.

From Madrid, dialing a subscriber in Copenhagen requires 9 digits, Brussels 10, near London (Croydon) 10, Harlow (England) outskirts 11, Harlow center 10, and New York City 11 (not including the international prefix). This raises the concepts of uniform and nonuniform numbering as well as some ambiguity.

The CCITT defines *uniform numbering* as a numbering scheme in which the length of the subscriber numbers is uniform inside a given numbering area. It defines *nonuniform numbering* as a scheme in which the subscriber numbers vary within a given numbering area. With uniform numbering, each subscriber, by using the same number of digits, can be reached inside a numbering area and from one numbering area to another inside national boundaries. Theoretically, this is true for North America (north of the Rio Grande River), where a subscriber can dial seven digits and always seven digits for that subscriber to reach any other subscriber inside the calling area. Ten digits is required to reach any subscriber outside the numbering area. Note that in North America the trunk code is called the *area code*. This arrangement is shown in the following formula as it was used at the end of 1973, before the introduction of "interchangeable codes" discussed briefly further in this section.

Area code	Telephone number (subscriber number)
$N\frac{0}{1}X$*	$NNX\text{-}XXXX$

where X is any number from 0 to 9, N is any number from 2 to 9, and 0/1 is the number 0 or 1.

Following a plan set up as in this formula, ATT (the major North American telephone administration prior to 1983) provided for an initial arrangement of 152 area codes, each with a capacity of 540 exchange codes. Remember that there are 10 digits involved in dialing between areas including all the 50 United States, Canada, Puerto Rico, and parts of Mexico.

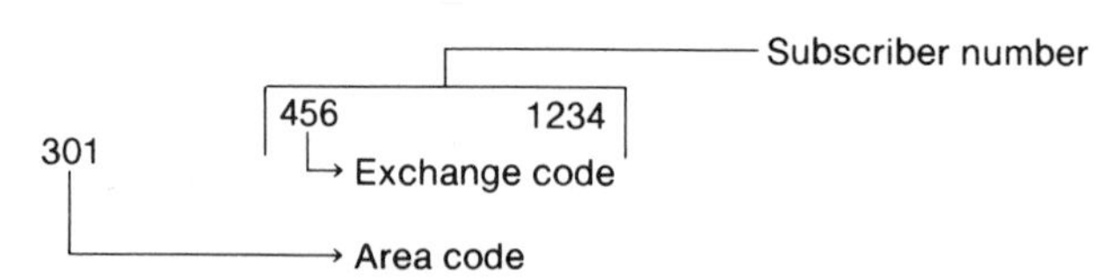

Subsequently, the 540 exchange codes were expanded to 640 when ANC (all number calling) was introduced. This simply means that the use of letters was eliminated and digit combinations 55, 57, 95, and 97 could be utilized where names and resulting letters could not be structured from such combinations, and the code group $NN0$ was also added. This provided needed number relief. This relief was not sufficient to meet telephone growth on the continent, and further code relief had to be provided. This was carried out by realigning code areas, introducing interchangeable codes with code areas where required, and splitting existing code areas and introducing new area codes [13].

It should be appreciated that a switch must be able to distinguish between receipt of calls bound both out of and within the calling area. The introduction of "interchangeable" codes precludes the ability of a switch to determine whether a seven-digit number (bound for a subscriber inside the same calling area) or a 10-digit number (for a call bound for a subscriber in another calling

*Excludes the combination $N11$.

area) can be expected. This was previously based on the presence of a "0" or "1" in the second digit position. Now the switch must use either of two different methods to distinguish seven-digit numbers: the "timing method," where the exchange is designed so that the equipment waits for 3–5 s after receiving seven digits to distinguish between seven- and ten-digit calls (if one or more digits are received in the waiting period, the switch then expects a ten-digit call) or the "prefix method," which utilizes the presence of either a "1" or "0" prefix that identifies the call as having a ten-digit format. This, of course, is the addition of an extra digit.

The prefix method described here for ATT is the "trunk prefix" defined previously. Such a trunk prefix is used in Spain, for example. In the automatic service Spain uses either six or seven digits for the subscriber number. Numbering areas with high population density and high telephone growth have seven digits, such as Madrid, Barcelona, and Valencia. For numbering areas of low telephone density, six digits are used. However, when dialing between numbering areas, the subscriber always dials nine digits. This is also referred to as a *uniform system*. Inside numbering areas the subscriber number is made up of six or seven digits. Area codes (trunk codes) consist of one or two digits. The ninth digit is used for toll access. To dial Madrid, 91 + 7 digits is used, and to dial Huelva (a small province in southwestern Spain), 955 + 6 digits is used. The United Kingdom presents an example of nonuniform numbering where subscriber numbers in the same numbering area may be five, six, or seven digits. Dialing from one area to another often involves different procedures.

16.3 Factors Affecting Numbering

16.3.1 General. In telecommunications network design there are numerous trade-offs between economy and operability. Operability covers a large realm, one aspect of which is the human interface. The subscriber must use his/her telephone, and its use should be easy to undersand and apply. Uniform numbering and number length notably improve operability.

Number assignment should leave as large a reserve of numbers as possible for growth. The simplest method to accomplish this tends toward longer numbers and nonuniform numbers. Another goal is to reduce switching costs. One way is to reduce number analysis, that is, the number of digits to be analyzed by a switch for proper routing and charging. We find that an economical analysis becomes increasingly necessary as the network becomes more complex with more and more direct routes.

16.3.2 Routing. Consider the scheme for number analysis in routing shown in Figure 3.24. Only one-digit analysis is required to route any call from exchange X to any station in the network. The first digit selects the required outlet. However, if a direct route was established between X and Z_1, two-digit analysis would be required. If a direct route was established to Z_2, three-digit

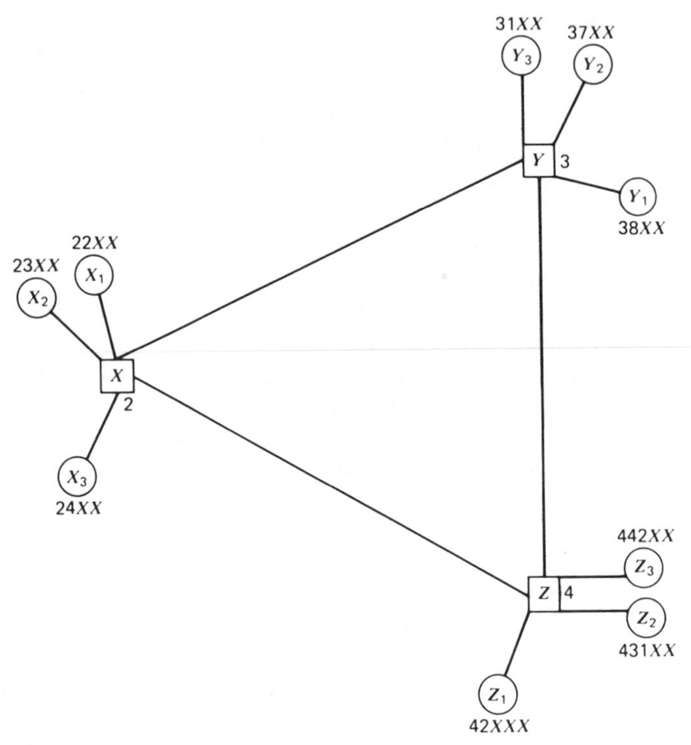

Figure 3.24 Number analysis in routing (X = any digit from 0 to 9).

analysis would be necessary. Note that some freedom of number assignment has been lost because all numbers beginning with digit 2 home on X, 3 on Y, and 4 on Z. With the loss of such freedom, we have gained simplification of switches. We could add more digit analysis capability at each exchange and gain more freedom of digit assignment.

This brings up one additional point, geographical significance and the definition of numbering areas. When dealing with trunk codes, "geographical significance" means that neighboring call areas are assigned digits beginning with the same number. In Spain there is geographical significance; in the United States there is not. In the northwestern corner of Spain all trunk codes begin with 8. Aside from Barcelona all codes in northeastern Spain begin with 7. In Andalusia all begin with 5, and around Valencia the codes begin with 6; Alicante is 65 and Murcia is 68. In the United States, New York City is 212 and 718, just to the north is 914, across the Hudson River is 301, and Long Island is 516.

However, both countries try to use political administrative boundaries to coincide with call-area boundaries. In the United States in many cases we have a whole state or a grouping of counties, but never crossing state boundaries. In Spain we are dealing with provinces, with each province bearing its own call-area number (although some provinces share the same number). Nevertheless, boundaries do coincide with political demarcations. This eases subscriber understanding and simplifies tariffing procedures.

16.3.3 Tariffs (Charging). The billing of telephone calls is automated in the automatic service. There are two basic methods for charging, bulk billing and detailed billing (see Section 14 of this chapter). Detailed billing is essentially a North American practice. Detailed billing includes in the subscriber's bill a listing of each toll call made, number called, date and time, charge time, and individual charge entry for each toll call. Such a form of billing requires extensive number analysis, which is usually carried out by data-processing equipment in centralized locations.

When charges are determined typically by accumulated meter steps, billing is defined as bulk billing. The subscriber is periodically presented with a bill indicating the total of steps incurred for the period at a certain money rate per step. Bulk billing is much more economical to install and administer. It does lead to much greater subscriber misunderstanding and, in the long run, to dissatisfaction.

With bulk billing, stepping meters are part of the switching equipment. The switching equipment determines the tariff and call duration and actuates the meters accordingly. For bulk billing the relationship between numbering and billing is much closer than in the case of detailed billing. Ideally, tariff zones should coincide with call areas (routing areas). Care should be taken in the compatibility of charging for tariff zones and the numerical series used in numbering. It follows that the larger the tariff zone, the less numerical analysis required.

16.3.4 Size of Numbering Area and Number Length. Size relates not only to geographical dimension but also to the number of telephones encompassed in the area. Areas of very large geographical size present problems in network and switch design. Principally, large areas show a low level of community of interest between towns. In this case switching requires more digit analysis, and tariff problems may be complicated. If the area has a large telephone population, a larger number of digits may be required, with implied longer dialing times and more postdial delay for analyses. However, large areas are more efficient for number assignment.

With smaller numbering areas, short-length subscriber numbers may be used for those with a higher community of interest. Smaller digit storage would be required for intra-area calls with shorter holding times for switch control units. Smaller areas do offer less flexibility, particularly in the future when uneven growth takes place and forecasts are in error. In the mid-1970s North America faced a situation requiring reconfiguration and the addition of new areas and also the use of an extra digit or access code.

One report [10] recommends, as a goal, that a numbering area be no greater than 70,000 km^2 nor have less than 100,000 subscribers at the end of a numbering-plan validity period. Subscribers should be able to dial a shorter number when calling other subscribers inside their own numbering area. All other subscribers would be reached by dialing the national number. This is done by adding a simple trunk code in front of the subscriber number. The use of a trunk prefix has advantages in switch design. This should be a single digit, preferably 0. Zero is recommended because few if any subscriber numbers start with zero. Thus when a switch receives the initial digit as zero, it is prepared to receive the longer toll number for interarea dialing, whereas if it receives any other digit, it is prepared for receipt of a subscriber number for intra-area dialing. In the United Sates, the initial digit 1 is often used to indicate a toll call.

For dialed international calls, the CCITT recommends number length no greater than 12 digits. A little research will show that few international calls require 12 digits. Number length, of course, deals with number of subscribers, code blocks reserved, immediate and future spare capacity, call-area size, and trunk prefix assignment. Uniform and nonuniform numbering are also factors. Other factors are 40-year forecast accuracy, subscriber habits, routing and translation facility availability, switching system capabilities, and existing numbering scheme.

16.4 In-Dialing

Numbers, particularly in uniform numbering systems, must be set aside for PABXs with in-dialing capability. This is particularly true when the PABX numbering scheme is built into the national scheme for in-dialing. For example, suppose that a PABX main number is 543-7000, with an extension to that PABX of 678. To dial the PABX extension directly from the outside, we would dial 543-7678, and the 543-7*XXX* block would be lost for use except for 999 extension possibilities for that PABX. If it were a small installation, the numbering block loss could be reduced by 543-7600 (PABX main number), extension 78—543-7678. The PABX would be limited to 99 (or 98) direct in-dialed extensions. If the PABX added digits onto the end of the main number, there would be no impact on national or area numbering. For example:

Main PABX number:	543-7000
PABX extension number:	678
Extension dialed directly from outside:	543-7000678

Here we face the CCITT limitation of 12 digits for international dialing; 10 of the 12 have been used. We must also consider digit storage in switches handling such calls. Of course, the realm of nonuniform numbering has been entered. A third method suggested is shown in the following example:

Main PABX number:	53-87654
PABX extension number:	789
Extension dialed directly from outside:	503-8789

Here the first and third digits locate the local exchange on which the PABX is a subscriber. That local exchange must then analyze the fourth, fifth, and possibly the sixth digits. Disadvantages are that translation is always required and that there is some loss to the numbering system.

17 TELEPHONE TRAFFIC MEASUREMENT

Statistics on telephone traffic is mandatory both at the individual exchange level and for the network as a whole. Traffic measurements provide the required statistics when carried out in accordance with a well-organized plan. The statistics include traffic intensity (erlangs or ccs) and its distribution by type of subscriber and service on each route and circuit group and its variation daily, weekly, and seasonally. In the process of traffic measurement, congestion is indicated, if present, and switch and network efficiency can be calculated. The most common measurement of "efficiency" is grade of service.

Congestion and its causes tell the systems engineer whether an exchange is overdimensioned, underdimensioned, or of improper traffic balance. Traffic statistics provide a concrete base or starting point from which to forecast growth. It gives past traffic evolution of subscriber-generated traffic growth by type of subscriber and class of service. It also provides a prediction of the evolution of local traffic between exchanges of national long-distance (toll) traffic and international traffic. Traffic measurements (plus forecasts) supply the data necessary to dimension new exchanges and for the extension of existing ones. They are especially important when a new exchange is to replace an existing exchange or when a new exchange will replace several existing exchanges.

Traffic measurement involves a number of parameters, including seizures (call attempts), completed calls, traffic intensity (involving holding times), and congestion. The term "seizures" indicates the number of times a switching unit or groups are seized without taking into account holding time. Seizures may be equated to call attempts and give an expression of how much exchange control equipment is being worked. The number of completed calls is of interest in the operation and administration of an exchange. Of real interest are statistics on uncompleted calls that are not attributed to busy lines or lack of answer or that can be attributed to the specified grade of service. However, at a particular exchange a completed call means only that the switch in question has carried out its function, which does not necessarily imply that a connection has been established between two subscribers. The intensity of traffic or traffic volume, a most important parameter, directly provides a measure of usage of a circuit. It is especially useful in those switching units involved directly in the speech network. On the other hand, it is not directly indicative of grade of service. Approximate grade of service should be taken from the traffic tables used to dimension the exchange by using measured traffic intensity as an input. The approximation is most accurate when traffic intensity approaches dimensioned intensity.

Congestion involves three characteristics: "all circuits busy," overflow, and dial-tone delay. "All circuits busy" is an indication of the number of times, and eventually the duration, where all units of a switching group are handling traffic simultaneously. Thus it represents an index of the real grade of service. Its use is particularly effective for those switching groups that are operating near maximum capacity. For an overdimensioned exchange, the "all circuits busy" index is useless and tells us nothing. Overflow is an index of the number of call attempts that have not been able to proceed due to congestion. For networks with alternative routing, overflow tells us the number of offered calls not handled by specific switching equipment group. Dial-tone delay is directly indicative of overall grade of service that an exchange provides its subscribers, particularly in the preselection switching stages. Dial-tone delay is normally expressed in the time required in getting dial tone compared to a fixed time, usually 3 s, as a percentage of total calls.

Traffic measurements should be made through the busy hour (BH), which can be determined by reading amperage of exchange battery over the estimated period of occurrence, say, 9:30 A.M. to 12:30 P.M., every 10 min (or by peg count usage devices, microprocessors on lines, equipment or trunk circuits). These measurements should be done daily for at least three weeks. Traffic measurements should be carried out for the work week, one week per month for an entire year. The means of measurement depends on available equipment and exchange type. This may range from simple observation to fully automatic traffic-measuring equipment with recorders on magnetic or paper tape. The engineer may have to resort to the use of electromechanical counters and/or electromechanical or electronic hunting devices with which many exchanges, particularly those with common control, are equipped.

18 DIAL-SERVICE OBSERVATION

Dial-service observation provides an index or measurement of the telephone service provided to the customer (see Chapter 1, Section 13). It gives an administration or telephone company a sample measure of maintenance required (or how well it is being carried out), load balance, and adequacy of installed equipment. The number of customer-originated calls sampled per day is of the order of 1–1.5% of the total calls originated; 200 observations per exchange per day is a minimum [20].

Traditionally, service observation has been done manually, requiring the presence of an observer. Service observation positions have some forms of automation, such as automatic recording of a calling number and/or the mark sensing or keypunching to be employed for computer-processing input. Tape recorders may be placed at exchanges to automatically record calls on selected lines. The tapes are then periodically sent to the service observation desk. Tapes usually have a "time hack" to record the time and the duration of calls. A service observation desk usually serves many exchanges or an entire local area. Tandem exchanges are good candidates because they concentrate the service function. The results of service observation are intended to represent average service; thus care should be taken to ensure that subscribers selected represent the average customer. The following is one list of data to be collected per line observed:

- Total call attempts (local and long distance).
- Completed calls.
- Ineffective calls due to calling subscriber: incomplete call, late dialing, unavailable number dialed, call abandoned prematurely.
- Ineffective calls due to called subscriber: busy, no answer.
- Ineffective calls due to equipment: wrong number, congestion, no ringing, busy tone, no answer, interruptions.
- Transmission quality (level, distortion, noise, etc.).
- Wrong number dialed by subscriber.
- Dial-tone delay.
- Postdial delay.

Results of service observation are computerized and summary tables published monthly or quarterly. Individual tables are often made for each exchange, relating it to the remainder of the network. Quality of route can be determined since observations identify the called local exchange. A similar set of observations may be made for the long-distance (toll) service exclusively.

There is a tendency to fully automate service observation without any human intervention. The major weakness to this approach is the lack of subjective observation of factors such as type of noise and its effect on a call, subscriber behavior to certain stimuli, and so forth.

REVIEW QUESTIONS

1. List at least four user requirements with regard to switching.
2. Why do we use switching in the first place? Explain the rationale of switching versus transmission.

3. Give two basic functions of a telephone number.

4. What is a *serving area*?

5. Without digit blocking, how many distinct entities can be served by a three-digit number? a four-digit number?

6. What are the two basic functions of a local switch?

7. In a local area a certain traffic relation is 12 erlangs. What would be the most economical form of routing: direct or tandem?

8. There are eight basic functions that a conventional generic switch carries out. Name at least six.

9. What does the term *supervision* mean in switching?

10. In the concentration stage of a local switch, are there more or less inputs than outputs?

11. In the expansion stage of a local switch, the number of outputs equals the number of______.

12. Define a trunk group using the term *traffic relation*.

13. There are two basic categories of electromechanical space division switches. Name them and give an example switch type for each.

14. What does *degeneration* tell us about a particular switch?

15. Differentiate between *limited availability* and *full availability*.

16. Define *grading*.

17. What is the function of latching on a crossbar switching matrix?

18. Differentiate between direct progressive control and register progressive control.

19. Based on the definition in this text, what major advance does *common control* have over a progressive control system?

20. In a crossbar switch, what unit carries out the common-control function?

21. What is the function of a *junctor* on crossbar switches?

22. Why would a marker need more than one register?

23. Why is it advantageous to use *equipment number* (EN) rather than *dial number* (DN) in a switch?

24. SPC simply means that a switch is ______-controlled.

25. Give at least three major advantages of SPC over hard-wired control switches.

26. What is the meaning of *holding time* and *delay* with regard to SPC systems?

27. Give at least two advantages and one disadvantage of using centralized control (versus distributed control).

28. What is the one major difference between a *concentrator* and a *satellite*?

29. Describe the differences between European and North American charging methods. Bring in the acronyms *CAMA* and *LAMA*.

30. Give at least five transmission impairments encountered in an analog switch.

31. Give the CCITT and North American definitions of 0 TLP.

32. Discuss the importance of longitudinal balance on a wire pair trunk.

33. Give the two different characteristic impedances that may be encountered on local exchange VF switch ports.

34. Differentiate between uniform and nonuniform numbering.

35. With switch complexity in mind, discuss the importance of numbering on routing.

36. Describe at least two methods of measuring telephone traffic through a switch. Describe one indirect method that is fairly accurate.

37. Give at least three traffic parameters that are measured at a switch that give valuable insight to planners, system design engineers, and telephone company/administration managers.

38. In what period during a weekday is traffic measurement most important?

39. Give at least five parameters to be included in dial-service observation.

REFERENCES

1. *Switching Systems*, American Telephone and Telegraph Company, New York, 1961.
2. Marvin Hobbs, *Modern Communication Switching Systems*, Tab Books, Blue Ridge Summit, Pa., 1974.
3. T. H. Flowers, *Introduction to Exchange Systems*, Wiley, New York, 1976.
4. Amos F. Joel, "What is Telecommunication Circuit Switching?" *Proc. IEEE*, **65** (9) (September 1977).
5. *Fundamental Principles of Switching Circuits and Systems*, American Telephone and Telegraph Company, New York, 1963.
6. International Telephone and Telegraph Corporation, *Reference Data for Radio Engineers*, 6th ed., Howard W. Sams, Inc., Indianapolis, 1976.

7. J. G. Pearce, *Electronic Switching*, Telephony Publishing Company, Chicago, 1968.
8. J. P. Dartois, "Metaconta L Medium Size Local Exchanges," *Electr. Commun.*, **48** (3) (1973).
9. J. G. Pearce, "The New Possibilities of Telephone Switching," *Proc. IEEE*, **65** (9) (September 1977).
10. "Numbering," Telecommunication Planning, ITT Laboratories (Spain), Madrid, 1973.
11. CCITT, Q Recommendations (Rec. Q), Red Books, VIII Plenary Assembly, Malaga-Torremolinos, Spain, October 1984.
12. L. F. Goeller, *Design Background for Telephone Switching*, Lees ABC of the Telephone, Geneva, Ill., 1977.
13. *Notes on the Network 1980*, American Telephone and Telegraph Company, New York, 1980.
14. Bruce E. Briley and Wing N. Toy, "Telecommunication Processors," *Proc. IEEE*, **65** (9) (September 1977).
15. T. H. Flowers, "Processors and Processing in Telephone Exchanges," *Proc. IEEE*, **119** (3) (March 1972).
16. Enn Aro, "Stored Program Control-Assisted Electromechanical Switching—An Overview," *Proc. IEEE*, **65** (9) (September 1977).
17. P. J. Hiner, "TXE4 and the New Technology," *Telecommunications* (January 1977).
18. USITA Symposium, April 1970, Open Questions 18–37.
19. Roger L. Freeman, *Telecommunication Transmission Handbook*, 2nd ed., Wiley, New York, 1981.
20. L. Alvarez Mazo and M. Poza Martinez, "Dial Service Observation," Telecommunication Planning Symposium, STC (SA) Boksburg, South Africa, June 1972.
21. L. Alvarez Mazo, "Network Traffic Measurements," Telecommunication Planning Symposium, STC (SA) Boksburg, South Africa, June 1972.
22. Robin Deese, "A History of Switching," *Telecommunications* (February 1984).
23. *Notes on the BOC Intra-LATA Networks—1986*, TR-NPL-OO275, Bellcore, Livingston, N.J. April 1986. Bell Communications Research Technical Reference.

4

SIGNALING FOR ANALOG TELEPHONE NETWORKS

1 INTRODUCTION

In a switched telephone network signaling conveys the intelligence needed for one subscriber to interconnect with any other in that network. Signaling tells the switch that a subscriber desires service and then gives the local switch the data necessary to identify the required distant subscriber and hence to route the call properly. It also provides supervision of the call along its path. Signaling also gives the subscriber certain status information, such as dial tone, busy tone (busy back), and ringing. Metering pulses for call charging may also be considered a form of signaling.

There are several classifications of signaling:

1. General.
    a. Subscriber signaling.
    b. Interswitch signaling.
2. Functional.
    a. Audible–visual (call progress and alerting).
    b. Supervisory.
    c. Address signaling.

Figure 4.1 shows a more detailed breakdown of these functions.

It should be appreciated that on many or even on most telephone calls more than one switch is involved in call routing. Therefore switches must interchange information among switches in fully automatic service. Address information is provided between modern switching machines by interregister signaling and the supervisory function by line signaling. The audible–visual category of signaling functions inform the calling subscriber regarding call *progress*, as shown in Figure 4.1. The *alerting* function informs the called

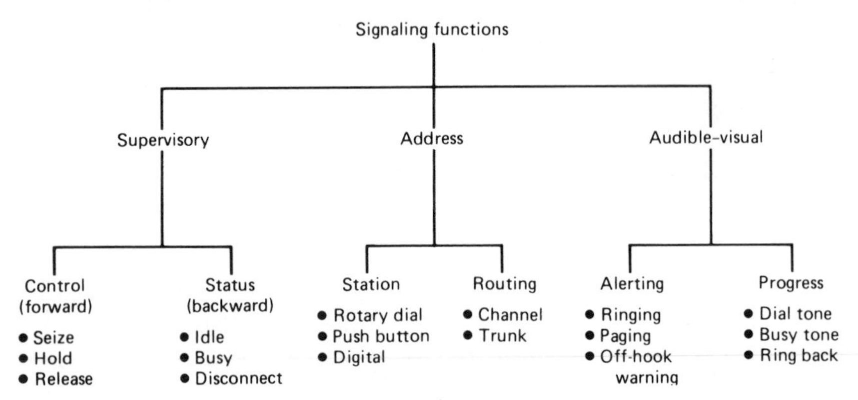

Figure 4.1 A functional breakdown of signaling.

subscriber of a call waiting or an extended "off-hook" condition of his or her handset. Signaling information can be conveyed by a number of means from subscriber to switch or between (and among) switches. Signaling information can be transmitted by means such as

- Duration of pulses (pulse duration bears a specific meaning).
- Combination of pulses.
- Frequency of signal.
- Combination of frequencies.
- Presence or absence of a signal.
- Binary code.
- For dc systems, the direction or level of transmitted current.

2 SUPERVISORY SIGNALING

Supervisory signaling provides information on line or circuit condition and indicates whether a circuit is in use. It informs the switch and interconnecting trunk circuits information on line condition, such as calling party "off hook," calling party "on hook," called party "off hook," and called party "on hook." The terms "on hook" and "off hook" were derived from the position of the receiver of an old-fashioned telephone set in relation to the mounting, in this case a hook, provided for it. If a subscriber subset is on hook, the conductor (subscriber loop) between the subscriber and his local exchange is open and no current is flowing. For the reverse or off-hook condition, there is a dc shunt across the line, and current is flowing. These terms have also been found convenient for designating the two signaling conditions of a trunk (junction). Usually, if a trunk is not in use, the on-hook condition is indicated toward

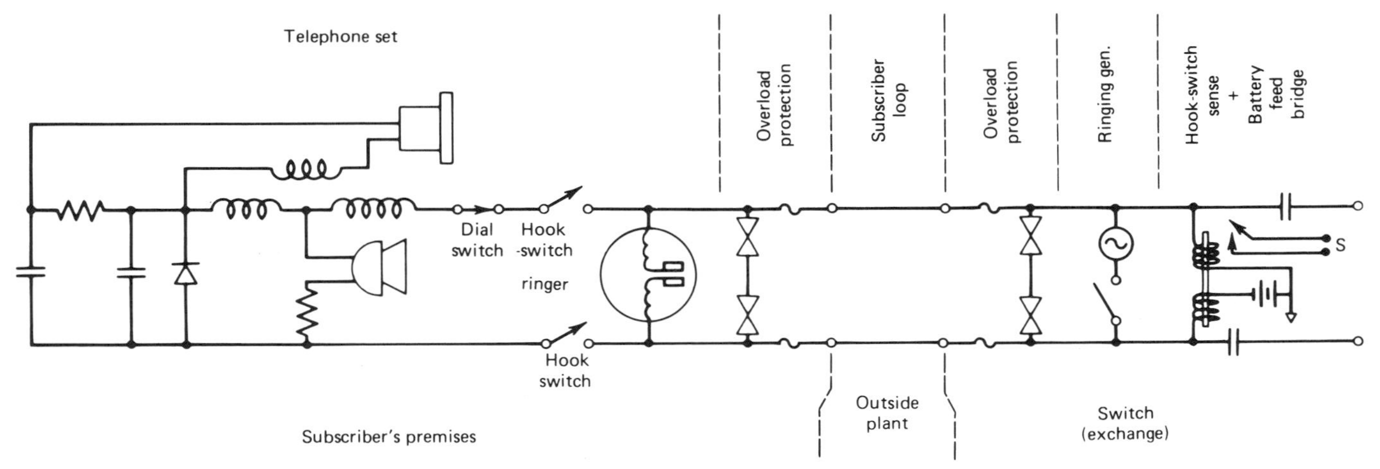

Figure 4.2 Signaling with a conventional telephone subset. Note functions of hook switch, dial, and ringer.

both ends. Seizure of the trunk at the calling end initiates an off-hook signal transmitted toward the called end.

The reader must appreciate that supervisory information–status must be maintained end to end on every telephone call. It is necessary to know when a calling subscriber lifts his telephone off hook, thereby requesting service. It is equally important that we know when the called subscriber answers (i.e., lifts her telephone off hook), for that is when we may start metering the call to establish charges. It is also important to know when the called and calling subscribers return their telephones to the on-hook condition. Charges stop, and the intervening trunks comprising the talk path as well as the switching points are then rendered idle for use by another pair of subscribers. During the period of occupancy of a talk path end to end, we must know that this particular path is busy (is occupied) so that no other call attempt can seize it.

Dialing of a subscriber line is merely interruption of the subscriber loop's off-hook condition, often called "make and break." The "make" is a current flow condition (or off-hook), and the "break" is the no-current condition (or on-hook). How do we know the difference between supervisory and dialing? Primarily by duration—the on-hook interval of a dial pulse is relatively short and is distinguishable from an on-hook disconnect signal (subscriber hangs up), which is transmitted in the same direction for a longer duration. Thus the switch is sensitized to duration to distinguish between supervisory and dialing of a subscriber loop. Figure 4.2 is a simplified diagram of a subscriber loop showing its functional signaling elements.

2.1 E and M Signaling

Probably the most common form of trunk supervision is E and M signaling. Yet it only becomes true E and M signaling where the trunk interfaces with the switch (see the following diagram).

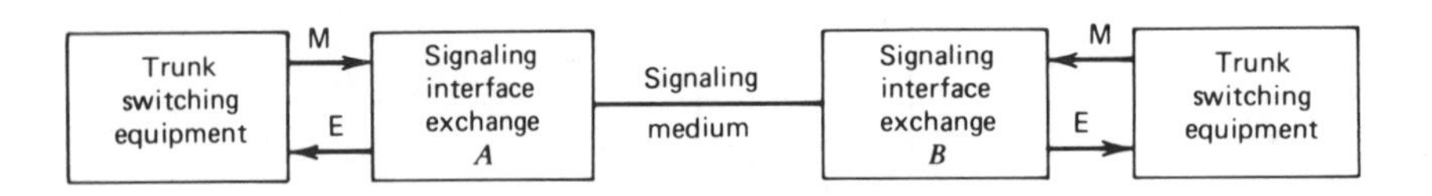

E-lead and M-lead signaling systems are semantically derived from historical designation of signaling leads on circuit drawings covering these systems. Historically, the E and M signaling interface provides two leads between the switch and what we may call *trunk-signaling equipment* (signaling interface). One lead is called the "E-lead," which carries signals *to* the switching equipment. Such signal directions are shown in the preceding diagram, where we see that signals from switch A and switch B leave A on the M-lead and are delivered to B on the E-lead. Likewise, from B to A, supervisory information leaves B on the M-lead and is delivered to A on the E-lead.

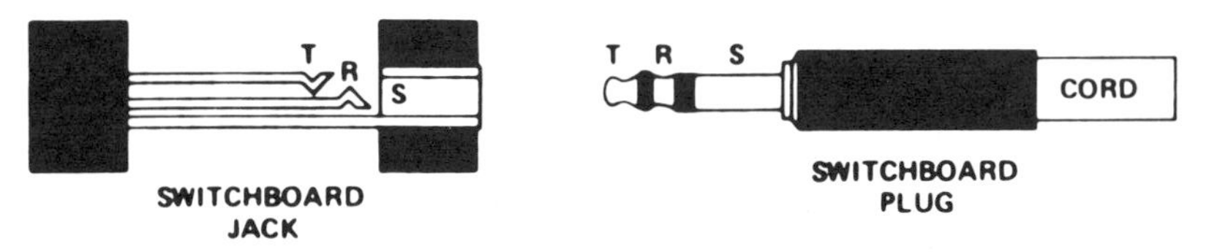

Figure 4.3 Switchboard plug with corresponding jack (R, S, and T are ring, sleeve, and tip, respectively).

For conventional E and M signaling (referring to electromechanical exchanges), the following supervisory conditions are valid:

Direction		Condition at *A*		Condition at *B*	
Signal *A* to *B*	Signal *B* to *A*	M-Lead	E-Lead	M-Lead	E-Lead
On hook	On hook	Ground	Open	Ground	Open
Off hook	On hook	Battery	Open	Ground	Ground
On hook	Off hook	Ground	Ground	Battery	Open
Off hook	Off hook	Battery	Ground	Battery	Ground

2.2 Reverse Battery Signaling

Another method of supervisory signaling used on metallic pair trunks is *reverse battery signaling*. The on-hook and off-hook conditions are given by the polarity across the loop (e.g., direction of current flow). Polarity in the case of dc trunk loops refers to battery and ground state. Terminology in signaling often refers back to manual switchboards or, specifically, to the plug used with these boards and its corresponding jack, as shown in Figure 4.3. The off-hook condition places the battery on the tip and ground on the ring of the plug. The on-hook condition has reverse current conditions with ground on the tip and battery on the ring.

3 AC SIGNALING

3.1 General

Up to this point we have reviewed two of the most employed means of supervisory trunk signaling (or line signaling). Direct current signaling has notable limits on distance because it cannot be applied directly to carrier systems (Chapters 5 and 9) and is limited on metallic pairs due to the IR drop of the lines involved.

There are many ways to extend these limits, but from a cost-effectiveness standpoint there is a limit that we cannot afford to exceed. On trunks exceeding dc capabilities some form of ac signaling will be used. Traditionally, ac signaling systems are divided into three categories: low frequency, in-band, and out-band (out-of-band) systems.

3.2 Low-Frequency ac Signaling Systems

An ac signaling system operating below the limits of the conventional voice channel (i.e., < 300 Hz) are termed *low frequency*. Low-frequency signaling systems are one-frequency systems, typically 50 Hz, 80 Hz, 135 Hz, or 200 Hz. It is impossible to operate such systems over carrier-derived channels (see Chapter 5) because of the excessive distortion and band limitation introduced. Thus low-frequency signaling is limited to metallic pair transmission systems. Even on these systems, cumulative distortion limits circuit length. A maximum of two repeaters may be used, and, depending on the type of circuit (open wire, aerial cable, or buried cable) and wire gauge, a rough rule of thumb is a distance limit of 80–100 km.

3.3 In-Band Signaling

In-band signaling refers to signaling systems using an audio tone, or tones inside the conventional voice channel, to convey signaling information. In-band signaling is broken down into three categories: (1) one frequency (SF or single frequency), (2) two frequency (2VF), and (3) multifrequency (MF). As the term implies, in-band signaling is where signaling is carried out directly in the voice channel. As the reader is aware, the conventional voice channel as defined by the CCITT occupies the band of frequencies from 300 Hz to 3400 Hz. Single-frequency and two-frequency signaling systems utilize the 2000–3000-Hz portion, where less speech energy is concentrated.

3.3.1 Single-Frequency Signaling. Single-frequency signaling is used almost exclusively for supervision. In some locations still it is used for interregister signaling, but the practice is diminishing in favor of more versatile methods such as MF signaling. The most commonly used frequency is 2600 Hz, particularly in North America. On two-wire trunks 2600 Hz is used in one direction and 2400 Hz in the other. A diagram showing application of SF signaling on a four-wire trunk is shown in Figure 4.4.

3.3.2 Two-Frequency Signaling. Two-frequency signaling is used for both supervision (line signaling) and address signaling. We often associate SF and 2VF supervisory signaling systems with carrier (FDM) operation. Of course, when we discuss such types of line signaling (supervision), we know that the term "idle" refers to the on-hook condition and "busy" to the off-hook condition. Thus, for such types of line signaling that are governed by audio tones of which SF and 2VF are typical, we have the conditions of "tone on when idle" and "tone on when busy." The discussion holds equally well for in-band and out-of-band signaling methods.

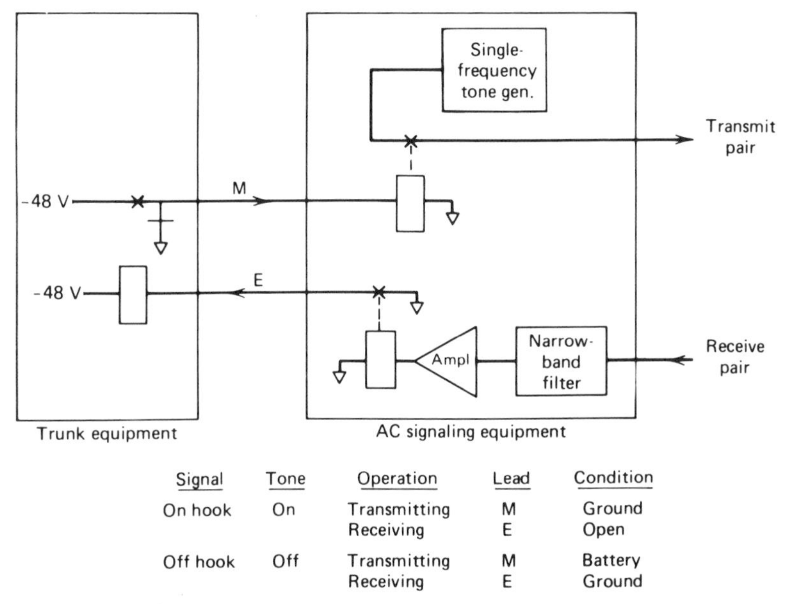

Signal	Tone	Operation	Lead	Condition
On hook	On	Transmitting	M	Ground
		Receiving	E	Open
Off hook	Off	Transmitting	M	Battery
		Receiving	E	Ground

Figure 4.4 Functional block diagram of a single-frequency signaling circuit (*Note*: Wire pairs, "receive" and "transmit," derive from carrier-equipment "receive" and "transmit" channels.)

However, for in-band signaling supervision is by necessity carried out only when the call is being set up and when the call is being taken down or terminated ("gone on hook"). A major problem with in-band signaling is the possibility of "talk-down," which refers to the premature activation or deactivation of supervisory equipment by an inadvertent sequence of voice tones through normal usage of the channel. Such tones could simulate the SF tone. Chances of simulating a 2VF tone set are much less likely. To avoid the possibility of talk-down on SF circuits, a time-delay circuit or slot filters to bypass signaling tones may be used. Such filters do offer some degradation to speech unless they are switched out during conversation.

Thus it becomes apparent why some administrations and telephone companies have turned to the use of 2VF for supervision. For example, a typical 2VF line signaling arrangement is the CCITT No. 5 code, where f_1 (one of the two frequencies) is 2400 Hz and f_2 is 2600 Hz. 2VF signaling is also used widely for address signaling (see Section 4.1 of this chapter).

3.4 Out-of-Band Signaling

With out-of-band signaling, supervisory information is transmitted out of band (e.g., > 3400 Hz). In all cases it is a single-frequency system. Some out-of-band systems use "tone on when idle," indicating the on-hook condition, whereas others use "tone off." The advantage of out-of-band signaling is that either system, tone on or tone off, may be used when idle. Talk-down cannot occur because all supervisory information is passed out of band, away from the speech-information portion of the channel.

The preferred CCITT out-of-band frequency is 3825 Hz, whereas 3700 Hz is commonly used in the United States. It also must be kept in mind that out-of-band signaling is used exclusively on carrier systems, not on wire trunks. On the wire side, inside an exchange, its application is E and M signaling. In other words, out-of-band signaling is one method of extending E and M signaling over a carrier system.

In the short run out-of-band signaling is attractive in terms of both economy and design. One drawback is that when channel patching is required, signaling leads have to be patched as well. In the long run the signaling equipment required may indeed make out-of-band signaling even more costly because of the extra supervisory signaling equipment and signaling lead extensions required at each end and at each time that the carrier (FDM) equipment demodulates to voice. The major advantage of out-of-band signaling is that continuous supervision is provided, whether tone on or tone off, during the entire telephone conversation. In-band SF signaling and out-of-band signaling are illustrated in Figure 4.5. An example of out-of-band signaling is the CCITT R-2 System (CCITT Rec. Q.351) (see Table 4.1).

4 ADDRESS SIGNALING: INTRODUCTION

Address signaling originates as dialed digits from a calling subscriber, whose local switch accepts these digits and, using that information, directs the telephone call to the desired distant subscriber. If more than one switch is involved in the call setup, signaling is required between switches (both address and supervisory). Address signaling between switches in conventional systems is called *interregister signaling*.

The paragraphs that follow discuss various more popular standard ac signaling techniques such as 2VF, MF pulse, MF tone, and the more advanced common-channel signaling. Although interregister signaling is stressed where appropriate, some supervisory techniques are also reviewed.

4.1 Two-Frequency Pulse Signaling

Two-frequency signaling is commonly used as an interregister mode of signaling employing the speech band for the transmission of information. It may also be used for line signaling.* There are various methods of using two-voice frequencies to transmit signaling information. For example, CCITT No. 4 uses

*Supervision (interswitch).

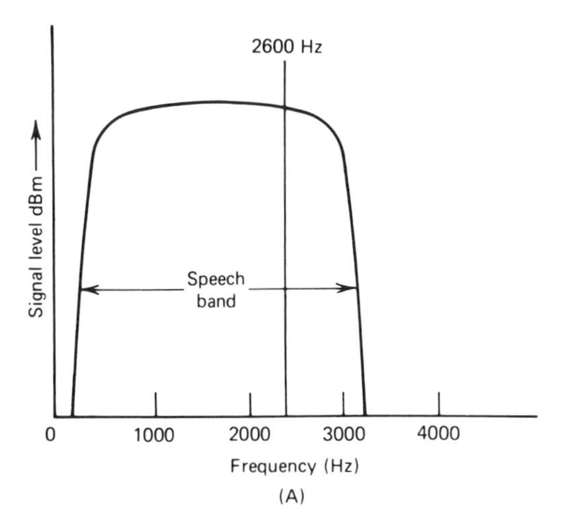

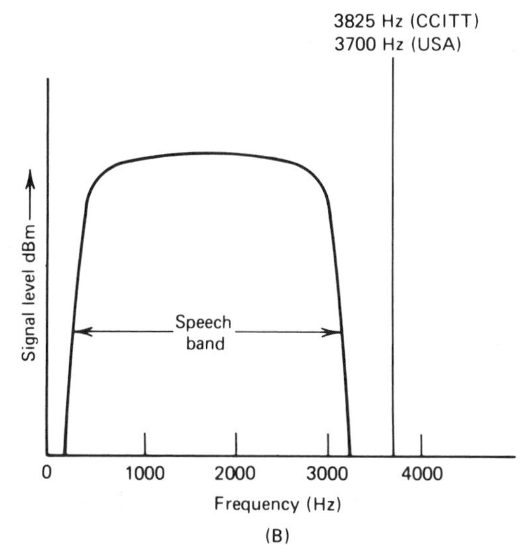

Figure 4.5 Single-frequency signaling: (A) in-band; (B) out-of-band.

TABLE 4.1 R-2 Line Signaling (3825 Hz)

	Direction	
Circuit State	Forward (Go)	Backward (Return)
Idle	Tone on	Tone on
Seized	Tone off	Tone on
Answered	Tone off	Tone off
Clear back	Tone off	Tone on
Release	Tone on	Tone on or off
Blocked	Tone on	Tone off

TABLE 4.2 CCITT Signal Code System No. 4

		Combination Elements			
Digit	Number	1	2	3	4
1	1	*y*	*y*	*y*	*x*
2	2	*y*	*y*	*x*	*y*
3	3	*y*	*y*	*x*	*x*
4	4	*y*	*x*	*y*	*y*
5	5	*y*	*x*	*y*	*x*
6	6	*y*	*x*	*x*	*y*
7	7	*y*	*x*	*x*	*x*
8	8	*x*	*y*	*y*	*y*
9	9	*x*	*y*	*y*	*x*
0	10	*x*	*y*	*x*	*y*
Call operator code 11	11	*x*	*y*	*x*	*x*
Call operator code 12	12	*x*	*x*	*y*	*y*
Spare code (see CCITT Rec. Q.104)	13	*x*	*x*	*y*	*x*
Incoming half-echo suppressor required	14	*x*	*x*	*x*	*y*
End of pulsing	15	*x*	*x*	*x*	*x*
Spare code	16	*y*	*y*	*y*	*y*

Sending duration of binary elements 35 ± 7 ms. Sending duration of blank elements between binary elements 35 ± 7 ms. Element x is 2040 Hz; element y is 2400 Hz.

2040 Hz and 2400 Hz to represent binary 0 and 1, respectively. It uses a four-element code, permitting 16 different coded characters, as shown in Table 4.2. With the CCITT No. 4 code both interregister and line signaling utilize the 2VF technique. Interregister signaling in this case is a pulse-type signaling, and line signaling utilizes the combination of the two frequencies and the duration of signal to convey the necessary supervisory information, as shown in Table 4.3.

As we mentioned, line signaling with the CCITT No. 4 code is based on signal duration as well as frequency. The line-signaling format uses both tone frequencies, 2040 Hz and 2400 Hz. Each line signal consists of an initial *prefix* (P) signal followed by a control signal element, called a *suffix*. The P signal consists of both frequencies (2VF), and the suffix signal consists of one frequency, where x is 2040 Hz and y is 2400 Hz (see Table 4.3). Now consider the durations of the following signal elements used for line signaling:

$$P = 150 \pm 30 \text{ ms}$$
$$x, y = 100 \pm 20 \text{ ms (each)}$$
$$xxyy = 350 \pm 70 \text{ ms (each)}$$

This set of values refers to transmitted signal duration, that is, as transmitted by the signaling sender.

Let us see how these signals are used in the line (interswitch supervision) signaling (see Table 4.3).

TABLE 4.3 CCITT No. 4 Line Signaling

Forward Signals	
Terminal seizing	*Px*
Transit seizing	*Py*
Numerical signals	As in Table 4.2
Clear forward	*Pxx*
Forward transfer	*Pyy*
Backward Signals	
Proceed to send	
Terminal	*x*
International transit	*y*
Number received	*p*
Busy flash	*pX*
Answer	*pY*
Clear back	*Px*
Release guard	*Pyy*
Blocking	*Px* (congestion)
Unblocking	*Pyy*

The supervisory functions "clear forward" and "release guard" do the reverse of "call setup." They take the call down or disconnect, readying the circuit for the next user. "Clear back" is another example.

4.2 Multifrequency Signaling

Multifrequency signaling is in wide use today for interregister signaling. It is an in-band method utilizing five or six tone frequencies, two at a time. Multifrequency signaling works equally well over metallic and carrier (FDM) systems. Four commonly used MF signaling systems follow with a short discussion for each. Tables 4.10 and 4.11 show the North American MF push-button codes and subscriber audible tones, respectively.

4.2.1 SOCOTEL. SOCOTEL is an interregister signaling system used principally in France, areas of French influence, and Spain with some modifications. The frequency pairs and their digit equivalents are shown in Table 4.4. Line signaling used with SOCOTEL may be dc, 50 Hz, or 2000 Hz. The same frequencies are used in both directions.

4.2.2 Multifrequency Signaling in North America: The R-1 Code. The MF signaling system principally used in the United States and Canada is recognized by the CCITT as the R-1 code. It is a two-out-of-five frequency-pulse system. Additional signals for control functions are provided by combinations using a sixth frequency. Table 4.5 shows digits and other applications and their corresponding frequency combinations as well as a brief explanation of "other applications."

TABLE 4.4 Basic SOCOTEL MF Signaling Code

Tone Frequencies (Hz)[a]	Digit
700 + 900	1
700 + 1100	2
900 + 1100	3
700 + 1300	4
900 + 1300	5
1100 + 1300	6
700 + 1500	7
900 + 1500	8
1100 + 1500	9
1300 + 1500	0

[a]The 1700-Hz frequency is also used for signaling system check and when more code groups are required by a telephone company or national administration. When 1700 Hz is used for coding, 1900 Hz is used for checking the system; 1700 Hz and / or 1900 Hz may also be used for control purposes.

TABLE 4.5 The R-1 Code[a] (North American MF)

Digit	Frequency Pair (Hz)	
1	700 + 900	
2	700 + 1100	
3	900 + 1100	
4	700 + 1300	
5	900 + 1300	
6	1100 + 1300	
7	700 + 1500	
8	900 + 1500	
9	1100 + 1500	
10 (0)	1300 + 1500	
Use	**Frequency Pair**	**Explanation**
KP	1100 + 1700	Preparatory for digits
ST	1500 + 1700	End of pulsing sequence
STP	900 + 1700	Used with TSPS (traffic service position system)
ST2P	1300 + 1100	
ST3P	700 + 1700	
Coin collect	700 + 1100	Coin control
Coin return	1100 + 1700	Coin control
Ringback	700 + 1700	Coin control
Code 11	700 + 1700	Inward operator (CCITT No. 5)
Code 12	900 + 1700	Delay operator
KP1	1100 + 1700	Terminal call
KP2	1300 + 1700	Transit call

[a]Pulsing of digits is at the rate of about seven digits per second with an interdigital period of 68 $\pm$ 7 ms. For intercontinental dialing for CCITT No. 5 code compatibility, the R-1 rate is increased to 10 digits per second. The KP pulse duration is 100 ms.

TABLE 4.6 CCITT No. 5 Code[a] Showing Variations with R-1 Code

Signal	Frequencies (Hz)	Remarks
KP1	1100 + 1700	Terminal traffic
KP2	1300 + 1700	Transit traffic
1	700 + 900	
2	700 + 1100	
3 – 0	Same as Table 4.5	
ST	1500 + 1700	
Code 11	700 + 1700	Code 11 operator
Code 12	900 + 1700	Code 12 operator

[a]Line signaling for CCITT No. 5 code is 2VF, with f_1 2400 Hz and f_2, 2600 Hz. Line-signaling conditions are shown in Table 4.7.

TABLE 4.7 Line-Signaling Code CCITT No. 5

Signal	Direction	Frequency	Sending Duration	Recognition Time (ms)
Seizing	⟶	f_1	Continuous	40 ± 10
Proceed to send	⟵	f_2	Continuous	40 ± 10
Busy flash	⟵	f_2	Continuous	125 ± 25
Acknowledgment	⟶	f_1	Continuous	125 ± 25
Answer	⟵	f_1	Continuous	125 ± 25
Acknowledgment	⟶	f_1	Continuous	125 ± 25
Clear back	⟵	f_2	Continuous	125 ± 25
Acknowledgment	⟶	f_1	Continuous	125 ± 25
Forward Transfer	⟶	f_2	850 ± 200 ms	125 ± 25
Clear forward	⟶	$f_1 + f_2$	Continuous	125 ± 25
Release guard	⟵	$f_1 + f_2$	Continuous	125 ± 25

4.2.3 CCITT No. 5 Signaling Code. Interregister signaling with the CCITT No. 5 code is very similar in makeup to the North American R-1 Code. Variations with R-1 are shown in Table 4.6. The CCITT No. 5 line signaling code is shown in Table 4.7.

4.2.4 The R-2 Code. The R-2 code is listed by CCITT (Rec. Q.361) as a European regional signaling code. Taking full advantage of combinations of two-out-of-six tone frequencies, 15 frequency-pair possibilities are available. This number is doubled in each direction by having meaning groups I and II in the forward direction and groups A and B in the backward direction (see Table 4.8).

Groups I and A are said to be of primary meaning and groups II and B, secondary. The change from primary to secondary meaning is commanded by the backward signal A-3 or A-5. Secondary meanings can be changed back to primary meanings only when the original change from primary to secondary

TABLE 4.8 European R-2 System

Index No. for Groups I / II and A / B	Frequencies (Hz)						
	1380	1500	1620	1740	1860	1980	Forward Direction I / II
	1140	1020	900	780	660	540	Backward Direction A / B
1	x	x					
2	x		x				
3		x	x				
4	x			x			
5		x		x			
6			x	x			
7	x				x		
8		x			x		
9			x		x		
10				x	x		
11	x					x	
12		x				x	
13			x			x	
14				x		x	
15					x	x	

was made by the use of the A-5 signal. Referring to Table 4.8, the 10 digits to be sent in the forward direction in the R-2 system are in group I and are index numbers 1 through 10 in the table. The index 15 signal (group A) indicates "congestion in an international exchange or at its output." This is a typical backward information signal giving circuit status information. Group B consists of nearly all "backward information" and in particular deals with subscriber status.

The R-2 line-signaling system has two versions: the one used on analog networks is discussed here; the other, on digital (PCM) networks, is briefly covered in Chapter 9. The analog version is an out-of-band tone-on-when-idle system. Table 4.9 shows the line conditions in each direction, forward and backward. Note that the code takes advantage of a signal sequence that has six characteristic operating conditions.

Let us consider several of these conditions.

TABLE 4.9 Line Conditions for R-2 Code

Operating Condition of the Circuit	Signaling Conditions	
	Forward	Backward
1. Idle	Tone on	Tone on
2. Seized	Tone off	Tone on
3. Answered	Tone off	Tone off
4. Clear back	Tone off	Tone on
5. Release	Tone on	Tone on or off
6. Blocked	Tone on	Tone off

Seized. The outgoing exchange (call-originating exchange) removes the tone in the forward direction. If seizure is immediately followed by release, removal of the tone must be maintained for at least 100 ms to ensure that it is recognized at the incoming end.

Answered. The incoming end removes the tone in the backward direction. When another link of the connection using tone-on-when-idle continuous signaling precedes the outgoing exchange, the "tone-off" condition must be established on the link as soon as it is recognized in this exchange.

Clear Back. The incoming end restores the tone in the backward direction. When another link of the connection using tone-on-when-idle continuous signaling precedes the outgoing exchange, the "tone-off" condition must be established on this link as soon as it is recognized in this exchange.

Clear Forward. The outgoing end restores the tone in the forward direction.

Blocked. At the outgoing exchange the circuit stays blocked as long as the tone remains off in the backward direction.

TABLE 4.10 North American Push-Button Codes

Digit	Dial Pulses (Breaks)	Multifrequency Push-button Tones
0	10	941, 1336 Hz
1	1	697, 1209 Hz
2	2	697, 1336 Hz
3	3	697, 1474 Hz
4	4	770, 1209 Hz
5	5	770, 1336 Hz
6	6	770, 1477 Hz
7	7	852, 1209 Hz
8	8	852, 1336 Hz
9	9	852, 1477 Hz

TABLE 4.11 Audible Tones Commonly Used in North America

Tone	Frequencies (Hz)	Cadence
Dial	350 + 440	Continuous
Busy (station)	480 + 620	0.5 s on, 0.5 s off
Busy (network congestion)	480 + 620	0.2 s on, 0.3 s off
Ring return	440 + 480	2 s on, 4 s off
Off-hook alert	Multifreq. howl	1 s on, 1 s off
Recording warning	1400	0.5 s on, 15 s off
Call waiting	440	0.3 s on, 9.7 s off

4.2.5 Subscriber Tones and Push-Button Codes (North America). Subscriber subsets in many places in the world are either dial or push button. The push-button type is more versatile, and more rapid dialing can be accomplished by a subscriber. Table 4.10 compares digital dialed, dial pulses (breaks), and MF push-button tones. Table 4.11 shows the audible tones commonly used in North America. Functionally, these are the call-progress tones presented to the subscriber.

5 COMPELLED SIGNALING

In many of the signaling systems discussed thus far, signal element duration is an important parameter. For instance, in a call setup an initiating exchange sends a 100-ms seizure signal. Once this signal is received at the distant end, the distant exchange sends a "proceed to send" signal back to the originating exchange; in the case of the R-1 system, this signal is 140 ms or more in duration. Then, on receipt of "proceed to send" the initiating exchange spills all digits forward. In the case of R-1 each digit is an MF pulse of 68 ms duration with 68 ms between each pulse. After the last address digit an ST (end-of-pulsing) signal is sent. In the case of R-1 the incoming (far-end) switch register knows the number of digits to expect. Thus there is an explicit acknowledgment that the call setup has proceeded satisfactorily. Thus R-1 is a good example of noncompelled signaling.

A fully compelled signaling system is one in which each signal continues to be sent until an acknowledgment is received. Thus signal duration is not significant and bears no meaning. The R-2 and SOCOTEL are examples of fully compelled signaling systems. Figure 4.6 shows a fully compelled signaling sequence. Note the small overlap of signals, causing the acknowledging (reverse) signal to start after a fixed time on receipt of the forward signal. This is because of the minimum time required for recognition of the incoming signal. After the initial forward signal, further forward signals are delayed for that short recognition time (see Figure 4.6). Recognition time is normally less than 80 ms.

Fully compelled signaling is advantageous in that signaling receivers do not have to measure duration of each signal, thus making signaling equipment simpler and more economical. Fully compelled signaling adapts automatically to the velocity of propagation, to long circuits, to short circuits, to metallic pairs, or to carrier, and is designed to withstand short interruptions in the transmission path. The principal drawback of compelled signaling is its inherent lower speed, thus requiring more time for setup. Setup time over space-satellite circuits with compelled signaling is appreciable and may force the system engineer to seek a compromise signaling system.

There is also a partially compelled type of signaling, where signal duration is fixed in both forward and backward directions according to system specifications; or the forward signal is of indefinite duration and the backward is of fixed duration. The forward signal ceases once the backward signal has been

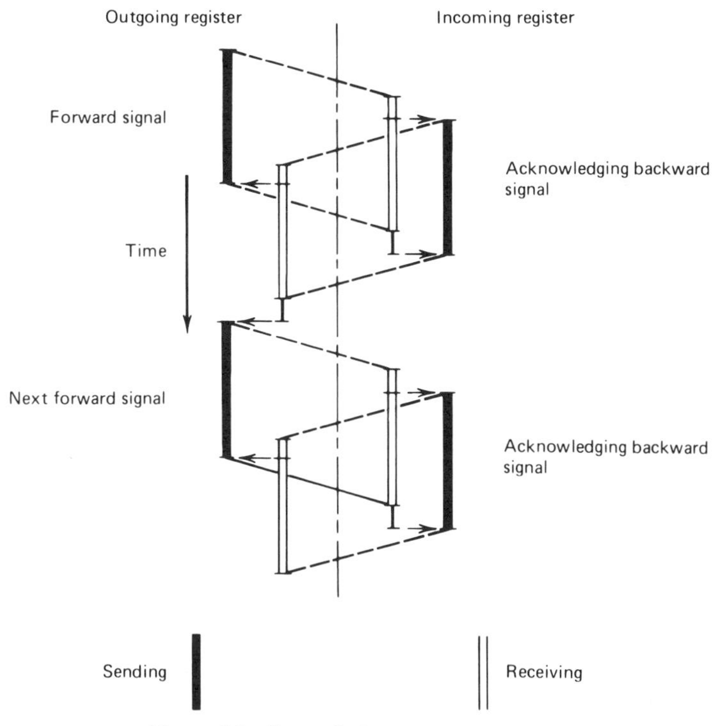

Figure 4.6 Compelled signaling procedure.

received correctly. The CCITT No. 4 is a variation of partially compelled signaling.

6 LINK-BY-LINK VERSUS END-TO-END SIGNALING

An important factor to be considered in switching system design that directly affects both signaling and customer satisfaction is postdialing delay. This is the amount of time it takes after the calling subscriber completes dialing until ring-back is received. Ring-back is a backward signal to the calling subscriber telling her that her dialed number is ringing. Postdialing delay must be made as short as possible.

Another important consideration is register occupancy time for call setup as the setup proceeds from originating exchange to terminating exchange. Call-setup equipment, that equipment used to establish a speech path through a switch and to select the proper outgoing trunk, is expensive. By reducing register occupancy per call, we may be able to reduce the number of registers (and markers) per switch, thus saving money.

Link-by-link and end-to-end signaling each affect register occupancy and postdialing delay, each differently. Of course, we are considering calls involving one or more tandem exchanges in a call setup, as this situation usually occurs on long distance or toll calls. Link-by-link signaling may be defined as a signaling system where *all* interregister address information must be transferred to the subsequent exchange in the call-setup routing. Once this information is received at this exchange, the preceding exchange control unit (register) releases. This same operation is carried on from the originating exchange through each tandem (transit) exchange to the terminating exchange of the call. The R-1 system is an example of link-by-link signaling.

End-to-end signaling abbreviates the process such that tandem (transit) exchanges receive only the minimum information necessary to route the call. For instance, the last four digits of a seven-digit telephone number need be exchanged only between the originating exchange (e.g., the calling subscriber's local exchange or the first toll exchange in the call setup) and the terminating exchange in the call setup. With this type of signaling, fewer digits are required to be sent (and acknowledged) for the overall call-setup sequence. Thus the signaling process may be carried out much more rapidly, decreasing postdialing delay. Intervening exchanges on the call route work much less, handling only the digits necessary to pass the call to the next exchange in the sequence.

The key to end-to-end signaling is the concept of "leading register." This is the register (control unit) in the originating exchange that controls the call routing until a speech path is set up to the terminating exchange before releasing to prepare for another call setup. For example, consider a call from subscriber X to subscriber Y.

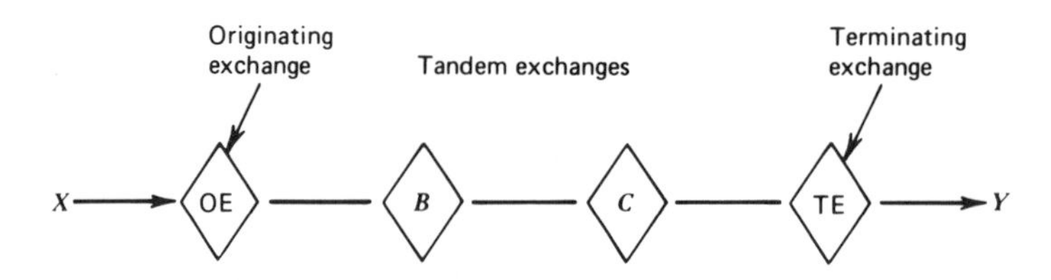

The telephone number of subscriber Y is 345-6789. The sequence of events is as follows using end-to-end signaling:

- A register at exchange OE receives and stores the dialed number 345-6789 from subscriber X.
- Exchange OE analyzes the number and then seizes a trunk (junction) to exchange B. It then receives a "proceed-to-send" signal indicating that the register at B is ready to receive routing information (digits).
- Exchange OE then sends digits 34, which are the minimum necessary to effect correct transit.
- Exchange B analyzes the digits 34 and then seizes a trunk to exchange C. Exchanges OE and C are now in direct contact and exchange B's register releases.

- Exchange OE receives the "proceed-to-send" signal from exchange *C* and then sends digits 45, those required to effect proper transit at *C*.
- Exchange *C* analyzes digits 45 and then seizes a trunk to exchange TE. Direct communication is then established between the leading register for this call at OE and the register at TE being used on this call setup. The register at *C* then releases.
- Exchange OE receives the "proceed-to-send" signal from exchange TE, to which it sends digits 5678, the subscriber number.
- Exchange TE selects the correct subscriber line and returns to *A* ring-back, line busy, out of order, or other information after which all registers are released.

Thus we see that a signaling path is opened between the leading register and the terminating exchange. To accomplish this, each exchange in the route must "know" its local routing arrangements and request from the leading register those digits it needs to route the call further along its proper course.

Again, the need for backward information becomes evident, and backward signaling capabilities must be nearly as rich as forward signaling capabilities when such a system is implemented.

R-1 is a system inherently requiring little backward information (interregister). The little information that is needed, such as "proceed to send," is sent via line signaling. The R-2 system has major backward information requirements, and backward information and even congestion and busy signals are sent back by interregister signals.

7 THE EFFECTS OF NUMBERING ON SIGNALING

Numbering, the assignment and use of telephone numbers, affects signaling as well as switching. It is the number or the translated number, as we found out in Chapter 3, that routes the call. There is "uniform" numbering and "nonuniform" numbering. How does each affect signaling? Uniform numbering can simplify a signaling system. Most uniform systems in the nontoll or local-area case are based on seven digits, although some are based on six. The last four digits identify the subscriber. The first three digits (or the first two in the case of a six-digit system) identify the exchange. Thus the local exchange or transit exchanges know when all digits are received. There are two advantages to this sort of scheme:

1. The switch can proceed with the call once all digits are received because it "knows" when the last digit (either the sixth or seventh) has been received.

2. "Knowing" the number of digits to expect provides inherent error control and makes "time out"* simpler.

For nonuniform numbering, particularly on direct distance dialing in the international service, switches require considerably more intelligence built in. It is the initial digit or digits that will tell how many digits are to follow, at least in theory.

However, in local or national systems with nonuniform numbering, the originating register has no way of knowing whether it has received the last digit, with the exception of receiving the maximum total used in the national system. With nonuniform numbering an incompletely dialed call can cause a useless call setup across a network up to the terminating exchange, and the call setup is released only after time out has run its course. It is evident that with nonuniform numbering systems national (and international) networks are better suited to signaling systems operating end to end with good features of backward information, such as the R-2 system.

8 COMMON-CHANNEL SIGNALING

8.1 General

There are two types of common-channel signaling (CCS) in existence today, European CCITT No. 6 and the North American CCIS (common-channel interoffice signaling). Another system is CCITT No. 7, which will probably be implemented worldwide by the year 2000. This system is discussed in Chapter 14. In our previous discussion all signaling systems covered were associated channel systems, meaning that each voice trunk carried its own signaling, both supervisory (line) and interregister. Signaling was associated with the channel, whether in band, out of band, pulse or MF, or MF pulse.

Common-channel signaling separates signaling from its associated speech path by placing the signaling from a group (or several groups) of voice trunks on a separate path dedicated to signaling only. The signaling information is transmitted by means of serial binary data. Serial binary data transmission is discussed further in Chapter 8.

The basic distinction between conventional associated channel signaling and CCS is shown in Figure 4.7. From Figure 4.7 we see that CCS is feasible only between processor-controlled exchanges, more commonly known as *stored-program control* (SPC) exchanges. The advantages of CCS become obvious between switch processors. Signaling on the telephone network is essentially digital for both line and interregister signaling. Why, then, translate into a complex analog mode rather than just leaving it digital where it

*"Time out" is the resetting of call-setup equipment and return of dial tone to subscriber as a result of incomplete signaling procedure, subset left off hook, and so forth.

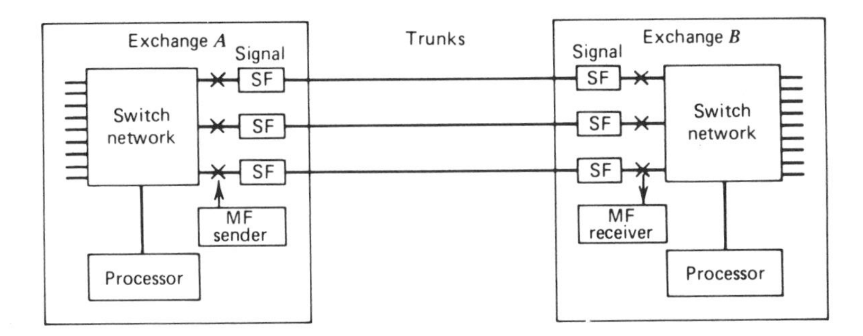

CONVENTIONAL SF-MF
(Associated Channel Signaling)

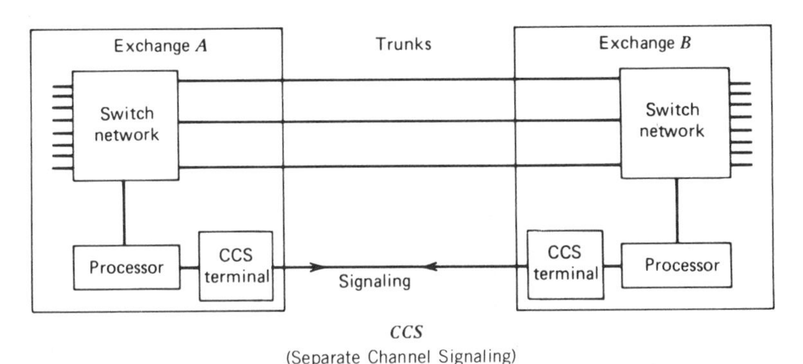

CCS
(Separate Channel Signaling)

Figure 4.7 Conventional analog versus CCS signaling techniques. (*Note*: Signaling in upper drawing accompanies voice paths; signaling in lower drawing is conveyed on a separate circuit.)

belongs? Common-channel signaling does just this. Of course, on analog networks the interprocessor digital information has to be conditioned for analog voice channels; this is done by the data modem.

Figure 4.8 illustrates the basic functional components of CCS. The diagram is valid for both CCITT No. 6 and CCIS operation. Essentially, then, leaving aside error control features, a CCS link consists of the voice-frequency channel (four-wire), two signaling terminals, and two modems. The signaling terminals store both incoming signaling information awaiting processing and outgoing signaling information awaiting transmission. The terminals also perform error control in the North American CCIS system.

With conventional signaling the signal path and the speech path or voice channel occupy the same media; if signaling is effected, there is continuity of the voice or speech path. Because CCS does not pass signaling over the speech trunks that are to be set up and "supervised," call-path continuity must be checked once a call is set up. This is done with tone transceivers that are connected at the time of setup to ensure path continuity. With CCIS the

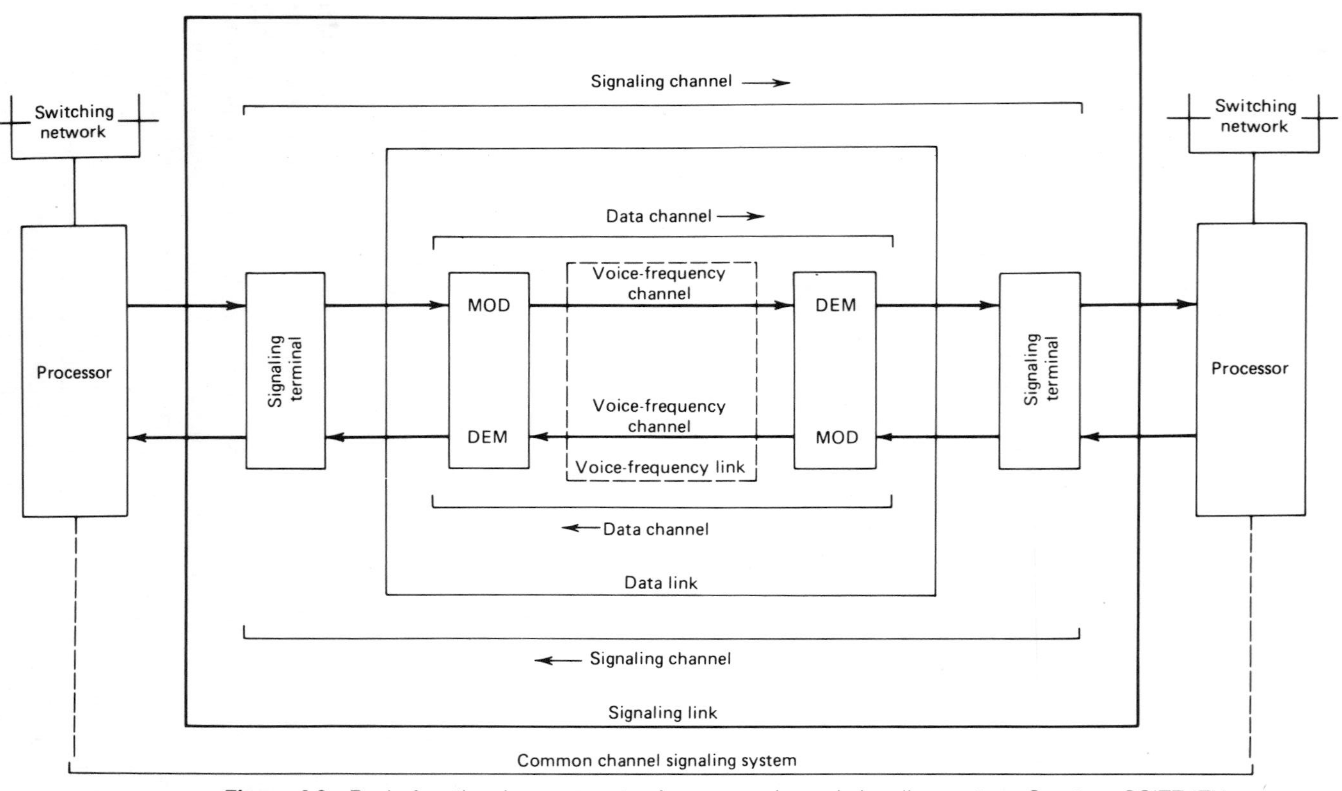

Figure 4.8 Basic functional components of common-channel signaling system. Courtesy CCITT-ITU.

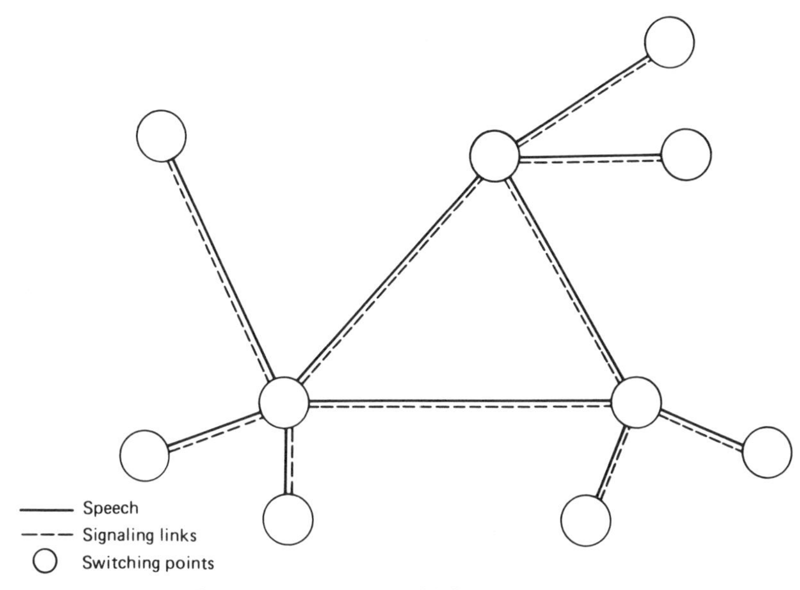

Figure 4.9 Associated CCIS signaling mode.

transceivers operate at 2010 Hz on four-wire trunks. On two-wire trunks, 1780 Hz is transmitted by the originating exchange, and a 2010-Hz tone is returned by the terminating exchange. When no continuity is achieved, a second attempt is made and the failed trunk is blocked. The blocked trunk is then retested. If it continues to fail continuity with CCIS, a trunk failure message is printed out at the maintenance position.

The CCIS (North America) is designed to operate in two basic signaling modes of operation, referred to as *associated* and *nonassociated*. In the associated mode a separate voice channel (or channels) carries signaling information, and this channel is routed with the speech channels it serves. Such a concept is represented in Figure 4.9. Compare this to Figure 4.10, which depicts the concept of fully nonassociated signaling, where the signaling information traverses a routing entirely separate from the voice path that it controls.

Signal-transfer points (STPs)* are used in North America with nonassociated signaling. A signal-transfer point consists of a processor with signaling terminals and data modems (see Chapter 8) on either side. In effect, an STP is a data-message transfer point or switching point, if you will. Signal-transfer points allow the concentration of signaling for a large number of trunks and will also provide for improved circuit reliability by allowing alternative routing of the CCIS signaling path.

*Signal-transfer points are also used in the nonassociated mode with CCITT No. 6; however, North American practice is discussed here.

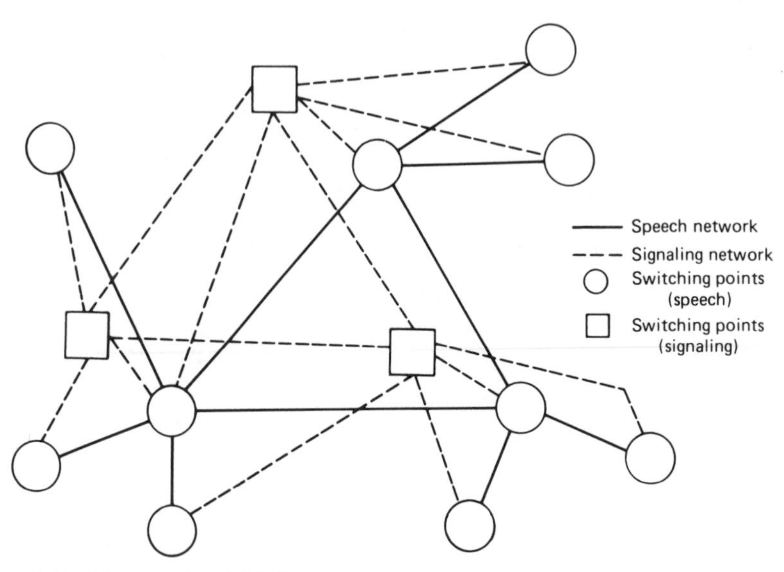

Figure 4.10 Nonassociated CCIS signaling mode, also called fully disassociated mode.

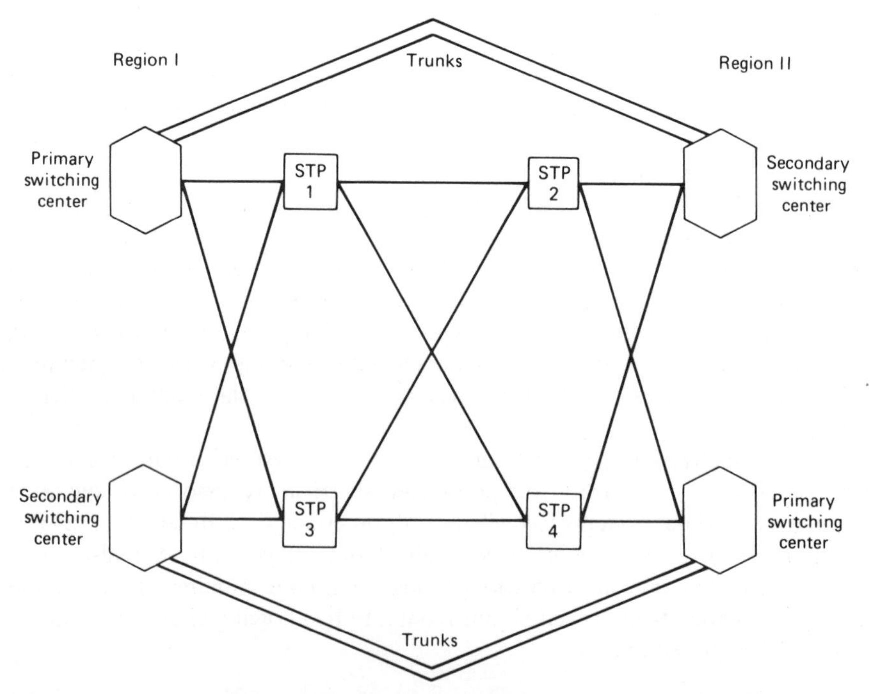

Figure 4.11 Nonassociated CCIS mode, according to concept of signal-transfer point (STP) (CCITT hierarchy) (single solid lines = STP signaling links). Copyright © American Telephone and Telegraph Co., 1975.

To improve reliability of communications, all CCIS signaling paths in North America are duplicated and fully redundant. In addition, all STPs are fully interconnected to provide several signaling-path possibilities. The STPs are overdimensioned, so that either STP can carry the full signaling load if one STP fails. Such a network (see Figure 4.11) has a hierarchical structure, with primary and secondary centers. Of course, it is fully nonassociated.

For cost-effectiveness, associated signaling is generally used with large trunk groups. The concentration aspects of nonassociated signaling would make it more attractive for small trunk groups in the long-distance (toll) network. It is also undesirable to have many STPs in tandem because of the added delay. Each STP processes signal units from its input link to its output link, each adding incremental delay. The limit of STPs in tandem is set at two in the normal or primary path by ATT and four under failure conditions.

8.2 Common-Channel Interoffice Signaling Format

Both CCIS and CCITT No. 6 signaling systems carry signaling information in a serial binary format. Thus we are dealing with digital data transmission, which is introduced in Chapter 8.

The basic (data) word in the CCIS system is the signal unit (SU). A signal unit is 28 bits long, with the last 8 bits used for error checking. Thus the signaling information is actually contained in the first 20 bits of the basic "word." Signal units are grouped into blocks of 12 for transmission; thus a *block* contains 12 × 28 or 336 serial bits. The last signal unit of each block is the acknowledgment signal unit (ACU).

Common-channel interoffice "messages" can be one or more SUs long. Length, as we can imagine, depends on the amount of information to be sent. There are single-unit messages, called *lone signal unit* (LSU), and *multiunit messages* (MUMs). The LSU is generally used for specific control information (e.g., "answer"), whereas MUMs are generally used for passing address information such as digits. Figure 4.12 shows the format of an LSU and a MUM. Note the trunk label, which is used to identify the trunk being served. The trunk label is subdivided into two fields: (1) the band number, one or more of which is associated with a trunk group and is used to determine the routing of the message in the signaling network, and (2) the trunk number, which identifies a specific trunk (see Figure 4.12). The number of trunk labels handled by a link gives a measure of CCS link capacity.

The MUM shown in Figure 4.12 has an ISU (initial signal unit) and a single SSU (subsequent signal unit). The ISU has a unique heading code identifying it as the message start (ISU) of a MUM. The ISU then has a length indicator, (②) in Figure 4.12B, which indicates the number of SSUs to follow. The SSU starts with two data fields. The first, a unique heading code, is used to identify the SSU (i.e., to differentiate it from LSU, ISU, etc.). The second field gives message category to identify the type of MUM. This is shown in Table 4.12.

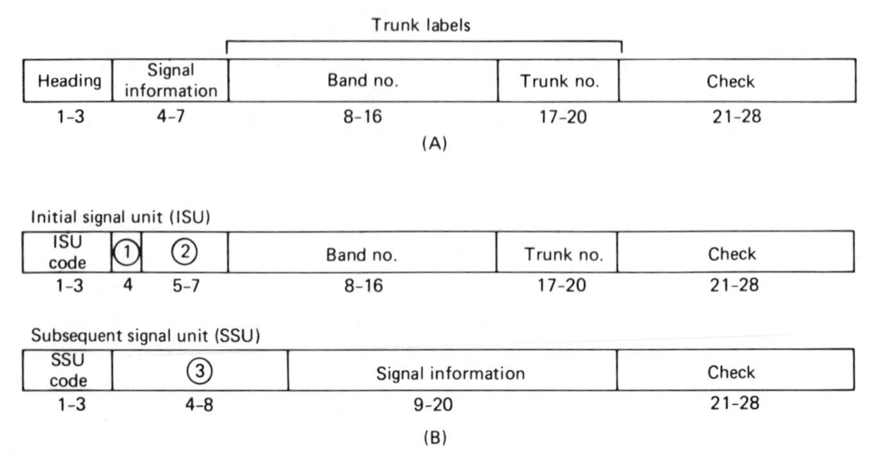

Figure 4.12 Message formats of CCIS signals: (A) lone-signal unit (LSU); (B) multiunit message (MUM) [① ISU-type indicator, ② length indicator, ③ message category]. Copyright © American Telephone and Telegraph Co., 1975.

The heading field of three bits (binary digits) provide eight different combinations of 1's and 0's (see Chapter 8) to identify signal groups. These combinations of 1's and 0's can be called a *code*. These eight combinations are shown in Table 4.12.

Only the first 11 SUs of a signaling message block carry signaling information. The twelfth SU of the block is an ACU, which is coded to indicate the number of blocks being acknowledged, and the acknowledgment bits, indicating whether each of the 11 signal units of the block being acknowledged was received without error.

Figure 4.13 illustrates the IAM (initial address message) and LSUs associated with a routine 10-digit call and the actions performed at the originating

TABLE 4.12 Binary Combinations

Heading Code	Signal Unit Type
000	Lone signal unit (LSU)—telephone signals
001	Lone signal unit (LSU)—telephone signals
010	Lone signal unit (LSU)—telephone signals
011	Acknowledgment signal unit (ACU)
100	Lone signal unit (LSU)—telephone signals
101	Initial signal unit (ISU)
110	Subsequent signal unit (SSUs)
111	Lone signal units (LSU)
	Telephone signals
	System control signals
	Management signals
	Maintenance signals

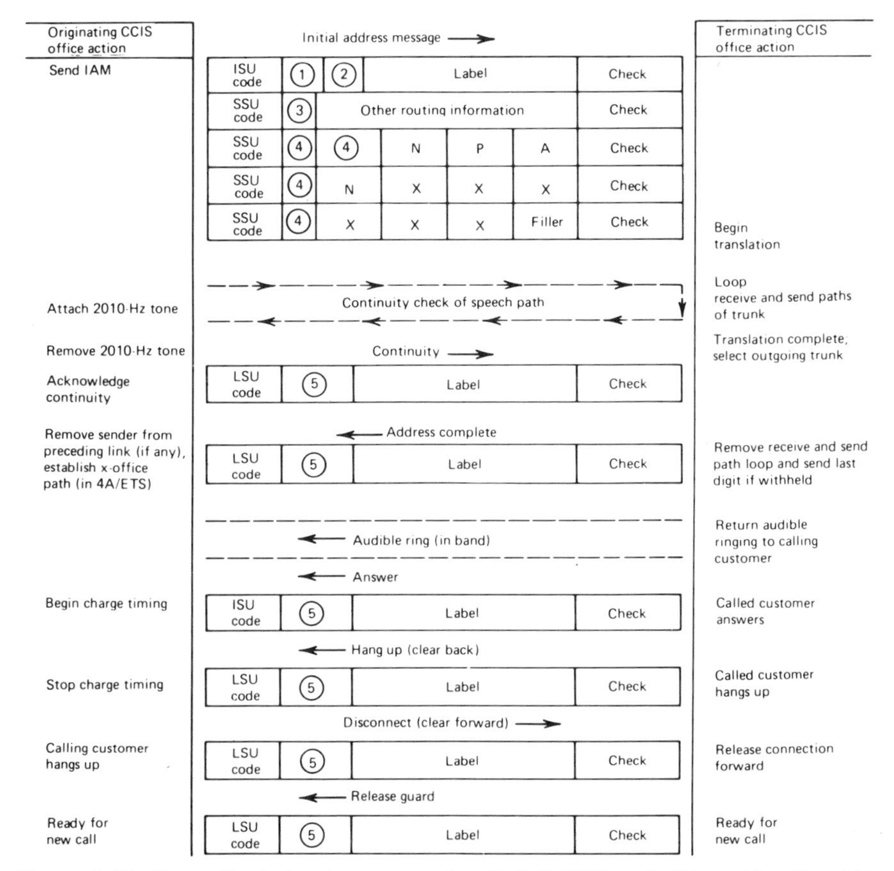

Figure 4.13 Generalized signal sequence for 10-digit CCIS call. [*Note*: All units within dotted lines are in-band; ① ICU-type indicator (IAM), ② length indicator, ③ fully routing information indicator, ④ no information; all zeros, ⑤ signal information ("answer," "disconnect," etc.).]

and terminating CCIS exchanges. Speech-path continuity checks are also shown in their proper chronological order in Figure 4.13. Error detection of CCIS signaling links is achieved by redundant coding. A data-carrier failure detector complements the bit-error detection and is helpful for longer error bursts. Error-free messages are processed on receipt while a retransmission (a form of ARQ (automatic repeat request) described in Chapter 8) is requested of those found in error.

8.3 CCITT No. 6 Code

The CCITT developed a CCS code very similar to the North American CCIS. The No. 6 code is embodied in CCITT Rec. Q.251 through Rec. Q.267. However, the variances are too great for the two codes to interoperate (see

TABLE 4.13 Variance between CCITT No. 6 and CCIS[a]

	CCITT No. 6	CCIS
Heading	5 bits	3 bits
$ Signal information	4 bits	4 bits
$ Band no.	7 bits	9 bits
Circuit no.–trunk no.	4 bits	4 bits
Check (error)	8 bits	8 bits
	28 bits	28 bits
Label capacity	$2^{11^a} = 2048$	$2^{13^a} = 8192$

[a] The exponents 11 and 13 are derived from the sum of bits with $ indicator; thus 7 + 4 = 11 and 9 + 4 = 13, respectively. Thus CCIS label capacity is four times greater than CCITT No. 6. There are other format variations as well. Although international and North American signaling requirements differ, spare capacity was allocated in the CCITT No. 6 code to allow for national variants such as the CCIS.

Table 4.13). The CCITT No. 6 format is 28-bit for a signal unit (SU), with the last 8 bits of an SU for error check; however, compare this with an LSU. Both systems use a loop-back technique on four-wire speech paths for continuity using a nominal 2000-Hz tone (CCIS uses a 2010-Hz tone).

REVIEW QUESTIONS

1. Give the three generic signaling functions, and explain the purpose of each.
2. Differentiate between line signaling and interregister signaling.
3. There are seven ways to transmit signaling information; one is frequency; name five others.
4. How does a switch know whether a particular talk path is busy or idle?
5. A most common form of line signaling is E and M signaling. Describe how it works in three sentences or less.
6. Compare in-band and out-of-band supervisory signaling regarding tone-on idle/busy, advantages, and disadvantages.
7. What is the most common form of in-band supervisory signaling in North America (analog)?
8. What is the standard CCITT out-of-band signaling frequency for North America?
9. Give the principal advantage of 2VF over SF supervisory signaling.
10. Compare CCITT No. 5, R-1, and R-2 supervisory signaling systems.

11. List three types of interregister signaling using MF.

12. Clearly distinguish compelled and noncompelled signaling. Give advantages and disadvantages of each.

13. On connections involving communication satellites, what would be more attractive regarding postdial delay, compelled or noncompelled signaling?

14. Describe and compare link-by-link and end-to-end signaling. Associate R-1 and R-2 with each.

15. Describe at least four types of backward information.

16. Describe the effects of uniform and nonuniform numbering on signaling/switching. Why is uniform numbering more advantageous to signaling/switching?

17. Distinguish associated channel signaling from separate channel signaling.

18. What is "nonassociated channel" signaling? Give some examples.

19. What signaling system has more capacity, CCS (i.e., CCITT No. 6) or CCIS? What is the measurement of capacity in these cases?

20. Compare CCS/CCIS with associated (common-) channel signaling, giving advantages and disadvantages of each.

21. How is talk path continuity verified on CCS/CCIS?

REFERENCES

1. C. A. Dahlbom and C. Breen, "Signaling Systems for Control of Telephone Switching," *BSTJ*, **39** (November 1960).
2. J. D. Sipes, "Common Channel Interoffice Signalling—International Field Trial," IEEE International Switching Symposium, June 1972.
3. *Notes on Distance Dialing*, American Telephone and Telegraph Company, New York, 1975.
4. *Signalling*, from Telecommunication Planning Documents, ITT Laboratories (Spain), Madrid, November 1974.
5. *National Networks for the Automatic Service*, CCITT-ITU, Geneva, 1964.
6. M. Den Hertog, "Inter-register Multifrequency Signalling for Telephone Switching in Europe," *Electr. Commun.*, **38** (1) (1972).
7. "USITA Minutes Attachments," May 18, 1972, "CCIS" and "Common Channel Interoffice Signaling" memorandum.
8. International Telephone and Telegraph Corporation, *Reference Data for Radio Engineers*, 6th ed, ITT Howard W. Sams, Indianapolis, 1976.
9. R. Freeman, *Telecommunication Transmission Handbook*, 2nd ed. Wiley, New York, 1981.

10. *Lenkurt Demodulator—World-Wide E & M Signalling*, July–August 1977; *Common Channel Signalling Systems*, April 1973; *A Glossary of Signalling Terms*, April 1974, GTE Lenkurt, San Carlos, Calif.
11. CCITT Recommendations, *Orange Books*, Vol. V1. ITU Geneva 1976 (Sixth Plenary Assembly CCITT).
12. J. Gordon Pearce, "The CCITT No. 6 Signaling System," *Telephony* (May 17, 1971).
13. Karl F. Steinhawer, "International Signalling with CCITT No. 6 System," *Telephony* (May 17, 1971).
14. *Access Area Switching and Signaling: Concepts, Issues, and Alternatives*, NTIA, Boulder, Colo., May 1978.
15. C. A. Dahlbom, "Signaling Systems and Technology," *Proc. IEEE*, **65**, 1349–1353 (1977).
16. CCITT Recommendations, *Red Book*, Vol. VI (Fascicles), VI.1 through VI.6. VIIIth Malaga-Torremolinas, Oct. 1984 (Seventh Plenary Assembly).

5

INTRODUCTION TO TRANSMISSION FOR TELEPHONY

1 GENERAL

The basic building block for transmission is the telephone channel or voice channel. "Voice channel" implies spectral occupancy, whether the voice path is over wire, radio, or optical fiber. If a pair of wires of a simple subscriber loop is extended without loading, we could expect to see the spectral content of the signal deriving from the average talker with frequencies as low as 20 Hz and as high as 20 kHz if the transducer of the telephone set was at all efficient across this band. Our ear, at least in younger people, is sensitive to frequencies from 30 Hz to as high as 30 kHz. However, the primary content of a voice signal (energy plus emotion) will occupy a much narrower band of frequency (~ 100–4000 Hz). Considering these and other factors, we say that the *nominal* voice channel occupies the band 0–4 kHz. The CCITT voice channel occupies the band 300–3400 Hz.

There are essentially four other parameters we can use to define the voice channel:

1. Attenuation distortion.
2. Phase distortion.
3. Level (signal power level).
4. Noise and signal-to-noise ratio.

Singing and stability, echo, and reference equivalent are other important parameters but are more applicable to a voice channel operating in a specific network. Each parameter is treated in its appropriate place further on in the discussion.

2 THE FOUR ESSENTIAL PARAMETERS

2.1 Attenuation Distortion

A signal transmitted over a voice channel suffers various forms of distortion. That is, the output signal from the channel is distorted in some manner such that it is not an exact replica of the input. One form of distortion is called *attenuation distortion* and is the result of imperfect amplitude–frequency response. Attenuation distortion can be avoided if all frequencies within the passband are subjected to exactly the same loss (or gain). Whatever the transmission medium, however, some frequencies are attenuated more than others. For example, on loaded wire-pair systems higher frequencies are attenuated more than lower ones. On carrier equipment (see Section 4 of this chapter), band-pass filters are used on channel units, where, by definition, attenuation increases as the band edges are approached. Figure 5.1 is a good example of the attenuation characteristics of a voice channel operating over carrier equipment.

Attenuation distortion across the voice channel is measured against a reference frequency. The CCITT specifies 800 Hz as a reference, which is universally used in Europe, Africa, and parts of Hispanic America, whereas 1000 Hz is the common reference frequency in North America. Let us look at some ways attenuation distortion may be stated. For example, one European requirement may state that between 600 Hz and 2800 Hz the level will vary no more than $-1 + 2$ dB, where the plus sign means more loss and the minus sign means less loss. Thus if a signal at -10 dBm is placed at the input of the channel, we would expect -10 dBm at the output at 800 Hz (if there were no overall loss or gain), but at other frequencies we could expect a variation between -1 and $+2$ dB. For instance, we might measure the level at the output at 2500 Hz at -11.9 dBm and at 1100 Hz at -9 dBm.

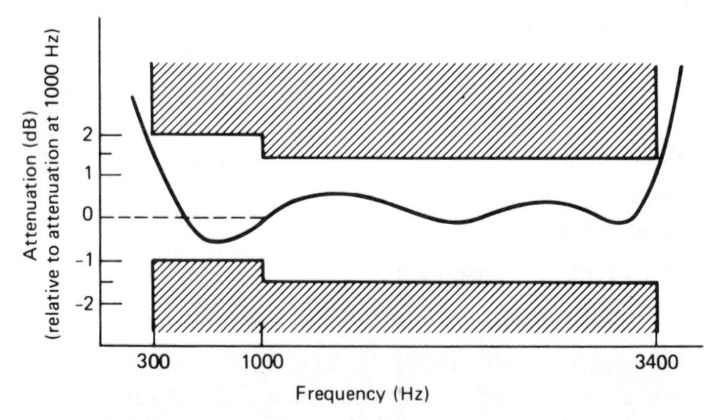

Figure 5.1 Typical attenuation–frequency response (attenuation distortion) for a voice channel. Cross-hatched areas show specified limits.

2.2 Phase Distortion

A voice channel may be regarded as a band-pass filter. A signal takes a finite time to pass through a filter. This time is a function of the velocity of propagation, which varies with the medium involved. This velocity also tends to vary with frequency because of the electrical characteristics associated with it; it tends to increase toward band center and decrease toward band edge, as shown in Figure 5.2.

The finite time during which a signal takes to pass through the total extension of a voice channel or any network is called *delay*. Absolute delay is the delay a signal experiences while passing through the channel at a reference frequency. But we see that the propagation time is different for different frequencies with the wavefront of one frequency arriving before the wavefront of another in the passband. Thus we can say that the phase has shifted or has been distorted. A modulated signal will not be distorted on passing through the channel if the phase shift changes uniformly with frequency, whereas if the phase shift is nonlinear with respect to frequency, the output signal is distorted when compared to the input.

In essence, therefore, we are dealing with the phase linearity of a circuit. If the phase–frequency relationship over a passband is not linear, distortion will occur in the transmitted signal. This phase distortion is often measured by a parameter called *envelope delay distortion*. Mathematically, envelope delay is the derivative of the phase shift with respect to frequency. The maximum

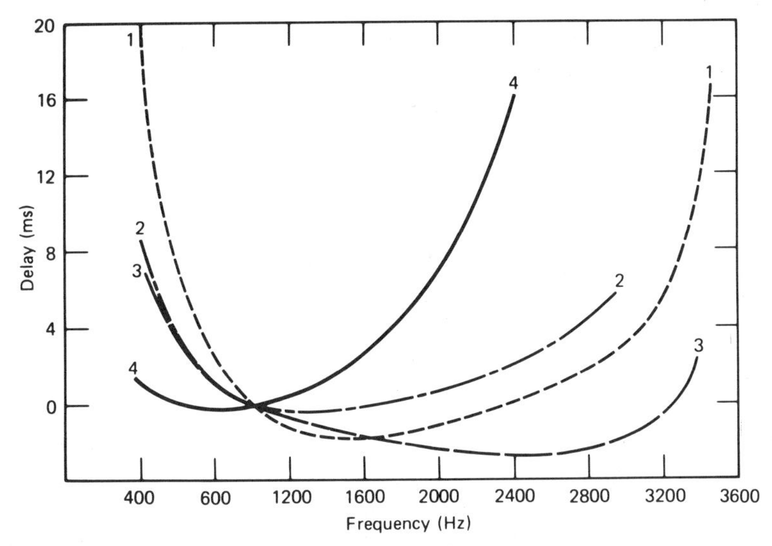

Figure 5.2 Comparison of envelope delay in some typical voice channels. Curves 1 and 3 represent delay of several thousand miles of a toll-quality carrier system. Curve 2 shows delay produced by 100 mi of loaded cable. Curve 4 shows delay in 200 mi of heavily loaded cable. Courtesy of GTE Lenkurt Demodulator, San Carlos, Calif.

difference in the derivative over any frequency interval is called envelope delay distortion (EDD). Therefore EDD is always a difference between the envelope delay at one frequency and that at another frequency of interest in a passband. The EDD unit is milli- or microseconds. Note that envelope delay is often defined the same as group delay, that is, the rate of change, with angular frequency, of the phase shift between two points in a network [11].

2.3 Level

When referring to level, the connotation is either signal intensity or noise intensity. In most telecommunication systems "level" means power level measured in dBm, dBW, or other power units such as picowatts. One notable exception is video, which uses voltage, usually dBmV. Level is an important system parameter. If levels are maintained at too high a point, amplifiers become overloaded, with resulting increases in intermodulation products or crosstalk. If levels are too low, customer satisfaction may suffer.

System levels are used for engineering a communication system. These are usually taken from a level chart or reference system drawing made by a planning group or as a part of an engineered job. On the chart a 0 TLP (zero test-level point) is established. A test-level point is a location in a circuit or system at which a specified test-tone level is expected during alignment. A 0 TLP is a point at which the test-tone level should be 0 dBm.

From the 0 TLP other points may be shown using the unit dBr (decibel reference). A minus sign shows that the level is so many decibels below reference and a positive sign, above. The unit dBm0 is an absolute unit of power in dBm referred to the 0 TLP. The dBm can be related to dBr and dBm0 by the following formula:

$$\text{dBm} = \text{dBm0} + \text{dBr} \tag{5.1}$$

For instance, a value of -32 dBm at a -22-dBr point corresponds to a reference level of -10 dBm0. A -10 dBm0 signal introduced at the 0-dBr point (0 TLP) has an absolute signal level of -10 dBm.

2.4 Noise

2.4.1 General. Noise, in its broadest definition, consists of any undesired signal in a communication circuit. The subject of noise and noise reduction is probably the most important single consideration in transmission engineering. It is the major limiting factor in system performance. For this discussion noise has been broken down into four categories: (1) thermal noise, (2) intermodulation noise, (3) crosstalk, and (4) impulse noise.

2.4.2 *Thermal Noise.* Thermal noise occurs in all transmission media and all communication equipment. It arises from random electron motion and is characterized by a uniform distribution of energy over the frequency spectrum with a Gaussian distribution of levels.

Every equipment element and the transmission medium proper contribute thermal noise to a communication system if the temperature of that element of medium is above absolute zero. Thermal noise is the factor that sets the lower limit of sensitivity of a receiving system and is often expressed as a temperature, usually given in units referred to absolute zero. These units are Kelvins.

Thermal noise is a general expression referring to noise based on thermal agitations. The term "white noise" refers to the average uniform spectral distribution of energy with respect to frequency. Thermal noise is directly proportional to bandwidth and temperature. The amount of thermal noise to be found in 1 Hz of bandwidth in an actual device is

$$P_n = kT(\mathrm{W/Hz}) \tag{5.2}$$

where k is Boltzmann's constant, equal to $1.3803(10^{-23})\mathrm{J/°K}$ and T is the absolute temperature (°K) of the circuit (device). At room temperature, $T = 17°\mathrm{C}$ or 290°K; thus

$$P_n = 4.00(10^{-21})\mathrm{W/Hz} \text{ of bandwidth}$$

or

$$= -204 \mathrm{\ dBW/Hz} \text{ of bandwidth}$$
$$= -174 \mathrm{\ dBm/Hz} \text{ of bandwidth.}$$

For a band-limited system (i.e., a system with a specific bandwidth), $P_n = kTB(\mathrm{W})$, where B refers to the so-called noise bandwidth in Hertz. Thus at 0 K $P_n = -228.6$ dBW/Hz of bandwidth and for a system with a noise bandwidth measured in Hertz (B) and its noise temperature is T:

$$P_n = -228.6 \mathrm{\ dBW} + 10\log T + 10\log B \tag{5.3}$$

2.4.3 *Intermodulation Noise.* Intermodulation (IM) noise is the result of the presence of intermodulation products. If two signals with frequencies F_1 and F_2 are passed through a nonlinear device or medium, the result will be IM products that are spurious frequency components. These components may be present either inside or outside the band of interest for the device. Intermodulation products may be produced from harmonics of the signals in question, either as products between harmonics or as one of the signals or both signals themselves. The products result when the two (or more) signals beat together or "mix." Look at the mixing possibilities when passing F_1 and F_2 through a nonlinear device. The coefficients indicate first, second, or third harmonics.

- Second-order products $F_1 \pm F_2$.
- Third-order products $2F_1 \pm F_2$; $2F_2 \pm F_1$.
- Fourth-order products $2F_1 \pm 2F_2$; $3F_1 \pm F_2 \ldots$.

Devices passing multiple signals simultaneously, such as multichannel radio equipment, develop intermodulation products that are so varied that they resemble white noise.

Intermodulation noise may result from a number of causes:

- Improper level setting. If the level of input to a device is too high, the device is driven into its nonlinear operating region (overdrive).
- Improper alignment causing a device to function nonlinearly.
- Nonlinear envelope delay.
- Device malfunction.

To summarize, intermodulation noise results from either a nonlinearity or a malfunction that has the effect of nonlinearity. The cause of intermodulation noise is different from that of thermal noise. However, its detrimental effects and physical nature can be identical with those of thermal noise, particularly in multichannel systems carrying complex signals.

2.4.4 Crosstalk. *Crosstalk* refers to unwanted coupling between signal paths. There are essentially three causes of crosstalk: (1) electrical coupling between transmission media, such as between wire pairs on a VF cable system, (2) poor control of frequency response (i.e., defective filters or poor filter design), and (3) the nonlinear performance in analog (FDM) multiplex systems. There are two types of crosstalk:

1. *Intelligible*, where at least four words are intelligible to the listener from extraneous conversation(s) in a 7-s period.
2. *Unintelligible*: crosstalk resulting from any other form of disturbing effects of one channel on another.

Intelligible crosstalk presents the greatest impairment because of its distraction to the listener. Distraction is considered to be caused by either fear of loss of privacy or primarily by the user of the primary line consciously or unconsciously trying to understand what is being said on the secondary or interfering circuits; this would be true for any interference that is syllabic in nature.

Received crosstalk varies with the volume of the disturbing talker, the loss from the disturbing talker to the point of crosstalk, the coupling loss between the two circuits under consideration, and the loss from the point of crosstalk to the listener. The most important of these factors for this discussion is the coupling loss between the two circuits under consideration. Talker levels have been discussed elsewhere in this text. Also, we must not lose sight of the fact that the effects of crosstalk are subjective, and other factors have to be considered when crosstalk impairments are to be measured. Among these factors are the type of people who use the channel, the acuity of listeners, traffic patterns, and operating practices.

2.4.5 Impulse Noise. Impulse noise is noncontinuous, consisting of irregular pulses or noise spikes of short duration and of relatively high amplitude. These spikes are often called "hits." Impulse-noise degrades voice telephony only marginally, if at all; however, it may seriously degrade error rate on data or other digital circuits. Impulse noise is treated in greater depth when we discuss data and other digital communications.

2.5 Signal-to-Noise Ratio

When dealing with transmission engineering, signal-to-noise ratio is perhaps more frequently used than any other criterion when designing a telecommunication system. Signal-to-noise ratio expresses in decibels the amount by which a signal level exceeds the noise within a specified bandwidth.

As we review several types of material to be transmitted, each will require a minimum signal-to-noise ratio to satisfy the customer or to make the receiving instrument function within certain specified criteria. We might require the following signal-to-noise ratios (S/N) with the corresponding end instruments:

- Voice: 40 dB } based on customer satisfaction.
- Video: 45 dB }
- Data: ~ 15 dB, based on a specified error rate and modulation type.

In Figure 5.3* a 1000-Hz signal has a signal-to-noise ratio of 10 dB. The level of the noise is +5 dBm and the signal, +15 dBm. Thus

$$\left(\frac{S}{N}\right)_{\text{dB}} = \text{level}_{(\text{signal in dBm})} - \text{level}_{(\text{noise in dBm})} \tag{5.4}$$

*Assume a nominal 4-kHz bandwidth in the example.

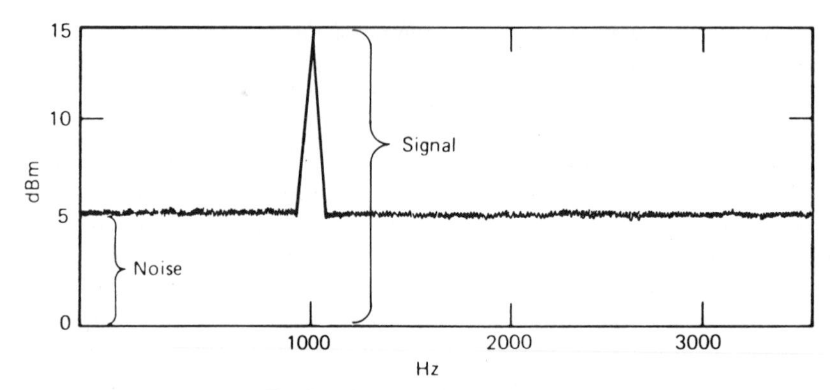

Figure 5.3 Signal-to-noise ratio.

3 TWO-WIRE – FOUR-WIRE TRANSMISSION

3.1 Two-Wire Transmission

A telephone conversation inherently requires transmission in both directions. When both directions are carried on the same pair of wires, it is called *two-wire transmission*. The telephones in our homes and offices are connected to a local switching center (exchange) by means of two-wire circuits. A more proper definition for transmitting and switching purposes is that when oppositely directed portions of a single telephone conversation occur over the same electrical transmission channel or path, we call this *two-wire operation*.

3.2 Four-Wire Transmission

Carrier and radio systems require that oppositely directed portions of a single conversation occur over separate transmission channels or paths (or use mutually exclusive time periods). Thus we have two wires for the transmit path and two wires for the receive path, or a total of four wires for a full-duplex (two-way) telephone conversation. For almost all operational telephone systems, the end instrument (i.e., the telephone subset) is connected to its intervening network on a two-wire basis.

Nearly all long-distance (toll) telephone connections traverse four-wire links. From the near-end user the connection to the long-distance network is two-wire or via a two-wire link. Likewise, the far-end user is also connected to the long-distance (toll) network via a two-wire link. Such a long-distance connection is shown in Figure 5.4. Schematically, the four-wire interconnection is shown as if it were a single channel wire-line with amplifiers; however, it would more likely be a multichannel carrier on cable and/or multiplex on radio. However, the amplifiers in Figure 5.4 serve to convey the ideas that this section considers. As shown in Figure 5.4, conversion from two-wire to four-wire operation is carried out by a terminating set, more commonly

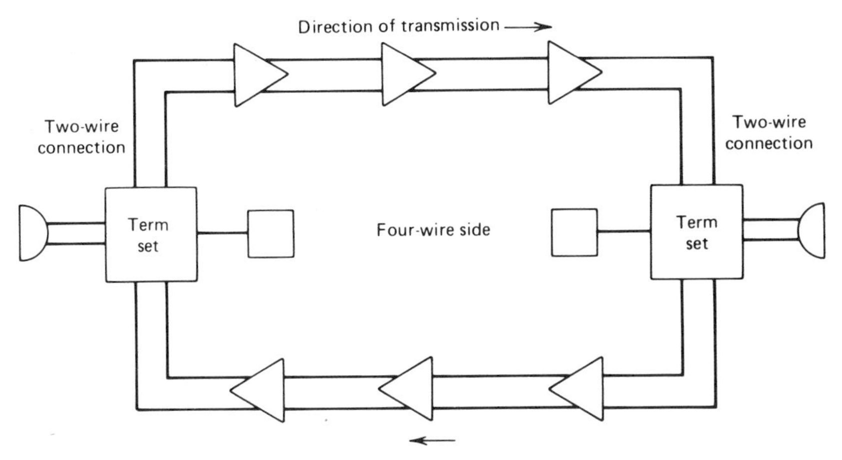

Figure 5.4 A typical long-distance (toll) telephone connection.

referred to in the industry as a *term set*, which contains a four-port balanced transformer (a hybrid) or, less commonly, a resistive network.

3.3 Operation of a Hybrid

A hybrid, in terms of telephony (at voice frequency), is a transformer. For a simplified description, a hybrid may be viewed as a power splitter with four sets of wire-pair connections. A functional block diagram of a hybrid device is shown in Figure 5.5. Two of the wire-pair connections belong to the four-wire path, which consists of a transmit pair and a receive pair. The third pair is the connection to the two-wire link that is eventually connected to the subscriber subset via one or more switches. The last wire pair of the four connects the hybrid to a resistance–capacitance balancing network, which electrically balances the hybrid with the two-wire connection to the subscriber subset over the frequency range of the balancing network. An artificial line may be used for this purpose.

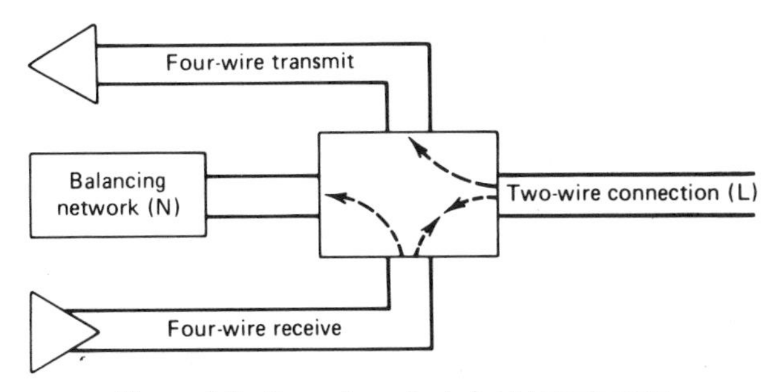

Figure 5.5 Operation of a hybrid transformer.

Signal energy entering from the two-wire subset connection divides equally, half of it dissipating in the impedance of the four-wire side receive path and the other half going to the four-wire side transmit path, as shown in Figure 5.5. Here the *ideal* situation is that no energy is to be dissipated by the balancing network (i.e., there is a perfect balance). The balancing network is supposed to display the characteristic impedance of the two-wire line (subscriber connection) to the hybrid. Signal energy entering from the four-wire side receive path is also split in half in the ideal situation where there is perfect balance. Half of the energy is dissipated by the balancing network (N) and half at the two-wire port (L) (see Figure 5.5).

The reader notes that in the description of the hybrid, in every case, ideally half of the signal energy entering the hybrid is used to advantage and half is dissipated or wasted. Also keep in mind that any passive device inerted in a circuit, such as a hybrid, has an insertion loss. As a rule of thumb, we say that the insertion loss of a hybrid is 0.5 dB. Thus there are two losses here that the reader must not lose sight of:

Hybrid insertion loss	0.5 dB
Hybrid dissipation loss	3.0 dB (half of the power)
	3.5 dB (total)

As far as this section is concerned, any signal passing through a hybrid suffers a 3.5-dB loss. This is a good design number for gross engineering practice. However, some hybrids used on short subscriber connections purposely have higher losses, as do special resistance-type hybrids.

4 FREQUENCY DIVISION MULTIPLEX

4.1 General

Frequency division multiplex (FDM) is a method of allotting a unique band of frequencies in a comparatively wide band frequency spectrum of the transmission medium to each communication channel on a continuous time basis. The communication channel may be a voice channel 4 kHz wide, a 120-Hz telegraph channel (for example), a 15-kHz broadcast channel, a 48-kHz data channel, or a 4.2-MHz television channel.

Before launching into multiplexing, keep in mind that all multiplex systems work on a four-wire basis. The transmit and receive paths are separate (see Figure 5.6). For the discussion that follows, it is assumed that the reader has some background on how an SSB (single sideband) signal is developed and how a carrier is suppressed in SSBSC systems.

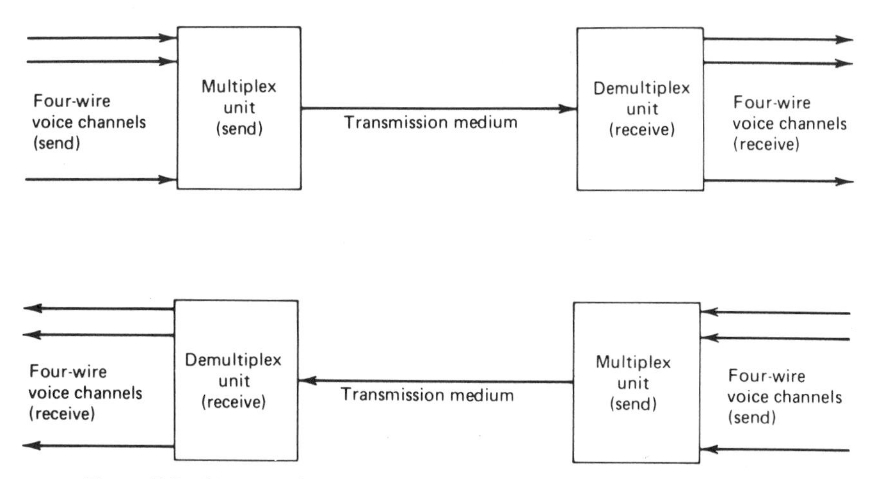

Figure 5.6 Simplified block diagram of a frequency division multiplex link.

4.2 Mixing

The heterodyning or mixing of signals of frequencies A and B is shown in the following diagram. What frequencies may be found at the output of the mixer?

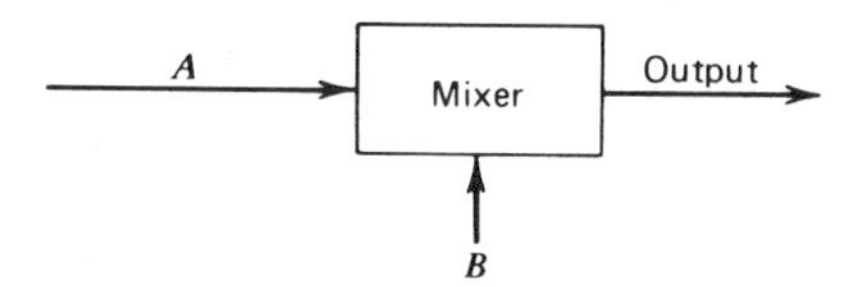

Both of the original signals will be present, as well as signals representing their sum and difference in the frequency domain. Thus signals of frequency A, B, $A + B$, and $A - B$ will be present at the output of the mixer.* Such a mixing process is repeated many times in FDM equipment.

Let us look at the boundaries of the nominal 4 kHz voice channel. These are 300 Hz and 3400 Hz. Let us further consider these frequencies as simple tones of 300 Hz and 3400 Hz. Now consider the mixer below and examine the possibilities at its output. First, the output may be summed, such that

$$\begin{array}{r} 20{,}000 \text{ Hz} \\ +300 \text{ Hz} \\ \hline 20{,}300 \text{ Hz} \end{array} \qquad \begin{array}{r} 20{,}000 \text{ Hz} \\ +3{,}400 \text{ Hz} \\ \hline 23{,}400 \text{ Hz} \end{array}$$

*Of course, if the mixer is a balanced mixer, only signals $A + B$ and $A - B$ will be present.

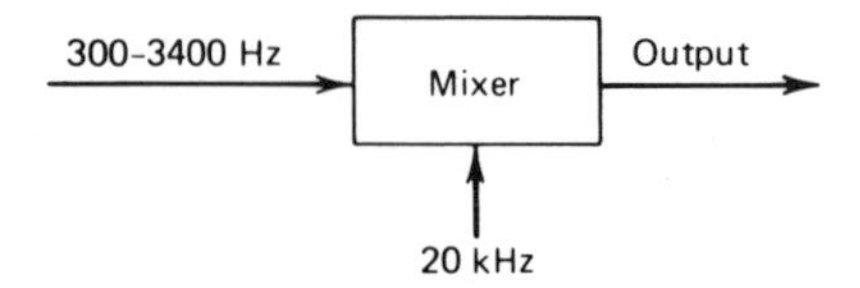

A simple low-pass filter could filter out all frequencies below 20,300 Hz. Now imagine that instead of two frequencies we have a continuous spectrum of frequencies between 300 Hz and 3400 Hz (i.e., we have the CCITT voice channel). We represent the spectrum as a triangle:

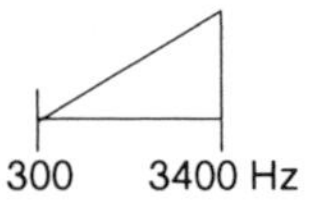

As a result of the mixing process (translation), we have another triangle:

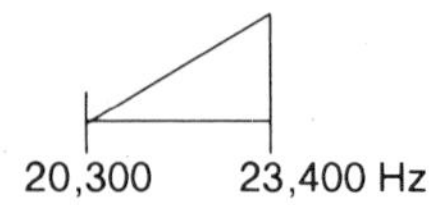

When we take the sum, as we did previously, and filter out all other frequencies, we say we have selected the upper sideband. Thus we have a triangle facing to the right, termed an *upright* or *erect* sideband.

We can also take the difference, such that

$$\begin{array}{r} 20{,}000 \text{ Hz} \\ -300 \text{ Hz} \\ \hline 19{,}700\ Hz \end{array} \qquad \begin{array}{r} 20{,}000 \text{ Hz} \\ -3{,}400 \text{ Hz} \\ \hline 16{,}600\ Hz \end{array}$$

and we see that in the translation (mixing process) we have had an inversion of frequencies. The higher frequencies of the voice channel become the lower frequencies of the translated spectrum, and the reverse occurs when the difference is taken. We represent this by a right triangle facing the other direction (left):

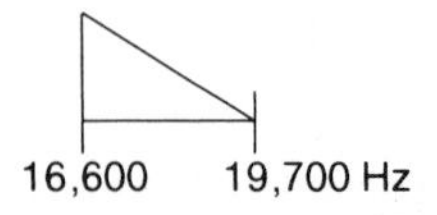

This is called an *inverted sideband*. To review, when we take the sum, we get an erect sideband. When we take the difference in the mixing process, frequencies invert and we have an inverted sideband represented by a triangle facing left.

4.3 CCITT Modulation Plan

4.3.1 Introduction. A modulation plan sets forth the development of a band of frequencies called the *line frequency* (i.e., ready for transmission on the line or other transmission medium). The modulation plan usually is a diagram showing the necessary mixing, local oscillator insertion frequencies, and the sidebands selected by means of the triangles described previously, in a step-by-step process from voice-channel input to line-frequency output. The CCITT has recommended a standardized modulation plan with a common terminology allowing large telephone networks, in both national and multinational systems, to interconnect. In the paragraphs that follow the reader is advised to be careful with terminology.

4.3.2 Formation of Standard CCITT Group. The standard *group* as defined by CCITT occupies the frequency band 60–108 kHz and contains 12 voice channels. Each voice channel is the nominal 4-kHz voice channel occupying the 300–3400-Hz spectrum. The group is formed by mixing each of the 12 voice channels with a particular carrier frequency associated with the channel. Lower sidebands are then selected. Figure 5.7 shows the basic (and preferred) approach to the formation of the standard CCITT group. It should be noted that the 60–108-kHz band voice channel 1 occupies the highest frequency segment by convention, between 104 kHz and 108 kHz. The layout of the standard group is shown in Figure 5.8. For more information refer to CCITT Rec. G.232.

Single sideband suppressed carrier (SSBSC) modulation techniques are used in all cases. The CCITT recommends that the carrier leak be down to at least −26 dBm0, referred to the zero relative level point (0 TLP) (see Section 2.3 of this chapter).

4.3.3 Formation of Standard CCITT Supergroup. A supergroup contains 5 standard CCITT groups, equivalent to 60 voice channels. The standard supergroup occupies the frequency band 312–552 kHz. Each group making up the supergroup is translated in frequency to the supergroup band by mixing with the proper carrier frequency. The carrier frequencies are 420 kHz for group 1, 468 kHz for group 2, 516 kHz for group 3, 564 kHz for group 4, and 612 kHz for group 5. In the mixing process the difference is taken (lower sidebands are selected). This translation process is shown in Figure 5.9.

4.3.4 Formation of Standard CCITT Basic Mastergroup and Supermastergroup. The basic mastergroup contains 5 supergroups (300 voice channels) and occupies the spectrum 812–2044 kHz. It is formed by translating the five standard supergroups, each occupying the 312–552-kHz band, by a process similar to that used to form the supergroup from five standard CCITT groups. This process is shown in Figure 5.10.

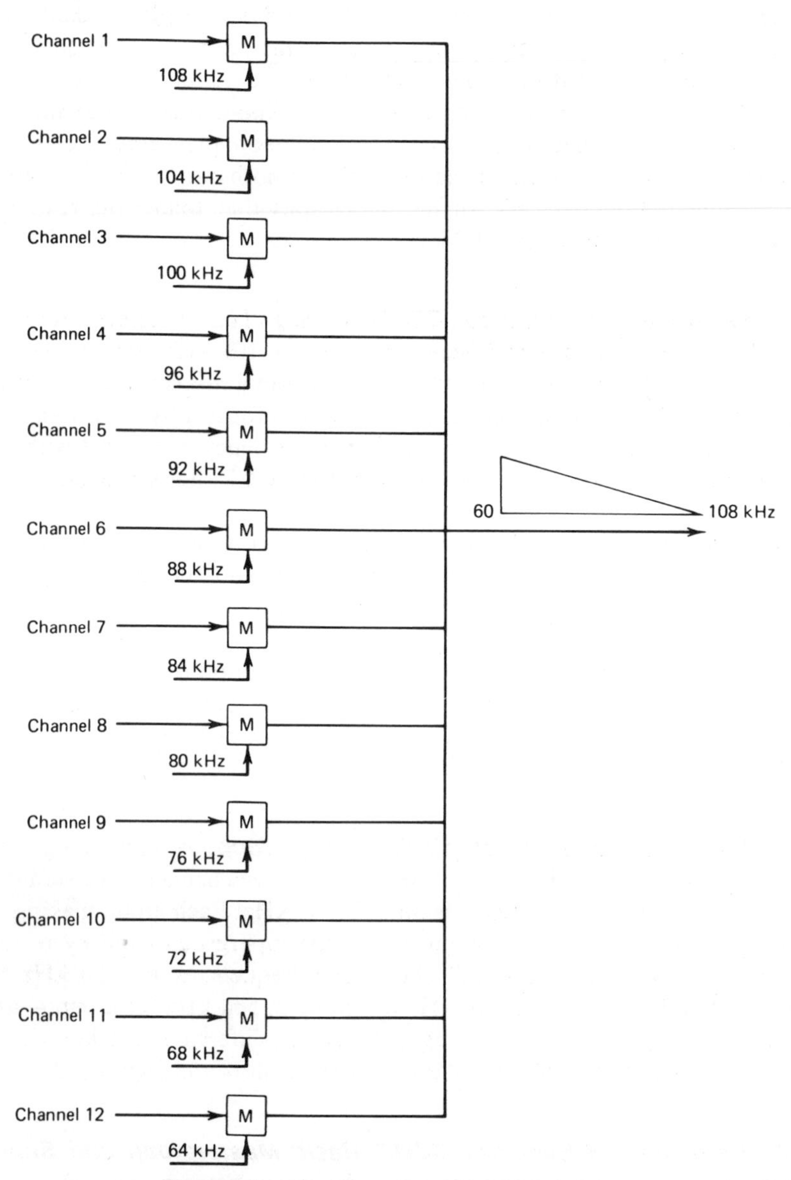

Figure 5.7 Formation of standard CCITT group.

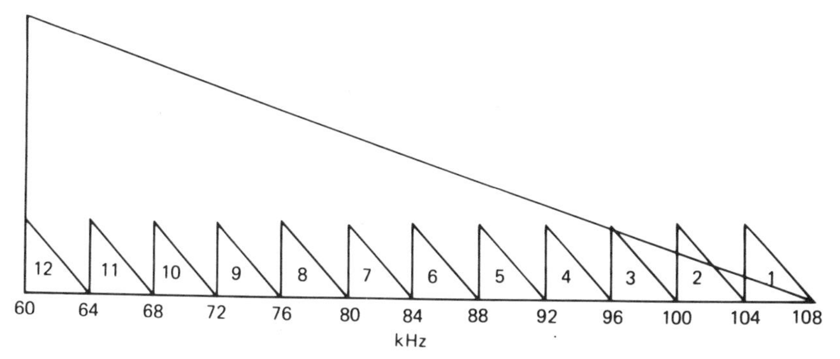

Figure 5.8 Layout of standard CCITT group.

The basic supermastergroup contains three mastergroups and occupies the band 8516–12,388 kHz. The formation of the supermastergroup is shown in Figure 5.11.

4.3.5 Line Frequency. The band of frequencies that the multiplex applies to the line, whether the line is a radiolink, coaxial cable, wire pair, or open wire, is called the *line frequency*. Another expression is HF (high frequency), which is ambiguous and often confused with HF radio transmission.

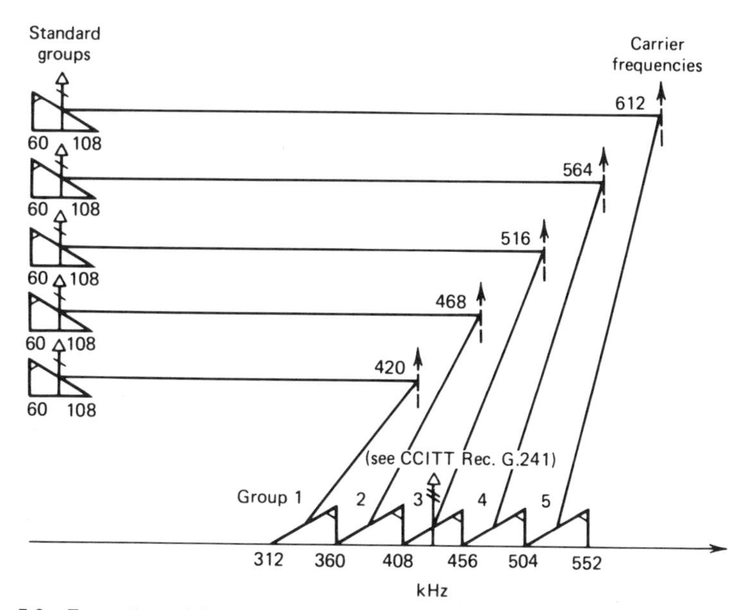

Figure 5.9 Formation of the standard CCITT supergroup. Vertical arrows with solid lines are level-regulating pilot tones; arrows with dashed lines are translation carrier frequencies. Courtesy of the International Telecommunication Union–CCITT.

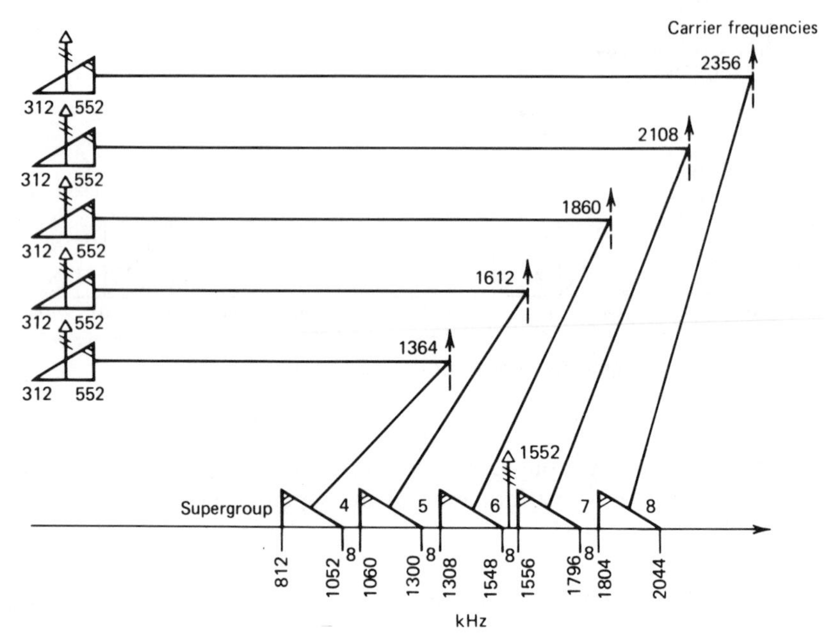

Figure 5.10 Formation of standard CCITT mastergroup. Courtesy of the International Telecommunication Union–CCITT.

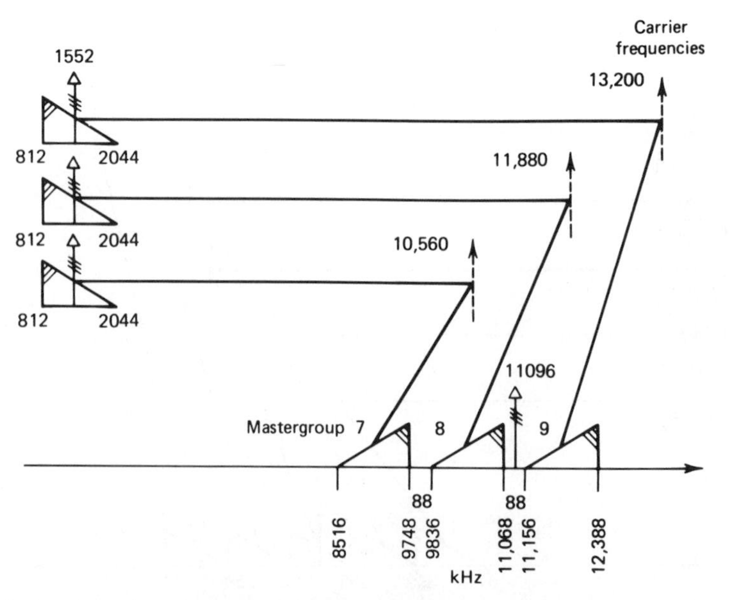

Figure 5.11 Formation of standard CCITT supermastergroup. Courtesy of the International Telecommunication Union–CCITT.

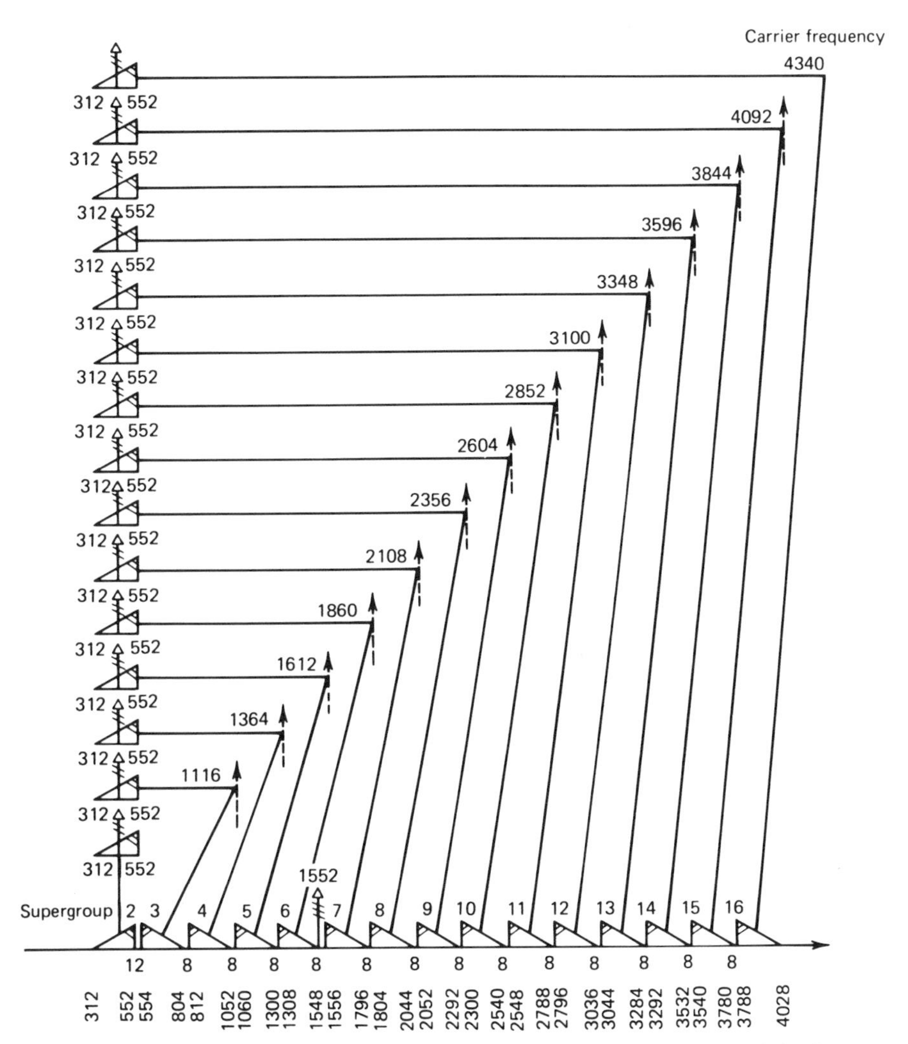

Figure 5.12 Makeup of basic CCITT 15-supergroup assembly. Courtesy of the International Telecommunication Union–CCITT.

The line frequency in any one particular case may be just the direct application of a group or supergroup to the line. However, a final translation stage more commonly occurs, particularly in high-density systems. Two of the recognized CCITT configurations are shown in Figures 5.12 and 5.13. Figure 5.12 shows the makeup of the standard CCITT 15-supergroup assembly and Figure 5.13, the standard 15-supergroup assembly No. 3 deriving from the basic 15-supergroup assembly shown previously.

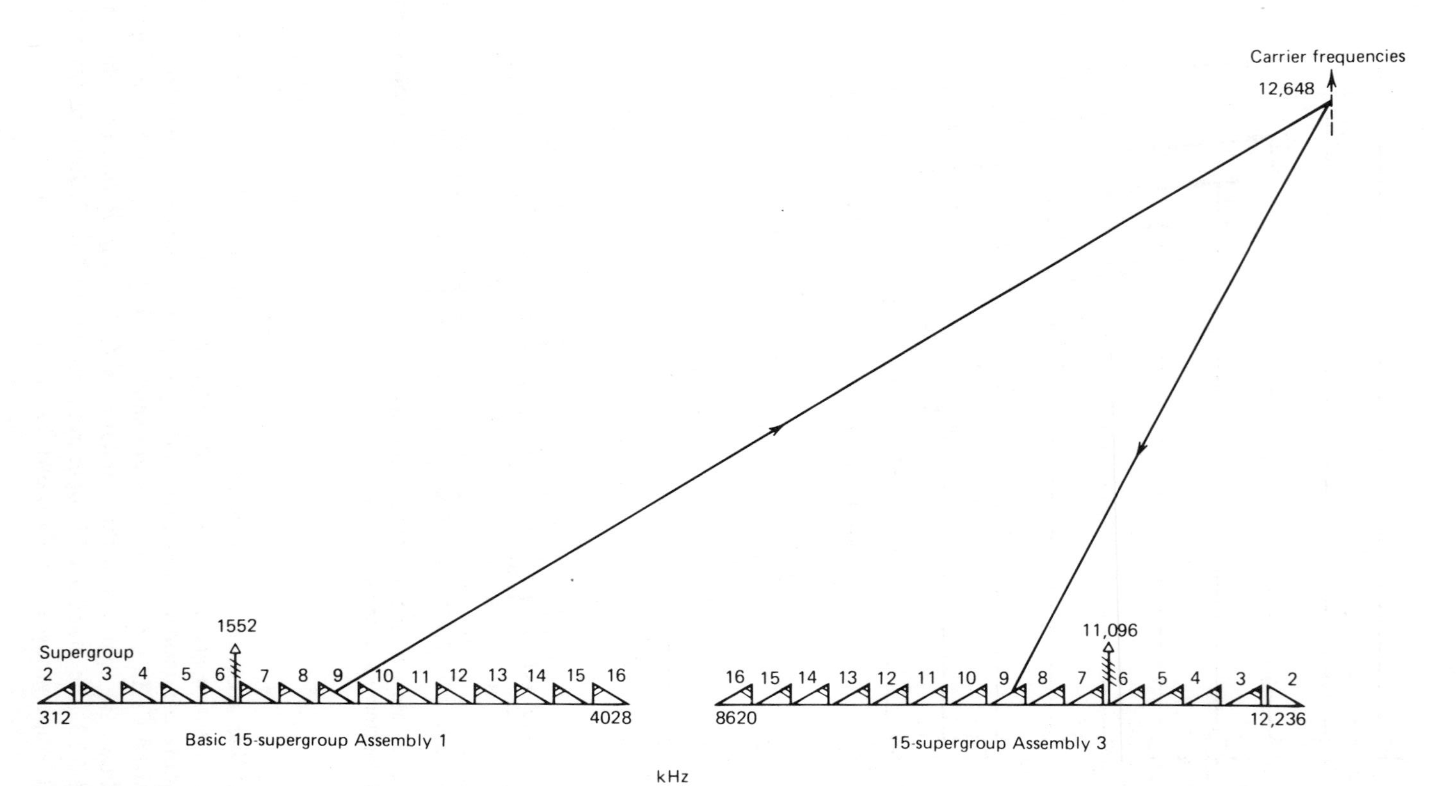

Figure 5.13 Makeup of standard CCITT 15-supergroup assembly No. 3 as derived from basic 15-supergroup assembly. Courtesy of the International Telecommunication Union–CCITT.

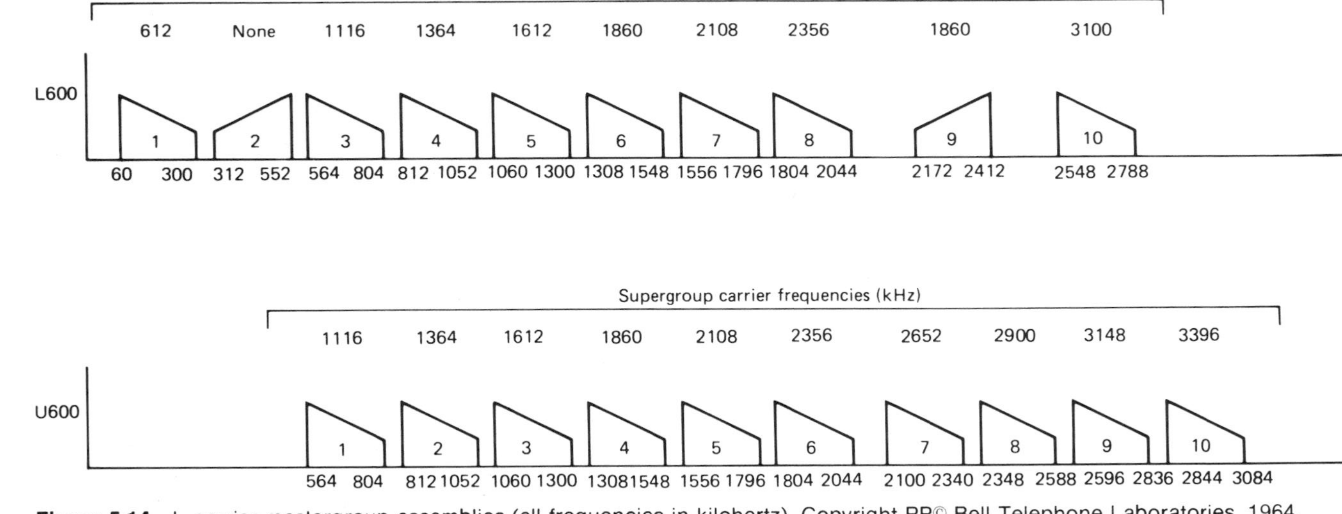

Figure 5.14 L-carrier mastergroup assemblies (all frequencies in kilohertz). Copyright PP© Bell Telephone Laboratories, 1964.

TABLE 5.1 L-Carrier and CCITT Comparison Table

Item	ATT L-Carrier	CCITT
Level		
Group		
Transmit	−42 dBm	−37 dBm
Receive	−5 dBm	−8 dBm
Supergroup		
Transmit	−25 dBm	−35 dBm
Receive	−25 dBm	−30 dBm
Impedance		
Group	130 Ω balanced	75 Ω unbalanced
Supergroup	75 Ω unbalanced	75 Ω unbalanced
VF channel	200–3300 Hz	300–3400 Hz
Response	+1.0 to −1.0 dB	+0.9 to −3.5 dB
Channel carrier		
Levels	0 dBm	Not specified
Impedances	130 Ω balanced	Not specified
Signaling	2600 Hz in-band	3825 Hz out-of-band
Group pilot		
Frequencies	92 or 104.08 kHz	84.08 kHz
Relative levels	−20 dBm0	−20 dBm0
Supergroup carrier		
Levels	+19.0 dBm per mod or demod	Not specified
Impedances (ohm)	75 unbalanced	Not specified
Supergroup pilot frequency	315.92 kHz	411.92 kHz
Relative supergroup pilot levels	−20 dBm0	−20 dBm0
Frequency synchronization	Yes, 64 kHz	Not specified
Line pilot frequency	64 kHz	308, 12,435 kHz
Relative line pilot level	−14 dBm0	−10 dBm0
Regulation		
Group	Yes	Yes
Supergroup	Yes	Yes

4.3.6 North American L-Carrier. "L-carrier" is the generic name given by the Bell System of North America to their long-haul SSB carrier system. Its development of the basic group and supergroup assemblies is essentially the same as that of the CCITT described previously. There is a variance in levels and pilot tone frequencies. The basic mastergroup differs, however. It consists of 600 VF channels (i.e., 10 standard supergroups). The L600 configuration occupies the band 60–2788 kHz and the U600 configuration, the band 564–3084 kHz. The relevant mastergroup assemblies are shown in Figure 5.14. The Bell System (ATT) identifies specific long-haul line-frequency configurations by adding a simple number after the letter "L." For example, the L3 carrier, which is used on coaxial cable and one type of ATT microwave (TH), has three mastergroups (see Figure 5.14) plus one supergroup, comprising 1800 VF channels on line occupying the band 312–8284 kHz. Table 5.1 compares L-carrier with CCITT carrier (FDM) standards. Subsequent paragraphs discuss the parameters and their meanings.

4.4 Loading of Multichannel Frequency-Division-Multiplex Systems

4.4.1 Introduction. Most of the FDM (carrier) equipment in use today carries speech traffic, which is sometimes misnamed "message traffic" in North America. In this context we refer to full-duplex conversations by telephone between two "talkers." However, the reader should remember that there is a marked increase in the use of these same intervening talker facilities for the transmission of data and fascimile.

For this discussion, the problem essentially concerns human speech and how multiple users may load a carrier system. If we load a carrier system too heavily—if the input levels are too high—IM noise and crosstalk would become intolerable, eventually leading to the breakdown of the system. If we do not load the system sufficiently, the signal-to-noise ratio will suffer. The problem is fairly complex because speech amplitude varies:

1. With talker volume.
2. At a syllabic rate.
3. At an audio rate.
4. With varying circuit losses as different loops and trunks are switched into the same channel bank voice-channel input.

4.4.2 Loading. For the loading of multichannel FDM systems, CCITT Rec. G.223 recommends:

> It will be assumed for the calculation of intermodulation below the overload point that the multiplex signal during the busy hour can be represented by a uniform spectrum [of] random noise signal, the mean absolute power level of which, at a zero relative level point—(in dBm0):
>
> $$P_{av} = -15 + 10 \log N^* \tag{5.5}$$
>
> when N is equal to or greater than 240 and
>
> $$P_{av} = -1 + 4 \log N \tag{5.6}$$
>
> When N is equal to or greater than 12 and less than 240...

where N is the number of voice channels (all logs to the base 10).

4.4.3 Single-Channel Loading. Many telephone administrations have attempted to standardize on -16 dBm0 for single-channel speech input to multichannel FDM equipment. With this input, peaks in speech level may reach -3 dBm0. Tests indicate that such peaks will not be exceeded more than 1% of the time. However, the conventional value of average power per

*For North American practice use $-16 + 10 \log N$.

voice channel allowed by the CCITT is −15 dBm0 (see CCITT Rec. G.223). This assumes a standard deviation of 5.8 dB and the traditional activity factor of 0.25 (see Section 4.4.5). Average talker level is assumed to be at −11.5 VU (volume units). We must turn to the use of standard deviation because we are dealing with talker levels that vary with each talker and thus with the mean or average.

4.4.4 Loading with Constant-Amplitude Signals. Speech on multichannel systems has a low duty cycle or activity factor. We exemplify a simplified derivation of the traditional activity factor of 0.25%. Certain other types of signals transmitted over multichannel equipment have an activity factor of 1. This means that they are transmitted continuously, or continuously over fixed time frames. They are also usually characterized by constant amplitude. The following are some examples of these types of signal:

- Telegraph tone or tones.
- Signaling tone or tones.
- Pilot tones.
- Data signals [particularly PSK (phase shift keying) and FSK (frequency shift keying) modulation].
- Facsimile (in digital or FM mode).

To provide a safety factor on overload the tendency would be to reduce level. Here again, if we reduce level too much, the signal-to-noise ratio and hence the error rate will suffer.

Using the loading formulas in Section 4.4.2 may permit the use of several data or telegraph channels without serious consequence. But if any wide use is to be made of the system for data–telegraph–facsimile, then other loading criteria must be used. For typical constant-amplitude signals, traditional (CCITT) transmit levels as seen at the input of the channel modulator of FDM carrier equipment are as follows:

- Data: −13 dBm0.
- Signaling [single-frequency supervision, tone on when idle (see Chapter 4)]: −20 dBm0.
- Composite telegraph: −8.7 dBm0.

For one FDM system now on the market with 75% speech loading and 25% data loading with more than 240 channels (total), the manufacturer recommends the following:

$$P_{\mathrm{rms}} = -11 + 10 \log N \tag{5.7}$$

(units in dBm0) using −5 dBm0 per channel for data-input levels and −8 dBm0 for the composite telegraph signal level. However, the manufacturer hastens to mention that all VF channels may be loaded with data signals (or telegraph) at −8 dBm0 level per channel. But for the −5-dBm0 signal level only two channels per group may be assigned to this level, with the remainder speech, or the group must be "deloaded," which means that a certain number of channels must remain idle.

4.4.5 Activity Factor. Speech is characterized by large variations in amplitude, ranging from 30 dB to 50 dB. We often use the VU (volume unit) to measure speech levels. The VU can be equated to the dBm for a simple sine wave in the VF range across 600 Ω. For a complex signal such as our speech signal, the average power measured in dBm of a typical single talker is

$$P_{\mathrm{dBm}} = V_{\mathrm{VU}} - 1.4 \tag{5.8}$$

or that a 0-VU talker has an average power of −1.4 dBm. Empirically, the peak power is about 18.6 dB higher than average power for a typical talker. The peakiness of speech level means that carrier (FDM) equipment must be operated at a low average power to withstand voice peaks to avoid overloading and distortion. These can be related to an activity factor T_a, which is defined as that proportion of the time that the rectified speech envelope exceeds some threshold. If the threshold is about 20 dB below the average power, the activity dependence on threshold is fairly weak. The preceding equation for average talker power can now be rewritten in relation to the activity factor as follows:

$$P_{\mathrm{dBm}} = V_{\mathrm{VU}} + 10 \log T_a \tag{5.9}$$

If $T_a = 0.725$, the results will be the same for the equation relating VU to dBm.

Now consider adding a second talker at a different frequency segment on the same equipment, but independent of the first talker. Of course, we are describing here the operation of FDM equipment discussed previously. With the second talker added, the system average power will increase 3 dB as expected. If we have N talkers, each on a different frequency segment, the average power developed will be

$$P_{\mathrm{dBm}} = V_{\mathrm{VU}} - 1.4 + 10 \log N \tag{5.10}$$

where P_{dBm} is the power developed across the frequency band occupied by all talkers.

Empirically, it has been found that the peakiness or peak factor of many talkers over a multichannel analog system reaches the characteristics of random noise peaks when the number of talkers, N, exceeds 64. When $N = 2$, the peaking factor is 18 dB; when $N = 10$, it is 16 dB; when $N = 50$, it is 14 dB; and so forth.

An activity factor—a term analogous to a "duty cycle" of 1—which we have been using in the preceding argument, is quite unrealistic. If it is 1, it means that somebody is talking on each channel all the time. The traditional figure for activity factor accepted by the CCITT and used in North American practice is 0.25. Follow the argument in the next paragraph on how to reach this lower figure.

The multichannel FDM equipment cannot be designed for N callers and no more. If this were true, a new call would have to be initiated every time a call terminated, or calls would have to be turned away because of congestion. Thus the equipment must be "overdimensioned" for BH service. For this dimensioning problem, we drop the activity factor from 1 to 0.70. Other causes will reduce this figure even more. For instance, circuits are essentially inactive during call setup as well as during pauses for thinking during a conversation. The 0.70 figure now must be dropped to 0.50. This latter figure is divided in half because of the talk–listen condition. If we disregard cases of "double-talking," it is obvious that while one end is talking on a full-duplex telephone circuit, the other is listening. Thus a circuit (in one direction) is idle half the time during the "listen" period. The resulting activity factor is 0.25.

4.5 Pilot Tones

In FDM equipment pilot tones are used primarily for level regulation and secondarily for fault alarms. The nature of speech, particularly in varying amplitude, makes it a poor prospect as a reference for level control. Ideally, simple single-sinusoid constant-amplitude signals with 100% duty cycles provide simple control information for level-regulating equipment. Frequency division multiplex level regulators operate very much in the same manner as automatic-gain control circuits on radio systems, except that their dynamic range is considerably smaller. Modern FDM carrier systems initiate a level-regulating pilot tone on each group at the transmit end of the system. Individual level-regulating pilots are also inserted on all supergroups and mastergroups. The intent is to regulate system level within ± 0.5 dB.

Pilots are assigned frequencies that are part of the transmitted spectrum yet do not interfere with voice-channel operation. They usually are assigned a frequency appearing in the guard band between voice channels or are residual carriers (i.e., partially suppressed carriers). The CCITT recommends one of the following as group-regulation pilots:

$$84.080 \text{ kHz (at a level of } -20 \text{ dBm0)}$$

$$84.140 \text{ kHz (at a level of } -25 \text{ dBm0)}$$

The Defense Communications Agency of the US Department of Defense recommends 104.08 kHz for group regulation and alarm. For CCITT group pilots, the maximum level of interference permissible in the voice channel is

−73 dBm0p. CCITT pilot filters have a bandwidth at the 3-dB points of 50 Hz (see CCITT Rec. G.241 for further information on other CCITT pilot frequencies and levels). The operating range of level control equipment activated by pilot tones is usually about ±4 or ±5 dB. If the incoming level of a pilot tone in the multiplex receive equipment drops outside the level regulating range, an alarm will be indicated (if such an alarm is included in the system design); CCITT Rec. G.241 suggests such an alarm when the incoming level varies 4 dB up or down from the nominal.

4.6 Noise and Noise Calculations in FDM Carrier Systems

4.6.1 General. Carrier equipment is the principal contributor of noise on coaxial cable systems and other metallic transmission media. On radiolinks (line-of-sight microwave) it makes up about 25% of the total noise on the overall system link. The traditional approach is to consider noise with respect to a hypothetical reference circuit. Several methods are possible, depending on the application. These include the CCITT method, which is based on a 2500-km hypothetical reference circuit, and the US Department of Defense method used in specifying communication systems. Such military systems are based on a 6000-nautical-mile reference circuit with 1000-mi links and 333-mi sections.

4.6.2 CCITT Approach. CCITT Rec. G.222 states:

> The mean psophometric* power, which corresponds to the noise produced by all modulating (multiplex) equipment... shall not exceed 2500 pW† at a zero relative level point. This value of power refers to the whole of the noise due to various causes (thermal, intermodulation, crosstalk, power supplies, etc.). Its allocation between various equipments can be to a certain extent left to the discretion of design engineers. However, to ensure a measurement agreement in the allocation chosen by different administrations, the following values are given as a guide to the target values:
>
> | | |
> |---|---|
> | for 1 pair of channel modulators | 200 pW |
> | for 1 pair of group modulators | 80 pW |
> | for 1 pair of supergroup modulators | 60 pW |
>
> The following values are recommended on a provisional basis:
>
> | | |
> |---|---|
> | for 1 pair of mastergroup modulators | 60 pW |
> | for 1 pair of supermastergroup modulators | 60 pW |
> | for 1 pair of 15-supergroup assembly modulators | 60 pW |

*See Section 5 of this chapter.
†1 pW = 1×10^{-9} mW.

Experience has shown that these target figures can often be improved considerably. The CCITT has purposely loosened the noise value allotted to channel modulators so that other modulation schemes from voice channel to group can be used rather than the direct modulation approach that we have described here. For instance, one solid-state FDM equipment now on the market, when operated with CCITT loading, has the following characteristics:

1 pair of channel modulators	31 pWp*
1 pair of group modulators	50 pWp
1 pair of supergroup modulators	50 pWp

Using another solid-state equipment and increasing the loading to 75% voice, 17% telegraph tones, and 8% data, the following noise information is applicable:

1 pair of channel modulators	322 pWp
1 pair of group modulators	100 pWp
1 pair of supergroup modulators	63 pWp

If this equipment is used on a real circuit with heavier loading, the sum for noise for channel modulators, group modulators, and supergroup modulator pairs is 485 pWp. Thus by simple arithmetic we see that such a system would be permitted to demodulate to voice only 5 times over a 2500-km route if we were to adhere to CCITT recommendations (i.e., $5 \times 485 = 2425$ pWp). This leads to the use of through-group and through-supergroup techniques discussed in Section 4.7. Figure 5.15 shows a typical application of this same equipment using CCITT loading.

4.6.3 US Military Approach. The following values are set forth in US Military Standard 188-100 for loaded noise:

Channel modulator pair	31 pWp0
Group modulator pair	50 pWp0
Supergroup modulator pair	50 pWp0
Through-group equipment	10 pWp0
Through-supergroup equipment	50 pWp0
Multiplex noise of FDM reference voice bandwidth link	131 pWp0

and it should be noted that the design objective is 100% data loading.

*Picowatts psophometrically weighted. See Section 5 of this chapter.

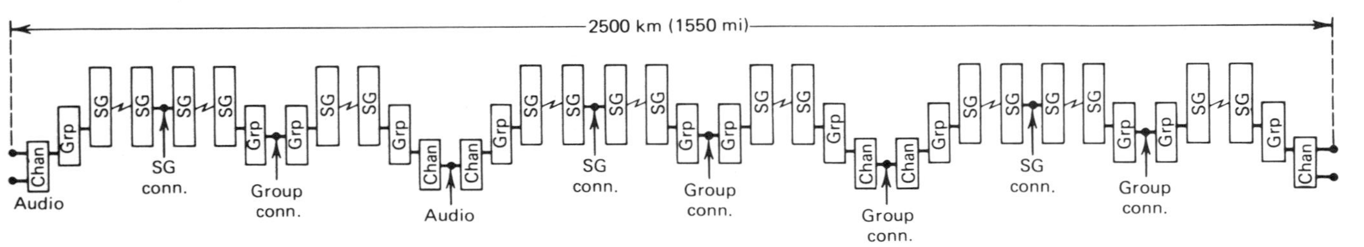

Figure 5.15 Typical CCITT reference-system noise calculations. These calculations are for typical equipment using CCITT loading Courtesy of GTE Lenkurt, Inc., San Carlos, Calif.

Multiplex		pWp0	dBrnCO
3 complete sets of channel bank equipment (XMT and REC)	at 110 pW	330	26
6 complete sets of group bank equipment	at 45 pW	270	25
9 complete sets of supergroup bank equipment	at 58 pW	522	28
3 sets of group connector equipment	at 5 pW	15	12.6
3 sets of supergroup connector equipment	at 16 pW48	17.6	
Total 46A3 Carrier Noise		1185	31.5
Microwave			
9 complete radio sections of 6 hops each			
78A3 radio (600 ch.) at 500 pWp0/section		4500	37.3
75A3 radio (960 ch.) at 400 pWp0/section		3600	36.4
Overall System Noise			
Using 78A3 radio (600 channels)	(4500 + 1185) 5685		
Using 75A3 radio (960 channels)	(3600 + 1185) 4785		

Note: Conversion from pWp0 to dBrnC0 is dBrnC0 = (10 log pWp0) + 0.8

Note: Calculations are for a typical equipment with CCITT loading.
Courtesy GTE Lenkurt Incorporated, San Carlos, Calif.

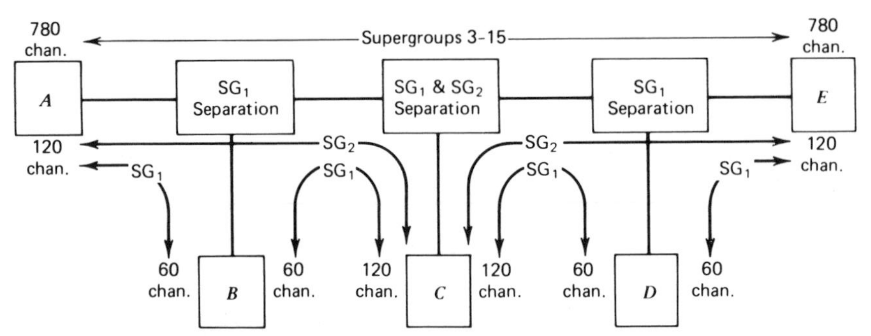

Figure 5.16 Typical drop and insert of supergroups.

4.7 Through-Group and Through-Supergroup Techniques

In Section 4.6 we indicated that modulation–translation steps in long FDM carrier systems must be limited to avoid excessive noise accumulation. One method used widely is to employ group connectors and through-supergroup devices. Figure 5.16, which illustrates this technique, shows that supergroup 1 is passed directly from point A to point B while supergroup 2 is dropped at C, a new supergroup is inserted for onward transmission to E, and so forth. At the same time supergroups 3 to 15 are passed directly from A to E on the same line frequency (baseband).

The expression "drop and insert" is used in carrier-system terminology to indicate that at some point a number of channels are "dropped" to voice (if you will) and an equal number are "inserted" for transmission back in the opposite direction. For full-duplex operation, if channels are dropped at B from A, B necessarily must insert the same number of channels going back again to A.

Through-group and through-supergroup techniques are used especially on long trunk routes where excessive noise accumulation can be a problem. Such route plans can be very complex. However, the savings on equipment and reduction of noise accumulation should be obvious. When through-supergroup techniques are used, the supergroup pilot may be picked off and used for level regulation. Nearly all FDM carrier equipment manufacturers include level regulators as an option on through-supergroup equipment, whereas through-group equipment may not necessarily have the option.

5 SHAPING OF A VOICE CHANNEL AND ITS MEANING IN NOISE UNITS

The attenuation distortion or frequency response of a voice channel is termed "flat" because when it is tested from input to output, the response from, say, 300 Hz to 3400 Hz may vary by only several decibels. Of course, from the input of a voice channel modulator of an FDM carrier link to the output of the companion demodulator, flat response, which is indeed what we have, should vary no more than perhaps ± 0.5 dB.

Now connect a telephone handset transmitter (with appropriate talk battery) to the input of the voice channel modulator and handset receiver to the output of the companion voice channel demodulator and include the acuity of the "average" human ear. Now we see a "shaping" effect. The handset and the ear acuity "shape" the channel when the audio level is compared at various frequencies at the input of the acoustical–electrical transducer (handset transmitter) to the audio output from the electrical–acoustical transducer (the handset receiver).*

The frequency response measurement uses a reference frequency located at the point of minimum attenuation in the voice channel. In North America the reference frequency is 1000 Hz, whereas in Europe and many other locations in the world it is 800 Hz. All CCITT recommendations dealing with the voice channel use 800 Hz as reference.

For all voice-channel measurements affecting *speech*, we must note that certain frequencies in the voice channel are attenuated more than others. When transmission systems are tested, weighting networks are used to simulate these effects. Different types of telephone handset have different attenuation–frequency characteristics than others. In the literature we are liable to run into four different types, namely:

144	Line weighting (North America); seldom encountered; noise unit, dBrn
F1A	Line weighting (North America); being phased out; noise unit, dBa
C–Msg	*C* message weighting; presently applicable in North America; noise unit, dBrn*C*
CCIR–1951	Psophometric weighting; European and CCITT–CCIR; noise unit, pWp

Reference frequency was established where the reference signal level was just discernible by the human ear. This level, depending on the handset, is in the range −85 to −90 dBm. Thus the derived units would be positive numbers. Such units are used for noise measurement, given the zero reference and weighting characteristics referred to one of the four telephone handsets mentioned previously. These weighting characteristics are given in Figure 5.17.

The Western Electric 144 handset was an early model universally used in North America. The noise-measurement unit referred to the 144 was the dBrn (decibels reference noise), −90 dBm at 1000 Hz. Note that on the 144 curve in

*The acoustical properties of the typical human ear must also be taken into account because the design of a weighting network, using a particular handset, is based on the annoyance factor of a certain single frequency tone in the voice channel compared to the reference frequency.

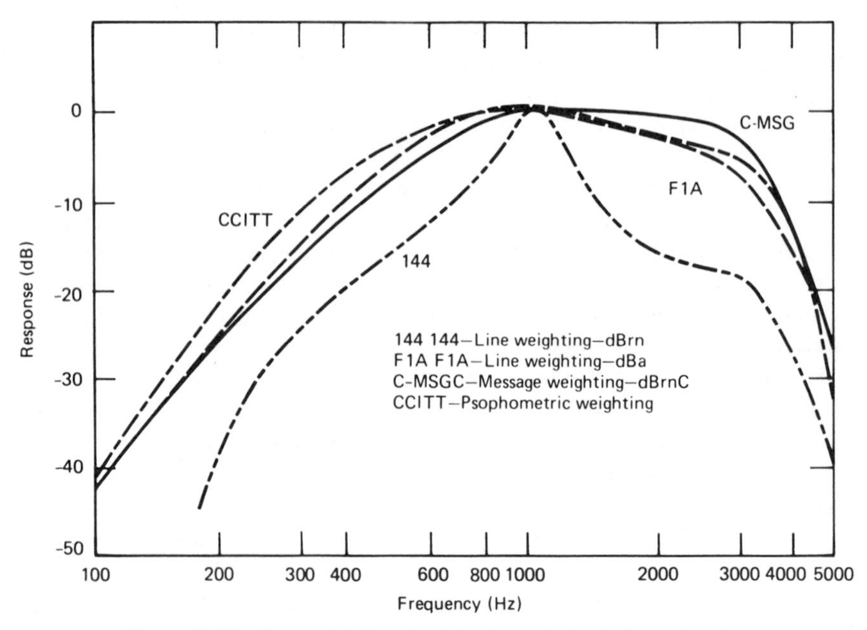

Figure 5.17 Line weightings for telephone (voice) channel noise.

Figure 5.17, a 500-Hz sinusoidal signal would need a 15-dB level increase to have the same interfering effect on the "average" listener over the 1000-Hz reference. A 3000-Hz signal requires an 18-dB increase to have the same interfering effect, 6 dB at 800 Hz, and so on. This curve, as shown in Figure 5.17, is called a *weighting curve*. Noise-measurement instruments have artificial filters made to simulate the response of the several handsets listed. Such response also includes the subjective effects of the average listener. These filters are called *weighting networks*.

Subsequent to the 144 handset, the Western Electric Company developed the F1A handset, which has a considerably broader response than the older 144 handset but was 5 dB less sensitive at 1000 Hz (as shown in Figure 5.17). The reference level for this type of handset was −85 dBm. Of course, the new weighting curve and associated weighting network were denoted "F1A." The noise measurement was dBa (dB adjusted).

A third, more sensitive handset is now in use in North America, giving rise to the *C*-message line-weighting curve and its companion noise-measurement unit, the dBrn*C*. In Figure 5.17 we see that it is 3.5 dB more sensitive at the 1000-Hz reference frequency than F1A and 1.5 dB less sensitive than the old 144 weighting. Rather than a new reference frequency power level (−88.5 dBm), the reference power level of −90 dBm was maintained.

One important weighting curve is the CCIR or CCITT psophometric weighting curve. The noise-measurement units associated with this curve are the dBmp (dBm psophometrically weighted) and the pWp (picowatts psophometrically weighted). The reference frequency in this case is 800 Hz rather than 1000 Hz.

The reader must be certain to differentiate between a voice channel where noise measurements are weighted because the eventual user of the channel is the human mouth–ear combination, and the associated handset as the transducer of electrical energy to speech or audio energy. We also talk about a flat channel that has a comparatively flat frequency response across the voice channel, and we generally define the voice channel today as occupying that spectrum from 300 Hz to 3400 Hz. Data–telegraph, facsimile, and video* transmission systems utilize "flat" channels, that is, where the amplitude–frequency response varies little between channel edges—say, of the order of ± 0.5 or ± 0.2 dB. On long networks with numerous points of modulation–demodulation the amplitude–frequency response can vary considerably more, but we still refer to the voice channel as being flat.

To convert from flat channel noise measurements in pW or dBm, the following excerpt from CCITT Rec. G.223 may be useful:

> If uniform-spectrum random noise is measured in a 3.1 kHz band with flat attenuation frequency characteristic, the noise level must be reduced 2.5 dB to obtain the psophometric power level. For another bandwidth, B, the weighting factor will be equal to:
>
> $$2.5 + \frac{(10 \log B)}{3.1} \text{ (dB)}$$
>
> when $B = 4$ kHz, for example, this formula gives a weighting factor of 3.6 dB.

REVIEW QUESTIONS

1. There are four basic transmission parameters for a telephone voice channel, three of which are impairments. Name and define each.
2. Define the specified CCITT voice channel.
3. What are the two reference frequencies for the voice channel? One is European and accepted by CCITT and the other is North American.
4. The reference test tone input to a telephone channel is -16 dBm and the output is $+7$ dBM. What range of values in dBm can be expected at the output if the attenuation distortion is -1 $+2.5$ dB from 600 Hz to 2600 Hz?
5. What causes phase distortion?

*Under certain conditions video signal-to-noise ratios are weighted [i.e., TASO (Television Allocations Studies Organization) weightings].

6. What test tone level should one expect at 0 TLP? A value of −10 dBm at the −12-dBr point corresponds to what dBm0 value?

7. Name the four basic categories of noise.

8. What thermal noise level would one expect with a receiver with 1000 K noise temperature and a bandwidth of 1 kHz?

9. Two frequencies of 1000 Hz and 1200 Hz appear at the input of a nonlinear device. Give values of third-order products.

10. Give at least two causes of intermodulation products.

11. Impulse noise generally does not affect speech telephony. What can it affect seriously?

12. The signal level coming out of a receiver is +7 dBm and the thermal noise level is −35 dBm. What is the signal-to-noise ratio in this case.

13. Define two-wire and four-wire transmission. Where would each be applied?

14. What are the two loss components across a model hybrid? In each case explain what causes the loss.

15. Channel 1 of the standard CCITT FDM group has an injection local oscillator carrier frequency of 108 kHz. What band does channel 1 occupy (3-dB points).

16. What frequency band does the CCITT group occupy? How many VF channels does it accommodate?

17. What is the standard activity factor for FDM telephony?

18. Considering question 17, what is the problem when voice channels carry conventional data signals on an FDM system? Discuss.

19. When an FDM system is loaded too heavily (i.e., levels are too high), what will result? It can be disastrous.

20. What are the primary and secondary functions of pilot tones on an FDM system?

21. What are the standard North American and CCITT test tone inputs on an FDM VF channel?

22. Why do we resort to through-groups and through-supergroups on FDM systems wherever possible?

23. Weighting of a VF telephone channel simulates three characteristics for speech. What are they?

24. *C*-message weighting noise uses what noise measurement units?

25. Express 1 pWp in watts and in milliwatts. Use the powers of 10.

26. What is the power level in dBm for 200 pWp?

REFERENCES

1. *Notes on Distance Dialing*, American Telephone and Telegraph Company, New York, 1975.
2. International Telephone and Telegraph Corporation, *Reference Data for Radio Engineers*, 6th ed., Howard W. Sams, Indianapolis, 1976.
3. *Lenkurt Demodulator*, Lenkurt Electric Company, San Carlos, Calif., December 1964. June 1965, September 1965.
4. *Transmission Systems for Communication*, 5th ed., rev., Bell Telephone Laboratories, 1982.
5. U.S. Military Standard Mil-Std-188-100, November 15, 1972, Department of Defense, Washington, D.C. 20301.
6. H. H. Smith, *Noise Transmission Level Terms in American and International Practice*, ITT Communication Systems, Paramus, N.J., 1964.
7. *CCITT Red Books*, VIII Plenary Assembly, Malaga-Torremolinos, 1984, G. Recommendations.
8. B. D. Holbrook and J. T. Dixon, "Load Rating Theory for Multichannel Amplifiers," *Bell. Syst. Tech. J.*, 624–644 (October 1939).
9. W. Oliver, *White Noise Loading of Multi-channel Communication Systems*, Marconi Instruments, St. Albans, Herts, UK, 1976.
10. Roger L. Freeman, *Telecommunication Transmission Handbook*, 2nd ed., Wiley, New York, 1981.
11. *IEEE Standard Dictionary of Electical & Electronic Terms*, 2nd ed., IEEE, New York, 1977.
12. *Notes on the Network*, American Telephone and Telegraph Co., New York, 1980.
13. *Notes on the Boc Intra-LATA Networks*, Bell Communications Research, 1986, Livingston, N.J., TR-NPL-000275.

6

LONG-DISTANCE NETWORKS

1 GENERAL

The design of a long-distance network involves basically three considerations: (1) routing scheme given inlet and outlet points and their traffic intensities, (2) switching scheme and associated signaling, and (3) transmission plan. In the design each criterion will interact with the others. In addition, the system designer must specify type of traffic, lost-call criteria or grade of service, a survivability criterion, forecast growth, and quality of service. The trade-off of all these factors with "economy" is probably the most vital part of initial planning and downstream system design.

Consider transcontinental communications in the United States. Service is now available for people in New York to talk to people in San Francisco. From the history of this service, we have some idea of how many people wish to talk, how often, and for how long. These factors are embodied in traffic intensity and calling rate. There are also other cities on the West Coast to be served and other cities on the East Coast. In addition, there are existing traffic nodes at intermediate points such as Chicago and St. Louis. An obvious approach would be to concentrate all traffic into one transcontinental route with drops and inserts at intermediate points.

Again, we must point out that switching enhances the transmission facilities. From an economic point of view, it would be desirable to make transmission facilities (carrier, radio, and cable systems) adaptive to traffic load. These facilities taken alone are inflexible. The property of adaptivity, even when the transmission potential for it has been predesigned through redundancy, cannot be exercised except through the mechanism of switching in some form. It is switching that makes transmission adaptive.

The following requirements for switching ameliorate the weaknesses of transmission systems: concentrate light, discretely offered traffic from a multiplicity of sources and thus enhance the utilization factor of transmission trunks; select and make connections to a statistically described distribution of destinations per source; and restore connections interrupted by internal or external disturbances, thus improving reliabilities (and survivability) from the levels of the order of 90% to 99% to levels of the order of 99% to 99.9% or better. Switching cannot carry out this task alone. Constraints have to be iterated or fed back to the transmission systems, even to the local area. The transmission system must not excessively degrade the signal to be transported; it must meet a reliability constraint expressed in MTBF (mean time between failures) and availability and must have an alternative route scheme in case of facility loss, whether switching node or trunk route. This latter may be termed *survivability* and is only partially related to overflow (e.g., alternative routing).

The single transcontinental main traffic route in the United States suggested earlier has the drawback of being highly vulnerable. Its level of survivability is poor. At least one other route would be required. Then why not route that one south to pick up drops and inserts? Reducing the concentration in the one route would result in a savings. Capital, of course, would be required for the second route. We could examine third and fourth routes to improve reliability–survivability and reduce long feeders for concentration at the expense of less centralization. In fact, with overflow, one to the other, dimensioning can be reduced without reduction of overall grade of service.

2 THE DESIGN PROBLEM

The same factors enter into long-distance network routing decisions as were discussed for the local area in Chapter 2. The first step is exchange placement. Here we follow North American practice and call the exchange in the long-distance network a "toll exchange." Rather than base the placement decision on subscriber density distribution and their calling rates, the basic criterion is economy, the most cost-effective optimum. Toll-center placement is discussed in Section 5 of this chapter.

Having chosen toll-center locations, the design procedure is the familiar traffic matrix, where cost ratio studies are carried out to determine whether routing will be direct or tandem. The tendency is tandem (or "transit" exchanges) working and direct routes with overflow. The economic decision arises to balance switching against transmission (considering our arguments in Section 1). Compare local versus long-distance networks:

	Switching Cost per Circuit	Transmission Cost per Circuit	Favored Network
Local network	Relatively high	Low	Mesh
Toll network	Relatively low	High	Star

For the long-distance network we can nearly always assume a hierarchical structure with three, four, or even five levels. Ideally, the highest levels should be connected in mesh for survivability.

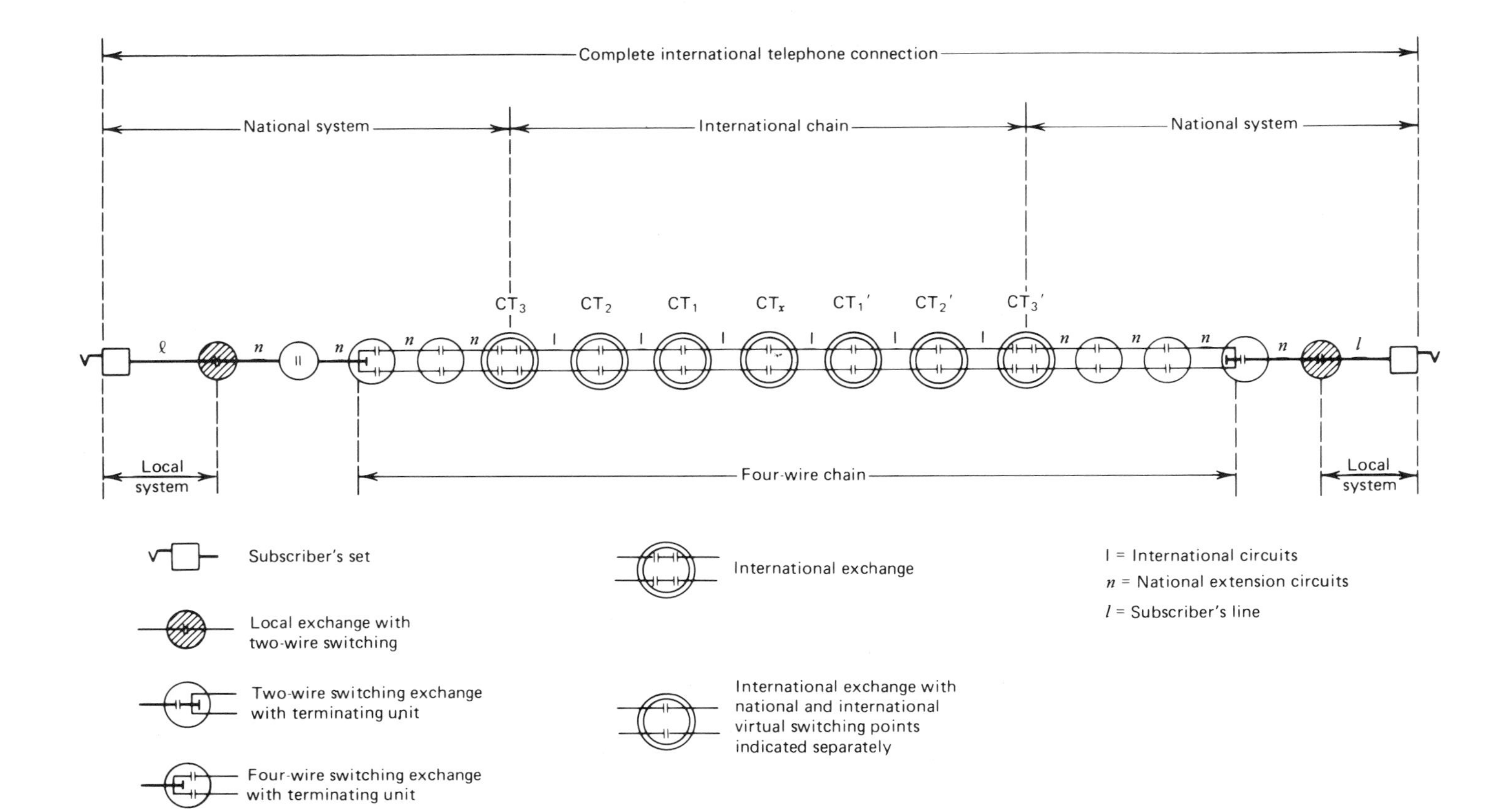

Figure 6.1 An international connection to illustrate the nomenclature adopted and the maximum number of links in tandem for an international connection (CCITT Recs. G.101 and Q.40). Courtesy CCITT-ITU.

In local network design, particularly in metropolitan areas, we could generally assume a mesh connection. There might be an exceptional case where tandem working would be economical, where traffic flows were 20 erlangs or less. Because of low traffic flows in long-distance network design, star connections can be assumed at the outset, and we would then proceed to determine cases where direct links may be justified with or without alternative routing.

3 LINK LIMITATION

It is stated in CCITT Rec. Q.40, para. 3:

> For reasons of transmission quality and the efficient operation of signalling, it is desirable to limit as much as possible the number of circuits connected in tandem.
>
> The apportionment between national and international circuits in such a chain may vary.
>
> The maximum number of circuits to be used for an international call is 12 with up to a maximum of four of the circuits being international.
>
> In exceptional cases and for a low number of calls, the total number of circuits may be 14, but even in this case the maximum number of international circuits is four.

If twelve circuits in tandem is the absolute maximum and we subtract four for the international portion, eight are left for the national portions. We then allow four maximum for each national network. This limit is crucial in the national network design. The concept is illustrated in Figure 6.1.

4 INTERNATIONAL NETWORK

Before 1980 the CCITT routing plan was based on a network with a hierarchical structure with descending levels called CT1, CT2, CT3, and CTX. Since 1980 CCITT has made a radical change in its international routing plan. The new plan might be called a "free routing structure." It assumes that national administrations (telephone companies) will maintain national hierarchical networks. Obviously the change was brought about by the long reach of satellite communications with which international high-usage (HU) trunks can terminate practically anywhere in the territory of a national administration.

The CCITT International Telephone Routing Plan is contained in CCITT Rec. E.171 (Ref. 19) and is reviewed below.

In practice, the large majority of international telephone traffic is routed on direct circuits (i.e., no intermediate switching point) between international switching centers (ISCs). It should be noted that it is the rules governing routing of connections consisting of a number of circuits in tandem that this recommendation primarily addresses. These connections have an importance in the network because:

- they are used as alternate routes to carry overflow traffic in busy periods to increase network efficiency
- they can provide a degree of service protection in the event of failure of other routes
- they can facilitate network management when associated with ISCs having temporary alternative routing capabilities.

This plan replaces the previous one established in 1964. Rec. E.171 continues under "Principles":

The Plan preserves the freedom of administrations: a) to route their originating traffic directly or via any transit administration they choose; b) to offer transit capabilities to as wide a range of destinations as possible in accordance with the guidelines it provides. . . .

The governing features of this plan are:
a) it is *not* hierarchical,
b) administrations are free to offer whatever transit capabilities they wish, providing they conform to the Recommendation,
c) direct traffic should be routed over final (fully provided) or high usage circuit groups,
d) no more than 4 international circuits in tandem between the originating and terminating ISCs,
e) advantage should be taken of the non-coincidence of international traffic by the use of alternative routings and provide route diversity (Rec. E.523),
f) the routing of transit switched traffic should be planned to avoid possibility of circular routings,
g) when a circuit group has both terrestrial and satellite circuits, the choice of routing should be governed by:

- the guidance given in Rec. G.114 (e.g., no more than 400 ms one-way propagation time),
- the number of satellite circuits likely to be utilized in the overall connection,
- the circuit which provides the better transmission quality and overall service quality,

h) the inclusion of two or more satellite circuits in the same connection should be avoided in all but exceptional cases. Regarding (h), reference should be made to Annex A of Recs. E.171 and Q.14.

5 EXCHANGE LOCATION

The design of the toll network is closely related to the layout of toll areas; thus the system designer would start by placing a toll exchange in each toll area, probably in or near a large city in the area. Tariffs (tolls) for long-distance calls are usually based on the crow-fly distance traversed by the call. For instance, there may be fixed charges for calls over distances 0–20 km, 20–50 km, 50–100 km, 100–200 km, 200–300 km, and so on. It is expensive to measure route distance for every call. Hence a country or telephone serving area (in the macro sense) is usually divided into tariff areas and charged the same amount for calls between two different areas, no matter what the exact origin and destination in each of the areas. Tariff areas should not be made too small, as this would result in numerous areas with costly charging and billing equipment. When tariff areas are too large, tariffs may tend to be inequitable to subscribers. An area 50 km in diameter may be used as a guide for desirable size of tariff area. Areas will be considerably larger in sparsely populated regions. After tentatively placing a toll exchange in each toll area, the system designer should then examine adjacent pairs of toll exchanges to determine whether one exchange could serve both areas.

The next step is to examine assignments of toll exchanges regarding numbering. In Chapters 3 and 4 we saw how numbering routes a call and inputs call accounting equipment or call metering. Numbering may entail consideration of more than one toll exchange in geographically large tariff areas. Another consideration is maximum size of a toll exchange. Depending on expected long-distance calling rate and holding times, we might suppose 0.003 erlangs* per subscriber line; thus a 4000-line toll exchange could serve just under a million subscribers. The exchange capacity should be dimensioned to the forecast long-distance traffic load 10 years after installation. If the system undergoes a 15% expansion in long-distance traffic volume per year, it will grow to 4 times its present size in 10 years. Exchange location in the toll area is not very sensitive to traffic; it is more important to make maximum use of existing plant.

Hierarchy is another essential aspect. One important criterion is establishing the number of hierarchical levels in a national network. The tendency today is to reduce the number of hierarchical levels. Factors leading to more than two levels (e.g., the United States has five levels) are:

- Geographical size.
- Telephone density, usually per 100 inhabitants.
- Toll traffic trends.

*There is a marked tendency of toll traffic growth per subscriber in highly developed countries.

The trend toward greater use of direct HU routes may also force the use of less hierarchical levels. Once the number of hierarchical levels has been established, the number of fan outs must be considered to establish the number of long-distance exchanges in the network. Fan outs of six to eight are desirable. Thus with a two-stage hierarchy with a fan out of 6 and then 8, there would be 48 of the lowest-level long-distance (toll) exchanges. A three-level hierarchy, using the same rules, would have $48 \times 8 = 384$, a formidable number.

As can be seen, there are many choices open to the system engineer to establish the route-plan hierarchy. For example, if there are 24 long-distance exchanges in an area, the network will initially be a star connection, either three-stage, two-stage with low (initial) fan out, or two-stage with high initial fan out. Here "fan out" refers to the highest level and works downward. Figure 6.2 illustrates these principles. Figure 6.2A shows an area with 26 exchanges and part of a larger area with perhaps 100 or more exchanges, with the principal city in the upper left-hand corner. Three choices of fan outs are shown. Figure 6.2B is a three-level hierarchy with a four-to-five fan out at each stage. For a two-level hierarchy, two possibilities are suggested. Figure 6.2C has low initial fan out and 6.2D, a high one. The choice between 6.2C and 6.2D may depend on traffic intensity between nodes or availability of routes. For national networks, the fan out in Figure 6.2D may be most economical since traffic is brought to a common point more quickly, leaving the individual branches to be least traffic efficient.

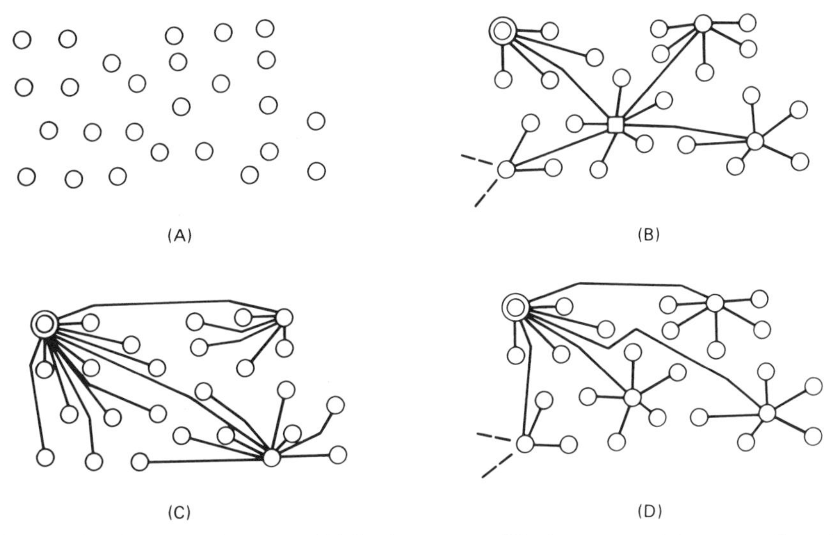

Figure 6.2 Choice of fan outs: (A) basic pattern; (B) three-stage interconnection; (C) two-stage interconnection with low initial fan out; (D) two-stage interconnection with high initial fan out.

6 NETWORK DESIGN PROCEDURES

The attempt to attain a final design of an optimum national network is a major "cut-and-try" process. It lends itself well to computer techniques such as a program developed by ITT, "Optimization of Telephone Trunking Networks with Alternate Routing." A second program complements the series, namely, "Optimization of Telephone Networks with Hierarchical Structure" [10, 11].

In every case the design must first take into account the existing network. Major changes in that network require large expenditure. The network also represents an existing investment that should be amortized over time. Elements of the system, such as switches, are of varying age; some switches remain in service for up to 40 years, and others have been recently installed. Removal of a switch with only several years of service would not be economical. These switches also have specific signaling characteristics and are interconnected by a trunk network.

To simplify the design process, visualize a group of local areas. That is, the geographical and demographic area of interest in which a national network is to be designed is made up of contiguous local areas (Chapter 2). There are now three bases to work from:

1. There are existing local areas, each of which has a toll exchange.
2. There is one or more ISCs placed at the top of the network hierarchy.
3. There will be no more than four links in tandem (Section 3) on any connection to reach an ISC.

Point 1 may be redefined as a toll area made up of a grouping of local areas probably coinciding with a numbering (plan) area. This is illustrated in a very simplified manner in Figure 6.3, where T, in European (CCITT) terminology, is a primary center, or a class 4 exchange in North American terminology. Center T, of course, is a tandem exchange with a fan out of four; these are four local exchanges, *A*, *B*, *C*, and *D* homing on T. The entire national

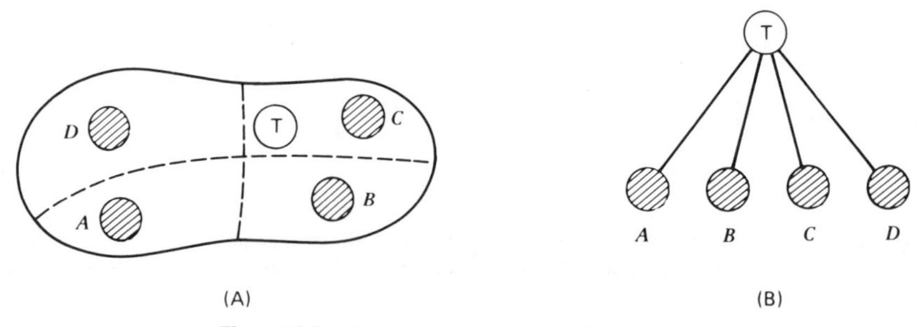

Figure 6.3 Areas and exchange relationships.

TABLE 6.1 Toll Traffic Matrix (Sample) (in Erlangs)

From Exchange	To Exchange									
	1	2	3	4	5	6	7	8	9	10
1		57	39	73	23	60	17	21	23	5
2	62		19	30	18	26	25	2	9	6
3	42	18		28	17	31	19	8	10	12
4	70	31	23		6	7	5	8	4	3
5	25	19	32	5		22	19	31	13	50
6	62	23	19	8	20		30	27	19	27
7	21	30	17	40	16	32		15	16	17
8	21	5	12	3	25	19	17		18	29
9	25	10	9	1	16	22	18	19		19
10	7	8	7	2	47	25	13	30	17	

geographic area will be made up of small segments, as shown in Figure 6.3, and each may be represented by a single exchange such as T.

The next step is to examine traffic flows to and from (originating and terminating at) each T. This information is organized and tabulated on a traffic matrix. A simplified example is shown in Table 6.1. Care must be taken in the preparation and subsequent use of such a table. The convention used here is that values are read *from* the exchange in the left-hand column *to* the exchange in the top row. For example, traffic from exchange 1 to exchange 5 is 23 erlangs, and traffic from exchange 5 to exchange 1 is 25 erlangs. It is often useful to set up a companion matrix of distances between exchange pairs. The matrix (Table 6.1) immediately offers candidates for high-usage routes. Nonetheless, this step is carried out after a basic hierarchical structure is established.

The highest level of the structure has been set forth initially. The CCITT hierarchy was shown in Figure 1.14. The lowest level of the network was established with local areas. The fact that Figure 1.14 has a five-level hierarchy does not imply the use of five levels. For instance, if the country in question is large, four links are permitted on a five-level hierarchy. A medium-size country might require the total links to the top of the hierarchy to be three for a four-level hierarchy, and a smaller country, two links for a three-level hierarchy, and so forth. We are, of course, working with maximums on final routes.

The outline of a five-level hierarchy is shown in Figure 6.4 with high-usage (HU) routes. Note that the lowest level is not included in the figure, that of the local exchange. High-usage routes ameliorate the problem of excessive links in tandem on the great majority of calls, thereby meeting the intent of CCITT Rec. Q.40.

Suppose, for example, that a country had four major population centers and could be divided into four areas around each center. Each of the four major population centers would have a tertiary center assigned, one of which would be the ISC. Each tertiary center would have one or several secondary

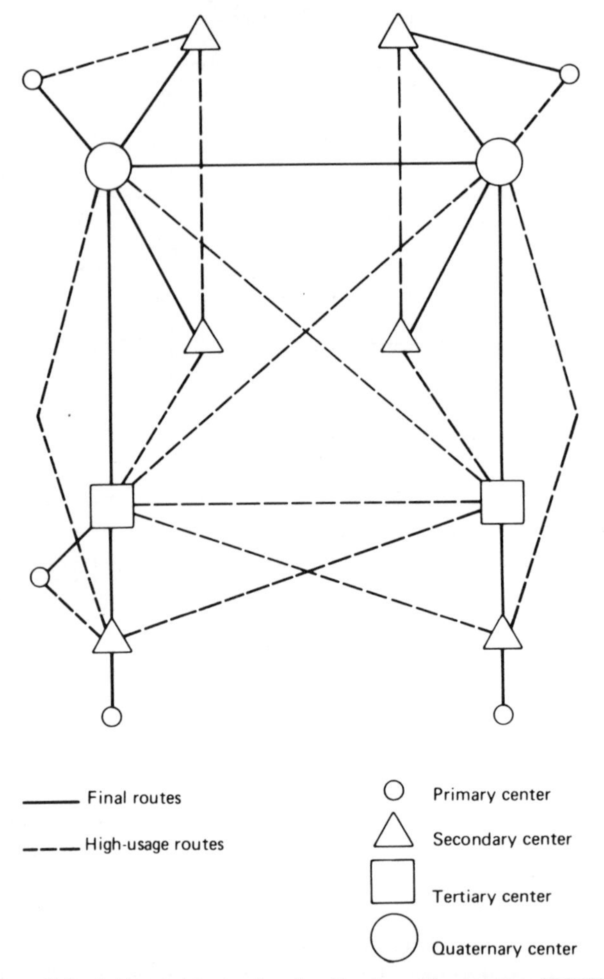

Figure 6.4 A hierarchical network with alternative routing (CCITT).

centers homing to it, and a number of primary centers would home to the secondary centers. This procedure is illustrated in Figure 6.5 and is represented systematically in Figure 6.6, thus establishing a hierarchy and setting out the final routes. In this case one of the tertiary exchanges would be an ISC. We define a final route as a route from which no traffic can overflow to an alternative route. It is a route that connects an exchange immediately above or below it in the network hierarchy and there is also connection of the two exchanges at the top level of the network. Final routes are said to make up the "backbone" of a network. Calls that are offered to the backbone but cannot be completed are lost calls.

A *high-usage route* is defined as any route that is not a final route; it may connect exchanges at a level of the network hierarchy *other than* the top level,

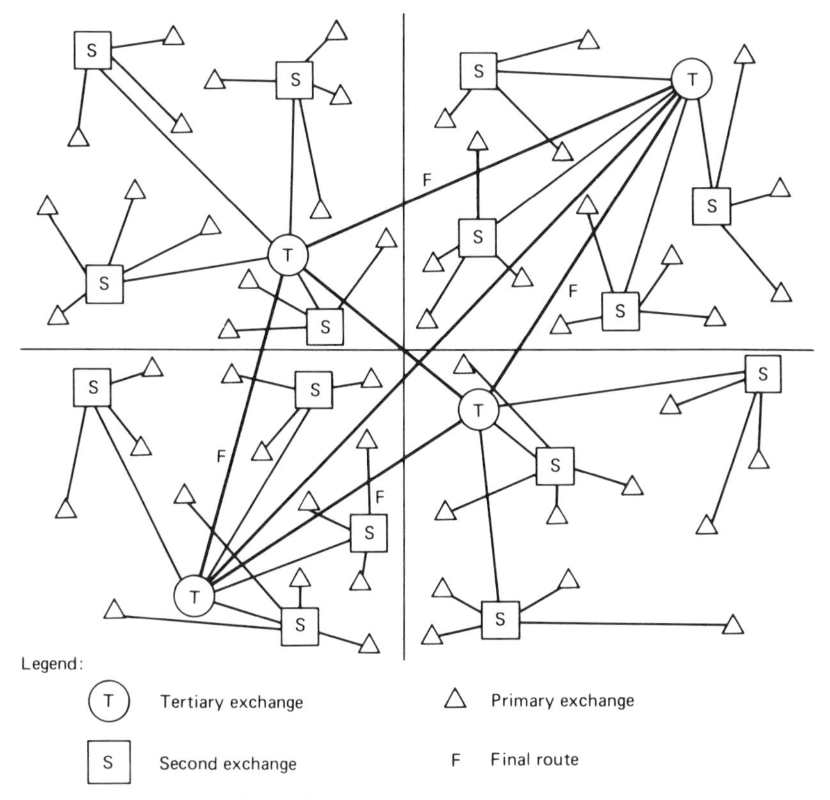

Figure 6.5 A sample network design.

such as between T_1 and T_2 in Figure 6.6. It may also be a route between exchanges on different hierarchical levels when the lower-level exchange does not home on the higher level. A *direct route* is a special type of high-usage route connecting exchanges of the lowest rank in the hierarchy. Figure 6.7 illustrates these two definitions. High-usage routes are between exchanges 1 and 2 and 3 and 2. The direct route is also between exchanges 1 and 2, with exchanges 1 and 2 the lowest level in the hierarchy.

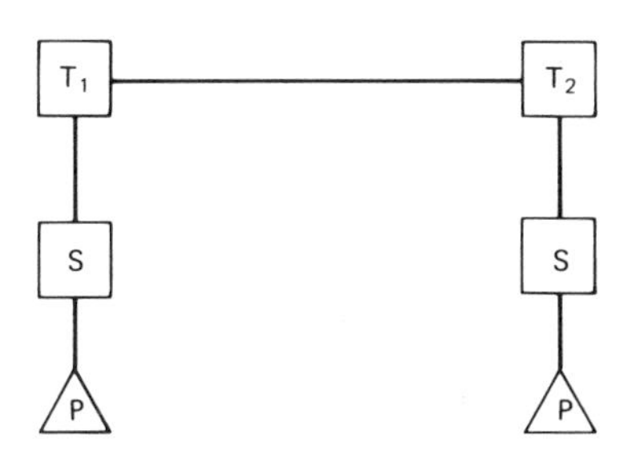

Figure 6.6 Hierarchical representation showing final routes.

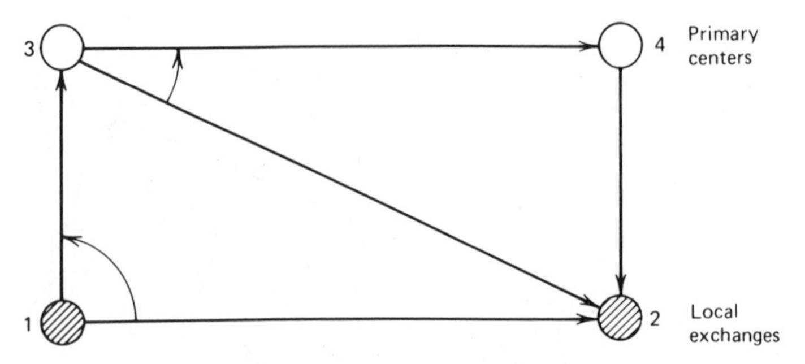

Figure 6.7 Hierarchical network segment.

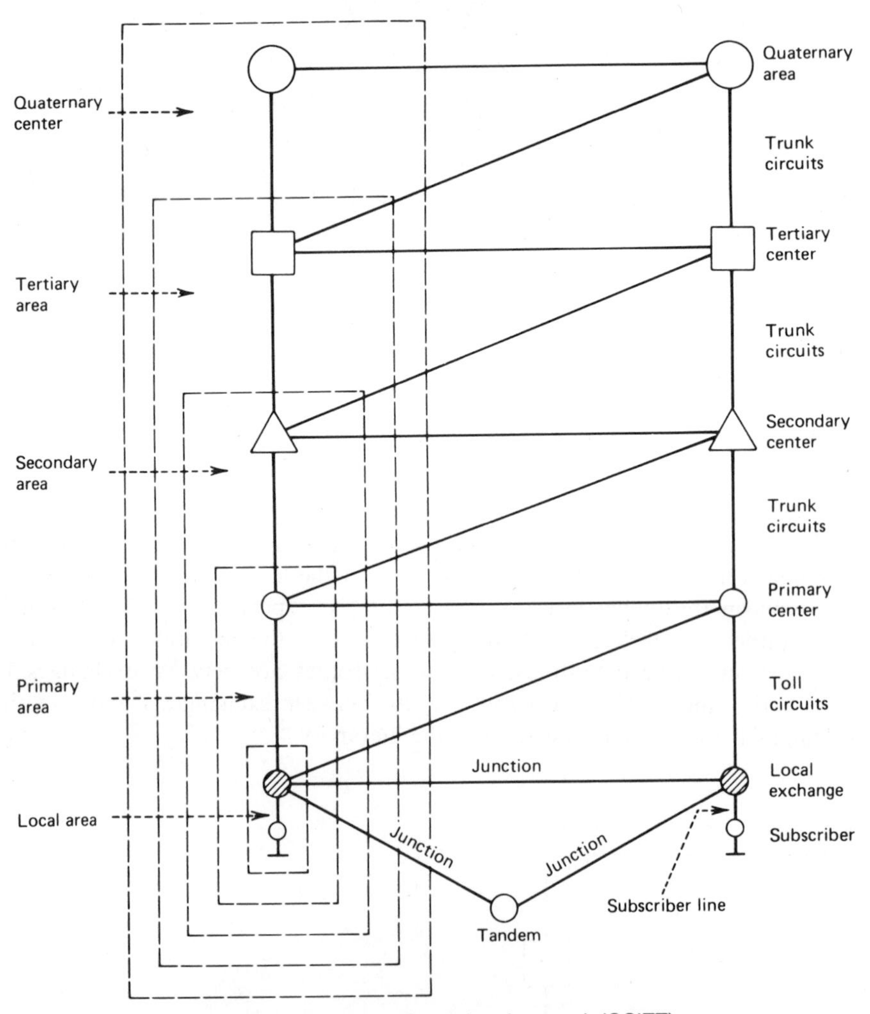

Figure 6.8 Structure of a national network (CCITT).

Before final dimensioning can be carried out, a grade of service p must be established, usually at no greater than 1% per link on the final route during the busy hour (BH). If the maximum number of links in an international call is established at 12 (in some cases 14), the very worst grade of service would be $12 \times 1\%$; however, on most calls the overall grade of service would be significantly better. These would be figures for direct-dialed international calls (see CCITT Recs. Q.13 and Q.95 bis). If there are three links in tandem on final choice routes, there would be up to 3% grade of service under worst conditions and with four links, 4%. These latter figures are for national connections. The use of HU connections reduces tandem operation and tends to improve overall grade of service.

The next step is to lay out high-usage routes. As mentioned earlier, this is done with the aid of the traffic matrix. One method of dimensioning with overflow (alternative routing) was discussed in Section 8.1 of Chapter 1 and the discussion continued in Section 9 of that chapter. Reference should also be made to Section 10 of Chapter 2. Trunks are costly. The exercise is to optimize the number of trunks and maintain a given grade of service. The methodologies given in Chapters 1 and 2 will help carry out this exercise. However, an increasing number of network designers are using computers to carry out this function. Two such computer programs are referenced at the beginning of this subsection. Traffic intensities used in the traffic matrices should be those taken from a 10-year forecast.

In much of this discussion CCITT terminology has been used. It would be helpful if the reader consulted Figure 6.8, which shows the standard structure of a national network according to the CCITT.

7 TRUNKING DIAGRAMS

After toll exchange locations have been established (Section 5 of this chapter), a hierarchical structure and routing scheme set out (Section 6), and trunks dimensioned, the trunking diagrams must be prepared for each toll exchange. A trunking diagram not only assures the designers of the network of a dimensioning interface on trunks, but in another context assures them of that all-important interface between switching and transmission. For instance, if there are 12 outgoing trunks at exchange A for exchange B, then at exchange B there must be provision for 12 inlets from A. The diagram must also indicate the number of one-way circuits and two-way circuits. Likewise, if exchange A is designed for CCITT No. 5 signaling, inlets at exchange B proceeding from A must be prepared to accept CCITT No. 5 interregister signaling and supervision, or conversion must be made for compatible signaling A to B.

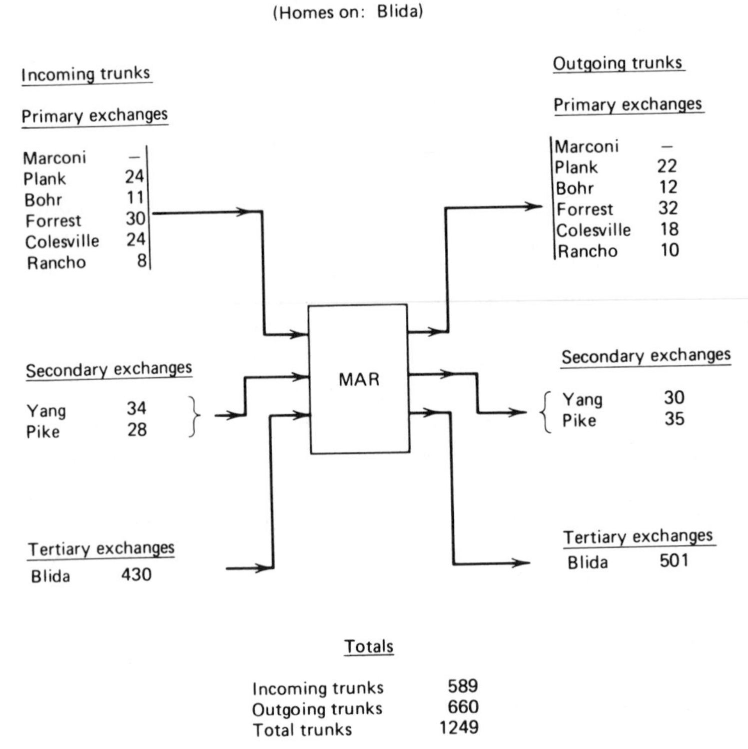

Figure 6.9 A typical trunking diagram for a transit exchange.

A trunking diagram for dimensioning a network contains the name of the exchange and the exchange on which it homes and is dependent within the hierarchy. An example is shown in Figure 6.9, where incoming trunks are on the left and outgoing trunks on the right. All circuits are one-way. In Europe two-way circuits are seldom in the toll plant but are found in rural areas. Two-way circuits or a mix of one- and two-way circuits are quite broadly used in North America. The Americans argue that use of circuit groups is more efficient. Europeans argue that the additional expense for plant mix and additional equipment to ensure no double seizure are not worth any additional marginal efficiencies. The tendency toward longer circuit routes in North America may justify attempts to optimize efficiency, particularly when the tendency to traffic peaks in one direction at one time during the day and then peaks in the other direction at another time.

There is still another trunking diagram required to properly engineer trunk groups, that ensuring proper signaling interface. It is common to use a circle associated with each trunk group. The circle is divided into three sections, as shown in the following diagram:

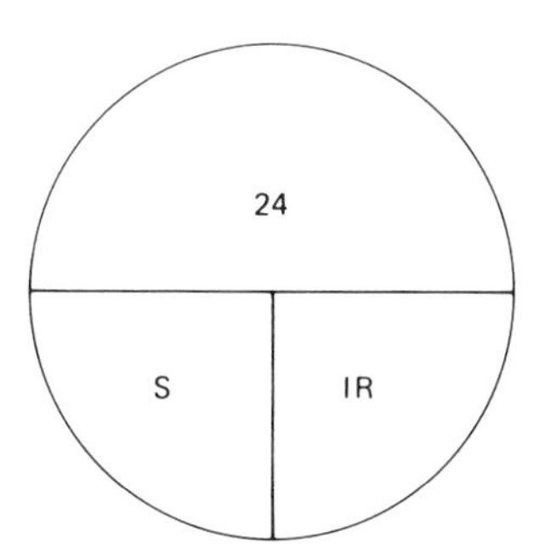

The upper half of the diagram shows the number of trunks, the lower left shows the type of supervision (S), and the lower right shows the interregister (IR) type of signaling. One- and two-way trunks must be separated, and a circle such as the preceding one must be used for one-way in each direction and for the two-way. Refer to Chapter 4 for a description of the various signaling and supervision types and methods.

8 TRANSMISSION FACTORS IN LONG-DISTANCE TELEPHONY

8.1 Introduction

Long-distance analog communication systems require some method to overcome losses. As a wire-pair telephone circuit is extended, there is some point where loss accumulates so as to attenuate signals to such a degree that the far-end subscriber is dissatisfied. The subscriber cannot hear the near-end talker sufficiently well. Extending the wire connection still further, the signal level can drop below the noise level. For good received level, a 40-dB signal-to-noise ratio is desirable (see Chapter 5, Section 2.5). To overcome the loss, amplifiers can be added; in fact, amplifiers are installed on many wire-pair trunks. Early North American transcontinental circuits were on open-wire lines using amplifiers quite widely spaced. However, as BH demand increased to thousands of circuits, such an approach was not cost-effective.

System designers turned to wideband radio and coaxial cable systems where each bearer or pipe* carried hundreds (and now thousands) of simultaneous telephone conversations. Carrier (frequency division) multiplex techniques made this possible (see Chapter 5). Frequency division multiplex (FDM) requires separation of transmit and receive voice paths. In other words, the circuit must convert from two-wire to four-wire transmission. This is normally carried out by a hybrid transformer, or resistive hybrid. Figure 6.10 is a

*On a pair of coaxial cables or a pair of radio frequency carriers, one coming and one going.

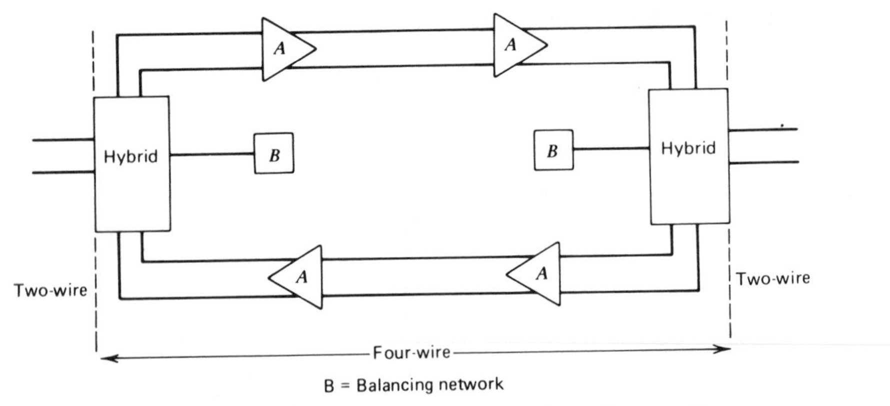

Figure 6.10 Simplified schematic of two-wire/four-wire operation.

simplified block diagram of a telephone circuit with transformation from two-wire to four-wire operation at one end and conversion back to two-wire operation at the other end. This concept was introduced in Chapter 5, Section 3.

The two factors that must be considered that greatly affect transmission design are "echo" and "singing." These are related, as we see in Sections 8.2 and 8.3

8.2 Echo

As the name implies, echo in telephone systems is the return of a talker's voice. To be an impairment, the returned voice must suffer some noticeable delay. Thus we can say that echo is a reflection of the voice. Analogously, it may be considered as that part of the voice energy that bounces off obstacles in a telephone connection. These obstacles are impedance irregularities, more properly called *impedance mismatches*. Echo is a major annoyance to the telephone user. It affects the talker more than the listener. Two factors determine the degree of annoyance of echo: its loudness and the length of its delay.

8.3 Singing

Singing is the result of sustained oscillations due to positive feedback in telephone amplifiers or amplifying circuits. Circuits that sing are unusable and promptly overload multichannel carrier equipment (FDM; see Chapter 5).

Singing may be regarded as echo that is completely out of control. This can occur at the frequency at which the circuit is resonant. Under such conditions the circuit losses at the singing frequency are so low that oscillation will continue, even after cessation of its original impulse.

8.4 Causes of Echo and Singing

Echo and singing can generally be attributed to the mismatch between the balancing network of the hybrid and its two-wire connection associated with the subscriber loop. It is at this point that the major impedance mismatch usually occurs and an echo path exists. To understand the cause of the mismatch, remember that we always have at least one two-wire switch between the hybrid and the subscriber. Ideally, the hybrid-balancing network must match each subscriber line to which it may be switched. Obviously, the impedances of the four-wire trunks (lines) may be kept fairly uniform. However, the two-wire subscriber lines may vary over a wide range. The subscriber loop may be long or short, may or may not have inductive loading, and may or may not be carrier derived. The hybrid imbalance causes signal reflection or signal "return." The better the match, the more the return signal is attenuated. The amount that the return signal (or reflected signal) is attenuated is called the *return loss*, and it is expressed in decibels. The reader should remember that any four-wire circuit may be switched to hundreds or even thousands of different subscribers; if not, it would be a simple matter to match the four-wire circuit to its single subscriber through the hybrid. This is why the hybrid to which we refer has a compromise balancing network rather than a precision network. A compromise network is usually adjusted for a compromise in the expected range of impedance (Z) encountered on the two-wire side.

Let us now consider the problem of match. If the impedance match is between the balancing network (N) and the two-wire line (L) (see Figures 5.4 and 5.5), then

$$\text{Return loss dB} = 20 \log_{10} \frac{Z_N + Z_L}{Z_N - Z_L} \tag{6.1}$$

If the network perfectly balances the line, then $Z_N = Z_L$, and the return loss would be infinite.

Return loss may also be expressed in terms of reflection coefficient, or

$$\text{Return loss dB} = 20 \log_{10} \frac{1}{\text{Reflection coefficient}} \tag{6.2}$$

where the reflection coefficient is equal to the reflected signal/incident signal.

We use the term "balance return loss" (see CCITT Rec. G.122) and classify it as two types:

1. Balance return loss from the point of view of echo.* This is the return loss across the band of frequencies from 300 Hz to 3400 Hz.
2. Balance return loss from the point of view of stability. This is the return loss between 0 and 4000 Hz.

*Called *echo return loss* (ERL) in via net loss (VNL) (North American practice; Section 8.7) but uses a weighted distribution of level.

The band of frequencies most important in terms of echo for the voice channel is that from 300 Hz to 3400 Hz. A good value for echo return loss for toll telephone plant is 11 dB, with values on some connections dropping to as low as 6 dB. For further information, the reader should consult CCITT Recs. G.122 and G.131.

Echo and singing may be controlled by:

- Improved return loss at the term set (hybrid).
- Adding loss on the four-wire side (or on the two-wire side).
- Reducing the gain of the individual four-wire amplifiers.

The annoyance of echo to a subscriber is also a function of its delay. Delay is a function of the velocity of propagation of the intervening transmission facility. A telephone signal requires considerably more time to traverse 100 km of a voice-pair cable facility, particularly if it has inductive loading, than it requires to traverse 100 km of radio facility (as low as 22,000 km/s for a loaded cable facility and 240,000 km/s for a carrier facility). Delay is measured in one-way or round-trip propagation time measured in milliseconds. The CCITT recommends that if the mean round-trip propagation time exceeds 50 ms for a particular circuit, an echo suppressor or echo canceler should be used. Practice in North America uses 45 ms as a dividing line. In other words, where echo delay is less than that stated previously here, echo can be controlled by adding loss.

An echo suppressor is an electronic device inserted in a four-wire circuit that effectively blocks passage of reflected signal energy. The device is voice operated with a sufficiently fast reaction time to "reverse" the direction of

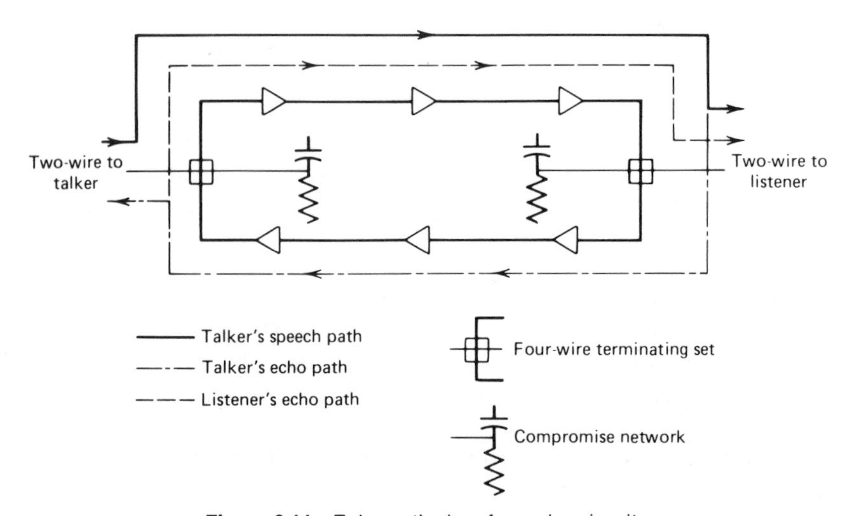

Figure 6.11 Echo paths in a four-wire circuit.

transmission, depending on which subscriber is talking at the moment. The blocking of reflected energy is carried out by simply inserting a high loss in the return four-wire path. Figure 6.11 shows the echo path on a four-wire circuit. An echo canceler generates an echo-canceling signal.

8.5 Transmission Design to Control Echo and Singing

As stated previously, echo is an annoyance to the subscriber. Figure 6.12 relates echo path delay to echo path loss. The curve in Figure 6.12 traces a group of points at which the average subscriber will tolerate echo as a function of its delay. Remember that the longer the return signal is delayed, the more annoying it is to the telephone talker (i.e., the more the echo signal must be attenuated). For instance, if the echo path delay on a particular circuit is 20 ms, an 11-dB loss must be inserted to make echo tolerable to the talker. The careful reader will note that the 11 dB designed into the circuit will increase the end-to-end reference equivalent by that amount, which is quite undesirable. The effect of loss design on reference equivalents and the trade-offs available are discussed in the paragraphs that follow.

If singing is to be controlled, all four-wire paths must have some loss. Once they go into a gain condition, and we refer here to overall circuit gain, positive feedback will result and the amplifiers will begin to oscillate or "sing." North American practice calls for a 4-dB loss on all four-wire circuits to ensure against singing. The CCITT recommends a minimum loss for a national network of 10 dB (CCITT Rec. G.122, p. 2).

Almost all four-wire circuits have some form of amplifier and level control. Such amplifiers are often embodied in the channel banks of the carrier (FDM) equipment.

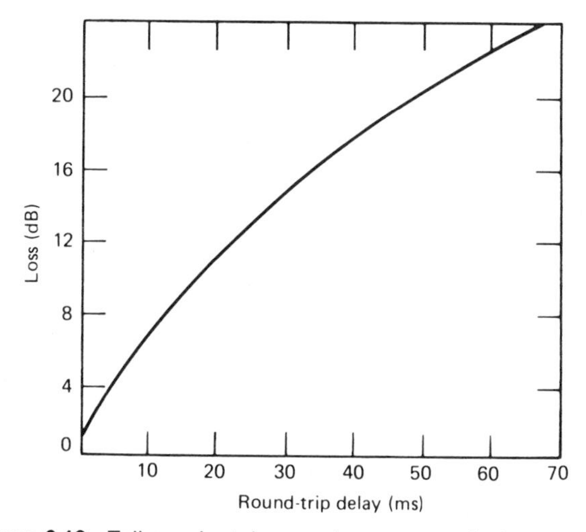

Figure 6.12 Talker echo tolerance for average telephone users.

8.6 Introduction to Transmission-Loss Engineering

One major aspect of transmission system design for a telephone network is to establish a transmission-loss plan. Such a plan, when implemented, is formulated to accomplish three goals:

1. Control singing (stability).
2. Keep echo levels within limits tolerable to the subscriber.
3. Provide an acceptable overall reference equivalent to the subscriber.

For North America the via net loss (VNL) concept embodies the transmission plan idea (VNL is covered in Section 8.7).

From our preceding discussions we have much of the basic background necessary to develop a transmission-loss plan. We know the following:

1. A certain minimum loss must be maintained in four-wire circuits to ensure against singing.
2. Up to a certain limit of round-trip delay, echo is controlled by loss.
3. It is desirable to limit these losses as much as possible to improve reference equivalent.

National transmission plans vary considerably. Obviously, length of circuit is important, as well as the velocity of propagation of the transmission media. Two approaches are available in the preparation of a loss plan, variable-loss plan (i.e., VNL) and fixed-loss plan (i.e., as used in Europe). A national transmission-loss plan for a small country (i.e., small in geographic area) such as Belgium could be quite simple. Assume that a 4-dB loss is inserted in all four-wire circuits to prevent singing. Consult Figure 6.12, where 4 dB allows for 4 ms of round-trip delay. If we assume carrier transmission for the entire length of the connection and use 105,000 mi/s for the velocity of propagation, we can satisfy Belgium's echo problem. The velocity of propagation used comes out to 105 mi/ms (169 km/ms). By simple arithmetic, we see that a 4-dB loss on all four-wire circuits will make echo tolerable for all circuits extending 210 mi (338 km) (i.e., 2×105). This is an application of the fixed-loss type of transmission plan. In the case of small countries or telephone companies operating over a small geographic extension, the minimum loss inserted to control singing controls echo as well for the entire country.

Let us try another example. Assume that all four-wire connections have a 7-dB loss. Figure 6.12 indicates that 7 dB permits an 11-ms round-trip delay. Assume that the velocity of propagation is 105,000 mi/s. Remember that we are dealing with round-trip delay. The talker's voice reaches the far-end hybrid and is then reflected back. This means that the signal traverses the system twice, as shown in Figure 6.13. Thus 7 dB of loss for the given velocity of propagation allows about 578 mi of extension or, for all intents and purposes,

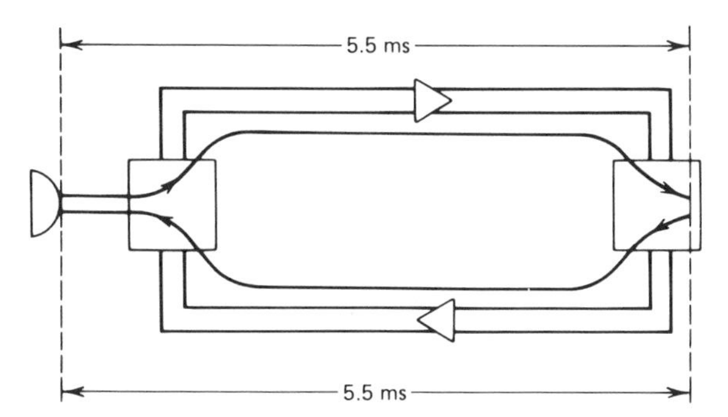

Figure 6.13 Example of echo round-trip delay (5.5 + 5.5 = 11-ms round-trip delay).

the distance between subscribers and will satisfy the loss requirements for a country with a maximum extension of 578 mi.

It has become evident by now that we cannot continue increasing losses indefinitely to compensate for echo on longer circuits. Most telephone companies and administrations have set a 45- or 50-ms round-trip delay criterion, which sets a top figure above which echo suppressors are to be used. One major goal of the transmission-loss plan is to improve overall reference equivalent or to apportion more loss to the subscriber plant so that subscriber loops can be longer or to allow the use of less copper (i.e., smaller-diameter conductors). The question arises as to what measures can be taken to reduce losses and still keep echo within tolerable limits. One obvious target is to improve return losses at the hybrids. If all hybrid return losses are improved, the echo tolerance curve shifts; this is because improved return losses reduce the intensity of the echo returned to the talker. Thus the talker is less annoyed by the echo effect.

One way of improving return loss is to make all two-wire lines out of the hybrid look alike, that is, have the same impedance. The switch at the other end of the hybrid (i.e., on the two-wire side) connects two-wire loops of varying length, thus causing the resulting impedances to vary greatly. One approach is to extend four-wire transmission to the local office such that each hybrid can be better balanced. This is being carried out with success in Japan. The US Department of Defense has its Autovon (automatic voice network), in which every subscriber line is operated on a four-wire basis. Two-wire subscribers connect through the system via PABXs (private automatic branch exchanges).

Let us return to standard telephone networks using two-wire switches in the subscriber area; suppose that balance return loss could be improved to 27 dB. Thus minimum loss to ensure against singing could be reduced to 0.4 dB. Now suppose that we distribute this loss across four four-wire circuits in tandem. Thus each four-wire circuit would be assigned a 0.1-dB loss. If we have gain in the network, singing will result. The safety factor between loss and gain is 0.4 dB. The loss in each circuit or link is maintained by amplifiers. It is difficult to

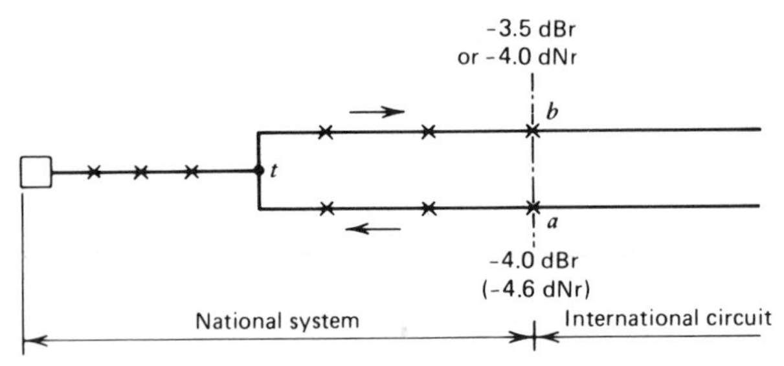

Figure 6.14 Definition of points *a–t–b* (CCITT Rec. G.122); X indicates a switch. Courtesy of the International Telecommunication Union-CCITT.

adjust the gain of an amplifier to 0.1 dB, much less keep it there over long periods, even with good automatic regulation. *Stability* or *gain stability* is the term used to describe how well a circuit can maintain a desired level. Of course, in this case we refer to a test tone level. In the preceding example it would take only one amplifier to shift 0.4 dB, two to shift in the positive direction 0.2 dB, and so forth. The importance of stability, then, becomes evident.

The stability of a telephone connection depends on three criteria:

1. The variation of transmission level with time.
2. The attenuation–frequency characteristics of the links in tandem.
3. The distribution of balance return loss.

Each criterion becomes magnified when circuits are switched in tandem. To handle the problem properly, we must talk about statistical methods and standard distributions. In the case of criteria 1 and 2, we refer to the tandem nature of the four-wire circuits. Criterion 3 refers to switching subscriber loops–hybrid combinations that will give a poorer return loss than will the 11 dB stated earlier. Return losses on some connections can drop to 3 dB or less.

Stability is discussed in CCITT Recs. G.122, G.131, and G.151C. In essence, the loss through points *a–t–b* in Figure 6.14 should have a value not less than $(10 + N)$ dB for established connections, where N is the number of four-wire circuits in the national chain. Thus the minimum loss is stated (CCITT Rec. G.122), and Rec. G.131 is quoted in part as follows:

> The standard deviation of transmission loss among international circuits routed in groups equipped with automatic regulation is 1 dB.... This accords with... the tests... [which] indicate that this target is being approached in that 1.1 dB was the standard deviation of the recorded data.
>
> It is also evident that those national networks which can exhibit no better stability balance return loss than 3 dB, 1.5 dB standard deviation, are unlikely to

> seriously jeopardize the stability of international connections as far as oscillation is concerned. However, the near-singing [rain-barrel effect] distortion and echo effects that may result give no grounds for complacency in this matter.

Stability requirements in regard to North American practice are embodied in the VNL concept discussed in the next section.

8.7 Via Net Loss

Via net loss (VNL) is a concept or method of transmission planning that permits a relatively close approach to an overall zero transmission loss in the telephone network (lowest practicably attainable) and maintains singing and echo within specified limits. The two criteria that follow are basic to VNL design:

1. Customer–customer talker echo should be satisfactorily low on more than 99% of all telephone connections that encounter the maximum delay likely to be experienced.
2. The total amount of overall loss is distributed throughout the trunk segments of the connection by allocation of loss to the echo characteristics of each segment.

One important concept in the development of the discussion on VNL is that of echo return loss (ERL) (see subsection 8.2). For this discussion, we consider ERL as a single-valued weighted figure of return losses in the frequency band 500–2500 Hz. Echo return loss differs from return loss in that it takes into account a weighted distribution of level versus frequency to simulate the nonlinear characteristics of the transmitter and receiver of the telephone instrument. By using ERL measurements, it is possible to arrive at a basic design factor for the development of the VNL formula. This design factor states that the average return loss at class 5 offices (local exchanges) is 11 dB, with a standard deviation of 3 dB. Considering a standard distribution curve and the one, two, or three σ points on the curve, we could thus expect practically all measurements of ERL to fall between 2 dB and 20 dB at class 5 offices (local exchanges). Via net loss also considers that reflection occurs at the far end in relation to the talker where the four-wire trunks connect to the two-wire circuits (i.e., at the far-end hybrid).

The next concept in the development of the VNL discussion is overall connection loss (OCL), which is the value of one-way trunk loss between two end (local) offices (not subscribers). Consider that

$$\text{Echo path loss} = 2 \times \text{Trunk loss (one-way)} + \text{Return loss (hybrid)}$$

where all units are in decibels. Now let us consider the average tolerance for a particular echo path loss. Average echo tolerance is taken from the curve in

Figure 6.12. Therefore

$$\text{OCL} = \frac{\text{Average echo tolerance (loss)} - \text{Return loss}}{2} \tag{6.3}$$

where all units are in decibels. Return loss in this case is the average echo return loss that must be maintained at the distant local exchange—the 11 dB given earlier.

An important variability factor not considered in the formula is trunk stability, which determines how close assigned levels are maintained on a trunk. Via net loss practice dictates trunk stability to be maintained with a normal distribution of levels and a standard deviation of 1 dB in each direction. For a round-trip echo path the deviation is taken as 2 dB. This variability applies to each trunk in a tandem connection. If there are three trunks in tandem, this deviation must be applied to each.

The reader will recall that the service requirement in VNL practice is satisfactory echo performance for 99% of all connections. This may be considered a cumulative distribution, or 2.33 standard deviations, summing from negative infinity toward the positive direction. The OCL formula may now be rewritten as follows:

$$\text{OCL} = \frac{\text{Average echo tolerance} - \text{Average return loss} + 2.33D}{2} \tag{6.4}$$

where D is the composite standard deviation of all functions and all units are in decibels. The derivation of D, the composite standard deviation of all functions, is as follows:

$$D = \sqrt{D_{\mathrm{t}}^2 + D_{\mathrm{r1}}^2 + ND_1^2} \tag{6.5}$$

where D_{t} is the standard deviation of distribution of echo tolerance among a large group of observers, given as 2.5 dB; D_{r1} is the standard deviation of distribution of return loss, given as 3 dB; D_1 is the standard deviation of distribution of the variability of trunk loss for a round-trip echo path, given as 2 dB; and N is the number of trunks switched in tandem to form a connection class 5 office to class 5 office. Now consider several trunks in tandem; it can be calculated that at just about any given echo path delay, the OCL increases approximately 0.4 dB for each trunk added. With this simplification, once we have the OCL for one trunk, all that is needed to compute the OCL for additional trunks is to add 0.4 dB times the number of trunks added in tandem. This loss may be regarded as an additional constant needed to compensate for variations in trunk loss in the VNL formula.

Figure 6.15 relates echo path delay (round-trip) to overall connection loss (OCL for one trunk, then for a second trunk in tandem, and for four and six trunks in tandem). Although the straight-line curve has been simplified, the

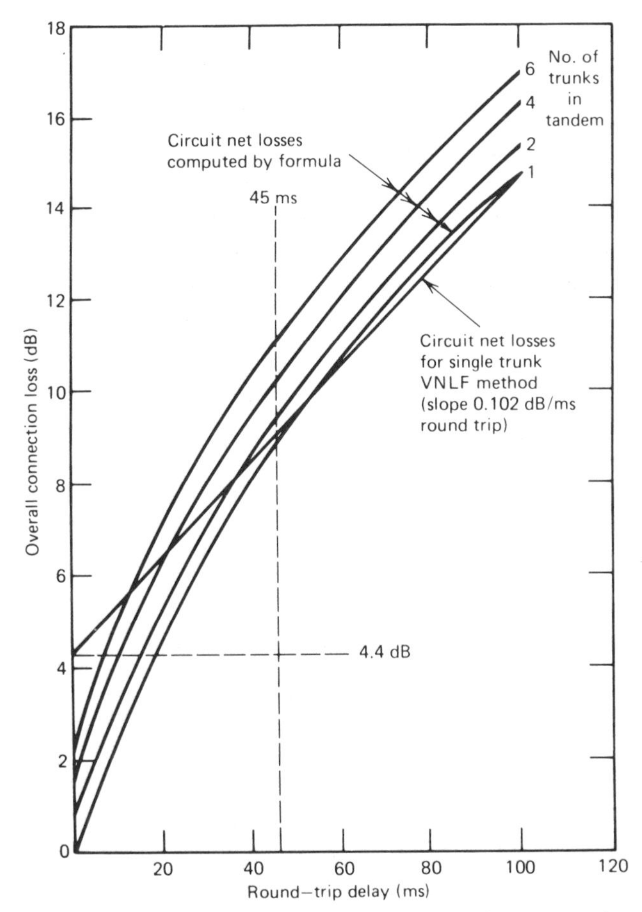

Figure 6.15 Approximate relationship between round-trip echo delay and overall connection loss (OCL).

approximation is sufficient for engineering VNL circuits. Note that the straight-line curve in Figure 6.15 cuts the Y axis at 4.4 dB, where round-trip delay is 0. This 4.4 dB is based on two conditions, namely, that all trunks have a minimum of 4 dB to control singing and that there is 0.4 dB protection against negative variation of trunk loss. Another important point to be defined on the linear curve in Figure 6.15 is a round-trip delay of 45 ms, which corresponds to an OCL of 9.3 dB. Empirically, it has been determined that echo suppressors must be used for delays greater than 45 ms. From this same linear curve the following formula for OCL may be derived:

$$\text{OCL} = (0.102)\,(\text{Path delay in ms}) + (0.4\text{ dB})\,(\text{Number of trunks in tandem}) + 4\text{ dB} \qquad (6.6)$$

Usually the 4 dB* as shown in the preceding OCL equation is applied to the extremity of each trunk network, namely, to the toll connecting trunks, 2 dB to each.

Overall connection loss deals with the losses of an entire network consisting of trunks in tandem, whereas VNL deals with the losses assigned to one trunk. The VNL formula follows from the OCL formula. The key here is the round-trip delay on the trunk in question. The delay time for a transmission facility employing only one particular medium is equal to the reciprocal of the velocity of propagation of the medium multiplied by the length of the trunk. To obtain round-trip time, this figure must be multiplied by 2; thus

$$\text{VNL} = 0.102 \times 2 \times \frac{1}{\text{Velocity of propagation}} \times (\text{One-way length of the trunk}) + 0.4\ \text{dB} \quad (6.7)$$

Another term is often introduced to simplify the equation, the via net loss factor (VNLF):

$$\text{VNL} = \text{VNLF} \times (\text{One-way length of trunk in miles}) + 0.4\ \text{dB} \quad (6.8)$$

$$\text{VNLF} = \frac{2 \times 0.102}{\text{Velocity of propagation of the medium}} (\text{dB/mi}) \quad (6.9)$$

The velocity of propagation of the medium used here must be modified by such things as delays caused by repeaters, intermediate modulation points, and facility terminals.

Via net loss factors for loaded two-wire facilities are 0.03 dB/mi, with H-88 loading on 19-gauge wire and increased to 0.04 dB/mi on B-88 and H-44 facilities. On four-wire carrier and radio facilities, the factor improves to 0.0015 dB/mi. For connections with round-trip delay times in excess of 45 ms, the standard VNL approach must be modified. As mentioned previously, these circuits use echo suppressors that automatically switch about 50 dB into the echo return path, and the switch actuates when speech is received in the "return" path, thus switching the pad into the "go" path.

Via net loss practice in North America treats long delay circuits with up to a maximum of 45 ms of delay in the following manner (see Figure 6.16). Here the total round-trip delay is arbitrarily split into two parts for connections involving regional intertoll trunks. If the regional intertoll delay exceeds 22 ms, echo suppressors are used. If the figure is 22 ms or less, echo is controlled by VNL design. Thus allow 22 ms for the maximum delay for the regional intertoll segments to the connection. This leaves 23-ms maximum delay for the other segments (45 − 22). Now apply the VNL formula for a delay of 22 ms. Thus

$$\text{VNL} = 0.102 \times 22 + 0.4 = 2.6\ \text{B}$$

*This value in equation 6.6 has been changed to 5 dB or 2.5 dB at each end applied to the toll-connecting trunks. The additional 0.5 dB added to each end accommodates nominal switch loss at the extreme ends of the circuit.

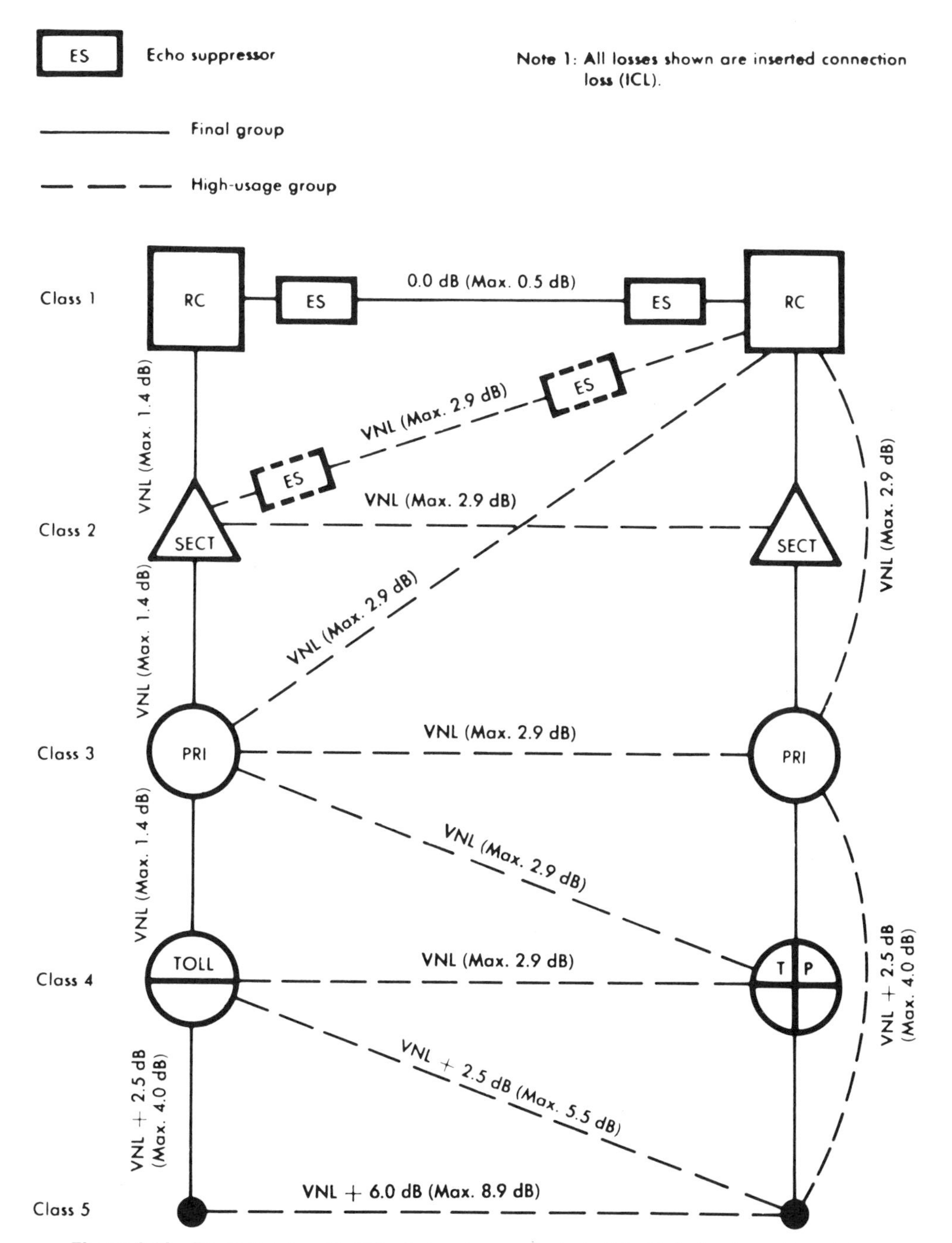

Figure 6.16 Trunk losses with VNL design. Losses include 0.4-dB design loss allowed for maintenance. Copyright © American Telephone and Telegraph Co., 1977.

This loss is equivalent to the maximum length of an intertoll trunk without an echo suppressor. What is that length?

$$\text{Length (one-way)} = \frac{\text{VNL} - 0.4}{\text{VNLF}}\,\text{mi}$$

$$= \frac{0.102 \times 22 + 0.4 - 0.4}{0.0015} = 1498\ \text{mi}^* \qquad (6.10)$$

In summary, in VNL design we have three types of loss that may be assigned to a trunk:

Type	Loss
Toll-connecting trunk	VNL + 2.5 dB
Intertoll trunk (no echo suppressor)	VNL
Intertoll trunk (with echo suppressor)	0 dB†

8.8 Transmission-Loss Plan for Japan (An Example)

An example of the current transmission-loss plan for Japan is presented in Figure 6.17. It is a fixed-loss plan. The ORE for 97% of national connections is only 26 dB. One way of achieving this level, as suggested previously, is to have four-wire switching extend through primary centers and four-wire transmission to local exchanges. The breakdown of the ORE is as follows:

0.5 dB for local switches
12.5 dB TRE for subscriber loop
2.5 dB RRE for subscriber loop
10.0 dB four-wire system, local exchange to local exchange.
26 dB total (= ORE)

The reference equivalents are based on the Japanese 600P telephone sets with a 1200-Ω subscriber loop and a 0.4-mm nonloaded cable fed with −48 V. The plan assumes a mean echo balance return loss of 11 dB. Loss stability is given as 1 dB of standard deviation per four-wire circuit, with a singing probability of 0.1%.

8.9 Future Fixed-Loss Plan for the United States

Reference 18 describes the rationale for a future fixed-loss plan in the United States once a switched digital network (SDN) is implemented. The VNL design plan described in Section 8.7 is not well suited for the SDN because the VNL plan requires loss to be inserted in each VF trunk. This would require

*The VNLF in this equation indicates carrier and/or radio for the whole trunk.
†See discussion on echo suppressor for explanation of the 0-dB figure.

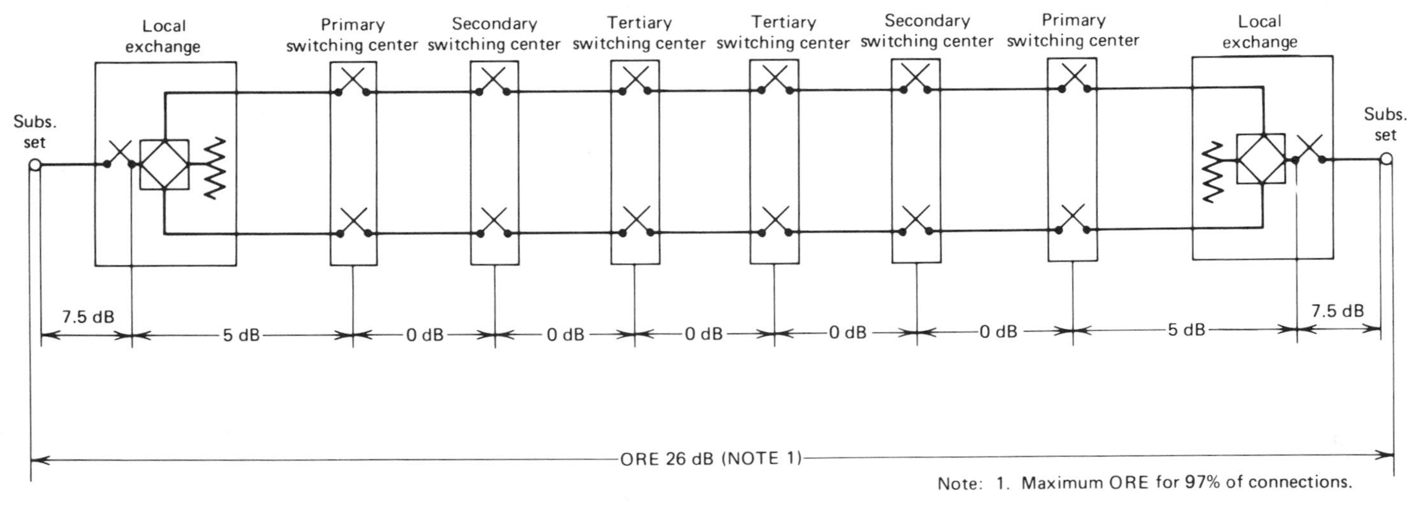

Figure 6.17 National loss distribution for the Japanese network. (Ref. 6)

that the digital signal be decoded (see Chapter 9) to an analog signal, loss inserted, and the signal recoded or that the encoded signal level be changed by some digital processing technique so that when it is decoded, a lower signal level will result. Either of these techniques would add cost and introduce transmission impairments.

The switching of digits without processing at higher-class offices (class 4 and above) requires end-to-end connections on purely digital facilities that have the required loss for control of talker echo. This loss is inserted at the local office (switch) ends of the connections by digital processing before digital-to-analog conversion, by electrical loss after conversion, or by processes associated with conversion. Since it is impractical to insert different losses on end-to-end connections having different mileages, a fixed amount of loss has been selected for all connections, according to Ref. 18.

A fixed-loss network is feasible for the digital network but impractical for the analog network because of the different delay and noise characteristics [4]. Connection delay in the digital network is the sum of the propagation delay, the delay of the terminals at the class 5 offices (local switches), and the delay through the switching systems. The delay of digital terminals and switching systems is analogous to the component of the delay introduced by each trunk of the analog network; however, the amount of delay will be less. The noise on a digital system depends on the number of encodings and decodings, not on length. For the digital network with only one encoder and decoder, the noise is less in comparison to the analog network and has a constant value for all connection lengths. Furthermore, the talker echo performance of overall connections will be improved relative to the present analog network due to the increased use of toll-connecting trunks (TCTs) and their improved echo return loss (ERL).

Reference 18 states that a loss of 6 dB for all lengths of toll connections was selected as the best compromise. In comparison with the VNL design, this plan has better loss-noise grade of service for all connection lengths but particularly for longer connections due to the reduced noise.

REVIEW QUESTIONS

1. What are the three basic underlying considerations in the design of a long-distance (toll) network?

2. What is the fallacy of providing just one high-capacity trunk across the United States to serve all major population centers by means of tributaries off that main trunk?

3. How can the utilization factor of trunks be improved?

4. For long-distance (toll) switching centers, what is the principal factor involved in the placement of such exchanges (differing from local ex-

change placement substantially)—economy, subscriber density, altitude, or what?

5. How are the highest levels of a national hierarchical network connected, and why is this approach used?

6. On a long-distance (toll) connection, why must the number of links in tandem be limited?

7. Describe in one short sentence the current approach to structure of the international network as recommended by CCITT.

8. What type of routing is used on the majority of international connections?

9. Why do we limit the number of satellites used on an international full-duplex speech telephone connection?

10. Name three principal factors used in deciding how many and where toll (long-distance) exchanges will be located in a given geographic area.

11. Discuss the impact of fan outs on the number of hierarchical levels in a national network.

12. Name the three principal bases required at the outset for the design of a toll (long-distance) network.

13. In the design of a long-distance (toll) network, once the hierarchical levels have been established, what is assembled next?

14. Define a final route.

15. Define high-usage (HU) route and direct route.

16. A grade of service no greater then ______% per link is recommended on a final route?

17. When assembling a traffic matrix, at the ends of what forecast period are the traffic intensities valid for?

18. What do conventional trunking diagrams tell us (three things)?

19. What is the principal cause of echo in a telephone network?

20. What causes singing in the telephone network?

21. Differentiate balance return loss and echo return loss.

22. What is the return loss of a particular device displaying 600-Ω characteristic impedance terminating in a toll switch port that displays the specified characteristic impedance? (The switch operates in North America.)

23. There are four methods given for the control of echo and singing. Give three of them.

24. What are the two basic methods of loss design?

25. Why is it advantageous to use fixed-loss design on an all digital network? Discuss the pros and cons.

26. Why must we limit loss at some maximum point in a loss design plan and resort to the use of echo suppressors or cancelers? Bring reference equivalent into the discussion.

27. The stability of a telephone connection depends on three criteria. Name two of the criteria.

28. A VNL value is a function of two factors. What are they?

29. What is the conventional design loss value of an analog space division switch? of a hybrid?

30. Noise on digital systems depends solely on the number of ______ and ______.

REFERENCES

1. International Telephone and Telegraph Corporation, *Reference Data for Radio Engineers*, 6th ed., Howard W. Sams, Indianapolis, 1976.
2. *National Networks for the Automatic Service*, International Telecommunications Union, Geneva, 1964.
3. *Overall Communication System Planning*, Vols. I–III, IEEE North Jersey Section Seminar, 1964.
4. *Notes on the Network*,—1980 American Telephone and Telegraph Company, New York, 1980.
5. *Switching Systems*, American Telephone and Telegraph Company, New York, 1961.
6. *Telecommunication Planning*, ITT Laboratories (Spain), Madrid, 1973 (in particular, Section 2, "Networks").
7. CCITT Recommendations, Vols. III and VI, *Red Books*, Malaga-Torremolinos, 1984.
8. J. E. Flood, *Telecommunication Networks*, IEEE Series, London, 1974.
9. *Telecommunication Planning Symposium*, ITT Laboratories (Spain), presented to SAPO, Boksburg, South Africa, 1972.
10. "Optimization of Telephone Networks with Hierarchical Structure" (computer program), ITT Laboratories (Spain), Madrid, 1973.
11. "Optimization of Telephone Trunking Networks with Alternate Routing" (computer program), ITT Laboratories (Spain), Madrid, 1973.

12. Roger L. Freeman, *Telecommunication Transmission Handbook*, 2nd ed. Wiley, New York, 1981.
13. F. T. Andrews and R. W. Hatch, "National Telephone Network Planning in the ATT," *IEEE Com. Tech. J.* (June 1971).
14. Ramses R. Mina, *Introduction to Teletraffic Engineering*, Telephony Publishing Corporation, Chicago, 1974.
15. *Theory of Telephone Traffic: Tables and Diagrams*, Siemens, Berlin-Munich, 1971, Part 1.
16. M. A. Clement, "Transmission," reprint from *Telephony* (magazine), Telephony Publishing Corporation, Chicago, 1969.
17. *Transmission Systems for Communication*, 5th ed., Bell Telephone Laboratories, Holmdel, N.J., 1982.
18. *Notes on the BOC Intra-LATA Networks—1986*, Bell Communications Research Technical Reference TR-NPL-000275, issue 1, Bellcore, Livingston, N.J., April 1986.
19. CCITT Recommendation E.171, "International Routing Plan," *Red Book*, Vol. II, Fascile II.2, Malaga-Torremolinas 1984 (Eighth Plenary Assembly CCITT).

7

THE DESIGN OF LONG-DISTANCE LINKS

1 INTRODUCTION

In Chapter 6 we proposed a methodology for the design of a long-distance network. The network may be defined as a group of switching nodes interconnected by links. We may refer to a link as a transmission highway between switches carrying one or more traffic relations. The link could appear as that in the following diagram, where switches A, B, and C are connected to switches X, Y, and Z over a link as shown. The discussion that follows introduces the essentials of transmission design of such links.

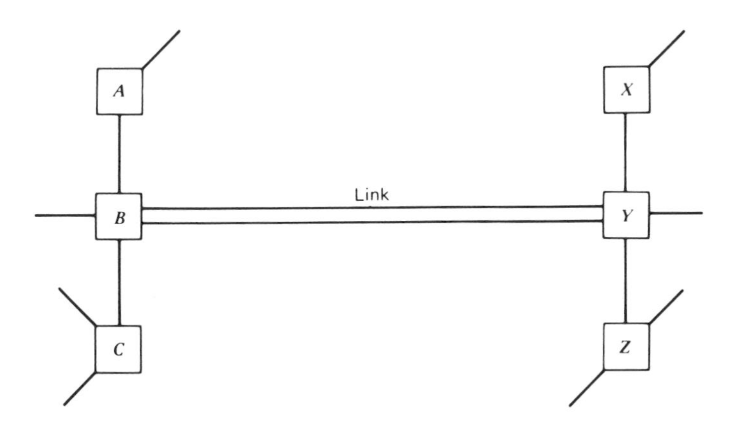

2 THE BEARER

British telephone engineers are fond of the term "bearer," which is quite descriptive. The bearer is what carries the signal(s). It could be a pair of wires or two pairs on a four-wire basis, a radio carrier in each direction, a coaxial cable, or a fiber optics cable. The wire pair could be open wire line, aerial cable, or buried cable. In the text that follows it is assumed that a number of telephone channels are transmitted by each bearer in an FDM configuration as discussed in Chapter 5.

Modern long-distance links use one of three types of bearer: radio, coaxial cable, or fiber optics. The decision of which to use is basically economic. For instance, if a link, as defined in Section 1, is to carry 5000 or more circuits (telephone channels) at the end of a 10-year planning period, coaxial or fiber optics cable may be strongly recommended as the bearer. If less than 5000 circuits are to be carried, radio may be the more feasible medium. If the link is digital with high forecast growth, then optical fiber may be the alternative. Other considerations are discussed in Chapter 9, Sections 5 and 8.

3 INTRODUCTION TO RADIO TRANSMISSION

Unlike wire, cable, and fiber as a transmission medium, which are designed for that function, the radio medium is apt to be quite far from optimum, and its characteristics are not always fully understood. Thus much of radio system design is devoted to calculating the probable behavior of a given path and to finding modulation and signal-processing techniques that will overcome defects of the medium. Radio circuits may be characterized by their carrier frequency, which largely determines behavior of the path.

As we look at the electromagnetic spectrum progressing from the lowest to the highest frequencies, certain generalities can be made regarding behavior and application. The *lowest frequencies* are VLF (very low frequency, 3–30 kHz) and LF (low frequency, 30–300 kHz) and are useful for very long-range communications but have very limited information bandwidths and require very high power. Propagation in this frequency range is worldwide; thus frequencies can essentially be assigned only once (i.e., they cannot be assigned again in another part of the world). The MF band (300–3000 kHz) is traditionally used for broadcast and marine, with fairly limited information capacity, requiring power of the order of kilowatts and with an effective daytime range in the hundreds of miles. Basic propagation is by ground wave (i.e., follows the curvature of the earth). The HF (high frequency) band (3–30 MHz) is the old, traditional long-haul point-to-point communication band. Propagation at long distance is by one or more reflections from ionospheric layers and is therefore variable, as the ionosphere varies with sunspot conditions and time of day. Since the ionosphere is multilayered and irregular in motion and since fairly large areas of these surfaces are involved in the propagation path, the received signal exhibits multipath effects and is subject to statistical fading. On VLF = very low frequency (3–30 kHz) LF = low frequency (30–300 kHz) HF = high frequency (3–30 MHz) communication paths, communication can be effective for circuits a few hundred miles long, and even worldwide coverage can be obtained. Nearly 90% path reliability on the longer term can be expected for well-designed HF circuits. To accommodate the very high demand for HF frequencies, modulation bandwidths are limited by law to 12 kHz (maximum), the equivalent of four 3-kHz telephone channels.

Above approximately 30–50 MHz, radio signals tend to pass through the ionosphere, rather than reflect or refract sufficiently for use far beyond the visible horizon. These higher frequencies are useful for line-of-sight communication, troposcatter, diffraction, or satellite relay. All three radio systems offer advantages and disadvantages for the design of telephone transmission links.

Line-of-sight microwaves (radiolinks) in the 150-MHz, 450-MHz, and 900-MHz bands provide multichannel transmission capability of 12 to 120 nominal 4-kHz voice channels in an FDM configuration over line-of-sight paths (FDM is discussed in Chapter 5, Section 4). Above 2 GHz, line-of-sight systems transmit up to 1800 and in some cases up to 2700 FDM telephone channels per radio carrier frequency. Modulation is often FM on older systems. On new systems some form of digital modulation should be considered. This point is discussed further in this section and in Chapter 9.

Radio waves in line-of-sight systems travel in a straight line and are limited by the horizon due to the curvature of the earth. Radio waves propagated line of sight (ca. > 50 MHz) are usually bent or defracted beyond the optical horizon, the one that limits our vision beyond a certain point. The optical horizon may be approximated by the following formula:

$$d = \sqrt{\frac{3h}{2}} \tag{7.1A}$$

and the radio horizon:

$$d = \sqrt{2h} \tag{7.1B}$$

where d is the distance to the optical horizon in miles from antenna and h is the height in feet of antenna above the earth's surface.

The distance to the radio horizon varies with the index of refraction. Some designers generalize and say it is $\frac{4}{3}$ the distance to the optical horizon. However, such generalization may be overly optimistic to use in certain circumstances. The concept of optical and radio horizon is shown in Figure 7.1. Microwave radio paths over several miles long may suffer from fading. The longer the path, the more prone it is to fading. Fading is the variation of a received radio signal level with time. In microwave line-of-sight systems, fading is usually caused by atmospheric refractive changes and ground and water reflections in the propagation path. When using frequencies above 10 GHz, rainfall attenuation must also be taken into account. The most commonly used line-of-sight microwave frequency bands are 2 GHz, 4 GHz, 6 GHz, and 7 GHz. All bands 4 GHz and above generally will carry up to 1800 FDM voice channels by national regulation and CCIR recommendations with

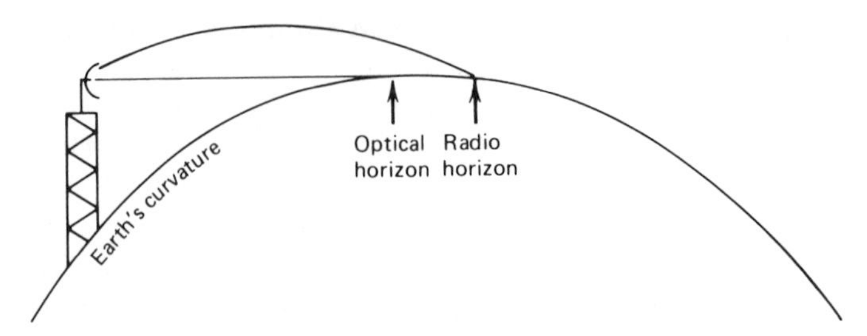

Figure 7.1 Radio and optical horizon.

portions of the 6-GHz and 7-GHz band, where 2700-channel operation is permitted per radio carrier frequency.

Geostationary satellites may be considered as radio frequency (RF) repeaters for up to 960, 1200, or 2400 voice channels (per transponder) in an FDM configuration or, as now developing, in a time-division configuration for digital systems. The 6-GHz band transmits to the satellite that converts and amplifies the received signal to the 4-GHz bands. Both bands are shared with terrestrial line-of-sight microwave services. Due to crowded conditions in the 4-GHz and 6-GHz bands, more and more services are turning to the 11-GHz and 14-GHz bands, especially for domestic and intracontinental communications. Satellite systems are used for very long links and up to about 1972 were almost exclusively used for intercontinental communications. Today there is a veritable rush to what might be familiarly termed "domestic systems." As the technology advances, the cost per voice channel has markedly reduced and can compete or be more economical than line-of-sight terrestrial systems for links over 300 miles (500 km) or to provide comparatively thin line service to rural areas. Figure 7.2 illustrates a simplified application of satellite communication for intercontinental service, where three properly placed geostationary satellites can provide nearly 100% earth coverage. By definition, geostationary satellites are 22,300 statute miles above the earth's surface with a concurrent propagation delay of about 0.25 s earth-satellite-earth and a round-trip delay of about 0.5 s. This comparatively long propagation time must be taken into account, particularly in the design of speech telephone and signaling and certain data circuits.

Tropospheric (tropo) *scatter* is an over-the-horizon microwave communication technique, operating in the 400-MHz and 900-MHz bands and in the 2-GHz and 4-GHz bands, handling 12 to 240 FDM telephone channels or up to 120 digital channels. Tropospheric scatter takes advantage of the refraction and reflection phenomena in a section of the earth's atmosphere called the *troposphere*. This is shown functionally in Figure 7.3. With such systems UHF radio signals can be transmitted beyond line of sight on single hops up to 640 km (400 mi). Tropo systems are expensive. Transmitters emit 1 kW or 10 kW; antennas use parabolic reflectors 5 m, 10 m, or 20 m in diameter, and receiving

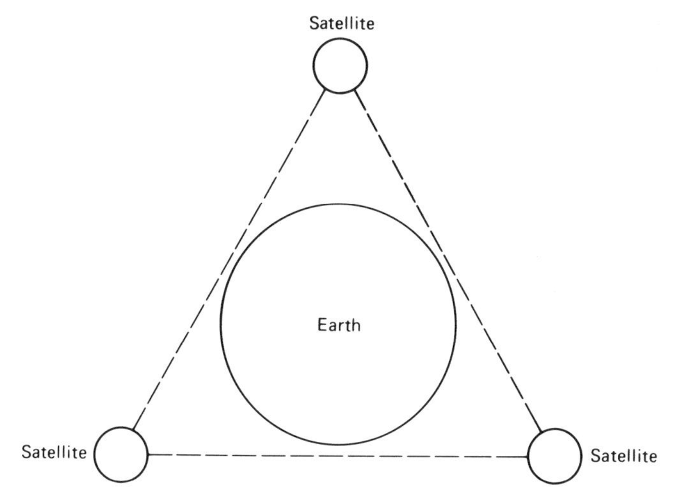

Figure 7.2 Three geostationary satellites properly placed can provide 100% earth coverage.

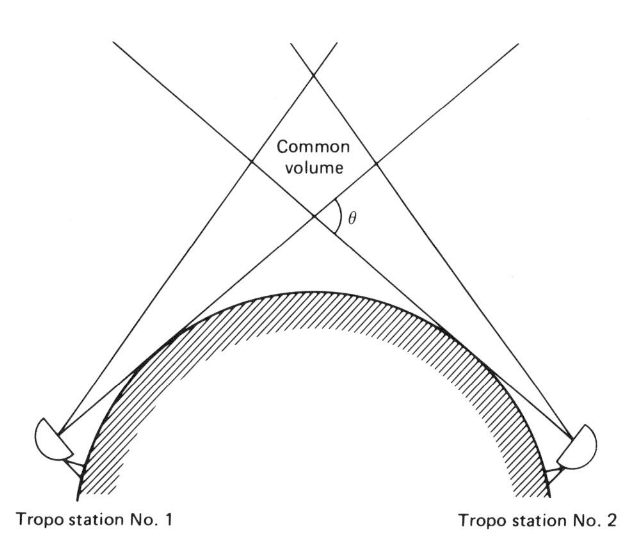

Figure 7.3 Tropospheric scatter model (θ = scatter angle.)

systems are quadruple diversity with low noise receivers. Tropo is used for comparatively thin line links to reach over difficult terrain or over water and into rural regions.

4 DESIGN ESSENTIALS FOR RADIOLINKS

4.1 General

For this discussion, radiolink and line-of-sight microwave are considered synonymous. The greater portion of long-distance communication links or "connections" use line-of-sight microwave. A radiolink is made up of terminal radios and often one or more repeaters separated approximately 20–50 mi (35–85 km). Of course, some "hops" may be shorter than 20 mi or longer than 50 mi. Only frequency-modulated systems are considered for this discussion. Digital radio-transmission techniques are discussed in Chapter 9, Section 8.5.

Frequency modulation (FM) is so widely used because of noise-improvement factors, namely, the trade-off of bandwidth for improved signal-to-noise ratio above a certain "noise" threshold. This threshold is called the *FM improvement threshold*. The improvement over amplitude-modulation systems is of the order of 20 dB. The bandwidth (Bw) of an FM radio system is defined by the following formula:

$$\mathrm{Bw} = 2(F_{\mathrm{dev}} + BB) \quad \text{(Carson's rule)}$$

If Bw is in megahertz, then F_{dev} (the peak frequency deviation) is in megahertz, and BB (the maximum baseband frequency) must also be in megahertz. The design of a radiolink hop includes the determination of tower heights by path profiling and path calculations. From the latter, the system designer derives the necessary equipment parameters.

4.2 Site Placement and Tower Heights

As mentioned previously, radiolinks have terminal sites and repeater sites. At terminal sites all RF carriers are demodulated to baseband; the resulting baseband signal is demultiplexed* to individual voice-frequency channels. Figure 7.4A illustrates the concept pictorially, and Figure 7.4B is a simplified functional block diagram taken from Figure 7.4A. Radiolink repeaters are discussed at length further on. However, it can be seen here that baseband repeating is carried on at the simple repeater site. At exchange *B* there will be straight-through circuit groups from exchanges *A* to *C*. This would most likely be done by using through-group or through-supergroup techniques. This means that not all channels are demultiplexed to voice channels. The "through" channels are bunched in groups of 12 or supergroups of 60. Through-grouping has two advantages: (1) less multiplex equipment is required; (2) there is less

*This, of course, refers to frequency division multiplex. See Chapter 5, Section 4.

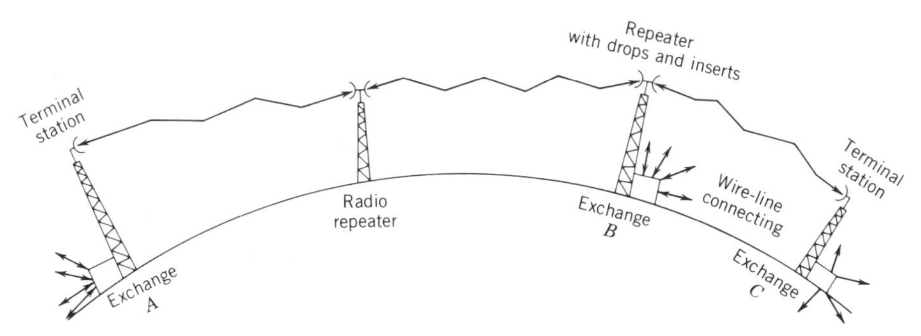

Figure 7.4A Sketch of a microwave (LOS) radio relay system.

noise accumulation on through circuits. Figures 7.4A and 7.4B show terminal and relay sites colocated with an exchange. In many cases this may not be feasible. We would want as much concentration as possible of traffic entering (inserts) and leaving (drops) the radio relay system. There are several trade-offs to be considered:

1. Bringing traffic in by wire from several exchanges traded off by elimination of drop and insert at relay site (savings on multiplex equipment).
2. Siting due to propagation constraints (or advantages) versus colocation with exchange or distant from any exchange (savings in land and access problems by colocation).
3. Method of feeding (feeders):* by light-route radio, coaxial cable (multiplex at exchange), or wire pairs, multiplex on wire (aerial or buried cable).

In gross system design, exchange location, particularly tandem-toll exchanges, must be considered in light of probable radio and cable routes. Another consideration is RF interference or electromagnetic compatibility in general. Midcity relay or terminal sites have the following advantages:

- Colocation with toll exchange.
- Use of tall buildings as natural towers.

And they have the following disadvantages:

- Wave reflections (multipath) off buildings.
- Electromagnetic compatibility (EMC) problems, particularly from other nearby emitters and industrial emission.
- Low-grade labor market.

*These are feeders to the main-line radio system.

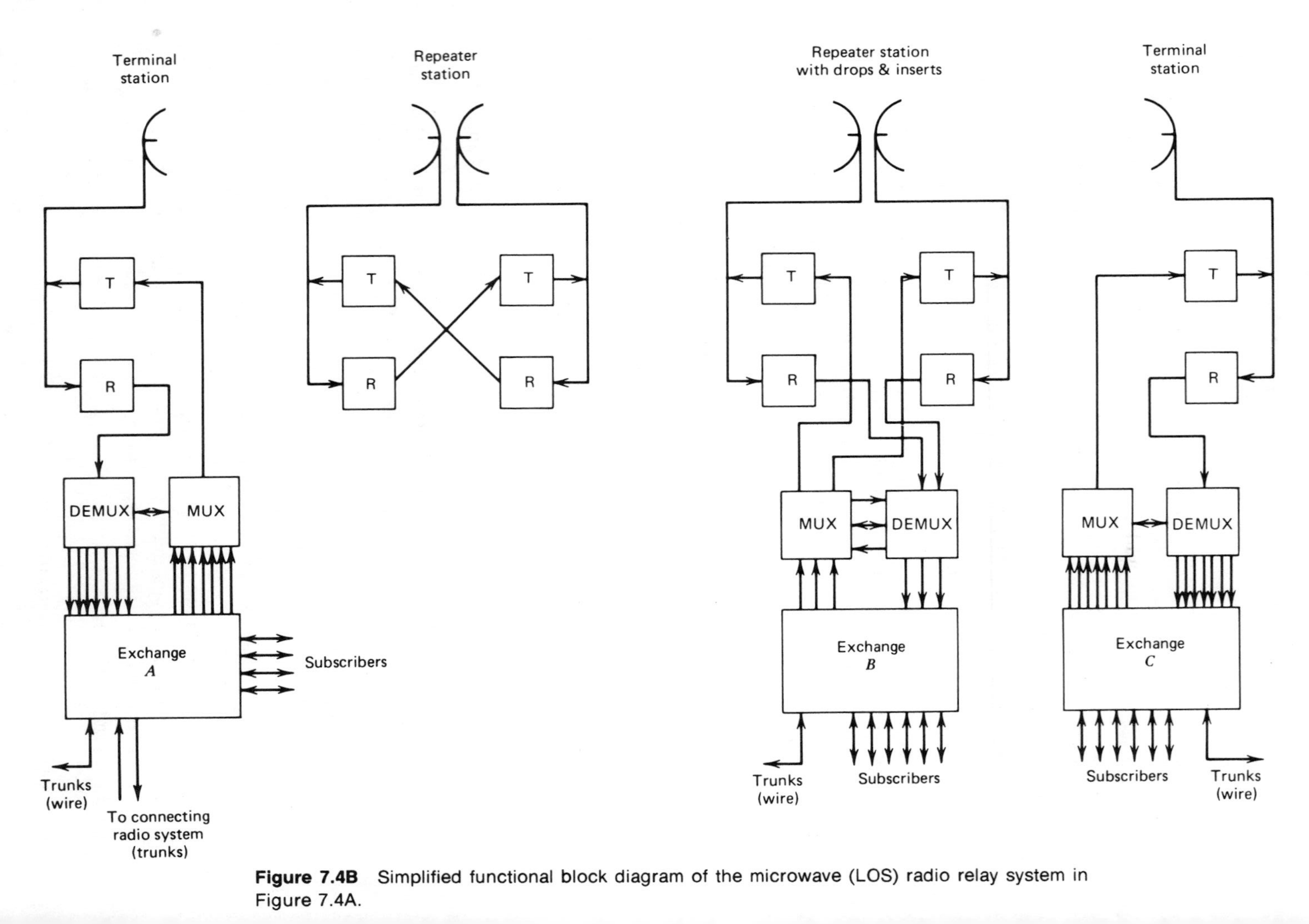

Figure 7.4B Simplified functional block diagram of the microwave (LOS) radio relay system in Figure 7.4A.

Sitings in the country have fewer EMC problems and usually a better labor force (for operators, other operational personnel, and technicians), and right-of-way for cable is easier.

We are now led to propagation constraints. Terminal sites will be in or near heavily populated areas and preferably colocated with a toll exchange. The tops of modern large office buildings, if properly selected, are natural towers. Relay sites are heavily influenced by intermediate terrain. Accessible hilltops or mountain tops are good prospective locations. Draw a line along the path of the desired route. Sites would zigzag along the line with optical or "radio" separation distances. If tower costs were $300 per foot ($900 per meter), 300-ft or 100-m towers might be the height limit for economic reasons. If hilltop or mountain top sites are well selected, towers that high may never have to be considered. High towers are the rule over flat country. The higher the tower, the longer the line-of-sight distance. Thus, on a given link, fewer repeaters would be required if towers could be higher. Hence there is a trade-off between tower height and number of repeaters.

4.3 Calculation of Tower Heights

Assume now that sites along a microwave radio relay route have been carefully selected. The next step in engineering is the determination of tower heights. The objective is to keep the tower height as low as possible and still maintain effective communication. The towers must just be high enough for the radio beam to surmount obstacles in the path. As the discussion proceeds, the term "high *enough*" is carefully defined. What obstacles might there be in the path? To name some, there are terrain such as mountains, ridges, hills, and earth curvature—which is highest at midpath—and buildings, towers, grain elevators, and so on.

All obstacles along the path must be scaled on graph paper in an exercise called *path profiling*. Good topographical maps are required of the region. Ideally, such maps should be 1 : 24,000, although 1 : 62,500 maps are acceptable. A straight line is drawn between the sites in question and then on linear graph paper scaled to 1 in. for 2 mi on the horizontal or 1 cm for 1 or 2 km. Vertical scales depend on the rate of change of elevation along the path. An ideal scaling is 100 ft per inch or 1 cm for 10 m and over hilly country, 1 in. equivalent to 200 ft or 1 cm equivalent to 20 m. In mountainous country the vertical scale may have to be as much as 1 in. equivalent to 1000 ft or 1 cm equivalent to 100 m. Each obstacle encountered must be identified with a letter or number on the horizontal scale. The next step is to establish a point directly above, giving altitude above mean sea level. The bottom of the chart need not be mean sea level; it may be mean sea level plus so many meters. Once the reference altitude has been established, we must give several additional clearances. If the obstacle is terrain with vegetation, especially trees, a clearance for trees and growth must be established. If no other values are available, use 40 feet and 10 feet, respectively.

To the altitude or height of each obstacle must be added "earth bulge," the number of feet or meters an obstacle is raised higher in elevation (into the path) as a result of earth curvature or "earth bulge." The amount of earth bulge at any point in the path may be calculated by the formula(s):

$$h = 0.677 d_1 d_2 \qquad (h \text{ in feet; } d \text{ in miles}) \tag{7.2A}$$

$$h = 0.078 d_1 d_2 \qquad (h \text{ in meters; } d \text{ in km}) \tag{7.2B}$$

where d_1 is the distance from the near end of the hop to the obstacle in question and d_2 is the distance from the far end of the hop to the obstacle in question. Equation (7.2) is for a ray beam that is a straight line (i.e., no bending). Atmospheric refraction may cause the beam to be bent either toward or away from the earth. This bending effect is handled by adding the factor K to equation (7.2), where

$$K = \frac{\text{Effective earth radius}}{\text{True earth radius}}$$

such that

$$h_{\text{ft}} = \frac{0.667 d_1 d_2}{K} \qquad (d \text{ in miles}) \tag{7.3A}$$

$$h_{\text{m}} = \frac{0.078 d_1 d_2}{K} \qquad (d \text{ in km}) \tag{7.3B}$$

If the factor K is greater than 1, the ray beam is bent toward the earth and the radio horizon is greater than the optical horizon. If K is less than 1, the radio horizon is less than the optical horizon. For general system planning purposes, $K = \frac{4}{3}$ may be used. However, for specific path engineering, K must be selected with care. The value of h or earth curvature corrected for K from equation (7.3) must be added to obstacle height in the path-profile exercise for each obstacle.

Still another factor must be added to obstacle height, namely, Fresnel zone clearance. This factor derives from the electromagnetic wave theory that a wave front, which our ray beam is, has expanding properties as it travels through space. These expanding properties result in reflections and phase transitions as the wave passes over an obstacle. The outcome is an increase or a decrease in received signal level. The amount of additional clearance over obstacles that must be allowed to avoid problems of the Fresnel phenomenon (diffraction) is expressed in Fresnel zones. The first Fresnel zone radius may be calculated from the following formula:

$$R_{\mathrm{ft}} = 72.1\sqrt{\frac{d_1 d_2}{FD}} \tag{7.4A}$$

where F is the frequency in gigahertz, d_1 is the distance from transmit antenna to obstacle (statute miles), d_2 is the distance from path obstacle to receive antenna (statute miles), and $D = d_1 + d_2$. For metric units:

$$R_{\mathrm{m}} = 17.3\sqrt{\frac{d_1 d_2}{FD}} \tag{7.4B}$$

where F is the frequency in gigahertz and d_1, d_2, and D are the same as in equation (7.4A), but d and D are in kilometers and R in meters.

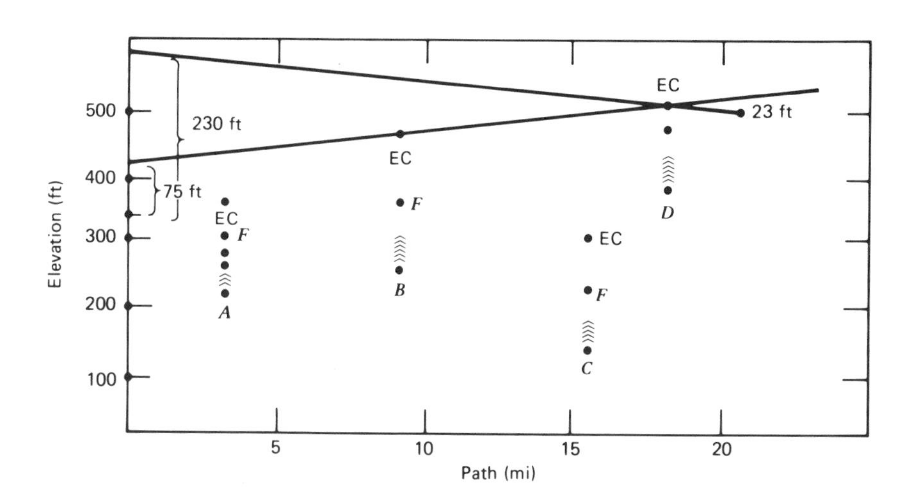

	Obstacle	d_1	d_2	Basic Height (ft)	F (Fresnel) (ft)	E.C. (ft)	T and G (ft)	Adjusted Total Height (ft)
Tree conditions: 40 + 10 ft growth (T and G)	A	3.5	19.0	220	30	49	50	349
Frequency band: 6 GHz	B	10	12.5	270	41	91.7	50	452.7
Midpath Fresnel (0.6)	C	17	5.5	160	36	68.6	50	314.6
= 42 ft	D	20	2.5	390	25.2	36.6	50	501.8

Figure 7.5 Practice path profile (x in miles, y in feet; assume that $K = 0.9$).

Previously, a clearance of 0.6 Fresnel zone [0.6 the value of R in equation (7.4)] was considered sufficient. A new rule of thumb is evolving, namely, when $K = \frac{2}{3}$, at least 0.3 Fresnel zone clearance is required, and 1.0 Fresnel zone clearance must be allowed when $K = \frac{4}{3}$. At points near the ends of a path, Fresnel zone clearances should be at least 6 m or 20 ft [5].

The three basic increment factors that must be added to obstacle heights are now available: vegetation height and its growth, earth bulge corrected for K factor, and Fresnel zone clearance. These are marked as indicated previously on our path-profile chart. A straight line is drawn from right to left, just clearing the obstacle points as corrected for the three factors. Another line is then drawn from left to right. A sample profile is shown in Figure 7.5. Some balance is desirable so at one extreme we have a very tall tower and at the other, a little stubby tower. This is true but for one exception: when a reflection point exists at an inconvenient spot along the path.

4.3.1 Reflection Point. Possible reflection points may be obtained from the profile. The objective is to adjust tower heights such that the reflection point is adjusted to fall on land area where the reflected energy will be broken up and scattered. Bodies of water and other smooth surfaces cause reflections that are undesirable. Figure 7.6 can facilitate calculations for the adjustment of the reflection point. It uses a ratio of tower heights, h_1/h_2, and the shorter tower height is always h_1. The reflection area lies between a K factor of grazing ($K = 1$) and a K factor of infinity. The distance expressed is always from h_1, the shorter tower. The reflection point can be moved by adjusting the ratio h_1/h_2.

For a path that is highly reflective for much of its length, space-diversity operation may minimize the effects of multipath reception.

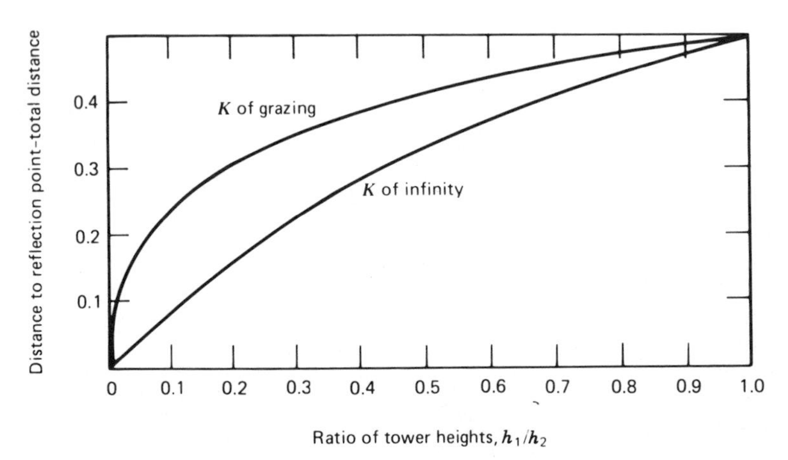

Figure 7.6 Calculation of reflection points.

4.4 Path Calculations

4.4.1 Introduction. Once the path profile has been completed and checked by a path survey, the next exercise in radiolink design is path calculation. The profile gave us tower heights. Now we want to assign certain parameters to the radio equipment we wish to install. It was assumed at the outset that the band of frequencies had been selected. Among the other parameters we must work with are:

1. Path loss in decibels.
2. Operating bandwidth and peak deviation (we assume frequency modulation).
3. Receiver thermal noise and FM improvement thresholds.
4. A desired, unfaded signal-to-noise ratio in the telephone channel.
5. A fade margin ensuring a certain noise specification for 99%, 99.9%, or 99.99% (or some intermediate value) of the time for the worst month or for the year.

From these we will determine parabolic antenna diameters. Generally, reflector diameters greater than 12 feet are impractical for these applications. If we cannot reach the noise objectives in items 4 and 5, the designer may have to resort to:

1. More transmitter power.
2. More sensitive receiver front ends.
3. Use of diversity reception.
4. Reduction of hop length (as a last resort).

As an introduction to the overall problem, Figure 7.7 graphically gives an idea of the gains and losses for a single hop in a radiolink system. The gains and losses as shown in Figure 7.7 are summed algebraically. Hence antenna sizes are adjusted to meet a noise objective.

4.4.2 Path Loss. For all intents and purposes, path loss up to about 10 GHz can be considered "free-space loss." To introduce the reader to the problem, consider an isotropic antenna, that is, an antenna that radiates uniformly in all directions. If the isotropic radiator is fed by a transmitted power P_t, it radiates $P_t/4\pi d^2$ (W/m^2) at a distance d, and if a radiator has a gain G_t, the power flow is enhanced by the factor G_t. Finally, the power intercepted by an antenna of effective cross section A (related to the gain by $G_r = 4\pi A/\lambda^2$) is $P_t G_t G_r(\lambda/4\pi d)^2$. The term $(\lambda/4\pi d)^2$ is known as the free-space loss and represents the steady decrease of power flow (in W/m^2) as the wave propagates. From this we can derive the more common formula for

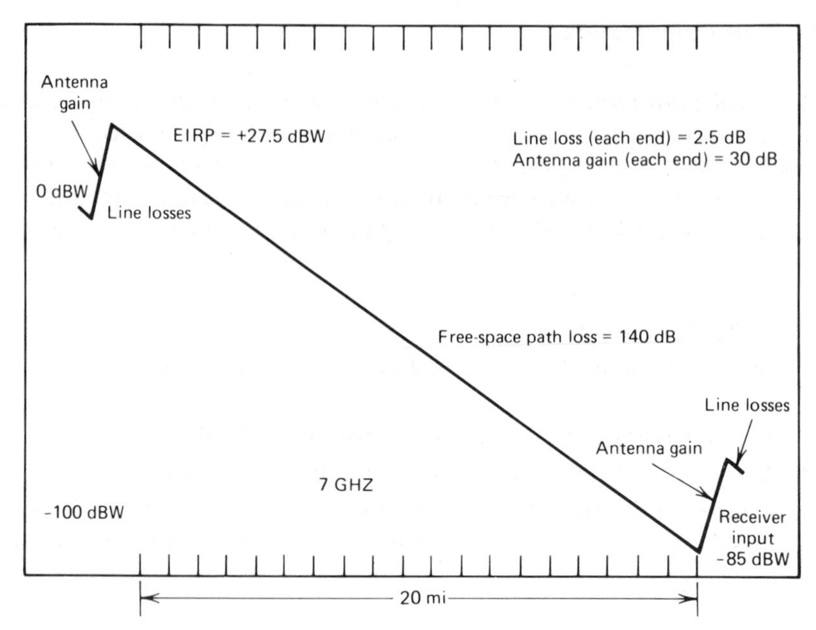

Figure 7.7 Radiolink gains and losses (simplified) (transmitter output = 0 dBW).

free-space path loss, which reduces to

$$L = 96.6 + 20 \log_{10} F + 20 \log_{10} D \tag{7.5}$$

where L is the free-space attenuation between isotropic antennas in dB, F is the frequency in GHz, and D is the path distance in statute miles. In the metric system

$$L_{\mathrm{dB}} = 92.4 + 20 \log_{10} F_{\mathrm{GHz}} + 20 \log_{10} D_{\mathrm{km}}$$

Consider the problem from a different aspect. It requires 22 dB to launch a wave to just 1 wavelength (1λ) distant from an antenna. Thus for an antenna emitting +10 dBW, we could expect the signal one wavelength away to be 22 dB down, or −12 dBW. Whenever we double the distance, we incur an additional 6 dB of loss. Hence at 2λ from the +10-dBW radiator, we would find −18 dBW; at 4λ, −24 dBW; 8λ, −30 dBW; and so on. Now suppose that we have an emitter where $F = 1$ GHz; what is the path loss at 1 statute mile?

$$L = 96.6 + 20 \log_{10} 1 + 20 \log_{10} 1 = 96.6 \text{ dB}$$

For rough calculations, the 6-dB relationship is worthwhile and also gives insight in that if we have a 20-mi path and shorten or lengthen it by a mile, our signal level will be affected little.

4.4.3 Receiver Thermal Noise Threshold: The Starting Point. In this step in path calculations the objective is to calculate the thermal noise level of the receiver to be used at the distant end of the path. If we are given a transmitter with a known output, an attenuation of the signal calculated in Section 4.4.2, and a receiver at the far end, the exercise is to find the proper signal level input to the receiver equal to the thermal noise. The problem is shown diagrammatically as follows:

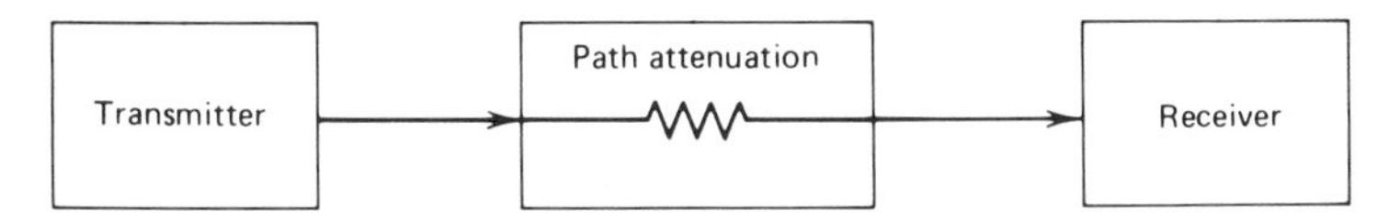

Receiver thermal noise level can be calculated from the following formula using Boltzmann's constant:

$$N_{\text{dBW}} = 10 \log_{10} kTB \tag{7.6}$$

where T is the receiver noise temperature in degrees kelvin; B is the noise bandwidth in hertz; and k is Boltzmann's constant (1.3803×10^{-23} J/°K). By converting Boltzmann's constant to dBW, we have

$$N_{\text{dBW}} = -228.6 \text{ dBW/Hz} + 10 \log T + 10 \log B_{\text{if}} \tag{7.7}$$

where B_{if} may be calculated according to Carson's rule as $B_{\text{if}} = 2$(peak FM deviation) + 2(highest modulating frequency). Of course, if we were transmitting only supergroup 2 in an FDM configuration (Chapter 5), the highest modulating frequency would be 552 kHz.

In equation (7.7) the term -228.6 dBW/Hz represents the thermal noise of a receiver with a 1-Hz bandwidth, unadjusted for noise temperature. The term T, or $10 \log T$, adjusts the receiver to its real noise temperature, and B_{if} adjusts the calculation to its design intermediate-frequency (IF) bandwidth. For the more common radiolink application, noise temperature may be converted to the more familiar noise figure by

$$NF_{\text{dB}} = 10 \log_{10} \left(1 + \frac{T_e}{290}\right) \tag{7.8}$$

where the effective noise temperature of the receiver, T_e, is compared to room temperature, 290K.

The noise figure is a measure of the noise produced by a practical network compared to an ideal network (i.e., one that is noiseless). For a linear system, noise figure (NF) is expressed by

$$NF = \frac{S/N_{\text{in}}}{S/N_{\text{out}}} \tag{7.9}$$

where S/N is the signal-to-noise ratio.

For NF expressed in decibels:

$$NF_{\text{dB}} = 10 \log_{10} NF \qquad (7.10)$$

The receiver noise threshold formula may be simplified still further, assuming that the receiver operates at room temperature (290K or ~ 17°C):

$$N_{\text{dBW}} = -228.6 \text{ dBW/Hz} + 10 \log 290\text{K} + NF_{\text{dB}} + 10 \log B_{\text{if}}$$

or

$$N_{\text{dBW}} = -204 \text{ dBW/Hz} + NF_{\text{dB}} + 10 \log B_{\text{if}} \qquad (7.11)$$

If a receiver has a 10-MHz IF bandwidth and a noise figure of 10 dB, what is the noise threshold?

$$\begin{aligned} N_{\text{dBW}} &= -204 \text{ dBW/Hz} + 10 \text{ dB} + 10 \log_{10} 10^7 \\ &= -124 \text{ dBW} \end{aligned}$$

Let us now examine such a receiver on the distant end of a 20-mi hop operating at 6 GHz. First, we would calculate the free-space attenuation of the hop, or

$$\begin{aligned} \text{Path loss} &= 96.6 + 20 \log_{10} 6 + 20 \log_{10} 20 \\ &= 138.24 \text{ dB} \end{aligned}$$

Look again at the simplified network diagram, and calculate the transmitter output (assuming isotropic antennas and no other losses) so that the incoming signal at the receiver front end just equals the thermal noise level of the receiver.

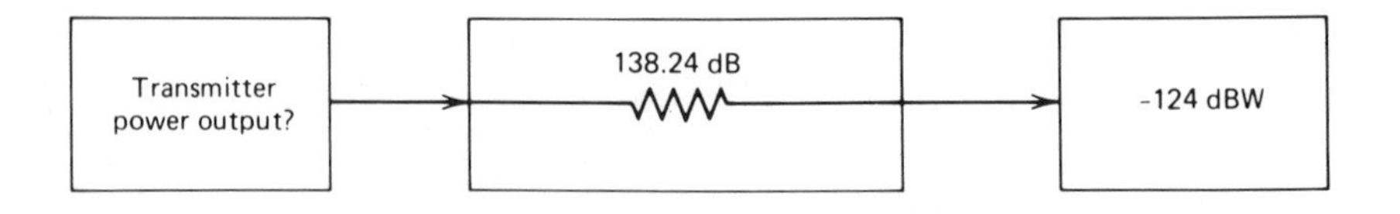

Thus

$$138.24 \text{ dB} - 124 \text{ dBW} = +14.24 \text{ dBW}$$

Hence the transmitter power output would have to be at least +14.24 dBW or 17.4 W.

To make the same exercise more realistic, assume a transmitter with 0 dBW output, 2 dB of loss for transmission lines from the transmitter to its antenna, and 2 dB of loss from the receiver antenna to the receiver front end. What antenna gains will be required to reach a 10-dB carrier-to-noise ratio at the receiver front end?

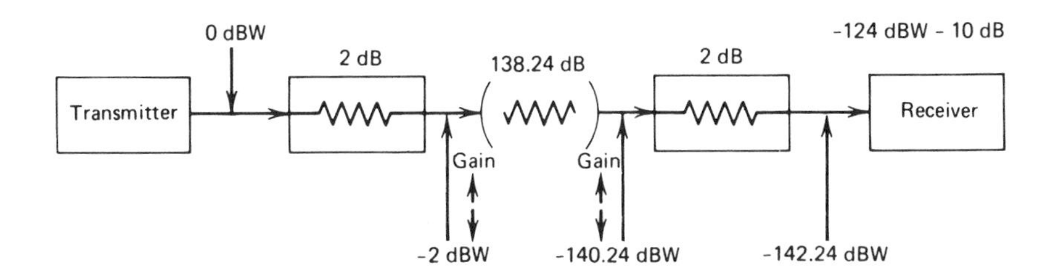

Without any antenna gain (i.e., 0 dBi), we have a level of -142.24 dBW at the receiver front end, but as we require $(-124 + 10)$ dBW, an antenna gain of $-114 + 142.24$ dB or 28.24 dB is necessary. The antennas (there are two) should have at least 28.24 dB gain between them, or 14.12 dB gain each. We round off upward; thus 15 dB gain is required for each antenna.

4.4.4 Parabolic Antenna Gain. At a given frequency the gain of a parabolic antenna is a function of its effective area and may be expressed by the formula

$$G = 10 \log_{10} (4\pi A\eta/\lambda^2) \tag{7.12}$$

where G is the gain in decibels relative to an isotropic antenna, A is the area of antenna aperture, η is the aperture efficiency, and λ is the wavelength at the operating frequency. Commercially available parabolic antennas with a conventional horn feed at their focus usually display a 55% efficiency or somewhat better. With such an efficiency, gain (G, in decibels) is then

$$G = 20 \log_{10} D + 20 \log_{10} F + 7.5 \tag{7.13}$$

where F is the frequency in gigahertz and D is the parabolic diameter in feet.

In metric units, we have

$$G = 20 \log_{10} D + 20 \log_{10} F + 17.8 \tag{7.14}$$

where D is measured in meters and F in gigahertz.

What size antenna would be required in the preceding example? Let $G = 15$ dB and $F = 6$ GHz; then

$$\begin{aligned} 15 &= 20 \log_{10} D + 20 \log_{10} 6 + 7.5 \\ 20 \log D &= +15 - 20 \times 0.7782 - 7.5 \\ &= -\left(\frac{8.06}{20}\right) \\ D &= 0.395 \text{ ft} \end{aligned}$$

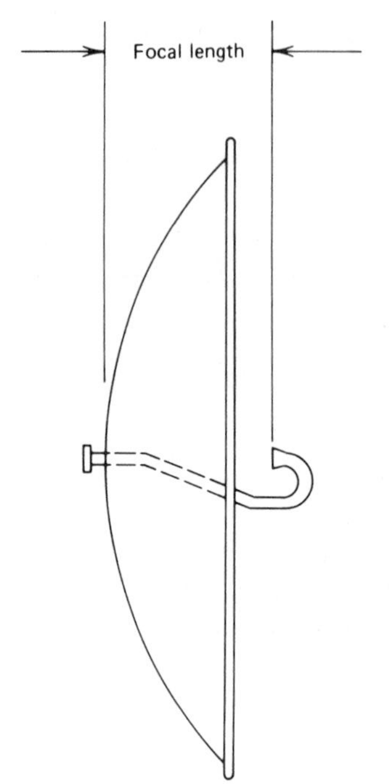

Figure 7.8 Typical parabolic antenna with front feed.

Parabolic dish antennas, with waveguide (horn) feeds (see Figure 7.8), are probably the most economic antennas for radiolinks operating from 3 GHz upward. From 900 MHz to about 3 GHz, coaxial feeds are used. Coaxial cable transmission lines deliver the RF energy from/to transmitter/receiver to the antenna in this range. Above 3 GHz coaxial cable becomes too lossy and waveguide is more practical.

To every rule there are exceptions. One method of delivering the RF signal to/from the point of radiation is by the so-called periscopic method. In this case the antenna is mounted on the radio equipment building or shelter, and a plane reflector is placed at the point of radiation on the tower. If the antenna is directly below the reflector, the reflector would be oriented at 45° to permit the radio beam to be emitted on a straight line parallel to level earth, as in Figure 7.9.

For path calculations, there may be a gain or loss of up to several decibels on each end using periscopic techniques. The gain or loss depends on frequency, antenna size, reflector cross area, and distance from antenna to reflector [3, 5–7]. It should be noted that in the United States new periscopic installations are not generally authorized.

Other types of antennas may also be used, such as the "cornucopia," horn, and spiral. Besides cost and gain, other features are front-to-back ratio, side lobes, and efficiency. For instance the "cornucopia," called such because it looks like "the horn of plenty," has efficiencies in excess of 60% and improved side-lobe discrimination but is more costly.

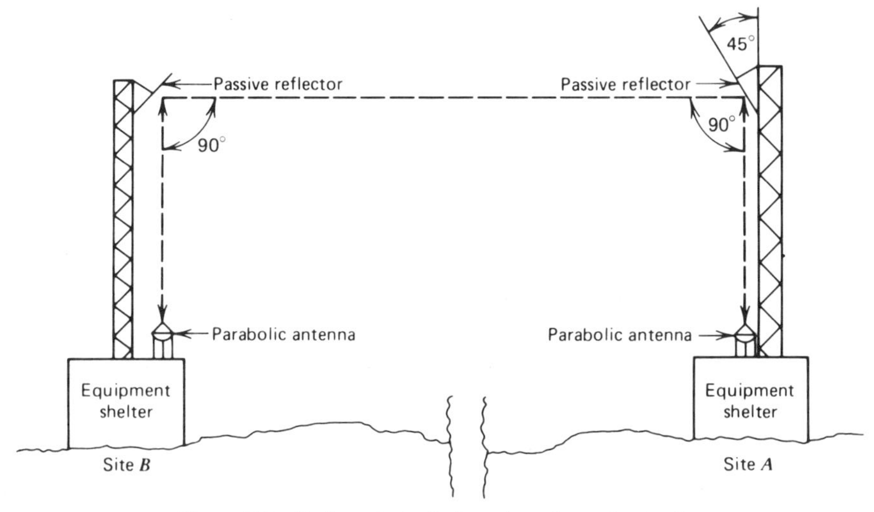

Figure 7.9 Periscopic technique (passive reflectors).

4.4.5 Effective Isotropically Radiated Power. Effective isotropically radiated power (EIRP) is a term used by the radio engineer to conveniently describe the power in the radio beam compared to the isotropic antenna. Remember that the isotropic radiator is "a hypothetical antenna radiating or receiving equally in all directions" (IEEE definition). The IEEE dictionary goes on to add that "an isotropically radiating antenna does not exist physically but represents a convenient reference for expressing the directive properties of actual antennas." It has a gain of 1, expressed as 0 dB.

The EIRP is the algebraic sum of the transmitter output (expressed in dBm or dBW) plus the gains and losses of the transmitting antenna system. The system includes all antenna and transmission-line elements from the transmitter output to the antenna feed.

If a transmitter has a +10 dBW output and there are 2 dB of transmission-line losses and the antenna has a 20 dB gain, the EIRP is

$$\begin{aligned} \text{EIRP}_{\text{dBW}} &= +10\ \text{dBW} - 2\ \text{dB} + 20\ \text{dB} \\ &= +28\ \text{dBW} \end{aligned}$$

4.4.6 Fades and Fade Margin. In Section 4.4.2 we showed how path attenuation can be calculated. This was a fixed loss and can be simulated in the laboratory with a transmission-line attenuator. On short radiolink paths below about 10 GHz, the signal level impinging on the distant end receiving antenna can be calculated to less than 1 dB. If the transmitter continues to give the same output, that receive level will remain the same over long periods of time, for years. As the path is extended, the calculated level will tend to decrease once in awhile. These drops in level may last for seconds, minutes, or even longer. This is the phenomenon of fading. The radiolink design must take fading into account, first on a system basis and then on a per-hop basis. The system designer is concerned with accumulated noise contribution in the derived voice channel. As the signal fades, the signal-to-noise ratio decreases and the system noise increases in the derived voice channel.

The system is designed for specified signal levels. Microwave receiver AGC (automatic gain control) and FDM level regulation maintain these levels regardless of the signal-to-noise ratio. Thus in the system design, noise level in the derived voice channel is specified not to exceed a certain value for a percentage of the time, such as 1%, 0.1%, or 0.01%. Or we can say that noise will be *less* than a certain value for 99%, 99.9%, or 99.99% of the time. That is, a system would not meet a predetermined noise criterion for 8.8 hr per year if we assigned a 99.9% path reliability criteria and would meet that criterion for the remainder of the year.

If a radiolink system has 10 hops and the system noise criterion is established for 99.9% of the time, the *worst*-case per-hop criterion must be 99.99%. Fades will not occur at the same time on each hop in the system, so the hop criterion can be relaxed considerably and still maintain the system noise time percentage. But for the worst case the designer often assigns such a noise criterion per hop because noise power is the summation of the noise powers of individual hops.

The CCIR recommends 3-pWp noise accumulation per kilometer, adding a fixed noise power quantity of 200 pWp for systems from 50 km to 840 km long, 400 pWp from 840 km to 1670 km, and 600 pWp from 1670 km to 2500 km. For instance, a system 2000 km long would be 600 pWp + 2000 × 3 pWp or 6600 pWp accumulated noise for the system. These, of course, are unfaded figures. The CCIR further states that the 1-min mean power of 47,500 pWp shall not be exceeded 0.1% of the time over a 2500-km reference circuit.

Two approaches may be taken to establish a fade margin. The first, developed by Bullington of Bell Telephone Laboratories [9], is shown in Figure 7.10. Using Figure 7.10, a 6-GHz path fade margin may be derived from path length for 99%, 99.9%, and 99.99% of the time. For instance, a path 25 mi long with a 99.9% path reliability must have a 30-dB fade margin. Another approach is to assume that the fading follows a Rayleigh distribution. Here a 8-dB fade margin would be required for a 90% path reliability.* A Rayleigh distribution has a 10-dB slope for each full order of improvement, or:

Propagation or Path Reliability	Required Fade Margin
90%	8 dB
99%	18 dB
99.9%	28 dB
99.99%	38 dB
99.999%	48 dB

*This is often called *time availability*.

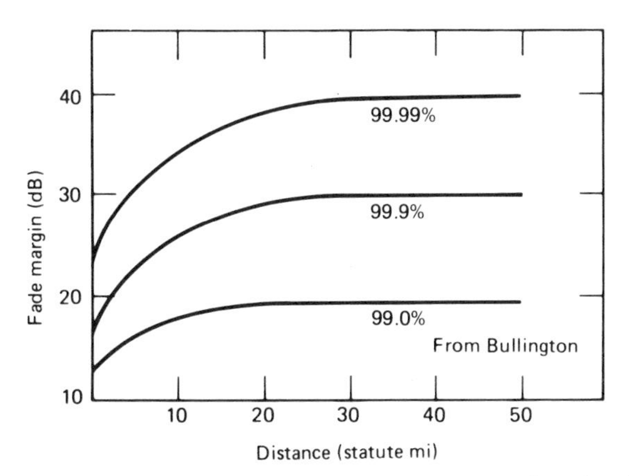

Figure 7.10 Nomogram for calculation of fade margins as a function of path length for the 6-GHz band [3, 6].

A major cause of fading derives from reflections from a stratified atmosphere or from surface or land conditions along the path. This type of fading is called *multipath* and causes destructive and constructive interference in the level of the incoming signal. Of course, the adjustment of the reflection point as discussed in Section 4.3.1 can reduce or eliminate ground reflections in many cases. Likewise, space diversity can reduce the effects of multipath fading (see Section 4.4.7).

Noise accumulation following North American (ATT) practice states that analog trunks should display no more than for the trunk lengths indicated [28]:

Short Haul (3 dB per Double Distance)	Long Haul (3 dB per Double Distance)
20–60 mi: 28 dBrnC	250 mi: 28 dBrnC
61–120 mi: 31 dBrnC	1000 mi: 34 dBrnC
121–240 mi: 34 dBrnC	4000 mi: 40 dBrnC

When referring to trunks, in this case total noise accumulation is implied from all sources. It is the sum of radio noise (or cable noise), switch noise, and multiplex noise, among others. The figures are for BH measurement, unfaded. To relate dBrnC0 to pWp:

$$\text{dBrnC0} = 10 \log \text{pWp} + 0.8 \text{ dB} \tag{7.15}$$

To meet system noise criterion, a per-hop fade margin must be established. This means the hop is overbuilt; or, in other words, more gain is engineered in the hop than would be necessary under no-fade conditions. Overbuilding costs money and must be limited to just the proper amount necessary to meet system specifications.

Fading varies with path length and frequency. This is a simplification of a very complex phenomenon. Let us make the assumption first and then discuss the factors that will affect the design. Multipath fading not only varies with path length and frequency as in our simplification but also is a function of climate and terrain. For example, in dry, windy, mountainous areas the multipath phenomenon may be nonexistent. Flat terrain along a path tends to increase the incidence of fading. In hot, humid coastal regions a very high incidence of fading would be expected.

4.4.7 Diversity. "Diversity" here refers to simultaneous reception of a radio signal over several "paths." The signal "paths" are combined by predetection or postdetection combiners in the radio equipment so that the composite signal is less affected by fades. In fact, on well-designed radio diversity systems the frequency of fades (i.e., the number fades per time interval) and the depth of fades (in decibels) are notably less. The simplest form of diversity is space diversity. Such a configuration is shown in Figure 7.11.

The two diversity paths in space diversity are derived at the receiver end from two separate receivers with a combined output. Each receiver is connected to its own antenna, separated vertically on the same tower. The separation distance should be at least 70 wavelengths and preferably 100 wavelengths. In theory, fading will not occur on both paths simultaneously.

Frequency diversity is more complex and more costly than space diversity. It has advantages as well as disadvantages. Frequency diversity requires two transmitters at the near end of the link. The transmitters are modulated simultaneously by the same signal but transmit on different frequencies. Frequency separation must be at least 2%, but 5% is preferable. Figure 7.12 is an example of a frequency-diversity configuration. The two diversity paths are derived in the frequency domain. When a fade occurs on one frequency, it will probably not occur on the other frequency. The more one frequency is separated from the other, the less chance there is that fades will occur simultaneously on each path.

Frequency diversity is more expensive, but there is greater assurance of path reliability. It provides full and simple equipment redundancy and has the great operational advantage of two complete end-to-end electrical paths. In this case failure of one transmitter or one receiver will not interrupt service,

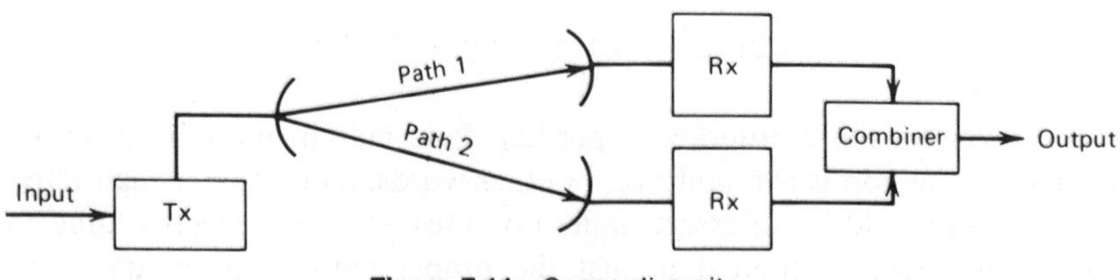

Figure 7.11 Space diversity.

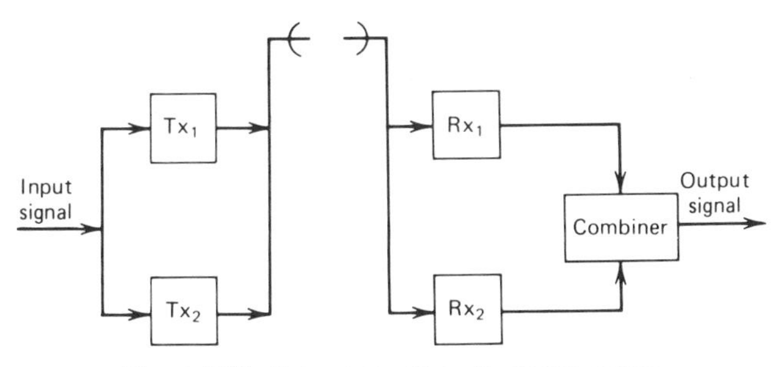

Figure 7.12 Frequency-diversity configuration.

and a transmitter and/or a receiver can be taken out of service for maintenance. The primary disadvantage of frequency diversity is that it doubles the amount of frequency spectrum required in this day and age when spectrum is at a premium. In many cases it is prohibited by national licensing authorities. For example, the US Federal Communications Commission (FCC) does not permit frequency diversity for industrial users. It also should be appreciated that it will be difficult to get the desired frequency spacing.

The full equipment redundancy aspect is very attractive to the system designer. Another approach to achieve diversity improvement in propagation plus reliability improvement by fully redundant equipment is to resort to the "hot standby" technique. On the receive end of the path, a space-diversity configuration is used. On the transmit end a second transmitter is installed as in Figure 7.12, but the second transmitter is on "hot standby." This means that the second transmitter is on but its signal is not radiated by the antenna. On a one-for-one basis the second transmitter is on the same frequency as the first transmitter. On failure of transmitter 1, transmitter 2 is switched on automatically.

One-for-*N* hot-standby is utilized on large radiolink systems employing several radio carriers, where the cost for duplicate equipment for each channel may be prohibitive. In this case one full set of spare equipment in the "on" condition serves to replace one of several operational channels, and the spare equipment is assigned its own frequency. On the receive side there is just one extra receiver. Relay cut-over must be provided, and no space-diversity improvement is afforded. Likewise, there is no paralleling of inputs on the transmit side; thus the switching from the operational pair to the standby pair is much more complex. On such multi-RF channel arrangements it is customary to assign some sort of priority arrangement. Often the priority channel enjoys the advantage of frequency diversity, whereas the other RF channels do not. On failure of one of the other channels, the diversity improvement is lost on the priority channel, with the diversity pair switched to carry the traffic on the failed pair. In another arrangement the standby channel carries low-priority traffic and does not operate in a frequency-diversity arrangement, while providing protection for perhaps two or three other channels carrying the higher-priority traffic. On failure of a high-priority channel, the lower-priority channel drops its traffic, replacing one of the RF channels carrying the more important traffic flow. Once this occurs, the remaining channels operate without standby equipment protection.

Diversity Improvement. Propagation reliability improvement can be exemplified as follows. If a 30-mi path required a 51-dB fade margin to achieve a 99.999% reliability on 6.7 GHz without diversity, with space diversity on the same path, only a 33-dB fade margin would be required for the same propagation reliability, namely, 99.999% (Vigants IEEE Trans. Com., December 1968 and Ref. 22). For frequency diversity in the nondiversity condition, assuming Rayleigh fading, a 30-dB fade margin would display something better than a 99.9% path reliability. But under the same circumstances with frequency diversity, with only a 1% frequency separation, propagation reliability on the same path would be improved to 99.995% [23].

4.4.8 Path Calculations: Conclusions. The exercise of carrying out path calculations basically involves algebraically summing the gains and losses in the system to reach a specified noise criterion that is valid for a time percentage. The basic contributors to loss, such as path loss and transmission-line loss, have been discussed. Researching the literature still further will uncover yet other losses that must be considered in any well-engineered system. Gain is achieved by selecting the proper antennas and applying diversity when needed. There are other gains as well, such as preemphasis gain, and, when required, the use of low-noise front ends and improved antenna feeds among other tools available to the design engineer to optimize radiolink design.

4.5 Noise Considerations: System Loading

For this discussion, there are two types of noise to be considered in radiolink systems, thermal noise and intermodulation (IM) noise. The calculation of thermal noise was discussed in Section 4.4.3. Once the receiver noise figure is given, thermal noise can be calculated. The receiver noise figure is given by the equipment manufacturer.

Intermodulation noise is also characteristic of the equipment and can be determined either from manufacturer specifications or from actual measurements on the system itself. Under multichannel loading, with more than approximately 60 voice channels, IM noise resembles thermal noise, and both can be considered "white noise." The IEEE dictionary defines white noise as "noise, of either random or impulsive type, that has a flat frequency spectrum at the frequency range of interest." This means that the noise-power distribution is equal throughout the frequency spectrum of the demodulated signal.

The important fact is that the IM noise increases with load (i.e., increased traffic and/or signal level), and when a certain "breakpoint" of load-handling capability is exceeded, IM noise becomes excessively high. On the other hand, residual (thermal) noise is not affected by the amount of traffic or traffic load of the system. Frequency-modulation radio overcomes much of the residual noise that appears in a radio system by distributing it over a wide radio bandwidth. Such an exchange of bandwidth for lower noise is also proportional to the signal level appearing at the system receiver. At periods of low signal level (i.e., during a radio fade), thermal or residual noise becomes dominant.

4.5.1 Noise-Power Ratio. Noise-power ratio (NPR) has come into wide use in radiolink systems to describe IM noise performance. This ratio gives an excellent indication of performance of IM noise when measured under standard fixed conditions. It is measured in decibels, and we can say that a radiolink transceiver combination has a specified NPR under certain load conditions. Noise-power ratio, which is a meaningful measurement for radiolink systems that carry multichannel FDM telephone baseband information, can be defined as the ratio, expressed in decibels, of the noise in a test channel with all channels loaded with white noise to the noise in the test channel with all channels except the test channel fully noise loaded.

When NPR is measured on a baseband–baseband basis, a radiolink transmitter is connected back to back with a receiver using proper waveguide attenuators to simulate real-path conditions. A white noise generator is connected to the transmitter baseband input. This generator produces a noise spectrum that approximates a spectrum produced by a multichannel (FDM) system. The output noise level from the generator is adjusted to a desired *composite noise baseband power*. A notched filter is then switched in to clear a narrow slot in the spectrum of the noise signal, and a noise analyzer is connected at the output of the system. The analyzer is used to measure the ratio of the noise in the illuminated (noise-loaded) section of the baseband to the noise power in the cleared slot. The slot noise level is equivalent to the total noise (residual plus intermodulation) that is present in the slot bandwidth. Slot bandwidths are the width of a standard voice channel and are taken at the upper, middle, and lower portions of the baseband.

The composite noise power is taken from one of the following formulas for N telephone channels in an FDM (SSBSC) configuration (see Chapter 5):

$$P(\text{dBm0}) = -1 + 4\log_{10} N \qquad \text{(CCIR)} \tag{7.16A}$$

when $N < 240$ channels;

$$P(\text{dBm0}) = -15 + 10\log_{10} N \qquad \text{(CCIR)} \tag{7.16B}$$

when $N > 240$ channels;

$$P(\text{dBm0}) = -16 + 10\log_{10} N \qquad \text{(ATT)} \tag{7.16C}$$

$$P(\text{dBm0}) = -10 + 10\log_{10} N \tag{7.16D}$$

for certain US military systems with heavy data usage, and

$$\text{NPR} = \text{Composite power (dB)} - \text{Noise power in slot (dB)} \qquad (7.17)$$

A good guide for NPR for high-capacity radiolink systems should be 55 dB; for IF repeaters, 59 dB.

4.5.2 Derived Signal-to-Noise Ratio. Given the NPR of a system, we can then compute test-tone-to-noise ratio for each voice channel.

$$\frac{S}{N} = \text{NPR} + \text{BWR} - \text{NLR} \qquad (7.18)$$

$$\text{BWR (bandwidth ratio)} = 10 \log \frac{\text{Occupied-baseband bandwidth}}{\text{Voice-channel bandwidth}} \qquad (7.19)$$

NLR (noise load ratio) = P which is taken from the load equation (7.16). The signal-to-noise ratio as given is unweighted. For psophometric weighting, add 3 dB (when reference is 1000 Hz).* For an 800-Hz reference,* add 2.5 dB.

Total noise in a derived FDM or voice channel is the sum of the intermodulation noise component, which is derived from NPR, thermal noise, and intermodulation noise from antenna feeder distortion. It should be noted that the noise values from each contributor are converted to power equivalents, then summed and reconverted to decibel values in the case of dBrnC0.

4.5.3 Conversion of Signal-to-Noise Ratio to Channel Noise. Using the signal-noise-ratio calculated in equations (7.18) and (7.19),

$$\text{Noise power in dBa0} = 82 - \frac{S}{N} \qquad (7.20\text{A})$$

$$\text{Noise-power ratio in dBrnC} = 88.5 - \frac{S}{N} \qquad (7.20\text{B})$$

$$\text{Noise-power ratio in pW} = \frac{10^9}{\text{antilog } S/N} \qquad (7.20\text{C})$$

$$\text{Noise-power ratio in pWp} = \frac{10^9 \times 0.56}{\text{antilog } S/N} \qquad (7.20\text{D})$$

4.6 Radiolink Repeaters

Up to this point the only radiolink repeaters discussed have been baseband repeaters. Such a repeater fully demodulates the incoming RF signal to baseband. In the most simple configuration this demodulated baseband is used

*These are voice-channel test-tone reference signals (tones); 1000 Hz is used in North America, whereas 800 Hz is used in Europe and is recommended by the CCITT (see Chapter 5, Section 5).

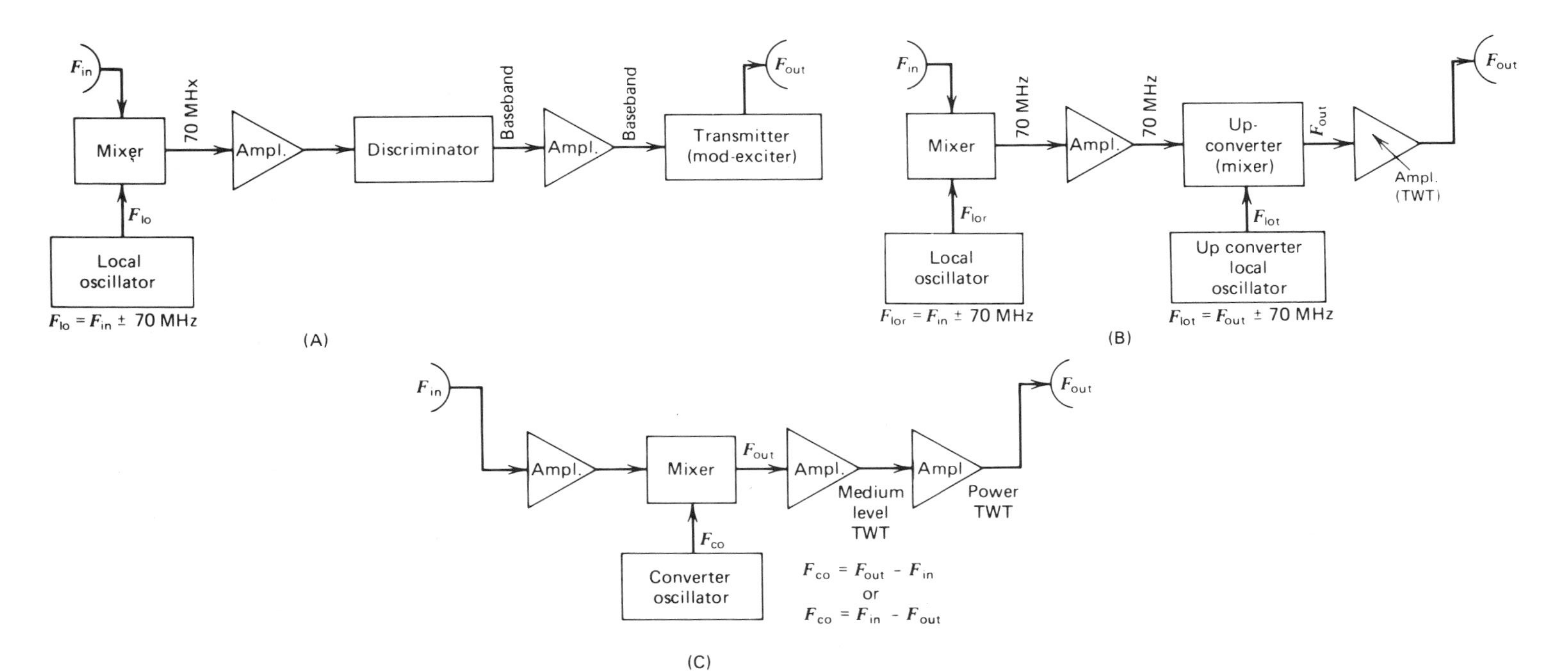

Figure 7.13 Radiolink repeaters: (A) baseband repeater; (B) IF heterodyne repeater; (C) RF heterodyne repeater; F_{in}, input frequency to receiver; F_{out}, output frequency of transmitter; F_{co}, output frequency of the converter local oscillator; F_{lor}, frequency of receiver local oscillator; F_{lot}, frequency of transmitter local oscillator; F_{lo}, frequency of the local oscillator; twt, traveling-wave tube.

to modulate the transmitter used in the next link section. This type of repeater also lends itself to dropping and inserting voice channels, groups, and supergroups. It may also be desirable to demultiplex the entire baseband down to voice channel for switching and insert and drop a new arrangement of voice channels for multiplexing. The new baseband would then modulate the transmitter of the next link section. A simplified block diagram of a baseband repeater is shown in Figure 7.13A. Two other types of repeater are also available: the IF heterodyne repeater (Figure 7.13B) and the RF heterodyne repeater (Figure 7.13C). The IF repeater is attractive for use on long backbone systems where noise and/or differential phase and gain should be minimized.

Generally, a system with fewer modulation–demodulation stages or steps is less noisy. The IF repeater eliminates two modulation steps. The repeater simply translates the incoming signal to IF with appropriate local oscillator and a mixer, amplifies the derived IF, and then up-converts it to a new RF frequency. The up-converted frequency may then be amplified by a traveling-wave tube (TWT) amplifier. An RF heterodyne repeater is shown in Figure 7.13C. With this type of repeater amplification is carried out directly at RF frequencies. The incoming signal is translated to a different frequency, amplified, usually by a TWT, and then reradiated. Radio frequency repeaters are troublesome in design in things such as sufficient selectivity, limiting and automatic gain control, and methods to correct envelope delay. However, some RF repeaters are now available, particularly for operation below 6 GHz.

4.7 Frequency Planning

4.7.1 General. To derive maximum performance from a radiolink system, the system engineer must set out a frequency-usage plan that may or may not have to be approved by the local administration.

The problem has many aspects. First, the useful RF spectrum is limited from above dc to about 150 GHz. The frequency range of discussion for radiolinks is essentially from the VHF band at 150 MHz (overlapping) to the millimeter region of 30 GHz. Second, the spectrum from 150 MHz to 30 GHz often must be shared with other services, such as radar, navigational aids,

TABLE 7.1 Some Radiolink Frequency Bands

450– 470 MHz	5,925– 6,425 MHz
890– 960 MHz	7,300– 8,400 MHz
1,710–2,290 MHz	10,550–12,700 MHz
2,550–2,690 MHz	14,400–15,250 MHz
3,700–4,200 MHz	13,300–19,700 MHz

Source: Ref. 1.

research (i.e., space), meteorological, and broadcast. Some radiolink frequency bands are shown in Table 7.1.

Although many of the allocated bands are wide, some up to 500 MHz in width, FM inherently is a wide-band form of emission. It is not uncommon to have $B_{rf} = 25$ (B_{rf} = RF bandwidth) or 30 MHz for just one RF carrier. Guard bands must also be provided. These are a function of frequency drift of transmitters as well as "splatter" or out-of-band emission, which in some areas is not well specified.

Occupied bandwidth has been specified in Section 4.4.3, according to Carson's rule. This same rule is followed by the FCC.

4.7.2 Radio Frequency Interference. On planning a new radiolink system or on adding RF carriers to an existing installation, RF interference of the existing (or planned) emitters in the area must be carefully considered. Usually the governmental authorizing agency has information on these and their stated radiation limits. Limits are established by national authorities. Equally important is antenna directivity and side-lobe radiation. Not only must the radiation of other emitters be examined from this point of view but also the capability of the planned antenna to reject unwanted signals. The radiation pattern of all licensed emitters should be known. The side-lobe level should be converted in the direction of the planned installation to EIRP in dBW. This should be done for all interference candidates within interference frequency range. For each emitter's EIRP, a path loss should be run to the planned installation to determine interference. Such a study could well affect a frequency plan or antenna design.

Nonlicensed emitters should also be considered. Many such emitters may be classified as industrial noise sources, such as heating devices, electronic ovens, electric motors, and unwanted radiation from privately owned and other microwave installations (i.e., radar harmonics). In the 6-GHz band a coordination contour should be carried out to verify interference from earth stations (see CCIR Rep. 448 and Rec. 359). For general discussion on the techniques for calculating interference noise in radiolink systems, see CCIR Rep. 388, p. 1.

4.7.3 Overshoot. Overshoot interference may occur when radiolink hops in tandem are in a straight line. Consider stations A, B, C, and D in a straight line or that a straight line on a map drawn between A and C also passes through B and D. Link A–B has frequency F_1 from A to B, and F_1 is reused in direction C–D. Care must be taken that some of the emission F_1 on the A–B hop does not spill into the receiver at D. Reuse may even occur on an A, B, and C combination, so F_1 at A–B may spill into a receiver at C tuned to F_1. This can be avoided if stations are removed from the straight line. In this case the station at B should be moved to the north of a line A–C, for example.

4.7.4 Transmit–Receive Separation. If a transmitter and receiver are operated on the same frequency at a radiolink station, the loss between them must be at least 120 dB. One way to ensure the 120-dB figure is to place all "go" channels in one-half of an assigned band and all "return" channels in the other. The terms "go" and "return" are used to distinguish the two directions of transmission.

4.7.5 Basis of Frequency Assignment. "Go" and "return" channels are assigned as in the preceding section. For adjacent RF channels in the same half of the band, horizontal and vertical polarizations are used alternately; thus we may assign, as an example, horizontal polarization (H) to the odd-numbered channels in both directions on a given section and vertical polarization (V) to the even-numbered channels. The order of isolation between polarizations is of the order of 35 dB.

5 COAXIAL CABLE TRANSMISSION LINKS

5.1 Introduction

A coaxial cable is simply a transmission line consisting of an unbalanced pair made up of an inner conductor surrounded by a grounded outer conductor, which is held in a concentric configuration by a dielectric. The dielectric can be of many different types, such as solid "poly" (polyethylene or polyvinyl chloride), foam, Spirafil, air, or gas. In the case of air–gas dielectric, the center conductor is kept in place by spacers or disks.

Systems have been designed to use coaxial cable as a transmission medium with a capability of transmitting an FDM configuration ranging from 120 to 10,800 voice channels. Community antenna television (CATV) systems use single cables for transmitted bandwidths of the order of 300 MHz.

Frequency division multiplex was developed originally as a means to increase the voice-channel capacity of wire systems. At a later date the same techniques were applied to radio. Then for the 20 years after World War II radio systems became the primary means for transmitting long-haul toll-telephone traffic. Coaxial cable had been making a strong comeback in this area until the advent of fiber optics (Section 6).

One advantage of coaxial cable systems is reduced noise accumulation when compared to radiolinks. For point-to-point multichannel telephony the FDM line frequency (see Chapter 5) configuration can be applied directly to the cable without further modulation steps, as required in radiolinks, thus substantially reducing system noise. In most cases radiolinks will prove more economical than coaxial cable. Nevertheless, because of the congestion of centimetric radio wave (radiolink) systems, coaxial cable is a good alternative. Coaxial cable should be considered in lieu of radiolinks using the following general guidelines:

- In areas of heavy microwave (including radiolink) RF (radio frequency interference).
- On high-density routes where coaxial cable may be more economical than radiolinks. (Consider a system that will require 5000 or more circuits at the end of 10 years.)
- On long national or international backbone routes where the system designer is concerned with noise accumulation.

Coaxial cable systems may be attractive for the transmission of television or other video applications. Some activity has been noted in the joint use of TV and FDM telephone channels on the same conductor. Another advantage in some circumstances is that system maintenance costs may prove to be less than for equal-capacity radiolinks.

One deterrent to the implementation of coaxial cable systems, as with any cable installation, is the problem of getting right-of-way for installation, and

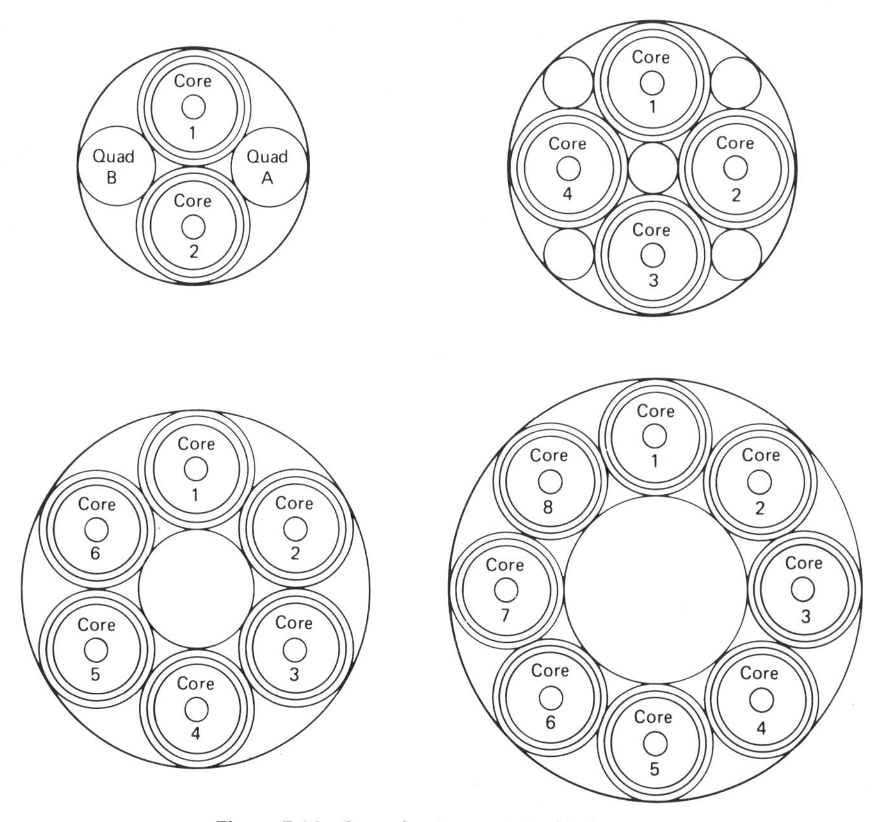

Figure 7.14 Some basic coaxial cable lay-ups.

its subsequent maintenance (gaining access), particularly in urban areas. Another consideration is the possibility of damage to the cable once it is installed. Construction crews may unintentionally dig up or cut the cable.

5.2 Basic Construction Design

Each coaxial line is called a "tube." A pair of these tubes is required for full-duplex long-haul application. One exception is the CCITT small-bore coaxial cable system where 120 voice channels, both "go" and "return," are accommodated in one tube. For long-haul systems, more than one tube is carried in a sheath. In the same sheath filler pairs or quads are included, sometimes placed in the interstices, depending on the size and lay-up of the cable. The pairs and quads are used for order wire and control purposes as well as for local communication. Some typical cable lay-ups are shown in Figure 7.14. Coaxial cable is usually placed at a depth of 90–120 cm, depending on frost penetration, along the right-of-way. Tractor-drawn trenchers or plows normally are used to open the ditch where the cable is placed, using fully automated procedures.

Cable repeaters are spaced uniformly along the route. Secondary or "dependent" repeaters are often buried. Primary power feeding or "main" repeaters are installed in surface housing. Cable lengths are factory cut so that the splice occurs right at repeater locations.

5.3 Cable Characteristics

For long-haul transmission standard cable sizes are as follows:

Inches	Millimeters
0.047/0.174	1.2/4.4 (small diameter)
0.104/0.375	2.5/9.5

The fractions express the outside diameter of the inner conductor over the inside diameter of the outer conductor. For instance, for the large-bore cable the outside diameter of the inner conductor is 0.104 in. and the inside diameter of the outer conductor is 0.375 in. This is shown in Figure 7.15. As can be seen from equations (7.21) and (7.22), the ratio of the diameters of the inner and outer conductors has an important bearing on attenuation. If we can achieve a

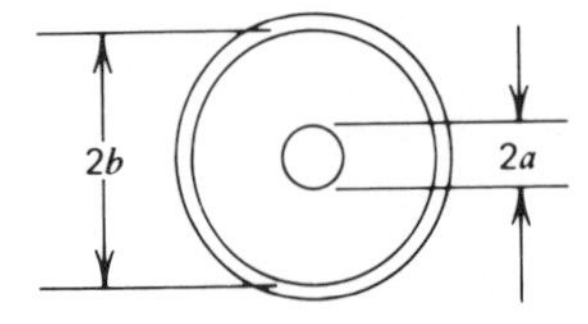

Figure 7.15 Basic electrical characteristics of coaxial cable. See equations (7.21) and (7.22).

ratio of $(b/a) = 3.6$, a minimum attenuation per unit length will result.

ε is the dielectric constant and
for air dielectric cable pair, $\varepsilon = 1.0$
Outside diameter of inner conductor $= 2a$
Inside diameter of outer conductor $= 2b$
Attenuation constant dB/mi,

$$\alpha = 2.12 \times 10^{-5} \frac{\sqrt{f}\,[(1/a) + (1/b)]}{\log b/a} \tag{7.21}$$

where a is the radius of inner conductor and b is the radius of outer conductor. Characteristic impedances (Ω) are

$$Z = \left(\frac{138}{\sqrt{\varepsilon}}\right) \log \frac{b}{a} = 138 \log \frac{b}{a} \text{ in air} \tag{7.22}$$

The characteristic impedance of coaxial cable is $Z_0 = 138 \log(b/a)$ for an air dielectric. If $b/a = 3.6$, then $Z_0 = 77\ \Omega$. Using a dielectric other than air reduces the characteristic impedance. If we use the disks mentioned in Section 5.1 to support the center conductor, the impedance lowers to 75 Ω.

Figure 7.16 is a curve giving attenuation per unit length in decibels versus frequency for the two most common types of coaxial cable discussed in this chapter. Attenuation increases rapidly as a function of frequency and is a function of the square root of frequency, as shown in equation (7.21). The

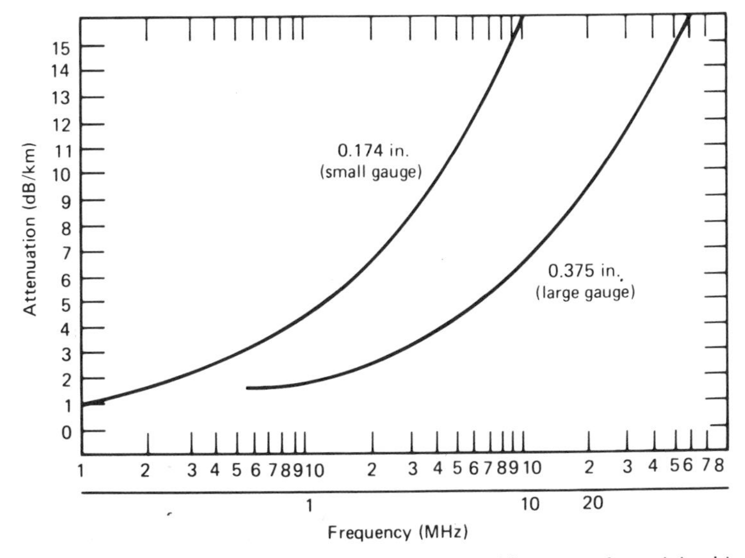

Figure 7.16 Attenuation–frequency response per kilometer of coaxial cable.

transmission system engineer is basically interested in how much bandwidth is available to transmit an FDM line-frequency configuration. For instance, the 0.375-in. cable has an attenuation of about 5.8 dB/mi at 2.5 MHz and the 0.174-in. cable, 12.8 dB/mi. At 5 MHz the 0.174-in. cable has about 19 dB/mi and the 0.375-in. cable, 10 dB/mi. Attenuation is specified for the highest frequency of interest.

Coaxial cable can transmit signals down to dc, but in practice, frequencies below 60 kHz are not used because of difficulties of equalization and shielding. Some engineers raise the lower limit to 312 kHz. The high-frequency limit of the system is a function of the type and spacing of repeaters as well as cable dimensions and the dielectric constant of the insulating material. It will be appreciated from equation (7.21) that the gain-frequency characteristics of the cable follow a root-frequency law, and equalization and "preemphasis" should be designed accordingly.

5.4 System Design

Figure 7.17 is a simplified application diagram of a coaxial cable system in long-haul point-to-point multichannel telephone service. To summarize system operation, an FDM line frequency (Chapter 5) is applied to the coaxial cable system via a line-terminal unit. Dependent repeaters are spaced uniformly along the length of the cable system. These repeaters are fed power from the cable itself. In the ITT design [14] the dependent repeater has a plug-in automatic level-control unit. In temperate zones where cable is laid to sufficient depth and where diurnal and seasonal temperature variations are "normal" (a seasonal swing of $\pm 10°C$), a plug-in level-control (regulating) unit is incorporated in every fourth dependent amplifier (see Figure 7.18). We use the

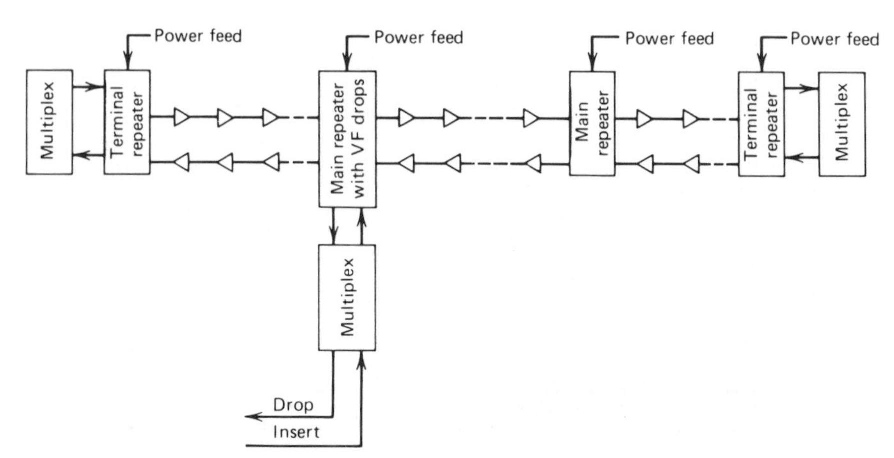

Figure 7.17 Simplified application diagram of a long-haul coaxial cable system for multichannel telephony.

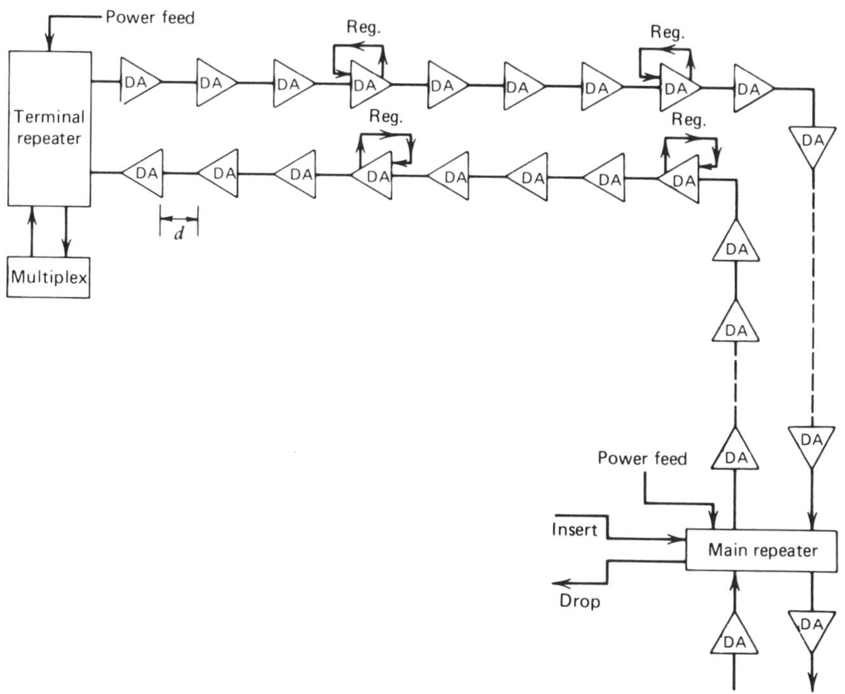

Figure 7.18 Detail of application diagram: DA, dependent amplifier (repeater); Reg, regulation circuitry; *d*, distance between repeaters.

word "dependent" for the dependent repeater (DA in Figure 7.18) because it depends on a terminal or main repeater for power and also provides fault information to the terminal or main repeater.

Let us examine Figures 7.17 and 7.18 at length. Assume we are dealing with a nominal 12-MHz system on a 0.375-in. (9.5-mm) cable. Up to 2700 voice channels can be transmitted. To accomplish this, two tubes are required, one in each direction. Most lay-ups, as shown in Figure 7.14, have more than two tubes. Consider Figure 7.17 from left to right. Voice channels in a four-wire configuration connect with the multiplex equipment in both the "go" and "return" directions. The output of the multiplex equipment is the line frequency (baseband) to be fed to the cable. Various line-frequency configurations are shown in Figures 5.12 and 5.13. The line signal is fed to the terminal repeater, which performs the following functions:

- Combines the line-control pilots with the multiplex line frequency.
- Provides "preemphasis" to the transmitted signal, distorting the output signal such that the higher frequencies get more gain than the lower frequencies (see Figure 7.16).

- Equalizes the incoming wide-band signal.
- Feeds power to dependent repeaters.

The output of the terminal repeater is a preemphasized signal with required pilots along with power feed. In the ITT design this is a dc voltage up to 650 V with a stabilized current of 110 mA. A main (terminal) repeater feeds, in this design, up to 15 dependent repeaters in each direction. Thus a maximum of 30 dependent repeaters appear in a chain for every main or terminal repeater. Other functions of a main repeater are to equalize the wide-band signal and to provide access for drop and insert of telephone channels by means of through-group filters. Figure 7.18 is an enlargement of a section of Figure 7.17 showing each fourth repeater with its automatic level regulation circuitry. Distance *d* between "DA" (dependent amplifier) repeaters is 4.5 km or 2.8 mi for a nominal 12-MHz system (0.375-in. cable). Amplifiers have gain adjustments of ± 6 dB, equivalent to varying repeater spacing ± 570 m.

As can be seen from the preceding discussion, the design of coaxial cable systems for both long-haul multichannel telephone service and CATV systems has become, to a degree, a "cookbook" design. Basically, system design involves the following:

- Repeater spacing as a function of cable type and bandwidth.
- Regulation of signal level.
- Temperature effects on regulation.
- Equalization.
- Cable impedance irregularities.
- Fault location or the so-called supervision.
- Power feed.

Other factors are, of course, right-of-way for the cable route with access for maintenance and the laying of the cable. With these factors in mind, consult Tables 7.2 and 7.3, which review the basic parameters of the CCITT approach (Table 7.2) and the Bell System approach (Table 7.3) to standard coaxial cable systems. For the 0.375-in. coaxial cable systems, practical noise accumulation is less than 1 pWp/km, whereas radiolinks allocate 3 pWp/km. These are good guideline numbers to remember for gross system considerations. Noise in coaxial cable systems derives from the active devices in the line (e.g., the repeaters) as well as the terminal equipment, both line conditioning and multiplex. Noise design of these devices is a trade-off between thermal and intermodulation (IM) noise. Intermodulation noise is the principal limiting parameter forcing the designer to install more repeaters per unit length with less gain per repeater.

Refer to Chapter 5 for CCITT-recommended FDM line-frequency configurations, in particular Figures 5.12 and 5.13, which are valid for 12-MHz systems.

TABLE 7.2 Characteristics of CCITT Specified Coaxial Cable[a] Systems (Large-Diameter Cable)

Item	Nominal Top Modulation Frequency				
	2.6 MHz	4 MHz	6 MHz	12 MHz	60 MHz
CCITT Rec.	G.337A	G.338	G.337B	G.332	G.333
Repeater type	Tube	Tube	Tube	Transistor	Transistor
Video capability	No	Yes	Yes	Yes	Not stated
Video + FDM capability	No	No	No	Yes	Not stated
Nominal repeater spacing	6 mi (9 km)	6 mi (9 km)	6 mi (9 km)	3 mi (4.5 km)	1 mi (1.55 km)
Main line reg. pilot (kHz)	2604	4092	See CCITT Rec. J.72	12,435	12,435/4287
Auxiliary reg. pilot(s) (kHz)		308 and 60	See Rec. J.72	4287 and 308	61,160, 40,920, and 22,372

[a]Cable type for all systems, 0.104/0.375 in. = 2.6/9.5 mm.

TABLE 7.3 Characteristics of "L" Coaxial Cable Systems

Item	"L" System Identifier[a]			
	L_1	L_3	L_4	L_5
Maximum design line length (mi)	4000	4000	4000	4000
Number of 4-kHz FDM VF channels[b]	600	1860	3600	10,800[c]
TV NTSC	Yes	Yes plus 600 VF	No	Not stated
Line frequency (kHz)	60–2788	312–8284	564–17,548	1590–68,780
Nominal repeater spacing (mi)	8	4	2	1
Power feed points	160 mi or every 20 rptrs	160 mi or 42 rptrs	160 mi or every 80 rptrs	75 mi or every 75 rptrs

[a]Cable type of all "L" systems, 0.375 in.
[b]Number of VF channels expressed per pair of tubes, one tube "go" and one tube "return."
[c]L5E ~ 13,200 VF channels.

5.5 Repeater Considerations: Gain and Noise

Consider a coaxial cable system 100 km long using 0.375-in. cable capable of transmitting up to 2700 voice (VF) channels in an FDM configuration, in this case a 12-MHz system. At the highest modulation frequency, 12 MHz, the cable attenuation per kilometer is approximately 8.3 dB (from Figure 7.16). The total loss at 12 MHz for the 100-km cable section is $8.3 \times 100 = 830$ dB. Thus one approach the system design engineer might take would be to install a

830-dB amplifier at the front end of the 100-km section. This approach is rejected out of hand.

Another approach would be to install a 415-dB amplifier at the front end and another at the 50-km point. Suppose the signal level was −15 dBm composite at the originating end. Thus −15 dBm + 415 dB = +400 dBm or +370 dBW. Remember that +60 dBW is equivalent to a megawatt; otherwise, we would have an amplifier with an output of 10^{37} W or 10^{31} MW. Still other approaches would be to have 10 amplifiers with 83-dB gain, each spaced at 10-km intervals, or to install 20 amplifiers or (830/20) = 41.5 dB each, or to install 30 amplifiers at (830/30) = 27.67 dB, each spaced at 3.33-km intervals. As we see later, the latter approach begins to reach an optimum from a noise standpoint, keeping in mind that the upper limit for noise accumulation is 3 pWp/km. The gain most usually encountered in coaxial cable amplifiers is 30–35 dB.

If we remain with the 3-pWp/km criterion, in nearly all cases radiolinks will be installed because of their economic advantage. Assuming 10 full-duplex RF channels per radio system at 1800 VF channels per RF channel, the radiolink can transmit 18,000 full-duplex channels, which is probably more economical on an installed cost basis. On the other hand, if we can show noise accumulation less on coaxial cable systems, these systems will prove in at some number of channels less than 18,000 if the reduced cumulative noise is included as an economic factor. There are other considerations, such as maintenance and reliability, but let us discuss noise further. In the basic design of coaxial cable repeaters noise consists of two major components, thermal noise (white noise) and intermodulation (IM) noise. To reach a goal of 1 pWp/km of noise accumulation, coaxial cable amplifier design must walk a "tightrope" between thermal and IM noise. It is also very sensitive to overload, with its consequent impact on intermodulation noise. For a deeper analysis of repeater design, the reader should consult Refs. 3 and 4.

5.6 Powering the System

Power feeding of buried repeaters in a typical coaxial cable link such as the ITT system permits the operation of 15 dependent repeaters from each end of a feed point (12-MHz cable). Thus up to 30 dependent repeaters can be supplied power between power feed points. A power feed unit at the power feed point provides up to 650 V dc between center conductor and ground using 110-mA stabilized direct current. Power feed points may be spaced as far apart as 140 km (87 mi) on large-diameter cable.

5.7 60-MHz Coaxial Cable Systems

Wide-band coaxial cable systems have been implemented because of the ever-increasing demand for long-haul toll-quality telephone channels. Such systems are designed to carry 10,800 nominal 4-kHz FDM channels per pair of

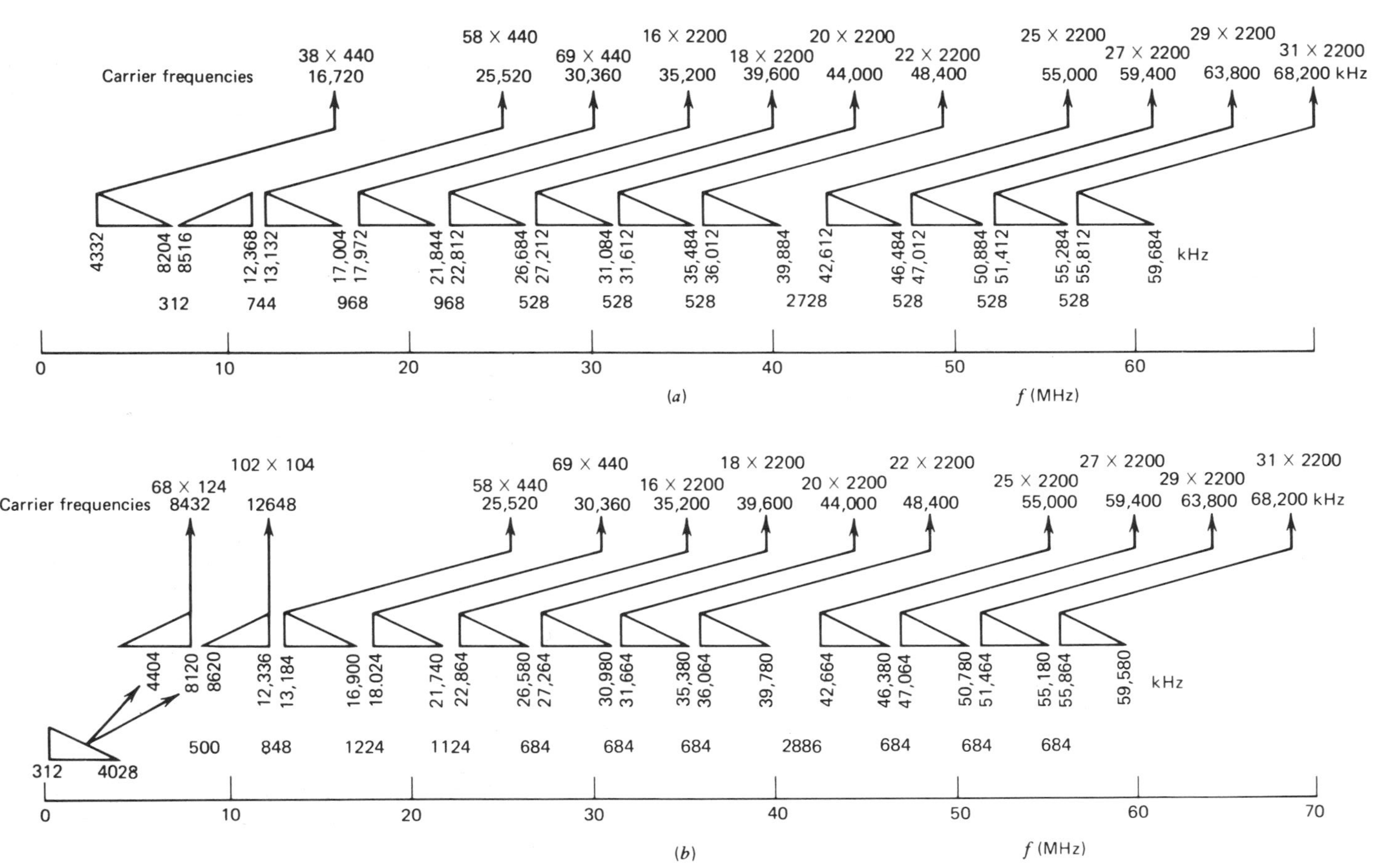

Figure 7.19 Line-frequency allocation for 40-MHz and 60-MHz systems on 2.6/9.5 mm coaxial cable pairs using (A) plan 1 and (B) plan 2 from CCITT Rec. G.333. Courtesy of the International Telecommunication Union–CCITT.

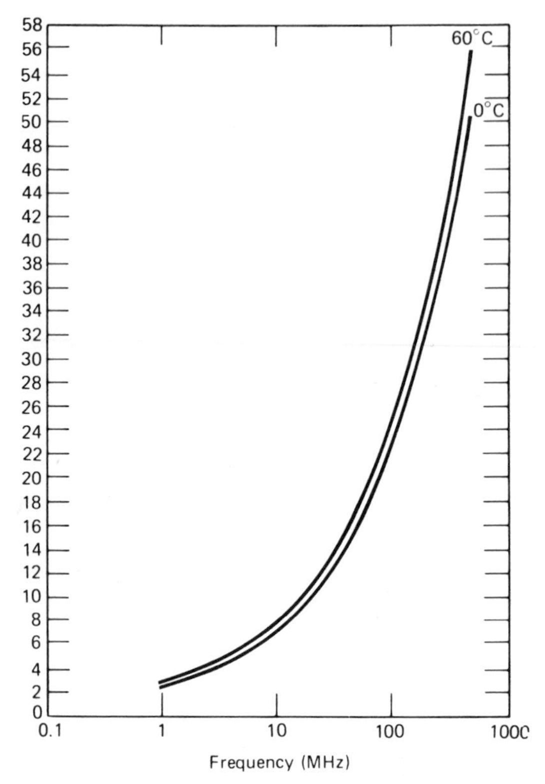

Figure 7.20 Attenuation of large coaxial cable (0.375 in.).

tubes. The line-frequency configuration for such a system, as recommended in CCITT Rec. G.333, is shown in Figure 7.19. For long-haul noise objectives, the large-diameter cable is recommended (e.g., 2.6 mm/9.5 mm).

When expanding a coaxial cable system, a desirable objective is to use the same repeater locations as with the old cable and to add additional repeaters at intervening locations. For instance, if we have 4.5-km spacing for a 12-MHz system and our design shows that we need three times the number of repeaters for an equal-length 60-MHz system, repeater spacing should be at 1.5-km intervals.

The ITT 12-MHz system uses 4.65-km spacing. Thus its 60-MHz system will use 1.55-km (0.95-mi) spacing with a mean cable temperature of 10°C. The attenuation characteristic of the large-gauge cable is shown in Figure 7.20, which is an extension of Figure 7.16. Repeater gain for the ITT system is nominally 28.5 dB at 60 MHz and can be varied ±1.5 dB. Line build-out networks allow still greater tolerance. The overload point, following CCITT Rec. 223, is taken at +20 dBm with a transmit level of −18 dBm. System pilot frequency is 61.160 MHz for regulation. A second pilot frequency of 4.287 MHz corrects the level of the lower-frequency range. Pilot regulation repeaters are installed at from 7 to 10 nonregulated repeaters, with deviation equalization at every 24th repeater. All repeaters have temperature control (controlled by the buried ambient). Power feeding is carried out at every 100 km (63 mi). Thus 64 repeaters will be fed remotely using constant direct current feed over the conductors. Each repeater will tap off about 15 V, requiring 2 W. Thirty-two repeaters at 2 × 15 V each will require 960 V. An additional 120-V dc is required for pilot-regulated repeaters plus one repeater with deviation equalization. Added to this is the 50-V IR drop on the cable. The total feed voltage adds to 1226 V dc. Fault location is similar to that for the 12-MHz ITT system.

5.8 L5 Coaxial Cable Transmission System

A good example of a 60-MHz coaxial cable transmission system that is presently operational and carrying traffic is the L5 system operating on a transcontinental route in North America (see Table 7.3). In its present lay-up it consists of 22 tubes, of which 20 are on line and 2 are spare. Each tube has the capacity to transmit 10,800* VF channels in one direction. For full-duplex operation two tubes are required for 10,800 VF channels, or the total system capacity is $[(22 - 2)/2] \times 10{,}800$, or 108,000 VF channels. The system is designed for a 40-dBrnC0 (8000-pWp) noise objective in the worst VF channel at the end of a 4000-mi (6400-km) system. This is a noise accumulation of 8000/6400 or 1.25 pWp/km. The system design is such that it is second-order intermodulation and thermal noise limited. Repeater overload is at the 24-dBm point.

The modulation plan is an extension of that shown in Figure 5.14. The basis of the plan is the development of the "jumbo group" (JG) made up of six mastergroups (Bell System FDM hierarchy). Keep in mind that the basic mastergroup consists of 600 voice channels (in this case) or 10 standard supergroups and occupies the band 564–3084 kHz. The basic jumbo group occupies the band 564–17,548 kHz with a level-control pilot at 5888 kHz. The three jumbo groups are assigned the following line frequencies:

JG1	3,124 to 20,108 kHz
JG2	22,068 to 39,056 kHz
JG3	43,572 to 60,556 kHz

Equalizing pilots are at 2976 kHz, 20,992 kHz, and 66,048 kHz as transmitted to the line. There is a temperature pilot at 42,880 kHz. The basic JG frequency generator is built around an oscillator that has an output of 5.12 MHz. This oscillator has a drift rate of less than 1 part in 10^{10}/day after aging and a short-term stability of better than 1 part in 10^{8}/ms. Excessive frequency offset is indicated by an alarm. Automatic protection of the 10 operating systems is

*The ATT L5E system has a channel capacity of 13,200 voice channels.

afforded by the LPSS (line-protection switching system) on a 1 : 10 basis. A maximum length of switching span is 150 mi. Power feeds are at 150-mi intervals, feeding power in both directions. Thus a power span is 75 mi long, or has 75 repeaters. Power is 910 mA on each cable, +1150 V and −1150 V operating against ground. The basic repeater is a fixed-gain amplifier, spaced at 1-mi intervals. Typically, every fifth repeater is a regulating repeater, and this regulation is primarily for temperature compensation.

6 FIBER OPTICS COMMUNICATION LINKS

6.1 Application

Fiber optics as a transmission medium has a comparatively unlimited bandwidth. It has excellent attenuation properties, as low as 0.25 dB/km. A major advantage fiber has when compared to coaxial cable is that no equalization is necessary. Also, repeater separation is of the order of 10 to 100 times that of coaxial cable for equal transmission bandwidths. Other advantages are:

- Electromagnetic immunity.
- Ground loop elimination.
- Security.
- Small size and lightweight.
- Expansion capabilities requiring change out of electronics only, in most cases.

Fiber has analog transmission application, particularly for video/TV. However, for this discussion we will be considering only digital applications, principally as a PCM highway or "bearer."

Fiber optics transmission is used for links under 1 ft in length all the way up to and including transoceanic undersea cable. In fact, all transoceanic cables presently being installed and planned for the future are based on fiber optics.

Fiber optics technology was developed by physicists, and, following the convention of optics, wavelength rather than frequency is used to denote the position of light emission in the electromagnetic spectrum. The fiber optics of today uses three wavelength bands: around 800 nm, 1300 nm, and 1600 nm or near-visible infrared. This is shown in Figure 7.21.

This section includes an overview of how a fiber optics link works, including types of fiber, a discussion of sources, detectors, and connectors and splices. More detail is provided in Chapter 9, Section 8.4, regarding digital transmission over fiber.

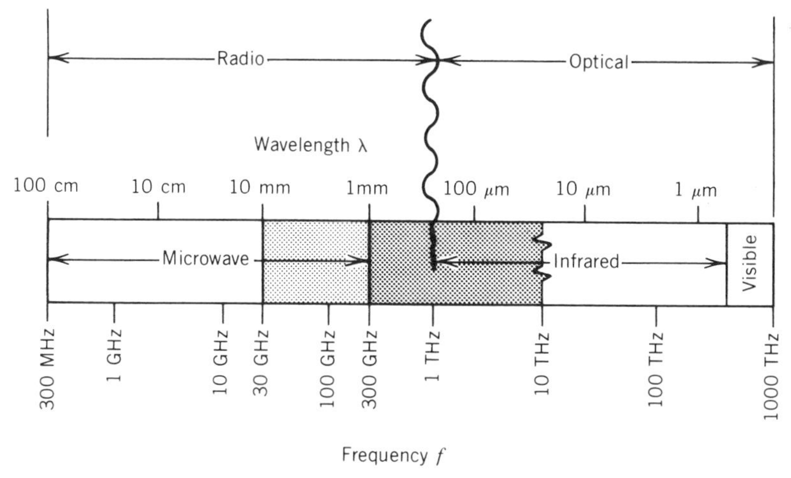

Figure 7.21 Frequency spectrum above 300 MHz.

6.2 Introduction to Optical Fiber as a Transmission Medium

The practical propagation of light through an optical fiber may best be explained using ray theory and Snell's law. Simply stated, we can say that when light passes from a medium of higher refractive index (n_1) into a medium of lower refractive index (n_2), the refracted ray is bent away from the normal. For instance, a ray traveling in water and passing into an air region is bent away from the normal to the interface between the two regions. As the angle of incidence becomes more oblique, the refracted ray is bent more until finally the refracted ray emerges at an angle of 90° with respect to the normal and just grazes the surface. Figure 7.22 shows various incidence angles. Figure 7.22b illustrates what is called the critical angle, where the refracted ray just grazes the surface. Figure 7.22c is an example of total reflection. This occurs when the angle of incidence exceeds the critical angle. A glass fiber, when utilized as a medium for transmission of light, requires total internal reflection.

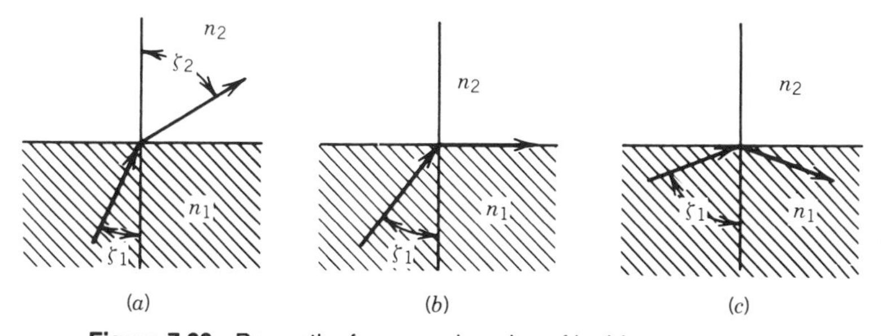

Figure 7.22 Ray paths for several angles of incidence, $n_1 > n_2$.

Another property of the fiber for a given wavelength λ is the normalized frequency V; then

$$V = \frac{2\pi a}{\lambda}\sqrt{n_1^2 - n_2^2} \tag{7.23}$$

where a is the core radius and n_2 for unclad fiber equals 1. In equation (7.23) the term $\sqrt{n_1^2 - n_2^2}$ is called the numerical aperture. In essence the numerical aperture is used to describe the light-gathering ability of fiber. In fact, the amount of optical power accepted by a fiber varies as the square of its numerical aperture. It is also interesting to note that the numerical aperture is independent of any physical dimension of the fiber.

As shown in Figure 7.23, there are three basic elements in an optical fiber transmission system: the light source, the fiber link, and the optical detector. Regarding the fiber link, there are two basic design parameters that can limit the length of such a link without resorting to repeaters or that can limit the distance between repeaters. These parameters are loss, usually expressed in decibels per kilometer, and dispersion, usually expressed as bandwidth per unit length or megahertz per kilometer. A particular link may be power limited or dispersion limited.

Dispersion, manifesting itself in intersymbol interference at the far end, is brought about by two factors, material dispersion and modal dispersion. Material dispersion is the change in the refractive index of the material with frequency. If the fiber waveguide supports several modes, we have modal dispersion. Since different modes have different phase and group velocities, energy in the respective modes arrives at the detector at different times. Consider that most optical sources excite many modes, and if these modes propagate down the fiber waveguide, delay distortion (dispersion) will result.

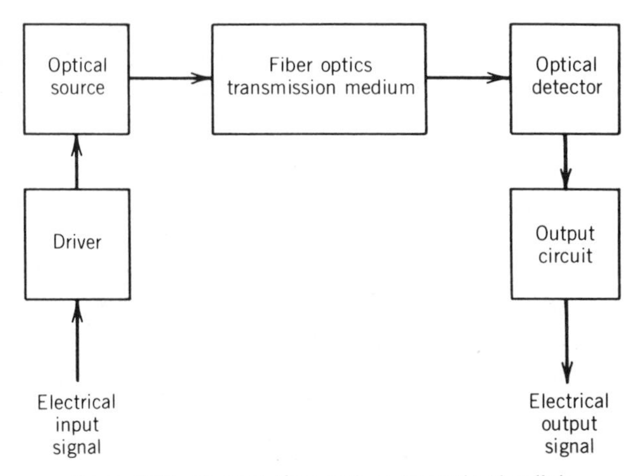

Figure 7.23 Typical fiber optic communication link.

The degree of distortion depends on the amount of energy in the various modes at the detector input.

One way of limiting the number of propagating modes in the fiber is in the design and construction of the waveguide itself. Return again to equation (7.23). The modes propagated can be limited by increasing the radius a and keeping the ratio n_1/n_2 as small as practical, often 1.01 or less.

We can approximate the number of modes N that a fiber can support by applying equation (7.23). If $V = 2.405$, only one mode will propagate (HE_{11}). If V is greater than 2.405, more than one mode will propagate, and when a reasonably large number of modes propagate,

$$N = (1/2)V^2 \tag{7.24}$$

6.3 Types of Fiber

There are three categories of optical fiber as distinguished by their modal and physical properties:

- Single mode.
- Step index (multimode).
- Graded index (multimode).

Single-mode fiber is designed such that only one mode is propagated. To do this, V must equal 2.405. Such a fiber exhibits no modal dispersion at all (theoretically). Typically we might encounter a fiber with indices of refraction of $n_1 = 1.48$ and $n_2 = 1.46$. If the optical source wavelength is 820 nm, for single-mode operation the maximum core diameter would be 2.6 μm, a very small diameter indeed. (*Note*: 1 micron = 1 micrometer.)

Step-index fiber is characterized by an abrupt change in refractive index, and graded-index fiber is characterized by a continuous and smooth change in refractive index (i.e., from n_1 to n_2). Figure 7.24 shows the fiber construction and refractive index profile for step-index fiber (Figure 7.24a) and graded-index fiber (Figure 7.24b).

Step-index fiber is more economical than graded-index fiber. For step-index fiber the distance–bandwidth product, the measure of dispersion discussed, is of the order of 10–150 MHz/km. With repeater spacings of the order of 12 km, only a few megahertz of bandwidth is possible.

Graded-index fiber is more expensive than step-index fiber, but it is one alternative for improved distance–bandwidth products. When a laser diode source is used, values of from 400 MHz/km to over 1000 MHz/km are possible. If an LED source is used with its much broader emission spectrum, distance–bandwidth products with graded-index fiber can be achieved up to about 400 MHz/km. In this case, material dispersion is what principally limits the usable bandwidth.

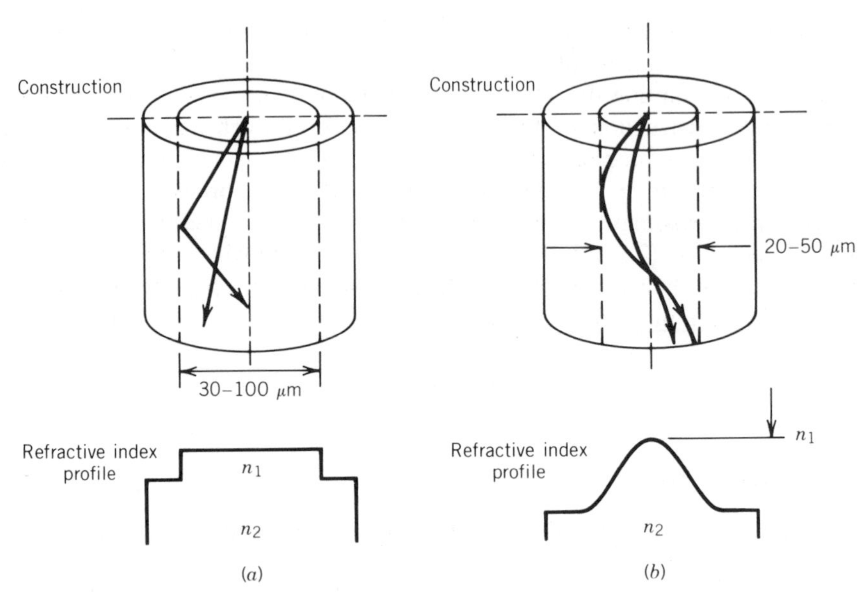

Figure 7.24 Construction and refractive index properties for (a) step-index fiber, (b) graded-index fiber.

There are two additional factors that the fiber optics communication system designer must take into account, minimum bending radius and fiber strength. Radiation losses at fiber waveguide bends are usually quite small and may be neglected in system design unless the bending radius is smaller than that specified by the fiber manufacturer. Minimum bending radii vary from about 2 cm to 10 cm, depending on the cable characteristics, or, as a rule of thumb, about 10 times the cable diameter. Fiber cable strength is also specified by the manufacturer. For example, one manufacturer for a specific cable type specifies a maximum pulling tension of 1780 N (400 lb), at 20°C, a maximum permissible compression load of 655 N/cm (375 lb/in.) flat plate, and a maximum permissible impact force of 280 N/cm (160 lb-in.).

Figure 7.25 shows a typical five-fiber cable for direct burial.

6.4 Splices and Connectors

Optical fiber cable is commonly available in 1-km sections; it is also available in longer sections, in some types up to 10 km or more. In any case there must be some way of connecting the fiber to the source and to the detector as well as connecting the reels of cable together, whether in 1 km or more lengths, as required. There are two methods of connection, splicing or using connectors. The objective in either case is to transfer as much light as possible through the coupling. A good splice couples more light than the best connectors.

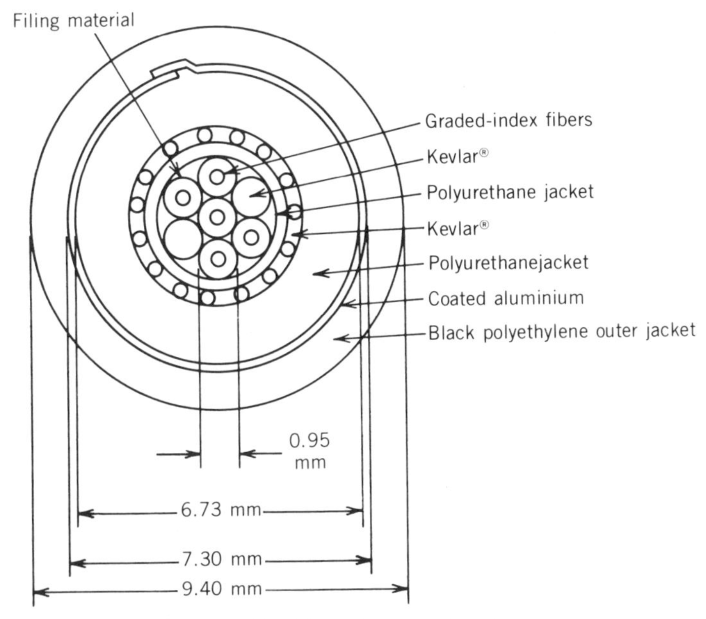

Figure 7.25 Direct-burial optical fiber.

A good splice can have an insertion loss as low as 0.2 dB, whereas connectors, depending on the type and how well they are installed, can have insertion losses as low as 0.7 dB and some as high as 2.5 dB or more.

An optical fiber splice requires highly accurate alignment and an excellent end finish to the fibers. There are three causes of loss at a splice:

1. Lateral displacement of fiber axes.
2. Fiber end separation.
3. Angular misalignment.

Splice loss also varies directly with the numerical aperture of the fiber in question.

There are two types of splice now available, the mechanical splice and the fusion splice. With a mechanical splice an optical matching substance is used to reduce splicing losses. The matching substance must have a refractive index close to the index of the fiber core. A cement with similar properties is also used, serving the dual purpose of refractive index matching and fiber bonding. The fusion splice, also called a hot splice, is where the fibers are fused together. The fibers to be spliced are butted together and heated with a flame or electric arc until softening and fusion occur. Fusion splices show 0.2–0.3 dB insertion loss.

Splices require special splicing equipment and trained technicians. Thus it can be seen that splices are generally hard to handle in a field environment such as a cable manhole. Connectors are much more amenable to field connecting. However, connectors are lossier and can be expensive. Repeated mating of a connector may also be a problem, particularly if dirt or dust deposits occur in the area where the fiber mating takes place.

6.5 Light Sources

A light source, perhaps more properly called a photon source, has the fundamental function in a fiber optics communication system to convert efficiently electrical energy (current) into optical energy (light) in a manner that permits the light output to be effectively launched into the optical fiber. The light signal so generated must also accurately track the input electrical signal so that noise and distortion are minimized.

The two most widely used light sources for fiber optics communication systems are etched-well surface light-emitting diode (LED) and injection laser diode (ILD). LEDs and ILDs are fabricated from the same basic semiconductor compounds and have similar heterojunction structures. They do differ considerably in their performance characteristics. LEDs are less efficient than ILDs but are considerably more economical. The spatial intensity distribution of an LED is Lambertian (cosine), whereas a laser exhibits a relatively high degree of waveguiding and thus for a given acceptance angle can couple more power into a fiber than the LED. In other words, the LED has a comparatively broad output spectrum and the ILD has a narrow spectrum, of the order of 1 or 2 nm wide. This is shown in Figure 7.26. Figure 7.26a shows the spectral distribution of an LED, and Figure 7.26b, that of an ILD.

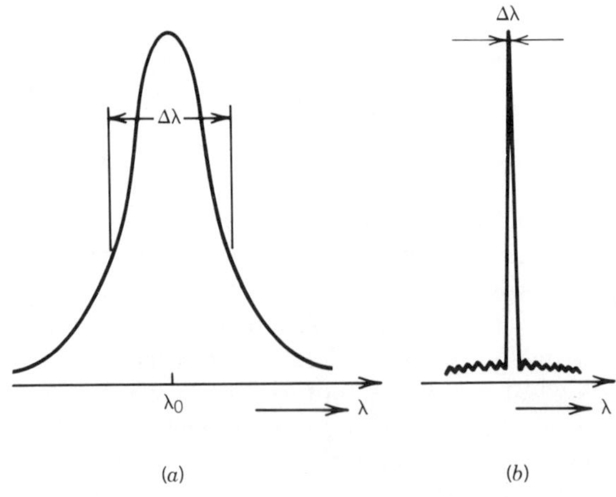

Figure 7.26 Spectral distribution of the emission from (a) an LED and (b) an ILD. λ-optical wavelength.

With present technology the LED is capable of launching about 100 μW (−10 dBm) or better of optical power into the core of a fiber with a numerical aperture of 0.2 or better. A laser diode with the same input power can couple up to 7 mW (+8.5 dBm) into the same cable. The coupling efficiency of an LED is of the order of 2%, whereas the coupling efficiency of an ILD is better than 50%, with one manufacturer reporting about 70%.

Methods of coupling a source into an optical fiber vary, as do coupling efficiencies. To avoid ambiguous specifications on source output powers, such powers should be stated at the "pigtail." A pigtail is a short piece of optical fiber coupled to the source at the factory and, as such, is an integral part of the source. Of course, the pigtail should be the same type of fiber as that specified for the link.

Component lifetimes for LEDs are of the order of 100,000 hr (MTBF) with up to a million hours reported in the literature. Many manufacturers guarantee an ILD for 20,000 hr and more. One-hundred fifty thousand hours are expected from ILDs after stressing and the culling of unstable units. Such ILDs are used in the TAT8 cable connecting the United States to Great Britain.

It should be noted that the life expectancy of an ILD is reduced when it is overdriven to derive more coupled output power (such as more than 7 mW).

The ILD is a temperature-dependent device. Its threshold current increases nonlinearly with temperature. Rather than attempt to control the device's temperature, a negative feedback circuit is used whereby a portion of the emitted light is sampled, detected, and fed back to control the drive current. Such circuits are similar to the familiar AGC circuits used in radio receivers.

Fiber optics communication systems operate in the nominal wavelength regions of 820 nm, 1330 nm, and 1550 nm. If we examine the attenuation versus wavelength curve in Figure 7.27, we see that for the 820-nm region about the best attenuation is 3 dB/km. As the wavelength increases, going to the right on the curve, we see another valley around 1330 nm, where the loss per unit length is 0.5 dB/km, and there is another valley at 1550 nm, where the loss is 0.25 dB/km or a little less. There is a comparatively mature technology available for all three wavelength regions.

6.6 Light Detectors

The most commonly used detectors (receivers) for fiber optics communication systems are photodiodes, either PIN or APD. The terminology *PIN* derives from the semiconductor construction of the device where an intrinsic (I) material is used between the p–n junction of the diode.

A photodiode can be considered a photon counter. The photon energy E is a function of frequency and is given by

$$E = hv \qquad (7.25)$$

where h is Planck's constant (W/s^2) and v is the frequency in hertz. E is measured in watt-seconds or kilowatt-hours.

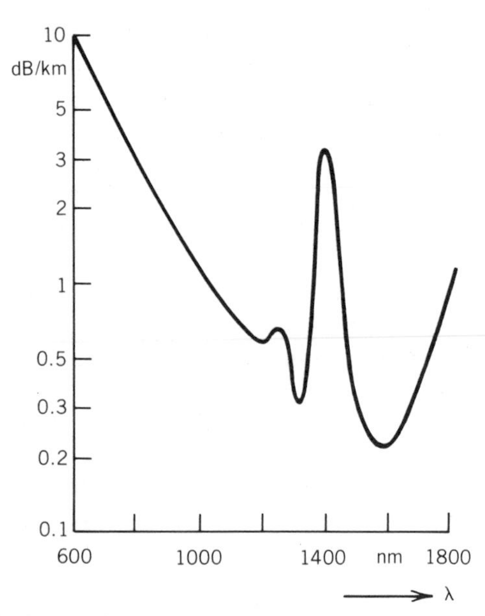

Figure 7.27 Attenuation per unit length versus wavelength of glass fiber.

The receiver power in the optical domain can be measured by counting, in quantum steps, the number of photons received by a detector per second. The power in watts may be derived by multiplying this count by the photon energy, as given in equation (7.25).

The efficiency of the optical-to-electrical power conversion is defined by a photodiode's *quantum efficiency*, which is the average number of electrons released by each incident photon. A highly efficient photodiode would have a quantum efficiency of 1, and decreasing from 1 indicates progressively poorer efficiencies. The quantum efficiency, in general, varies with wavelength and temperature.

For the fiber optic communication system engineer *responsivity* is a most important parameter when dealing with photodiode detectors. Responsivity is expressed in amperes per watt or volts per watt and is sometimes called *sensitivity*. Responsivity is the ratio of the root mean square (rms) value of the output current or voltage of a photodetector to the rms value of the incident optical power. In other words, responsivity is a measure of the amount of electrical power we can expect at the output of a photodiode, given a certain incident light power signal input. For a photodiode the responsivity R is related to the wavelength λ of the light flux and to the quantum efficiency η, the fraction of the incident photons that produce a hole-electron pair. Thus

$$R = \frac{\eta\lambda}{1234}(A/W) \tag{7.26}$$

with λ measured in nanometers.

TABLE 7.4 Receiver (APD) Sensitivities

Bit Rate (Mbps)*	Threshold (dBm)
1.5	−67
6	−64
12	−61
45	−54
90	−51
170	−48

For a BER = 1×10^{-9}

The avalanche photodiode (APD) is a gain device displaying gains of the order of 15–20 dB. The PIN diode is not a gain device. Table 7.4 summarizes APD sensitivities for the 850-nm region for the standard BER of 1×10^{-9} for bit rates in common use for the North American PCM industry [3]. Noise equivalent power (NEP) is commonly used as the figure of merit of a photodiode. NEP is defined as the rms value of optical power required to produce a unit signal-to-noise ratio (i.e., signal-to-noise ratio = 1) at the output of a light-detecting device. NEPs vary for specific diode detectors between 1×10^{-13} W/Hz$^{1/2}$ and 1×10^{-14} W/Hz$^{1/2}$.

Of the two types of photodiodes discussed here, the PIN is more economical and requires less complex circuitry than its APD counterpart. The PIN diode has peak responsivity from about 800 nm to 900 nm for silicon devices. These responsivities range from 300 μA/mW to 600 μA/mW. The overall response time for the PIN diode is good for about 90% of the transient but sluggish for the remaining 10%, which is a "tail." The power response of the tail portion of a pulse may limit the net bit rate on digital systems.

The PIN detector does not display gain, whereas the APD does. The response time of the APD is far better than that of the PIN diode, but the APD displays certain temperature instabilities where responsivity can change significantly with temperature. Compensation for temperature is usually required in APD detectors and is often accomplished by a feedback control of bias voltage. It should be noted that bias voltages for APDs are much higher than for PIN diodes, and some APDs require bias voltages as high as 200 V. Both the temperature problem and the high-voltage bias supply complicate repeater design.

6.7 Link Design

The design of a fiber optics communication link involves several steps. Certainly the first consideration is to determine the feasibility of such a transmission system for a desired application. There are two aspects of this

*We use "bps" for bits per second, "Mbps" for megabits per second and "kbps" for kilobits per second.

decision, economical and technical. Analog applications are for wide-band transmission of such information as video, particularly CATV trunks, studio-to-transmitter links, and multichannel FDM mastergroups. As mentioned previously, we stress digital applications for fiber optics transmission and in particular for:

- On-premises data bus; LANs.
- Higher-level PCM or CVSD configurations for telephone trunks.
- Radar data links.
- Conventional data links well in excess of 19.2 kbps.
- As a bus for an analog format such as in radio system IFs.
- Digital video.

The present trend continues for the cost of fiber cable and components to go down. Fiber optics repeaters are more costly than their metallic PCM counterparts, but much fewer repeaters are required for a given distance, and the powering of these repeaters is more involved, particularly if the power is to be taken off the cable itself. In this case a metallic pair (or pairs) must be included in the cable sheath if the power is to be provided from trunk terminal points. Another approach is to supply power locally at the repeater site with a floating battery backup.

6.7.1 Design Approach. The first step in designing a fiber optic communication system is to establish the basic input system parameters. Among these we would wish to know:

- Signal to be transmitted.
- Link length.
- Growth requirements (i.e., additional circuits, increased bit rates).
- Tolerable signal impairment levels stated as signal-to-noise ratio or BER at the output of the terminal detector.

Throughout the design procedure, when working with trade-offs, the designers must establish whether they are working in a power-limited domain or in a dispersion- (bandwidth-) limited domain. For instance, with low bit rates (bandwidth) such as T1 (DS1) carrier (1.544 Mbps), we would expect to be power limited in almost all circumstances. Once we get into higher bit rates, say, above 45 Mbps, the proposed system (if it uses multimode fiber) must be tested (by calculation) to determine if the system rise time is not bit rate limited. The designers then select the most economic alternatives and trade-offs among the following:

- Fiber parameters: single mode or multimode; if multimode, step index or graded index; number of fibers, cable makeup, and strength.
- Transmission wavelength: 820 nm, 1330 nm, or 1550 nm.
- Source type: LED or ILD.
- Detector type: PIN or APD.
- Repeaters, if required, and how they will be powered.
- Modulation type and waveform (code format, such as Manchester code), impact on design.

Permanently installed systems, such as with telephone companies and administrations, would probably use splices. Temporarily installed systems, such as military tactical systems, would tend to use connectors. The cost of fiber is dropping dramatically. At some point, fiber becomes the system cost driver, and on shorter links the electronics is the cost driver. Of course, for short links (such as on-premises data bus systems) low cost cable, possibly plastic, and low cost components such as LED/PIN combinations would be more cost effective.

When we build long systems, over about 5 km, we may have to resort to more expensive components, such as glass fiber, and over 10–20 km, ILD–APD combinations may have to be considered. Systems transmitting more than a gigabit are becoming more common with repeater separations from 20 km to 150 km, depending on the bit rate.

6.7.2 Loss Design. As a first step, let's assume that the system is power limited. Nearly all systems today can stay in the power-limited regime if monomode fiber is used with ILDs. That may be an easy way out and add extra burden financially on first cost and present value of annual charges. Multimode fiber is less expensive and may meet requirements, especially for the lower bit rates. This can hold true for comparatively long links. It is a high bandwidth product that drives the designer to monomode operation. The bandwidth product is measured in MHz/km and was discussed above. Another guideline is that an RZ waveform requires twice the bandwidth of an NRZ waveform. (See Chapter 8 for a discussion of RZ and NRZ waveforms.)

Link margin is another factor. Fiber optics links are overbuilt by a certain number of decibels to allow for component aging, temperature variation, and reel-to-reel loss variation in fiber. Margin often is a function of link length. Margins from 2 dB to 6 dB are common.

The system designer develops a power budget that is somewhat similar to a microwave link budget but certainly less complex. For a first-cut design, there are two source types, LED and ILD. LEDs display around −10 dBM output power; ILDs around 0 to +7 dBm. There are two types of detectors, PIN diodes and APDs. For long links the APD may be the choice. Again, for long links the longer wavelengths are much more attractive. However, for short links LEDs at 820 nm may well suffice and, as a bonus, provide considerably greater MTBF.

Consider the following detectors for a 1×10^{-9} BER threshold:

Detector Type	7 Mbps	45 Mbps
PIN (typical)	−47 dBm	−34 dBm
APD	−64 dBm	−54 dBm

For the sources, LED = −12 dBM; ILD = +3 dBM.

Taking these in pairs (i.e., source and detector), first with the LED source and a PIN detector at 7 Mbps, the total link budget loss is −47 dBm − (−12 dBm) = 35 dB. This 35 dB is allotted to:

- Fiber loss.
- Splice or connector loss.
- Margin.

By observation we see that the link must be short, and therefore the margin can be reduced to about 3 dB, leaving 32 dB for actual link losses. If we splice rather than use connectors, using a multimode cable at 4 dB/km, a 7-km link is possible assuming 1-km sections of cable. Using 3-dB/km graded-index cable, we could have a 9-km link. The trade-off is cost. If we used connectors at 1 dB per connector and 3 dB/km cable, we would be back at a 7-km cable again.

For 45 Mbps operation, which would support US DS3 operation (see Chapter 9), using an ILD source and APD detector, we would have 57 dB available for the link budget (i.e., the difference between −54 dBm and +3 dBm). By observation, we would be dealing with a longer link where a margin of 5 dB may be appropriate. Thus we are left with 52 dB to apportion over link losses. Assume again that splices are used with 0.25 dB per splice. The fiber is monomode at 1550 nm with 0.25-dB/km loss with 10-km reels. The link could be 188 km without repeaters or $(188 \times 0.25) + (20 \times 0.25)$, which is 52 dB. Remember that there is a splice at each end connecting to the pigtail, thus 19 segments with 20 splices. However, for ease of changing out electronics, connectors are often used on pigtails, those short lengths of fiber connecting the source and detector.

The reader should consult Ref. 3 for a discussion of dispersion-limited links.

7 INTRODUCTION TO EARTH STATION TECHNOLOGY

7.1 General Application

Satellite communication is another method of establishing a communication link. For network planning, satellite trunks for telephony may be economically the optimum for a variety of applications, including the following:

1. On international high-usage trunks country to country.
2. On national trunks, between switching nodes that are well separated in distance (i.e., > 200 miles) in highly developed countries. Again, the tendency is high usage to serve in addition to radiolink–coaxial cable–fiber optics.
3. In areas under development where satellite links replace HF and a high growth is expected to be eventually supplemented by radiolink and coaxial cable.
4. In sparsely populated, highly rural, "out-back" areas where it may be the only form of communication. Northern Canada and Alaska are good examples.
5. On final routes for overflow on a demand-assignment basis. Route length again is a major consideration.
6. In many cases, on international connections reducing such connections to one link.
7. On private and industrial networks.
8. On specialized common carriers.

7.2 Definition

A number of world bodies, including the CCIR and the US Federal Communications Commission (FCC), have now accepted the term "earth sation" as a radio facility located on the earth's surface that communicates with satellites. A "terrestrial station" is a radio facility on the earth's surface that communicates with other similar facilities on the earth's surface. Section 4 of this chapter dealt with one form of terrestrial stations. The term "earth station" as used today has come more to mean a radio station operating with other stations on the earth via an orbiting satellite relay.

Nearly all commercial communication satellites are geostationary. Such satellites orbit the earth in a 24-hr period. Thus they appear stationary over a particular geographic location on earth. For a 24-hr synchronous orbit the altitude of a geostationary satellite is 22,300 statute miles or 35,900 km above the earth's equator.

7.3 The Satellite

A communication satellite is an RF repeater, which may be represented in its most simple configuration as that shown in Figure 7.28. Theoretically, as mentioned earlier, three such satellites properly placed in equatorial synchronous orbit could provide communication from one earth station to any other located anywhere on the earth's surface (see Figure 7.29).

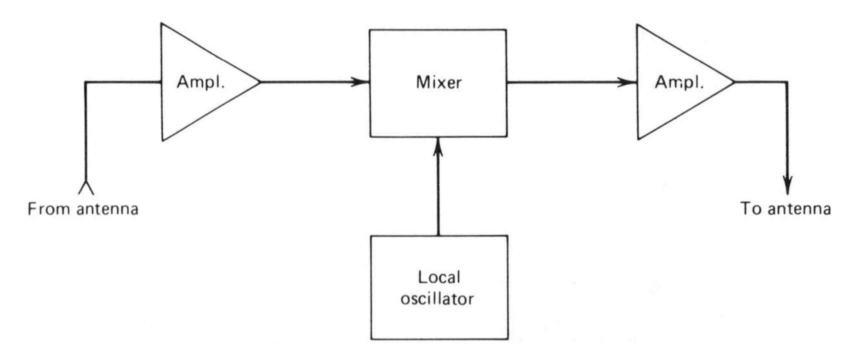

Figure 7.28 Simplified functional block diagram of radio relay portion of a typical communication satellite.

7.4 Three Basic Technical Problems

As the reader can appreciate, satellite communication is nothing more than radiolink (microwave line-of-sight) communication using one or two RF repeaters located at great distances from the terminal earth stations, as shown in Figure 7.29. Because of the distance involved, consider the slant range from earth antenna to satellite to be the same as the satellite altitude. This would be true if the antenna were pointing at zenith to the satellite. Distance increases as the pointing angle to the satellite decreases (elevation angles).

We thus are dealing with very long distances. The time required to traverse these distances, namely, earth station to satellite to another earth station, is of

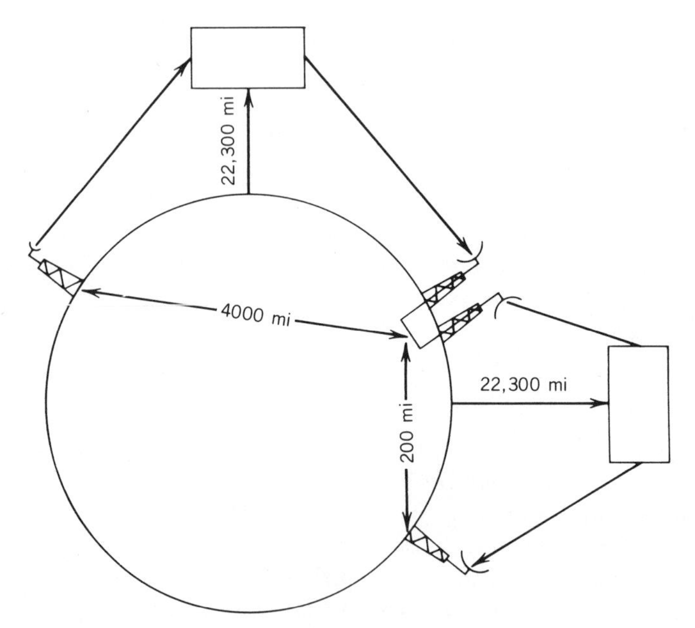

Figure 7.29 Distances involved in SatCom.

the order of 250 ms. Round-trip delay will be 2 × 250 or 500 ms. These propagation times are much greater than those encountered on conventional terrestrial systems. So one major problem is propagation time and resulting echo on telephone circuits. It influences certain data circuits in delay to reply for block or packet transmission systems and requires careful selection of telephone signaling systems, or call-setup time may become excessive.

Naturally, there are far greater losses. For radiolinks we encounter free-space losses possibly as high as 145 dB. In the case of a satellite with a range of 22,300 mi operating on 4.2 GHz, the free-space loss is 196 dB and at 6 GHz, 199 dB. At 14 GHz the loss is about 207 dB. This presents no insurmountable problem from earth to satellite, where comparatively high power transmitters and very high gain antennas may be used. On the contrary, from satellite to earth the link is power limited for two reasons: (1) in bands shared with terrestrial services such as the popular 4-GHz band to ensure noninterference with those services, and (2) in the satellite itself, which can derive power only from solar cells. It takes a great number of solar cells to produce the RF power necessary; thus the down-link, from satellite to earth, is critical, and received signal levels will be much lower than on comparative radiolinks, as low as −150 dBW. A third problem is crowding. The equatorial orbit is filling with geostationary satellites. Radio frequency interference from one satellite system to another is increasing. This is particularly true for systems employing smaller antennas at earth stations with their inherent wider beamwidths. It all boils down to a frequency congestion of emitters.

7.5 Frequency Bands: Desirable and Available

The most desirable frequency bands for commercial satellite communication are in the spectrum 1000–10,000 MHz. These bands are:

3700–4200 MHz (satellite-to-earth or down-link)
5925–6425 MHz (earth-to-satellite or up-link)
7250–7750 MHz* (down-link)
7900–8400 MHz* (up-link)

These bands are preferred by design engineers for the following primary reasons:

- Less atmospheric absorption than higher frequencies.
- Rainfall loss not a concern.
- Less noise, both galactic and man-made.
- A well-developed technology.
- Less free-space loss compared to the higher frequencies.

*These two bands are intended mainly for military application.

There are two factors contraindicating application of these bands and pushing for the use of higher frequencies:

- The bands are shared with terrestrial services.
- There is orbital crowding (discussed earlier).

Higher-frequency bands for commercial satellite service are:

11.7–12.2 GHz (down-link)
14.0–14.5 GHz (up-link)
17.7–20.2 GHz (down-link)
27.5–30.0 GHz (up-link)

Above 10-GHz rainfall attenuation and scattering and other moisture and gaseous absorption must be taken into account. The satellite link must meet a 10,000-pWp total VF channel noise criterion at least 99.9% of the time. One solution is a space-diversity scheme where we can be fairly well assured that one of the two antenna installations will not be seriously affected by the heavy rainfall cell affecting the other installation. Antenna separations of 1 km to 10 km are being studied. Another advantage with the higher frequencies is that requirements for down-link interference are less; thus satellites may radiate more power. This is often carried out on the satellite using spot-beam antennas rather than general-coverage antennas.

7.6 Multiple Access of a Satellite

Multiple access is defined as the ability of a number of earth stations to interconnect their respective communication links through a common satellite. Satellite access is classified (1) by assignment, whether quasipermanent or temporary, namely, (a) preassigned multiple access or (b) demand-assigned multiple access (DAMA), and (2) according to whether the assignment is in the frequency domain or the time domain, namely, (a) frequency-division multiple access (FDMA) or (b) time-division multiple access (TDMA). On comparatively heavy routes, ($\geq$ 10 erlangs) preassigned multiple access may become economical. Other factors, of course, must be considered, such as whether the earth station is "INTELSAT" standard as well as the space-segment charge that is levied for use of the satellite. In telephone terminology, "preassigned" means dedicated circuits. Demand-assigned multiple access is useful for low-traffic multipoint routes where it becomes interesting from an economic standpoint. Also, an earth station may resort to DAMA as a remedy to overflow for its FDMA circuits.

7.6.1 Frequency-Division Multiple Access. Historically, FDMA has the highest usage and application of the various access techniques. The several RF bands available (from Section 7.5) have a 500-MHz bandwidth. A satellite

TABLE 7.5 INTELSAT V, VA, and VI Global Beam: Voice-Channel Capacity versus Frequency Assignments (Partial listing) (Ref. 18)

Carrier capacity (number of voice channels)	24.0	60.0	96.0	132.0	252.0	432.0	792.0
Top baseband frequency (kHz)	108.0	252.0	408.0	552.0	1052.0	796.0	4028.0
Allocated satellite bandwidth (MHz)	1.25	2.5	5.0	5.0	7.5	10.0	25.0
Occupied bandwidth (MHz)	1.1	2.1	3.3	4.3	6.3	8.3	21.6

contains a number of transponders, each of which cover a frequency segment of the 500-MHz bandwidth. One method of segmenting the 500-MHz is by utilizing 12 transponders, each with a 36-MHz bandwidth. Sophisticated satellites such as INTELSAT V segment the 500-MHz available with transponders up to 77-MHz bandwidth at 4–6 GHz and at 11–14 GHz have one transponder with a 241-MHz bandwidth.

With FDMA operation, each earth station is assigned a segment or a portion of a segment. For a nominal 36-MHz transponder, 14 earth stations may access in an FDMA format, each with 24 voice channels in a standard CCITT modulation plan (Chapter 5). The INTELSAT VI assignments for a 36-MHz transponder are shown in Table 7.5, where it can be seen that when larger channel groups are used, fewer earth stations can access the same transponder.

Consider the following hypothetical example of how FDMA works. Frequency translation is via a 2225-MHz local oscillator in the satellite, and the difference mode is used in the mixer; for example:

Satellite Receive Frequency (MHz)		Mixer Frequency (MHz)		Satellite Transmit Frequency (MHz)
5925	−	2225	=	3700
6425	−	2225	=	4200
If a transponder had the 6262 – 6298-MHz segment, then:				
6262	−	2225	=	4037
6298	−	2225	=	4073

In the segment, assign three RF carriers from three locations in common view of an Atlantic satellite. Each carrier will be frequency modulated with two supergroups, as in Table 7.5. Subgroup A, the remaining 12 channels of the 132 VF channel total, are spare and are disregarded. Let us assign Spain

(Buitrago) the transponder subsegment 6262–6267 MHz; Etam, West Virginia (USA), 6272–6277 MHz; and Longoville, Chile, 6282–6287 MHz. These are the up-link frequencies. When converted in the satellite, they are as follows:

	Up-link (MHz)	Down-link (MHz)
Buitrago	6262–6267 MHz	4037–4042 MHz
Etam	6272–6277 MHz	4047–4052 MHz
Longoville	6282–6287 MHz	4057–4062 MHz

The United States would transmit one carrier at Etam to communicate with both Buitrago and Longoville. The carrier contains two supergroups, as shown in the following diagram.

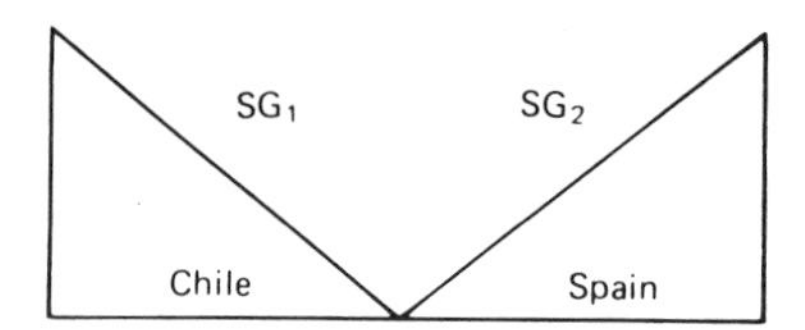

The United States would receive two carriers at Etam for the receive end, one from Buitrago and one from Longoville, shown in the following diagram, and would pick off supergroup 1 (as in our example) from the Chile carrier and supergroup 2 from the Spanish carrier. A similar exercise can be carried out for the Buitrago and Longoville situations. Note that one transmitter can access many locations, but to receive multiple locations, a receiver chain is required for each distant location to be received. If a single up-link carrier occupied an entire transponder (Table 7.5), it could serve (36/1.25) or 28 down-link separate locations, provided that each location required no more than 24 VF channels.

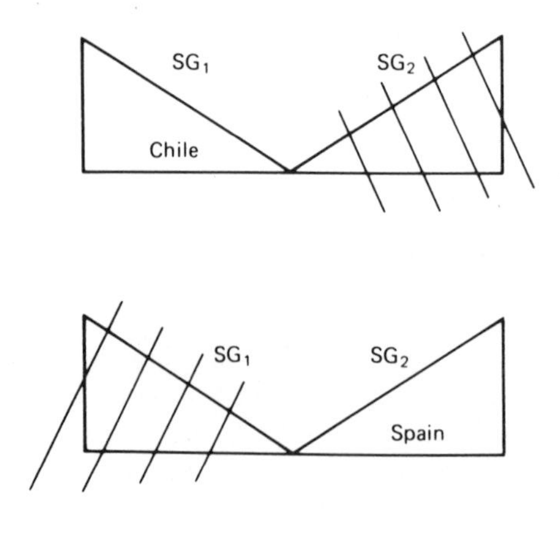

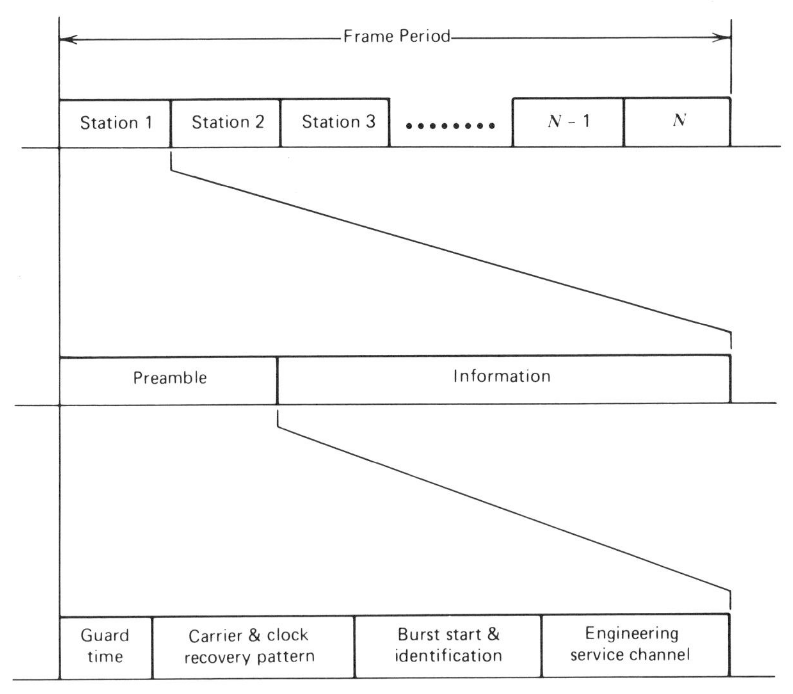

Figure 7.30 Example of TDMA burst format.

7.6.2 Time-Division Multiple Access. Time-division multiple access (TDMA) operates in the time domain. Use of the satellite transponder is on a time-sharing basis. Individual time slots are assigned to earth stations in a sequential order. Each earth station has full and exclusive use of the transponder bandwidth during its time-assigned segment. Depending on the bandwidth of the transponder, bit rates of 10–100 Mbps (megabits per second) are used.

With TDMA operation earth stations use digital modulation and transmit with bursts of information. The duration of a burst lasts for the time period of the slot assigned. Timing synchronization is a major problem.

A frame, in digital format, may be defined as a repeating cycle of events. It occurs in a time period containing a single digital burst from each earth station and the guard periods or guard times between each burst. A sample frame is shown in Figure 7.30 for earth stations 1, 2, and 3 to N. Typical frame periods are 750 μs for INTELSAT and 250 μs for the Canadian Telesat.

The reader will appreciate that timing is crucial to effective TDMA operation. The greater N becomes (i.e., the more stations operating in the frame period), the more clock timing affects the system. The secret lies in the "carrier and clock (timing) recovery pattern" as shown in Figure 7.30. One way to ensure that all stations synchronize to a master clock is to place a sync burst as the first element in the format frame. The INTELSAT V does just this. The burst carries 44 bits, starting with 30 bits carrier and bit timing recovery, 10 bits for the "unique word," and 4 bits for the station identification code.

Why use TDMA in the first place? It lies in a major detraction of FDMA. Satellites use traveling-wave tubes (TWTs) in their transmitter final amplifiers. A TWT has the undesirable property of nonlinearity in its input–output (I/O) characteristics when operated at full power. When there is more than one carrier accessing the transponder simultaneously, high levels of intermodulation products are produced, thus increasing noise and crosstalk. When a transponder is operated at full power output, such noise can be excessive and intolerable. Thus input must be backed off (i.e., level reduced) by ≥ 3 dB. This, of course, reduces the EIRP and results in reduced efficiency and reduced information capacity. Consequently, each earth station's up-link power must be carefully coordinated to ensure proper loading of the satellite. The complexity of the problem increases when a large number of earth stations access a transponder, each with varying traffic loads.

On the other hand, TDMA allows the transponder's TWT to operate at full power because only one earth-station carrier is providing input to the satellite transponder at any one instant.

To summarize, consider the following advantages and disadvantages of FDMA and TDMA. The major advantages of FDMA are:

- No network timing is required.
- Channel assignment is simple and straightforward.

The major disadvantages of FDMA are:

- Up-link power levels must be closely coordinated to obtain efficient use of transponder RF output power.
- Intermodulation difficulties require power back-off as the number of RF carriers increase with inherent loss of efficiency.

The major advantages of TDMA are:

- There is no power sharing and IM product problems do not occur.
- The system is flexible with respect to user differences in up-link EIRP and data rates.
- Accesses can be reconfigured for traffic load in almost real time.

The major disadvantages of TDMA are:

- Accurate network timing is required.
- There is some loss of throughput due to guard times and preambles.
- Large buffer storage may be required if frame lengths are long.

7.6.3 Demand-Assignment Multiple Access. The demand-assignment multiple access (DAMA) method has single voice channels allocated to an earth station on demand. A pool of idle channels is available, and assignments from the pool are made on request. When a call has been completed on the channel, the channel is returned to the idle pool for reassignment.

The DAMA method is analogous to a telephone switch. When a subscriber goes off hook, a line is seized; on dialing, a connection is made; and when the call is completed and there is an on hook condition, the voice path through the switch is returned to "idle" and is ready for use by another subscriber. There are three methods available for handling DAMA in a satellite system:

- Polling.
- Random access–central control.
- Random access–distributed control.

The polling method is fairly self-explanatory. A master station "polls" all other stations in the system sequentially. When a positive reply is received, a channel is assigned accordingly. As the number of stations increases, the polling interval becomes longer and the system tends to become unwieldy. With the random access–central control method, status of channels is coordinated by a central control computer, which is usually located at a "master" earth station. Call requests (call attempts in switching) are passed to the central processor via digital order wire (digitally over the radio service channel), and a channel is assigned if available. Once the call is completed and the subscriber goes on hook, the speech path is taken down and the channel used is returned to the demand-access pool. According to the system design there are various methods to handle blocked calls ["all trunks busy" (ATB)], such as queuing and other repeat attempts.

The distributed control–random access method utilizes a processor control at each earth station in the system. All earth stations in the network monitor the status of all channels where channel status is continuously updated via a digital order-wire circuit. When an idle circuit is seized, all users are informed and the circuit is removed from the pool. Similar information is transmitted to all users when the circuit returns to the idle condition. The same problems arise regarding blockage (ATB) as in the central control system. Distributed control is more costly, particularly in large systems with many users. It is attractive in the international environment, as it eliminates the "politics" of a master station.

Many systems use a mix of preassigned channels on an FDMA basis and DAMA channels. The DAMA concept uses only transponder "space" when in use, and all DAMA stations in the system can *directly* access each other. On low-usage routes and at earth stations with a low traffic volume, DAMA is attractive, as proven on low space-segment-usage costs. DAMA employs the technique of single channel per carrier (SCPC), whereas TDMA and FDMA

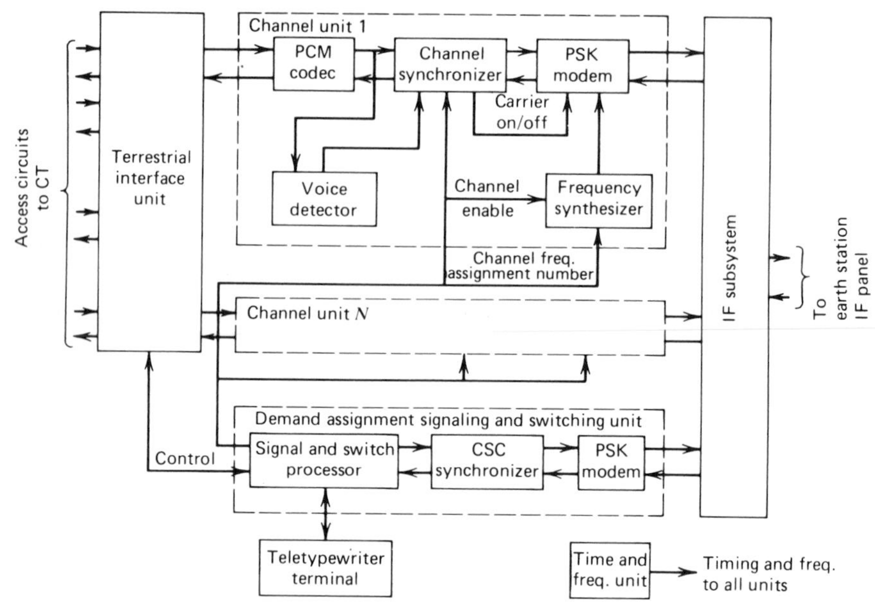

Figure 7.31 Block diagram of a SPADE terminal.

systems are multichannel. SCPC may use FM or phase-shift keying (PSK) modulation. Frequency modulation systems operate on an analog basis, usually with preemphasis, threshold extension of the carrier, and syllabic companding. With PSK systems the voice is digitized in a PCM or delta-modulation format. Modulation rates for PCM are 64 kbps per channel and with delta-modulation, 32 or 40 kbps (see Chapter 9).

A typical distributed control DAMA system (SPADE) used by INTELSAT is shown in Figure 7.31. This system uses PCM on the voice circuit and four-phase modulation. A typical SCPC FM system is illustrated in Figure 7.32.

7.7 Earth Station Link Engineering

7.7.1 Introduction. Up to this point we have discussed basic earth communication problems such as access and coverage. This section reviews some of the link engineering problems associated with earth station system engineering. The approach used to introduce the reader to essential path engineering expands on the basic principles previously discussed (Section 4) in this chapter on radiolinks. As we saw in Section 7.3, an earth station is a distant RF repeater. By international agreement the repeater's EIRP is limited because nearly all bands are shared by terrestrial services. The limit for satellites transmitting in the 4-GHz band is $-142\ \text{dBW/m}^2$ (flux density) in a 4-kHz bandwidth and $-140\ \text{dBW/m}^2$ for the 11-GHz band.

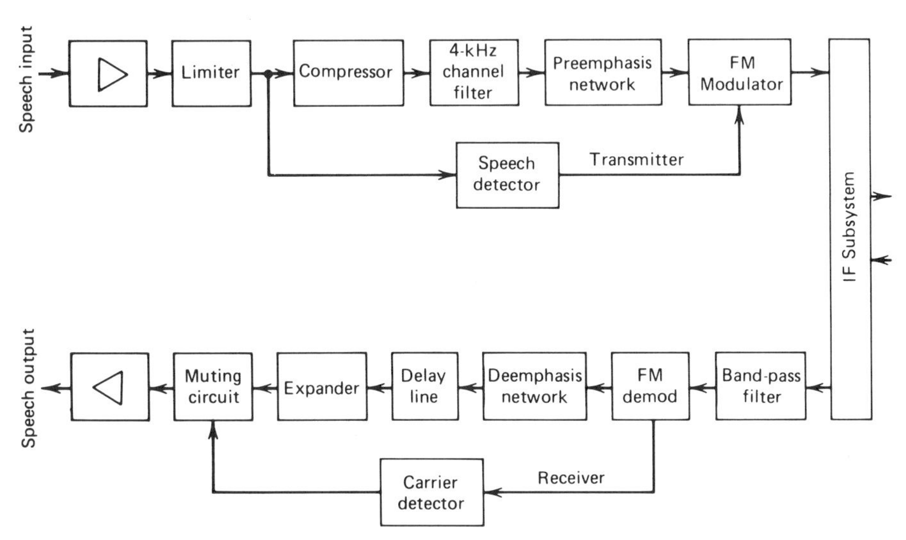

Figure 7.32 Block diagram of a typical SCPC FM station unit.

7.7.2 Earth Station Receiving System Figure of Merit, G/T. The figure of merit of an earth station receiving system, G/T, has been introduced into the technology to describe the capability of an earth station or a satellite to "*receive*" a signal. It is also a convenient tool in the link budget analysis. A link budget is used by the system engineer to size components of earth stations and satellites, such as RF output power, antenna gain and directivity, and receiver front end characteristics.

G/T can be written as a mathematical identity:

$$G/T = G_{\text{dB}} - 10\log T_{\text{sys}} \tag{7.27}$$

where G is the net antenna gain up to an arbitrary reference point or reference plane in the down-link receive chain (for an earth station). Conventionally, in commercial practice the reference plane is taken at the input of the low noise amplifier (LNA). Thus G is simply the gross gain of the antenna minus all losses up to the LNA. These losses include feed loss, waveguide loss, bandpass filter loss, and where applicable, directional coupler loss, waveguide switch insertion loss, radome loss, and transition losses.

T_{sys} is the effective noise temperature of the receiving system and

$$T_{\text{sys}} = T_{\text{ant}} + T_{\text{recvr}} \tag{7.28}$$

T_{ant} or the antenna noise temperature includes all noise-generating components up to the reference plane. The components include sky noise (T_{sky}) plus the thermal noise generated by ohmic losses created by all devices inserted into the system up to the reference plane, including the radome. A typical

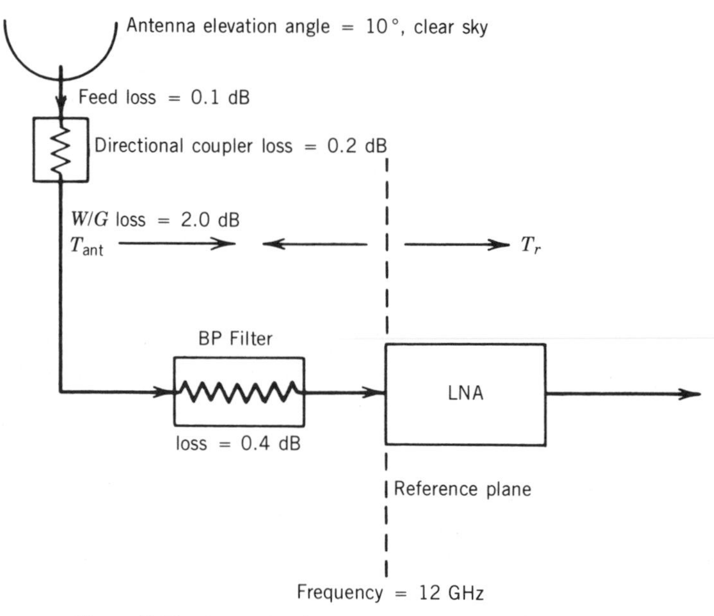

Figure 7.33 Example of an earth station receiving system.

earth station receiving system is shown in Figure 7.33 for an 11-GHz down-link. For a 4-GHz down-link the minimum elevation angle would be 5°. The elevation angle is that angle measured from the horizon (0°) to the antenna main beam when pointed at the satellite. Antenna noise (T_{ant}) is calculated by the following formula:

$$T_{\text{ant}} = \frac{(l_a - 1)290 + T_{\text{sky}}}{l_a} \tag{7.29}$$

where l_a is the numeric equivalent of the sum of the ohmic losses up to the reference plane. l_a is calculated by

$$l_a = \log_{10}^{-1} \frac{L_a}{10} \tag{7.30}$$

where L_a is the sum of the losses in decibels.

Sky noise varies with frequency and elevation angle. Typical values of sky noise are (from CCIR Rec. 720):

Frequency (GHz)	Elevation Angle	Sky Noise (K)
4.0	5°	25
7.5	5°	35
11.7	10°	30
20.0	10°	110

An earth station operating at 12 GHz with a 10° elevation angle will typically display an antenna noise temperature of 150°K [22].

Using values given in Figure 7.33 and T_{sky} of 30°K, we can calculate T_{ant} for a typical earth station where the down-link operating frequency is 12 GHz.

Sum the losses:	Feed	0.1 dB
	Directional coupler	0.2 dB
	Waveguide	2.0
	Bandpass filter	0.4 dB
	Total	2.7 dB = L_a

Using equation (7.30):

$$l_a = \log_{10}^{-1}(2.7/10)$$
$$= 1.86$$

Now use the value for l_a and T_{sky} in equation (7.29):

$$T_{ant} = [(1.86 - 1)290 + 30]/1.86$$
$$= 150°\text{K}$$

The noise figure for a typical commercial LNA or down-converter is 4 dB at 12 GHz. Convert the 4-dB value to equivalent noise temperature. Noise figure can be related to noise temperature (T_e) by the following formula:

$$F_{dB} = 10 \log(1 + T_e/290) \qquad (7.31)$$

For the sample problem, $T_e = T_{recvr}$

$$4 \text{ dB} = 10 \log(1 + T_e/290)$$
$$0.4 = \log(1 + T_e/290)$$

Take the antilog of 0.4, and

$$2.51 = 1 + T_e/290$$
$$T_e = 438°\text{K} = T_{recvr}$$
$$T_{sys} = T_{ant} + T_{recvr}$$
$$= 150 + 438$$
$$= 588°\text{K}$$

Suppose that the antenna in Figure 7.33 had a 47-dB gross gain. What, then, is the G/T of the receiving system at a 10° elevation angle? Calculate net antenna gain.

$$\begin{aligned} G_{\text{net}} &= 47 - 2.7 \\ &= 44.3 \text{ dB} \\ G/T &= 44.3 - 10 \log T_{\text{sys}} \\ &= 44.3 - 10 \log 588 \\ &= 44.3 - 27.7 \\ &= 16.7 \text{ dB/°K} \end{aligned}$$

7.7.3 Station Margin. One major consideration in the design of radiolink and tropo systems is the fade margin, which is the additional signal level added in the system calculations to allow for fading. This value is often of the order of 20–50 dB. In other words, the receive signal level system in question was overbuilt to provide 20–50 dB above threshold to overcome most fading conditions or to ensure that noise would not exceed a certain norm for a fixed time frame.

As we saw in Section 4, fading is caused by anomalies in the intervening medium between stations or by the reflected signal, thus causing interference to the direct ray signal. There would be no fading phenomenon on a radio signal being transmitted through a vacuum. Thus satellite earth station signals are subject to fade only during the time they traverse the atmosphere. For this case, most fades, if any, may be attributed to rainfall or very low elevation angle refractive anomolies.

Margin or station margin is an additional design advantage that compensates for deteriorated propagation conditions or fading. The margin designed into an LOS system is large and is achieved by increasing antenna size, improving the receiver noise figure, or increasing transmitter output power. The station margin of a satellite earth station in comparison is small, of the order of 4–6 dB. Typical rainfall attenuation exceeding 0.01% of a year may be from 1 dB to 2 dB (in the 4-GHz band) without a radome on the antenna and when the antenna is at 5° elevation angle. As the antenna elevation increases to zenith, the rain attenuation notably decreases because the signal passes through less atmosphere. The addition of a radome could increase the attenuation to 6 dB or greater during precipitation. Receive station margin for an earth station, those extra decibels on the down-link, is sometimes achieved by use of threshold-extension demodulation techniques. Up-link margin is provided by using larger transmitters and by increasing power output when necessary. A G/T ratio in excess of the minimum required for clear sky conditions at the 5° elevation angle will also provide margin but may prove expensive to provide.

7.7.4 Typical Down-Link Power Budget. A link budget is a tabular method of calculating space communication system parameters. The approach is similar to that used on line-of-sight microwave links (radiolinks) (see Section

4.4). We start with the EIRP of the satellite for the down-link or the EIRP of the earth station for the up-link. The bottom line is C/N_0 and link margin (in decibels). C/N_0 is the carrier-to-noise level in 1 Hz of bandwidth at the input of the LNA. (*Note*: RSL or receive signal level and C are synonymous.) Expressed as an equation:

$$\frac{C}{N_0} = \text{EIRP} - \text{FSL}_{\text{dB}} - (\text{other losses}) + G/T_{\text{dB/K}} - k \qquad (7.32)$$

where FSL is the free-space loss to the satellite for the frequency of interest and k is Boltzmann's constant expressed in decibel-watts. "Other losses" may include (where applicable):

- Polarization loss (0.5 dB).
- Pointing losses, terminal and satellite (0.5 dB each).
- Off-contour loss (depends on satellite antenna characteristics).
- Gaseous absorption loss (varies with frequency, altitude, and elevation angle).
- Excess attenuation due to rainfall (for systems operating above 10 GHz).

The loss values in parentheses are conservative estimates and should be used only if no definitive information is available.

The off-contour loss refers to spacecraft antennas that provide a spot or zone beam with a footprint on a specific geographical coverage area. There are usually two contours, one for G/T (up-link) and the other for EIRP (down-link). Remember that these contours are looking from the satellite down to the earth's surface. Naturally, an off-contour loss would be invoked only for earth stations located outside of the contour line. This must be distinguished from satellite pointing loss, which is a loss value to take into account that satellite pointing is not perfect. The contour lines are drawn as if the satellite pointing were "perfect."

Gaseous absorption loss (or atmospheric absorption) varies with frequency, elevation angle, and altitude of the earth station. As one would expect, the higher the altitude, the less dense the air and thus the less loss. Gaseous absorption losses vary with frequency and inversely with elevation angle. Often, for systems operating below 10 GHz, such losses are neglected. Reference 23 suggests a 1-dB loss at 7.25 GHz for elevation angles under 10° and for 4 GHz, 0.5 dB below 8° elevation angle.

Example of Link Budget. Assume the following: a 4-GHz down-link, FDM/FM waveform, 5° elevation angle, EIRP +30 dBW; satellite range 25,573 statute miles (sm), and terminal G/T +20.0 dB/K. Calculate down-link C/N_0.

First calculate the free-space loss. Use equation (7.5):

$$\begin{aligned} L_{\mathrm{dB}} &= 96.6 + 20 \log F_{\mathrm{GHz}} + 20 \log D_{\mathrm{sm}} \\ &= 96.6 + 20 \log 4.0 + 20 \log 25{,}573 \\ &= 96.6 + 12.04 + 88.16 \\ &= 196.8 \text{ dB} \end{aligned}$$

Example Link Budget: Down-Link

EIRP of satellite	+30 dBW
Free space loss	−196.8 dB
Satellite pointing loss	−0.5 dB
Off-contour loss	0.0 dB
Excess attenuation rainfall	0.0 dB
Gaseous absorption loss	−0.5 dB
Polarization loss	−0.5 dB
Terminal pointing loss	−0.5 dB
Isotropic receive level	−168.8 dBW
Terminal G/T	+20.0 dB/K
Sum	−148.8 dBW
Boltzmann's constant (dBW)	−(−228.6 dBW)
C/N_0	79.8 dB

On repeatered satellite systems, sometimes called "bent-pipe satellite systems" (those that we are dealing with here), the link budget is carried out only as far as C/N_0, as we did above. It is calculated for the up-link and for the down-link separately. We then calculate an equivalent C/N_0 for the system (i.e., up-link and down-link combined). Use the following formula to carry out this calculation:

$$\left(\frac{C}{N_0}\right)_{(s)} = \frac{1}{1/(C/N_0)_{(u)} + 1/(C/N_0)_{(d)}} \tag{7.33}$$

EXAMPLE. Suppose that an up-link has a C/N_0 of 82.2 dB and its companion down-link has a C/N_0 of 79.8 dB. Calculate the C/N_0 for the system $(C/N_0)_s$. First calculate the equivalent numeric value (NV) for each C/N_0 value.

$$\begin{aligned} \mathrm{NV}(1) &= \log^{-1}(79.8/10) = 95.5 \times 10^6 \\ \mathrm{NV}(2) &= \log^{-1}(82.2/10) = 166 \times 10^6 \\ C/N_0 &= 1/\left[(10^{-6}/95.5) + (10^{-6}/166)\right] \\ &= 1/(0.016 \times 10^{-6}) = 62.5 \times 10^6 = 77.96 \text{ dB} \end{aligned}$$

This is the carrier-to-noise ratio in 1 Hz of bandwidth. To derive C/N for a particular RF bandwidth, use the following formula:

$$C/N = C/N_0 - 10 \log BW_{\mathrm{Hz}} \tag{7.34}$$

Suppose the example system had a 1.2-MHz bandwidth with the C/N_0 of 77.96 dB, what is the C/N?

$$\begin{aligned} C/N &= 77.96 \text{ dB} - 10 \log(1.2 \times 10^6) \\ &= 77.96 - 60.79 \\ &= 17.17 \text{ dB} \end{aligned}$$

7.7.5 Up-Link Considerations. A typical specification for INTELSAT states that the EIRP per voice channel must be +61 dBW (example); thus to determine the EIRP for a specific number of voice channels to be transmitted on a carrier, we take the required output per voice channel in dBW (the above) and add logarithmically $10 \log N$, where N is the number of voice channels to be transmitted.

For example, consider the case for an up-link transmitting 60 voice channels; thus

$$+61 \text{ dBW} + 10 \log 60 = 61 + 17.78 = +78.78 \text{ dBW}$$

If the nominal 100-ft (30-m) antenna has a gain of 63 dB (at 6 GHz) and losses typically of 3 dB, the transmitter output power, P_t required is

$$\mathrm{EIRP}_{\mathrm{dBW}} = P_t + G_{\mathrm{ant}} - \text{line losses}_{\mathrm{dB}} \tag{7.35}$$

where P_t is the output power of the transmitter (in decibel-watts) and G_{ant} is the antenna gain (in decibels) (up-link). Then in the example:

$$\begin{aligned} +78.78 \text{ dBW} &= P_t + 63 - 3 \\ P_t &= +18.78 \text{ dBW} \\ &= 75.6 \text{ W} \end{aligned}$$

7.8 Domestic and Regional Satellite Systems

7.8.1 Introduction. Domestic or regional satellite communication systems are attractive to regions of the world with a large community of interest. These regions may be just one country with a comparatively large geographical expanse (United States, Canada, Indonesia, Mexico, Brazil, or the Soviet Union) or a group of countries with common interests (Europe), a common culture (Arab countries), or a common language (Hispanic America). Such systems also serve areas that are sparsely populated to bring in quality telephone service and TV programming; Canada's TeleSat is a good example.

Still another family of systems serves the business community, providing basically a digital offering. There are two distinct approaches to these latter business systems:

1. DTU or direct-to-user (sometimes called bypass because such systems bypass the local telephone company or administration).
2. Trunking systems, where a city may have a "teleport" or central earth station. The earth station or teleport is connected to each business premise by the standard local telephone connection, usually by wire pair.

Businesses and other institutions such as universities are establishing private networks by leasing all or a portion of a satellite transponder.

7.8.2 Rationale. The basic underlying guideline for any communication system is cost-effectiveness. As a general rule, to achieve a cost-effective system for a low-population terminal segment, less investment is placed in the space segment (the spacecraft) and more in the terminal segment. On the other hand, for a high-population terminal segment, more investment is placed in the space segment, allowing the cost of each terminal in the terminal segment to be reduced.

Generally, domestic and regional satellite systems fall into the latter category, where the terminal segment is fairly highly populated to very highly populated. One cogent example is direct broadcast satellite (television), where it is hoped that every home will have a terminal.

To be cost-effective, first cost must be reduced as well as present value.

7.8.3 Approaches to Cost Reduction. Terminal cost reduction can be achieved by:

- Reducing performance.
- Eliminating connecting links.
- Optimizing bandwidth.
- Augmenting the space segment.
- Mass producing terminals.

There are a number of measures of performance. There are those that have a serious impact on customer satisfaction and, of course, should then remain unmodified. There are others that will have less customer impact, such as link availability, small reductions in S/N ratio and possibly BER, small increase in postdial delay, and increases in blockage probability.

Link performance can be reduced without affecting subscribers when connecting links are eliminated. This is certainly the case with DTU systems, where connecting links to the nearest earth station or teleport are eliminated.

As an example, INTELSAT links (i.e., ground–satellite–ground) are built to a 10,000-pWp specification for analog voice channel service. This noise value is based on the CCIR 2500-km hypothetical reference circuit (HRC). The HRC value is based on the CCITT reference connection of no more than 12 links in tandem. Here there is one, possibly two links in tandem; 50,000 pWp may then be a more appropriate value. An analogous reasoning can be shown for BER in the case of a digital connectivity.

If we consider the INTELSAT Standard A earth station as a model, then the major cost item is the antenna system with its autotracking feature. Reduction of antenna aperture can reduce cost by a power law relationship. As aperture size decreases, beamwidth increases. Depending on station keeping of the satellite in question, there is some point where autotracking can be eliminated entirely, and the station operator need only trim up the antenna pointing periodically—weekly, monthly, annually, and with VSATs possibly never. At this point we can see that dramatic savings on earth station first cost can be achieved.

Without compromising performance, antenna aperture reductions can be realized in several ways. The Standard A earth station must accommodate the entire 500-MHz bandwidth of the satellite. Suppose that a particular domestic earth station is required to accommodate only a 36-MHz transponder rather than the entire 500-MHz bandwidth encompassing all transponders. This can permit a direct decibel-for-decibel reduction. A rule of thumb is that if we reduce gain requirements of antenna by 6 dB, reflector diameter can be cut in half; by 3 dB, 25%, etc. In this case the gain reduction is $10\log(500/36)$ or 11.4 dB. The 30-m dish can be reduced to 8-m.

A similar rationale can be used by beefing up the space segment. For every decibel of improvement in satellite G/T, earth station EIRP can be reduced by the same amount for a given performance criterion. Likewise, for every decibel of improvement of satellite EIRP, earth station G/T can be reduced by the same amount. Of course, satellite EIRP has a limit if we are to follow CCIR recommendations.

7.8.4 Sample Applications. A good example of industrial application is the SBS (Satellite Business Systems). This system covers the continental United States with two spot beams. One spot beam is superimposed on the other; the narrower one covers the east central states where rainfall will affect path reliability more, and the wider one covers nearly all the contiguous states. The narrow beam has an EIRP of +42 dBW and the wide beam, +38 dBW. The satellite keeps station with a total excursion of 0.07°, as seen from an optimally pointed 7-m dish. At maximum satellite excursion, up-link additional loss is about 1.2 dB and for that critical down-link, 0.8 dB. Operation for the system is in the 12-GHz and 14-GHz bands (see Section 7.5); thus rainfall is a major consideration. The system availability goal is 99.5%. The down-link margin is 5 dB. Margin, as discussed in Section 7.7.3, is the additional gain added to the system to ensure the specified reliability and availability.

Earth stations will have 5-m antennas in the region of coverage for the narrow beam and 7-m antennas for the remainder of the coverage area. Autotracking is not required. The estimated system noise temperature (T_e) is 225°K; G/T for the 5-m antenna is 30.4 dB/°K and for the 7-m antenna, 33.3 dB/°K. Some of the rain margin in the narrow beam region is made up from decreased losses due to satellite excursion. This is because smaller-diameter parabolic antennas (5 m) have wider beamwidths. The two different antenna configurations have beamwidths as follows:

Beamwidth	Receive	Transmit
5 m	0.37°	0.31°
7 m	0.27°	0.22°

Full redundancy in the satellite will help ensure the 99.95% total-availability goal figures. With a bit error rate (Chapter 8) of 1×10^{-4}, the availability is 99.5%. The operational mode is TDMA and demand access, using burst transmissions with bit rates as high as 50 Mbps. Although we indicated previously that satellite communications was down-link limited, SBS is partially up-link limited. A greater noise contributions is allowed the up-link so that earth-station transmitter size (and resulting cost) may be reduced. Maximum transmitter output for any earth station is 4 kW. In other systems the transmitter available power is sufficient for the up-link to have little effect as a noise contributor when compared to the down-link. Individual transmitter output will be a function of the bit rate to be transmitted. A 38-Mbps burst will require about 300 W; 45 Mbps, 1 kW; and 50 Mbps, about 4 kW.

Considering that SBS and many other use Ku band (11–12 GHz down-link), a down-link analysis would be revealing. The signal level impinging on the antennas will be:

SAT EIRP: Free-Space Loss (12 GHz)

5 m	+42 dBW − 206 dB = −164 dBW
7 m	+38 dBW − 206 dB = −168 dBW

Signal Level at Receiver Front End

Level	Gain (dB)
5 m	−164 dBW + 52.5 = −111.5 dBW
7 m	−168 dBW + 56.5 = −111.5 dBW

With a system noise temperature of 225K and an IF of approximately 60

MHz, the bandwidth noise threshold is N_{th}:

$$
\begin{aligned}
N_{th} &= -228.6 \text{ dBW} + 10 \log K + 10 \log B_{if} \\
&= -228.6 \text{ dBW} + 10 \log 225 + 10 \log 60 \times 10^6 \\
&= -127.18 \text{ dBW} \\
C/N &= 111.5 - (-127.18) \\
&= 15.68 \text{ dB}
\end{aligned}
$$

where 15.68 dB for C/N (carrier-to-noise ratio) provides a good margin to achieve the minimum acceptable bit-error rate (BER) for digital modulation. This minimum BER is specified as 1×10^{-4} 99.5% of the time (per year). With the use of forward error correction (FEC) data transmitted over an SBS link will have a BER better than 1×10^{-7}. Modulation has been assumed as QPSK, or quaternary phase shift keying. [*Note*: Digital transmission (PCM) is discussed in Chapter 9, data transmission in Chapter 8, and data network operation in Chapter 11.]

7.8.5 Very Small Aperture Terminal. Very small aperture terminals (VSATs) carry the rationale one step further. Aperture, of course, refers to antenna aperture. In this case aperture and parabolic dish diameter are synonymous. VSATs have dishes with apertures in the range of 1–2 m and all are DTU facilities where the equipment is on the customer premises.

The application of VSATs is generally for personal computer (PC) connectivity via a considerably larger hub station. Hub stations have dishes in the range of 5–8 m. VSAT bit rates are usually 2400* bps with up-link transmit power in the range of 0–10 dBW. Hub transmit power ranges to +20 dBW or more. The 12–14-GHz frequency band is generally favored for VSAT operation where greater satellite EIRP is permitted compared to the 4–6-GHz band. The 12–14-GHz band suffers excess attenuation due to rainfall sufficient to disrupt operations during a downpour. Some systems try mitigation techniques; others wait until the rain passes, usually no more than 20 min or so.

Typical applications of VSAT systems are:

- Short computer transactions for large chains of businesses covering major geographical areas.
- Reservation systems.
- Credit card verifications and transactions.

All traffic must flow through the hub because the system, with its disadvan-

*Some VSAT systems operate up to the DS1 rate of 1.544 Mbps requireing somewhat larger antennas.

taged users (VSATs) can operate only with the enhancement the hub provides. Direct VSAT-to-VSAT connectivity is not feasible. In some cases a single hub station can serve thousands of 2400-bps VSAT accesses.

REVIEW QUESTIONS

1. What transmission medium should be considered as a principal candidate for a digital waveform that would carry more than 5000 full-duplex telephone circuits at the end of the forecast period?
2. What are the advantages of using the RF bands from 2 GHz to 10 GHz for trunk telephony? Name at least two.
3. Frequencies above 10 GHz have an additional major impairment that must be taken into account in radiolink design. What is the cause of this impairment?
4. Discuss the problem of delay in speech telephone circuits traversing a satellite. Can you think of any problems in telephone signaling or in data transmission?
5. Give the two basic procedure steps in the design of line-of-sight radiolinks.
6. What are the advantages of "through-grouping" in FDM systems?
7. Name three basic planning considerations in siting radiolinks.
8. Name two advantages and at least two disadvantages to a radiolink terminal sited in the middle of an urban area?
9. Why do we limit the heights of radio towers serving radiolink (line-of-sight) systems?
10. Describe how earth bulge varies from one end of a radiolink hop to the other. Where is earth bulge maximum?
11. When a K factor of $\frac{4}{3}$ is used, does the microwave ray beam bend toward or away from the earth?
12. In the path profile, what are the three basic increment factors that must be added to obstacle height?
13. Name at least five of the basic factors that a radiolink design engineer must deal with when carrying out the path calculation (analysis) stage of the design effort.
14. Calculate the free-space loss in decibels of a radiolink hop 24 statute miles long operating at 4100 MHz.
15. Express Boltzmann's constant in decibel-watts.

16. If the signal-to-noise ratio into a device is 31 dB and the signal-to-noise ratio at the device output is 28 dB, what is the device noise figure?

17. A receiving system operates at room temperature; its noise figure is 11 dB, and its bandwidth is 2000 kHz. What is the thermal noise threshold of the system?

18. What are the three factors that affect the gain of a parabolic dish antenna?

19. What is the standard efficiency (%) used with line-of-sight microwave parabolic dish antennas?

20. Calculate the EIRP of a radiolink transmitting system where the transmitter has 1.0-W output, 125 ft of waveguide with a loss of 0.015 dB/ft, and an antenna with 35-dB gain.

21. What is the most common type of fading encountered on a line-of-sight radiolink? Discuss its causes and cures.

22. On a 20-mi radiolink hop, a path propagation reliability of 99.5% is desired. Based on the Rayleigh fading criterion, what fade margin is required (in decibels)?

23. Name the two types of voice channel noise weighting units in use today.

24. A picowatt equals how many watts?

25. There are two common types of diversity used on line-of-sight radiolinks to mitigate the effects of fading. What are they? Discuss the pros and cons of each.

26. What type of noise does NPR quantify?

27. A radiolink does not meet its noise specifications. What steps can be taken, in ascending order of cost (descending order of desirability), to meet specifications?

28. What is the loading of a radio system by an FDM configuration a function of?

29. Name three types of analog system radiolink repeaters. Discuss each and its application.

30. Total noise in a derived FDM voice channel over a radiolink is the sum of what three sources of noise?

31. Discuss radio frequency interference in the planning of a radiolink installation.

32. Give two applications of coaxial cable systems. Discuss trade-offs with radiolinks.

33. On a coaxial cable system, as the baseband frequency increases, what increases exponentially?

34. Coaxial cable design has been called a "cookbook" design. There are seven identified principal design factors involved in this design. Name five of them.

35. What makes radiolink design not a "cookbook" design?

36. Name at least five advantages of fiber optics links compared to radiolink and coaxial cable.

37. Identify the three principal fiber optics wavelength bands used for communications. Relate each comparatively regarding loss per unit length.

38. Name the three basic components of a fiber optics link.

39. The parameter "numerical aperture" describes what capability of a fiber?

40. What are the two basic parameters that limit the length of a fiber optics link (without repeaters)?

41. What are the basic types of optical fiber used for communications?

42. An optical fiber consists of what two concentric components?

43. Discuss applications of splices and connectors and the advantages and disadvantages of each.

44. What are the two basic light sources discussed in the text? Discuss advantages and disadvantages of each.

45. What is a "pigtail," and what does it do?

46. There are two basic types of light detectors covered in the text. Identify each and discuss advantages and disadvantages.

47. A fiber optic link uses an ILD with +4-dBm output and an APD with a threshold of −41 dBm. The bit rate is 500 Mbps. Design a link around these parameters. What is the maximum link length achievable without repeaters? Show rationale.

48. Give five applications of fiber optics links.

49. What is the standard BER (bit error rate) for a fiber optics link?

50. What are the three factors affecting link margin for a fiber optics link?

51. How does range (free-space loss) vary with elevation angle of a geostationary communication satellite system?

52. What is the free-space loss to a geostationary satellite for a range of 24,500 sm at 4100 MHz?

53. There are two basic types of access to a communication satellite discussed in the text. Define each and give their advantages and disadvantages.

54. Give two principal applications of SCPC DAMA systems (i.e., where/when would they be used?).

55. An earth station has a net antenna gain on the down-link referenced to the input of the LNA of 40 dB and the receiving system noise temperature is 200°K. What is the G/T?

56. What are the principal noise components in a satellite receiving system that contribute to T_{sys}?

57. T_{sky} varies with what two factors?

58. Besides free-space loss, name at least four other loss components in a link budget.

59. C/N_0 is given as 97 dB. The noise bandwidth is 8.0 MHz. What is C/N?

60. Where do VSAT systems find application?

61. Why do VSAT systems trend more for 12–14-GHz operation than for 4–6-GHz operation?

REFERENCES

1. International Telephone and Telegraph Corporation, *Reference Data for Radio Engineers*, 6th ed., Howard W. Sams, Indianapolis, 1976.
2. *Overall Communication Systems Planning*, Vol. 3, IEEE North Jersey Section, 1964.
3. Roger L. Freeman, *Telecommunication Transmission Handbook*, 2nd ed., Wiley, New York, 1981.
4. *Transmission Systems for Communications*, 4th ed., Bell Telephone Laboratories, Holmdel, N.J., 1973.
5. *Engineering Considerations for Microwave Communication Systems*, GTE Lenkurt, San Carlos, Calif., 1975.
6. *Jerrold Path Calculations*, Jerrold Electronics Corporation, Philadelphia, 1967.
7. Philip F. Panter, *Communication System Design—Line-of-Sight and Tropo-Scatter Systems*, McGraw-Hill, New York, 1972.
8. J. Jasik, *Antenna Engineering Handbook*, 2nd ed., McGraw-Hill, New York, 1981.
9. K. Bullington, "Radio Propagation Fundamentals," *Bell Syst. Tech. J.* (May 1957).
10. *A Survey of Microwave Fading Mechanisms, Remedies and Applications*, ESSA Technical Report ERL-69-WPL4 Boulder, Colo., March 1968.
11. R. G. Medhurst, "Rainfall Attenuation of Centimeter Waves: Comparison of Theory and Measurement," *IEEE Transact. Antennas Propag.*, July 1965.

12. Technical Note 100, National Bureau of Standards, January 1965.
13. M. J. Tant, *Multichannel Communication Systems and White Noise Testing*, Marconi Instruments, St. Albans, Herts (UK), 1974.
14. P. J. Howard, M. F. Alarcon, and S. Tronsli, "12-Megahertz Line Equipment," *Electr. Commun.*, **48** (1973).
15. William A. Rheinfelder, *CATV System Engineering*, TAB Books, Blue Ridge Summit, Pa., 1970.
16. CCITT, *Red Books*, VIII Plenary Assembly, Malaga-Torremolinos, 1984, particularly Vol. III.
17. CCIR, *Green Books*, XV Plenary Assembly, Dubrovnik, 1986.
18. INTELSAT Earth Station Standards IESS-302, INTELSAT, Washington, D.C., Dec. 1986.
19. J. A. Lawlor, *Coaxial Cable Communication Systems, Management Overview*, ITT, New York, February 1972 (technical memorandum).
20. F. J. Herr, "The L5 Coaxial System—Transmission System Analysis," *IEEE Trans. Commun.* (February 1974).
21. L. Becker, "60-Megahertz Line Equipment," *Electr. Commun.*, **48** (1973).
22. *Satellite Communication Reference Data Handbook*, Defense Communication Agency (NTIS), Washington, D.C., July 1972.
23. Roger L. Freeman, *Radio System Design for Telecommunications*, Wiley, New York, 1987.
24. B. Edelson, "Cost Effectiveness in Global Satellite Communications," *IEEE Commun. Soc. Mag.* (January 1977).
25. R. B. Marsten, "Service Needs and Systems Architecture in Satellite Communications," *IEEE Commun. Soc. Mag.* (May 1977).
26. E. E. Basch, Ed. *Optical Fiber Transmission*, Howard W. Sams, Indianapolis, 1987.
27. *Small Earth Stations 1976–1986*, ComQuest Corporation, Palo Alto, Calif., 1977.
28. *Transmission Systems for Communications*, 5th ed., Bell Telephone Laboratories, Holmdel, N.J., 1982.
29. *INTELSAT V Satellite Specification, Exhibit A*, ComSat, Washington, D.C., BG-23-10E W/9/76, 1976.
30. Roger L. Freeman, "An Approach to Earth Station Technology," *Telecommun. J. (ITU)* (June 1971).
31. R. G. Gould and Y. F. Lum, *Communication Satellite Systems: An Overview of the Technology*, IEEE Press, New York, 1976.
32. *Determining Site Coordinates*, GTE Lenkurt Demodulator, San Carlos, Calif., January–February 1978.

8

THE TRANSMISSION OF OTHER INFORMATION OVER THE TELEPHONE NETWORK

1 INTRODUCTION

Up to this point we have discussed voice communications over telephone systems. The word "telephone" implies sound, particularly "articulate speech." The existing telephone network covers the entire world. It is gigantic, with an annual investment in a new plant in the many thousands of million dollars. This worldwide network, in part or in whole, can be used to transmit information other than speech. Such information includes data, facsimile, and video. The bottleneck in the telephone network would appear to be the limitations of a wire pair. There are wire pairs connecting subscribers to switches; the conventional analog switch is built around wire pairs, and wire pairs make up the greater portion of local trunk connections. We find that even with the limitations of bandwidth and envelope delay, wire pairs may be used as is for the transmission of data, telegraph, and facsimile and may even be used over short runs for video. To improve transmission characteristics, wire pairs can be conditioned for wider bandwidths and improved delay characteristics. The real bottleneck that restricts transmission capabilities and places a firm top on data rates or facsimile speed in the conventional telephone network is carrier (multiplex) equipment.

The objective of this chapter is to introduce the principles of data, telegraph, and facsimile over the telephone network. No distinction is made between data and telegraph communication from the transmission viewpoint. Both are binary, and often the codes used for one serve equally well for the other. The transmission problems of data apply equally to telegraph communication, which is usually in message format, transmitted at rates less than 110 words per minute (wpm), and asynchronous. Data transmission is most often synchronous (but not necessarily so) and usually is an alphanumeric mix of information primarily destined for computers, work stations, and other data processing equipment. Data are often transmitted at rates in excess of 110 wpm.

Besides data, there are other forms of information that can be transmitted over the telephone network, such as telemetry, video, and facsimile. Because of its forecast rapid growth and our references to it in Chapter 11, we elected to provide a short description of current facsimile technology at the end of this chapter.

2 ANALOG – DIGITAL

Up to this point the text has dealt essentially with analog signals. An analog transmission system has an output at the far end that is a continuously variable quantity representative of the input. With analog transmission, the signal containing the information is continuous; with digital transmission, the signal is discrete. The simplest form of digital transmission is binary, where an information element is assigned one of two states. There are many binary situations in real life where only one of two possible values can exist; for example, a light may be either on or off, an engine is running or not, and a person alive or dead.

An entire number system has been based on two values, which by convention have been assigned the symbols 1 and 0. This is the binary system, and its number base is 2. Our everyday number system has a base of 10 and is called the *decimal system*. Another system has a base of 8 and is called the *octal system*, and still another is the *hexadecimal* representation, with a number base of 16. The basic information element of the binary system is called the *bit*, which is an acronym for "binary digit." The bit may have the value of either 1 or 0. A number of discrete bits can be encoded to identify a larger piece of information, which we may call a *character*. A code is defined by the IEEE as "a plan for representing each of a finite number of values or symbols as a particular arrangement or sequence of discrete conditions or events."

Binary coding of written information has been in existence for a long time. An example of teleprinter service (i.e., that used in the transmission of a telegram). The majority of computers now in operation operate in binary languages; thus binary transmission fits in well with computer communication, whether terminal to computer or computer to computer. The facilities used to effect this communications make up the elements of a data transmission system (see Chapter 11).

3 THE BIT – BINARY CONVENTION

In a binary transmission system the smallest unit of information is the bit, which we know to be either one of two states. We call one state a *mark* and

TABLE 8.1 Equivalent Binary Designations: Summary of Equivalence

Symbol 1	Symbol 0
Mark or marking	Space or spacing
Current on	Current off
Negative voltage	Positive voltage
Hole (in paper tape)	No hole (in paper tape)
Condition *Z*	Condition A
Tone on (amplitude modulation)	Tone off
Low frequency (frequency shift keying)	High frequency
Inversion of phase	No phase inversion (differential phase shift keying)
Reference phase	Opposite to reference phase

Source: CCITT Recs. V.1, V.10, and V.11.

the other, a *space*. These states may be indicated electrically by the presence or absence of current flow. Unless some rules are established, an ambiguous situation would exist. Is the "1" condition a mark or a space? Does the "no current" condition mean transmission of a 0 or a 1? To avoid confusion and to establish a unique identity to binary conditions, CCITT Rec. V.1 recommends equivalent binary designations. These are shown in Table 8.1. If this table is adhered to universally, no confusion should exist as to which is a mark, which is a space, which is the active condition, which is the passive condition, which is 1, and which is 0. Table 8.1 defines the sense of transmission so that the mark and space, the 1 and 0, respectively, will not be inverted. Data-transmission engineers often refer to such a table as a "table of mark–space convention."

4 CODING

4.1 Introduction to Binary-Coding Techniques

Written information must be coded before it can be transmitted over a digital system. The discussion that follows covers binary codes only. Not only does this simplify our argument, but certainly binary codes are by far the most widely used codes in data and telegraph networks.

One of the first questions that arises is in regard to the size or extent of a binary code. The answer involves yet another question, specifically, how much information is to be transmitted. One binary digit (bit) carries little information; it has only two possibilities. If two binary digits are transmitted in sequence, there are four possibilities:

00 10
01 11

or four pieces of information. Suppose 3 bits are transmitted in sequence. Now there are eight possibilities:

000	100
001	101
010	110
011	111

We can now see that for a binary code, the number of distinct information characters available is equal to two raised to a power equal to the number of elements or bits per character. For instance, the last example was based on a three-element code giving eight possibilities or information characters, or 2^3.

Another more practical example is the CCITT ITA No. 2 teleprinter code (Figure 8.1), which has 5 bits or information elements per character. Therefore the number of different graphics* and characters available are $2^5 = 32$. The American Standard Code for Information Interchange (ASCII) has seven information elements per character, or $2^7 = 128$; thus it has 128 distinct combinations of marks and spaces that are available for assignment as characters or graphics.

The number of distinct characters for a specific code may be extended by establishing a bit sequence (a special character assignment) to shift the system or machine to uppercase (as is done with a conventional typewriter). Uppercase is a new character grouping. A second distinct bit sequence is then assigned to revert to lowercase. For example, the CCITT ITA No. 2 code (Figure 8.1) is a five-unit code with 58 letters, numbers, graphics, and operator sequences. The additional characters and graphics (additional above $2^5 = 32$) originate from the use of uppercase. Operator sequences appear on a keyboard as "space" (spacing bar), "figures" (uppercase), "letters" (lowercase), "carriage return," "line feed" (spacing vertically), and so on. When we refer to a 5-unit, 6-unit, or 12-unit code, we refer to the number of information units or elements that make up a single character or symbol. That is, we refer to those elements assigned to each character that carry information and that distinguish it from all other characters or symbols of the code.

4.2 Hexadecimal Representation and BCD Code

The hexadecimal system is a numeric representation in the number base 16. This number base uses 0 through 9 as in the decimal number base, and the letters A through F to represent the decimal numbers 10 through 15. The

*In this context a graphic is a printing character other than a letter or number. Typical graphics are asterisks, punctuation, parentheses, dollar signs, and so forth.

Characters				Code Elements[a]						
Letters Case	Communications	Weather	CCITT #2[b]	START	1	2	3	4	5	STOP
A	-	↑			X	X				X
B	?	⊕			X			X	X	X
C	:	○				X	X	X		X
D	$	↗	WRU		X			X		X
E	3	3			X					X
F	1	→	Unassigned		X		X	X		X
G	&	↘	Unassigned			X		X	X	X
H	STOP[c]	↓	Unassigned				X		X	X
I	8	8				X	X			X
J	'	↙	Audible signal		X	X		X		X
K	(	←			X	X	X	X		X
L	)	↖				X			X	X
M	.	.					X	X	X	X
N	,	⦶					X	X		X
O	9	9						X	X	X
P	θ	θ				X	X		X	X
Q	1	1			X	X	X		X	X
R	4	4				X		X		X
S	BELL	BELL	,		X		X			X
T	5	5							X	X
U	7	7			X	X	X			X
V	;	⦶	=			X	X	X	X	X
W	2	2			X	X			X	X
X	/	/			X		X	X	X	X
Y	6	6			X		X		X	X
Z	"	+	+		X				X	X
BLANK		-								X
SPACE							X			X
CAR. RET.								X		X
LINE FEED						X				X
FIGURE					X	X		X	X	X
LETTERS					X	X	X	X	X	X

[a] Blank, spacing element; crosshatched, marking element.

[b] This column shows only those characters that differ from the American "communications" version

[c] Figures case H(COMM) may be STOP or +.

Figure 8.1 Communication and weather codes, CCITT International Alphabet No. 2 (ITA2).

"hex" numbers can be translated to the binary base as follows:

Hex	Binary	Hex	Binary
0	0000	8	1000
1	0001	9	1001
2	0010	A	1010
3	0011	B	1011
4	0100	C	1100
5	0101	D	1101
6	0110	E	1110
7	0111	F	1111

Two examples of the hexadecimal notation are

Number Base 10	Number Base 16
21	15
64	40

The BCD (binary-coded decimal) is a compromise code assigning 4-bit binary numbers to the digits between 0 and 9. The BCD equivalents to decimal digits appear as follows:

Decimal Digit	BCD Digit	Decimal Digit	BCD Digit
0	1010	5	0101
1	0001	6	0110
2	0010	7	0111
3	0011	8	1000
4	0100	9	1001

To cite some examples, consider the number 16; it is broken down into 1 and 6. Thus its BCD equivalent is 0001 0110. If it were written in straight binary notation, it would appear as 10000. The number 25 in BCD combines the digits 2 and 5 above as 0010 0101.

4.3 Some Specific Binary Codes for Information Interchange

In addition to the ITA No. 2 code (Figure 8.1), some of the more commonly used binary codes are the American Standard Code for Information Interchange (ASCII) (Figure 8.3), the CCITT No. 5 (Figure 8.4) Code, and the EBCDIC (Figure 8.5). The ASCII, CCITT No. 5, and EBCDIC codes are referred to as "eight-level" code sets because a character is expressed with 8 binary digits or bits. The levels are numbered from 1 to 8, starting from the right or least significant binary digit.

The term "level" originates from perforated-paper-tape technology and implies a channel on the tape. On a paper tape a 1 is expressed with a punched

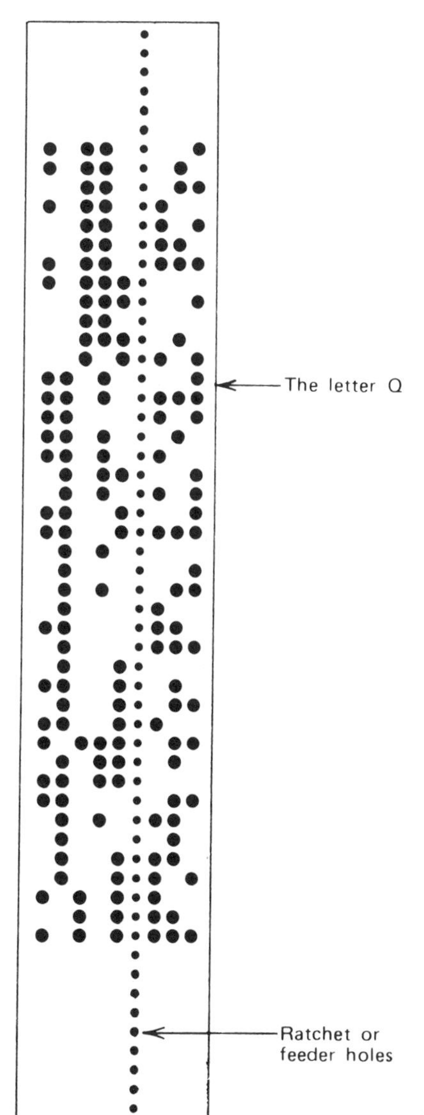

Figure 8.2 Perforated paper tape with the ASCII code. (*Note*: Even parity; parity bits are in far-left column, characters are top to bottom: 1234567890: QWERTYUIOPASDFGHJKL; ZX-CVBNM, etc.)

hole (Table 8.1), and a 0 is expressed with no hole punched in the appropriate channel. With magnetic tape, 1's and 0's are represented by changes in magnetic flux. An example of punched paper tape is shown in Figures 8.2. The small holes offset just to the right of the tape center are the feed sprocket holes. The ASCII, CCITT No. 5, and EBCDIC are stated as eight-level codes. However, only EBCDIC* is a true eight-level code; ASCII and CCITT No. 5

*EBCDIC—Extended Binary Coded Decimal Interchange Code.

codes are actually 7-bit codes with an extra bit added to each character for parity. Here we mean that the 7 bits (levels) are used for information, whereas EBCDIC uses all 8 bits for information.

Parity or parity checks provide a means for determining whether a character contains an error after transmission. We speak of "even parity" and "odd parity." On a system using an odd-parity check, the total count of 1's or marks has to be an odd number per character (e.g., it carries 1, 3, 5, or 7 marks or 1's). To explain parity and parity checks a little more clearly, let us look at some examples. Consider a seven-level code with an extra parity bit. By system convention, even parity has been established; suppose a character is transmitted as 1111111. There are seven marks, so to maintain even parity, we would need an even number of marks. Thus an eighth bit is added, which must be a mark (1). Now consider another bit pattern, 1011111. Here there are six marks, even; thus the eighth (parity) bit must be a space. Still another example would be 0001000. To obtain even parity, a mark must be added on transmission and the character transmitted would be 00010001, maintaining even parity. Suppose that, as a result of some sort of signal interference, one signal element was changed on reception. No matter which element was changed, the receiver would indicate an error because we would no longer have even parity. If two elements were changed, however, the error could be masked. This would happen in the case of even or odd parity if two marks were substituted for two spaces or vice versa at any element location in the character.

The American Standard Code for Information Interchange (see Figure 8.3), better known as the ASCII, is the latest effort on the part of the North American industry and common-carrier systems, backed by the US Standards Institute (now the American National Standards Institute), to produce a universal common language code. The ASCII is 7-unit code with all 128 combinations available for assignment. Here again, the 128 bit patterns are divided into 2 groups fo 64. One group is assigned to a subset of graphic printing characters. The second subset of 64 is assigned to control characters. An eighth bit is added to each character for parity check. The ASCII is widely used in North America and has received considerable acceptance in Europe and Hispanic America.

A seven-level code is recommended in CCITT Rec. V.4 as an international standard for information interchange. It is not intended as a substitute for the CCITT No. 2 code. The CCITT No. 5, or the new alphabet No. 5, as the seven-level code is more commonly referred to, is basically intended for data transmission. Although the CCITT No. 5 is considered a seven-level code, CCITT Rec. V.4 advises that an eighth bit may be added for parity. Under certain circumstances odd parity is recommended; otherwise, even parity.

Figure 8.4 shows the CCITT No. 5 code, where b_1 is the first signal element in serial transmission, and b_7 is the last element of a character. Like the ASCII code, the CCITT No. 5 does not normally need to shift out (i.e., uppercase or lowercase, as in CCITT No. 2). However, like ASCII, it is

Bit Number b_7	b_6	b_5		0 0 0	0 0 1	0 1 0	0 1 1	1 0 0	1 0 1	1 1 0	1 1 1
$b_4 b_3 b_2 b_1$			Row \ Column	0	1	2	3	4	5	6	7
0 0 0 0			0	NUL	DLE	SP	Ø	@	P	`	p
0 0 0 1			1	SOH	DC1	!	1	A	Q	a	q
0 0 1 0			2	STX	DC2	"	2	B	R	b	r
0 0 1 1			3	ETX	DC3	#	3	C	S	c	s
0 1 0 0			4	EOT	DC4	$	4	D	T	d	t
0 1 0 1			5	ENQ	NAK	%	5	E	U	e	u
0 1 1 0			6	ACK	SYN	&	6	F	V	f	v
0 1 1 1			7	BEL	ETB	'	7	G	W	g	w
1 0 0 0			8	BS	CAN	(	8	H	X	h	x
1 0 0 1			9	HT	EM	)	9	I	Y	i	y
1 0 1 0			10	LF	**SUB**	*	:	J	Z	j	z
1 0 1 1			11	VT	ESC	+	;	K	[	k	{
1 1 0 0			12	FF	FS	,	<	L	\	l	|
1 1 0 1			13	CR	GS	–	=	M	]	m	}
1 1 1 0			14	SO	RS	.	>	N	^	n	~
1 1 1 1			15	SI	US	/	?	O	–	o	DEL

Figure 8.3 American Standard Code for Information Interchange (ASCII). Columns 2, 3, 4, and 5 indicate printable characters in DCS Autodin (US Defense Communications System Automatic Digital Network); columns 6 and 7 fold over into columns 4 and 5, respectively, except DEL.

Bit Number									
b_7 →		0	0	0	0	1	1	1	1
b_6 →		0	0	1	1	0	0	1	1
b_5 →		0	1	0	1	0	1	0	1
$b_4 b_3 b_2 b_1$	Row \ Column	0	1	2	3	4	5	6	7
0 0 0 0	0	NUL	(TC_r)DLE	SP	0	(@)	P	‘	p
0 0 0 1	1	(TC_1)SOH	DC_1	!	1	A	Q	a	q
0 0 1 0	2	(TC_2)STX	DC_2	”	2	B	R	b	r
0 0 1 1	3	(TC_3)ETX	DC_3	£	3	C	S	c	s
0 1 0 0	4	(TC_4)EOT	DC_4	$	4	D	T	d	t
0 1 0 1	5	(TC_5)ENQ	(TC_8)NAK	%	5	E	U	e	u
0 1 1 0	6	(TC_6)ACK	(TC_9)SYN	&	6	F	V	f	v
0 1 1 1	7	BEL	(TC_{10})ETB	’	7	G	W	g	w
1 0 0 0	8	FE_0(BS)	CAN	(	8	H	X	h	x
1 0 0 1	9	FE_1(HT)	EM	)	9	I	Y	i	y
1 0 1 0	10	FE_2(LF)	SUB	*	:	J	Z	j	z
1 0 1 1	11	FE_3(VT)	ESC	+	;	K	([)	k	•
1 1 0 0	12	FE_4(FF)	IS_4(FS)	,	<	L	•	l	•
1 1 0 1	13	FE_5(CR)	IS_3(GS)	–	=	M	(])	m	•
1 1 1 0	14	SO	IS_2(RS)	.	>	N	^	n	‾
1 1 1 1	15	SI	IS_1(US)	/	?	O	–	o	DEL

Figure 8.4 CCITT No. 5 Code for Information Interchange (CCITT Recs. V.4 and X.4). Courtesy of the ITU-CCITT, Geneva.

B			4	0	0	0	0	0	0	0	0	1	1	1	1	1	1	1	1
I			3	0	0	0	0	1	1	1	1	0	0	0	0	1	1	1	1
T			2	0	0	1	1	0	0	1	1	0	0	1	1	0	0	1	1
		S	1	0	1	0	1	0	1	0	1	0	1	0	1	0	1	0	1
8	7	6	5																
0	0	0	0	NUL				PF	HT	LC	DEL								
0	0	0	1					RES	NL	BS	IL								
0	0	1	0					BYP	LF	EOB	PRE			SM					
0	0	1	1					PN	RS	UC	EOT								
0	1	0	0	SP										¢	.	<	(	+	|
0	1	0	1	&										!	$	*	)	;	¬
0	1	1	0	-	/									^	,	%	—	>	?
0	1	1	1											:	#	@	'	=	"
1	0	0	0		a	b	c	d	e	f	g	h	i						
1	0	0	1		j	k	l	m	n	o	p	q	r						
1	0	1	0			s	t	u	v	w	x	y	z						
1	0	1	1																
1	1	0	0		A	B	C	D	E	F	G	H	I						
1	1	0	1		J	K	L	M	N	O	P	Q	R						
1	1	1	0			S	T	U	V	W	X	Y	Z						
1	1	1	1	0	1	2	3	4	5	6	7	8	9						¤

PF – Punch Off
HT – Horiz. Tab
LC – Lower Case
DEL – Delete
SP – Space
UC – Upper Case
RES – Restore
NL – New Line
BS – Backspace
IL – Idle
PN – Punch On
EOT – End of Transmission
BYP – Bypass
LF – Line Feed
EOB – End of Block
PRE – Prefix
RS – Reader Stop
SM – Start Message

Figure 8.5 Extended Binary-Coded Decimal-Interchange Code (EBCDIC).

provided with an escape,* 1101100. Few differences exist between the ASCII and CCITT No. 5 codes.

The ASCII and CCITT No. 5 codes are known as "computable codes," where the letters of the alphabet and all other characters and graphics are assigned values in continuous binary sequence; thus these codes are in the native binary language of today's common digital computers. The CCITT No. 2 (ITA No. 2) is not a computable code and when used with a computer, often requires special processing.

The EBCDIC (extended binary-coded decimal-interchange code) is similar to the ASCII but is a true 8-bit code. The eighth bit is used as an added bit to "extend" the code, providing 256 distinct code combinations for assignment. Figure 8.5 illustrates the EBCDIC code.

*An "escape" is a code sequence indicating that the succeeding sequences have an interpretation that differs from the conventional meanings of the code in use. In other words, an escape has been made from the code.

5 ERROR DETECTION AND ERROR CORRECTION

5.1 Introduction

In data transmission one of the most important design goals is to minimize the error rate. Error rate may be defined as the ratio of the number of bits incorrectly received to the total number of bits transmitted. On many data circuits the design objective is an error rate better than one error in 1×10^5 (often expressed 1×10^{-5}), and for telegraph circuits, one error in 1×10^4.

One method for minimizing the error rate would be to provide a "perfect" transmission channel, one that will introduce no errors in the transmitted information at the output of the receiver. However, that perfect channel can never be achieved. Besides improvement of the channel transmission parameters themselves, error rate can be reduced by forms of a systematic redundancy. In old-time Morse code, words on a bad circuit were often sent twice; this is redundancy in its simplest form. Of course, it took twice as long to send a message; this is not very economical if the number of useful words per minute received is compared to channel occupancy.

This illustrates the trade-off between redundancy and channel efficiency. Redundancy can be increased such that the error rate could approach zero. Meanwhile, the information transfer on efficiency across the channel would also approach zero. Thus unsystematic redundancy is wasteful and merely lowers the rate of useful communication. On the other hand, maximum efficiency could be obtained in a digital transmission system if all redundancy and other code elements, such as "start" and "stop" elements, parity bits, and other "overhead" bits were removed from the transmitted bit stream. In other words, the channel would be 100% efficient if all bits transmitted were information bits. Obviously, there is a trade-off of cost and benefits somewhere between maximum efficiency on a data circuit and systematically added redundancy (see Chapter 11, Sections 6 and 7).

5.2 Throughput

Throughput of a data channel is the expression of how much data are put through. In other words, throughput is an expression of channel efficiency. The term gives a measure of *useful* data put through the communication link. These data are directly useful to the computer or DTE (data-terminal equipment).

Therefore, on a specific circuit throughput varies with the raw-data rate, is related to the error rate and the type of error encountered (whether burst or random), and varies according to the type of error detection and correction system used, the message-handling time, and the block length from which we must subtract overhead bits such as parity, flags, and cyclic redundancy checks. Throughput and the operational features of data circuits are described in detail in Chapter 11.

5.3 The Nature of Errors

In data transmission an error is a bit that is incorrectly received; for instance 1 is transmitted in a particular time slot and the element received in that slot is interpreted as a 0. Bit errors occur either as single random errors or bursts of error. For instance, lightning or other forms of impulse noise often cause bursts of error, where many contiguous bits show a very high number of bits in error. The IEEE defines error burst as "a group of bits in which two successive erroneous bits are always separated by less than a given number of correct bits."

5.4 Error Detection and Correction Defined

The data system engineer differentiates between error detection and error correction. Error detection identifies that a symbol has been received in error. As discussed earlier, parity is primarily used for error detection. However, parity bits add redundancy and thus decrease channel efficiency or throughput.

Error correction corrects the detected error. Basically, there are two types of error-correction technique: forward acting (FEC) and two-way error correction [automatic repeat request (ARQ)]. This latter system uses a return channel (backward channel). When an error is detected, the receiver signals this fact to the transmitter over the backward channel, and the block* or packet of information containing the error is transmitted again. Forward-acting error correction utilizes a type of coding that permits a limited number of errors to be corrected at the receiving end by means of software (or hardware) implemented at both ends.

5.4.1 Error Detection. There are various arrangements or techniques available for the detection of errors. All error-detection methods involve some form of redundancy, those additional bits or sequences that can inform the system of the presence of error or errors. The parity discussed in Section 4.3 was character parity, and its weaknesses were presented. Commonly the data system engineer refers to such parity as *vertical redundancy checking* (VRC).

Another form of error detection utilizes longitudinal redundancy checking (LRC), which is used in block transmission where a data message consists of one or more blocks. Remember that a block is a specific group of digits or data characters sent as a "package." An LRC character, often called a BCC (block check character) or frame check sequence is appended at the end of each block. The BCC verifies the total number of 1's and 0's in the columns of the block (vertically). The receiving end sums the 1's (or the 0's) in the block, depending on the parity convention for the system. If that sum does not

*A "block" is a group of digits (data characters) transmitted as a unit over which a coding procedure is usually applied for synchronization and error-control purposes.

correspond to the BCC, an error (or errors) exists in the block. The LRC ameliorates much of the problem of undetected errors that could slip through with VRC, if used alone. It is simply the extension of parity in two dimensions. The LRC method is not foolproof, however, as it uses the same thinking of VRC. Suppose errors occur such that two 1's are replaced by two 0's in the second and third bit positions of characters 1 and 3 in a certain block. In this case the BCC would read correctly at the receive end and the VRC would pass over the errors as well. A system using both LRC and VRC is obviously more immune to undetected error than either system implemented alone. A more effective method of error detection is CRC (cyclic redundancy check), which is based on a cyclic code and is used in block transmission with a BCC. In this case the transmitted BCC represents the remainder of a division of the message block by a "generating polynomial."

Mathematically, a message block can be treated as a function, such as

$$a_n X^n + a_{n-1} X^{n-1} + a_{n-2} X^{n-2} + \cdots + a_1 X + a_0$$

where coefficients a are set to represent a binary number. Consider the binary number 11011, which is represented by the polynomial

$$\begin{matrix} 1 & 1 & 0 & 1 & 1 \\ a_4 & a_3 & a_2 & a_1 & a_0 \end{matrix}$$

and then becomes

$$X^4 + X^3 + X + 1$$

Or consider another example

$$\begin{matrix} 0 & 1 & 1 & 0 & 1 \\ a_4 & a_3 & a_2 & a_1 & a_0 \end{matrix}$$

which becomes

$$X^3 + X^2 + 1$$

The CRC character used as the BCC is the remainder of a data polynomial divided by the generating polynomial.

More specifically, if a data polynomial $D(X)$ is divided by the generating polynomial $G(X)$, the result is a quotient polynomial $Q(X)$ and a remainder polynomial $R(X)$ or

$$\frac{D(X)}{G(X)} = Q(X) + \frac{R(X)}{G(X)}$$

The CRC character in most applications is 16 bits in length or two 8-bit bytes.

At present there are three standard generating polynomials commonly in use:

1. CRC-16 (ANSI): $X^{16} + X^{15} + X^2 + 1$.
2. CRC (CCITT): $X^{16} + X^{12} + X^5 + 1$.
3. CRC-12: $X^{12} + X^{11} + X^3 + X^2 + X + 1$.

Of course, again, if the computed BCC at the receive end differed from the BCC received from the transmit end, there would be an error (or errors) in the received block.

Reference 31 states that CRC-12 provides error detection of bursts of up to 12 bits in length. Additionally, 99.955% of error bursts up to 16 bits in length. The CRC-16 provides detection of bursts up to 16 bits in length. Additionally, 99.955% of error bursts greater than 16 bits can be detected.

5.5 Forward-Acting Error Correction

Forward-acting error correction (FEC) uses certain binary codes that are designed to be self-correcting for errors introduced by the intervening transmission media. In this form of error correction the receiving station has the ability to reconstitute messages containing errors.

The codes used in FEC can be divided into two broad classes: block codes and convolutional codes. In block codes information bits are taken k at a time, and c parity bits are added, checking combinations of the k information bits. A block consists of $n = k + c$ digits. The code consists of $2k$ words, each n digits long. When used for the transmission of data, block codes may be systematic. A systematic code is one in which the information bits occupy the first k positions in a block and are followed by the $(n - k)$ check digits.

Still another block code is the group code, where the modulo 2 sum of any two n-bit code words is another code word. Modulo 2 addition is denoted by the symbol $\oplus$. It is a binary addition without the "carry" or $1 + 1 = 0$, and we do *not* carry the 1. Summing 10011 and 11001 in modulo 2, we get 01010.

The minimum Hamming distance is a measure of the error-detection and-correction capability of a code. This "distance" is the minimum number of digits in which two encoded words differ. For example, to detect E digits in error, a code of a minimum Hamming distance of $(E + 1)$ is required. To correct E errors, a code must display a minimum Hamming distance of $(2E + 1)$. A code with a minimum Hamming distance of 4 can correct a single error *and* detect two digits in error.

A convolution(al) code is another form of coding used for error correction. As the word "convolution" implies, this is one code wrapped around or convoluted on another. It is the convolution of an input-data stream and the response function of an encoder. The encoder is usually made up of shift registers. Modulo 2 adders are used to form check digits each of which is a binary function of a particular subset of the informations digits in the shift register.

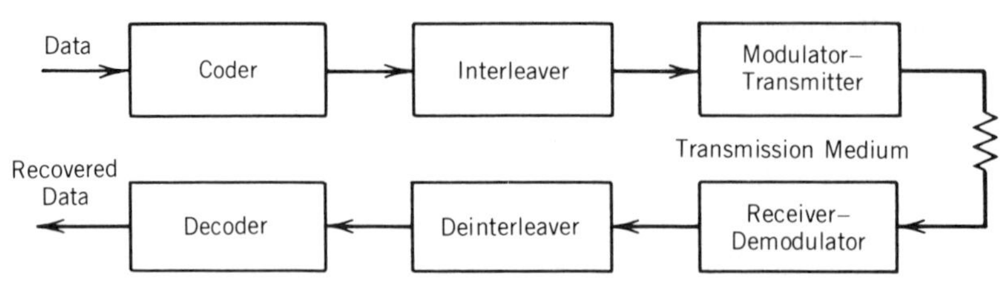

Figure 8.6 FEC scheme for a channel with burst errors.

Error performance can also be improved by the assistance of a microprocessor in the decoder. Mark or space decisions are not hard or irrevocable decisions; rather, these are called "soft" decisions. In this case a tag of 3 bits is attached to each received digit to indicate the confidence level of the decision in the demodulator before processing. After processing, when errors are indicated, the bits with the lowest confidence level are changed from 0 to 1 or 1 to 0, as the case may be.

The codes discussed up to this point are effective at detecting and correcting errors that are random, where the cause of error is due exclusively to perturbations by additive white Gaussian noise and limited signal power. On many transmission circuits burst noise is encountered where combatting such noise is by extending the duration of a bit. Such bursts, however, may have a duration from 10 ms to 100 ms. Typically, these are "hits" due to impulse noise or radio system fades. These fades may have durations from 1–15 s.

One forward-acting error-correcting code that can combat error bursts quite efficiently is the Hagelbarger code. Two requirements must be fulfilled with this code: the burst must be no longer than 8 digits in length, and there must be at least 91 correct digits between bursts. To pay for this capability, efficiency is reduced to 75% or, in other words, redundancy is added; 1 digit in 4 are check digits.

Another method that is now in wide use on military satellite data circuits is interleaving. Interleaving makes burst errors look like random errors. On the transmit side of a circuit an interleaver is inserted between the coder and the modulator, as shown in Figure 8.6. The interleaver shuffles the coded symbols around in a pseudorandom manner. At the distant end receiver a deinterleaver is inserted between the demodulator and the decoder. The deinterleaver uses the same pseudorandom sequence as at the transmit end. It shuffles the coded symbols back to their proper location.

5.6 Error Correction with Feedback Channel

Two-way or feedback error correction is used widely today on data and some telegraph circuits. Such a form of error correction is called ARQ. The letter sequence ARQ derives from the old Morse and telegraph signal, "automatic repeat request."

In most modern data systems block transmission is used, and the block is of a convenient length of characters sent as an entity. That "convenient" length is an important consideration, as we see in Chapter 1. One "convenient" number relates to the standard "IBM" card with 80 columns. With 8 bits per column, a block of 8 × 80 or 640 bits would be desirable as data text so we could transmit an IBM card in each block. In fact, one such operating system, Autodin, bases block length on that criterion, with blocks 672 bits long. The remaining bits, those in excess of 640, are overhead and check bits.

Optimal block length is a trade-off between block length and error rate, or the number of block repeats that may be expected on a particular circuit. Longer blocks tend to amortize overhead bits better (see Chapter 11) but are inefficient regarding throughput when an error rate is high. Under these conditions, long blocks tend to tie up a circuit with longer retransmission periods.

ARQ is based on the block or packet transmission concept. When a receiving station detects an error, it requests a repeat of the block in question from the transmitting station. That request is made on a "feedback" channel, which may be a channel especially dedicated for that purpose or may be the return side of a full-duplex link. Such an especially dedicated return channel is generally slow speed, often 75 bps, whereas the forward channel may be 2400 bps or better. For further discussion of ARQ, see Chapter 11.

6 THE dc NATURE OF DATA TRANSMISSION

6.1 Loops

Binary data are transmitted on a dc loop. More correctly, the binary data end instrument delivers to the line and receives from the line one or several dc loops. In its most basic form a dc loop consists of a switch, a dc voltage, and a termination. A pair of wires interconnects the switch and termination. The voltage source in data and telegraph work is called the *battery*, although the device is usually electronic, deriving the dc voltage from an ac power line source. The battery is placed in the line to provide voltage(s) consistent with the type of transmission desired. A simplified dc loop is shown in Figure 8.7.

6.2 Neutral and Polar dc Transmission Systems

Nearly all dc data and telegraph systems functioning today are operated in either a neutral or a polar mode. The words "neutral" and "polar" describe the manner in which battery is applied to the dc loop. On a "neutral" loop, following the mark–space convention of Table 8.1, battery is applied during marking (1) conditions and is switched off during spacing (0). Current therefore flows in the loop when a mark is sent and the loop is closed. Spacing is indicated on the loop by a condition of no current. Thus we have the two conditions for binary transmission, an open loop (no current flowing) and a

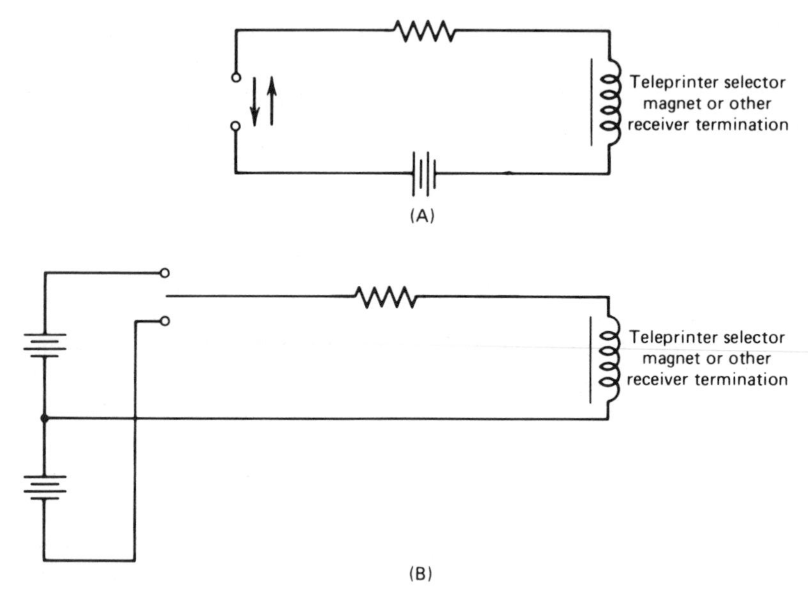

Figure 8.7 Simplified diagram of a dc loop with (A) neutral and (B) polar keying.

closed loop (current flowing). Keep in mind that we could reverse this, namely, change the convention (Table 8.1) and assign spacing to a condition of current flowing or closed loop and marking to a condition of no current or an open loop. This is sometimes done in practice and is called "changing the sense." Either way, a neutral loop is a dc loop circuit where one binary condition is represented by the presence of voltage and the flow of current, and the other by the absence of voltage–current. Figure 8.7A illustrates a neutral loop.

Polar transmission approaches the problem a little differently. Two batteries are provided, one "negative" and the other "positive." During a condition of marking, a positive battery is applied to the loop, following the convention of Table 8.1, and a negative battery is applied to the loop during the spacing condition. In a polar loop, current is always flowing. For a mark or binary 1 it flows in one direction and for a space or binary 0, it flows in the opposite direction. Figure 8.7B shows a simplified polar loop.

7 BINARY TRANSMISSION AND THE CONCEPT OF TIME

7.1 Introduction

Time and timing are most important factors in digital transmission. For this discussion, consider a binary end instrument sending out in series a continuous run of marks and spaces. Those readers who have some familiarity with the Morse code will recall that the spaces between dots and dashes told the

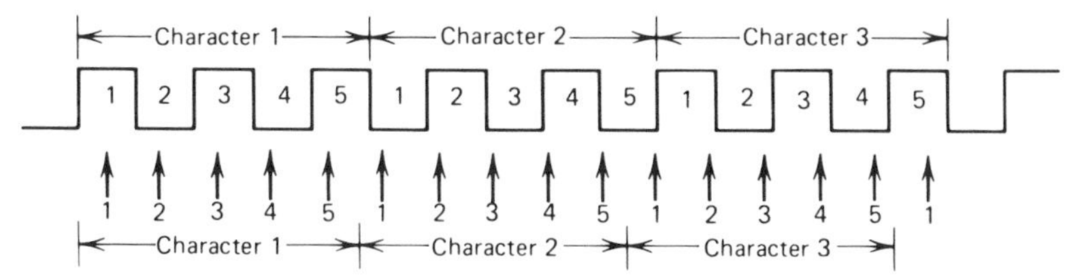

Figure 8.8 Five-unit synchronous bit stream with timing error.

operator where letters ended and where words ended. With the sending device or transmitter delivering a continuous series of characters to the line, each consisting of five, six, seven, eight, or nine elements (bits) per character, a receiving device that starts its print cycle when the transmitter starts sending and subsequently is perfectly in step with the transmitter can be expected to provide good printed copy and few, if any, errors at the receiving end.

It is obvious that when signals are generated by one machine and received by another, the speed of the receiving machine must be the same or very close to that of the transmitting machine. When the receiver is a motor-driven device, timing stability and accuracy are dependent on the accuracy and stability of the speed of rotation of the motors used. Most simple data–telegraph receivers sample at the presumed center of the signal element. It follows, therefore, that whenever a receiving device accumulates timing error of more than 50% of the period of one bit, it will print in error.

The need for some sort of synchronization is shown in Figure 8.8. A five-unit code is employed, and three characters transmitted sequentially are shown. Sampling points are shown in Figure 8.8 as vertical arrows. Receiving timing begins when the first pulse is received. If there is a 5% timing difference between the transmitter and receiver, the first sampling at the receiver will be 5% away from the center of the transmitted pulse. At the end of the tenth pulse or signal element the receiver may sample in error. The eleventh signal element will, indeed, be sampled in error, and all subsequent elements will be errors. If the timing error between transmitting machine and receiving machine is 2%, the cumulative error in timing would cause the receiving device to print all characters in error after the twenty-fifth bit.

7.2 Asynchronous and Synchronous Transmission

In the earlier days of printing telegraphy, "start–stop" transmission, or asynchronous operation, was developed to overcome the problem of synchronism. Here timing starts at the beginning of a character and stops at the end. Two signal elements are added to each character to signal the receiving device that a character has begun and ended.

For example, consider a five-element code such as CCITT No. 2 (see Figure 8.1). In the front of a character an element called a "start space" is added, and

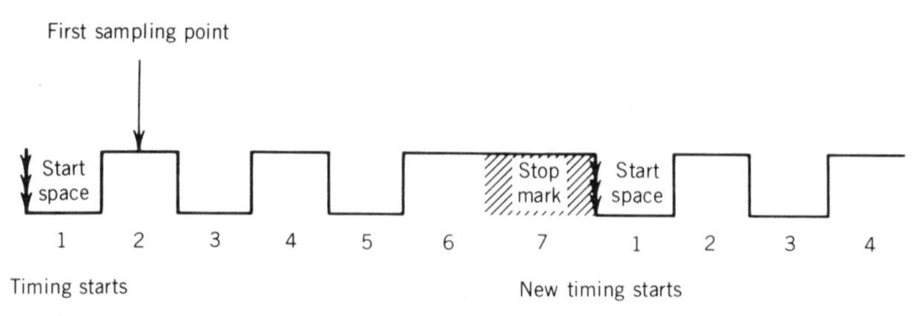

Figure 8.9 Five-unit start – stop stream of bits with a 1.5-unit stop element.

a stop mark is inserted at the end of each character. To send the letter Y in Figure 8.1, the receiving device starts its timing sequence on the first signal element, which is a space or 0, followed by 10101, which is the code sequence for the character Y, followed by a stop mark, which terminates the timing sequence, as shown in Figure 8.9. In such an operation, timing errors can accumulate only inside each character. Suppose the receiving device is again 5% slower or faster than its transmitting counterpart; now the fifth information element will be no more than 30% displaced in time from the transmitted pulse and well inside the 50% or halfway point for correct sampling to take place.

In start–stop transmission information signal elements are each of the same duration, which is the duration or pulse width of the start element. The stop element has an indefinite length or pulse width beyond a certain minimum. If a steady series of characters is sent, the stop element is always of the same width or has the same number of unit intervals. Consider the transmission of two Y's, 0101011010101111 → 11111. The start space (0) starts the timing sequence for six additional elements, which are the five code elements in the letter Y and the stop mark. Timing starts again on the mark-to-space transition between the stop mark of the first Y and the start of the second. Sampling is carried out at pulse center for most asynchronous systems. Note that a continuous series of marks is sent at the end of the second Y; thus the signal is a continuation of the stop element or just a continuous mark. It is the mark-to-space transition of the start element that tells the receiving device to start timing a character.

Minimum lengths of stop elements vary. The preceding example above shows a stop element of one-unit interval duration (1 bit). Some are 1.42-unit intervals, others are of 1.5- and 2-unit interval duration. The proper semantics of data–telegraph transmission would describe the code of the previous paragraph as a five-unit start–stop code with a one-unit stop element.

A primary objective in the design of telegraph and data systems is to minimize errors received or to minimize the error rate. There are two prime causes of errors, noise and improper timing relationships. With start–stop systems a character begins with a mark-to-space transition at the beginning of the start space. Then 1.5-unit intervals later the timing causes the receiving device to sample the first information element, which simply is a mark or space decision. The receiver continues to sample at one-bit intervals until the stop mark is received. In start–stop systems the last information bit is most susceptible to cumulative timing errors. Figure 8.9 is an example of a five-unit start–stop bit stream with a 1.5-unit stop element.

Another problem in start–stop systems is the mutilation of the start element. Once this happens, the receiver starts a timing sequence on the next mark-to-space transition it sees and then continues to print in error until, by chance, it cycles back properly on a proper start element.

Synchronous data–telegraph systems do not have start and stop elements but consist of a continuous stream of information elements or bits (see Figure 8.8). The cumulative timing problems eliminated in asynchronous (start–stop) systems are present in synchronous systems. Codes used on synchronous systems are often seven-unit codes with an extra unit added for parity, such as the ASCII or CCITT No. 5 codes. Timing errors tend to be eliminated because the exact rate at which the bits of information are transmitted is known.

If a timing error of 1% were to exist between transmitter and receiver, not more than 100 bits could be transmitted until the synchronous receiving device would be off in timing by the duration of 1 bit from the transmitter, and all bits received thereafter would be in error. Even if timing accuracy were improved to 0.05%, the correct timing relationship between transmitter and receiver would exist for only the first 2000 bits transmitted. It follows, therefore, that no timing error whatsoever can be permitted to accumulate since anything but absolute accuracy in timing would cause eventual malfunctioning. In practice, the receiver is provided with an accurate clock that is corrected by small adjustments, as explained in Section 7.3.

7.3 Timing

All currently used data-transmission systems are synchronized in phase and symbol rate in some manner. Start–stop synchronization has already been discussed. All fully synchronous transmission systems have timing generators or clocks to maintain stability. The transmitting device and its companion receiver at the far end of the circuit must maintain a timing system. In normal practice, the transmitter is the master clock of the system. The receiver also has a clock that in every case is corrected by some means to its transmitter's master clock equivalent at the far end.

Another important timing factor is the time it takes a signal to travel from the transmitter to the receiver. This is called *propagation time*. With velocities of propagation as low as 20,000 mi/s, consider a circuit 200 mi in length. The propagation time would then be 200/20,000 s or 10 ms. Ten milliseconds is the time duration of 1 bit at a data rate of 100 bps; thus the receiver in this case must delay its clock by 10 ms to be in step with its incoming signal. Temperature and other variations in the medium may also affect this delay, as well as variations in the transmitter master clock.

There are basically three methods of overcoming these problems. One is to provide a separate synchronizing circuit to slave the receiver to the transmitter's master clock. However, this wastes bandwidth by expending a voice channel or subcarrier just for timing. A second method, which was quite widely used until several years ago, was to add a special synchronizing pulse for groupings of information pulses, usually for each character. This method was similar to start–stop synchronization and lost its appeal largely because of the wasted information capacity for synchronizing. The most prevalent system in use today is one that uses transition timing, where the receiving device is automatically adjusted to the signaling rate of the transmitter by sampling the transitions of the incoming pulses. This type of timing offers many advantages, particularly automatic compensation for variations in propagation time. With this type of synchronization the receiver determines the average repetition rate and phase of the incoming signal transition and adjusts its own clock accordingly.

In digital transmission the concept of a transition is very important. The transition is what really carries the information. In binary systems the space-to-mark and mark-to-space transitions (or lack of transitions) placed in a time reference contain the information. In sophisticated systems, decision circuits regenerate and retime the pulses on the occurrence of a transition. Unlike decision circuits, timing circuits that reshape a pulse when a transition takes place must have a memory in case a long series of marks or spaces is received. Although such periods have no transitions, they carry meaningful information. Likewise, the memory must maintain timing for reasonable periods in case of circuit outage. Note that synchronism pertains to both frequency and phase, and that the usual error in high-stability systems is a phase error (i.e., the leading edges of the received pulses are slightly advanced or retarded from the equivalent clock pulses of the receiving device). Once synchronized, high-stability systems need only a small amount of correction in timing (phase). Modem internal timing systems may have a long-term stability of 1×10^{-8} or better at both the transmitter and receiver. At 2400 bps, before a significant timing error can build up, the accumulated time difference between transmitter and receiver must exceed approximately 2×10^{-4} s. Whenever the circuit of a synchronized transmitter and receiver is shut down, their clocks must differ by at least 2×10^{-4} s before significant errors take place. This means that the leading edge of the receiver-clock equivalent timing pulse is 2×10^{-4} in advance or retarded from the leading edge of the pulse received from the distant end. Often an idling signal is sent on synchronous data circuits during periods of no traffic to maintain the timing. Some high-stability systems need resynchronization only once a day.

Note that thus far in our discussion we have considered dedicated data circuits only. With switched (dial-up) synchronous circuits, the following problems exist:

- No two master clocks are perfect phase synchronization.
- The propagation time on any two paths may not be the same.

Thus such circuits will need a time interval for synchronization for each call setup before traffic can be passed.

To summarize, synchronous data systems use high-stability clocks, and the clock at the receiving device is undergoing constant but minuscule corrections to maintain an in-step condition with the received pulse train from the distant transmitter, which is accomplished by responding to mark-to-space and space-to-mark transitions. The important considerations of digital network timing are also discussed in Chapter 10.

7.4 Distortion

It has been shown that the key factor in data transmission is timing. Although the signal must be either a mark or space, this alone is not sufficient. The marks and spaces (or 1's and 0's) must be in a meaningful sequence based on a time reference.

In the broadest sense, distortion may be defined as any deviation of a signal in any parameter, such as time, amplitude, or wave shape, from that of the ideal signal. For data and telegraph binary transmission, distortion is defined as a displacement in time of a signal transition from the time that the receiver expects to be correct. In other words, the receiving device must make a decision as to whether a received signal element is a mark or a space. It makes the decision during the sampling interval, which is usually at the center of where the received pulse or bit should be; thus it is necessary for the transitions to occur between sampling times and preferably halfway between them. Any displacement of the transition instants is called "distortion." The degree of distortion suffered by a data signal as it traverses the transmission medium is a major contributor in determining the error rate that can be realized.

Telegraph and data distortion is broken down into two basic types, systematic and fortuitous. Systematic distortion is repetitious and is broken down into bias distortion, cyclic distortion, and end distortion, which is more common in start–stop systems. Fortuitous distortion is random and characterized by a displacement of a transition from the time interval in which it should have occurred. Distortion caused by noise spikes or other transients in the transmission medium may be included in this category. Characteristic distortion is still another type and is caused by transients in the modulation process that then appear in the demodulated signal.

Figure 8.10 shows some examples of distortion. Figure 8.10A is a binary signal without distortion, and Figure 8.10B shows the sampling instants, which should ideally occur in the center of the pulse to be sampled. From this we can see that the displacement tolerance is nearly 50%; namely, the point of sample could be displaced by up to 50% of a pulse width and still record the mark or space condition present without error. However, the sampling interval does require a finite amount of time thus in actual practice, the permissible displacement is somewhat less than 50%. Figures 8.10C, D show the two

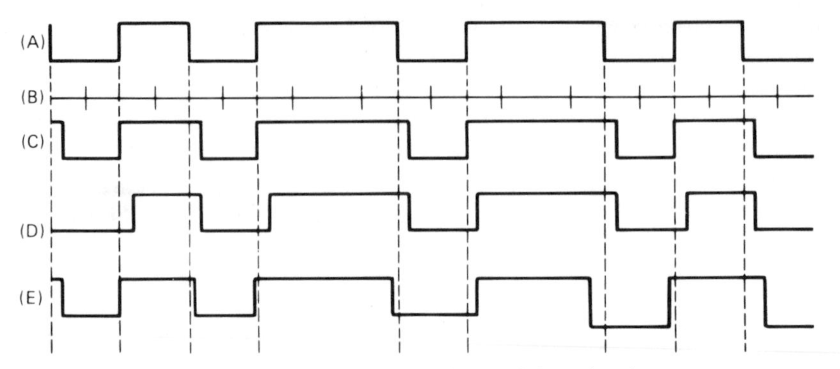

Figure 8.10 Three typical distorted data signals.

typical types of bias distortion. Spacing bias is shown in Figure 8.10C, where all the spacing impulses are lengthened at the expense of the marking impulses. Figure 8.10D shows marking bias, which is the reverse; the marking impulses are lengthened at the expense of the spaces. Figure 8.10E shows fortuitous distortion, which is random.

Figure 8.11 shows distortion that is more typical of start–stop transmission. Figure 8.11A is an undistorted start–stop signal. Figure 8.11B shows cyclic or repetitive distortion typical of mechanical transmitters. In this type of distortion the marking elements may increase in length for a period of time and then the spacing elements will increase in length. Figure 8.11C shows peak distortion. Identifying the type of distortion present on a signal often gives a clue to the source or cause of distortion. Distortion-measurement equipment measures the displacement of the mark-to-space transition from the ideal of the digital signal. If a transition occurs too near to the sampling point, the signal element is liable to be in error.

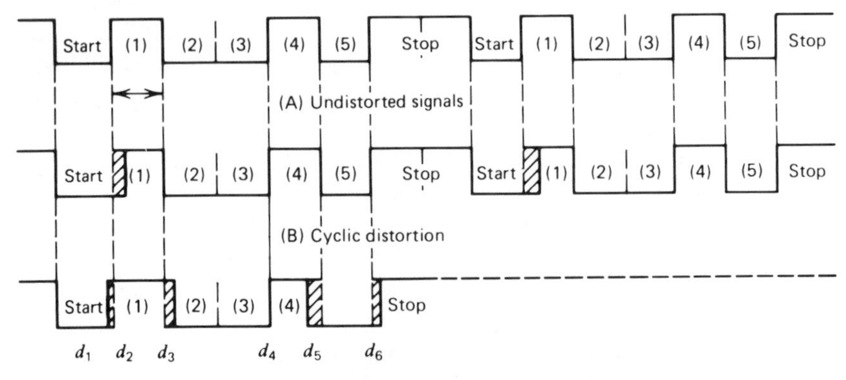

Figure 8.11 Distorted telegraph signals illustrating cyclic and peak distortion. Peak distortion appears at transition d_5.

7.5 Bits, Bauds, and Words per Minute

There is much confusion among transmission engineers in handling some of the semantics and simple arithmetic used in data and telegraph transmission, especially the terms "baud" and "bit." The bit has been defined previously. Now the term "words per minute" is introduced. A "word" in our telegraph and data language consists of six characters, usually five letters, numbers, or graphics and a space. All signal elements transmitted must be counted, such as "carriage return" and "line feed." Remember that the unit interval and the bit are synonymous. Let us look at some examples:

1. A channel is transmitting at 75 bps using a 5-unit start and stop code with a 1.5-unit stop element. Thus for each character there are 7.5 unit intervals (7.5 bits). Therefore the channel is transmitting at 100 wpm.

$$\frac{75 \times 60}{6 \times 7.5} = 100 \text{ wpm}$$

2. A system transmits in CCITT No. 5 code at 1500 wpm with parity. How many bits per second are being transmitted?

$$\frac{1500 \times 8 \times 6}{60} = 1200 \text{ bps}$$

The baud is the unit of modulation rate. In binary transmission systems, the number of bauds is equal to the number of bits per second. Thus a modem in a binary system transmitting to the line 110 bps has a modulation rate of 110 bauds. In multilevel (M-ary systems), the number of bauds is indicative of the number of transitions per second; it is synonymous with symbol rate. The baud is more meaningful to the transmission engineers concerned with the line side of a modem. This concept is discussed at greater length in Section 10.5. Table 8.2 also illustrates the concept.

The period or time duration of 1 bit is another parameter of interest. This is the inverse of the data rate in bits per second. A 75-bps system has a bit period of 1/75 or 0.01333 s. A 45.45-bps system has a bit period of 1/45.45 or 0.0220022 s, and 2400 bps has a bit period of 1/2400 s or 416.7 μsec. This simple technique is only valid for non-return-to-zero (NRZ) waveforms (see Section 7.6).

7.6 Digital Data Waveforms

Digital symbols may be represented in many different ways by electrical signals to facilitate data transmission. All these methods for representing (or coding) digital symbols assign electrical parameter values to the digital symbols. In binary coding, of course, these digital symbols are restricted to two

TABLE 8.2 Medium and High Data-Rate Modems for a Telephone Channel

Data Rate (bps)	Modulation Rate (Bauds)	Modulation	Bits per Hertz[a]	Bandwidth Required (Hz)[a]
1. 2400 Synchronous (e.g., Rec. V.26)	1200	Differential four-phase	2	1200
2. 4800 Synchronous (e.g., Rec. V.27)	1600	Differential eight-phase	3	1600
3. 3600 Synchronous	1200	Differential four-phase, two-level (combined PSK-AM)	3	1200
4. 2400 Synchronous	800	Differential eight-phase	3	800
5. 9600. Synchronous (e.g., Rec. V.29)	2400	Differential four-phase, two-level	4	2400[b]

[a]Theoretical values.
[b]Uses automatic equalizer.

states, space (0) and mark (1). The electrical parameters used to code digital signals are levels (or amplitudes), transitions between different levels, phases (normally 0° and 180° for binary coding), pulse duration, and frequencies or a combination of these parameters. There is a variety of coding techniques for different areas of application, and no particular technique has been found to be optimum for all applications, considering such factors as implementing the coding technique in hardware, type of transmission technique employed, decoding methods at the data sink or receiver, and timing and synchronization requirements.

In this section we discuss several basic concepts of *electrical* coding of binary signals. In this discussion reference is made to Figure 8.12, which graphically illustrates several line coding techniques.

Figure 8.12A shows what is still called by many today "neutral transmission." This was the principal method of transmitting telegraph signals until about 1960. In many parts of the world, neutral transmission is still widely used. First, this waveform is a non-return-to-zero (NRZ) format in its simplest form. "Non-return-to-zero" simply means that if a string of 1's (marks) are transmitted, the signal remains in the mark state with no transitions. Likewise, if a string of 0's is transmitted, there is no transition and the signal remains in the 0 state until a 1 is transmitted. As we can now see, with NRZ transmission, we can transmit information without transitions.

Figures 8.12B, D show the typical "return-to-zero" (RZ) waveform, where, when a continuous string of marks (or spaces) are transmitted, the signal level

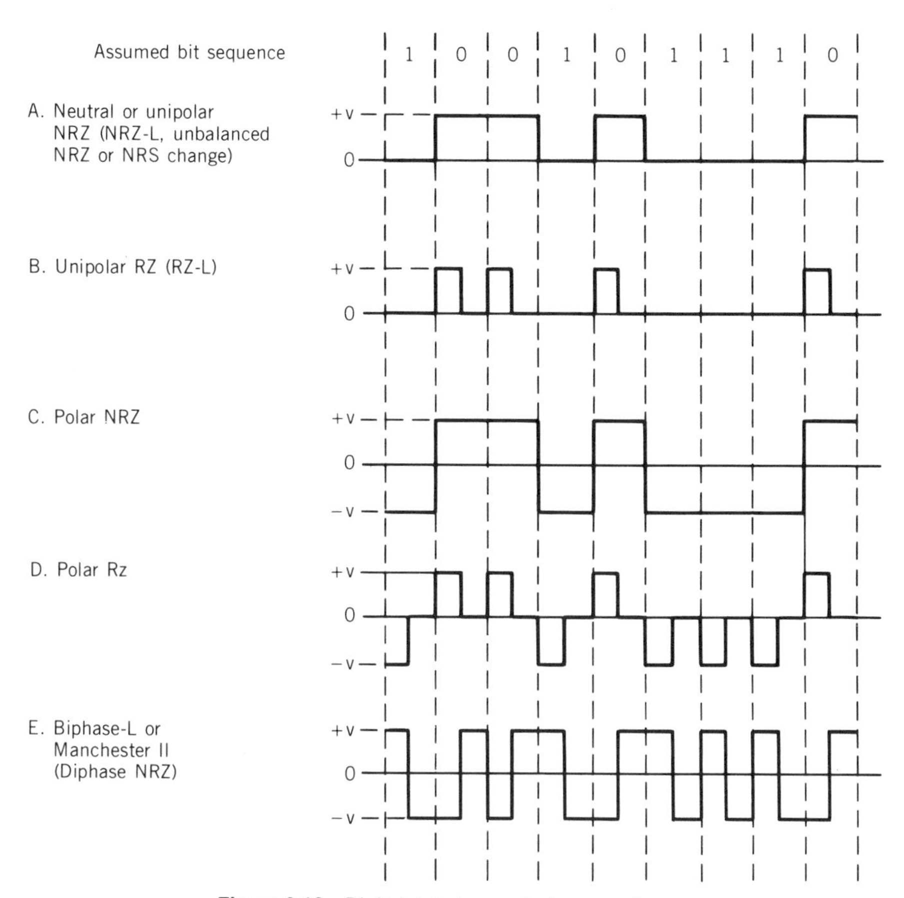

Figure 8.12 Digital data transmission waveforms.

(amplitude) returns to the zero voltage condition at each element or bit. Obviously RZ transmission is much richer in transitions than NRZ.

In section 6.2 we discussed neutral and polar dc transmission systems. Figure 8.12A shows a typical neutral waveform where the two state conditions are 0 V for the mark or 1 condition and some positive voltage for the space or 0 condition. On the other hand, in polar transmission, as shown in Figures 8.12C, D, a positive voltage represents a space and a negative voltage, a mark. With NRZ transmission, the pulse width is the same as the duration of a unit interval or bit. Not so with RZ transmission, where the pulse width is less than the duration of a unit interval. This is because we have to allow time for the pulse to return to the zero condition.

Bi-phase-L or Manchester coding (Figure 8.12E) is a code format that is being used ever more widely on digital systems such as wire pair, coaxial cable, and fiber optics. Here the binary information is carried in the transition. By convention a logic 0 is defined as a positive-going transition and a logic 1 as a negative-going pulse. Note that Manchester coding has a signal transition in the middle of each unit interval (or bit), Manchester coding is a form of phase coding.

The reader should be cognizant of and be able to differentiate between two sets or ways of classifying binary digital waveforms. The first set is *neutral* and *polar*. The second set is *NRZ* and *RZ*. Manchester coding is still another way to represent binary digital data where the transition takes place in the middle of the unit interval. In Chapter 9 one more class of waveform will be introduced: alternate mark inversion (AMI).

8 DATA INTERFACE (PHYSICAL LAYER)

With conventional analog telephony, a modem is a device that takes a digital data dc signal and makes it compatible with the standard 300–3400-Hz voice channel. Digital telephony circuits also require a device to condition data signals for transmission over the digital network. Once ISDN (Chapter 13) is implemented, these conditioning devices will no longer be required.

The modem or digital conditioning device is called *data communication equipment* (DCE). This equipment has two interfaces. The first, which we will discuss here, is on the user side, which is called *data terminal equipment* (DTE) and the applicable interface is the DTE-DCE interface. The second interface is on the line side, which is discussed in Section 10. It should be noted that the DTE-DCE interface is well defined; the interface DCE-line side is less well defined. Figure 8.13 illustrates these two interfaces.

The most well-known DTE-DCE standard has been developed by the (US) Electronics Industries Association (EIA) and is RS-232C, which in 1986 was superseded by RS-232D. The standard is a revision of RS-232C to bring it in line with international standards CCITT Recs. V.24 and V.28 and ISO IS2110. Several other modifications were made to RS-232C to make it more current.

RS-232D and most of the other standards discussed are applicable to the DTE-DCE interface employing serial binary data interchange. It defines signal

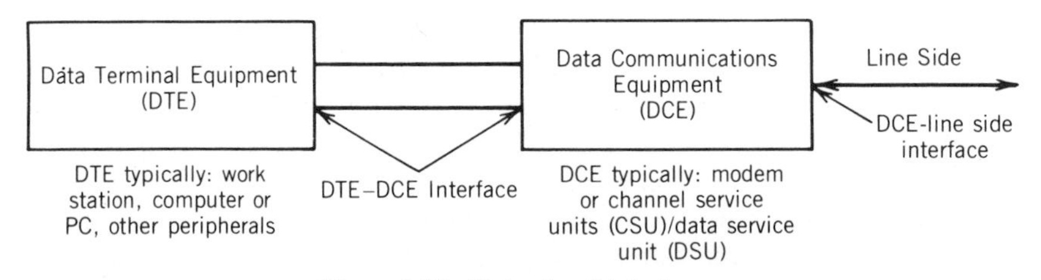

Figure 8.13 Data circuit interfaces.

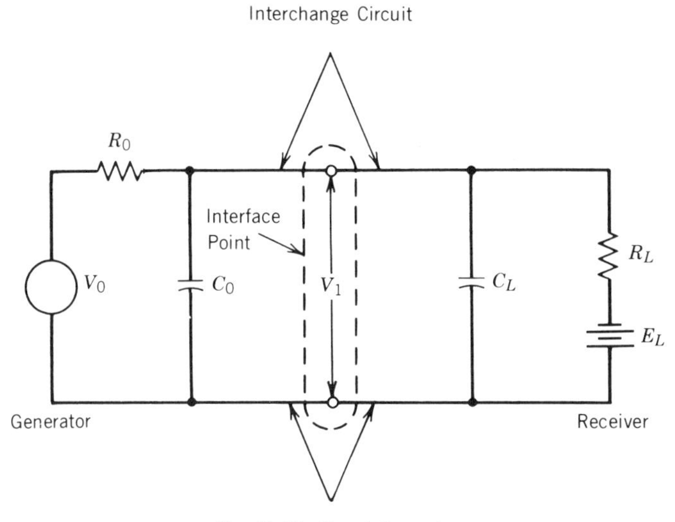

Figure 8.14 Interchange equivalent circuit. Courtesy of Electronics Industry Association, from EIA RS-232D.

V_0 is the open-circuit generator voltage.
R_0 is the generator internal dc resistance.
C_0 is the total effective capacitance associated with the generator, measured at the interface point and including any cable to the interface point.
V_1 is the voltage at the interface point.
C_L is the total effective capacitance associated with the receiver measured at the interface point and including any cable to the interface point.
R_L is the receiver load dc resistance.
E_L is the open-circuit receiver voltage (bias).

characteristics, interface mechanical characteristics, functional description of the interchange circuits, and some standard interfaces for selected communication system configurations. RS-232D is applicable for data signaling rates up to 20,000 bps and for synchronous/asynchronous serial binary data systems.

Section 2.1.3 is quoted from RS-232D which is crucial in the understanding of signal state convention and level:

> For data interchange circuits, the signal shall be considered in the marking condition when the voltage (V_1) on the interchange circuit, measured at the interface point [Figure 8.14], is more negative than minus three volts with respect to circuit AB (signal ground). The signal shall be considered in the spacing condition when the voltage V_1 is more positive than plus three volts with respect to circuit AB. . . . The region between plus three volts and minus three volts is defined as the transition region. The signal state is not uniquely defined when the voltage (V_1) is in this transition region.

During the transmission of data, the marking condition shall be used to denote the binary state *ONE* and the spacing condition shall be used to denote the binary state *ZERO*.

Figure 8.14 shows the interchange equivalent circuit.

Besides RS-232 there are many other interface standards issued by EIA, CCITT, US federal standards, US military standards, and ISO. Each define the DTE-DCE inteface. Several of these are briefly described.

The Electronics Industry Associated released RS-449 along with RS-422 and RS-423. The first describes the mechanical-functional interface of the DTE-DCE. The second two describe electrical interfaces.

Whereas RS-232D provides for a 25-pin interface, RS-449 provides for a 37-pin interface plus an additional 9-pin plug interface. RS-422 deals with a balanced electrical interface and RS-423 with an unbalanced electrical interface (i.e., one lead is grounded). The state transition region for these two standards is between +2 V and −2 V at the generator side. The sense is the same as RS-232D.

The applicable US military standard is MIL-STD-188-114. This is a potpourri of RS-232C/D and RS-422/423 parameters. The significant difference reflected in MIL-STD-188-114 is that the receiver has a balanced input even though it is used with an unbalanced generator. The choice of a balanced receiver was done deliberately for the following reasons:

1. Noise immunity and reducing problems of ground potential differences between generator and receiver(s).
2. Convenience of inverting mark and space signaling sense.
3. Uniformity of receivers for economic advantages in mass production.

CCITT has issued a number of recommendations for the DTE-DCE interface. The equivalent standard for RS-232D is Rec. V.24. The following are other pertinent CCITT recommendations:

Rec. V.10, "Electrical Characteristics for Unbalanced Double-Current Interchange Circuits for General Use with Integrated Circuit Equipment in the Field of Data Communications." The term *double-current* is synonymous with polar transmission discussed in Sections 6.2 and 7.6.

Rec. V.11, "Electrical Characteristics for Balanced Double-Current Interchange Circuits for General Use with Integrated Circuit Equipment in the Field of Data Communications."

Rec. V.28, "Electrical Characteristics for Unbalanced Double-Current Interchange Circuits."

Rec. V.31, "Electrical Characteristics for Single-Current Interchange Circuits Controlled by Contact Closure." Single-current refers to *neutral transmission*.

9 DATA INPUT - OUTPUT DEVICES

The following paragraphs are intended to give the reader a general familiarity with data-subscriber equipment, which we refer to as input–output (I/O) devices.* Such equipment converts user information (data or messages) into electrical signals and vice versa. The working human interface of a communication system is the I/O device. The data source is the input device, and the data sink is the output device.

Data input–output devices handle paper tape, punched cards, magnetic tape, drums, disks, visual displays, and printed page copy. Input devices may be broken down into the following categories:

Keyboard sending units.
Card readers.
Paper tape readers.
Magnetic tape (disk and drum) readers.
Optical character readers.

Output devices are as follows:

Printers.
Card punches.
Paper tape punches.
Magnetic tape recorders, magnetic cores, hard and floppy disks.
Visual display units (VDU) (CRT and plasma).

Further, these devices may be used as on-line or off-line devices. The latter are not connected directly to the communication system but serve as auxiliary equipment.

Off-line devices are used for tape, card, or floppy disk preparation for eventual transmission. In this case a keyboard is connected to a card punch for card preparation. Also, the keyboard may be connected to a paper tape punch or through a processor to a floppy disk drive. Once tape or cards are prepared, they are handled by on-line equipment, either tape readers or card readers. Intermediate equipment or line buffers supply timing, storage, and serial-to-parallel conversion to the input–output devices. The following table compares

*With other disciplines there may be an ambiguity here. For instance a computer can be a source or sink, which also may be considered I/O equipment.

the terminology of input–output devices for telegraph and data processing.

Data	Telegraph
Keyboard	Keyboard
Tape reader	Transmitter–distributor*
Printer	Teleprinter
Tape punch	Perforator, reperforator
Visual displays	–
Disk drives	–
Mass storage devices	–

The use of punched cards and paper tape as a medium for data storage and transmission is being phased out and being replaced by hard disk and floppy disks and other, more efficient storage devices, such as data storage derivatives of the compact disk (CD).

10 DIGITAL TRANSMISSION ON AN ANALOG CHANNEL

10.1 Introduction

Two fundamental approaches to the practical problem of data transmission are (1) to design and construct a complete, new network expressly for the purpose of data transmission and (2) to adapt the many existing telephone facilities for data transmission. The following paragraphs deal with the latter approach.

Analog transmission facilities designed to handle voice traffic have characteristics that hinder the transmission of dc binary digits or bit streams. To permit the transmission of data over voice facilities (i.e., the telephone network), it is necessary to convert the dc data into a signal within the voice frequency range. The equipment that performs the necessary conversion to the signal is generally called a *modem*, an acronym for *mo*dulator–*dem*odulator.

10.2 Modulation – Demodulation Schemes

A modem modulates and demodulates a carrier signal with digital data signals. The types of modulation used by present-day modems may be one or a combination of the following:

Amplitude modulation, double sideband (DSB).

Amplitude modulation, vestigial sideband (VSB).

Frequency shift modulation, commonly called frequency shift keying (FSK).

Phase shift modulation, commonly called phase shift keying (PSK).

*The distributor performs the parallel-to-serial equivalent conversion of data transmission.

10.2.1 Amplitude Modulation: Double Sideband. With the double-sideband (DSB) modulation technique, binary states are represented by the presence or absence of an audio tone or carrier. More often it is referred to as "on–off telegraphy." For data rates up to 1200 bps, one such system uses a carrier frequency centered at 1600 Hz. For binary transmission, amplitude modulation has significant disadvantages, which include (1) susceptibility to sudden gain change and (2) inefficiency in modulation and spectrum utilization, particularly at higher modulation rates (see CCITT Rec. R.70).

10.2.2 Amplitude Modulation: Vestigial Sideband. An improvement in the amplitude modulation double sideband (DSB) technique results from the removal of one of the information-carrying sidebands. Since the essential information is present in each of the sidebands, there is no loss of content in the process. The carrier frequency must be preserved to recover the dc component of the information envelope. Therefore digital systems of this type use VSB modulation in which one sideband, a portion of the carrier, and a " vestige" of the other sideband are retained. This is accomplished by producing a DSB signal and filtering out the unwanted sideband components. As a result, the signal takes only about 75% of the bandwidth required for a DSB system. Typical VSB data modems are operable up to 2400 bps in a telephone channel. Data rates up to 4800 bps are achieved using multilevel (M-ary) techniques. The carrier frequency is usually located between 2200 Hz and 2700 Hz.

10.2.3 Frequency Shift Modulation. Many data-transmission systems utilize frequency shift modulation (FSK). The two binary states are represented by two different frequencies and are detected by using two frequency-tuned sections, one tuned to each of the 2 bit-frequencies. The demodulated signal is then integrated over the duration of 1 bit, and a binary decision is based on the result.

Digital transmission using FSK modulation has the following advantages: (1) the implementation is not much more complex than an AM system; and (2) since the received signals can be amplified and limited at the receiver, a simple limiting amplifier can be used, whereas the AM system requires sophisticated automatic gain control for operation over a wide level range. Another advantage is that FSK can show a 3–4-dB improvement over AM in most types of noise environment, particularly at distortion threshold (i.e., at the point where the distortion is such that good printing is about to cease). As the frequency shift becomes greater (i.e., a greater frequency separation between the mark and space frequencies), the advantage over AM improves in a noisy environment.

Another advantage of FSK is its immunity from the effects of nonselective level variations, even when they occur extremely rapidly. Thus a major application is on worldwide high-frequency radio transmission where rapid fades are a common occurrence. In the United States FSK has nearly universal application for the transmission of data at the lower data rates (i.e., $\leq$ 1200 bps).

10.2.4 Phase Modulation. For systems using higher data rates, phase modulation becomes more attractive. Various forms are used, such as two-phase, relative phase, and quadrature phase. A two-phase system uses one phase of a the carrier frequency for one binary state and the other phase for the other binary state. The two phases are ideally 180° apart and are detected by a synchronous detector using a reference signal at the receiver that is of known phase with respect to the incoming signal. This known signal operates at the same frequency as the incoming signal carrier and is arranged to be in phase with one of the binary signals. In the relative-phase system a binary 1 is represented by sending a signal burst of the same phase as that of the previous signal burst sent. A binary 0 is represented by a signal burst of a phase opposite to that of the previous signal transmitted. The signals are demodulated at the receiver by integrating and storing each signal burst of 1-bit period for comparison in phase with the next signal burst. In the quadrature phase system (QPSK), two binary channels (2 bits) are phase multiplexed onto one tone by placing them in phase quadrature, as shown in the following sketch. An extension of this technique places two binary channels on each of several tones spaced across the voice channel of a typical telephone circuit.

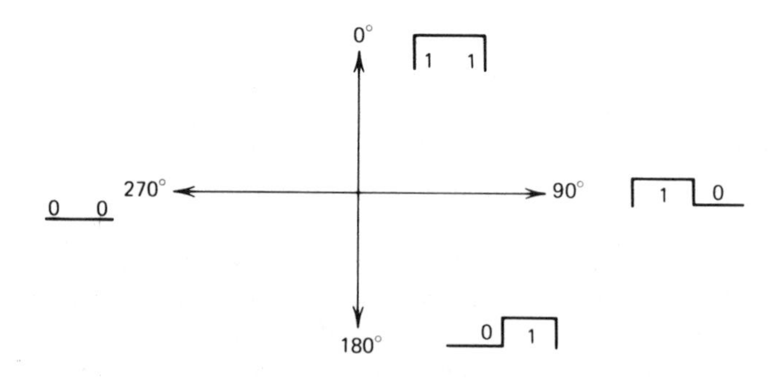

Some of the advantages of phase modulation are as follows:

1. All available power is utilized for intelligence conveyance.
2. The demodulation scheme has good noise-rejection capability.
3. The system yields a smaller noise bandwidth.

A disadvantage of such a system is the complexity of equipment required, compared to FSK systems.

10.3 Critical Parameters

The effect of the various telephone-circuit parameters on the capability of a circuit to transmit data is a most important consideration. The following discussion is intended to familiarize the reader with the problems most likely to be encountered in the transmission of data over analog circuits (e.g., the analog telephone network) and to make certain generalizations in some cases, which can be used to facilitate planning the implementation of data systems.

10.3.1 Phase Distortion. Phase distortion "constitutes the most limiting impairment to data transmission, particularly over telephone voice channels" [3]. When specifying phase distortion, the terms "envelope delay distortion" (EDD) and "group delay" are often used. The IEEE Standard Dictionary states that "envelope delay is often defined the same as group delay, that is the rate of change, with angular frequency, of the phase shift between two points in a network." (See Chapter 5, Section 2.2.)

The problem is that in a band-limited analog system, such as the typical telephone voice channel, not all frequency components of the input signal will propagate to the receiving end in exactly the same elapsed time, particularly on loaded cable circuits and FDM carrier systems. In carrier systems it is the cumulative effect of the many filters used in the FDM equipment. On long-haul circuits the magnitude of delay distortion is generally dependent on the number of carrier modulation stages that the circuit must traverse rather than the length of the circuit. Figure 8.15 shows a typical frequency–delay response curve in milliseconds of a voice channel due to FDM equipment only. For the voice channel (or any symmetrical passband, for that matter), delay increases toward band edge and is minimum around the center portion (around 1800 Hz).

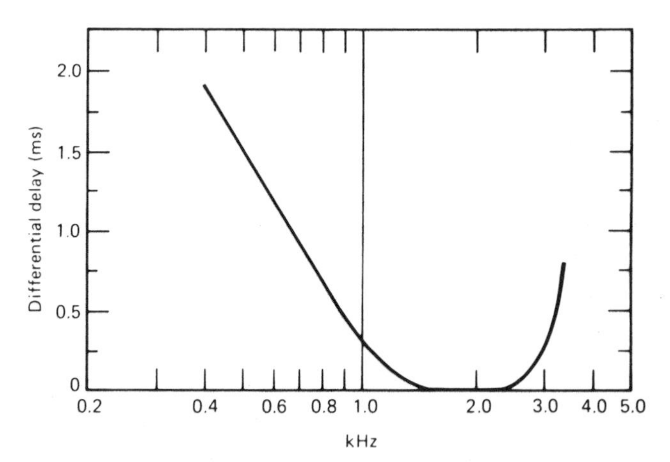

Figure 8.15 Typical differential delay across a voice channel, FDM equipment back to back.

Phase or delay distortion is the major limitation to modulation rate. The shorter the pulse width (the width of 1 bit in binary systems), the more critical will be the EDD parameters. As we discuss in Section 10.5, it is desirable to keep the delay distortion in the band of interest below the period of 1 bit.

10.3.2 Amplitude Response (Attenuation Distortion). Another parameter that seriously affects the transmission of data and can place very definite limits on the modulation rate is amplitude response, also called "attenuation distortion." Ideally, all frequencies across the passband of a channel of interest should undergo the same attenuation. For example, let a −10-dBm signal enter a channel at any frequency between 300 Hz, and 3400 Hz. If the channel has 13 dB of flat attenuation, we would expect an output of −23 dBm at any and all frequencies in the band. This type of channel is ideal but unrealistic in a real working system.

In Rec. G.132, the CCITT recommends no more than 9 dB of attenuation distortion relative to 800 Hz between 400 Hz and 3000 Hz. This figure, 9 dB, describes the maximum variation that may be expected from the reference level at 800 Hz. This variation of amplitude response is often called attenuation distortion. A conditioned channel, such as a Bell System C-4 channel, will maintain a response of −2 dB to +3 dB from 500 Hz to 3000 Hz and −2 dB to +6 dB from 300 Hz to 3200 Hz. Channel conditioning is discussed in Section 10.6.

Considering tandem operation, the deterioration of amplitude response is arithmetically cumulative when sections are added. This is particularly true at band edge in view of channel unit transformers and filters that account for the upper and lower cutoff characteristics. Figure 8.16 illustrates a typical example of amplitude response across FDM carrier equipment (see Chapter 5) con-

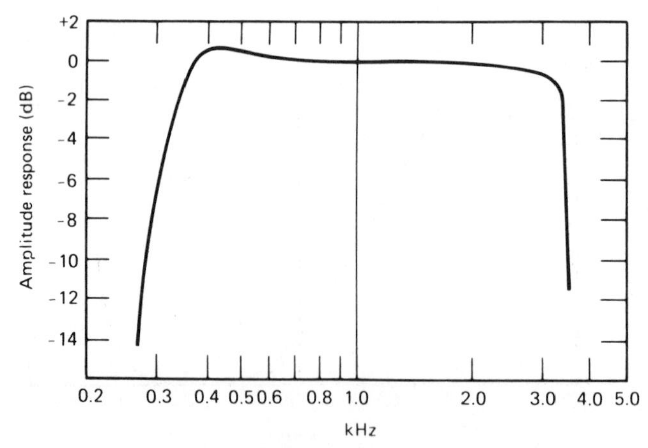

Figure 8.16 Typical amplitude-frequency response across a voice channel; channel modulator, demodulator back to back, FDM equipment.

nected back to back at the voice channel input–output. For additional discussion of attenuation distortion see Chapter 5, Section 2.1.

10.3.3 Noise. Another important consideration in the transmission of data is noise. All extraneous elements appearing at the voice channel output that were not due to the input signal are considered to be noise. For convenience, noise is broken down into four categories: (1) thermal, (2) crosstalk, (3) intermodulation, and (4) impulse. Thermal noise, often called "resistance noise," "white noise," or "Johnson noise," is of a Gaussian nature or completely random. Any system or circuit operating at a temperature above absolute zero inherently will display thermal noise. The noise is caused by the random motions of discrete electrons in the conduction path. Crosstalk is a form of noise caused by unwanted coupling from one signal path into another. It may be caused by direct inductive or capacitive coupling between conductors or between radio antennas. Intermodulation noise is another form of unwanted coupling, usually caused by signals mixing in nonlinear elements of a system. Carrier and radio systems are highly susceptible to intermodulation noise, particularly when overloaded. Impulse noise is a primary source of errors in the transmission of data over telephone networks. It is sporadic and may occur in bursts or discrete impulses called "hits." Some types of impulse noise are natural, such as that from lightning. However, man-made impulse noise is ever increasing, such as that from automobile ignition systems and power lines. Impulse noise may be of a high level in conventional telephone switching centers as a result of dialing, supervision, and switching impulses that may be induced or otherwise coupled into the data-transmission channel. The worst offender in the switching area is the step-by-step exchange.

For our discussion of data transmission, two types of noise are considered, random (or Gaussian) noise and impulse noise. Random noise measured with a typical transmission measuring set appears to have a relatively constant value. However, the instantaneous value of the noise fluctuates over a wide range of amplitude levels. If the instantaneous noise voltage is of the same magnitude as the received signal, the receiving detection equipment may yield an improper interpretation of the received signal and an error or errors will occur. Thus we need some way of predicting the behavior of data transmission in the presence of noise. Random noise or white noise has a Gaussian distribution and is considered representative of the noise encountered on the analog telephone channel (i.e., the voice channel). From the probability distribution curve for Gaussian noise shown in Figure 8.17, we can make some statistical predictions. It may be noted from this curve that the probability of occurrence of noise peaks that have amplitudes 12.5 dB above the rms level is 1 in 10^5. Hence, if we wish to ensure an error rate of 10^{-5} in a particular system using binary polar modulation, the rms noise should be at least 12.5 dB below the signal level [Ref. 3, p. 114]. This simple analysis is only valid for the type of modulation used (i.e., binary polar baseband modulation), assuming that no other factors are degrading the operation of the system and that a

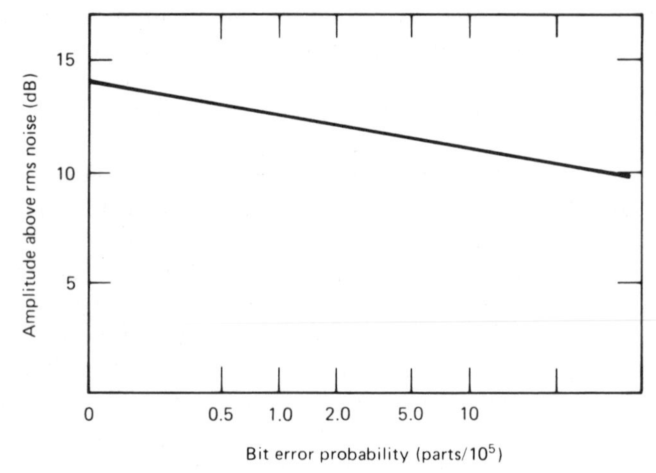

Figure 8.17 Probability of bit error in Gaussian noise; binary polar transmission with a Nyquist bandwidth.

cosine-shaped receiving filter is used. If we were to interject distortion such as EDD into the system, we could translate the degradation into an equivalent signal-to-noise ratio improvement necessary to restore the desired error rate. For example, if the delay distortion were the equivalent of one pulse width, the signal-to-noise ratio improvement required for the same error rate would be about 5 dB, or the required signal-to-noise ratio would now be 17.5 dB.

Let us assume a telephone system where the signal level is −10 dBm at the zero transmission-level point of the system; the rms noise measured at the same point would be −27.5 dBm to retain the error rate of 1 in 10^5. This figure can be significant only if it is related to the actual noise found in a channel. The CCITT recommends no more than 50,000 pW of noise psophometrically weighted (−43 dBmp) on an international connection made up of six circuits in a chain. However, the CCITT states [Rec. G.143(4)] that for data transmission at as high a modulation rate as possible without significant error rate, a reasonable circuit objective for maximum random noise would be −40 dBm0p for leased circuits (impulse noise not included) and −36 dBm0p for switched circuits without compandors. This figure obviously appears quite favorable when compared to the −27 dBm0 (−29.5 dBm0p) required in the preceding example. However, other factors developed later will consume much of the noise margin that appears to be available.

Unlike random noise, which is measured by its rms value when we measure level, impulse noise is measured by the number of "hits" or "spikes" per interval of time above a certain threshold. In other words, it is a measurement of the recurrence rate of noise peaks over a specified level. The word "rate" should not mislead the reader. The recurrence is not uniform per unit time, as the word "rate" may indicate, but we can consider a sampling and convert it to an average. The Bell System (circa 1982, Ref. 2) circuit objective on leased lines is 15 counts in 15 min at -69 dBm with equivalent C-message weighting. The CCITT (Rec. Q.45) states that "in any four-wire international exchange the busy hour impulsive noise counts should not exceed 5 counts in 5 min at a threshold level of -35 dBm0."

Remember that random noise has a Gaussian distribution and will produce peaks at 12.5 dB over the rms value (unweighted) 0.001% of the time on a data-bit stream for an equivalent error rate of 1×10^{-5}. It should be noted that some references use 12 dB, some use 12.5 dB, and others use 13 dB. The 12.5 dB above the rms random noise floor should establish the impulse noise threshold for measurement purposes. We should assume that in a well-designated data-transmission system traversing the telephone network, the signal-to-noise ratio of the data signal will be well in excess of 12.5 dB. Thus impulse noise may well be the major contributor to degradation of the error rate.

When an unduly high error rate has been traced to impulse noise, there are some methods for improving conditions. Noisy areas may be bypassed, repeaters may be added near the noise source to improve signal-to-impulse-noise ratio, or in special cases pulse smearing techniques may be used. This latter approach uses two delay-distortion networks that complement each other such that the net delay distortion is zero. By installing the networks at opposite ends of the circuit, impulse noise passes through only one network* and hence is smeared because of the delay distortion. The signal is unaffected because it passes through both networks.

Signal-to-noise ratio may be traded for the implementation of forward-acting error correction (FEC). This trade-off may be economically feasible or even mandatory, such as in certain digital satellite circuits, to reduce power output of a transmitter or reduce level out of a modem. Reducing power by 3 dB (Figure 8.17) on a circuit displaying a bit error rate of 1×10^{-5} would deteriorate error rate to some 5×10^{-2}. The error rate could be recovered by selecting the proper method of FEC, decoding algorithm and possibly implementing soft mark–space decisions. Soft decision decoding can improve error performance by an equivalent 2 dB. For additional discussion of FEC see Section 5 of this chapter.

10.3.4 Levels and Level Variations. The design signal levels of telephone networks traversing FDM carrier systems are determined by average talker levels, average channel occupancy, permissible overloads during busy hours, and so on. Applying constant amplitude digital data tone(s) over such an equipment of 0 dBm0 on each channel would result in severe overload and intermodulation within the system.

Loading does not affect (metallic) wire systems except by increasing crosstalk. However, once the data signal enters carrier multiplex (voice) equipment, levels must be carefully considered, and the resulting levels most

*This assumes that impulse noise enters the circuit at some point beyond the first network.

probably have more impact on the final signal-to-noise ratio at the far end than anything else. The CCITT (Rec. V.2) recommends that signal power shall not exceed −10 dBm0 in 10 Hz of bandwidth and −13 dBm0 when the portion of nonspeech circuits on an international carrier circuit exceeds 10% or 20%. For multichannel telegraphy −8.7 dBm0 for the composite level, or for 24 channels, each individual telegraph channel would be adjusted for −22.5 dBm0. Even this loading may be too heavy if a large portion of the voice channels is loaded with data. Depending on the design of the carrier equipment, cutbacks to −13 dBm0 or less may be advisable.

In a properly designed transmission system the standard deviation of the variation in level should not exceed 1.0 dB/circuit. However, data communication equipment should be able to withstand level variations in excess of 4 dB.

10.3.5 Frequency-Translation Errors. Total end-to-end frequency-translation error on a voice channel being used for data or telegraph transmission must be limited to 2 Hz (CCITT Rec. G.135); this is an end-to-end requirement. Frequency translations occur largely because of FDM carrier equipment modulation and demodulation steps. Frequency division multiplex carrier equipment widely uses single-sideband suppressed-carrier techniques. Nearly every case of error can be traced to errors in frequency translation (we refer here to deriving the group, supergroup, mastergroup, and its reverse process; see also Chapter 5) and carrier reinsertion frequency offset, where the frequency error is exactly equal to the error in translation and offset or the sum of several such errors. Frequency-locked (e.g., synchronized) or high-stability master carrier generators (1×10^{-8} or 1×10^{-9}, depending on the system), with all derived frequency sources slaved to the master source, are usually employed to maintain the required stability.

Although 2 Hz seems to be a very rigid specification, when added to the possible back-to-back error of the modems themselves, the error becomes more appreciable. Much of the trouble arises with modems that employ sharply tuned filters. This is particularly true of telegraph equipment. But for the more general case, some high-speed data modems occupying the whole voice channel can be designed to withstand greater carrier shifts than can their slower speed counterparts used for asynchronous data or printing telegraphy. On the other hand, V.22, V.27, and V.29 type modems standardized by CCITT require ±1 Hz stability.

10.3.6 Phase Jitter. The unwanted change in phase or frequency of a transmitted signal caused by modulation by another signal during transmission is defined as "phase jitter." If a simple sinusoid is frequency or phase modulated during transmission, the received signal will have sidebands. The amplitude of these sidebands compared to the received signal is a measure of the phase jitter imparted to it during transmission.

Phase jitter is measured in degrees of variation peak to peak for each hertz of transmitted signal. Phase jitter is manifest as unwanted variations in zero crossings of a received signal. It is the zero crossings that data modems use to distinguish marks from spaces. Thus the higher the data rate, the more jitter can affect error rate on the receive bit stream.

The greatest cause of phase jitter in the telephone network is FDM carrier equipment, where it is manifest as undesired incidental phase modulation. Modern FDM equipment derives all translation frequencies from one master frequency source by multiplying and dividing its output. To maintain stability, phase-lock techniques are used; thus the low-jitter content of the master oscillator may be multiplied many times. It follows, then, that there will be more phase jitter in the voice channels occupying the higher baseband frequencies.

Jitter most commonly appears on long-haul systems at rates related to the power line frequency (e.g., 60 Hz and its harmonics and submultiples) or is derived from 20-Hz ringing frequency. Modulation components that we define as "jitter" usually occur close to the carrier ± 300 Hz maximum.

10.4 Channel Capacity

A leased or switched voice channel represents a financial investment. Therefore one goal of the system engineer is to derive as much benefit as possible from the money invested. For the case of digital transmission, this is done by maximizing the information transfer across the system. This section discusses how much information in bits can be transmitted, relating information to bandwidth, signal-to-noise ratio, and error rate. These matters are discussed empirically in Section 10.5.

First, looking at very basic information theory, Shannon stated in his classic paper [13] that if input information rate to a band-limited channel is less than C (bps), a code exists for which the error rate approaches zero as the message length becomes infinite. Conversely, if the input rate exceeds C, the error rate cannot be reduced below some finite positive number.

The usual voice channel is approximated by a Gaussian band-limited channel (GBLC) with additive Gaussian noise. For such a channel, consider a signal wave of mean power of S watts applied at the input of an ideal low-pass filter that has a bandwidth of W (Hz) and contains an internal source of mean Gaussian noise with a mean power of N watts uniformly distributed over the passband. The capacity in bits per second is given by

$$C = W \log_2\left(1 + \frac{S}{N}\right)$$

Applying Shannon's "capacity" formula to an ordinary voice channel (GBLC) of bandwidth (W) 3000 Hz and a signal-to-noise (S/N) ratio of 1023, the

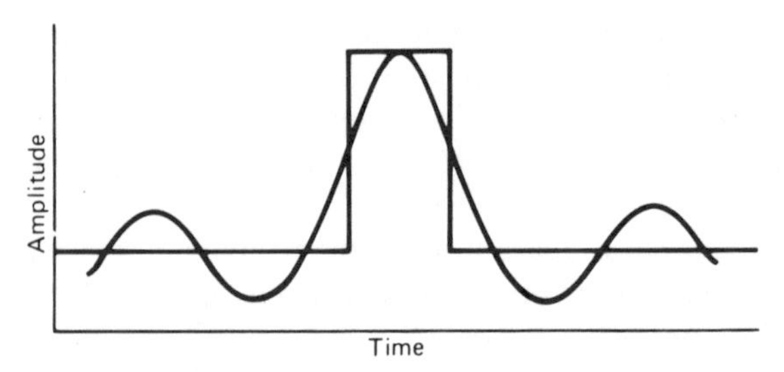

Figure 8.18 Pulse response through a Gaussian band-limited channel (GBLC).

capacity of the channel is 30,000 bps. (Remember that bits per second and bauds are interchangeable in binary systems.) Neither S/N nor W is an unreasonable value. Seldom, however, can we achieve a modulation rate greater than 3000 bauds. The big question in advanced design is how to increase the data rate and keep the error rate reasonable.

One important item not accounted for in Shannon's formula is intersymbol interference. A major problem of a pulse in a band-limited channel is that the pulse tends not to die out immediately, and a subsequent pulse is interfered with by "tails" from the preceding pulse (see Figure 8.18).

Nyquist (Ref. 12) provided another approach to the data-rate problem, this time using intersymbol interference (the tails in Figure 8.18) as a limit [30]. This resulted in the definition of the so-called Nyquist rate = $2W$ symbols/s, where W is the bandwidth (Hz) of a band-limited channel. In binary transmission we are limited to $2W$ bps, where a symbol is 1 bit. If we let $W = 3000$ Hz, the maximum data rate attainable is 6000 bps. Some refer to this as "the Nyquist 2-bit rule."

The key here is that we have restricted ourselves to binary transmission and are limited to $2W$ bps no matter how much we increase the signal-to-noise ratio. The Shannon GBLC equation indicates that we should be able to increase the information rate indefinitely by increasing the signal-to-noise ratio. The way to attain a higher C value is to replace the binary transmission system with a multilevel system, often termed an M-ary transmission system, with $M > 2$. An M-ary channel can pass $2W \log_2 M$ bps with an acceptable error rate. This is done at the expense of signal-to-noise ratio. As M increases (as the number of levels increases), so must S/N increase to maintain a fixed error rate.

Table 8.2 (page 340) shows some typical modem bit rates, corresponding baud rates and bandwidths.

10.5 Some Modem Selection Considerations

The critical parameters that affect data transmission have been discussed; these are amplitude–frequency response (sometimes called "amplitude distortion"), envelope delay distortion, and noise. Now we relate these parameters to the design of data modems to establish some general limits or "boundaries" for equipment of this type. The discussion that follows purposely avoids HF radio considerations.

As stated earlier in the discussion of envelope delay distortion, it is desirable to keep the transmitted pulse (bit) length equal to or greater than the residual differential EDD. Since about 1.0 ms is assumed to be reasonable residual delay after equalization (conditioning), the pulse length should be no less than approximately 1 ms. This corresponds to a modulation rate of 1000 pulses per second (binary). In the interest of standardization (CCITT Rec. V.22), this figure is modified to 1200 bps.

The next consideration is the usable bandwidth required for the transmission of 1200 bps. This figure is approximately 1800 Hz, using modulation methods such as phase shift (PSK), frequency shift (FSK) or double-sideband AM (DSB–AM), and somewhat less for vestigial-sideband AM (VSB–AM). Since delay distortion of a typical voice channel is at its minimum between 1700 Hz and 1900 Hz, the required band, when centered about these points, extends from 800 Hz to 2600 Hz or 1000 Hz to 2800 Hz. From the previous discussion and with reference to Figure 8.15, we can see that the EDD requirement is met easily over the range of 800–2800 Hz.

Bandwidth limits modulation rate. However, the modulation rate in bauds and the data rate in bits per second may not necessarily be the same. This is a very important concept. Suppose a modulator looked at the incoming serial bits stream 2 bits at a time rather than the conventional 1 bit at a time. Now let four discrete signal amplitudes be used to define each of four possible combinations of two consecutive bits such that

$$A_1 = 00$$
$$A_2 = 01$$
$$A_3 = 11$$
$$A_4 = 10$$

where A_1, A_2, A_3, and A_4 represent the four pulse amplitudes. This form of treating 2 bits at a time is called "di-bit coding" (see Section 10.2).

Similarly, we could let eight amplitude levels cover all the possible combinations of three consecutive bits so that with a modulation rate of 1200 bauds it is possible to transmit information at a rate of 3600 bps. Rather than vary amplitude to four or eight levels, phase can be varied. A four-phase system (PSK) could be coded as follows.

$$F_1 = 0^\circ = 00$$
$$F_2 = 90^\circ = 01$$
$$F_3 = 180^\circ = 11$$
$$F_4 = 270^\circ = 10$$

Again, with a four-phase system using di-bit coding, a tone with a modulation rate of 1200 bauds PSK can be transmitting 2400 bps. An eight-phase PSK system at 1200 bauds could produce 3600 bps of information transfer. Obviously, this process cannot be extended indefinitely. The limitation comes from channel noise. Each time the number of levels or phases is increased, it is necessary to increase the signal-to-noise ratio to maintain a given error rate.

Sufficient background has been developed to appraise the data modem for the voice channel. Now consider a data modem for a data rate of 2400 bps. By using quaternary phase shift keying (QPSK or 4-ary PSK) as described earlier, 2400 bps is transmitted with a modulation rate of 1200 bauds. Assume that the modem uses differential phase detection wherein the detector decisions are based on the change in phase between the last transition and the preceding one. Assume the bandwidth of the data modem under consideration to be 1800 Hz. It is now possible to determine whether the noise requirements can be satisfied. Figure 8.17 shows that a 12.5-dB signal-to-noise ratio (Gaussian noise) is required to maintain an error rate 1×10^{-5} for a binary polar (AM) system. It is well established that PSK systems has about 3-dB improvement. In this case only a 9.5-dB* signal-to-noise ratio would be needed, if all other factors were held constant (no other contributing factors). Assume the input from the line to be -10 dBm0 to satisfy loading conditions. To maintain the proper signal-to-noise ratio, the channel noise must be down to -19.5 dBm0. To improve the modulation rate without expense of increased bandwidth, quaternary phase-shift keying (four-phase) is used. This introduces a 3-dB noise degradation factor, bringing the required noise level down to -22.5 dBm0.

Now consider the effects of EDD. It has been found that for a four-phase differential system, this degradation will amount to 6 dB if the permissible delay distortion is one pulse length. This impairment brings the noise requirement down to -28.5 dBm0 of average noise power in the voice channel. Allow 1 dB for frequency-translation error or other factors, and the noise requirement is now down to -29 dBm0. If the transmission level were reduced by 3 dB to -13 dBm0 (see Section 10.3.4), the noise level must be reduced downward another 3 dB to -32.5 dBm0 (or 19.5 dB signal-to-noise ratio). Thus it can be seen that to achieve a certain error rate for a given modulation rate, several modulation schemes should be considered. It is safe to say that the noise requirement will fall somewhere between -25 dBm0 and -40 dBm0. This is well inside the CCITT figure of -43 dBmp referenced in Section 10.3.3.

10.6 Equalization

Of the critical circuit parameters mentioned in Section 10.3, two that have severely deleterious effects on data transmission can be reduced to tolerable

*Actually 9.6-dB theoretical ratio (E_b/N_0).

limits by *equalization*. These two are amplitude–frequency response (amplitude distortion) and EDD (delay distortion).

The most common method of performing equalization is the use of several networks in tandem. Such networks tend to flatten response and, in the case of amplitude response, add attenuation increasingly toward channel center and less toward its edges. The overall effect is one of making the amplitude response flatter. The delay equalizer operates in a similar manner. Delay increases toward channel edges parabolically from the center. To compensate, delay is added in the center much like an inverted parabola, with less and less delay added as the band edge is approached. Thus the delay response is flattened at some small cost to absolute delay, which has no effect in most data systems. However, care must be taken with the effect of a delay equalizer on an amplitude equalizer and conversely, of an amplitude equalizer on the delay equalizer. Their design and adjustment must be such that the flattening of the channel for one parameter does not entirely distort the channel for the other.

Another type of equalizer is the transversal type of filter, which is useful where it is necessary to select among, or to adjust, several attenuation (amplitude) and phase characteristics. The basis of the filter is a tapped delay line to which the input is presented. The output is taken from a summing network that adds or sums the outputs of the taps. Such a filter is adjusted to the desired response (equalization of both phase and amplitude) by adjusting the contributions for each tap.

If the characteristics of a line are known, a common method of equalization is predistortion of the output signal of the data set. Some devices use a shift register and a summing network. If the equalization needs to be varied, a feedback circuit from the receiver to the transmitter would be required to control the shift register. This type of dynamic predistortion is practical for binary transmission only.

A major drawback of all the equalizers discussed (with the exception of the latter one with the feedback circuit) is that they are useful only on dedicated or leased circuits where the circuit characteristics are known and remain fixed. Obviously, a switched circuit would require a variable automatic equalizer, or conditioning would be required on every circuit in the switched system that would be transmitting data.

Circuits are usually equalized on the receiving end. This is called *postequalization*. An equalizer must be balanced and must present the proper impedance to the line. Administrations may choose to condition trunks and attempt to eliminate the need to equalize station lines; the economy of considerably fewer equalizers is obvious. In addition, each circuit that would possibly carry high-speed data in the system would have to be equalized, and the equalization must be sufficient for any possible combination to meet the overall requirements. If equalization requirements become greater (i.e., parameters more stringent), the maximum number of circuits (trunks) in tandem may have to be restricted still further (i.e., < 12).

Equalization to meet amplitude–frequency response requirements is less exacting in the overall system than is envelope delay.* Equalization for envelope delay and its associated measurements are time consuming and expensive. In general, envelope delay is arithmetically cumulative. If there is a requirement of overall envelope delay distortion of 1 ms for a circuit between 1000 Hz and 2600 Hz, then for 3 links in tandem, each link must be better than 333 μs between the same frequency limits. For 4 links in tandem, each link would have to be at least 250 μs. In practice, accumulation of delay distortion is not entirely arithmetical, as it results in a reduction of requirements by about 10%. Delay distortion tends to be inversely proportional to the velocity of propagation. Loaded cables display greater delay distortion than do nonloaded cables. Likewise, with sharp filters a greater delay is experienced for frequencies approaching band edge than for filters with a more gradual cutoff.

In carrier multiplex systems, channel banks contribute more to the overall EDD than does any part of the system. Because channels 1 and 12 of the standard CCITT modulation plan (those nearest the group band edge) suffer additional delay distortion because of group filter effects and in some cases, supergroup filters, the system engineer should allocate channels for data transmission near group and supergroup center. On long-haul critical-data systems, the data channels should be allocated to through-groups and through-supergroups, minimizing as much as possible the steps of demodulation back to voice frequencies (channel demodulation).

Automatic equalization for both amplitude and delay are effective, particularly for switched data systems. Such devices are self-adaptive and require a short adaptation period after switching, of the order of < 1 s [27]. This can be carried out during synchronization. Not only is the modem clock being "averaged" for the new circuit on transmission of a synchronous idle signal, but the self-adaptive equalizers adjusts for optimum equalization as well. The major drawback of adaptive equalizers is cost.

10.7 Practical Modem Applications

10.7.1 Voice-Frequency Carrier Telegraph. Narrow-shifted FSK transmission of digital data is commonly referred to as voice-frequency telegraph (VFTG) and voice-frequency carrier telegraph (VFCT).

In practice, VFCT techniques handle data rates of up to 1200 bps by a simple application of FSK modulation. The voice channel is divided into segments or frequency-bounded zones or bands. Each segment represents a data or telegraph channel, each with a frequency-shifted subcarrier.

For proper end-to-end system interface, it is convenient to use standard modulation plans, particularly on international circuits. For the far-end demodulator to operate with the near-end modulator, the former must be tuned to the same center frequency and accept the same shift. Center frequency is

*Remember "envelope delay distortion" (EDD) is a measure of phase distortion.

that frequency in the center of the passband of the modulator–demodulator. The shift is the number of hertz that the center frequency is shifted up and down in frequency for the mark–space condition. From Table 8.1, by convention, the mark condition is the center frequency shifted downward and the space, upward. For modulation rates below 80 bps, bandpasses have either 170-Hz (CCITT Rec. R.39) or 120-Hz bandwidths, with frequency shifts of ± 42.5 Hz or ± 30 Hz, respectively. The CCITT recommends (R.31) the 120-Hz channels for operating at 50 bps and below; however, some administrations operate these channels at higher modulation rates.

The number of tone telegraph or data channels that can be accommodated on a voice channel depends partly on the usable voice-channel bandwidth. For high-frequency radio with a voice-channel limit on the order of 3 kHz, 16 channels may be accommodated using 170-Hz spacing (170 Hz between center frequencies). On the nominal 4-kHz voice channel 24 VFCT channels may be accommodated between 390 Hz and 3210 Hz with 120-Hz spacing, or 12 channels with 240-Hz spacing. This can easily meet standard telephone FDM carrier channels of 300–3400 Hz.

Some administrations use a combination of voice and telegraph data simultaneously on a telephone channel. This technique is commonly referred to as "voice plus" or $S + D$ (speech plus derived). There are two approaches to this technique. The first is recommended by CCITT and is used widely by INTELSAT order wires. It places five telegraph channels (channels 20 through 24) above a restricted voice band with a roofing filter* near 2500 Hz. Speech occupies a band between 300 Hz and 2500 Hz. Up to five 50-bps telegraph channels appear above 2500 Hz. The second approach removes a slot from the center of the voice channel into which up to two telegraph channels may be inserted. The slot is a 500-Hz band centered on 1275 Hz.

However, some administrations use a slot for telegraphy of frequencies 1680 Hz and 1860 Hz by either amplitude or frequency modulation (FSK) (see CCITT Rec. R.43). The use of speech plus should be avoided on trunks in large networks because it causes degradation to speech and also precludes the use of the channel for higher-speed data. In addition, the telegraph channels have to be removed before going into two-wire telephone service (i.e., at the hybrid or term set); otherwise, service drops to half-duplex on telegraph.

10.7.2 CCITT Recommended Medium Data Rate Modems. CCITT has issued a number of modem recommendations under the V series, in other words the recommendation number begins with the letter V. Some of the lower data rate modems are covered under the R recommendations.

In normal practice binary FSK is used for the transmission of data for data rates up to and including 1200 bps. For rates above 75 bps, one approach is to apply CCITT Rec. R.31. For instance, a 150-bps circuit would use 240-Hz spacing, a 300-bps circuit would use 480-Hz spacing, and so forth up to one

*This is a low-pass filter.

1200-bps channel. Of course, there would be concurrent increases for the frequency shift, such as ±60 Hz for the 150-bps channel and ±120 Hz for the 300-bps channel.

The first of the V recommendations of CCITT is V.21, which provides 300-bps service; two such channels can be accommodated on a standard VF channel. The mean frequency of each channel is 1080 Hz and 1750 Hz, and the frequency shift on each channel is ±100 Hz.

CCITT Rec. V.22 standardizes a 1200-bps modem that can operate either in the start–stop or synchronous mode. Two data rates are available—600 bps and 1200 bps—and the modulation rate is 600 bauds in either case. This modem is an exception to the rule stated about FSK. Rec. V.22 specifies a PSK waveform: BPSK for 600 bps operation and QPSK for 1200 bps operation. The modulation approach is similar to that described in Section 10.5 for PSK and QPSK operation; however, the phase values are different for the appropriate di-bit values. For instance, bit 00 is 90°. The recommendation also provides for a standardized scrambler/descrambler. Scrambling is not done for security but to randomize bit patterns. This allows for more uniform dispersal of energy and eliminates long strings of 1's or 0's when an NRZ format is used. Long strings of 1's or 0's give no transitions, which is an unfavorable condition for a data receiver that is looking for transitions to keep its timing circuit in sync with the far end transmitter.

A related CCITT recommendation is Rec. V.22 bis, which is for full-duplex 2400-bps operation in a frequency division mode on two-wire circuits. The "go" and "return" channels are separated by frequency division. Modulation is quadrature amplitude modulation utilizing four levels of phase and four levels of amplitude. A scrambler and an adaptive equalizer are also specified. The modulation rate in all cases is 600 bauds. For instance, for 2400-bps operation the data stream to be transmitted is divided into groups of 4 consecutive bits (quadbits). The first two bits of the quadbit are encoded as a phase quadrant change relative to the quadrant occupied by the preceding signal element.

Recommendation V.23 is titled "600/1200-Baud Modem Standardized for Use in the General Switched Telephone Network." It uses FSK and is compatible with synchronous and start–stop bit streams. Center frequencies are 1500 Hz for 600-bps operation and 1700 Hz for 1200-bps operation. Frequency shifts are ±200 Hz and ±400 Hz, respectively.

A 2400-bps operation modem is covered in CCITT Rec. V.26 which uses a conventional QPSK waveform. The center frequency is 1800 Hz. Eighteen-hundred hertz is chosen because for most VF channels it sits at about midpoint on the phase delay curve. Often we find envelope delay distortion specified for the band 1000–2600 Hz and, of course, 1800 Hz is just midpoint.

CCITT Rec. V.26 bis is similar in most respects to Rec. V.26. CCITT Rec. V.26 ter is also similar to Rec. V.26, but specified for half-duplex operation as well as full duplex. It provides for echo cancellation.

A 4800-bps capability modem is described in CCITT Rec. V.27, employing a manual equalizer for use on leased circuits. It uses 8-ary PSK on an 1800-Hz carrier frequency, and the modulation rate is 1600 bauds. In this case each transition carries 3 bits of information as shown below.

Tribit Value	Phase Change
001	0°
000	45°
010	90°
011	135°
111	180°
110	225°
100	270°
101	315°

As in many similar CCITT recommendations for data modems, there is provision for a 75-bps backward channel for ARQ commands such as ACK (acknowledgment a message received without error) and NACK (negative acknowledgement a result of a message received in error requesting repetition). There is provision for a scrambler having a generating polynomial of $1 + X^{-6} + X^{-7}$. Before data traffic exchange a synchronization signal is specified to sync the demodulator and to establish descrambler synchronization. The synchronization signal consists of a continuous 180° phase reversals on the line for about 9 ms followed by continuous 1's until full synchronization is achieved as indicated on circuit 106 (ready for sending). (Circuit 106, see CCITT Rec. V.24 or Ref. 26).

CCITT Rec. V.27 bis is similar in many respects to Rec. V.27. Some of the differences are that it has provision for 2400-bps operation using quadrature PSK as well as 4800 bps using 8-ary PSK. It is provided with an automatic equalizer, which uses rapid convergence techniques and thus requires a short training period. By a training period we mean the time it takes to adjust its constants in phase and amplitude, providing a best fit amplitude and phase response line equivalent of the automatic equalizer.

The third V.27 recommendation (CCITT Rec. V.27 ter) is a 4800/2400-bps modem standardized for use in the *general switched telephone network*. This modem is the Rec. V.27 bis equivalent for use on dial-up data connections.

Operation at 9600 bps over a standard VF channel is now covered by two CCITT recommendations: Recs. V.29 and V.32. Rec. V.29 is titled "9600 Bits per Second Modem Standardized for Use on Point-to-Point 4-Wire Leased Telephone-Type Circuits." Some of its principal characteristics are:

- Fallback rates of 7200 bps and 4800 bps.
- Capable of operating in duplex or half-duplex mode with continuous or controlled carrier.
- Combined amplitude and phase modulation with synchronous mode operation.

- Inclusion of an automatic adaptive equalizer.
- Optional inclusion of a multiplexer for combining data rates of 7200 bps, 4800 bps, and 2400 bps.

The carrier frequency is 1700 Hz, and the modulation is quadrature amplitude modulation (QAM) with 16 distinct phase-amplitude states. Thus the theoretical bit packing is 4 bits per hertz. As in other modems of similar type, a data scrambler/descrambler is provided based on a pseudorandom sequence generator.

CCITT Rec. V.32 is a recommendation for a family of two-wire duplex modems at data signaling rates up to 9600 bps for use on the general switched telephone network and on leased telephone-type circuits. The principal characteristics of these modems are:

- Duplex mode of operation on the GSTN (general switched telephone network) and two-wire point-to-point leased circuits.
- Channel separation by echo cancellation techniques.
- Quadrature amplitude modulation (QAM) for each channel with synchronous line transmission at 2400 bauds.
- Any combination of the following data rates may be implemented: 9600 bps, 4800 bps, and 2400 bps, all synchronous.
- At 9600 bps operation to alternative modulation schemes, one using 16 carrier states and one using trellis coding with 32 carrier states. It notes that modems providing the 9600 bps data rate shall be capable of interworking using the 16-state alternative.
- Exchange of rate sequences during start-up to establish the data rate, coding, and any other special facilities.

CCITT Rec. V.35 describes a wide-band modem for 48 kbps operation in the FDM group 60–108 kHz. CCITT Rec. V.36 is similar for access to a PCM 64 kbps channel with capabilities of extension over analog facilities. CCITT Rec. V.37 provides for data rates greater than 72 kbps using the FDM group 60–108 kHz.

10.8 Serial-to-Parallel Conversion for Transmission on Impaired Media

The transmission medium, in most cases the voice channel, often cannot support a high data rate, even with conditioning. The impairments may be due to poor amplitude–frequency response, EDD, or excessive impulse noise. One possible solution is to convert the high-speed serial bit stream at the dc level (e.g., demodulated) to a number of lower-speed parallel bit streams. A technique widely used is to divide a 2400-bit serial stream into 16 parallel streams, each carrying 150 bps. If each slower stream is di-bit coded (2 bits at a time; discussed in Section 10.5) and applied to a QPSK tone modulator, the modulation rate on each subchannel is reduced in this case to 75 bauds. The equivalent period for a di-bit interval is 1/75 s or 13 ms.

There are two obvious advantages to this technique. First, each subchannel has a comparatively small bandwidth and thus looks at a small and tolerable segment of the total delay across the channel. The EDD impairment impacts less on slower-speed channels. Second, there is less chance of a noise burst or hit of impulse noise smearing the subchannel signal beyond recognition. If the duration of noise burst is less than half the pulse width, the data pulse can be regenerated and the pulse will not be in error. The longer the pulse width, the less chance of disturbance from impulse noise. In this case the symbol interval or pulse width has had an equivalent lengthening by a factor of 32.

10.9 Parallel-to-Serial Conversion for Improved Economy of Circuit Usage

Long, high-quality (conditioned) toll telephone circuits are costly to lease or are a costly investment. The user is often faced with a large number of slow-speed circuits (50–300 bps) that originate in one general geographic location, with a general destination to another common geographic location. If we assume these to be 75-bps circuits (100 wpm), which are commonly

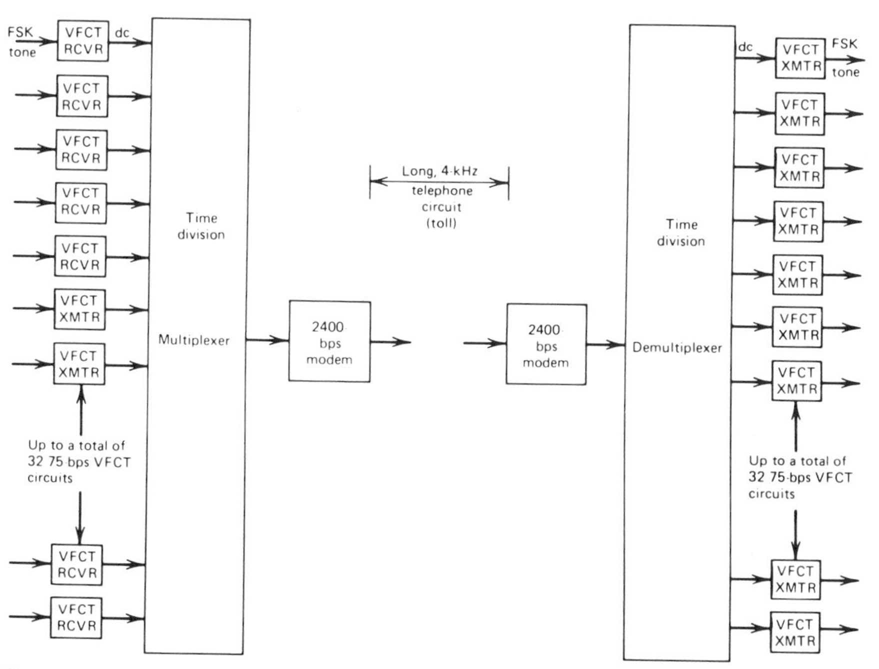

Figure 8.19 Typical application of parallel-to-serial conversion: VFCT, voice frequency carrier telegraph; rec., receiver (converter); xmt, transmitter (keyer); bps, bits per second; FSK, frequency shift keying.

encountered in practice, then only 18 to 24 telegraph channels can be transmitted on a high-grade telephone channel by conventional voice-frequency telegraph techniques (see Section 10.7.1).

Circuit economy can be affected using a data–telegraph time-division multiplexer. A typical application of this type is illustrated in Figure 8.19, which shows one direction of transmission only. Here incoming, slow-speed VFCT channels are converted to equivalent dc bit streams. Up to 32 of these bit streams serve as input to a time-division multiplexer in the application illustrated in Figure 8.19. The output of the multiplexer is a 2400-bit synchronous serial bit stream. This output is fed to a conventional 2400-bit modem. At the receiving end the 2400 bps serial stream is demodulated to dc and fed to the equivalent demultiplexer. The demultiplexer breaks the serial stream down back to the original 75-bps circuits.

Figure 8.19, which illustrates the concept of parallel-to-serial conversion, does not show clocking or other interconnecting circuitry. By use of a time-division mutiplexer, the circuit utilization can be increased by a factor of nearly 2. Whereas only about 18 75-bps circuits can be transmitted on a good telephone channel by conventional VFCT means, an equivalent of up to 32 such circuits can be transmitted on the same channel by means of the time-division multiplexer.

11 FACSIMILE

11.1 Background

Facsimile (fax) is a method of graphics communication. It permits the electrical transmission of printed or pictorial matter from one location to another with a reasonably faithful copy permanently recorded at the receiving end. Facsimile transmission over telephone lines has had widespread application since the 1920s. Before World War II its principal application was the transmission of weather maps and "wirephoto," which is the transmission of pictures for newspapers.

Today facsimile is showing new vigor, particularly in the office environment. Of course, it is still used for the transmission of weather maps and news photographs. In the commercial field facsimile is used for the transmission of way bills by trucking firms, signature verification by banks, and payment notices. In publishing it is used to expedite graphic communication, to dispatch news copy from satellite offices or bureaus to a newspaper's main newsroom, and to eliminate duplication of typesetting effort between separate printing facilities. It is also used in engineering and production for the transmission of engineering drawings. In law enforcement fax is used to transmit mugshots, and high-resolution fax is used for the transmission of fingerprints. "Electronic mail" is yet another application. Indeed fax is becoming ubiquitous in the office environment.

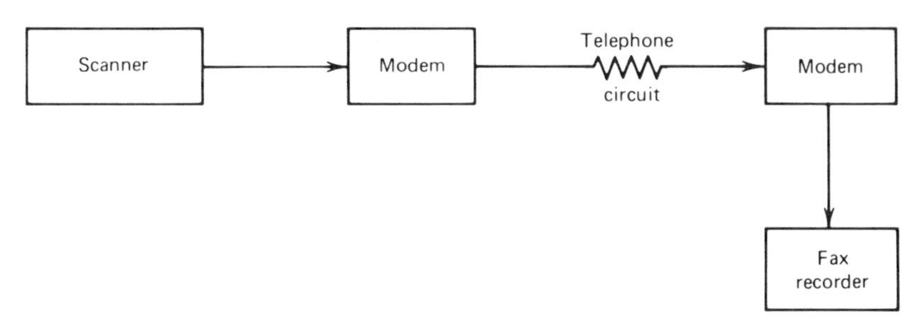

Figure 8.20 A facsimile (fax) system.

It may be said that data–telegraph methods of transmitting graphics are more rapid than fax. Whereas a standard printed page may take only 30 sec to transmit by digital data techniques (transmitting 300 characters per second), that same page may take up to 1 min to transmit by facsimile. Many fax circuits can transmit a standard printed page under 30 s. However, in the case of data transmission, we often forget that to the 30 sec of transmit time we must add the operator keyboard time to compose the page from original copy. With fax, the original copy is only inserted into the machine. Further, fax is less error prone—considerably less, in fact. However, once we enter the domain of computer-to-computer or terminal-to-computer operations, such communication should be left to the techniques of data transmission explored earlier in this chapter.

11.2 Introduction to Facsimile Systems

A fax system consists of some method of converting graphic copy on paper to an electrical equivalent suitable for transmission on a telephone pair, transmission and connection of the pair–telephone circuit to the desired distant user, and the recording–printing of the copy by that user. Basically, we are dealing with analog technology. However, there is a marked trend toward digital techniques. Figure 8.20 is a simplified block diagram of a facsimile system.

11.3 Scanning

The purpose of scanning is to produce an electrical analog signal representing the graphic copy to be transmitted. Scanning can be carried out by (1) a spot of light scanning a fixed copy or (2) the copy moving across a fixed spot of light. In either method the light is reflected from the copy to a photoelectric cell that senses the total variations as a function of the mirror reflection from the copy. Thus a scanner is simply a photoelectric transducer.

There are two approaches to lighting the copy on the scanner, spot, and flood. Both employ the technique of "bouncing" (reflecting) light off the graphic material to be transmitted. The spot type of scanning projects a tiny spot of light onto the surface of the printed copy. The reflection of the spot is then picked up directly by a photoelectric cell or transducer. With flood projection the copy is illuminated with diffuse light in the area of scan. The reflected light is then optically projected through a very small aperture onto the cathode of a photoelectric transducer.

Scanning in modern facsimile systems is electromechanical, either with cylinder scanning or flat-bed scanning. In the former case the copy to be transmitted is wrapped around a cylinder. The cylinder is rotated such as to effect a continuous helical scan of the entire copy. Flat-bed scanners use a feed mechanism. Copy to be transmitted is fed into a slot where the mechanism takes over, slowly advancing the copy through the machine. In another flat-bed method a flying spot is used, where the copy remains at rest and the spot does all the movement. Optical transducers such as those used on facsimile scanners are based on the photoelectric cell or the photomultiplier tube. Most electronic scanners require a highly stabilized dc source.

11.4 Recording

11.4.1 Introduction. Recording is the reproduction from an electric signal of visual copy of graphic material. Like scanning at the transmit end of a fax circuit, recording at the receive end is nearly always electromechanical—drum or flat bed. There are four basic electromechanical processes of recording in use today: (1) electrolytic, (2) electrothermal, (3) electropercussive, and (4) electrostatic. There is one well-established recording process that is not electromechanical. It is photographic and is used in the facsimile reproduction of newspapers.

11.4.2 Electrolytic Recording. Electrolytic recording, the oldest and one of the most popular recording processes, requires a special type of paper that is saturated with an electrolyte. When an electric current is passed through the paper, it tends to discolor. The amount of discoloration or darkness is a function of the current passing through the paper.

For facsimile recorders the electrolytic paper is passed between two electrodes. One is a fixed electrode, a backplate or platen on the machine; the other is the moving stylus. Horizontal lines of varying darkness appear on the paper and the stylus sweeps across the sheet. As each recorded horizontal line is displaced one line width per sweep of the stylus, a pattern begins to take shape. This pattern is a facsimile of the original pattern transmitted from the distant-end scanner.

This older, conventional stylus–backplate recording method is now giving way to what is called "helix-blade" recorders. The concept is basically the same, except that a special drum containing a helix at the rear and a stationary blade in front are used as the two electrodes. The drum-helix makes one complete revolution per scanning line. The rotating drum moving the helix carries out the same function as the moving stylus of the more conventional electrolytic recorders.

11.4.3 Electrothermal Recording. This process of facsimile recording is called "thermal" only because it appears to be so. It is similar to the electrolytic process in that the recording paper is interposed between two electrodes. The recorded pattern is made by arcing electric current through the paper. The arc gives the appearance of burning or a thermal process. A major characteristic of electrothermal recording is the high contrast achieved.

11.4.4 Electropercussive Recording. Electropercussive recording in facsimile is similar to the recording of audio on a record. An amplified facsimile signal is fed to an electromagnetic transducer that actuates a stylus in response to electrical signal variation. If a sheet of carbon paper is interposed between the stylus and a sheet of plain paper, a carbon impression is made on the paper by the vibrations of the stylus in accordance with the signal variation. The amount or intensity of darkness varies in proportion to the variations in strength of the picture signal. Electropercussive recording is also known by the terms *pigment transfer*, *impact*, or *impression recording*.

11.4.5 Electrostatic Recording. There are essentially two types of electrostatic recording, both of which are based on printing the image from a cathode ray tube (CRT): (1) transfer xerography and (2) a direct copy requiring special recording paper. It must be pointed out that the image tube need not necessarily be a CRT. Other image devices with the proper scanning mechanism can be used, such as the crater tube.

11.5 Technical Requirements for Facsimile Transmission

11.5.1 Phasing and Synchronization. Phasing and synchronization of the far-end fax receiver with the transmitting scanner permit the reassembly of picture elements in the same spatial order as when the picture was scanned by the transmitter. Phasing and synchronization must be distinguished. Phasing ensures that the receiving recorder stylus coincides in time and position on the copy at the start of transmission. Synchronization keeps the two this way throughout the transmission of a single graphic copy.

In many systems phasing is carried out by what may be termed a "stop–start" technique, where the receiving end is not really stopped, only retarded until a start-of-stroke signal is received from the transmitting scanner. This is the phasing signal, which usually is a pulse at the scan line rate of fixed amplitude and duration. A phasing sequence often lasts for several seconds before scanning on one end and printing on the other commence. Thus phasing may be called the "retarding of recording stroke until the start of recording stroke coincides with the start of scan stroke." Of course, once the recorder is retarded to permit coincidence, it must be allowed to speed up or "catch up" with the scanner.

Synchronization ensures that the fax recorder remains in perfect step with the transmit scanner. There are three methods of synchronization: (1) common ac power-source frequency, (2) individual stabilized frequency power sources, and (3) transmission of sync signal during picture transmission. Methods 1 and 2 use 50-Hz (in Europe) or 60-Hz (in North America) synchronous motors for scanning and recording. Frequency stability of the ac power grid or the individual ac stabilized frequency power sources should be 1×10^{-5} per day as a minimum. Any degradation in stability from this figure will cause skew of the received picture.

"Slaving" the fax recorder to the distant-end transmitter essentially circumvents the stability problem inherent in the first two methods of synchronization. There are two ways of accomplishing sync. The first uses a sync tone, usually a multiple of 50 Hz or 60 Hz, that is transmitted above or below the picture signal in the voice-frequency passband. Effective filtering is required to separate picture from sync signal. The second method is to transmit pulses in the amplitude domain where the sync pulses are below the white level [or the black level, depending on signal sense (polarity)].

Another method of pulse synchronization that can be continuous throughout picture transmission uses short sync pulses that provide a check on the speed of the recorder drive motor. The pulses are transmitted at the scan line rate and do not interfere with the picture signal by transmitting them during line flyback or at start of line.

11.5.2 Index of Cooperation. The index of cooperation is defined by the IEEE as the product of scan density measured in lines per inch (LPI) times the effective stroke length. This is a basic standard for facsimile transmission. If a scanner–transmitter and receiver–recorder have the same index of cooperation, they are considered to be compatible. This only means that the received copy is a faithful reproduction of the transmitted copy. It does not mean, however, that they are the same size.

To convert IEEE standards to the CCITT index of cooperation, multiply the IEEE index by 0.318. CCITT (Rec. T.1) recommends an index of cooperation of 352 or, alternatively, 264. These are equivalent to the IEEE indices of 1105 and 829, respectively. The World Meteorological Organization (WMO) specifies 576 and 288, which are equivalent to IEEE 1809 and 904, respectively. The EIA recommends IEEE 829, which is equivalent to CCITT 264.

11.5.3 Transmission Methods and Impairments

General. The output of a fax scanner contains electrical transitions representing reflectance changes of scanned copy. These transitions have a frequency component from subaudio, very nearly dc, up through the lower audio range. The transmission problem that the telecommunication engineer must face is to condition or convert this equivalent frequency spectrum so it can pass over a telephone circuit. We remember that the CCITT voice channel is encompassed in the band 300–3400 Hz. The critical fax frequencies are in the subaudio range, as low as 20 Hz.

Modulation. To overcome the frequency-response problem, simple carrier techniques have been adopted. With facsimile transmission the output of the scanner modulates an audio carrier using vestigial sideband modulation with an 1800-Hz carrier. When modulated by a fax signal, the carrier contains about 1300 Hz of information in the lower sideband, with the upper sideband vestiges extending to about 2300 Hz. With this approach the vital frequencies of the output of a facsimile scanner can be transmitted over a telephone voice channel.

Frequency-modulation techniques generally are more desirable than amplitude modulation (as described earlier) for transmission over a switched public telephone network or over a radio transmission media. The reason basically is that FM tends to be more impervious to noise. A common standard for FM transmission of fax is 1500 Hz for white and 2300 Hz for black.

Impairments. For good fax signal quality, a signal-to-noise ratio of 30 dB or better is required. This value is relative to maximum signal power level. Level should be maintained to at least 0.4 dB or better during transmission. Level variation has little effect on FM; this, then, is another good reason to use FM over most switched telephone circuits and on HF radio. Delay distortion in the audio range of interest should be maintained within ± 300 μs. This requirement may be difficult to meet on some switched telephone circuits.

11.6 Reduction of Redundancy

The transmission of facsimile is extremely redundant, particularly the transmission of white. With the advent of electronic data processing (EDP) and the microprocessor, methods of reducing redundancy have been developed. To apply this technology, the first step is to convert the analog facsimile signal to digital and then to process the digital information. One such method is called *run-length encoding*. It is based on the binary technique of viewing all digital elements as either black or white. In this system periodic sampling is carried out on the content of a scan stroke, determining whether black or white exists during a particular instant. A code is then transmitted indicating the color (e.g., black or white) and the length of time the color lasts. This information can be contained in only two code words. Another system uses a form of PCM to do this.

Yet another on-line digital system is multilevel, with the following binary notation:

11—White
10—Light gray
01—Dark gray
00—Black

For instance, 11 could be represented by a level of +2 V, 10 by +1 V, 01 by −1 V, and so forth.

CCITT Recs. T.4, T.5, and T.6 give some standardized facsimile run-length encoding techniques for 2400-bps and 4800-bps operation over the telephone network.

REVIEW QUESTIONS

1. What class of active devices in the analog telephone network is the major contributor to placing an upper limit on data rate?
2. ISDN provides 192-, 1544-, and 2048-kbps services to customer premises. Can these data rates be feasibly transmitted over a wire pair? What may have to be done to condition a typical wire-pair subscriber loop to accomplish these ISDN rates?
3. What is the most basic element of information in a binary data system? How much information does it contain? (Hint: How many distinct states does it represent?)
4. How does one extend the information content of the basic information element, for example, to construct an alphabet?
5. Give at least four synonymous terms for a binary 1 and for a binary 0.
6. How many distinct characters or symbols can be represented by a 4-unit binary code? a 7-unit binary code? an 8-unit binary code? (Hint: For this argument, consider that a unit is a bit.)
7. Name at least three nonprinting characters that might be encountered in a practical binary code.
8. How many information elements (bits) are in an ASCII character?
9. An ASCII character is represented as 1001011 with odd parity assumed. Give the value of the eighth bit. In another situation even parity is assumed and the ASCII character is 0101110. What is the value of the eighth bit?
10. Define the term *throughput*.

11. Describe the two common methods of correcting errors that occur on data links.

12. What is the most powerful method of error detection?

13. A frame check sequence (FCS) or BCC contains a 16- or 32-bit sequence. What is it and how is it derived?

14. When FEC (forward error correction) is implemented, what types of errors can be corrected? How can one use FEC to correct bursty errors?

15. Where and why would one use long message blocks (or packets) rather than short blocks? Why and where would short blocks be used? (The answers should not be application driven.)

16. Differentiate between neutral and polar transmission.

17. A data link transmits 2400 bps synchronously. There is a time base stability difference between transmitter and distant-end receiver of 20 ppm (parts per million). If the link can be considered error free except for timing, how many bits will the receiver receive correctly before printing in error? Assume, of course, that the receiver started copying the first bit out of the transmitter and that their two clocks started exactly in sync.

18. What is the duration of a start element? What is the duration of a stop element? Give the minimum duration of stop elements used in modern asynchronous data systems (three values).

19. Discuss how propagation time can affect a data link regarding timing.

20. Give two problems that are likely to be encountered in timing and synchronization on dial-up data circuits.

21. Regarding data transmission, define *distortion* in one short sentence.

22. What is the deleterious result of excessive distortion?

23. What are the two basic generic types of distortion?

24. Assuming an NRZ waveform, give the duration of 1 bit (i.e., 1 bit period) for the following bit rates: 75 bps, 110 bps, 4800 bps, and 19.200 kbps.

25. Using the ASCII code with parity operating at 2400 bps synchronous, how many words per minute are being transmitted?

26. Differentiate between baud and bit. For instance, a certain modem operates at 1200 bauds; it operates at 9600 bps. What are these terms telling us? Why is it important to know and appreciate the difference?

27. Give one advantage and one disadvantage of using an RZ waveform.

28. Draw a simple block diagram showing the location of the basic data transmission interfaces. (We say transmission meaning these are physical layer interfaces.)

29. Name at least three standards covering the DTE-DCE interface.

30. Based on RS-232D, at a transition the voltage rises to a +2 V peak. What information element does it represent? What information element do the following represent: −4 V, +6 V, and −6 V?

31. What is the overriding difference between RS-422 and RS-423?

32. There are three basic methods to modulate a carrier. What are they? What electrical device in the home uses all three simultaneously?

33. What is the advantage of quadrature PSK (QPSK) over binary PSK (BPSK)?

34. Give three *basic* impairments encountered in the telephone network that inhibit data rate, assuming a reasonable error rate?

35. Define *envelope delay distortion* (EDD).

36. Give four causes of errors on a data link that can be attributed to "transmission."

37. Voice channels have a useful bandwidth of about 3100 Hz. Give two basic techniques to allow us to transmit data in excess of 3100 bps.

38. What are the two parameters on which Shannon based his classic formula for channel capacity?

39. Why do most of the higher-speed modems use center tone frequencies in the range 1700–1800 Hz?

40. We usually equalize for two voice-channel impairments. What are these two basic impairments?

41. What type of modulation is used with conventional VFCT? Using conventional channel spacing (two types), how many 75-bps channels can the standard CCITT voice channel accommodate?

42. What type of modulation is commonly used with modems operating at 1200 bps and below? What type of modulation is used nearly universally for telephone circuit data modems operating at 2400 bps?

43. CCITT Rec. V.29 for 9600 bps modem for transmission over the telephone network recommends a complex waveform. Describe the waveform in two brief sentences, and give two reasons why we would resort to this more complex method of modulation.

44. Why should data be scrambled before transmission?

45. What are the operational advantages of facsimile over telex and electronic mail? Give at least one disadvantage.

46. Describe at least two facsimile scanning methods and two recording methods.

47. Define *index of cooperation*.

48. Give two modulation methods of transmitting facsimile over the telephone network.

49. Why is phasing and synchronization necessary? Describe how synchronization and phasing work for a facsimile link.

50. What is the primary advantage of digital facsimile typically using run-length encoding over its analog counterpart?

REFERENCES

1. International Telephone and Telegraph Corporation, *Reference Data for Radio Engineers*, 6th ed., Howard W. Sams, Indianapolis, 1976.
2. *Transmission Systems for Communication*, 5th ed., Bell Telephone Laboratories, Holmdel, N.J., 1982.
3. W. R. Bennett and J. R. Davey, *Data Transmission*, McGraw Hill, New York, 1965.
4. A. M. Rosie, *Information and Communication Theory*, Van Nostrand Reinhold, London, 1973.
5. Roger L. Freeman, *Reference Manual for Telecommunications Engineering*, Wiley, New York, 1985.
6. *Understanding Telegraph Distortion*, Stelma, Stamford, Conn., 1962.
7. *IEEE Standard Dictionary of Electrical and Electronic Terms*, IEEE Press, New York, 1977.
8. CCITT, *Red Books*, Vol. III, G. Recommendations, Malaga-Torremolinos, 1984.
9. CCITT, *Red Books*, Vol. VII, R. Recommendations, Malaga-Torremolinos, 1984.
10. CCITT, *Red Books*, Vol. VIII, V. Recommendations, Malaga-Torremolinos, 1984.
11. R. W. Lucky, J. Salz, and E. J. Weldon, *Principles of Data Communication*, McGraw-Hill, New York, 1968.
12. H. Nyquist, "Certain Topics in Telegraph Transmission Theory," *BSTJ*, 617–644 (April 1928).
13. C. E. Shannon, "A Mathematical Theory of Communication," *BSTJ*, 379–428 (July 1948); 623–656 (October 1948).
14. MIL-STD-188-114A, US Department of Defense, Washington, D.C., September 1985.
15. W. P. Davenport, *Modern Data Communications*, Hayden, New York, 1971.
16. S. Goldman, *Information Theory*, Dover, New York, 1968.

17. MIL-STD-188-100 with Notices 1–3, US Department of Defense, Washington, D.C., 1976.
18. M. P. Ristenhall, "Alternatives to Digital Communications," *Proc. IEEE*, **61** (6) (June 1973).
19. D. R. McGlynn, *Distributed Processing and Data Communications*, Wiley, New York, 1978.
20. C. L. Cuccia, "Subnanosecond Switching and Ultra-speed Communications," *Data Commun.* (November 1971).
21. D. R. Doll, "Controlling Data Transmission Errors," *Data Dynamics* (July 1971).
22. E. N. Gilbert, *Information Theory after Eighteen Years*, Bell Telephone Monograph, Bell Telephone Laboratories, Holmdel, N.J., 1965.
23. EIA Standard RS-232D, Electronic Industries Association, August 1986, Washington, D.C.
24. *Analog Parameters Affecting Voiceband Data Transmission—Description of Parameters*, Bell Systems Technical Reference Publication No. 41008, American Telephone and Telegraph Corporation, New York, October 1971.
25. W. Harper and R. Pollard, *Data Communications Desk Book: A Systems Analysis Approach*, Prentice-Hall, Englewood Cliffs, N.J., 1982.
26. R. L. Freeman, *Telecommunication Transmission Handbook*, 2nd ed., Wiley, New York, 1981.
27. K. Pahlavan and J. L. Holsinger, "Voice-Band Communication Modems: A Historical Review, 1919–1988," *IEEE Communications Magazine*, **26** (1) (January 1988).
28. MIL-STD-188C with Notice 1, US Department of Defense, Washington, D.C., June 1976.
29. D. R. Doll, *Data Communications: Facilities, Networks and Systems Design*, Wiley, New York, 1977.
30. W. R. Bennett, *Lecture Notes on Digital Communication Systems*, Michigan Univ. Press, East Lansing, July 1969.
31. John E. McNamara, *Technical Aspects of Data Communication* 2nd ed., Digital Equipment Corporation, Maynard, Mass., 1982.
32. *General Information: Binary Synchronous Communications*, IBM Report GA27-3004-1, December 1969.
33. K. Sherman, *Data Communications: A Users Guide*, Reston, Reston, Va., 1981.

9

DIGITAL TRANSMISSION SYSTEMS

1 DIGITAL VERSUS ANALOG TRANSMISSION

There are three notable advantages to digital transmission that make it extremely attractive to the telecommunication system engineer when compared to its analog counterpart. Dealing in generalities, we can say that:

1. Noise does not accumulate at repeaters and thus becomes a secondary consideration in system design, whereas in analog systems it is the primary consideration.
2. The digital format lends itself ideally to solid-state technology and in particular to integrated circuits.
3. It is inherently compatible with digital data, signaling, and computers.

The major portion of information to be transmitted in a common-carrier network is analog in nature, such as voice and video. Now convert these signals to a digital format, and we can take advantage of the three important features listed. Some readers justifiably will ask about the apparent data–telegraph disparity covered in the previous chapter. Is that not digital?

The apparent ambiguity stems from the input–output (I/O) devices. The telephone microphone generates an electrical-signal equivalent of the voice actuating the diaphragm, and this is analog in nature. On the other hand, the data–telegraph keyboard, tape reader, or computer delivers digital 1's and 0's to the line. To transmit this information over the telephone network, the digital signal is converted to an analog signal compatible with that network's facilities. The modem carries out this function.

The objective now is to do the reverse to the analog voice (telephone) signal, that is, to convert it to a digital signal that may be transmitted electrically. There are two different modulation methods commonly used to do this: pulse-code modulation (PCM), which is widely used for transmission in common-carrier communications, and delta modulation, which is finding broad application in military communications. Digital switches are now being implemented to accommodate these modulation types. The following paragraphs emphasize PCM because of its applicability to common-carrier communication. Delta modulation is reviewed in Section 7 of the chapter.

2 BASIS OF PULSE-CODE MODULATION

Pulse-code modulation is a method of modulation in which a continuous analog wave is transmitted in an equivalent digital mode. The cornerstone of an explanation of the functioning of PCM is the Nyquist sampling theorem, which states [Ref. 5, Section 21]:

> If a band-limited signal is sampled at regular intervals of time and at a rate equal to or higher than twice the highest significant signal frequency, then the sample contains all the information of the original signal. The original signal may then be reconstructed by use of a low-pass filter.

As an example of the sampling theorem, the nominal 4-kHz channel would be sampled at a rate of 8000 samples per second (i.e., 4000×2).

To develop a PCM signal from one or several analog signals, three processing steps are required: *sampling*, *quantization*, and *coding*. The result is a serial binary signal or bit stream, which may or may not be applied to the line without additional modulation steps. At this point a short review of Chapter 8 may be in order to clarify use of terminology such as mark, space, regeneration, and information bandwidth. One major advantage of digital transmission is that signals may be regenerated at intermediate points of links involved in transmission. The price for this advantage is the increased bandwidth required for PCM. Common systems in broad use require 16 times the bandwidth of their analog counterpart (e.g., a 4-kHz analog voice channel requires 16×4 or 64 kHz when transmitted by PCM), assuming 1 bit per Hz. Regeneration of a digital signal is simplified and particularly effective when the transmitted line signal is binary, whether neutral, polar, or bipolar. An example of a bipolar bit stream is shown in Figure 9.1.

Binary transmission tolerates considerably higher noise levels (i.e., degraded signal-to-noise ratios), when compared to its analog counterpart (i.e., FDM, Chapter 5). This fact, in addition to the regeneration capability, is a great step foreward in transmission engineering. The regeneration that takes place at each repeater by definition recreates a new digital signal; therefore, noise, as we know it, does not accumulate.

Error rate is another important factor in the design of PCM systems. If the error rate on a PCM system can be maintained end to end to 1 error in 10^5 bits, intelligibility will not be degraded. Even with an error rate of 1 bit in 10^3, intelligibility is fairly good. However, when errors exceed 1 in 10^2, intelligibil-

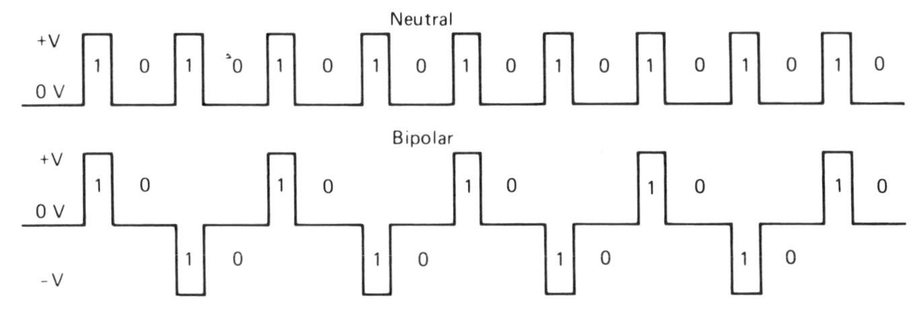

Figure 9.1 Neutral versus bipolar bit streams. The upper drawing illustrates alternate 1's and 0's transmitted in the neutral mode; the lower drawing illustrates the equivalent in a bipolar mode, which is also called AMI or alternate mark inversion.

ity is lost. Another factor imporant in the design of PCM cable installations is crosstalk, which can degrade error performance. This is crosstalk spilling from one PCM system to another or in the same system from the send path to the receive path inside the same cable sheath.

3 DEVELOPMENT OF A PULSE-CODE MODULATION SIGNAL

3.1 Sampling

Consider the sampling theorem given previously. If we now sample the standard CCITT voice channel, 300–3400 Hz (a bandwidth of 3100 Hz), at a rate of 8000 samples per second, we will have complied with the Nyquist sampling theorem and can expect to recover all the information in the original analog signal. Therefore a sample is taken every 1/8000 s, or every 125 μs. These are key parameters for our future argument.

Another example may be a 15-kHz program channel. Here the sampling rate would be 30,000 times per second. Samples would be taken at 1/30,000-s intervals, or at 33.3 μs.

3.2 The Pulse Amplitude Modulation Wave

With several exceptions (e.g., SPADE, Chapter 7, Section 7.6.3), practical PCM systems involve time division multiplexing. Sampling in these cases does not involve just one voice channel but several. In practice, one system (see following paragraph) samples 24 voice channels in sequence, and another samples 30 channels. The result of the multiple sampling is a PAM (pulse amplitude modulation) wave. A simplified PAM wave is shown in Figure 9.2, in this case a single sinusoid. A simplied diagram of the processing involved to derive a multiplexed PAM wave is shown in Figure 9.3.

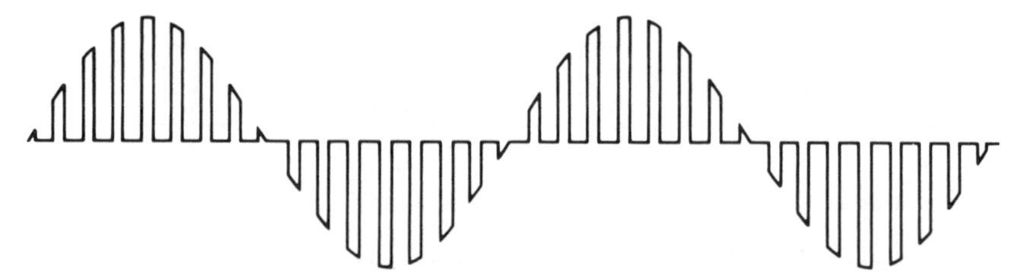

Figure 9.2 A PAM wave as a result of sampling a single sinusoid.

If the nominal 4-kHz voice channel must be sampled 8000 times per second and a group of 24 such voice channels are to be sampled sequentially to interleave them, forming a PAM multiplexed wave, this could be done by gating. The gate should be open for 5.2 μs (125/24) for each voice channel to be sampled successively from channels 1 through 24. This full sequence must be done in a 125-μs period ($1 \times 10^6/8000$). We call this 125-μs period a *frame*, and inside the frame all 24 channels are successively sampled once.

3.3 Quantization

It would appear that the next step in the process of forming a PCM serial bit stream would be to assign a binary code to each sample as it is presented to the coder.

Remember from Chapter 8 the discussion of code lengths, or what is more properly called coding "level." For instance, a binary code with four discrete elements (a four-level code) could code 2^4 separate and distinct meanings or 16 characters, not enough for the 26 letters in our alphabet; a five-level code would provide 2^5 or 32 characters or meanings. The ASCII is basically a seven-level code allowing 128 discrete meanings for each code combination ($2^7 = 128$). An eight-level code would yield 256 possibilities.

Another concept that must be keep in mind as the discussion leads into coding is that bandwidth is related to information rate (more exactly to modulation rate) or, for this discussion, to the number of bits per second transmitted. The goal is to keep some control over the amount of bandwidth necessary. It follows, then, that the coding length (number of levels) must be limited. As it stands, an infinite number of amplitude levels are being presented to the coder on the PAM highway. If the excursion of the PAM wave is 0 to +1 V, the reader should ask how many discrete values there are between 0 and 1. All values must be considered, even 0.0176487892 V.

The intensity range of voice signals over an analog telephone channel is of the order of 50 dB. The range −1 to 0 to +1 V of the PAM highway at the coder input may represent that 50-dB range. Further, it is obvious that the coder cannot provide a code of infinite length (e.g., an infinite number of coded levels) to satisfy every level in the 50-dB range (or a range from −1 V to +1 V). The key is to assign discrete levels from −1 V to +1 V (50-dB

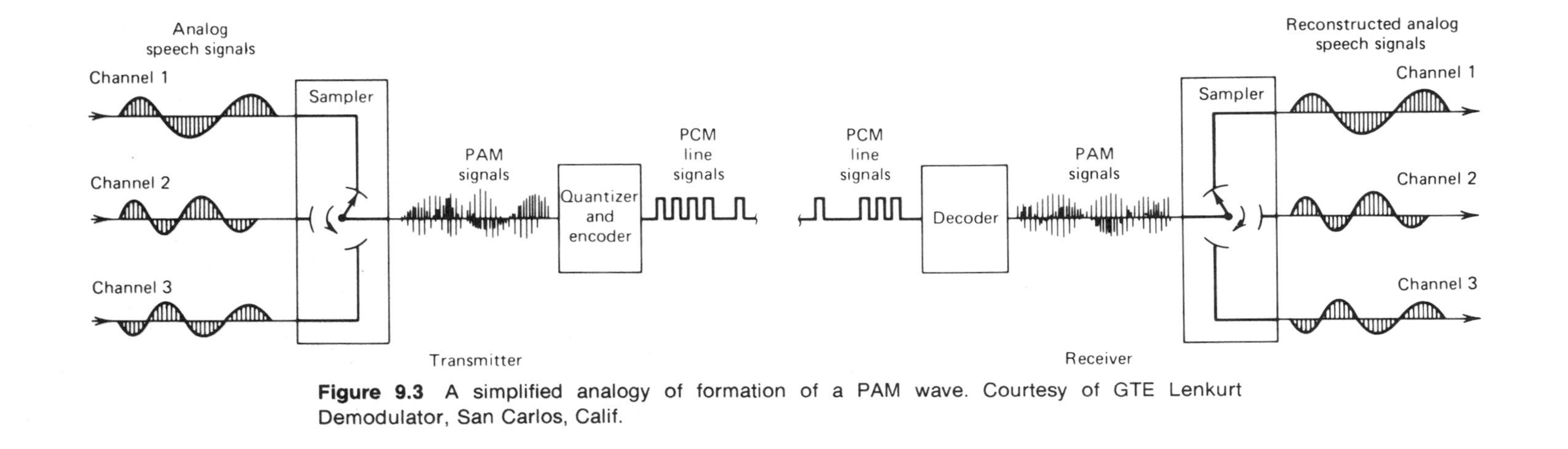

Figure 9.3 A simplified analogy of formation of a PAM wave. Courtesy of GTE Lenkurt Demodulator, San Carlos, Calif.

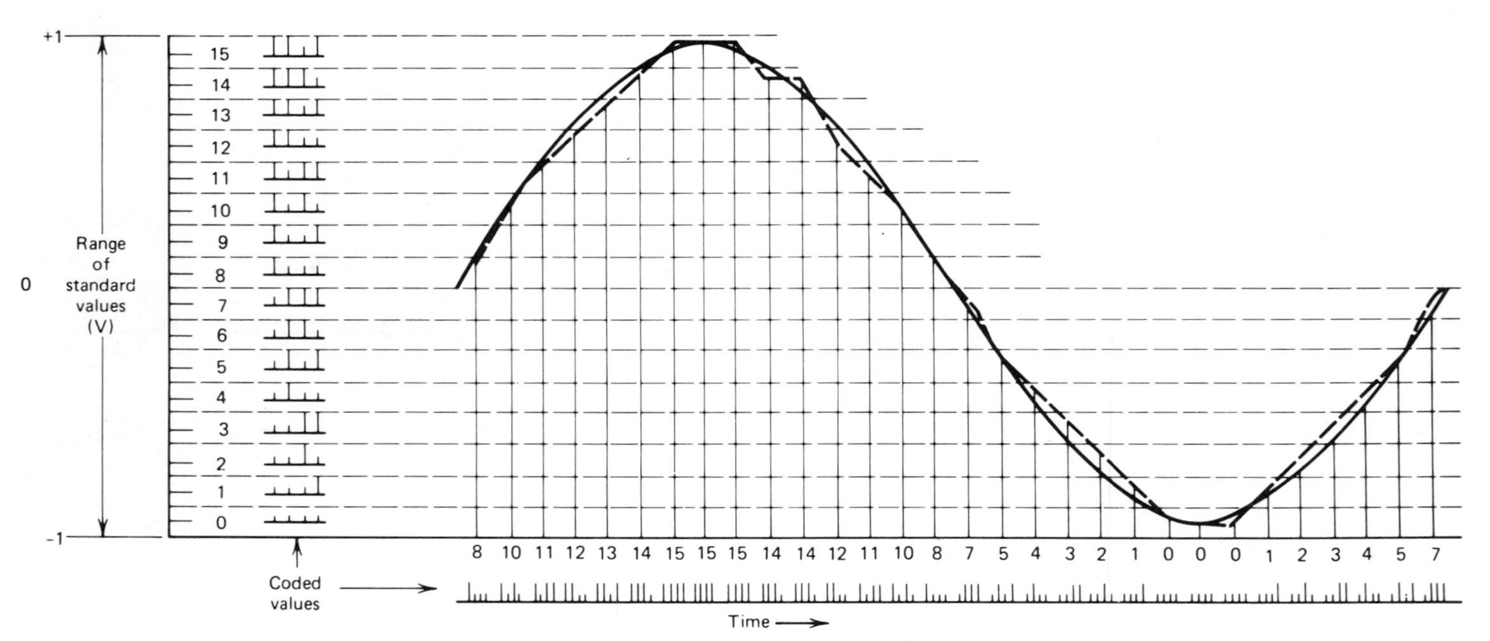

Figure 9.4 Quantization and resulting coding using 16 quantizing steps.

range). The assignment of discrete values to the PAM samples is called quantization. To cite an example, consider Figure 9.4.

Sixteen quantum steps exist between -1 V and $+1$ V and are coded as follows:

Step Number	Code	Step Number	Code
0	0000	8	1000
1	0001	9	1001
2	0010	10	1010
3	0011	11	1011
4	0100	12	1100
5	0101	13	1101
6	0110	14	1110
7	0111	15	1111

Examination of Figure 9.4 shows that step 12 is used twice. Neither time it is used is it the true value of the impinging sinusoid. It is a rounded-off value. These rounded-off values are shown with the dashed line in Figure 9.4, which follows the general outline of the sinusoid. The horizontal dashed lines show the point where the quantum changes to the next higher or next lower level if the sinusoid curve is above or below that value. Take step 14 in the curve, for example. The curve, dropping from its maximum, is given two values of 14 consecutively. For the first, the curve is above 14, and for the second, below. That error, in the case of 14, from the quantum value to the true value, is called *quantizing distortion*. This distortion is the major source of imperfection in PCM systems.

In Figure 9.4, maintaining the -1 -0- $+1$ V relationship, let us double the number of quantum steps from 16 to 32. What improvement would we achieve in quantization distortion? First determine the step increment in millivolts in each case. In the first case the total range of 2000 mV would be divided into 16 steps, or 125 mV/step. The second case would have 2000/32 or 62.5 mV/step. For the 16-step case, the worst quantizing error (distortion) would occur when an input to be quantized was at the half-step level, or in this case, 125/2 or 62.5 mV above or below the nearest quantizing step. For the 32-step case, the worst quantizing error (distortion) would again be at the half-step level, or 62.5/2 or 31.25 mV. Thus the improvement in decibels for doubling the number of quantizing steps is

$$20 \log \frac{62.5}{31.25} = 20 \log 2 \text{ or } 6 \text{ dB (approximately)}$$

This is valid for linear quantization only (see Section 3.6 of this chapter). Thus increasing the number of quantizing steps for a fixed range of input values reduces quantizing distortion accordingly. Experiments have shown that if

2048 uniform quantizing steps are provided, sufficient voice signal quality is achieved.

For 2048 quantizing steps, a coder will be required to code the 2048 discrete meanings (steps). Reviewing Chapter 8, we find that a binary code with 2048 separate characters or meanings (one for each quantum step) requires an 11-element code or $2^n = 2048$; thus $n = 11$. With a sampling rate of 8000 per second for each voice channel, the binary information rate per voice channel will be 88,000 bps. Consider that equivalent bandwidth is a function of information rate; thus the desirability of reducing this figure is obvious.

3.4 Coding

Practical PCM systems use seven- and eight-level binary codes, or

$$2^7 = 128 \text{ quantum steps}$$

$$2^8 = 256 \text{ quantum steps}$$

Two methods are used to reduce the quantum steps to 128 or 256 without sacrificing fidelity. These are nonuniform quantizing steps and companding before quantizing, followed by uniform quantizing. Keep in mind that the primary concern of digital transmission using PCM techniques is to transmit speech, as distinct from the digital transmission covered in Chapter 8, which dealt with the transmission of data and message information. Unlike data transmission, in speech transmission there is a much greater likelihood of encountering signals of small amplitudes than those of large amplitudes.

A secondary but equally important aspect is that coded signals are designed to convey maximum information, considering that all quantum steps (meanings or characters) will have an equally probable occurrence (i.e., the signal-level amplitude is assumed to follow a uniform probability distribution between 0 and $\pm$ the maximum voltage of the channel). To circumvent the problem of nonequiprobability of signal level for voice signals, specifically, that lower-level signals are more probable than higher-level signals, larger quantum steps are used for the larger-amplitude portion of the signal, and finer steps are used for the signals with low amplitudes. The two methods of reducing the total number of quantum steps can now be more precisely labeled:

- Nonuniform quantizing performed in the coding process.
- Companding (compression) before the signals enter the coder, which now performs uniform quantizing on the resulting signal before coding. At the receive end, expansion is carried out after decoding.

The first method is the one that is predominant in the industry today. An example of nonuniform quantizing could be derived from Figure 9.4 by changing the step assignment. For instance, 20 steps may be assigned between 0.0 and +0.1 V (another 20 between 0.0 and −0.1 V, etc.), 15 between 0.1 V and 0.2 V, 10 between 0.2 V and 0.35 V, 8 between 0.35 V and 0.5 V, 7 between 0.5 V and 0.75 V, and 4 between 0.75 V and 1.0 V.

Most practical PCM systems use companding to give finer granularity (more steps) to the smaller amplitude signals. This is instantaneous companding, as compared to the syllabic companding used in analog carrier telephony. Compression imparts more gain to lower amplitude signals. The compression and later expansion functions are logarithmic and follow one of two laws, the A law or the "mu" (μ) law. The curve for the A law may be plotted from the formula

$$\left(\frac{A|x|}{1 + \ln(A)}\right) 0 \leq |x| \leq \frac{1}{A}$$

$$\left(\frac{1 + \ln|Ax|}{1 + \ln(A)}\right) \frac{1}{A} \leq |x| \leq 1$$

where $A = 87.6$. The curve for the μ law may be plotted from the formula:

$$Y = \frac{\ln(1 + \mu|x|)}{\ln(1 + \mu)}$$

where x is signal imput amplitude and $\mu = 100$ for the original North American T1 system (now out dated) and 255 for later North American (DS1) systems and the CCITT 24-channel system (CCITT Rec. G.733). Note the use of the natural logarithms (ln) in these formulas. (Ref. 3).

A common expression used in dealing with the "quality" of a PCM signal is the signal-to-distortion ratio (expressed in decibels). Parameters A and μ determine the range over which the signal-to-distortion ratio is comparatively constant. This is the dynamic range. Using a μ of 100 can provide a dynamic range of 40 dB of relative linearity in the signal-to-distortion ratio.

In actual PCM systems the companding circuitry does not provide an exact replica of the logarithmic curves shown. The circuitry produces approximate equivalents using a segmented curve, and each segment is linear. The more segments the curve has, the more it approaches the true logarithmic curve desired. Such a segmented curve is shown in Figure 9.5. If the μ law were implemented using a seven (height)-segment linear approximate equivalent, it would appear as shown in Figure 9.5. Thus on coding, the first three coded digits would indicate the segment number (e.g., $2^3 = 8$). Of the seven-digit code, the remaining four digits would divide each segment into 16 equal parts to identify further the exact quantum step (e.g., $2^4 = 16$). For small signals,

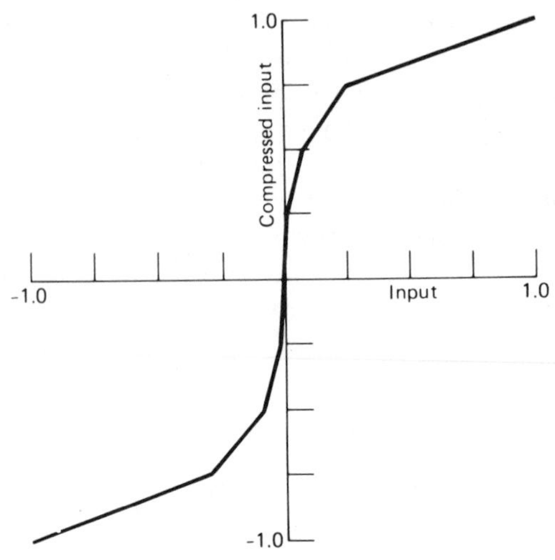

Figure 9.5 Seven-segment linear approximate of the logarithmic curve for μ law ($\mu = 100$) [6]. Copy © 1970 Bell Telephone Laboratories.

the companding improvement is approximately

$$
\begin{aligned}
A \text{ law:} &\quad 24 \text{ dB} \\
\mu \text{ law:} &\quad 30 \text{ dB}
\end{aligned}
$$

using a seven-level code. These values derive from the equation of companding improvement or

$$G_{\text{dB}} = 20 \log \frac{\text{Uniform (linear) scale}}{\text{Companded scale}}$$

Coding in PCM systems utilizes straightforward binary codes. Examples of such coding are shown in Figure 9.6, which is expanded in Figure 9.7, and in Figure 9.8, which is expanded in Figure 9.9 showing a number of example code levels.

The coding process is closely related to quantizing. In practical systems, whether the A law or the μ law is used, quantizing employs segmented equivalents of the companding curve (Figures 9.6 and 9.8), as discussed earlier. Such segmenting is a handy aid to coding. Consider the European 30 + 2 PCM system, which uses a 13-segment approximation of the A-law curve (Figure 9.6). The first code element indicates whether the quantum step is in the negative or positive half of the curve. For example, if the first code element were a 1, it would indicate a positive value (e.g., the quantum step is located above the origin). The following three-code elements (bits) identify the seg-

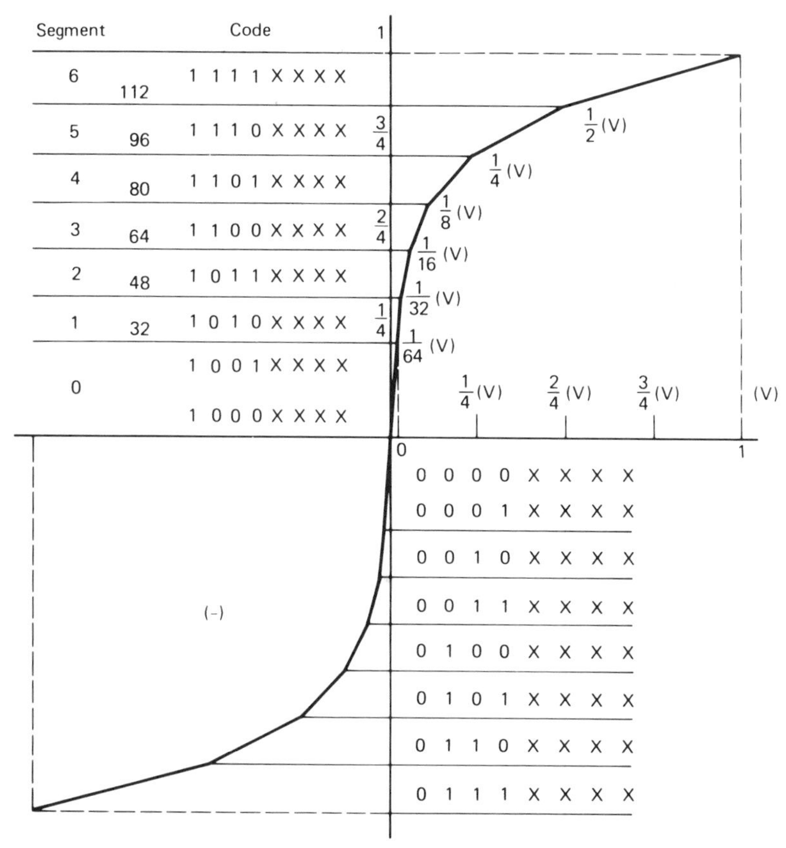

Figure 9.6 Quantization and coding used in the CEPT 30 + 2 PCM system.

ment, as there are seven segments above and seven segments below the origin (horizontal axis).

The first four elements of the fourth + segment are 1101. The first 1 indicates it is above the horizontal axis (e.g., it is positive). The next three elements indicate the fourth step or

0—1000 and 1001
1—1010
2—1011
3—1100
→4—1101
5—1110 etc.

Figure 9.7 shows a "blowup" of the uniform quantizing and subsequent straightforward binary coding of step 4. This is the final segment coding, the

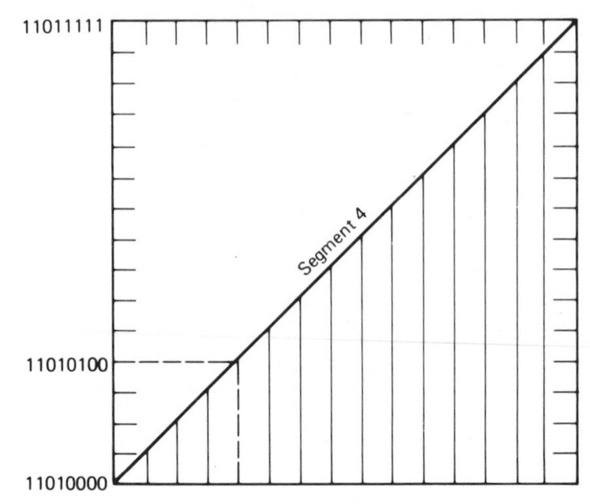

Figure 9.7 The CEPT 30 + 2 PCM system, coding of segment 4 (positive).

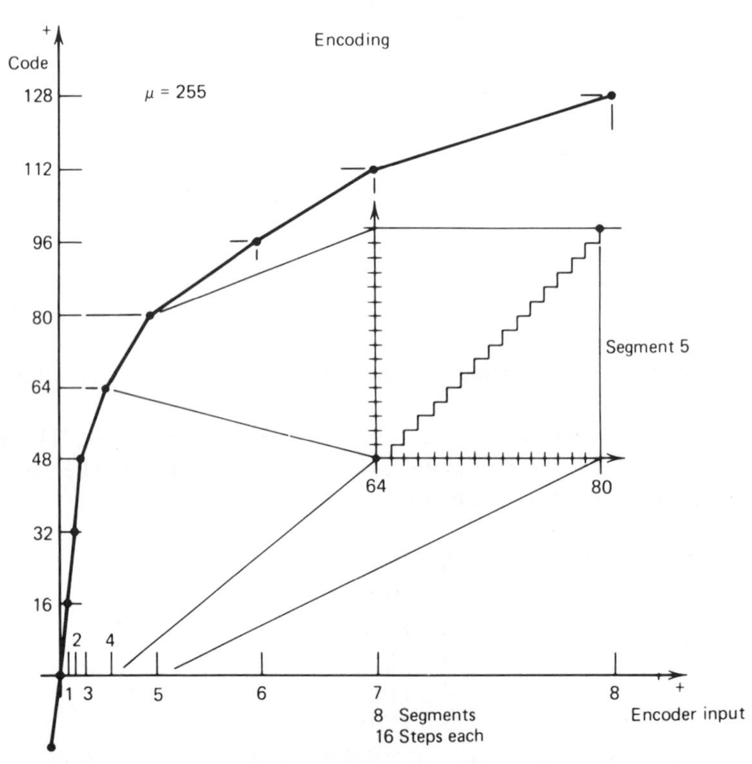

Figure 9.8 Positive portion of segmented approximation of μ law quantizing curve used in North American (ATT) DS1 PCM channelizing equipment. Courtesy of ITT Telecommunications, Raleigh, N.C.

Code Level		Digit Number							
		1	2	3	4	5	6	7	8
255	(Peak positive level)	1	0	0	0	0	0	0	0
239		1	0	0	1	0	0	0	0
223		1	0	1	0	0	0	0	0
207		1	0	1	1	0	0	0	0
191		1	1	0	0	0	0	0	0
175		1	1	0	1	0	0	0	0
159		1	1	1	0	0	0	0	0
143		1	1	1	1	0	0	0	0
127	(Center levels)	1	1	1	1	1	1	1	1
126	(Nominal zero)	0	1	1	1	1	1	1	1
111		0	1	1	1	0	0	0	0
95		0	1	1	0	0	0	0	0
79		0	1	0	1	0	0	0	0
63		0	1	0	0	0	0	0	0
47		0	0	1	1	0	0	0	0
31		0	0	1	0	0	0	0	0
15		0	0	0	1	0	0	0	0
2		0	0	0	0	0	0	1	1
1		0	0	0	0	0	0	1	0
0	(Peak negative level)	0	0	0	0	0	0	1*	0

*One digit is added to ensure that the timing content of the transmitted pattern is maintained.

Figure 9.9 Eight-level coding of North American (ATT) DS1 PCM system. Note that there are actually only 255 quantizing steps because steps 0 and 1 use the same bit sequence, thus avoiding a code sequence with no transitions (i.e., 0's only).

last four bits of a PCM code word for this system. Note the 16 steps in the segment, which are uniform in size.

The North American DS1 PCM system uses a 15-segment approximation of the logarithmic μ law. Again, there are actually 16 segments. The segments cutting the origin are colinear and counted as one. The quantization in the DS1 system is shown in Figure 9.8 for the positive portion of the curve. Segment 5, representing quantizing steps 64 through 80, is shown blown up in Figure 9.8. Figure 9.9 shows the DS1 coding. As can be seen in this figure, again the first code element, whether a 1 or a 0, indicates whether the quantum step is above or below the horizontal axis. The next three elements identify the segment, and the last four elements (bits) identify the actual quantum level inside that segment. Of course, we see that the DS1 is a basic 24-channel system using eight-level coding with μ-law quantization characteristic where $\mu = 255$.

3.5 The Concept of Frame

As shown in Figure 9.3, PCM multiplexing is carried out in the sampling process, sampling sources sequentially. These sources may be the nominal 4-kHz voice channels or other information sources, possibly data or video. The final result of the sampling and subsequent quantization and coding is a series of pulses, a serial bit stream (1's and 0's) that requires some indication or identification of the beginning of a scanning sequence. This identification is necessary at the far-end receiver so it will know exactly when each sampling sequence starts and ends; it times the receiver. Such identification is carried out by a *framing bit*, and a full sequence or cycle of samples is called a *frame* in PCM terminology.

Consider the framing structure of two widely implemented PCM systems: the North American DS1 and the European CEPT 30 + 2 systems. The North American DS1 system is a 24-channel PCM system using 8-level coding (e.g., $2^8 = 256$ quantizing steps or distinct PCM code words). Supervisory signaling is "in band" where bit 8 of every sixth frame is "robbed" for supervisory signaling. The DS1 signal format, shown in Figure 9.10, has one bit added as a framing bit called an "S" bit. The DS1 frame thus consists of

$$(8 \times 24) + 1 = 193 \text{ bits}$$

making up a full sequence or frame. By definition 8000 frames are transmitted per second, so the bit rate is

$$193 \times 8000 = 1{,}544{,}000 \text{ bps or } 1.544 \text{ Mbps.}$$

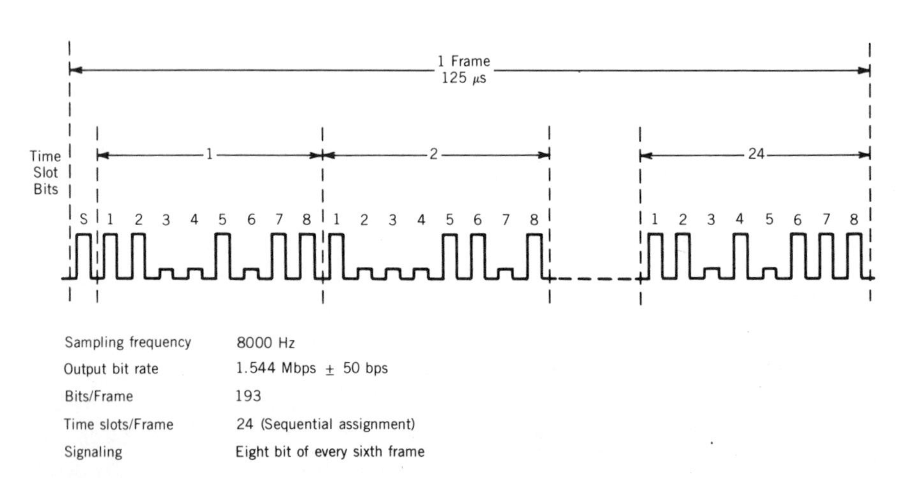

Sampling frequency	8000 Hz
Output bit rate	1.544 Mbps ± 50 bps
Bits/Frame	193
Time slots/Frame	24 (Sequential assignment)
Signaling	Eight bit of every sixth frame

The S-bit is time-shared between terminal framing (F_I) and signal framing (F_S).

Figure 9.10 DS1 signal format.

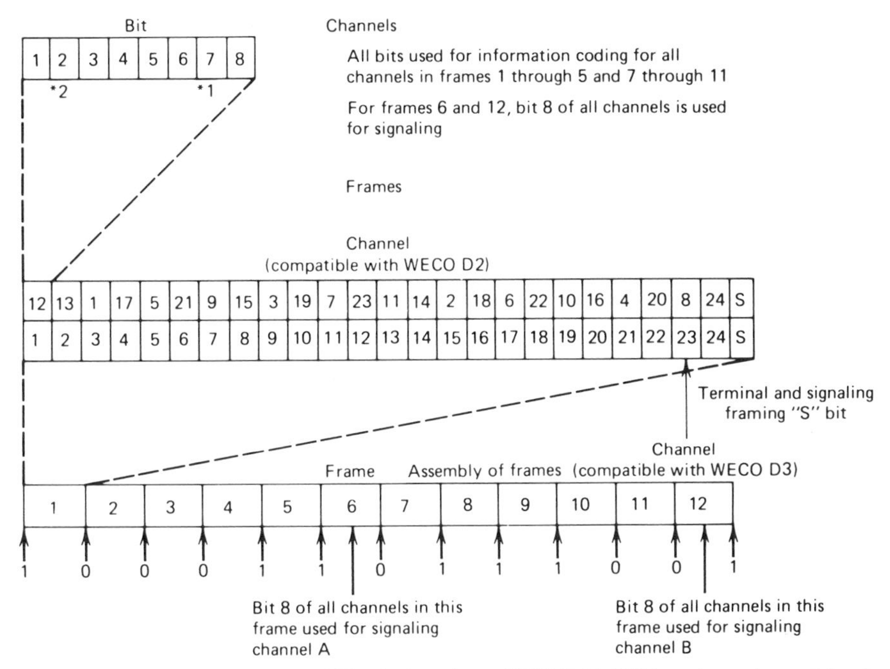

Figure 9.11 Frame structure of North American (ATT) DS1 PCM system channel bank. Note bit "robbing" technique used on each sixth frame to provide signaling information. Courtesy of ITT Telecommunications, Raleigh, N.C. [*Notes*: (1) If bits 1 to 6 and 8 are 0, then bit 7 is transmitted as 1; (2) bit 2 is transmitted as 0 on all channels for transmission of end-to-end alarm; (3) composite pattern 000110111001, etc.]

This frame structure is further clarified in Figure 9.11. The CEPT* 30 + 2 system is a 32-channel system where 30 channels transmit speech derived from incoming telephone trunks and the remaining 2 channels transmit signaling and synchronization information. Each channel is alloted a time slot (TS), and we can tabulate TS 0 through 31 as follows:

TS	Type of Information
0	Synchronizing (framing)
1–15	Speech
16	Signaling
17–31	Speech

In TS 0 a synchronizing code or word is transmitted every second frame, occupying digits 2 through 8 as follows:

0011011

In those frames without the synchronizing word, the second bit of TS 0 is

*This is the Conference Européene des Postes et Télécommunications.

frozen at a 1 so that in these frames the synchronizing word cannot be imitated. The remaining bits of time slot 0 can be used for the transmission of supervisory information signals (see Chapter 4).

Framing and basic timing should be distinguished. "Framing" ensures that the PCM receiver is aligned regarding the beginning (and end) of a sequence or frame; "timing" refers to the synchronization of the receiver clock, specifically, that it is in step with its companion (far-end) transmit clock. Timing at the receiver is corrected via the incoming mark-to-space (and space-to-mark) transitions. It is important, then, that long periods of no transitions do not occur. This point is discussed later in reference to line codes and digit inversion.

3.6 Quantizing Distortion

Quantizing distortion has been defined as the difference between the signal waveform as presented to the PCM multiplex (codec*) and its equivalent quantized value. For a linear codec with n binary digits per sample, the ratio of the full-load sine wave power to quantizing distortion power (S/D) is [6]

$$\frac{S}{D} = 6n + 1.8 \text{ dB}$$

where n is the number of bits per PCM word, the word expressing the sample. For instance, the older ATT D1 system uses a 7-bit word to express a sample (level), and the 30 + 2 and DS1 systems use essentially 8 bits. If we had a 7-bit word and uniform quantizing, S/D would be 43.8 dB. Each binary digit added to the PCM code word increases the S/D ratio 6 dB for linear quantization. Practical S/D values range in the order of 33–38 dB, depending largely on the talker levels (using 8-bit words).

4 PULSE-CODE MODULATION SYSTEM OPERATION

Pulse-code modulation (PCM) is four wire. Voice-channel inputs and outputs to and from a PCM multiplex are on a four-wire basis. The term "codec" is used to describe a unit of equipment carrying out the function of PCM multiplex and demultiplex and stands for *co*der-*dec*oder even though the equipment carries out more functions than just coding and decoding. A block diagram of a codec is shown in Figure 9.12.

A codec accepts 24 or 30 voice channels, depending on the system used; digitizes and multiplexes the information; and delivers a serial bit stream to the line of 1.544† Mbps or 2.048 Mbps. It accepts a serial bit stream at one or

*Codec is a term used in PCM meaning *co*der–*dec*oder and is analogous to the modem described in Chapter 8.

†This is the rate for the ATT DS1 24-channel bank.

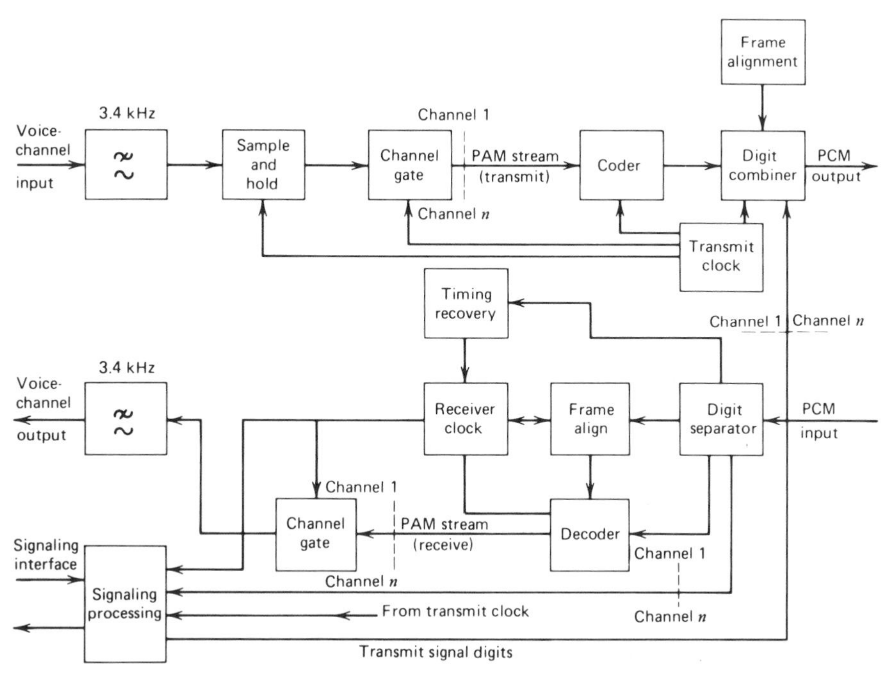

Figure 9.12 Simplified functional block diagram of a PCM codec.

the other modulation rate, demultiplexes the digital information, and performs digital to analog conversion. Output to the analog telephone network is the 24 or 30 nominal 4-kHz voice channels. Figure 9.12 illustrates the processing of a single analog voice channel through a codec. The voice channel to be transmitted is passed through a 3.4-kHz low-pass filter. The output of the filter is fed to a sampling circuit. The sample of each channel of a set of n channels (n usually equals 24 or 30) is released in turn to the pulse amplitude modulation (PAM) highway. The release of samples is under control of a channel gating pulse derived from the transmit clock. The input to the coder is the PAM highway. The coder accepts a sample of each channel in sequence and then generates the appropriate 8-bit signal character corresponding to each sample presented. The coder output is the basic PCM signal that is fed to the digit combiner where framing-alignment signals are inserted in the appropriate time slots, as well as the necessary supervisory signaling digits corresponding to each channel (European approach), and are placed on a common signaling highway that makes up one equivalent channel of the multiplex serial bit stream transmitted to the line. In North American practice supervisory signaling is carried out somewhat differently by "bit robbing," such as bit 8 in frame 6 and bit 8 in frame 12. Thus each equivalent voice channel carries its own signaling (see Figure 9.11).

On the receive side the codec accepts the serial PCM bit stream, in putting the digit separator where the signal is regenerated and split, delivering the

PCM signal to four locations to carry out the following processing functions: (1) timing recovery, (2) decoding, (3) frame alignment, and (4) signaling (supervisory). Timing recovery keeps the receive clock in synchronism with the far-end transmit clock. The receive clock provides the necessary gating pulses for the receive side of the PCM codec. The frame-alignment circuit senses the presence of the frame-alignment signal at the correct time interval, thus providing the receive terminal with frame alignment. The decoder, under control of the receive clock, decodes the code character signals corresponding to each channel. The output of the decoder is the reconstituted pulses making up a PAM highway. The channel gate accepts the PAM highway, gating the n-channel PAM highway in sequence under control of the receive clock. The output of the channel gate is fed in turn to each channel filter, thus enabling the reconstituted analog voice signal to reach the appropriate voice path. Gating pulses extract signaling information in the signaling processor and apply this information to each of the reconstituted voice channels with the supervisory signaling interface as required by the analog telephone system in question.

5 PRACTICAL APPLICATION

5.1 General

Pulse-code modulation has found wide application in expanding interoffice trunks (junctions) that have reached or will reach exhaust* in the near future. An interoffice trunk is one pair of a circuit group that connects two switching points (exchanges). Figure 9.13 sketches the interoffice trunk concept. Depending on the particular application, at some point where distance d is exceeded it will be more economical to install PCM on existing VF cable plant than to rip up streets and add more VF cable pairs. For the planning engineer the distance d where PCM becomes an economic alternative is called the "prove-in" distance. The distance d may vary from 8 km to 16 km (5 mi to 10 mi), depending on the location and other circumstances. For distances less than d, additional VF cable pairs should be used for expanding plant.

The general rule for measuring expansion capacity of a given VF cable is as follows:

- For ATT DS1/DS2 channelizing equipment, two VF pairs will carry 24 PCM channels.
- For the CEPT 30 + 2 system as configured by ITT, two VF pairs plus a phantom pair will carry 30 PCM speech channels.

*"Exhaust" is an outside-plant term meaning that the useful pairs of a cable have been used up (assigned) from a planning point of view.

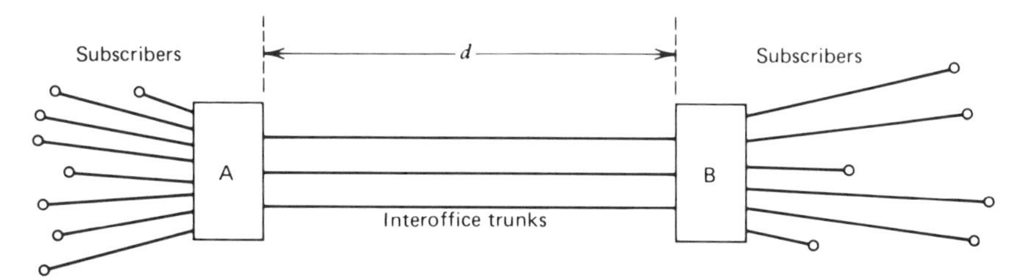

Figure 9.13 Simplified application diagram of PCM as applied to interoffice (interexchange) plant; *A* and *B* are switching centers.

All pairs in a VF cable may not necessarily be usable for PCM transmission, partly because there is a possibility of excessive crosstalk between PCM carrying pairs. The effect of high crosstalk levels is to introduce digital errors in the PCM bit stream. Error rate may be related on a statistical basis to crosstalk, which, in turn, is dependent on the characteristics of the cable and the number of PCM carrying pairs.

One method for reducing crosstalk and thereby increasing VF pair usage is to turn to two-cable working, rather than have the "go" and "return" PCM cable pairs in the same cable. Another factor that can limit cable pair usage is the incompatibility of FDM and PCM carrier systems in the same cable. On the cable pairs that will be used for PCM, the following should be taken into consideration:

- All load coils must be removed.
- Build-out networks and bridged taps must also be removed.
- No crosses, grounds, splits, high-resistance splices, or moisture are permitted.

The frequency response of the pair should be measured out to 1 MHz and considered as far out as 2.5 MHz. Insulation should be checked with a megger. A pulse reflection test using a radar test set is also recommended. Such a test will indicate opens, shorts, and high-impedance mismatches. A resistance test and balance test using a Wheatstone bridge may also be in order. Some special PCM test sets are available, such as the GTE Lenkurt 91100 PCM cable test set using pseudorandom PCM test signals and the conventional digital test eye pattern.

5.2 Practical System Block Diagram

A block diagram showing the elemental blocks of a PCM transmission link used to expand installed VF cable capacity is shown in Figure 9.14. Most telephone administrations (companies) distinguish between the terminal area of a PCM system and the repeatered line. The term "span" comes into play

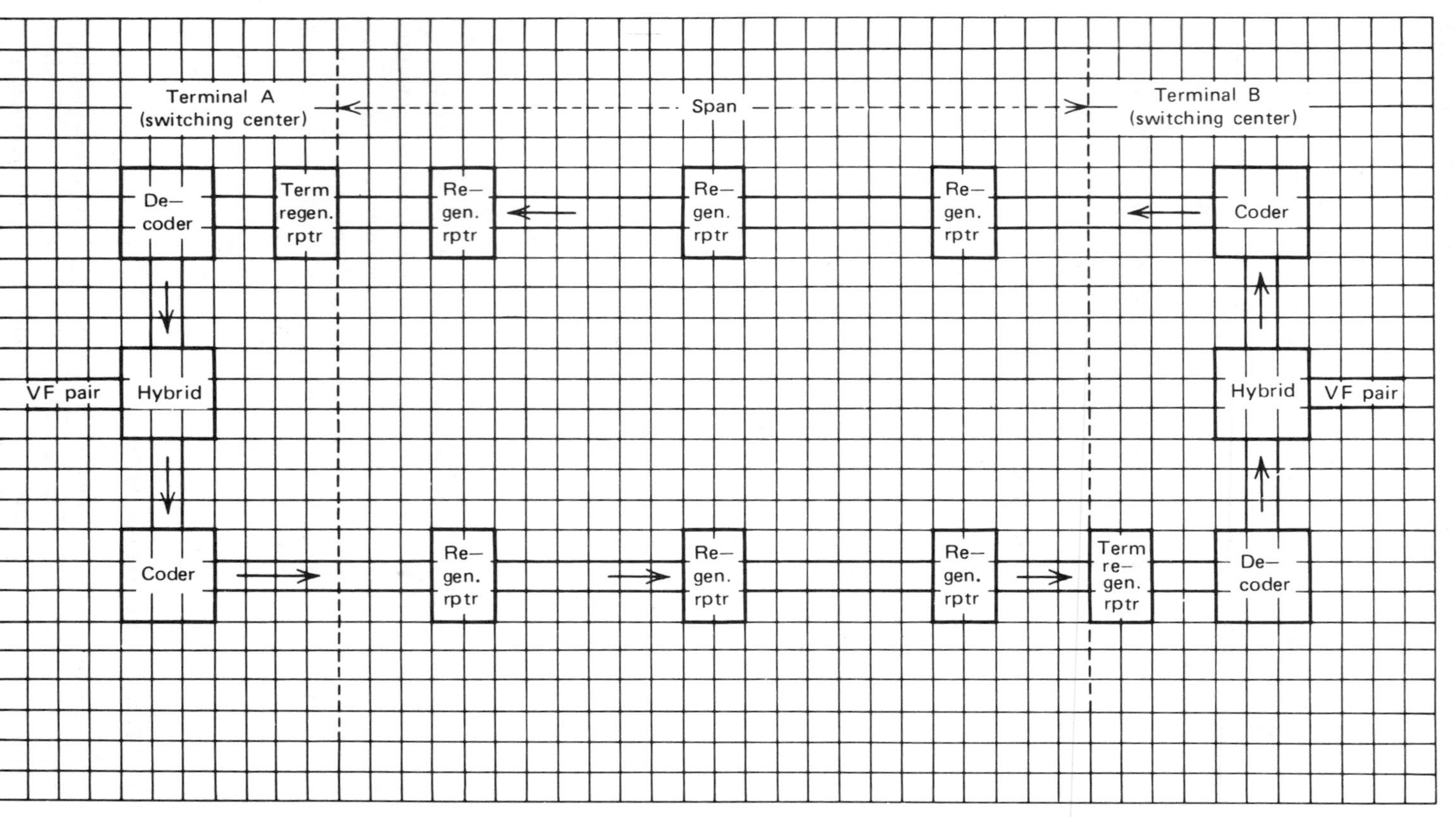

Figure 9.14 Simplified functional block diagram of a PCM link used to expand capacity of an existing VF cable (for simplicity, interface with only one VF pair is shown). Note the spacing between repeaters in the span line.

here. A span line is composed of a number of repeater sections permanently connected in tandem at repeater apparatus cases mounted in manholes or on pole lines along the span. A "span" is defined as the group of span lines that extend between two exchange (switching center) repeater points.

A typical span is shown in Figure 9.14. The spacing between regenerative repeaters is important. Section 5.1 mentioned the necessity of removing load coils from those trunk (junction) cable pairs that were to be used for PCM transmission. It is at these load points that the PCM regenerative repeater should be installed. On a VF line with H-type loading (see Chapter 2, Section 2.6), spacing between load points is normally about 6000 ft (1830 m). It will be remembered from Chapter 2 that the first load coil out from the exchange on a trunk pair is at half-distance or 3000 ft (915 m). This is provident, for a regenerative repeater also must be installed at this point. Such spacing is shown in Figure 9.14 (1 space = 1000 ft). The purpose of installing a repeater at this location is to increase the pulse level before entering the environment of an exchange area where the levels of impulse noise may be quite high. High levels of impulse noise introduced into the system may cause significant increases in digital error rate of the incoming PCM bit streams, particularly when the bit stream is of a comparatively low level. Generally, the amplitude of a PCM pulse output of a regenerative repeater is of the order of 3 V. Likewise, 3 V is the voltage on the PCM line cross connect field at the exchange (terminal area).

A guideline used by Bell Telephone Manufacturing Company (BTM) (Belgium) is that the maximum distance separating regenerative repeaters is that corresponding to a cable-pair attenuation of 36 dB at 1024 kHz at the maximum expected temperature. This frequency is equivalent to the half-bit rate for the CEPT systems (e.g., 2048 kbps). Actually, repeater design permits operation on lines with attenuations anywhere from 4 dB to 36 dB, allowing considerable leeway in placing repeater points. Table 9.1 gives some other practical repeater-spacing parameters for the CEPT-ITT-BTM 30 + 2 system. The maximum distance is limited by the maximum number of repeaters, which in this case is a function of power feeding and supervisory considerations. For instance, the fault-location (i.e., troubleshooting) system can handle up to a maximum of 18 tandem repeaters for the BTM (ITT) configuration.

TABLE 9.1 Line Parameters for ITT / BTM PCM Configuration

Pair Diameter (mm)	Loop Attenuation at 1 MHz (dB / km)	Loop Resistance (Ω/km)	Voltage Drop (V / km)	Maximum Distance[a] (km)	Total Repeaters	Maximum Distance System (km)
0.9	12	60	1.5	3	18	54
0.6	16	100	2.6	2.25	16	36

[a]Between adjacent repeaters.

Power for the BTM system is fed through a constant-current feeding arrangement over a phantom pair serving both the "go" and related "return" repeaters, providing up to 150 V dc at the power feed point. The voltage drop per regenerative repeater is 5.1 V; thus for a "go" and "return" repeater configuration the drop is 10.2 V. For example, let us determine the maximum number of regenerative repeaters in tandem that may be fed from one power feed point by this system, using 0.8-mm-diameter pairs with a 3-V IR drop in an 1830-m spacing between adjacent repeaters:

$$\frac{150}{(10.2 + 3)} = 11$$

Assuming power fed from both ends and an 1800-m "dead" section in the middle, the maximum distance between power feed points is approximately

$$(2 \times 11 + 1)1.8 \text{ km} = 41.4 \text{ km}$$

Fault tracing for the North American (ATT) T1 system is carried out by means of monitoring the framing signal, the 193d bit (Section 3.5). The framing signal (amplified) normally holds a relay closed when the system is operative. With loss of framing signal, the relay opens actuating alarms, and thus a faulty system is identified, isolated, and dropped from "traffic."

To locate a defective regenerator on the BTM (Belgium)-CEPT system, traffic is removed from the system and a special pattern generator is connected to the line. The pattern generator transmits a digital pattern with the same bit rate as does the 30 + 2 PCM signal, but the test pattern can be varied to contain selected low-frequency spectral elements. Each regenerator on the repeatered line is equipped with a special audio filter, each with a distinctive passband. Up to 18 different filters may be provided in a system. The filter is bridged across the output of the regenerator, sampling the output pattern. The output of the filter is amplified and transformer-coupled to a fault-transmission pair, which is normally common to all PCM systems on the route, span, or section. To determine which regenerator is faulty, the special test pattern is tuned over the spectrum of interest. As the pattern is tuned through the frequency of the distinct filter of each operative repeater, a return signal will derive from the fault-transmission pair at a minimum specified level. Defective repeaters will be identified by absence of return signal or a return level under specification. The distinctive spectral content of the return signal is indicative of the regenerator undergoing test.

5.3 The Line Code

Pulse-code-modulation signals are transmitted to the cable and are in the bipolar mode, as shown in Figure 9.1. The marks, or 1's, have only a 50% duty cycle. There are several advantages to this mode of transmission:

- No dc return is required; thus transformer coupling can be used on the line.
- The power spectrum of the transmitted signal is centered at a frequency equivalent to half the bit rate.

It will be noted in bipolar transmission that the 0's are coded as absence of pulses and 1's are alternately coded as positive and negative pulses, with the alternation taking place at every occurrence of a 1. This mode of transmission is also called *alternate mark inversion* (AMI).

One drawback to straightforward AMI transmission is that when a long string of 0's is transmitted (e.g., no transitions), a timing problem may arise because repeaters and decoders have no way of extracting timing without transitions. The problem can be alleviated by forbidding long strings of 0's. Codes have been developed that are bipolar but with N 0's substitution; they are called "BNZS" codes. For instance, a B6ZS code substitutes a particular signal for a string of six 0's.

Another such code is the HDB3 code (high-density binary 3), where the 3 indicates substitution for binary formations with more than three consecutive 0's. With HDB3, the second and third 0's of the string are transmitted unchanged. The fourth 0 is transmitted to the line with the same polarity as the previous mark sent, which is a "violation" of the AMI concept. The first 0 may or may not be modified to a 1 to ensure that the successive violations are of opposite polarity.

5.4 Signal-to-Gaussian-Noise Ratio on Pulse-Code Modulation Repeater Lines

As we mentioned earlier, noise accumulation on PCM systems is not an important consideration. However, this does not mean that Gaussian noise (or crosstalk or impulse noise) is unimportant. Indeed, it may affect error performance expressed as error rate (see Chapter 8). Errors are cumulative, as is the error rate. A decision in error, whether 1 or 0, made anywhere in the digital system is not recoverable. Thus such an incorrect decision made by one regenerative repeater adds to the existing error rate on the line, and errors taking place in subsequent repeaters further down the line add in a cumulative manner, thus tending to deteriorate the received signal.

In a purely binary transmission system, if a 20-dB signal-to-noise ratio is maintained, the system operates nearly error free. In this respect, consider Table 9.2.

As discussed in Section 5.3, PCM, in practice, is transmitted on-line with alternate mark inversion. The marks have a 50% duty cycle, permitting energy concentration at a frequency of half the transmitted bit rate. Thus it is

TABLE 9.2 Error Rate of a Binary Transmission System Versus Signal-to-rms-Noise Ratio

Error Rate	S/N (dB)	Error Rate	S/N (dB)
10^{-2}	13.5	10^{-7}	20.3
10^{-3}	16.0	10^{-8}	21.0
10^{-4}	17.5	10^{-9}	21.6
10^{-5}	18.7	10^{-10}	22.0
10^{-6}	19.6	10^{-11}	22.2

advisable to add 1 or 2 dB to the values shown in Table 9.2 to achieve a desired error rate in a practical system.

5.5 Regenerative Repeaters

As we are probably aware, pulses passing down a digital transmission line suffer attenuation and are badly distorted by the frequency characteristic of the line. A regenerative repeater amplifies and reconstructs such a badly distorted digital signal and develops a nearly perfect replica of the original at its output. Regenerative repeaters are an essential key to digital transmission in that we could say that the "noise stops at the repeater."

Figure 9.15 is a simplified block diagram of a regenerative repeater and shows typical waveforms corresponding to each functional stage of signal processing. As shown in the figure, the first stage of signal processing is amplification and equalization. Equalization is often a two-step process. The first is a fixed equalizer that compensates for the attenuation–frequency characteristic of the nominal section, which is the standard length of transmis-

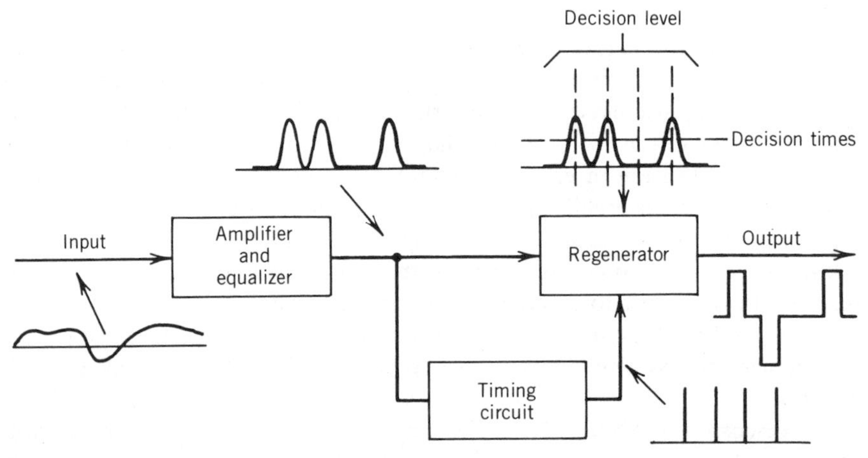

Figure 9.15 Simplified functional block diagram of a regenerative repeater for use on PCM cable systems.

sion line between repeaters (often 6000 ft). The second equalizer is variable and compensates for departures between nominal repeater section length and the actual length and loss variations due to temperature. The adjustable equalizer uses automatic line build-out (ALBO) networks that are automatically adjusted according to characteristics of the received signal.

The signal output of the repeater must be accurately timed to maintain accurate pulse width and space between the pulses. The timing is derived from the incoming bit stream. The incoming signal is rectified and clipped, producing square waves that are applied to the timing extractor, which is a circuit tuned to the timing frequency. The output of the circuit controls a clock-pulse generator that produces an output of narrow pulses that are alternately positive and negative at the zero crossings of the square wave input.

The narrow positive clock pulses gate the incoming pulses of the regenerator, and the negative pulses are used to run off the regenerator. Thus the combination is used to control the width of the regenerated pulses.

Regenerative repeaters are the major source of timing jitter in a digital transmission system. Jitter is one of the principal impairments in a digital network giving rise to pulse distortion and intersymbol interference. Jitter is discussed in more detail in Section 8.2.

Most regenerative repeaters transmit a bipolar (AMI) waveform (see Figure 9.1). Such signals can have one of three possible states in any instant in time, positive, zero, or negative, and are often designated +, 0, −. The threshold circuits are gated to admit the signal at the middle of the pulse interval. For example, if the signal is positive and exceeds a positive threshold, it is recognized as a positive pulse. If it is negative and exceeds a negative threshold, it is recognized as a negative pulse. If it has a value between the positive and negative thresholds, it is recognized as a 0 (no pulse).

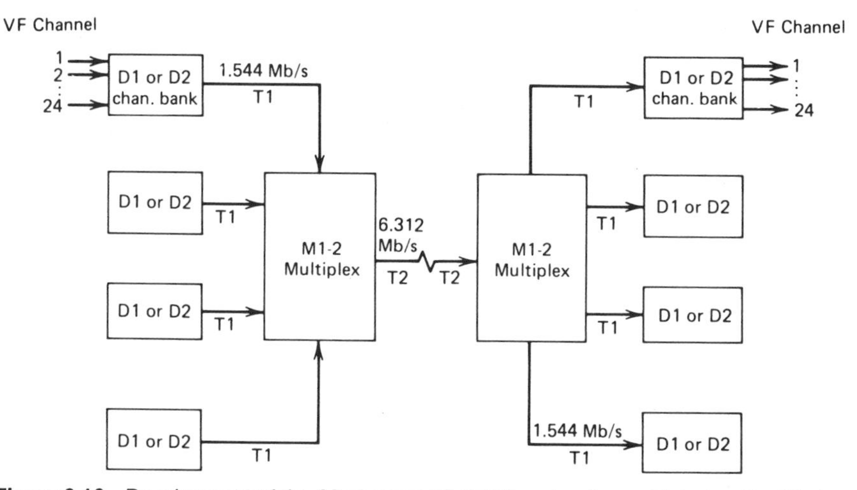

Figure 9.16 Development of the 96-channel T2 (ATT) system by multiplexing the 24-channel D1 channel bank outputs.

When either threshold is exceeded, the regenerator is triggered to generate a pulse of the appropriate duration, polarity, and amplitude. In this manner the distorted input signal is reconstructed as a new output signal for transmission to the next repeater [1].

6 HIGHER-ORDER PCM MULTIPLEX SYSTEMS

Using the 24-channel D1 channel bank as a basic building block, higher-order PCM multiplex systems are being developed in North America. For instance, four D1 channel banks are multiplexed by a M1–2 multiplexer, placing 6.312 Mbps on a single wire pair (T2 digital line). Figure 9.16 is a simplified block diagram of the first step in the development of a higher-order PCM multiplex configuration in North America.

The North American PCM multiplex hierarchy is shown in the following list and diagrammatically in Figure 9.17.

DS1	1.544 Mbps
DS1C	3.152 Mbps
DS2 (Output of multiplexer M1–2)	6.312 Mbps
DS3 (Output of multiplexer M2–3)	44.736 Mbps
DS4 (Output of multiplexer M3–4)	274.176 Mbps
DS432	432.000 Mbps (fiber)

As we have seen previously, there are basically two types of PCM system now in use in the world. The North American type is based on 24 voice channels. Japan follows a similar system inasmuch as its basic line rate is 1.544 Mbps

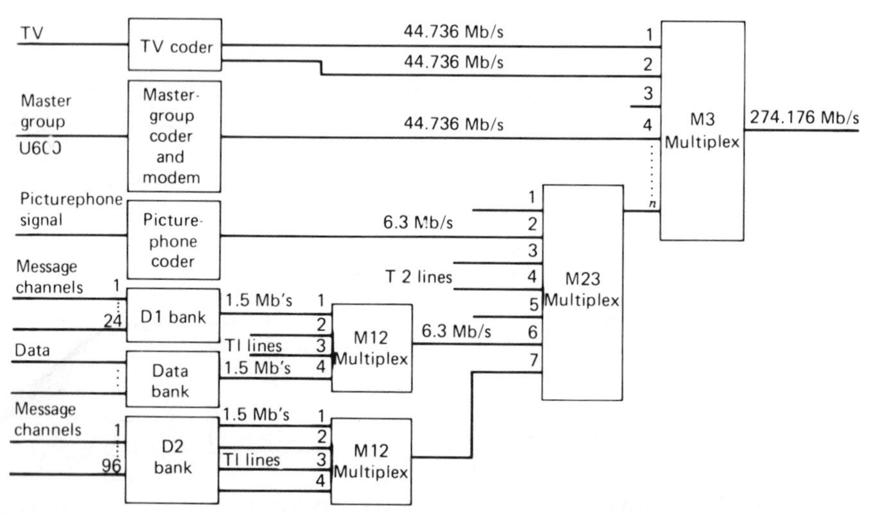

Figure 9.17 North American PCM hierarchy.

TABLE 9.3 Higher-Level PCM Multiplex Comparison

	Level				
System Type	1	2	3	4	5
North American T / D type	1	2	3	4	
Number of voice channels	24	96	672	4032	
Line bit rate (Mbps)	1.544	6.312	44.736	274.176	
Japan					
Number of voice channels	24	96	480	1440	5760
Line bit rate (Mbps)	1.544	6.312	32.064	97.728	400.352
Europe					
Number of voice channels	30	120	480	1920	7680
Line bit rate (Mbps)	2.048	8.448	34.368	139.264	560.0

based on 24 voice channels. The other system is European, based on 32 channels (30 channels of voice plus a signaling channel and a synchronization channel). The differences between these three systems for higher-level multiplex are shown in Table 9.3.

7 DELTA MODULATION

7.1 Introduction

Delta modulation is another method of transmitting an analog signal such as voice in a digital format. It is quite different from PCM in that coding is carried out before multiplexing and the code is far more elemental, actually coding at only 1 bit at a time. Delta modulation exploits the sample-to-sample redundancy typical in a speech or video waveform.

The delta modulation code is a one-element code and differential in nature providing 1 bit per sample of the difference signal. That single bit specifies the polarity of the difference sample. It thereby indicates whether the signal has increased in amplitude or decreased since the last sample.

An approximation of the input waveform is constructed in a feedback path by stepping up one quantization level when the difference is positive (1) and stepping down when the difference is negative (0). Here we mean that the derivative of the analog input is transmitted rather than the instantaneous amplitude as in PCM. This is done by integrating the analog input to decide which of the two has the larger amplitude. The polarity of the next digit placed on the line is either plus or minus to reduce the amplitude of the two waveforms [i.e., analog input and integrated digital output (previous digit)]. We thus see the delta encoder basically as a feedback circuit as shown in Figure 9.18.

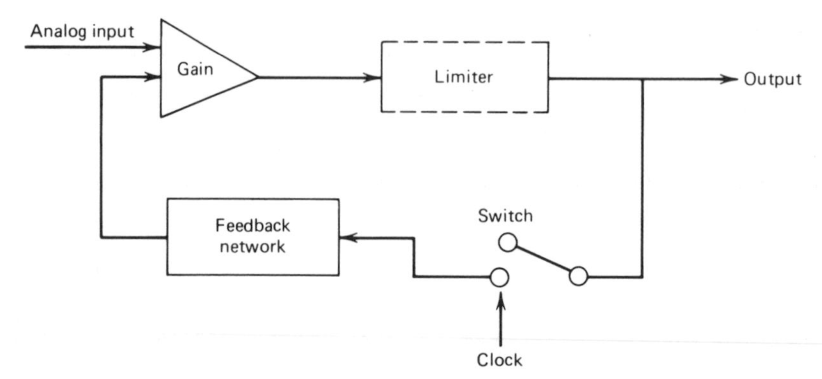

Figure 9.18 Basic electronic feedback circuit used in delta modulation.

Most delta encoders sample at a rate greater than the Nyquist rate, which, for a 4-kHz channel is 8000 samples per second. However, since each encoded sample contains relatively little information (i.e., 1 bit), delta modulation systems require a higher sampling rate than conventional PCM. These rates are typically 16 kbps and 32 kbps. Up to a certain point, the higher the sampling rate in this case, the better the signal-to-quantizing-noise ratio.

A simplified functional block diagram of a delta coder/decoder is shown in Figure 9.19 and Figure 9.20 shows a typical delta waveform superimposed in a simple sinusoid audio input signal.

From the Figure 9.19 block diagram we can see that the digital output signal of the delta coder is indicative of the slope of the analog input signal (its derivative of the slope)—1 for positive slope and 0 for a negative slope. But the 1 and 0 give no idea of the instantaneous or even semi-instantaneous steepness of the slope. This leads to the basic weakness found in delta modulation systems, namely, poor dynamic range or poor dynamic response given a satisfactory signal-to-quantizing-noise ratio.

With rapid changes of input level to the delta encoder, the output digital signal tends to lag behind these changes. This is called slope overload. Slope

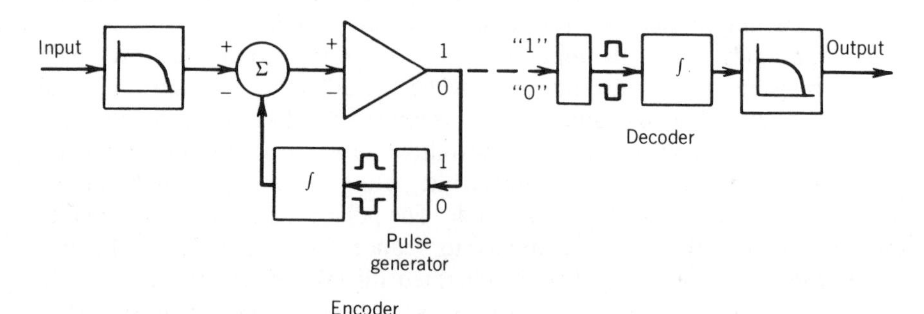

Figure 9.19 Simplied functional block diagram of a delta encoder/decoder.

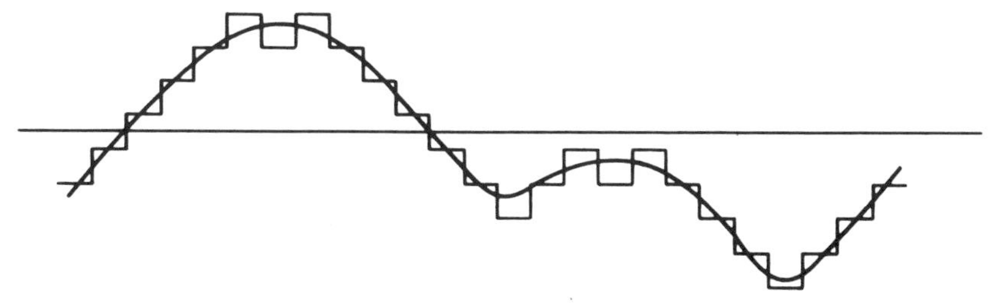

Figure 9.20 A delta encoding waveform.

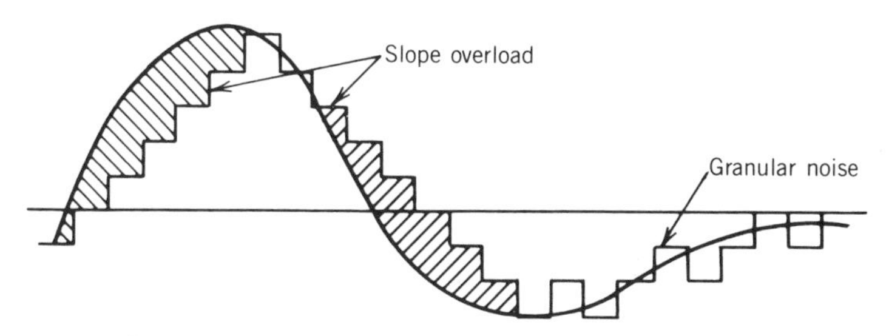

Figure 9.21 Typical slope overload of a delta modulator.

overload, shown diagrammatically in Figure 9.21, occurs when the rate of change of the input signal exceeds the maximum rate of change of the feedback loop. Thus a linear delta modulator (encoder) has severe dynamic range limitations.

7.2 Continuous Variable-Slope Delta Modulation

Continuous variable-slope delta modulation (CVSD) is a method of adaptive delta modulation using a form of digitally controlled companding (compression and expansion). It derives its step size from the transmit bit stream. As shown in Figure 9.22, adaptive logic monitors occurrence of three or four successive 1's or three or four successive 0's. A string of 1's indicates that the feedback is probably not rising as fast as the input, and conversely a string of 0's indicates that the feedback is not falling as fast as the input. This all 1's or all 0's condition enables control of a pulse generator increasing step size voltage. Through resistor leak-off, normal operation returns unless reenabled. Figure 9.22 shows a CVSD coder/decoder, and we see that the circuitry in the decoder is similar to that in the coder.

Turning to Figure 9.22, the analog input signal to the CVSD encoder is band limited by the input band-pass filter. The encoder compares the band-limited input signal S_i with the analog feedback approximation signal S_f, which is generated at the reconstruction integrator output. The digital output signal X_o of the encoder is the output of the first register in the "run-of-three"

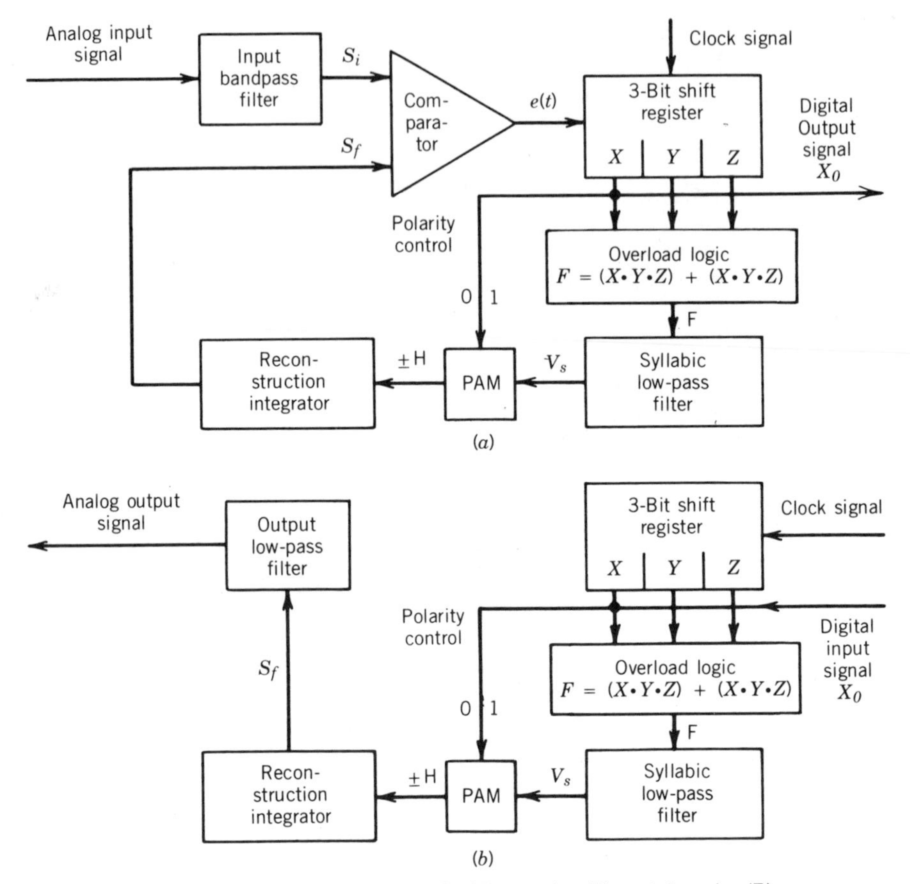

Figure 9.22 Block diagram of CVSD encoder (A) and decoder (B).

counter. The digital output signal is transmitted at the clock (sample) rate and will equal 1 if $S_i \geq S_f$ at the instant of sampling. For this value of X_o the pulse amplitude modulator (PAM) applies a positive feedback pulse $(+H)$ to the reconstruction integrator; otherwise, a negative pulse $(-H)$ is applied. This function is accomplished by the polarity control signal, which is equal to the digital encoder output X_o. The amplitude of the feedback pulse is derived by means of a 3-bit shift register, logic sensing for overload, and a syllabic low-pass filter. When a string of three consecutive 1's or 0's appears at the output of X_o, a discrete voltage level F is applied to the syllabic filter and the feedback pulse amplitude $(+H)$ increases until the overload string is broken. In such an event ground potential is fed to the filter by the overload logic, forcing a decrease in the amplitude of the slope voltage V_s.

The encoder and decoder, as mentioned, have identical characteristics, except for the comparator and filter functions. The CVSD decoder consists of the shift register, overload logic, syllabic filter, PAM, and reconstructed

TABLE 9.4 Signal-to-Quantizing-Noise Ratios for Typical Military CVSD Operation

Input Signal (dBm)	Minimum Test Signal-to-Quantizing-Noise Plus Idle Channel Noise Ratio (dB)	
	32 kbps	16 kbps
+2 to −3	22	15
−4 to −8	25	14
−9 to −13	26	14
−14 to −18	24	14
−19 to −23	22	14
−24 to −28	21	14
−29 to −33	19	13
−34 to −39	16	13

Source: From MIL-STD-188-200 (Ref. 13).

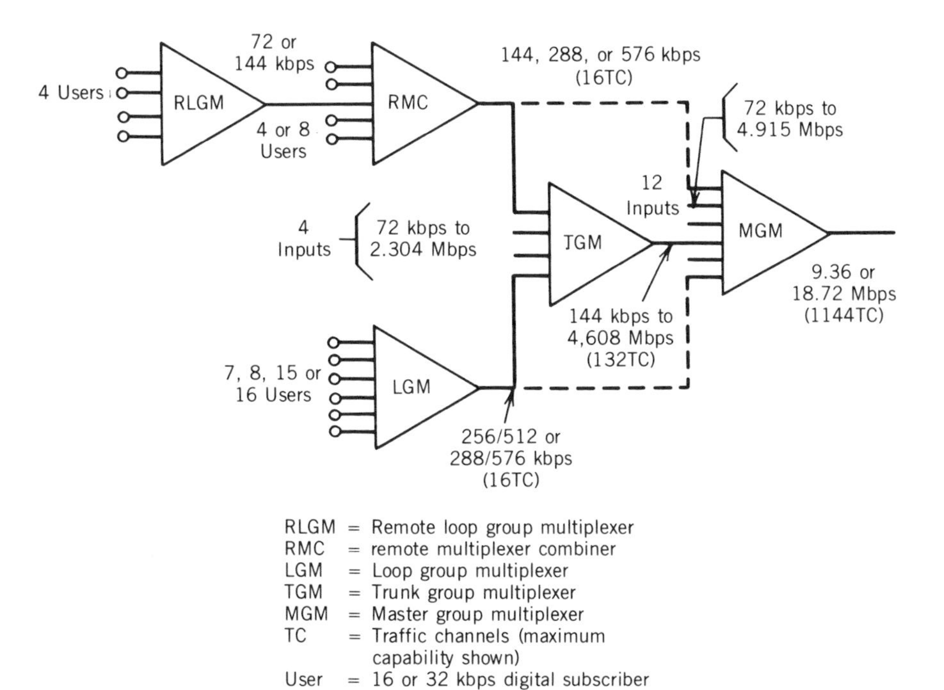

Figure 9.23 US Department of Defense TRI-TAC digital hierarchy. Courtesy of Raytheon Company (DGM data) (Ref. 18).

integrator used in the encoder, followed by a 4-kHz low-pass filter. The decoder performs the inverse function of the encoder and regenerates speech by passing the output signal S_f of the reconstruction integrator through the low-pass filter. Other characteristics optimize the CVSD technique for voice signals. These characteristics include:

- Changes in the slope of the analog input signal determine the step size changes of the digital output signal.
- The feedback loop is adaptive to the extent that the loop provides continuous or smoothly incremental changes in step size.
- Companding is performed at a syllabic rate to extend the dynamic range of the analog input signal.
- The reconstruction integrator of the CVSD device is of the exponential (leaky) type to reduce the effects of digital errors.

Table 9.4 shows typical signal-to-quantizing-noise ratio values. Figure 9.23 shows the US Department of Defense TRI-TAC digital multiplexer hierarchy where the user (subscriber) input signals are 16 kbps or 32 kbps CVSD waveforms.

8 LONG-DISTANCE DIGITAL TRANSMISSION

8.1 General

Binary digital transmission, with its capability of regeneration that essentially eliminates noise accumulation as the signal traverses the transmission media, has become the choice for long-distance (toll) transmission systems for backbone routes and tails. One driving factor for this choice is that the public switched telephone network (PSTN) is rapidly becoming all digital. Section 9 summarizes advantages and disadvantages of digital transmission. Here we cover several of the disadvantages and impairments that have an impact on toll system design. Here are two such factors we should not lose sight of:

- Bit errors accumulate as the digital signal traverses many nodes and regenerative repeaters.
- Digital transmission is not bandwidth efficient.

Accumulation of bit errors can be roughly approximated as a function of the number of nodes and repeater sections in tandem. Two measures can be taken to reduce bit error accumulation, which results in degraded error performance at end points. Obviously one way is to reduce the number of repeatered sections. The second measure is to specify improved BER performance per section (i.e., of the order of 1×10^{-9}).

Fiber optic cable transmission reduces the number of repeater sections by better than 10 : 1 over its metallic cable counterpart. Fiber also is an extremely wide bandwidth medium. Satellite circuits for high-usage routes are another alternative, but with considerably less information bandwidth than is available on optical fiber, for example.

Digital line-of-sight microwave, unless properly designed, can be wasteful bandwidth, approximately 16 : 1 compared to analog single sideband (SSB) techniques. SSB requires about 4 kHz per voice channel, whereas conventional PCM requires 64 kHz, assuming 1 bit per hertz. Modern digital LOS radiolinks use modulation waveforms with typical bit-packing ratios of 5–7 bits per hertz. This greatly alleviates the bandwidth conservation problem in exchange for more complex radio terminals.

This section briefly summarizes digital transmission impairments and discusses several long-distance digital system design approaches.

8.2 Digital Transmission Impairments

8.2.1 General. This section reviews five basic impairments to digital transmission: jitter, distortion, noise, crosstalk, and echo. The effects of these impairments on system performance drive the system and equipment design (selection of cable, repeater section length, repeater characteristics, equalizer design, and selection of waveform).

8.2.2 Jitter. Jitter is an important limitation of present-day technology on using PCM or delta modulation (DM) as a vehicle for long-haul transmission. A general definition of jitter is "the movement of zero crossings of a signal (digital or analog) from their expected time of occurrence." In Chapter 8 it was called "unwanted phase modulation or incidental FM." Such jitter or phase jitter affected the decision process of the zero crossing in a digital data modem. Much of this sort of jitter can be traced to the intervening FDM equipment between one end of a data circuit and the other.

Pulse-code modulation has no intervening FDM equipment, and jitter in PCM systems takes on different characteristics. However, the effect is essentially the same—uncertainty in a decision circuit as to when a zero crossing (transition) took place, or the shifting of a zero crossing from its proper location. In PCM it is more proper to refer to jitter as "timing jitter." The primary source of timing jitter is the regenerative repeater. In the repeatered line, jitter may be systematic or nonsystematic. Systematic jitter may be caused by offset pulses (i.e., where the pulse peak does not coincide with regenerator timing peaks or where transitions are offset), intersymbol interference (dependent on specific pulse patterns), and local clock threshold offset. Nonsystematic jitter may be traced to timing variations from repeater to repeater and to crosstalk.

In long chains of regenerative repeaters, systematic jitter is predominant and cumulative, increasing in rms value as $N^{1/2}$, where N is the number of repeaters in the chain. Jitter is also proportional to a repeater's timing filter bandwidth. Increasing the Q of these filters tends to reduce jitter of the regenerated signal, but it also increases error rate caused by sampling the incoming signal at nonoptimum times.

The principal effect of jitter on the resulting analog signal after decoding is to distort the signal. The analog signal derives from a PAM pulse train, which is then passed through a low-pass filter. Jitter displaces the PAM pulses from their proper location, showing up as undesired pulse-position modulation (PPM).

Because jitter varies with the number of repeaters in tandem, it is one of the major restricting parameters of long-haul high-bit-rate PCM systems. Jitter can be reduced in future systems by using an elastic store at each regenerative repeater (which is costly) and high-Q phase locked loops.

8.2.3 Distortion. On metallic transmission links, such as coaxial cable and wire-pair cable, line characteristics distort and attenuate the digital signal as it traverses the medium. There are three cable characteristics that create this distortion: loss, amplitude distortion (amplitude–frequency response), and delay distortion. Thus the regenerative repeater must provide amplification and equalization of the incoming digital signal before regeneration. There are also trade-offs between loss and distortion on the one hand and repeater characteristics and repeater section length on the other.

8.2.4 Noise. As in any electrical communication system, thermal noise, impulse noise, and crosstalk affect system design. Because of the nature of a digital system, these impairments need only be considered on a per-repeater-section basis because noise does not accumulate due to the regenerative process carried out at repeaters and nodes. Bit errors do accumulate, and this impairment family is one of several that create these errors. One way to limit error accumulation is to specify a stringent BER for each repeater section. Repeater sections are often specified with a median BER of 1 in 10^{-9}.

It is interesting to note that PCM provides reasonable voice performance for a BER as poor as 1 in 10^2. However, the worst tolerable BER is 1 in 10^3 at system end points. This value is required to ensure the correct operation of supervisory signaling. The reader should appreciate that such degraded BER values are completely unsuitable for data transmission.

8.2.5 Crosstalk. Crosstalk is a major impairment in PCM wire-pair systems, particularly when "go" and "return" channels are carried in the same cable sheath. The major offender of single cable operation is near-end crosstalk (NEXT). When the two directions of transmission are carried in separate cables or use shielded pairs in a common cable, far-end crosstalk (FEXT) becomes dominant.

One characteristic has been found to be a major contributor to poor crosstalk coupling loss. This is the capacitance imbalance between wire pairs. Stringent quality control during cable manufacture is one measure taken to ensure that minimum balance values are met.

8.2.6 Echo. Echo is caused by impedance discontinuities in the transmission line, including repeaters and terminations (MDFs, Codecs, switch ports). Good impedance match across the entire system eliminates the cause of echo or reduces its level. On a PCM transmission system there are many causes of echo, such as gas plugs and splices.

Gas plugs are used on cable systems to allow gas to be applied under pressure to the cable to prevent moisture buildup. The plug tends to add capacitance to the line. To compensate for this, repeater sections that incorporate plugs are made short to accommodate the added capacitance.

Other sources of echo are where gauge and insulation changes take place along the cable run. Bridged taps are still another potential source of mismatch.

8.3 Digital Metallic Transmission Systems

8.3.1 Scope. The earliest PCM systems were used to expand wire-pair cable capacity on local trunks. Today it is widely implemented on all types of terrestrial trunks on either wire-pair, coaxial, and fiber optic cable. Several typical North American PCM systems that operate on metallic media are described in the subsections that follow. System nomenclature derives from ATT and is listed below for reference to the corresponding PCM hierarchical level and bit rate.

Transmission System	Hierarchical Level	Bit Rate
T1 and T1/OS	DS1	1.544 Mbps
T1C	DS1C	3.152 Mbps
T2	DS2	6.312 Mbps
T4M	DS4	274.176 Mbps

8.3.2 The T1 System. The T1 system commonly uses wire-pair cable as the transmission medium, requiring two wire pairs per T1 system. The number of pairs carrying T1 must be limited due to crosstalk considerations. A variety of cables may be used for T1 systems, including pulp-insulated and polyethylene-insulated conductor (PIC) cables with copper pairs of 19 and 26 gauge and aluminum conductors of 17 and 20 gauge [17].

For those pairs assigned for T1 operation, all bridged taps, load coils and line build-out (LBO) networks must be removed. The maximum length of a T1 system on wire pair is that encompassed by 200 repeaters in tandem (about 200 mi). Several power spans can be accommodated within these performance limits. The signal waveform is the standard AMI (bipolar) waveform with a 50% duty cycle format and may contain no more than 15 consecutive 0's. Present power spans are 36 mi when using 22-gauge copper wire pairs. Repeaters are designed for optimum operation with cable loss of about 32 dB at 772 kHz, which is equivalent to a 6000-ft repeater section when using 22-gauge pulp-insulated cable or other commonly used cables [17].

Another type of T1 system is called the T1 Out State Digital Transmission System or T1/OS. System error rate is specified as 1×10^{-6}. Reference 17 states that transmission quality is ensured by application of a "design number" to each line section, span, or terminal-to-terminal system. The overall engineering design objective is that at least 95% of all properly engineered and installed systems have an error rate of 1 error in 10^6 bits. The probability of exceeding this error rate is allocated on a section-by-section basis. For T1/OS, 5% is allocated as the design number for the maximum allowable system length of 200 repeater sections. Hence each repeater section is allocated a design number of probability of $5/200 = 0.025\%$ of exceeding the error rate of 1×10^{-6}. The design number for a span or system is the sum of the design numbers for the component parts.

T1/OS also has improved fault locating capability and provides protection switching arrangements for better availability.

8.3.3 The T1C System. There are many similarities between T1C and T1 systems. Repeater spacings for T1 may be used for T1C, facilitating conversion or upgrading of T1. T1C also uses the same bipolar 50% duty cycle waveform of T1. T1C, however, provides twice the voice channel capacity of T1 or 48 voice channels and its bit rate is 3.152 Mbps.

Because of its higher bit rate, T1C requires different regenerative repeaters than those used on the T1 system. The equalizing network/amplifier can compensate for cable loss from about 10 dB to 53 dB at 1.576 MHz. The reader will note that this is the half-bandwidth value assuming again 1 bit/Hz. (i.e., $3.152/2 = 1.576$).

8.3.4 The T2 System. The T2 system has four times the capacity of the T1 system or 96 voice channels for transmission on tandem repeater links up to about 500 mi long. The T2 bit rate is 6.312 Mbps using a similar bipolar waveform as T1 but with B6ZS. In other words, for any successive six 0's there is an insertion of a special code (an AMI violation as described in Section 5.3). This ensures the presence of sufficient transitions to maintain correct repeater clock operation.

T2 requires a special low-capacitance cable that has been specifically designed for T2 operation. One such cable uses conductors that are dual-expanded polypropylene-insulated 22-gauge copper. Separate cables are used on the "go" and "return" directions to reduce crosstalk. These cables exhibit low capacitance, lower losses, and higher crosstalk coupling loss than conventional wire-pair cable such as PIC. A T2 repeater is made up of two regenerators, one for each direction of transmission. A selection of plug-in equalizers must be made to match the repeater gain to the cable loss in the preceding repeater section. Clock signal is extracted by a monolithic crystal filter with a very narrow passband. The timing circuit operates on the basis of the characteristics of the incoming pulse stream.

With the buried special low-capacitance cable, repeater spacing is 15,000 ft. A maximum of 250 repeaters in tandem is recommended [17], and on such a system the error rate should not exceed 1×10^{-7} on 95% of all lines. A maintenance span on the T2 system has up to 44 repeaters.

8.3.5 The T4M System. The T4M system is a high-capacity PCM transmission system designed for metropolitan and intercity circuits. It carries a DS4 signal at 274.176 Mbps rate, which accommodates 4032 VF channels. The line signal is unipolar (NRZ) and scrambled. T4M operates over standard 0.375-in. diameter coaxial cable. One such coaxial cable is required for the "go" direction and one for the "return" direction of transmission. Repeater spacings are up to 5700 ft.

The basic functions of the T4M repeater operation are similar to those described in Section 5.5. These functions include an equalizer and automatic line build-out (ALBO). A decision circuit recognizes the presence of positive-going or negative-going pulses and produces new undistorted pulses. These functions are carried out with the help of a timing and control circuit to provide accurate sampling of the pulse stream in the decision circuit. A quantized feedback circuit completes reshaping of the signal before the decision circuit. T4M has automatic protection switching to improve system availability. Service is switched to a protection line when the BER exceeds 1×10^{-6}. A protection line protects up to 10 working lines in one direction. Protection spans can be up to 111 mi long, and spans may be connected to form links up to 500 mi long.

8.4 Digital Transmission Over Fiber Optic Cables

Fiber optics offers many advantages for digital transmission. The two principal advantages are the very wide bandwidth available and the much greater distance permitted between repeaters. With fewer repeaters there is less jitter buildup and fewer errors will accumulate. These are some of the factors that drive the design engineer's decision to select fiber as the transmission medium rather than some form of metallic medium.

Single-mode fiber is now available with bandwidths in the range 25–50 GHz/km and attenuation as low as 0.25 dB/km. Repeater spacing on fiber optic cable systems is a function of cable attenuation, system bit rate/dispersion, and source detector characteristics. Multigigabit systems have been demonstrated.

As we mentioned in Section 8.1, reducing the number of repeaters for a given link distance reduces the accumulation of timing jitter and can provide a notable improvement in error performance. What fiber lacks is standardization. For instance, CCITT and US agencies have not issued recommendations or standards on high-bit-rate structures and format (i.e., above DS4 in North America). At this juncture manufacturers in concert with users seem to go their own way.

The US Electronics Industries Association (EIA) has made great strides in providing fiber cable standards and methods of testing. The US Department of Defense, working with EIA, likewise is starting to turn out fiber optic standards for cable, sources, detectors, and connectors. The concern here is system interfaces over 600 Mbps.

The major drawbacks of fiber optics are the same as for the related metallic media. These are obtaining right of way for the cable lay and accidental damage done to the cable after it is operational.

See Chapter 7 and Refs. 2 and 4 for general discussions of fiber optics transmission systems.

8.5 Digital Transmission by Radiolink

8.5.1 General. Digital transmission by radiolink is becoming increasingly important in both civilian and military communication. Civilian communication organizations such as telephone companies (administrations) and specialized common carriers have opted for PCM rather than DM, so our discussion from here onward stresses PCM.

The transmission of PCM by radiolink is a viable alternative to PCM VF cable pair transmission under the following circumstances:

- Where physical or natural obstructions make cable laying impractical.
- For relatively long metropolitan trunk groups under 6000 voice channels where cable laying is very costly.
- As an alternative routing of a cable system.
- Between PCM local switches to avoid requirements of A/D–D/A (analog-to-digital-to-analog) conversion, thus eliminating FDM as an economically viable possibility.

Consider a situation where a large number of trunk (junction) routes are presently equipped with PCM. An FDM–FM radiolink is contemplated, and the system engineer is faced with one or several of the preceding circumstances. The use of FM radiolinks with FDM will prove expensive. The existing PCM will have to be brought to VF (demultiplexed–demodulated) to interface with the new FDM equipment. Use of PCM eliminates the additional multiplex equipment cost. Further, PCM channelizing equipment, if we accept groups of 24 or 30 channels at a time, is less expensive on a per channel basis than is FDM equipment.

8.5.2 Modulation, RF Bandwidth, and Performance. As we have discussed previously, PCM requires large amounts of information bandwidth. Electromagnetic spectrum (i.e., radio spectrum) is at a premium, particularly where there is the greatest demand for radio facilities, namely, in industrially built-up areas. Consider a 672-voice-channel PCM radio system. Such a system transmits at 44.736 Mbps (Table 9.3). At 1 bit per hertz of RF bandwidth, about 45 MHz would then be required. On an FDM–FM conventional radiolink, about 6 MHz would be required. The system design engineer must resort to modulation methods that are bandwidth conservative or must resort to FDM–FM.

There are three basic methods of modulating a radio wave: by amplitude, by frequency, or by phase. With PCM we are dealing with binary conditions. The simplest approach would be to two-state modulate a carrier such that one of the states represents the 1 and the other the 0 of the PCM bit stream. Again, we arrive at a bandwidth of 1 bit per hertz. This is described in the literature as one symbol per hertz. For AM, this would represent a discrete level (amplitude) for the 1 and another for the 0; or for PSK, say, 0° phase for the 1 and 180° for the 0; or for FSK a discrete frequency for the 1 and another for the 0.

Suppose the carrier can take on four discrete states, coding the bit stream 2 bits at a time. A symbol represents a transition in state. In the case of phase 0° is assigned the symbol 00, 90° the symbol 01, 180° the symbol 10, and 270° the symbol 11. By this method we transmit 2 bits of information per symbol, or, if you will, change of state. Essentially, we have cut the required bandwidth in half. For higher logic level systems, this is done at the expense of signal-to-noise ratio (~ 3–4 dB) as the logic level is doubled [12]. Table 9.5 reviews modulation possibilities with the resulting bandwidth required.

Of course, the major advantage of digital modulation of radio systems is the ability to regenerate periodically the waveform at each repeater site. Another advantage is that it is little affected by traffic loading, and any mix of voice

TABLE 9.5 Digital Modulation Techniques versus Spectral Efficiency

Type of Modulation	Number of Logic Levels	Number of Bits per Symbol
Amplitude modulation	2	1
Frequency-shift keying	2	1
Two-phase shift keying	2	1
Four-phase shift keying	4	2
Eight-phase shift keying	8	3
Sixteen-phase shift keying	16	4
Quadrature AM (QAM)	16, 32, 64, 128, 256	4, 5, 6, 7, 8

and data traffic has no effect on system performance. Loading is a major problem on conventional FDM systems.

9 SUMMARY OF ADVANTAGES AND DISADVANTAGES OF DIGITAL TRANSMISSION

The following lists some advantages and disadvantages of digital transmission.

The advantages of digital transmission tend to far outweigh the disadvantages. This is being borne out with the rapid conversion of the trunk plant to all digital. Because of the large investment in the subscriber loop plant, conversion to all digital may take more time. However, this time may be shortened if ISDN (Chapter 13) implementation is accelerated.

Advantages

1. System noise is controlled by the design of the terminal (quantization noise) and is essentially independent of the length of the system or of line noise and distortion. This assumes that the full length of the system is digital.
2. The signal-to-distortion performance of the system increases linearly with the number of bits per sample, giving a more efficient noise/bandwidth trade-off than other *bandwidth expansion* techniques such as FM. This efficiency, combined with the ruggedness of digital transmission, gives better utilization of noisy media such as wire-pair cable.
3. Increases in device speed (i.e., ICs, VLSI) allow common circuit components to be shared by many channels, thus lowering the per-channel cost of a PCM terminal. This is one major factor that makes the per-channel cost of PCM more cost-effective than FDM.
4. Digital systems are insensitive to traffic loading up to their full capacity. FDM is highly sensitive to traffic loading.
5. Likewise, digital time division multiplex treats all channels alike in contrast to FDM regarding phase and amplitude distortion (and resulting noise degradation) suffered by channels at band edge.
6. There is no appreciable degradation incurred in multiplexing/demultiplexing so that facility arrangements do not need to take into account the number of previous multiplex/demultiplex operations.
7. Digital transmission gives complete freedom to multiplex digital data, voice, video, facsimile, etc. on the same facility, whereas analog transmission does not.
8. Digital systems are more efficient than analog in the transmission of digital data in that fewer voice channels must be displaced to obtain a given digital capacity.

9. Digital transmission provides the most economically possible interface to digital switching systems. Analog systems, on the other hand, require full demultiplexing/remultiplexing at switching nodes. Digital switches have inlets/outlets at the digital multiplex rates.
10. Fiber optics transmission systems tend to favor digital transmission in that its attenuation is relatively independent of frequency. This makes bandwidth expansion transmission techniques to reduce baseband noise particularly attractive since the added bandwidth on fiber optic systems is almost "free," and PCM is one of the most efficient of such schemes. In addition, light sources used in optical transmission exhibit nonlinearities that make them better suited to nonlinear modulation techniques such as PCM than to linear modulation methods such as AM.
11. Signaling is digital. Signaling on analog systems has to be converted to something compatible, such as a tone or multitone format. On digital systems only a bit has to be changed in state for supervisory signaling and a bit sequence for address signaling (SSN No. 7). A digital system is also compatible with digital processors used in SPC switches.

Disadvantages

1. Bit errors accumulate across a digital system. These are not recoverable unless we resort to an error-correction system such as ARQ or FEC, both of which require still additional bandwidth.
2. System timing is a major issue and is discussed in the following chapter.
3. Although digital terminals tend to be less expensive than their analog counterparts on a per-channel basis, digital transmission lines (metallic media) tend to be more expensive than their analog counterparts. With the cost of fiber optic systems dropping, digital transmission on fiber is less expensive than on metallic media. (See Ref. 17, Vol. 2.)

REVIEW QUESTIONS

1. Give the three principal advantages of digital transmission.

2. What are the three basic steps in the development of a PCM signal from an analog source, typically voice?

3. Following the Nyquist sampling theorem, what is the sampling rate of a 4-kHz voice channel? Of a 7.5-kHz program channel? of a 4.2 MHz video channel?

4. What is the polarity of a mark (1) and of a space (0) in AMI (bipolar) transmission?

5. For a 24-channel PCM system, calculate the period of one frame.

6. Define *quantization distortion*.

7. If the number of quantization steps is doubled in a particular PCM design using linear quantization, what improvement in quantization distortion (noise) is achieved? Express the answer in decibels.

8. Why is it desirable to reduce the size (number of bits or elements) of a PCM code word as much as possible and yet maintain reasonable voice quality?

9. There are only 16 quantization steps in a particular PCM system. What minimum length (bits) code word is required to accommodate these 16 discrete levels?

10. Identify the two distinct logarithmic companding laws used in modern PCM systems.

11. What key piece of information does the first significant bit in a PCM code word tell us?

12. Derive the value 1.544 Mbps in accordance with DS1 format.

13. Name at least five differences between European and North American PCM systems.

14. How is supervisory signaling carried out in the North American PCM system? In the European system? Argue pros and cons of each approach.

15. In the earliest implementations of PCM, where was it applied and why? (This does not imply that today it is not applied for the same reason in various situations.)

16. In the North American T1/DS1 system, what is the repeater spacing in feet? Going out from a switching node, why is there always a half-section rather than a full repeater section? For extra credit, what additional benefit can we get from that particular repeater spacing distance? (Hint: Turn back to subscriber/trunk loop design.)

17. Give at least two reasons why AMI (bipolar) waveform is used on T1/T1C (DS1/DS1C) cable transmission systems.

18. In a simplified functional block diagram of a regenerative repeater there are three basic functional blocks. Name and describe the function of each.

19. Give the two basic functions of overhead bits used in higher-order PCM multiplex. Explain each function.

20. Give two primary advantages and one principal disadvantage of delta modulation systems.

21. What causes slope overload in a delta modulator?

22. What is the advantage of CVSD compared to conventional delta modulation?

23. Why is the Nyquist sampling rate exceeded in delta/CVSD systems?

24. Describe how framing is carried out in the DS1 and in the CEPT 30 + 2 PCM systems.

25. Give at least four of the basic impairments affecting digital transmission. Describe the cause and effect of each. Concentrate on metallic transmission media.

26. Given the impairments in question 25, what are the two basic measures we can take to achieve end-to-end performance of a PCM system?

27. The radio frequency spectrum is a natural international resource. Wasteful bandwidth systems do not conserve this resource. With this background, compare bandwidth requirements of PCM/TDM versus some conventional analog techniques such as FDM–FM and SSB. (Hint: Compare on a VF channel basis.)

28. What are some techniques used to conserve bandwidth on digital radiolinks (LOS microwave)?

29. Considering only metallic transmission media, what limits T1 to 200 repeaters in tandem? T4M to about 465 repeaters in tandem?

30. Why must we limit the number of consecutive 0's in a PCM line code? Give the meaning of B6ZS.

31. T1 was designed primarily for what metallic medium? T2? T4M? This does not mean that their equivalent formats cannot be used on some other medium.

32. What cable parameter affects crosstalk more than any other?

33. With common cable transmission (both directions of transmission in one cable sheath), give one overriding reason why crosstalk considerations are so important.

34. Give the two principal advantages of digital transmission by fiber optics cable over its metallic counterparts.

35. Give advantages and disadvantages of digital fiber optics cable transmission versus digital radio.

36. What two key characteristics of fiber optics links make it so attractive for digital communications? (Give different answers than in questions 33 and 34.)

37. Considering CEPT 30 + 2 and the North American DS1, why would one say that the North American system gives the ISDN user a corrupted 64-kbps channel and the European does not? (Hint: What two things corrupt the DS1 bit stream, forcing the user down to 56 kbps rather than 64 kbps?)

38. If PCM transmission is performance limited as a function of the number of repeaters in tandem, then how can we get around this issue for trans- and intercontinental PCM transmission?

REFERENCES

1. *Transmission Systems for Communications*, 5th ed., Bell Telephone Laboratories, Holmdel, N.J., 1983.
2. Roger L. Freeman, *Reference Manual for Telecommunications Engineering*, Wiley, New York, 1985.
3. John Bellamy, *Digital Telephony*, Wiley, New York, 1983.
4. Roger L. Freeman, *Telecommunication Transmission Handbook*, 2nd ed., Wiley, New York, 1981.
5. International Telephone and Telegraph Corporation, *Reference Data for Radio Engineers*, 6th ed., Howard W. Sams, Indianapolis, 1976.
6. K. W. Catermole, *Principles of Pulse Code Modulation*, Illiffe, London, 1969.
7. *Digital Channel Bank, Requirements and Objectives*, Bell System Technical Reference, Pub. 43801, American Telephone and Telegraph, Basking Ridge, N.J., 1982.
8. *PCM: System Application 30 + 2TS*, BTM/ITT, Antwerp, Belgium 1976.
9. *Technical Manual: Operations and Maintenance Manual for T324 PCM Cable Carrier Systems*, ITT Telecommunications, Rayleigh, N.C., April 1973.
10. J. V. Marten and E. Brading, "30-Channel Pulse Code Modulation System," *Electr. Commun.*, **48** (1, 2) (1973).
11. CCITT, *Red Books*, Vol. III, VIII Plenary Assembly, Malaga-Torremolinos, 1984, in particular the G.700 recommendations.
12. Roger L. Freeman, *Radio System Design for Telecommunications*, Wiley, New York, 1987.
13. MIL-STD-188-200, US Department of Defense, Washington, D.C., June 1983.
14. GTE Lenkurt Demodulator, *PCM Update*, Parts 1 and 2, (GTE) Lenkurt Electric Company, San Carlos, Cal., February 1975.
15. *Notes on the BOC Intra-LATA Networks—1986*, Bell Communications Research, TR-NPL-000275.
16. P. Bylanski and D. G. W. Ingram, *Digital Transmission Systems*, IEE Telecommunications Series 4, Peter Peregrinus Ltd., Stevenage, Herts UK, 1976.
17. *Telecommunications Transmission Engineering*, ATT–Western Electric Co., Winston-Salem, N.C., 1977.

18. Technical Advisories, United States Independent Telephone Assoc., Washington, D.C., *Interconnection Specifications for Digital Cross-Connects*, No. 34, Issue 3 (1979); No. 42, Issue 2 (1981); No. 49; No. 50, Issue 5 (1979); No. 55, Issue 4 (April 1981); No. 56, Issue 1 (1980); No. 57, Issue 1 (January 1980).
19. Raytheon Company brochure: *Tri-Tac Digital Group Multiplex (DGM) Equipments: Characteristics, Specifications and System Interfaces*, Raytheon Company, Marlborough, Ma., 1987.
20. Roger L. Freeman, "An Overview of Digital Transmission and Multiplexing," a tutorial presentation to the MITRE Institute, Bedford, Ma., October 1987.

10

DIGITAL SWITCHING AND NETWORKS

1 INTRODUCTION

In Chapter 3 we dealt with analog space-division switching in which a metallic (conductive) path is set up between calling and called subscriber. "Space division" in this context refers to the fact that speech paths are physically separated (in space). Figure 10.1A illustrates this concept by showing a representative cross-point matrix. Time-division switching (Figure 10.1B) permits a common metallic path to be used by many simultaneous calls separated one from the other in the time domain. In this context, with time-division switching the speech or other information to be switched is digital in nature, either PCM or delta modulation (DM). Samples of each telephone call are assigned time slots, as described in Chapter 9. PCM or DM switching involves the distribution of these slots in sequence to the desired destination port(s) of the switch. Internal functional connectivities in the switch are carried out by digital "highways." The highways consist of sequential speech path time slots.

This chapter describes PCM switching in a simplified step-by-step fashion. It covers various generic switch architectures and explains basic functional operations. This is followed by a walk-through of the Northern Telecom DMS-100 digital switching family. Digital networks and their topology follow PCM switching. Then there is a discussion of network timing and synchronization. Finally, an overview is presented on digital network impairments, including error performance and "slips."

2 ADVANTAGES AND ISSUES OF PCM SWITCHING

There are both economic and technical advantages to digital switching; in this context we refer to PCM switching (of course, most of these same arguments hold for delta/CVSD switching as well). The economic advantages of time-

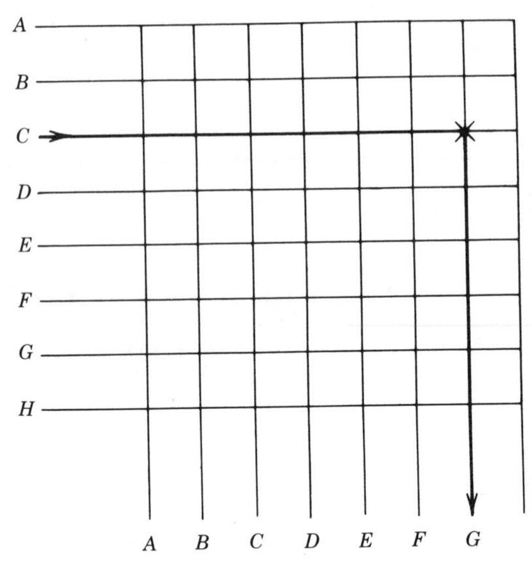

Figure 10.1A Space-division switch showing connectivity from user *C* to user *G*.

division PCM switching include:

- There are notably fewer equivalent cross-points for a given number of lines and trunks than in a space-division switch.
- A PCM switch is of considerably smaller size.
- It has more common circuitry (i.e., common modules).
- It is easier to achieve full availability within economic constraints.

The technical advantages include:

- It is regenerative (i.e., the switch does not distort the signal; in fact, the output signal is "cleaner" than the input).
- It is noise resistant.
- It is computer based and thus incorporates all the advantages of SPC.

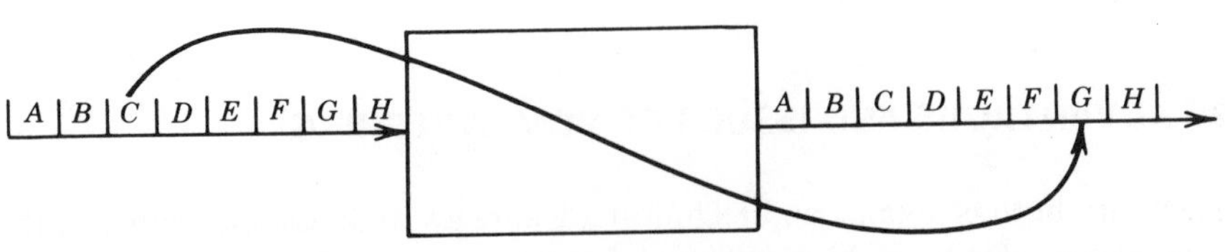

Figure 10.1B A time-division switch showing time slot interchange. Connectivity is from user *C* (in incoming time slot *C*) to user *G* (in outgoing time slot *G*).

- The binary message format is compatible with digital computers. It is also compatible with signaling.
- A digital exchange is lossless. There is no insertion loss as a result of a switch inserted in the network.
- It exploits the continuing cost erosion of digital logic and memory; LSI, VLSI, and VHSIC insertion.

Two technical issues may be listed as disadvantages:

- A digital switch deteriorates error performance of the system. A well-designed switch may only impact network error performance minimally, but it still does it.
- Switch and network synchronization and the reduction of wander and jitter can be gating issues in system design.

3 APPROACHES TO PCM SWITCHING

3.1 General

A digital switch's architecture is made up to two elements called T and S for time division switching (T) and space division switching (S) and can be made up of sequences of T and S. For example, the ATT No. 4 ESS is a TSSSST switch; No. 3 EAX is an SSTSS switch, and the Northern Telecom DMS-100 is TSTS folded. Some switches designed to serve local networks are simply TST switches. Now we want to examine these basic elements, the time and space switch networks.

3.2 Time Switch

In a most simplified way, Figure 10.1B is a time switch or time slot interchanger. From Chapter 9 we know that a time slot in PCM contains 8 bits and that a basic frame is 125 μs of duration. For the DS1 format the basic frame contains 24 time slots and for CEPT 30 + 2, 32 time slots. The time duration of an 8-bit time slot in each case is $125/24 = 5.2083$ μs for the DS1 case and $125/32 = 3.906$ μs for the CEPT 30 + 2 case. Time slot interchanging involves moving the data contained in each time slot from the incoming bit stream to an outgoing bit stream but with a different time slot arrangement in accordance with the destination of each time slot. What is done, of course, is to generate a new frame for transmission at the appropriate switch outlet.

Obviously, to accomplish this, at least one time slot must be stored in memory (read) and then called out of memory in a changed sequence position (write). The operations must be controlled in some manner and some of these control actions must also be kept in memory, such as the "idle" and "busy" condition of time slots. Now we can identify three of the basic blocks of a time switch:

1. Memory for speech (by time slots).
2. Memory for control.
3. Time slot counter or processor.

These three blocks are shown in Figure 10.2. The incoming time slots can be written into the speech memory either sequentially (i.e., as they appear in the incoming bit stream) (Figure 10.2A) or randomly (Figure 10.2B), where the order of appearance in memory is the same as the order of appearance in the outgoing bit stream. Now with the second version the outgoing time slots are read sequentially because they are in the proper order of the outgoing bit stream (e.g., random write, sequential read). This means that the incoming time slots are written into memory in the desired *output* order. The writing of incoming time slots into the speech memory can be controlled by a simple time slot counter and can be sequential (e.g., in the order in which they appear in the incoming bit stream) (Figure 10.2A). The readout of the speech memory is controlled by the control memory. In this case the readout is random where the time slots are read out in the desired output order. The memory has as many cells as there are time slots. For the DS1 example, there would be 24 cells. This time switch, as shown, works well for a single inlet–outlet switch. Consider a multiple port switch, which may be a tandem switch, for example,

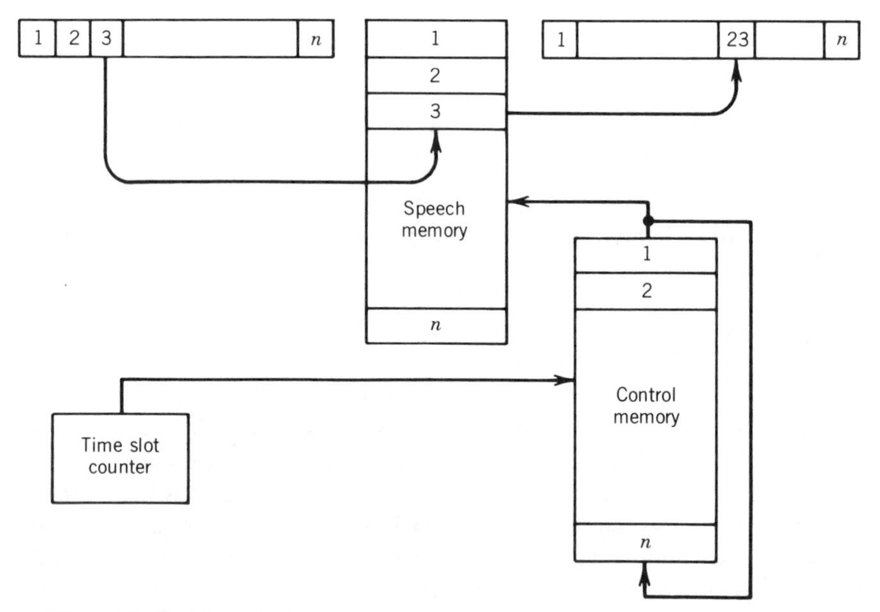

Figure 10.2A Time slot interchange: time switch. Sequential write, random read.

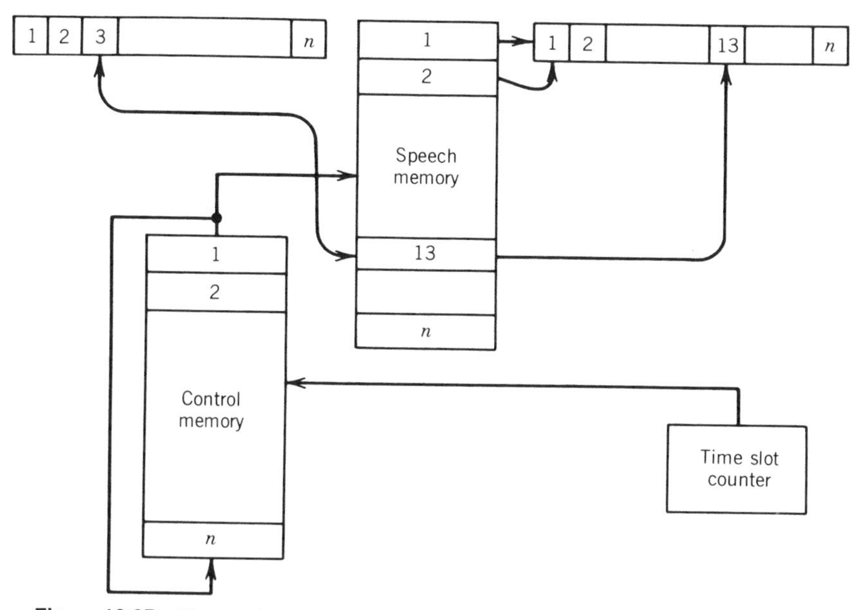

Figure 10.2B Time switch, time slot interchange. Random write, sequential read.

handling multiple trunks. Here time slots from an inlet port are destined for multiple outlet ports. Enter the space switch (S). Figure 10.3 is a simple illustration of this concept. For example, time slot B_1 is moved to the Z trunk into time slot Z_1 and time slot C_n is moved to trunk W into time slot W_n. However, we see that there is no change in time slot position.

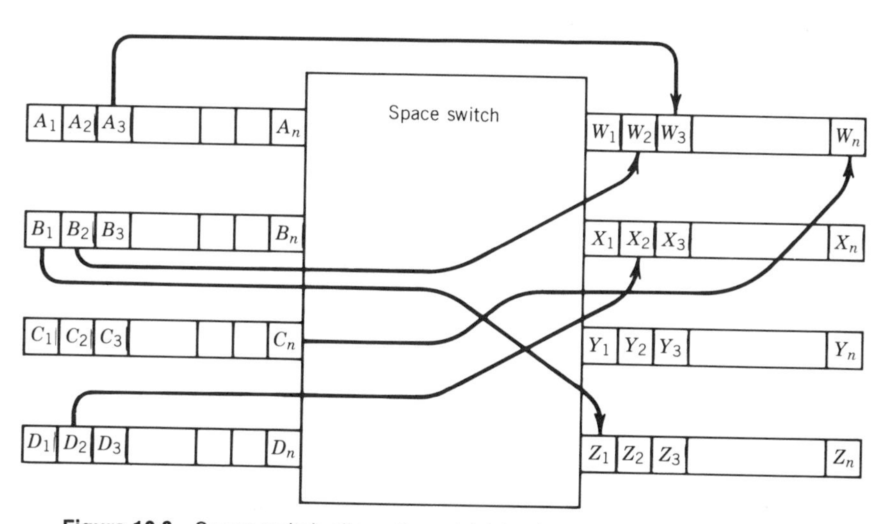

Figure 10.3 Space switch allows time slot interchange among multiple trunks.

3.3 Space Switch

A typical time-division space switch is shown in Figure 10.4. It consists of a cross-point matrix made up of logic gates that allow the switching of time slots in the PCM time slot bit streams in a pattern determined by the required network connectivity. The matrix consists of a number of input horizontals and a number of output verticals with a logic gate at each cross-point. The array, as shown in the figure, has m horizontals and n verticals, and we call it an $M \times N$ array. Sometimes other notation is used in the literature, such as N and K. If $M = N$, the switch is nonblocking. If $M > N$, the switch concentrates; and if $N > M$, the switch expands.

Return to Figure 10.4. The array consists of a number (M) input horizontals and (N) output verticals. For a given time slot the appropriate logic gate is enabled and the time slot passes from the input horizontal to the desired output vertical. The other horizontals, each serving a different serial stream of time slots, can have the same time slot (e.g., a time slot from time slots number 1–24, 1–30, or 1–n; for instance, time slot 7 on each stream) switched into other verticals by enabling their gates. In the next time slot position (e.g., time slot 8), a completely different path configuration could occur, again allowing time slots from horizontals to be switched to selected verticals. The selection,

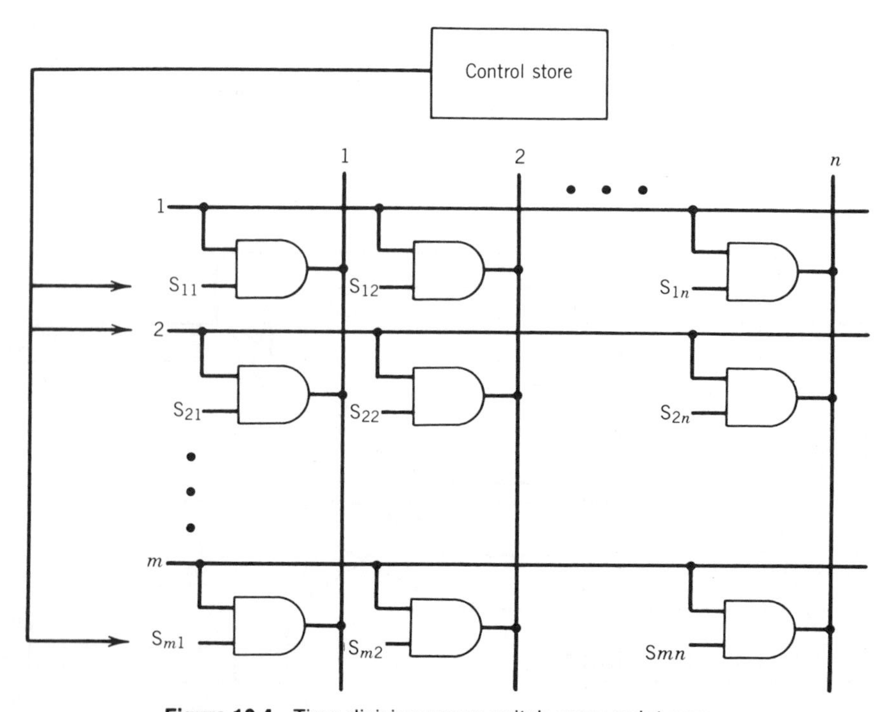

Figure 10.4 Time-division space switch cross-point array.

of course, is a function of how the traffic is to be routed at that moment for calls in progress or calls being set up.

The space array (cross-point matrix) does not switch time slots as does a time switch (time slot interchanger). This is because the occurrences of time slots are identical on the horizontal and on the vertical. The control memory in Figure 10.4 enables gates in accordance with its stored information.

If an array has m inputs and n outputs, m and n may be equal or unequal depending on the function of the switch on that portion of the switch. For a tandem or transit switch we would expect $m = n$. For a local switch requiring concentration and expansion, m and n would be unequal.

If, in Figure 10.4, it is desired to transmit a signal from input 1 (horizontal) to output 2 (vertical), the gate at that intersection would be activated by placing an enable signal on S_{12} during the desired time slot. Then the information bits of that time slot would pass through the logic gate onto the vertical. In the same time slot an enable signal on S_{m1} on the mth horizontal would permit that particular time slot to pass to vertical 1. From this we can see that the maximum capacity of the array during one time slot interval measured in simultaneous call connections is the smaller value of m or n. For example, if the array is 20×20 and a time slot interchanger is placed on each input (horizontal) line and the interchanger handles 32 time slots, the array then can serve $20 \times 32 = 640$ different time slots. The reader will note how the TSI (time slot interchanger) multiplies the call-handling capability of the array when compared to its analog counterpart.

3.4 Time-Space-Time Switch

Digital switches are composed of time and space switches in any order or in time switches only. We use the letter T to designate a time-switching stage and S to designate a space-switching stage. For instance, a switch that consists of a sequence of a time-switching stage, a space-switching stage, and a time-switching stage is called a TST switch. A switch consisting of a space-switching stage, a time-switching stage, and a space-switching stage is designated an STS switch. There are other combinations of T and S. The ATT No. 4 ESS switch is an example. It is a TSSSST switch.

Figure 10.5 illustrates the time-space-time (TST) concept. The first stage of the switch is the time slot interchanger (TSI) or time stage that interchanges information between external incoming channels and the subsequent space stage. The space stage provides connectivity between time stages at the input and output. We saw earlier that space stage time slots need not have any relation to either external incoming or outgoing time slots regarding number, numbering, or position. For instance, an incoming time slot 4 can be connected to outgoing time slot 19 via space network time slot 8.

If the space stage of a TST switch is nonblocking, blocking in such a switch occurs if there is no internal space stage time slot during which the link from

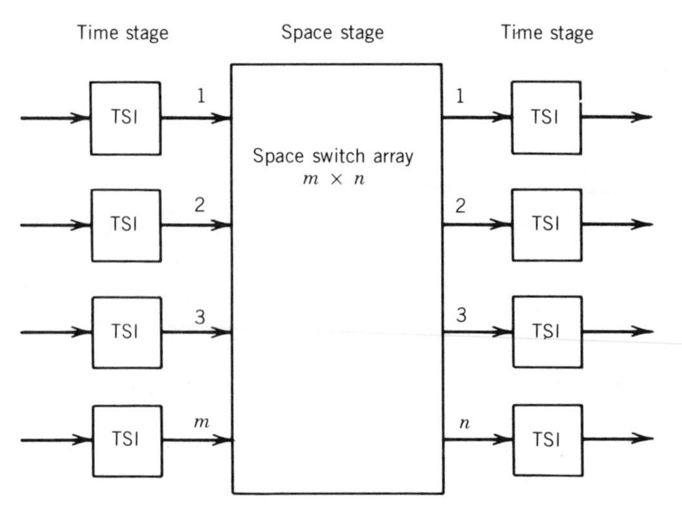

Figure 10.5 A time-space-time (TST) switch (TSI = time slot interchanger).

the inlet time stage and the link to the outlet time stage are both idle. The blocking probability can be minimized if the number of space stage time slots is large. A TST switch is strictly nonblocking if

$$l = 2c - 1 \tag{10.1}$$

where l is the number of space stage time slots and c is the number of external TDM time slots [3].

3.5 Space-Time-Space Switch

A space-time-space (STS) switch reverses the architecture of a TST switch. The STS switch consists of a space cross-point matrix at the input followed by an array of time slot interchangers whose ports feed another cross-point matrix at the output. Such a switch is shown in Figure 10.6. Consider this operational example with an STS. Suppose that an incoming time slot 5 must be connected to an output time slot 12. This requires a time slot interchanger that is available to interchange time slots 5 and 12. This can be accomplished by any of the n time slot interchangers in the time stage.

3.6 TST Compared to STS

Both TST and STS switches can be designed with identical call-carrying capacities and blocking probabilities. It can be shown that a direct one-to-one mapping exists between time-division and space-division networks [2].

The architecture of TST switching is more complex than STS switching with space concentration. The TST switch becomes more cost-effective because

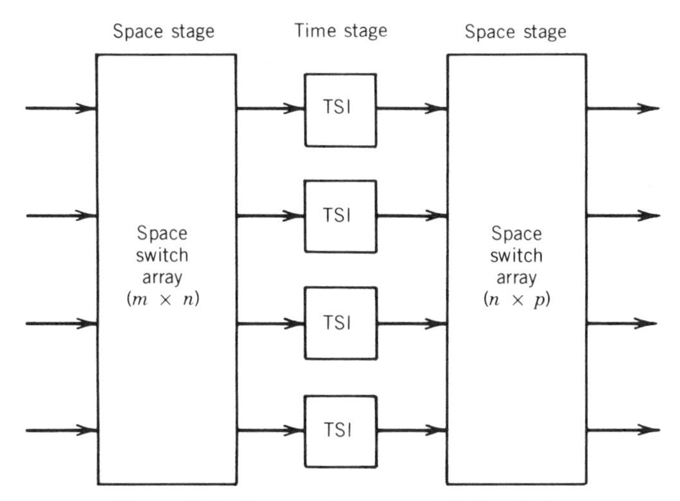

Figure 10.6 A space-time-space (STS) switch.

time expansion can be achieved at less cost than space expansion. Such expansion is required as link utilization increases because less concentration is acceptable as utilization increases.

It would follow, then, that TST switches have a distinct implementation advantage of STS switches when a large amount of traffic must be handled. Bellamy [3] states that for small switches STS is favored due to reduced implementation complexities. The choice of a particular switch architecture may be more dependent on such factors as modularity, testability, and expandability.

One consideration that generally favors an STS implementation is the relatively simpler control requirements. However, for large switches with heavy traffic loads, the implementation advantage of the TST switch and its derivatives is dominant. A typical large switch is the ATT No. 4 ESS, which has a TSSSST architecture and has the capability of terminating 107,520 trunks with a blocking probability of 0.5% and channel occupancy of 0.7.

4 NORTHERN TELECOM DMS-100 DIGITAL SWITCH

4.1 Introduction

In this section we describe a typical digital switch, the Northern Telecom DMS-100. This particular digital switch was selected because of its wide application accommodating both European and North American PCM formats. It can serve as a local switch. The DMS-200, which has a similar architecture and many similar modules, can serve as a tandem or transit (toll) switch and as a combined function switch, local–toll.

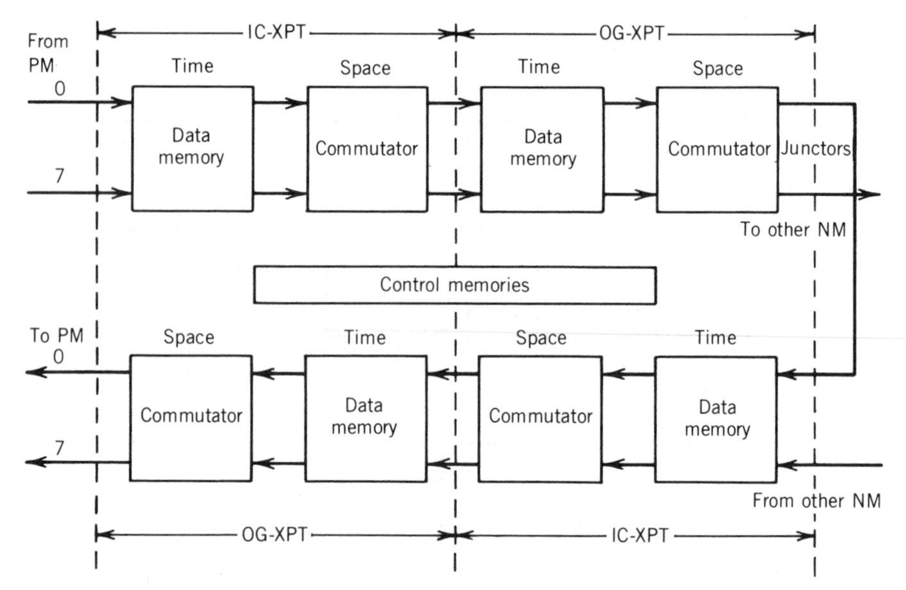

Figure 10.7 A simplified diagram showing the DMS-100/200 folded group architecture, network module (NM). PM = peripheral module.

4.2 Principal Parameters

The DMS-100 is a local switch with a capacity of from 1500 to 100,000 subscriber lines. Its maximum call capacity is 350,000 call attempts per average busy season busy hour (ABSBH). The DMS-200 is a toll (transit) switch that can serve from 400 to 60,000 trunks [4].

4.3 Basic Architecture

The DMS-100 is a TSTS folded group switch. A call routed through the switch traverses TSTS stages in a receive group that is connected by junctors to a transmit group that is also a TSTS architecture. This concept is shown in Figure 10.7.

4.4 Switching Network

The switching network employs four stages of time switching for each connection established between an originating peripheral module (PM) and a terminating peripheral module, as shown in Figures 10.7 and 10.8. The paths for each connection are assigned under the control of the central processing unit (CPU).

The network also distributes the control messages to and from the PMs and the CPU. The network is fully duplicated (i.e., plane 0 and plane 1) from the originating PM to the terminating PM to achieve the necessary reliability.

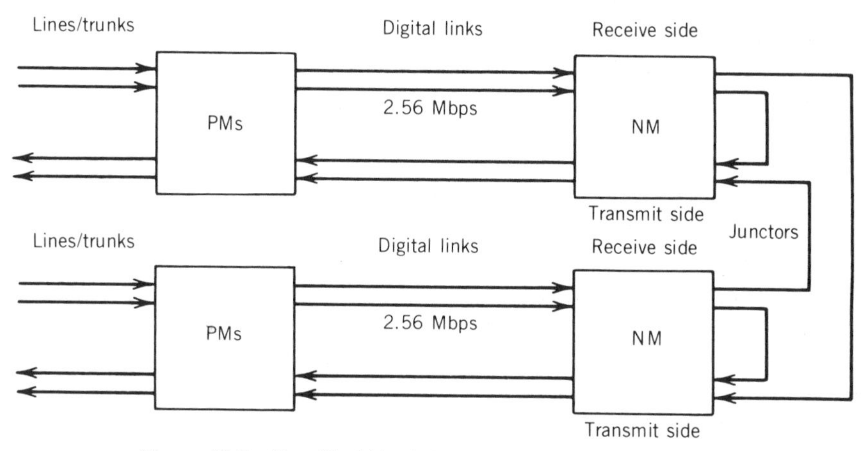

Figure 10.8 Simplified block diagram showing NM operation.

Plane 0 and plane 1 of the switching network each consist of a set of up to 32 network modules (NMs). These are identified as NM-0 through NM-31, each set of NMs forming an identical and independent half of the network. Digital signals between NMs are carried by *junctors*. The NM is the major building block of the plane and each NM has two sides:

- Receive side A (incoming paths from PMs).
- Transmit side B (outgoing paths to PMs).

These are illustrated in Figure 10.8. The separate receive and transmit paths give the network its inherent four-wire characteristics.

The basic building block of the network module is the time switch. In the network, 512 individual speech channels (i.e., 16 ports) are written into the readout of the time-switching memory of every frame time. Eight of these switching memories are used to create the 64 × 64 port matrix that forms one of the network module's two directions or sides. Four of the eight memories form the first stage of switching (time stage). The first stage is connected to an identical second stage via a commutator. The commutator is a space stage consisting of four 16 × 16 equivalent cross-point matrices to develop an equivalent 64 × 64 space array. The commutator gives programmable connectivity between each first stage memory and all second stage memories [5].

4.4.1 Network Module Structure. The network module has four stages of time switching. Each NM contains a network message controller (NMC). The function of the NMC is to communicate with the central control complex (CCC) and peripheral modules and control the network module. When control messages for the time switch and commutator are to be handled, the NMC translates CCC instructions into network connections. Figure 10.8 outlines the structure of a single plane of the network module.

Each NM contains eight cross-point cards. Each card contains two stages of 16 × 16 time switches. (A time switch consists of a time and a space stage.) The time switches are interconnected by means of interswitch links that can serve up to 512 simultaneous connections. Each NM has two sides, as mentioned. These are receive (side A) and transmit (side B), and each side contains two stages of time switches. These are shown in Figure 10.7, where a "stage" consists of a time switch carried out by the data memory and a space switch carried out by the commutator. Side A receives inputs from the peripheral module subsystem and passes switched outputs to side B of another or the same network module. Side B receives inputs from side A of another or the same NM and transmits switched outputs to the peripheral module subsystem. Each side has a peripheral face and a junctor face.

The network subsystem can grow from 1 to 32 NMs in steps of one. For each network subsystem size there is a particular junctor pattern for interconnecting junctor faces. Each junctor accepts 32 simultaneous connections (31 voice channels and 1 channel reserved for framing information). There are two types of junctors, internal junctor loop-back and regular junctor. The internal junctor loop-back eliminates junctor interfaces when intranetwork connections are required.

Ports are entry and exit points on the peripheral and junctor faces where parallel junctors (loop) are not used. Each port on the peripheral face carries 32 channels (30 voice, 1 signaling, and 1 unused) of PCM serial data. The sampling rate is 8 kHz and each sample consists of a 10-bit data word. For the 32 channels, therefore, the bit rate is 2.56 Mbps. Each 10-bit sample is composed of the standard 8-bit PCM voice sample, a parity bit, and an interperipheral signaling bit.

Within each time switch there is a data memory (DM) and a connection memory (CM). The DM is used to switch PCM speech samples between time slots, and the CM is used to store addresses of the speech samples to be moved into specific time slots by the switching process [5].

4.4.2 Operational Example of Time-Division Digital Network. An example of a typical time-switching operation is illustrated in Figure 10.9. The sequence of events is outlined below.

1. During the channel time for port 0, channel 2, a speech sample consisting of 10 bits of serial data enters the serial to parallel formatter. This serial data sample is converted to parallel format and written into the appropriate section of the DM assigned to port 0, channel 2 (i.e., data word location 32).

2. When the CCC selected an interswitch link bus, the link to the second stage time switch 0, channel 31 was found to be free and was assigned to the incoming call. The assignment was performed by the CCC, selecting the CM location associated with this interswitch link bus (i.e., CM word location 497).

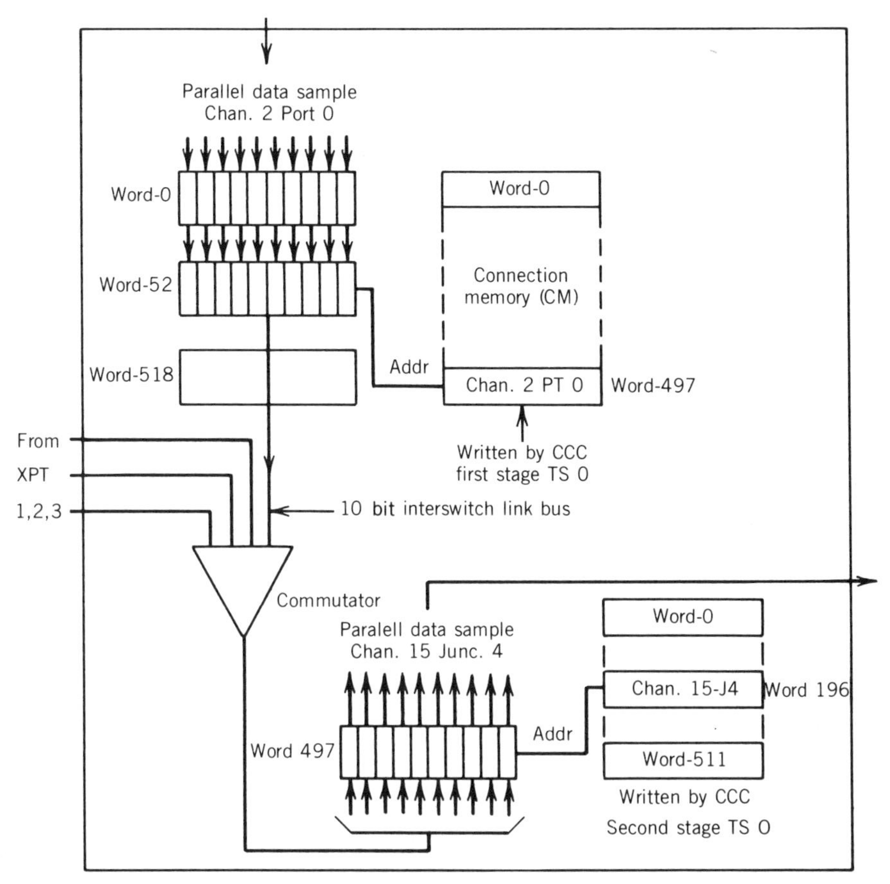

Figure 10.9 Operational example, time-switching function. Courtesy of Northern Telecom.

The address of the speech sample in DM word location 32 is channel 2, port 0. The CCC writes the code for this address into word location 497 in the CM, thus associating the speech sample with the assigned interswitch link bus. The CM also contains a 2-bit code that controls the multiplexer selection so that incoming time switch zero and outgoing time switch zero are connected.

3. All the CMs in the incoming time switch zero are then read cyclically from word zero through word 511 and when the time-slot corresponding to CM word 497 arrives, the timing of the multiplexer and DM access in the outgoing time switch zero is arranged to coincide with the establishment of the assigned interswitch link bus. The link bus runs from the DM of the incoming time switch zero through the zero segment of the multiplexer (commutator) to word 497 of the DM of the outgoing time switch.

4. All 10 bits of the speech sample are read out simultaneously during time slot 497 from DM location 32 of the first stage switch and written into DM location 497 of the second stage switch.

5. The speech sample has now been switched from time slot 32 to time slot 497. The same process has meanwhile been happening to the speech samples in the other DM time slots, with the samples being read out of the DM in order of the CM time slots and interswitch buses assigned to them by the CCC.

6. When the CCC selected a junctor for the speech sample to leave side A of the NM, junctor 4, channel 15 was found to be free and was assigned to the sample entered in DM location 497 of the outgoing switch. The assignment was performed by the CCC selecting the CM location associated with junctor 4, channel 15 (i.e., CM word 196). Word 196, not word 248, is used because timing is advanced on the junctor face by 26 bit times or 52 time slots. The address of the sample located in the DM 497 is link 127, incoming time switch zero. The CCC writes the code for this address into CM word 196, thus associating the speech sample in DM 497 with the assigned junctor and channel.

7. All the CM locations in the outgoing switch are read cyclically from word 0 to word 511, and when the time slot corresponding to CM word 196 arrives, the speech sample in DM 497 is read out and passed to the parallel to serial formatter. Here it is converted into serial data again after having undergone two stages of time switching, first from time slot 32 to time slot 497 and second from time slot 497 to time slot 196. The sample exits via the serial port interface, which transmits it over the assigned junctor and channel to the incoming junctor face of side B of another NM.

8. The same process has been occurring with the speech samples in the other DM time slots, with the samples being read out not in the order of their entry into the DM but in the order of the CM time slots representing the link buses or junctors and channels assigned to them by the CCC.

9. When the speech sample enters the incoming junctor face on the transmit side B of the NM, it undergoes the same process of decoding, serial to parallel conversion, entry into DM, etc., as just described for NM 0. Two more stages of time switching occur, with the second interswitch link bus being assigned by the CCC. The sample exits on the outgoing peripheral face of the NM 1 at the port and channel assigned by the CCC when the called party's digits were translated.

10. In this description the progress of only 1 speech sample through 1 channel out of a possible 1920 voice channels per NM has been covered. Actually, all the samples in all the DMs of all four stages of every time switch on all the channels are simultaneously in progress through the NMs.

When an NM is fully equipped, its traffic capacity is 50,000 ccs or 1388 erlangs, based on North American blocking criteria. Each port of an NM accepts a 2.560 Mbps bit stream consisting of 32 PCM channels, of which 30 are speech channels. A fully equipped NM handles 1920 speech channels or 64 ports times 30 speech channels per port. Internally it can also handle up to 128 control and signaling channels [5].

4.4.3 North American DS1 versus European CEPT 30 + 2. Five North American DS1 ports on the NM map into four DS30 links, whereas four CEPT 30 (+2) ports map into four DS30 links. In either case DS30 links are provisioned in increments of four. For instance, the DS1 mapping into DS30 links is shown below:

DS1 Ports	DS30 Links
0-4	0-3
5-9	4-7
10-14	8-11
15-19	12-15

4.5 Peripheral Modules

4.5.1 Introduction. The peripheral modules (PMs) are the elements of the DMS-100 that give the switch such versatility. First, the DMS-100 can interface with either CEPT 30 + 2 PCM format or North American DS1 format. Depending on peripheral module type and implementation, the DMS-100 can interface with lines and trunks, analog or digital. The switch can serve as a local switch or a toll (transit) switch, and as a toll/tandem switch, it has the nomenclature DMS-200 or as a local/tandem/transit switch, DMS-300. PMs can also serve as remote concentrators or satellites.

In other words, peripheral modules (PMs) serve as interfaces between the NM and the outside world, which is made up of digital carrier spans such as DS1, analog trunks, and subscriber loops. Under microprocessor control, the PMs carry out the following functions:

- Connect analog and digital facilities for conversion of data and signaling to internal DS30 format and vice versa.
- Transmit DS30 PCM signals and data to other PMs connected through the switching network.
- Scanning of lines for changes in circuit state.
- Performing timing functions for call processing; generating digital tones.
- Sending and receiving signaling and control information to and from the CCC.
- Providing integrity checking of the network.

For PMs that control subscriber line circuits, concentration occurs between terminals and links to the switching network. This is the case because line traffic is low (typical 4 ccs), whereas a DS30 link channel (and switching network port) can carry 36 ccs. Concentration makes efficient use of link and switching network port traffic capacities. The concentration is implemented by a time-division multiplexed switching matrix. Trunk PMs have no concentration since trunks must be capable of operating at a high traffic level (up to 36 ccs). Several representative PMs are described below.

4.5.2 Trunk Module. The trunk module (TM) provides an interface between analog trunks and NMs. It encodes and multiplexes 30 analog VF trunks into a standard digital format. The PCM information is then combined with internal control signals and trunk supervisory and control signals and transmitted at a rate of 2.560 Mbps over the speech links to the network (NM). In the opposite direction of transmission, the 30 digital speech channels and the 2 control channels from the NM are demultiplexed and decoded by the TM into 30 analog VF trunks with their respective speech and signaling information such as MF, SF, E & M, etc. The TM is not sensitive to traffic, so each trunk can carry up to 1 erlang (36 ccs) of traffic intensity.

4.5.3 Digital Trunk Controller. The digital trunk controller (DTC) is a two-shelf module capable of interfacing 20 DS1 or CEPT 30 + 2 lines with the switching network (NM). In the case of DS1, where each trunk has a data rate of 1.544 Mbps, the DTC links these lines with up to 16 DS30 speech links. Each speech link carries 30 voice channels plus 2 signaling channels at the internal 2.560-Mbps data rate. Because no concentration is performed in the DTC, each trunk is capable of handling 1 erlang or 36 ccs.

The DTC must convert the 8-bit standard PCM speech sample into the DMS-100 10-bit speech sample and vice versa. The mapping of the 8-bit format into the 10-bit format is shown in Figure 10.10.

4.5.4 Line Module. The line module (LM) provides an interface with the switching network (NM) for up to 640 analog VF subscriber loops and concentrates the voice and signaling information onto 2, 3, or 4 DS30 32-channel 2.560-Mbps speech links. Four speech links can accommodate 3700-ccs ABSBH (average busy season busy hour) per LM. LMs are installed in pairs within a double bay frame with one LM in each bay. Each LM has an associated controller, and access to the controller in the other bay and the controllers are reliability mates, each with separate battery feeds.

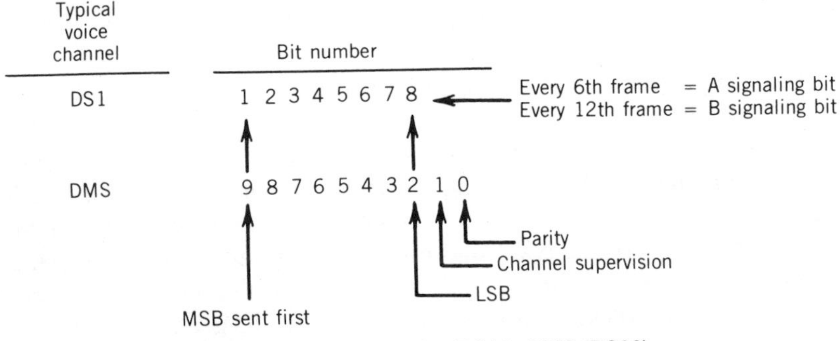

Figure 10.10 Bit mapping DS1 to DMS (DS30).

The LM can support subscriber loops with a maximum resistance of 1900 Ω, including the subscriber subset. There is also a long loop option of 2000 Ω or 4500 Ω.

4.5.5 Line Group Controller. The line group controller (LGC) is a two-shelf module that performs the medium-level processing tasks associated with both host and remote subscriber loop interfaces. The LGC supports line concentrating modules (LCMs), remote switching centers (RSCs), remote line concentrating modules (RLCMs) and outside plant modules (OPMs) to provide services to subscriber lines.

The LGC interfaces up to 20 DS30A (30 voice channels plus 2 signaling channels) speech links from LCMs or up to 20 DS1 (24 voice channels) serving RSCs, RLCMs or OPMs. The LGC provides up to 16 DS30 (30 voice channels plus 2 signaling channels) speech links to the network. As with the DTC, the LGC is provided with redundant processors, one on each shelf, for improved reliability.

4.5.6 Line Trunk Controller. The line trunk controller (LTC) combines most of the DTC and LGC functions and is a DTC/LGC hybrid capable of supporting such PMs as the LCM, RSC, and RLCM and digital trunk interface. The LTC interfaces up to 20 peripheral side ports from DS30A speech links and/or DS1 links to 16 network DS30 speech links.

4.5.7 Line Concentrating Module. The LCM in conjunction with remote units such as RSC, RLCM, and OPM, and the LGC or LTC provide an enhanced version of the line module (LM). The LCM serves as an interface between analog VF subscriber loops and DS30A PCM speech links at 2.560 Mbps data rate. When fully configured, the LCM can handle 640 subscriber loops and 6 DS30A links, 5390 ccs (150 erlangs) with 1.9% ABSBH terminating matching loss (TML) [4].

Each LCM shelf is equipped with a controller that provides line-related control functions such as scanning, ringing control, and channel assignment. The LCM controller is responsible for message handling to and from the LGC.

Two LCMs are accommodated within the line concentrating equipment (LCE) frame, providing a frame capacity of 1280 lines (2×640). The LCE is equipped with two ringing generators.

Five types of signaling, ringing, and special feature circuit card assemblies (CCAs) are available for use with the LCM providing versatile line interfaces.

4.5.8 Remote Switching Center. The remote switching center (RSC) interfaces subscriber lines at a location remote from the DMS-100. The RSC can connect up to 5760 subscriber loops and can serve as a community dial office (CDO) or PABX. It connects with the host DMS-100 with up to 16 DS1 links.

When fully configured, an RSC can accommodate 16,200 ccs (450 erlangs) ABSBH at 1.9% TML with 50% intracalling.

At the host DMS-100 up to 16 DS1 links from the RSC can serve as an interface to the switch through an LGC or LTC. All DS1 links from an RSC must terminate on the same LGC or LTC.

The major elements of the RSC are [4]:

- Line concentrator modules perform the line interface function.
- Remote cluster controller (RCC), a derivative of the LGC, performs the majority of the functions specific to the remote. These functions include the DS1/LCM interface, local switching within the remote through the intracalling feature, and local signaling and intelligence when in the emergency stand-alone (ESA) condition.

Some of the RSC optional features are [4]:

- Intracalling for calls that both originate and terminate within the RSC serving area. These calls require the use of DS1 links to the host for call setup and takedown. Once it is determined that a call is to terminate in an RSC serving area, channels internal to the RSC handle the call, thus reducing activity on the connecting DS1 links to the host.
- Remote maintenance module (RMM).
- Remote trunking, which provides support of RSC originating and terminating traffic for DS1 traffic off the RSC. Applications include trunking to a CDO colocated with the RSC or to PABX direct inward dialing (DID) trunks.

4.5.9 Remote Line Concentrating Module. The RLCM is essentially an LCM operating in a location remote from the controlling DMS-100 host through two to six DS1 links. An RLCM has a line capacity of 640 lines and can replace small CDOs and PABXs.

4.5.10 Outside Plant Module. The OPM is configured as an outside plant remote unit and can accommodate up to 640 subscriber loops over 2 to 6 DS1 links to the host DMS-100.

4.6 Central Control Complex

The central control complex (CCC) is composed at those modules that direct and control the operation and functions of the DMS-100 family of switches. It consists of a synchronized pair of central processing units (CPUs), each having dedicated program store (PS) and data store (DS) memories. Two central message controllers (CMC) also reside within the CCC, and each CMC is connected to both CPUs for improved reliability. The major units of the CCC are described below.

4.6.1 Central Message Controller. The central message controller (CMC) acts as the collector–distributor unit for signal interfacing, buffering, and routing between the CPU and peripheral equipment. As such it reduces the real-time load that the CPU would otherwise incur.

The CMC manages the collection and distribution of messages from up to 70 network and I/O controllers (peripherals) by converting messages from bit serial format to a 16-bit wide parallel format suitable for processing, and vice versa. It has a capacity for messages up to 256 bytes in length.

Two CMCs are normally on line, driven by the active CPU, and they share the load. An online CMC replies to both CPUs in the duplex mode and to the active CPU in the simplex mode. When a CMC is off line, it is driven by the inactive CPU. An off-line CMC replies to both CPUs in the duplex mode and to the inactive CPU only in the simplex mode. The CMC off-line mode is used for maintenance functions, such as loading an off-line inactive CPU from a peripheral device (e.g., a magnetic tape drive).

The CMC has serial interfacing to peripherals via ports using pairs of 2.56-Mbps ac-coupled unidirectional message links and parallel interfacing to the CPU via a pair of buses with 10 address lines, 16 data lines, and 9 control lines.

Messages in DMS-100 family of switches consist of primitives strung together. Each primitive is 1 byte long and represents a single telephony function, such as "collect digits" and "connect trunks." Messages are transmitted one at a time and conform to a standard message format.

Included in the CMC is the system clock that contains two independent synchronized timing sources derived from 10.24-MHz crystal oscillators. The clock is also the timing source for the network (NMs) and peripherals (PMs). The stability of the system clock is 1×10^{-9} when free running and therefore satisfies industry requirements for slip rates on digital trunks. During normal operation this clock is synchronized by means of software, either to an external reference frequency source, such as the digital network master clock (usually a cesium clock), or to an incoming DS1 link from a higher-ranking digital switch (slave operation).

4.6.2 Central Processor Unit. The CPU has access to programmed data memory and uses this information to decide what activity is required to satisfy the conditions of the network. The CPU is a high-speed 17-bit (16 data bits and 1 parity bit) processor with a microcycle of just over 100 ns. It has two independent parallel memory ports. One is the program port, which serves as the interface with external memory [program store (PS)]. The other port is the data port, which is used to gain access to the data store (DS) external memory and to serve as an interface for the redundant CMC. It should be noted that each CPU has its own dedicated PS and an associated DS.

The CPU uses a register stack to manipulate data internally and to shuttle information to and from the data port. The register stack contains frequently used data and is a high-speed bipolar store. When coupled with stack-oriented instructions, this register enhances the fast execution speed of the CPU. The CPU contains the microstore and microsequencing logic necessary to execute the program instructions. The instructions passed to the CPU are written in a modular high-level language called PROTEL. This language was developed specifically for the software programs required in a stored program control (SPC) switching system. In addition, the CPU contains many functions required for dual processor operations. These include matching, synchronization, intermachine communication, fault indication, and activity control. All manual and status indicators of the CPU are accessible or visible on the unit.

A 40-MHz free-running clock provides basic CPU timing, controls register gating and clock emergency timers, and interrupts logic. The approximate 100-ns microcycle is derived from this clock, and timing controls within each CPU ensure that synchronization is maintained.

4.6.3 Program Store Memory. The program store (PS) memory serves as a repository for program instructions required by the CPU for call processing, maintenance, and administration and the operating system. It interfaces the CPU through the program port. The PS uses the same memory cards as the DS. Each of these cards contains 1×10^6 17-bit words in memory. The maximum addressable PS is 8M words.

4.6.4 Data Store Memory. The data store (DS) memory contains transient per-call type information, customer data, and switch parameters. It interfaces the CPU through the data port. The DS uses the same memory cards as the PS, where each card contains 1×10^6 17-bit words. The maximum addressable DS is 16M words.

4.7 Block Diagrams

Figure 10.11 shows a simplified block diagram of the DMS-100, and Figure 10.12 shows a more detailed block diagram.

4.8 Transmission Characteristics of the DMS-100

The DMS-100 family of switches provides an essentially zero-loss network with the standard transmission level of 0 dBm. Switchable pads are provided for each interface to provide the correct transmitting levels for a fixed-loss national transmission plan. Thus the DMS-100 can operate as a toll (transit) switch (−3 dB total path loss for the North American plan) or as a local switch (0 dB total path loss) or any combination thereof.

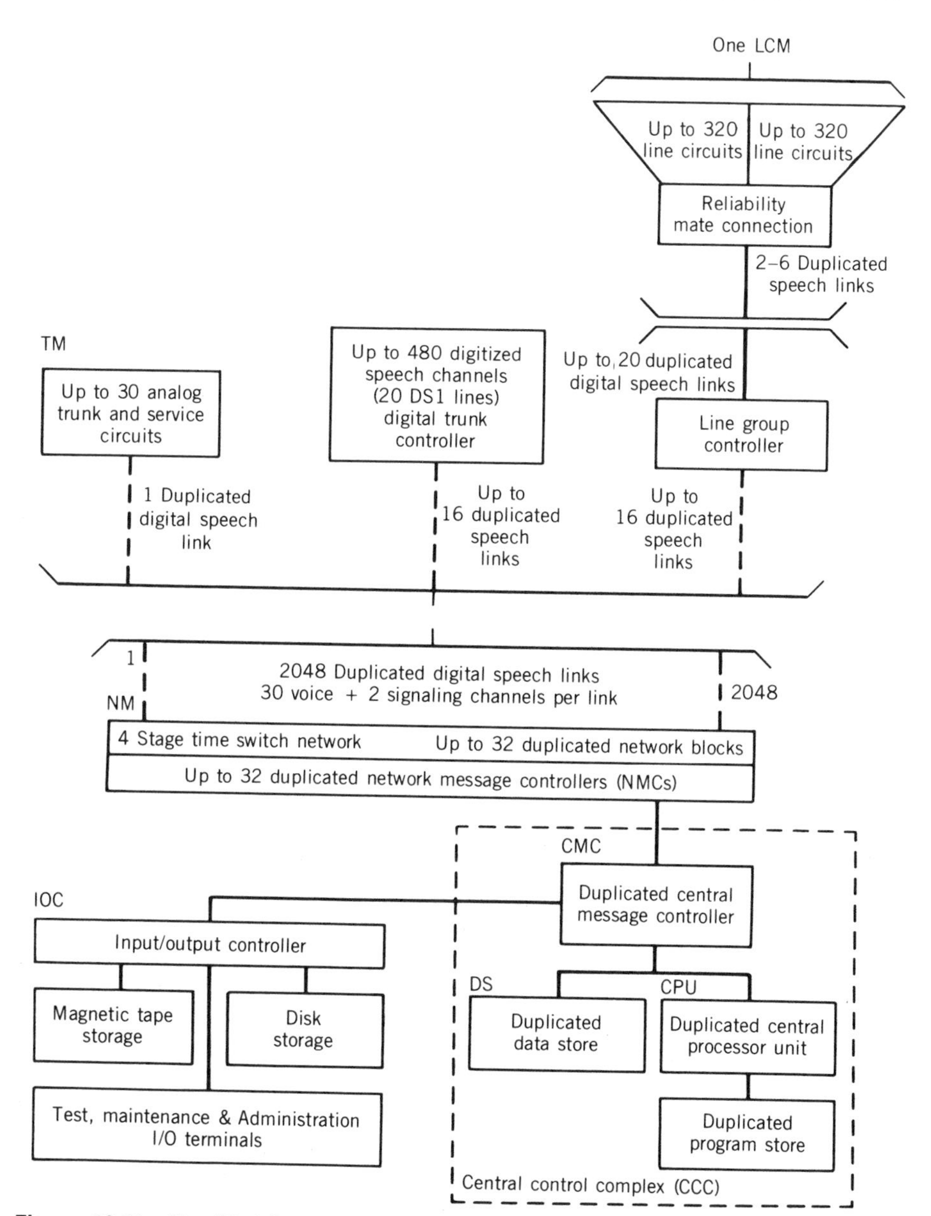

Figure 10.11 Simplified block diagram of a DMS-100 switch. Courtesy of Northern Telecom.

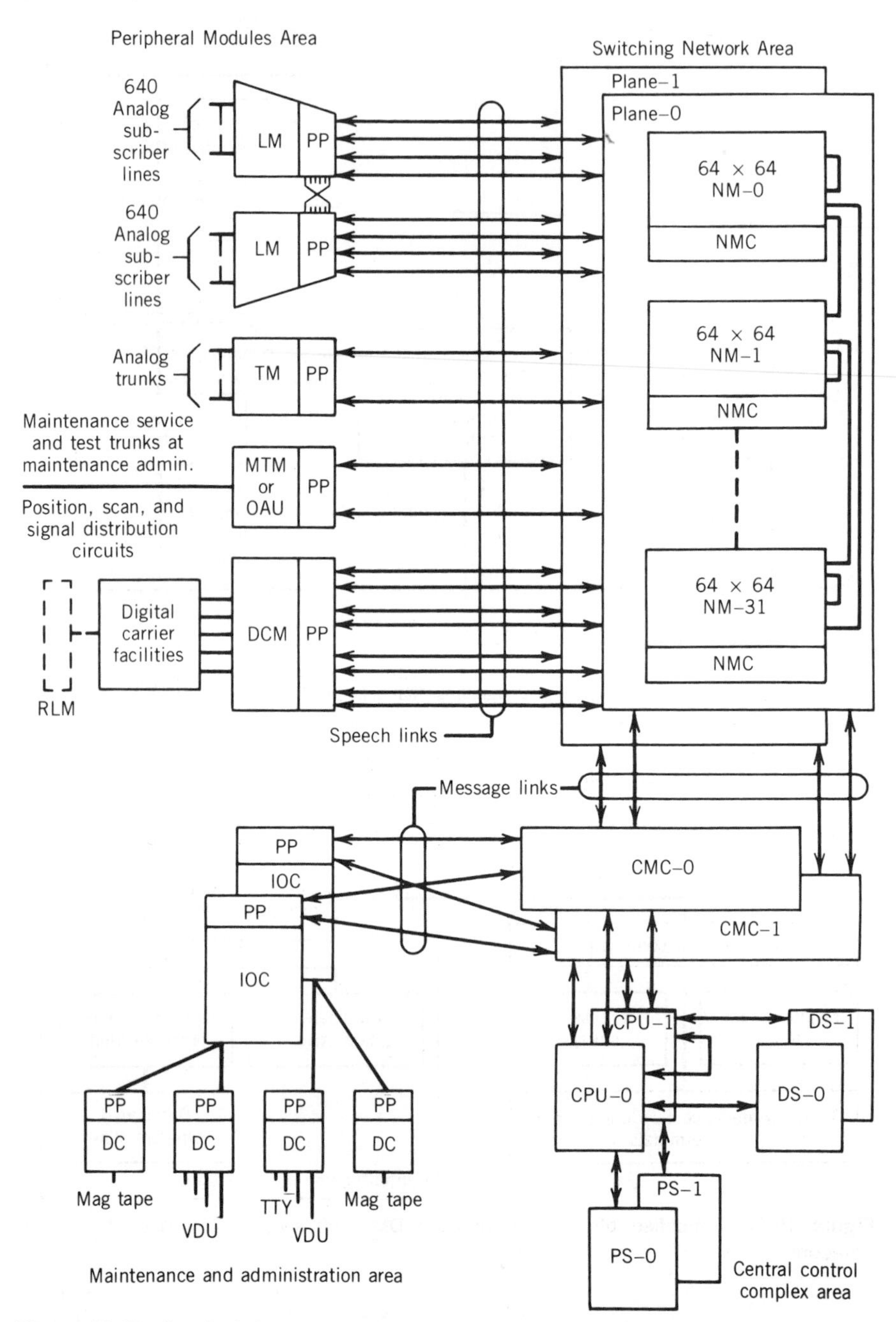

Figure 10.12 Detailed functional diagram of a DMS-100 switch. Courtesy of Northern Telecom.

Therefore the LM and LCM provide 0 dB transmission on intraoffice connections but insert the necessary losses to meet transmission requirements of local and toll (transit) trunk connections.

4.8.1 Digital Signal. Voice channels are companded and encoded according to North American DS1 format using a 15-segment companding law where $\mu = 255$ and standard 8-bit coding. These are byte interleaved before switching.

4.8.2 Return Loss. With a loop terminated in 900 $\Omega/2.16\ \mu F$, the following return loss values are specified for the DMS-100:

Type	Echo Return Loss	Singing Return Loss
Subscriber line	20 db	14 dB
Trunk circuit	26 dB	20 dB

4.8.3 Idle Channel Noise and Quantizing Noise. Idle channel noise will not exceed 23 dBrnC on standard configurations. The signal-to-noise ratio will not be less than 33 dB if a test tone of 1004 Hz with a level of −30 dBm0 is used.

4.8.4 Switching Loss. Figure 10.13 illustrates the switching losses for typical applications of the DMS-100 as integrated in the North American network.

4.9 DMS Supernode

The DMS supernode is a new product architecture that provides for the integration of previously separate network functions. For the DMS-100 family application it is an architecture upgrade that significantly increases its call attempt capacity. The DMS supernode incorporates state-of-the-art VLSI circuitry and software technology, uses increased distributed control, and retains the features of the DMS-100 software base.

The DMS supernode hardware architecture configuration is shown in Figure 10.14. The packaging of this architecture is as follows:

- DMS core. This consists of computing modules (CMs) and system load modules (SLMs).
- DMS bus. This consists of a series of high-speed packet buses that switch messages between various DMS components (DMS core, peripheral processors, application processors, and other system elements).

The DMS core performs the call management and system control functions. Speed and performance have been enhanced by use of the duplicated 32-bit

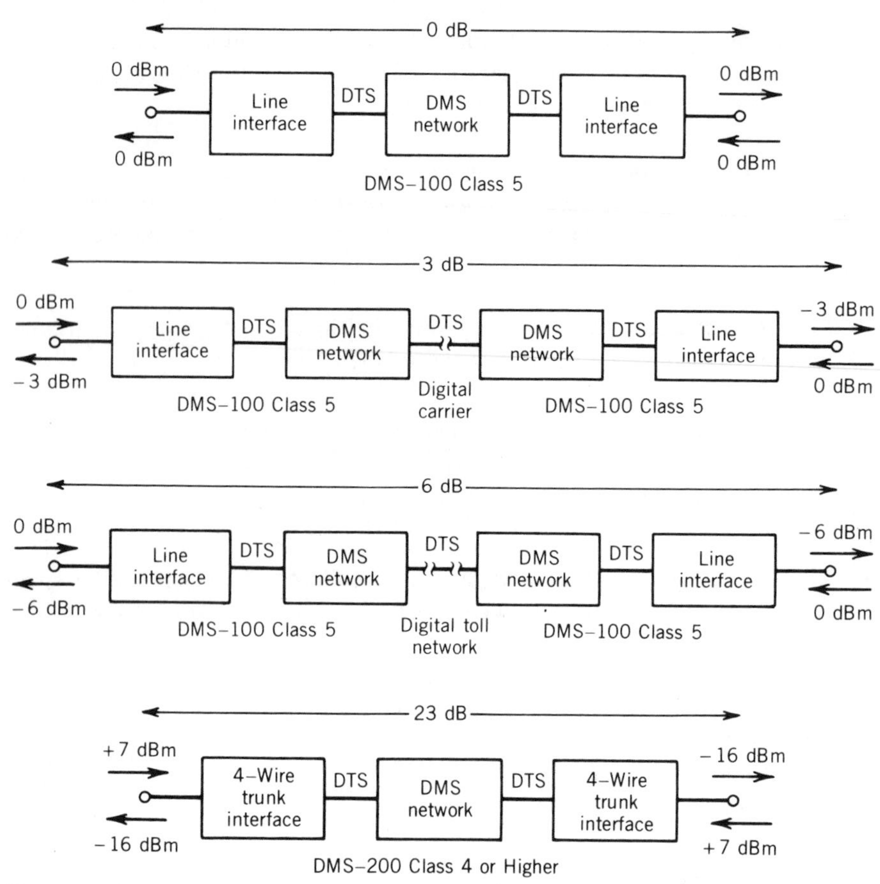

Figure 10.13 Typical transmission loss objectives for the DMS-100 and DMS-200. DTS = digital test sequence (1004 Hz, 0 dBm0 nominal). Courtesy of Northern Telecom.

Motorola MC68020 microprocessor. The DMS core provides 240 megabytes of call and program store per computing module.

The DMS bus includes microprocessors, supporting memory, and other bus components. It also employs the 32-bit Motorola MC68020 microprocessor. It is supported by 6 megabytes of memory and manages the overall performance of the DMS bus. The bus consists of one message switch (MS) configured in a modular, compact architecture that allows it to be economically configured from a few ports to a maximum of 1400 ports.

The DMS bus has a built-in flow control for managing buffers and preventing congestion. A simple message format is used: a 32-bit address header and a message body of variable length (to a maximum of 2 kilobytes). The high-speed transaction bus operates at 128 Mbps. A throughput of greater than 100,000 messages per second is possible with switching delays typically less than 100 μs. This capacity allows the DMS bus to be used for multiple network applications.

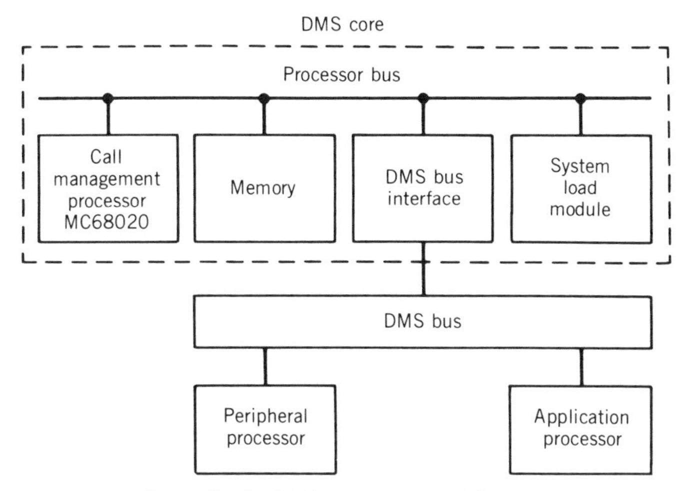

Figure 10.14 DMS supernode architecture.

5 THE DIGITAL NETWORK

5.1 Introduction

As digitization of the public switched telephone network (PSTN) progresses, the existing network topology and structure remains essentially unchanged. It is a three-, four-, or five-level hierarchical network. Most existing routes remain in place. Their augmentation in size and number is a function of forecast traffic intensity (i.e., growth) and the change in mix of voice and other services carried on the network, particularly data. The erosion of a national monopoly in the long-distance (toll) sector may be more of a factor for structure change. The impact of fiber optics transmission is another important factor.

The present network is thus left in place and is being converted to digital at an accelerating rate. The present conversion extends only to the local switch. Onward extension to the subscriber remains an issue. It is not a question of when. Our concern is how. The whole basis of ISDN is digital extension to the subscriber. This aspect is covered in Chapter 13. Here we consider interim measures; ISDN will bring the final solution. Now the argument surfaces: copper or fiber? ISDN is based on copper (i.e., existing copper pairs). Ideally it should be fiber for the bandwidth it offers. At this juncture it appears that an all-fiber outside plant is a long way off.

Likewise, ideally, the network structure should be reexamined. Idealism is tempered by the reality of current investment in existing plant yet to be amortized. This forces the macroplanner to gradual conversion rather than to rapid, radical change.

Our first consideration in this section is digital extension to the subscriber. We then cover other design issues of an all-digital PSTN. Among these issues are:

- Change in profile of services that digital brings about.
- Network performance and performance requirements.
- Synchronization and timing.
- International interface (DS1 versus CEPT 30 + 2).
- Signaling: CCITT SSN No. 7.
- Maintenance.

5.2 Extension to the Subscriber

The driving factor of digital extension directly to the subscriber is cost. The existing outside plant from the MDF to thousands of subscribers represents 30–50% of total plant investment. The straightforward approach is that each subscriber must have two pairs for four-wire operation needed for digital service. Today, in the analog world, it is one pair for two-wire operation. Methods have been suggested for two-wire full-duplex digital operation, such as bursting data in each direction alternatively. The complexity and cost of this scheme have dampened enthusiasm.

If we are to assume that we will remain with the present basic telephone subset, several other problems arise, deriving from the basic features that must be provided the subscriber. These are:

- Telephone alerting: bell, ringing voltage.
- Transmitter dc feed voltage; supervision.
- Subset sidetone.
- Overvoltage protection.

If we assume four-wire transmission to the telephone subset, then the hybrid in the subset will be removed and the subset will operate without sidetone. With no sidetone, the user will think that the set is not operational. Therefore some method must be devised to reinsert audio sidetone.

To ring the subset bell or other alerting device, a 70–75-V ringing voltage source is required at 15–25 Hz, requiring about 1 W of power. These characteristics are alien to solid-state devices such as a codec on an IC chip.

The carbon microphone requires, as a minimum 3–5 V dc for its operation at some 25–80 mA. This emf is derived from central battery (−48 V dc) at the host switch. Again such devices are alien to IC implementations.

Rather than extend directly to the subscriber, an alternative approach is much more feasible. This is the installation of a concentrator or outside plant module (OPM) in the vicinity of a number of subscribers. This concept is shown in Figure 10.15.

The OPM or concentrator accepts analog VF inputs from two-wire pairs connected through a cable route to subscriber subsets. The OPM/concentrator

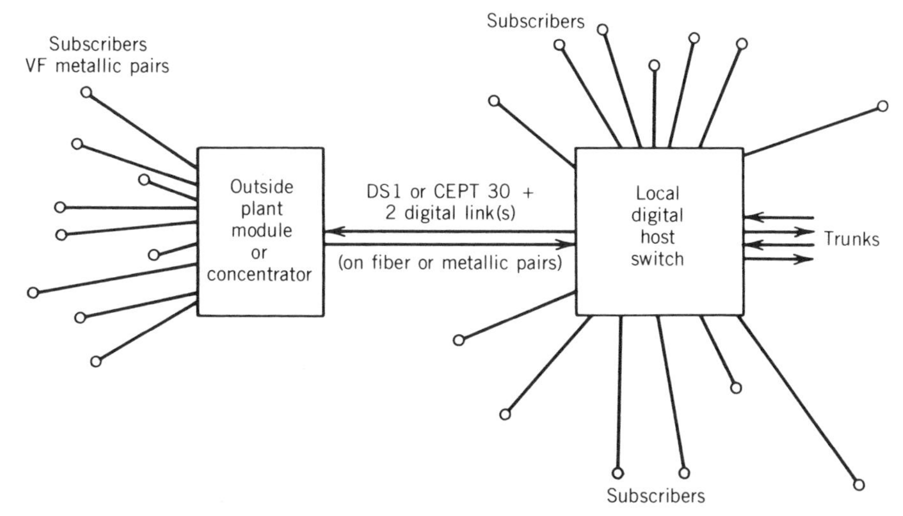

Figure 10.15 Remote concentrator or outside plant module performs SLIC functions.

carries out the SLIC (subscriber line interface circuit) functions. The functions of the SLIC are essentially those described before, such as −48 V dc battery, ringing voltage supply, overvoltage protection, and normal loop supervision. For PCM systems, it carries out the A/D and D/A conversion. The OPM/concentrator provides digital PCM links connecting to the host switch, such as the DMS-100.

There is another acronym commonly found in the literature describing these basic SLIC functions. The term is BORSCHT, where each letter covers a specific function as follows:

- B battery feed
- O overvoltage protection
- R ringing (ringing voltage source)
- S signaling (on-hook, off-hook supervision)
- C coding (A/D and D/A conversion to/from PCM format)
- H hybrid (two-wire to four-wire conversion)
- T test

Another approach is a satellite switch that is installed in a small community or settlement. A typical satellite switch is the Northern Telecom RSC discussed in Section 4.5.8. The satellite connects to the host switch, such as the DMS-100, by means of digital links, typically DS1 or CEPT 30 + 2.

Several advantages accrue from the use of these remote devices. Their use provides an interim solution of extending the digital network toward the subscriber. It also is a convenient means of loop extension, 100 miles or more if necessary. Such techniques also reduce the size of the intervening plant. When using the traditional analog wire-pair techniques, each subscriber pair required individual conditioning for long loops (e.g., loading). With the use of such digital plant extension techniques, the total pairs back to the host switch are reduced, 2 pairs for each 24 subscribers or 30 subscribers, depending on the system being employed. As an example, suppose there was a small community with 20 telephone lines some 25 miles from the host switch. Ordinarily this would require a 30-pair cable with the incumbent VF repeaters or carrier. Using the OPM/concentrator or remote satellite switch approaches, only two unconditioned pairs are required with repeaters every 6000 ft.

Of course it follows that a PCM-based PABX can carry out the same functions in addition to its capability of providing a rich variety of local switching functions and services.

5.3 Change of Profile of Services

Customers demand a greater variety of services than ever before, and that demand is growing. Speech telephony, however, remains the mainstay. Data communications is increasingly in demand with ever-increasing requirements for higher bit rates and special service offerings. Following our discussion in Section 5.2, the subscriber interface remains analog for the great majority of subscribers. These subscribers, then, are constrained by analog transmission. However, plant extension techniques, including the PABX, trend to shorter nonloaded subscriber loops. Such loops can, do, and will easily sustain 64 kbps, particularly if the pairs have good balance against ground. Links from digital PABXs to host exchanges commonly work at 1.544 or 2.048 Mbps. Thus the "analog constraint" is overstated.

Facsimile is still another service that is growing rapidly, and in nearly all cases the facsimile signal rides a standard voice channel through the PSTN. The trend is away from analog facsimile and toward digital facsimile (typically following CCITT Rec. T.3). Also the trend is toward faster operation, measured in page per unit time, where the slowest rate still in use is the 3-min page. This value is dropping to seconds per page. With the coming 64-kbps ubiquitous service, pages per second operation is within reach.

Other services are telemetry and video conferencing. Each service requires different bandwidths with specific BER and delay constraints.

5.4 Digital Transmission Network Models: CCITT

Digital transmission network models are hypothetical entities of defined length and composition for use in the study of digital transmission impairments, such as bit errors, jitter, wander, transmission delay, and slips. CCITT provides a standard *hypothetical reference connection* (HRX) based on an all-digital 64-kbps connection. The standard HRX is shown in Figure 10.16.

The implementation of the standard HRX for a particular application must be tailored. CCITT shows two additional models. One (Figure 10.17A) is for a moderate length circuit, and the second (Figure 10.17B) is also for a moderate

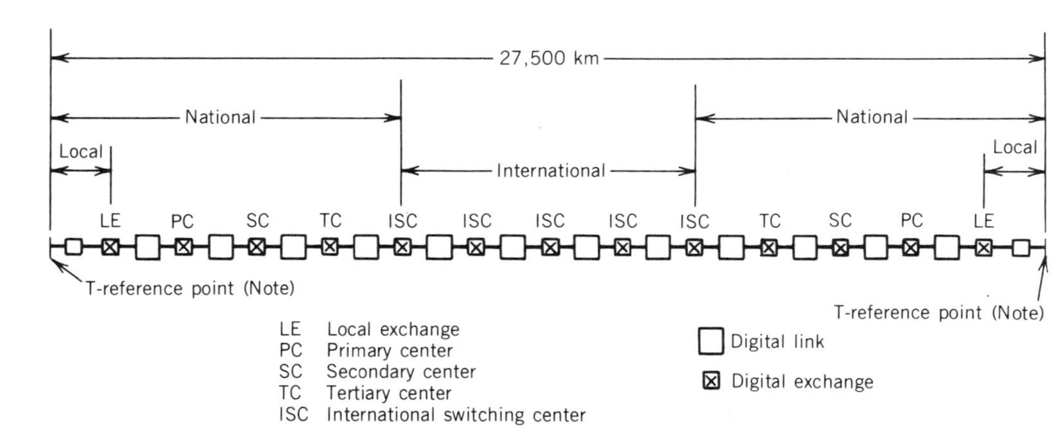

Figure 10.16 The standard CCITT hypothetical reference connection (HRX), longest length. From CCITT Rec. G.801 Figure 1, page 290, *Red Book*, Vol. 3, Fascicle III.3. Courtesy ITU-CCITT, Geneva. [*Note*: Rec. I.411 (applicable to ISDN only).]

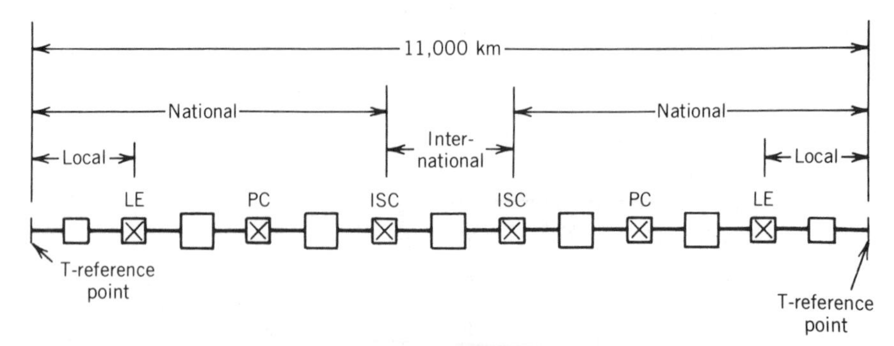

Figure 10.17A A standard CCITT HRX of moderate length. From CCITT Rec. G.801, Figure 2, Page 290, *Red Book,* Vol. 3, Fascicle III.3. Courtesy ITU-CCITT, Geneva. See Figure 10.16 for legend.

length connection but where a local exchange connects directly to an international switching center (ISC).

The HRX is broken down into hypothetical reference digital links (HRDLs) of 2500 km in length. A digital link is defined in CCITT Rec. G.701 as "the whole of the means of digital transmission of a digital signal of a specified rate between two digital distribution frames (or equivalent)." The HRDL is made up of hypothetical reference digital sections with lengths tentatively established as 50 km and 280 km.

5.5 Digital Network Synchronization

5.5.1 Need for Synchronization. When PCM pulses are transmitted over a communication link, there must be synchronization at three different levels. These are bit, time slot, and frame synchronization. Bit synchronization refers to the need for the transmitter (coder) and receiver (decoder) to operate at the

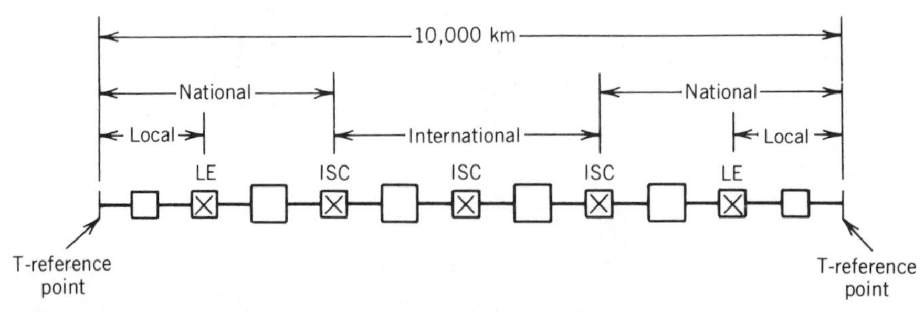

Figure 10.17B Standard CCITT HRX of moderate length connection where a local exchange connects directly to the international switching center (ISC). From CCITT Rec. G.801, Figure 3, Page 290, *Red Book,* Vol. 3, Fascicle III.3. Courtesy ITU-CCITT, Geneva. See Figure 10.16 for legend.

same bit rate so that the bits will not be misread by the receiver. Other levels of synchronization refer to the need of the transmitter and receiver to achieve proper phase alignment so that the beginning and the end of a time slot or frame can be readily identified for information retrieval.

Synchronous transmission is used for PCM transmission. The receiver (decoder) derives its timing from the incoming bit stream transistions to achieve bit synchronization; frame and time slot synchronization are achieved by the use of a frame aligning bit (in DS1) or separate channel alignment (for CEPT 30 + 2) and by using a fixed frame format to delineate separate time slots.

Network synchronization conveniently and economically achieves bit and time slot synchronization at all digital nodes within the network. Network synchronization can be accomplished by synchronizing all the clocks associated with these nodes so that transmissions from these nodes have the same average line bit rate. Buffer storage devices are judiciously placed at various transmission interfaces to absorb differences between the actual line bit rate and the average rate. Several methodologies used to achieve network synchronization are discussed in the Section 5.5.2. Slips, a major synchronization impairment, are discussed quantitatively in Section 5.6.3.

In most digital networks timing derives from the switch clock. Suppose a distant codec (channel bank) terminated at a switch port. The switch transmits a digital bit stream to the distant codec at bit rate $F(0)$, which is determined by the switch internal clock. The switch needs to receive each incoming bit stream at this same rate $F(0)$; otherwise its receiver buffer will eventually overflow or underflow. If an overflow or underflow occurs, an impairment called a *slip* is introduced into the bit stream.

Of course, this same situation can occur on trunks connecting switches. In this situation each switch transmits at a rate governed by its own internal clock. Unless the digital bit stream that is received at each switch arrives at the same nominal clock rate, slips will occur. To prevent these slips, it is necessary to force both switching systems to use the same common synchronized reference clock rate, $F(0)$.

When a slip occurs at a switch port buffer, it can be controlled to occur at frame boundaries. Such slips at buffers are termed controlled slips. Controlled slips occur for two basic reasons:

- Lack of frequency synchronization among clocks at various network nodes.
- Phase wander and jitter on the digital bit streams.

Thus, even if all the network nodes are operating in the synchronous mode, slips can still occur due to transmission impairments. An example of environmental effects that can produce phase wander of bit streams is the daily temperature variation affecting the electrical length of a digital transmission line such as T1 (DS1).

Slips due to wander and jitter can be prevented by adequate buffering. Therefore adequate buffer size at the digital line interfaces and synchronization of the network node clock rates are the basic means by which to achieve the network slip rate objective (from Ref. 7).

5.5.2 Methods of Network Synchronization. There are a number of methods that can be employed to synchronize a digital network. Six such methods are shown graphically in Figure 10.18.

Plesiochronous operation of a digital network (Figure 10.18a) is where each node has identical high-stability clocks operating at the same nominal rate. These clocks are free running. The accuracy and stability of each clock is such that there is almost complete coincidence in time keeping and the phase drift between many clocks is, in theory, avoided or the slip rate between network nodes is acceptably low. This requires that all nodes, no matter how small, have high-precision clocks. For commercial networks this causes a high cost burden. However, for military networks it is very attractive for survivability because there is no mutual network synchronization. Thus the loss of a node and its clock does not affect the rest of the network timing. CCITT also recommends (Rec. G.811; Ref. 11) such operation on transnational digital connectivities.

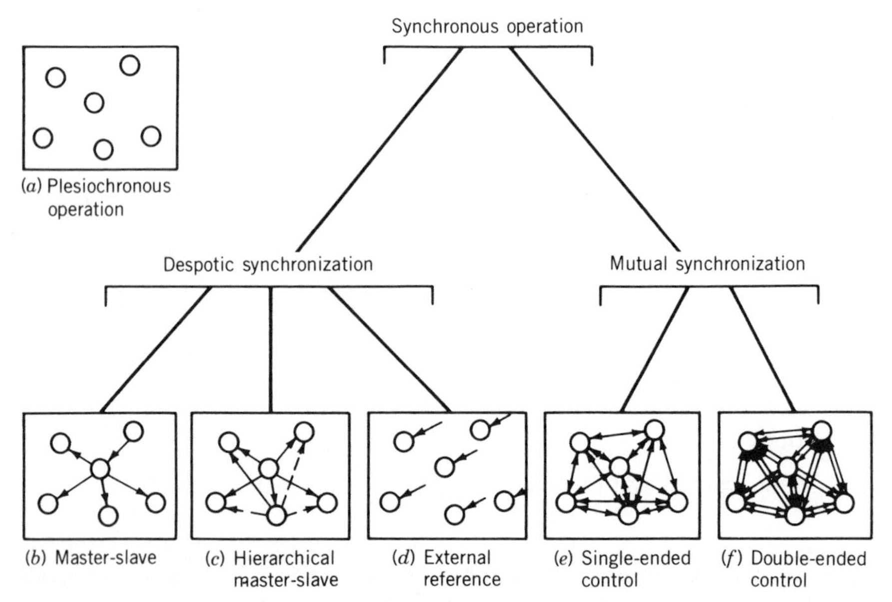

Figure 10.18 Digital network synchronization methods.

Another general synchronization scheme is mutual synchronization, which is shown in Figures 10.18e, f. Here all nodes in the network exchange frequency references, thereby establishing a common network clock frequency. Each node averages the incoming references and uses the result for its local transmitted clock. After an initialization period, the network aggregate clock normally converges to a single stable frequency.

The major advantage of mutual synchronization is survivability in spite of a clock failure at any node. The disadvantages are the uncertainties of the exact average frequency and the unknown transient behavior.

A number of military systems use external synchronization, as shown in Figure 10.18d. The external source of timing can be radio time dissemination systems, such as Transit Satellite, Geographical Positioning System (GPS), and MILSTAR. A number of terrestrial sources are also available, such as NBS's WWV/WWVB/WWVH and LORAN C and Omega.

5.5.3 The North American Synchronization Plan. The North American network uses a hierarchical timing distribution system, as shown in Figure 10.18c, with four levels called "strata." ATT divides the synchronization of clocks into two parts [7]: intrabuilding and interbuilding. We will call a "building" a switching node. Within a building or switching node a central clock is designated "building integrated timing supply" or BITS. It provides DS0 and DS1 synchronization references for all switching center clocks in one location—a building. This is *intrabuilding synchronization*. The BITS receives synchronization reference from remotely located BITSs. Such synchronization of BITSs between buildings is referred to as *interbuilding synchronization*.

To implement the ATT method, clocks within digital networks are divided into four stratum levels. The standard, known as the Bell System Reference Frequency (BSRF), is presently the stratum 1 clock, and it uses a cesium beam frequency standard. It is the primary frequency reference for all clocks in the synchronization network. All clocks in strata 2, 3, and 4 are synchronized with a reference traceable (frequency-locked) to the BSRF. In general, clocks in the different strata are distinguished by their free-running accuracies or by their stabilities during trouble conditions, such as during loss of all synchronization reference. Table 10.1 shows clock characteristics for each stratum.

For intrabuilding synchronization the highest stratum clock within the building is designated as the BITS for that building. All clocks within the building receive timing from the BITS, either at the DS1 or the DS0 rate. For example, if a stratum 2 clock resides in a building with one or more strata 3 and 4 clocks, the stratum 2 clock supplies timing reference for all strata 3 and 4 clocks in the building. The stratum 2 clock in the example is referred to as a stratum 2 BITS or BITS-2. The BITS, in turn, receives synchronization references from BITS in other buildings in a hierarchical manner. With this method frequency reference can be sent from one BITS to another BITS whose rate is regulated to follow the excursions of the transmitting BITS. The

TABLE 10.1 Clock Strata Requirements

Stratum	Minimum Accuracy[a]	Minimum Stability[b]	Pull-in Range[c]
1	$\pm 1\text{x}10^{-11}$	Not applicable	None
2	$\pm 1.6 \times 10^{-8}$ ($\pm$ 0.025 Hz @ 1.544 MHz)	1×10^{-10}/day	Must be capable of synchronizing to clock with accuracy of $\pm 1.6 \times 10^{-8}$
3	$\pm 4.6 \times 10^{-6}$ ($\pm$ 7 Hz @ 1.544 MHz)	3.7×10^{-7}/day	Must be capable of synchronizing to clock with accuracy of $\pm 4.6 \times 10^{-6}$
4	$\pm 32 \times 10^{-6}$ ($\pm$ 50 Hz @ 1.544 MHz)	Not applicable	Must be capable of synchronizing to clock with accuracy of $\pm 32 \times 10^{-6}$

[a]Minimum accuracy represents the maximum long-term (e.g., 20 years) deviation from the nominal frequency.
[b]Minimum stability or drift rate represents the maximum rate of change of the clock frequency with respect to time on loss of all frequency references.
[c]Pull-in range is a measure of the maximum input frequency deviation from the nominal clock rate that can be overcome by a clock to pull itself into synchronization with another clock.
Source: Ref. 8.

hierarchical level of the receiving BITS clock cannot be higher than the sending BITS clock. The receiving BITS clock in turn supplies reference frequency to BITSs of equal or lower hierarchical level. The flow of synchronization signals is such that no closed loops can form within the synchronization network. A closed loop is when a frequency source feeds back to itself through the network. Closed loops tend to cause frequency instability (from Ref. 7).

Figure 10.19 shows the ATT intrabuilding plan and Figure 10.20 illustrates the hierarchical interbuilding plan. Figure 10.21 shows the hierarchical structure envisioned for ISDN.

5.5.4 CCITT Synchronization Plans. CCITT Rec. G.811 deals with synchronization of international links. Plesiochronous operation is preferred (see Section 5.5.2). The recommendation states the problem at the outset:

> International digital links will be required to interconnect a variety of national and international networks. These networks may be of the following form:
>
> a) a wholly synchronized network in which the timing is controlled by a single reference clock.
> b) a set of synchronized subnetworks in which the timing of each is controlled by a reference clock but with plesiochronous operation between the subnetworks
> c) a wholly plesiochronous network (i.e., a network where the timing of each node is controlled by a separate reference clock).

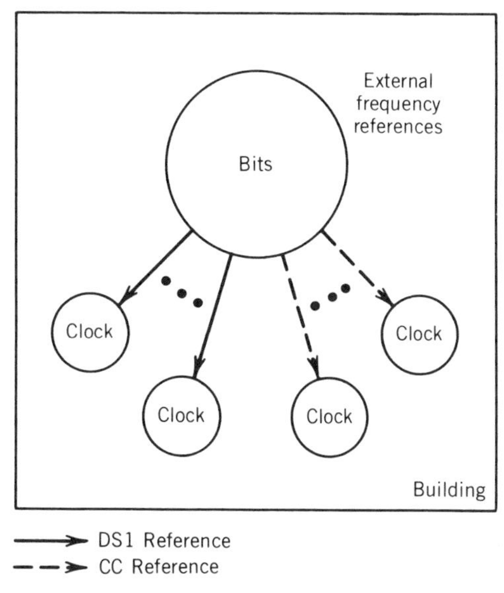

Figure 10.19 ATT intrabuilding plan.

Plesiochronous operation is the only type of synchronization that can be compatible with all three types listed. Such operation requires high-stability clocks. Thus Rec. G.811 states that all clocks at network nodes that terminate international links will have a long-term frequency departure of not greater than 1×10^{-11}. This is further described in what follows.

The theoretical long-term mean rate of occurrence of controlled frame or octet (time slot) slips under ideal conditions in any 64-kbps channel is consequently not greater than *1 in 70 days* per international digital link.

Any phase discontinuity due to the network clock or within the network node should result only in the lengthening or shortening of a time signal interval and should not cause a phase discontinuity in excess of one-eighth of a unit interval on the outgoing digital signal from the network node.

Rec. G.811 states that when plesiochronous and synchronous operation coexist within the international network, the nodes will be required to provide both types of operation. It is therefore important that the synchronization controls do not cause short-term frequency departure of clocks, which is unacceptable for plesiochronous operation. The magnitude of the short-term frequency departure should meet the requirements specified in Section 5.5.4.1

5.5.4.1 Time Interval Error and Frequency Departure. Time interval error (TIE) is based on the variation of ΔT, which is the time delay of a given timing signal with respect to an ideal timing signal, such as UTC (universal coordinated time). The TIE over a period of S seconds is defined to be the magnitude of difference between the time delay values measured at the end

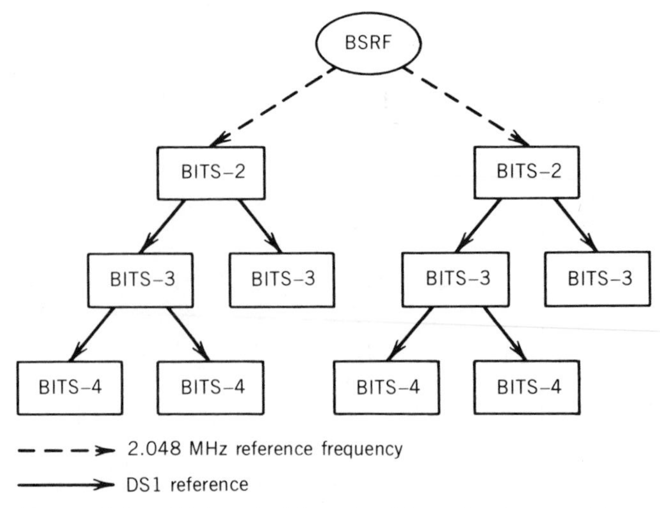

Figure 10.20 ATT interbuilding plan (hierarchical network) [7].

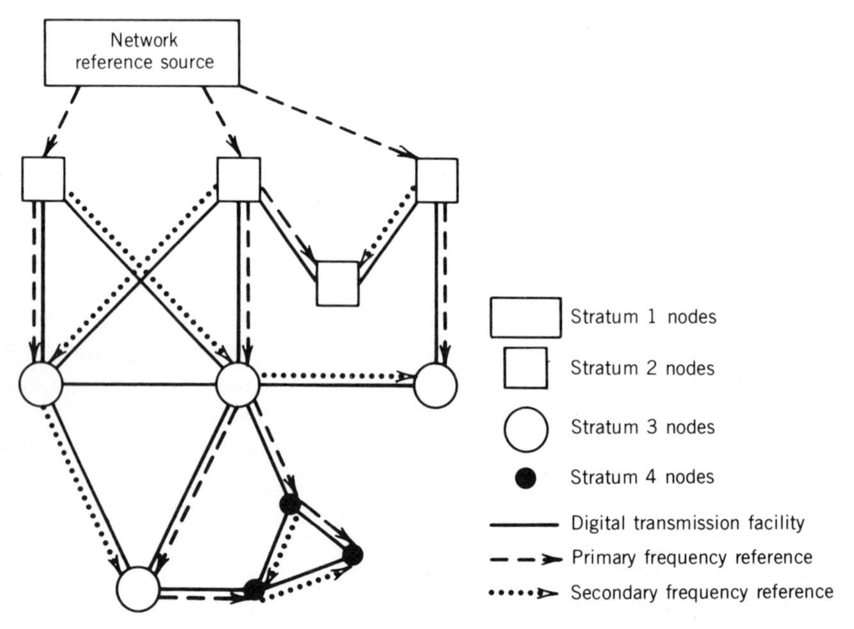

Figure 10.21 Hierarchical synchronization network for ISDNs (North America). From Ref. 8.

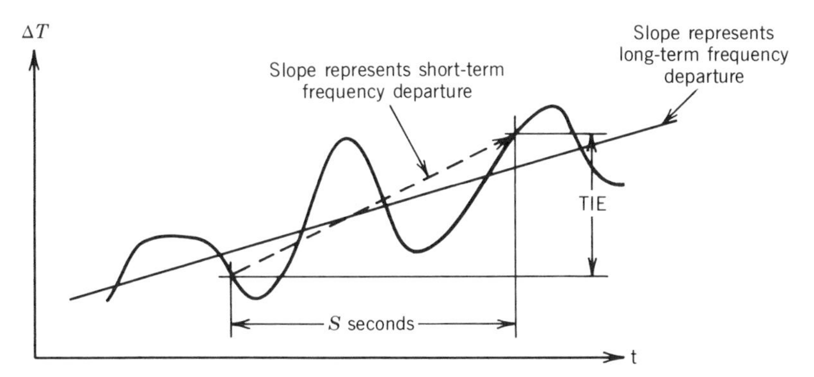

Figure 10.22 Definition of Time Interval Error (TIE). From CCITT Rec. G.811, page 298, *Red Book*, Vol. 3, Fascicle III.3. Courtesy ITU-CCITT, Geneva.

and at the beginning of the period or

$$\text{TIE}(S) = |\Delta T(t + S) - \Delta T(t)| \tag{10.2}$$

This is shown diagrammatically in Figure 10.22. The corresponding normalized frequency departure $\Delta f/f$ is the TIE divided by the duration of the period (i.e., S seconds).

The TIE at the output of a reference clock is specified in CCITT Rec. G.811 for three values of frequency as follows:

> The TIE over a period of S seconds shall not exceed the following limits:
>
> a) $(100S)$ ns + 1/8 unit interval. Applicable to S less than 5. These limits may be exceeded during periods of internal clock testing and rearrangements. In such cases the following conditions should be met: TIE over any period up to 2 UI (unit intervals) should not exceed 1/8 of a UI. For periods greater than 2 UI, the phase variation for each interval of 2 UI should not exceed 1/8 UI up to a total maximum TIE of 500 ns.
>
> b) $(5S + 500)$ ns for values of S between 5 and 500.
>
> c) $(10^{-2S} + 3000)$ ns for values of S greater than 500.

The allowance in (c) of 3000 ns is for component aging and environmental effects.

For clarification, CCITT defines UI in Rec. G.701 [9] as the nominal difference in time between consecutive significant instants of an isochronous signal. With NRZ coding we can think of a unit interval (UI) as the duration of 1 bit or the bit period. The bit period in NRZ is the inverse of the bit rate. In the case of AMI coding, the mark and space is usually not of equal duration.

TABLE 10.2 Maximum Permissible Degradation of Timing at a Network Node

Performance Category[a]	Frequency Departure $\left\|\frac{\Delta f}{f}\right\|$ of Node Timing[b]		Proportion of Time During Which Degradation May Occur, Referred to Total Time[d]	
	Local[c]	Transit	Local[c]	Transit
Nominal	See Section 5.5.4.1	See Section 5.5.4.1	$\geq 98.89\%$	$\geq 99.945\%$
(a)	$10^{-11} < \left\|\frac{\Delta f}{f}\right\| \leq 10^{-8}$	$10^{-11} < \left\|\frac{\Delta f}{f}\right\| \leq 2.0 \times 10^{-9}$	$\leq 1\%$	$\leq 0.05\%$
(b)	$10^{-8} < \left\|\frac{\Delta f}{f}\right\| \leq 10^{-6}$	$2.0 \times 10^{-9} < \left\|\frac{\Delta f}{f}\right\| \leq 5.0 \times 10^{-7}$	$\leq 0.1\%$	$\leq 0.005\%$
(c)	$\left\|\frac{\Delta f}{f}\right\| > 10^{-6}$	$\left\|\frac{\Delta f}{f}\right\| > 5.0 \times 10^{-7}$	$\leq 0.01\%$	$\leq 0.0005\%$

[a] The performance categories (b) and (c) correspond to (b) and (c) in Rec. G.822 while category (a) in Rec. G.822 corresponds to "Nominal" and (a) in Rec. G.811, combined.
[b] All values are provisional.
[c] The values for local nodes are given for guidance only, and administrations are free to adopt other performance levels provided the overall controlled slip performance objective of Rec. G.822 are met.
[d] These values are more stringent than would be strictly required by Rec. G.822 for a 64 kbps connection, to allow for the future introduction of services at higher bit rates that may require a better slip performance. They also allow a margin for possible network effects.
Source: From CCITT Rec. G.811, [11], page 298, *Red Book*, vol. 3, Fascicle III.3.

Table 10.2 shows the permissible degradation of the timing of a network node. The performance category refers to slip rate performance objectives defined briefly above and covered in Section 5.6.3.

5.6 Network Performance Requirements

5.6.1 Blocking Probability. A blocking probability of $B = 0.01$ on a trunk group is the engineering objective [8]. With alternative routing implemented, overall blocking probability of the network is closer to 0.005. Grade of service objectives for the public switched digital network is the same as for the public switched analog network.

5.6.2 Error Performance

5.6.2.1 North American Perspective. The error rate on a digital connection is the fraction of bits received that differ in binary value from the corresponding bits in the transmitted bit stream. Bit errors accumulate on a digital network from the source A/D conversion point to the sink D/A conversion point. An objective for end-to-end connections through the switched digital network (SDN) is that the error rate should be less than 1×10^{-6} on at least 95% of the connections [8].

Most digital facilities and connections have error rates lower than 1×10^{-6}. Where a higher error rate occurs on a connection, it is usually caused by a high error rate at one facility. The error rate objectives for digital facilities are therefore designed to control the probability that one or more facilities in a connection will have an error rate greater than 1×10^{-6}. The allocated objective of each digital facility specifies the percent of such facilities or the percentage of time for each facility for which the error rate should be less than 1×10^{-6}. The specified percentage is chosen so that on representative connections the total probability of having an error rate greater than 1×10^{-6} will be about 5% [8].

If we consider transmission of speech telephony only on a connection, PCM can withstand much greater error rates. Intelligibility is actually lost when the BER reaches about 1×10^{-2}. This error rate is unacceptable in the PSTN not because of intelligibility but because of supervisory signaling. Studies carried out by British Telecom indicate that a BER of 1×10^{-3} or better is required to maintain the supervisory signaling channel in its desired condition throughout the connection (i.e., idle or busy).

5.6.2.2 CCITT Perspective. CCITT Rec. G.821 [12] error performance objectives are based on a 64-kbps circuit-switched connection used for voice traffic or as a "bearer channel" for data traffic. The error performance parameters given refer to the HRX discussed in Section 5.4.

The CCITT error performance parameters are defined as follows (CCITT Rec. G.821): "The percentage of averaging periods each of time interval $T(0)$ during which the bit error rate (BER) exceeds a threshold value. The percentage is assessed over a much longer time interval $T(L)$." A suggested interval for $T(L)$ is 1 month.

It should be noted that total time $T(L)$ is broken down into two parts:

- Time that the connection is available.
- Time that the connection is unavailable.

The following BERs and intervals are used in CCITT Rec. G.821 in the statement of objectives [12]:

- A BER of less than 1×10^{-6} for $T(0) = 1$ min.
- A BER of less than 1×10^{-3} for $T(0) = 1$ s.
- Zero errors for $T(0) = 1$ s.

Table 10.3 gives the CCITT error performance objectives. Table 10.4 gives some guidelines for interpreting Table 10.3.

TABLE 10.3 CCITT Error Performance Objectives for International ISDN Connections

Performance Classification	Objective[c]
(a) (Degraded minutes)[a,b]	Fewer than 10% of one-minute intervals to have a bit error ratio worse than 1×10^{-6}[d]
(b) (Severely errored seconds)[a]	Fewer than 0.2% of one-second intervals to have a bit error ratio worse than 1×10^{-3}
(c) (Errored seconds)[a]	Fewer than 8% of one-second intervals to have any errors (equivalent to 92% error-free seconds)

[a]The terms "degraded minutes," "severely errored seconds," and "errored seconds" are used as a convenient and concise performance objective "identifier." Their usage is not intended to imply the acceptability, or otherwise, of this level of performance.
[b]The one-minute intervals mentioned in the table and in the notes are derived by removing unavailable time and severely errored seconds from the total time and then consecutively grouping the remaining seconds into blocks of 60. The basic one-second intervals are derived from a fixed time pattern.
[c]The time interval $T(L)$, over which the percentages are to be assessed has not been specified since the period may depend on the application. A period of the order of any one month is suggested as a reference.
[d]For practical reasons, at 64 kbps, a minute containing four errors (equivalent to an error ratio of 1.04×10^{-6}) is not considered degraded. However, this does not imply relaxation of the error ratio objective of 1×10^{-6}.
Source: From CCITT Rec. G.821 [12], page 302, *Red Book*, vol. III, Fascicle III.1.

5.6.3 Slips

5.6.3.1 Definition. A *slip* is a synchronization impairment of a digital network. No two free-running digital network clocks are exactly synchronized. Consider the simple case where two switches are connected by digital links. An incoming link to a switch terminates in an elastic store to remove transmission timing jitter. The elastic store at the incoming interface is *written into* at the recovered clock rate $R(2)$ but *read out* by the local clock rate $R(1)$. If the average clock rate of the recovered line clock $R(2)$ is different from $R(1)$, the elastic store will eventually underflow or overflow. When $R(2)$ is greater than $R(1)$, an overflow will occur, causing a loss of data. When the reverse happens, $R(1)$ greater than $R(2)$, an underflow occurs, causing extraneous data inserted into the bit stream. These disruptions of data are called *slips*.

Slips must be controlled to avoid loss of frame synchronization. The most common approach is to purposely add or delete a frame that does not affect framing synchronization.

Slips do not impair voice transmission per se. A slip to the listener sounds like a click. The impairments to signaling and framing depend on the system, whether it is DS1 or CEPT 30 + 2. Consider CEPT 30 + 2. It uses channel-associated signaling in channel 16 and may suffer from loss of multiframe alignment due to slip. Alignment may take up to 5 ms to be reacquired, and calls in the process of being setup may be lost. Common channel signaling such as CCITT SSN 7 (Chapter 14) is equipped with error detection and

TABLE 10.4 Guidelines for the Interpretation of Table 10.3

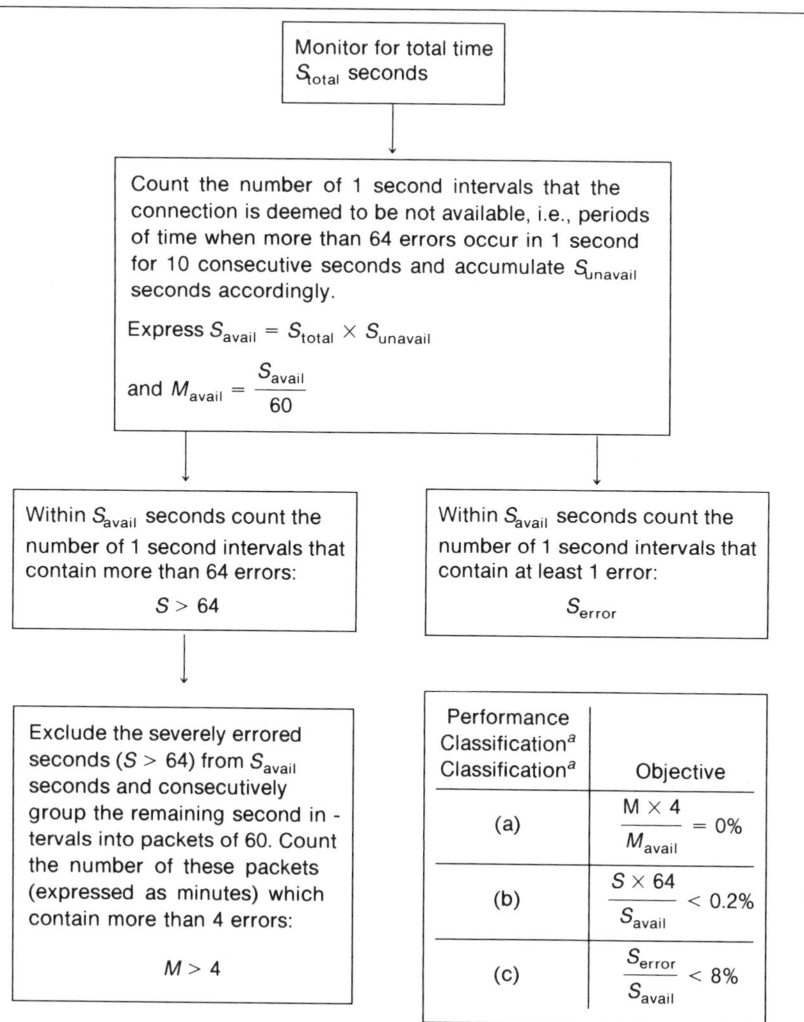

[a]See Table 10.3. From CCITT Rec. 6.811, Annex B, page 316, *Red Book*.

retransmission features so that a slip will cause increased retransmit activity, but the signaling function will be otherwise unaffected [1].

One can imagine the effect of slips on data transmission. A one frame slip causes a byte of extraneous data or a byte of errored data.

5.6.3.2 North American Slip Objectives. The North American end-to-end slip objective is 1 slip in 5 hours, where 1 slip in 10 hours (half of the end-to-end allocation) is assigned to transmission facilities between nodes [8].

The worst case, in which all reference clocks are lost, is 255 slips per day. However, Ref. 7 (ATT) states that during normal operation the network operates without slips and what slips do occur are traced to uncontrollable impairments, such as reference clock failures. However, most clocks are redundant (e.g., the DMS-100) so the probability of clock failure is very low.

5.6.3.3 CCITT Slip Criteria. The CCITT slip objective is 1 slip per 70 days per plesiochronous interexchange link using clocks specified in Section 5.5.4.

The CCITT HRX includes 13 nodes operating in the plesiochronous mode connected by 12 links. Thus the nominal slip performance of a connection is 70/12 or 1 slip in 5.8 days. CCITT Rec. G.822 [13] states that acceptable controlled slip rate performance on a 64-kbps international connection is less than 5 slips in 24 h for more than 98.9% of the time. Forty percent of the slip rate is allocated to the local network on each end of the connection, 6% for each national transit (toll) exchange, and 8% for the international transit portion (see CCITT Rec. G.822).

REVIEW QUESTIONS

1. Give a simple definition of time-division switching. Compare space-division switching with time-division switching.
2. Give at least five advantages of time-division switching compared to space-division switching.
3. Give two technical issues relating to time-division switching and digital networks. (Clue: These may be listed as disadvantages in relation to question 2.)
4. What are the two principal functional elements of digital switching? These elements carry out the actual switching of digital connections (e.g., not control).
5. Define the terms *read* and *write* as used in this chapter. Differentiate between sequential read–random write and random read–sequential write.
6. What are the three basic building blocks of a time switch?
7. What are the limitations of a time (T) switch? How are these solved by the addition of space (S) switching capability?
8. A space array has m horizontals and n verticals in its matrix. Relate m to n for nonblocking tandem switching, for local switch expansion, and for local switch concentration.
9. We have a 30×30 space array with a time slot interchange (TSI) at each input. The TSI is designed for DS1 operation. How many total time slots can the array handle?

10. Why is a TST switch more desirable than an STS switch? Where would an STS switch have application?

11. How can blocking probability be reduced in a TST switch?

12. What is the function of "junctors" in a switch (e.g., a DMS-100)?

13. What is the architecture of a DMS-100? (Clue: Use T and S.)

14. Considering the architecture of the DMS-100, what is the connotation of the word *folded*?

15. Why is the basic internal rate of the DMS-100 based on CEPT 30 + 2 rather than DS1 PCM format?

16. Give at least four interfaces feasible with the DMS-100. (Clue: Think basic telephony, both analog and digital.)

17. The network module (NM) is the heart of the DMS-100. What generic device provides the interface with the outside world? Name at least four specific devices in this generic device family, and describe in general terms the functions of each.

18. What is the basic function of a remote switching center (RSC), which is part of the DMS-100 family? How does it help to circumvent the major problem of extending digital service beyond a local switch toward users?

19. Briefly describe the generic processor and controller architecture of the DMS-100. Use words such as "centralized" and "distributed."

20. How is network clocking incorporated into the DMS-100?

21. What are the basic functions of the data store (DS) and the program store (PS) in the DMS-100?

22. What is the characteristic impedance of the DMS-100 line and trunk ports?

23. Give at least three switching loss capabilities of the DMS-100 and their respective network applications.

24. Relate at least three problems associated with extending digital telephone service to customer premises.

25. Give at least four functions of a SLIC. Define *BORSCHT*.

26. Why would it be advisable to use a PCM-based PABX rather than some other digital architecture or just straight PAM?

27. Give the basic reason why load coils must be removed from subscriber loops when converting to digital service.

28. Describe at least three different approaches to the synchronization of a digital network.

29. In the design of a digital network, of what value is a hypothetical reference connection (HRX)?

30. What is the CCITT rationale for recommending plesiochronous synchronization on international links?

31. Give the three levels of digital network synchronization as given in the text.

32. Where can one find the master clock for a local digital network?

33. What is the major impairment that derives from the network synchronization process? (Remember that the process goes on all the time.) What are the two causes of this impairment?

34. What drives specified BER performance in a digital network?

35. What is the most common boundary of a controlled slip? Why?

36. What keeps the designer from using high-precision clocks at *all* digital nodes? (Consider commercial practice only.)

37. What type of synchronization is used with the North American digital network?

38. What vehicle physically disseminates timing on the North American network? on the international CCITT network?

39. How will a "closed loop" affect timing on a hierarchical synchronization plan?

40. Define *unit interval* (UI).

41. What value BER is used in digital network planning?

42. How might a slip affect signaling?

43. What is the principal cause of slip? How can it be minimized?

44. Compare North American and CCITT basic slip rate criteria.

REFERENCES

1. John P. Ronadyne, *Introduction to Digital Communications Switching*, Howard W. Sams & Co., Indianapolis, Indiana, 1986.
2. John C. McDonald, ed., *Fundamentals of Digital Switching*, Plenum Press, New York, 1983.
3. John C. Bellamy, *Digital Telephony*, Wiley, New York, 1982.
4. *Planning Guide: DMS-100/200 Family*, Northern Telecom, Research Triangle, N.C., 1985.
5. *Network Frame*, General Specification, Northern Telecom, Research Triangle, N.C., 1983/1988.

7. *Technical Reference: Digital Synchronization Network Plan*, ATT Pub. 60110, New York, 1983.
8. *Notes on the BOC Intra-LATA Networks: 1986*, Bell Communications Research, Issue 1, TR-NPL-000275, Livingston, N.J., April 1986.
9. *Vocabulary of Digital Transmission and Multiplexing and Pulse Code Modulation (PCM) Terms*, CCITT Rec. G.701, Vol. 3, F.III.3, CCITT, Malaga-Torremolinos, 1984.
10. *Digital Transmission Models*, CCITT Rec. G.801, Vol. 3, F.III.3, CCITT, Malaga-Torremolinos, 1984.
11. *Timing Requirements at the Outputs of Reference Clocks and Network Nodes Suitable for Plesiochronous Operation of International Digital Links*, CCITT Rec. G.811, Vol. 3, F.III.3, CCITT, Malaga-Torremolinos, 1984.
12. *Error Performance of an International Digital Connection Forming Part of an ISDN*, CCITT Rec. G.821, Vol. 3, F.III.3, CCITT, Malaga-Torremolinos, 1984.
13. *Controlled Slip Rate Objectives of an International Digital Connection*, CCITT Rec. G.822, Vol. 3, F.III.3, CCITT, Malaga-Torremolinos, 1984.
14. Amos E. Joel, Jr., Ed. *Electronic Switching*: Digital Central *Office Systems of the World*, IEEE Press, New York, 1981.

11

DATA NETWORKS AND THEIR OPERATION

1 INTRODUCTION

Data communications continues to be the fastest growing segment in telecommunications. Its orientation is different from that of telephony. An outstanding difference is the human interface. Whereas in conventional telephony the human interface is transmission of voice through the mouth and reception by the ear, the data interface is transmission through the fingers (e.g., on a keyboard) and reception is through the eyes. In fact, some data operations have no direct human interface at all, or nearly none. We humans do, indeed, feel its effects. An example of the latter is automated control systems.

The major stimulus to advance data communications is the ever increasing use of electronic data processing, which is commonly referred to as EDP. Thus it is computer oriented, although in an increasing number of applications the computer may be nothing more than a microprocessor, as found in the PC (personal computer).

Unlike telephony, where the conventional telephone implies the spoken word, data communications handles the written word. It deals not only with *words* but with other symbols such as numbers, graphics, punctuation, or just bit sequences that have no direct meaning to us but act as a stimulus to a "machine" (or a network of machines) to bring about a desired reaction. The following is a list of examples where the communication of alphanumerics, graphics, and other bit sequences is required; it exemplifies some everyday problems solved by EDP and data communication networks.

1. Multipoint inventory control.
2. Airline, train, and hotel reservations.
3. Banking transactions, electronic funds transfer (EFT).

4. Truck, train, and ship cargo control.
5. Air traffic control.
6. Weather forecasting.
7. Defense: command, control, and communication networks, strategic and tactical, remote radar and other sensors, intelligence.
8. Research and development resource sharing.
9. Remote text composition (newspapers and magazines).
10. Police: anticrime EDP.
11. Factory automation and other industrial networks.
12. Office automation.
13. Finance (stock market).
14. Electronic mail, teletext.

Each application requires a distinct approach in the design of its associated data communications network. One major consideration is whether the data processing itself will be centralized or distributed processing. This and many other factors must be determined before proceeding in network design. The material in this chapter uses Chapter 8 as a base. The reader is advised to review Chapter 8 before continuing.

2 INITIAL DESIGN CONSIDERATIONS

2.1 General

Data communication networks can vary from an elemental two-terminal system, shown in Figure 11.1, to a multiterminal distributed processing network such as the US DARPA network, shown in Figure 11.2. Between these extremes there is a large variety of data networks regarding size, configuration, and capability. Some of the more important considerations entering into network design are:

- Data sources and destinations, locations, geographic dispersal.
- Type of network organization.

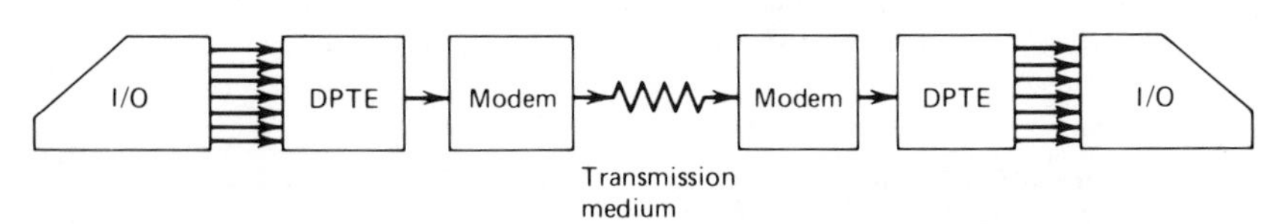

Figure 11.1 Elemental point-to-point data communication system (two PC-based terminals). I/O = input–output device; DPTE = data processing terminal equipment.

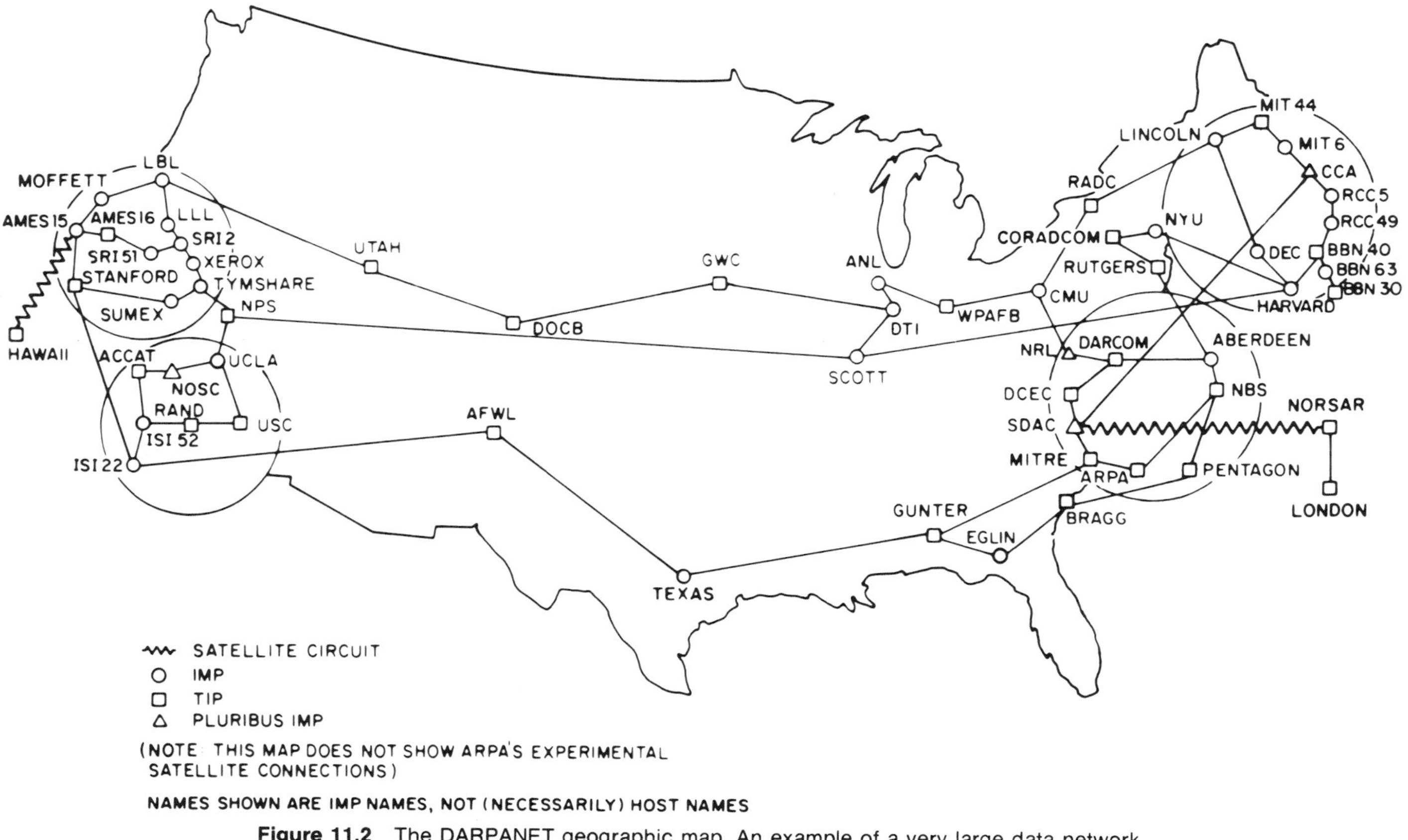

Figure 11.2 The DARPANET geographic map. An example of a very large data network. Courtesy of Bolt, Beranek and Newman (BBN).

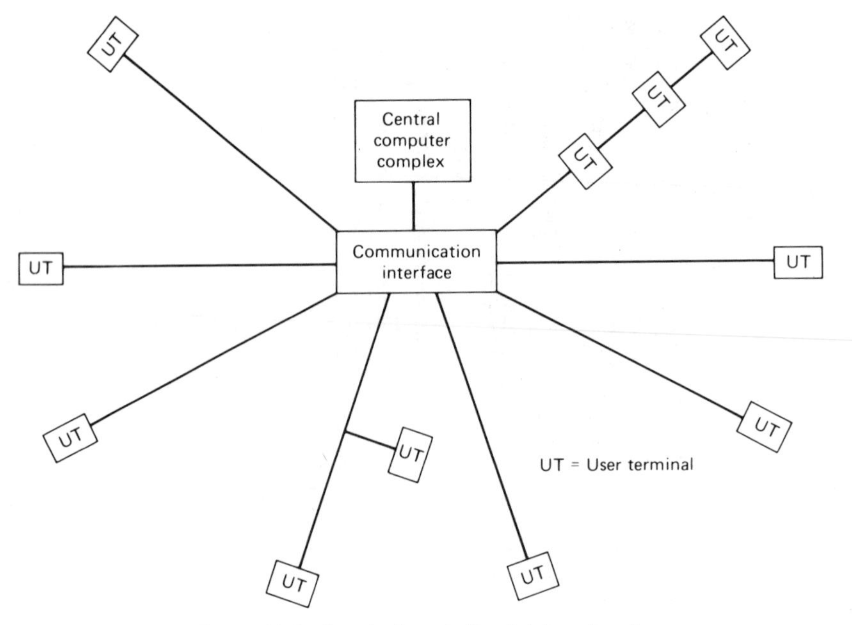

Figure 11.3 A typical centralized data network.

- Line routing.
- Service reliability and availability requirements.
- Tariffs and tariff structures.
- Type of communication service: switched lines, leased lines, private lines, or combinations thereof.
- Type of terminal equipment in place or planned.
- Protocols now used or planned.
- Traffic analysis, transactions per busy hour, peak minute; characterization of typical transaction length.
- Perishability of data (typical: inquiry/response).

There are two possibilities for organizing a data network: centralized and distributed. *Centralized* implies one main processing location, with all traffic between remote terminals and the single CPU (central processing unit). A centralized network is shown in Figure 11.3. In a *distributed* network major data processing capabilities are located in more than one location. Each of the several or multiple processing centers is often controlled by a different operating system. A typical distributed network is shown in Figure 11.4. There are arguments for and against each approach. With distributed processing, communication costs may be markedly lower, yet processing costs may rise.

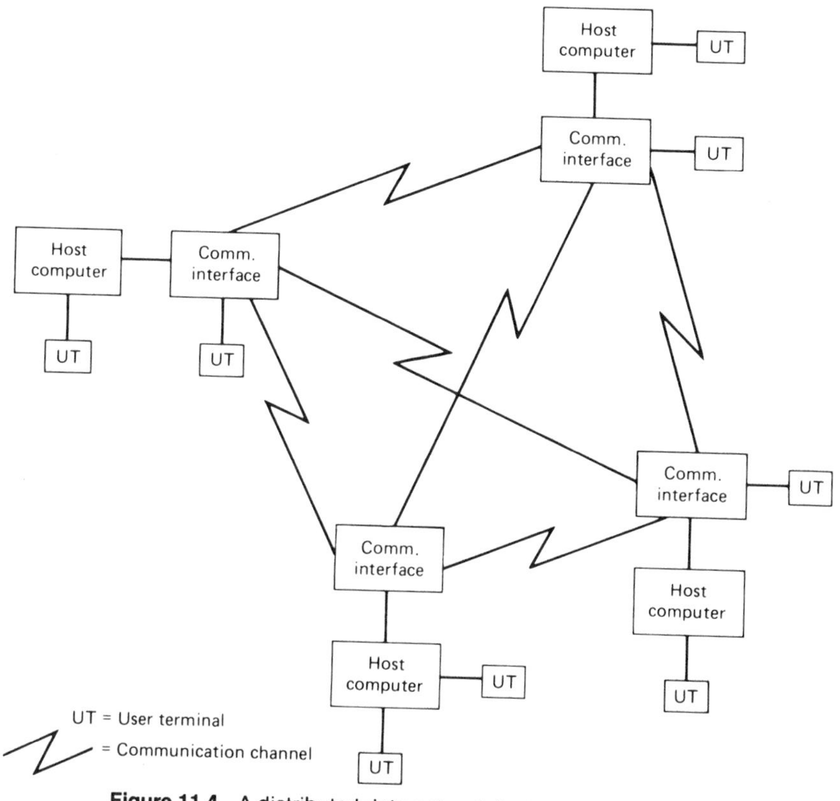

Figure 11.4 A distributed data network (note mesh connection).

This rise is due to not only the increase in processing equipment but in EDP administration as well. The corporate EDP manager may opt for the centralized approach. The telecommunications manager may show major savings in communication costs for the distributed network.

Many data networks today are fully distributed and use PCs or microcomputers at each access. These usually are connected to one or more mainframe computers (CPUs). The network can be further enhanced by adding mass storage devices and high-speed printers. In the local area a common method of networking these EDP assets is by means of a LAN (local area network). LANs are described in Chapter 12.

The *local area* may encompass a building or a group of buildings. A subset of the LAN concept is the metropolitan area network (MAN). In this case the network can cover a large city. When a network is to cover a still larger geographic area, a wide area network (WAN) is used. "Wide area" in this context can be a county, a state, a province, an entire nation, a community of countries, or the entire world. WANs can be set up in numerous ways. Several

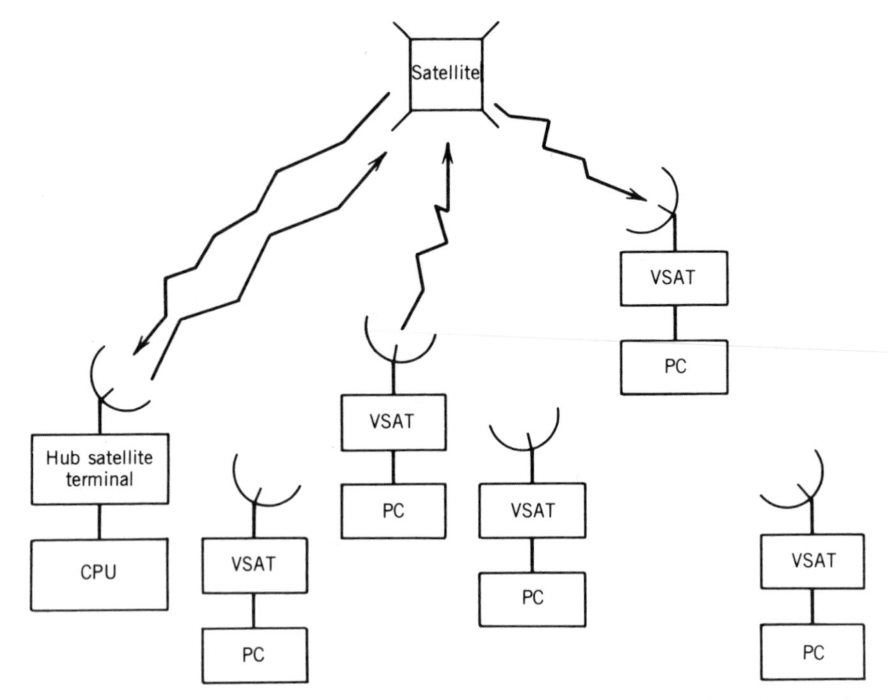

Figure 11.5 A typical VSAT network showing both centralized and distributed processing.

methods are:

- Private network using fiber optics, line-of-sight microwave radio, coaxial cable, or satellite or combinations of these media.
- Connectivity by means of a specialized common carrier and/or VAN (value-added network).
- Overlay of a common carrier or a specialized common carrier service such as ATT DDS (digital data system).
- VSAT (very small aperture satellite terminal) system such as that offered by Equatorial Communications.

Many turn to the alternative of leasing the transmission facilities and providing their own switching facilities.

It should be noted that some of the present VSAT networks are ideal examples of centralized and distributed processing. In this case user terminals operate from 2400 bps to 1.544/2.048 Mbps data rates. The VSAT accesses often are PC based, and the hub station connects to a mainframe computer.

Such an implementation is shown in Figure 11.5. VSATs were described in Chapter 7, Section 7.8.5.

2.2 User Requirements

A data network is built to fulfill a need. The technical specifications of the network must be based on the users' business application requirements. Before designing the initial network, we must know and quantify the following [1]:

1. Number and locations of processing sites.
2. Number and locations of remote terminals and their communication capabilities; processing capabilities.
3. Types of transactions to be processed.
4. Traffic intensity for each type of transaction by type of terminal.
5. Data perishability or the urgency of information to be transmitted (timeliness).
6. Patterns of traffic flow.
7. Acceptable error rates.
8. Required availability of the network.

First let us consider one of the simpler approaches to carrying out EDP requirements for a particular corporation. Let's assume that these requirements are accomplished by one centralized processing center. Magnetic tape will be prepared or cards will be punched at remote locations—a remote location may be down the hall or hundreds of miles away. The card decks or magnetic tapes are delivered by vehicle or express mail. The processing may be completed at the computer center within hours, but sometimes it takes days. The results, the output, are returned by the same means. Time or timeliness is no object. Payrolls, certain inventory operations, and sales analyses may fit into this category. It is simple, economical, and easy to administer. In other applications timeliness is very important. The data is "perishable." When the user needs faster response, say, in something less than "hours," electrical communications may well have to be considered. There is the old adage: "Time is money." Within the "urgency" requirements the total system cost should be optimized: communication costs versus processing costs. Again time can be related as a cost factor.

The next expedient is the telephone. Many telephone administrations permit either acoustic coupling or direct coupling to the public switched telephone network. Thus the user dials the telephone number of the computer center, connects a modem to the line, and dumps a serial data stream from 75 bps to 4800 bps to the processing center. Some minutes (or hours) later, the results may be returned to the remote user by the same means. The weaknesses to this approach are:

- There is no immediate assurance that the error rate is inside the user's specifications because of the half-duplex nature of the connection and because of the nonuniformity of quality of service from the data user viewpoint.
- There is a lack of high-speed duplex service, thus preventing interaction, or at least immediate interaction unless a second telephone line is dialed up to the center.
- The bit transfer rate is limited. Forced into the use of stop-and-wait ARQ.
- Telephone connection setup time may be a crucial factor. An all trunks busy (ATB) condition may be encountered.
- A fairly large amount of human intervention is required.
- The possibility exists of connection drop out (loss of connection in mid-transaction).

Figure 11.6 illustrates a typical "telephone connection."

The next level of increased "communication cost" is a leased-line system where, indeed, we configure our own network. The term "overlay network" may be used to describe a data communication system that utilizes leased lines. With a leased line the telephone company or administration provides full-period connection of a telephone circuit. There is no intervention of a switch. Connecting patches are installed at switch mainframes and are permanent. The data transmitted on a leased telephone line still suffer the limitations or transmission constraints on such a medium. For instance, the highest practical bit rate is 9600 bps, and in some cases up to 19.2 kbps can be

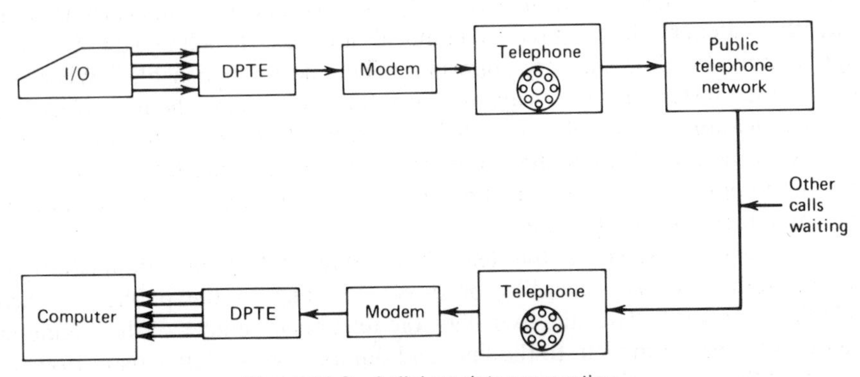

Figure 11.6 A dial-up data connection.

achieved (see Chapter 8). The leased service can be two-wire or four-wire. With two-wire service the user generally is limited to half-duplex service. With four-wire operation full-duplex service is achieved. With dial-up connections bit rates are generally limited to 4800 bps, but with leased service the 9600-bps rate can be used if the subscriber includes the necessary conditioning in the lease agreement. Selective ARQ, which further improves efficiency of usage, can be easily implemented.

The last and most expensive option is for the user to build a private network. In this case both the transmission media and switches belongs to the user. For WANs with large geographic dispersal, satellite transponder space may be leased. For WANs covering a smaller geographic area, LOS microwave or optical fiber or combinations of the two are utilized. Such networks should be digital, and consideration should be given to their compatibility with the public switched network in case an interconnect to that network is a future consideration.

2.2.1 Communication Needs: Capacities. A mainframe computer is an extremely high-speed device and a major capital investment. Its speed is measured in MIPS (millions of instructions per second) or MFLOPS (millions of floating point operations per second). To derive optimum benefit from a central computer with such speeds of operation, it would be uneconomical to permit processing capability to remain dormant or to be underemployed. With multi-MFLOPS capabilities on one hand and single communication accesses limited to 10–20 kbps on the other, we turn to methods of increasing I/O data rates of communication accesses to improve computer utilization.

Megabit LANs would appear to be one answer. Not quite. A LAN is meant to service many accesses, one at a time (see Chapter 12). Thus the apparent megabit/per second data rate over long periods is far less looking inward/outward from the computer port, assuming the data traffic is equipartitioned among the many accesses. On the other hand, if all traffic is destined to/from the mainframe, a LAN may serve a fair share of a computer's capacity.

To increase computer usage and improve accessibility of many low data rate users, concentrators, multiplexers, and port contention devices are used to connect low data rate users to a computer inlet/outlet. Each device makes it appear to the low data rate user (75–4800 bps) that he or she has the sole real-time use of the large CPU working in a time-sharing mode.

The PC or microcomputer workstation has revolutionized computer networks accessing large CPUs. Many computational and word processing transactions can be carried out by the PC alone. The mainframe computer is reserved only for complex transactions, large data base access, complex simulations, some CAD/CAM, and electronic mail. Now in many situations both the PC and the mainframe are underutilized.

PCs and other terminal or workstations with a basic common community of interest, such as a government agency, corporation, or university, are widely dispersed. Their effective connectivity among themselves and with other major computer assets becomes a problem to be solved by the telecommunication system engineer.

In the planning of such an installation, the starting point is user needs. Hence we turn to trade-offs of new equipment and facilities versus existing assets in computers and communications. Communication system engineers must design a cost-effective system meeting the needs. They must first quantify user requirements by location. Usually this is done by filling out a traffic flow matrix (transactions per hour and the average length of a transaction in bits or bytes). They then must consider:

- Data timeliness, urgency.
- Error performance requirements.
- Availability requirements.
- Existing system connectivity, capacity, and interfaces.
- Protocol issues.
- Candidate communication media and performance for WANs.
- Candidate communication media and performance for LANs and required gateways.

We have on one hand user needs and on the other hand a "toolbox" of candidate solutions to satisfy the needs. In the sections that follow we describe what is available in the toolbox and how to put the tools to work. Chapter 8 covered the transmission aspects of data. Here we cover network and operational issues.

3 DATA TERMINALS AND WORKSTATIONS

A key element of a data network is the data terminal or workstation. Generally, a network consists of many workstations and other computer assets such as a CPU, high-speed printer(s), and in some cases mass storage devices. Our interest in this section is how data terminals and workstations can drive the network design.

A workstation has a human operator. As a minimum it consists of a display, a keyboard or keypad input device, and a processor. To interface a data network, it must also have a communication interface device, which may or may not be physically a part of its processor, such as a plug-in card. It may also have a printer. For planning the system design engineers should determine the following:

- The type of protocol(s) supported.
- Allowance for protocol upgrades.

- Inclusion of automatic polling.
- Standard interface (OSI Layer 1): RS-232D, MIL-STD 188-114A, etc.
- Handling of varying data rates.
- Coprocessing and interrupt capabilities.
- LAN interface and NIU card slot.
- Processor's operating system.
- Buffer memory; can a portion be used for communications?
- Security device support, if any.
- Asynchronous/synchronous interface.

There are three ways of incorporating terminals into a WAN:

1. Stand-alone terminal with the necessary interface.
2. Cluster of terminals with communications controller.
3. LAN with interconnect through a gateway.

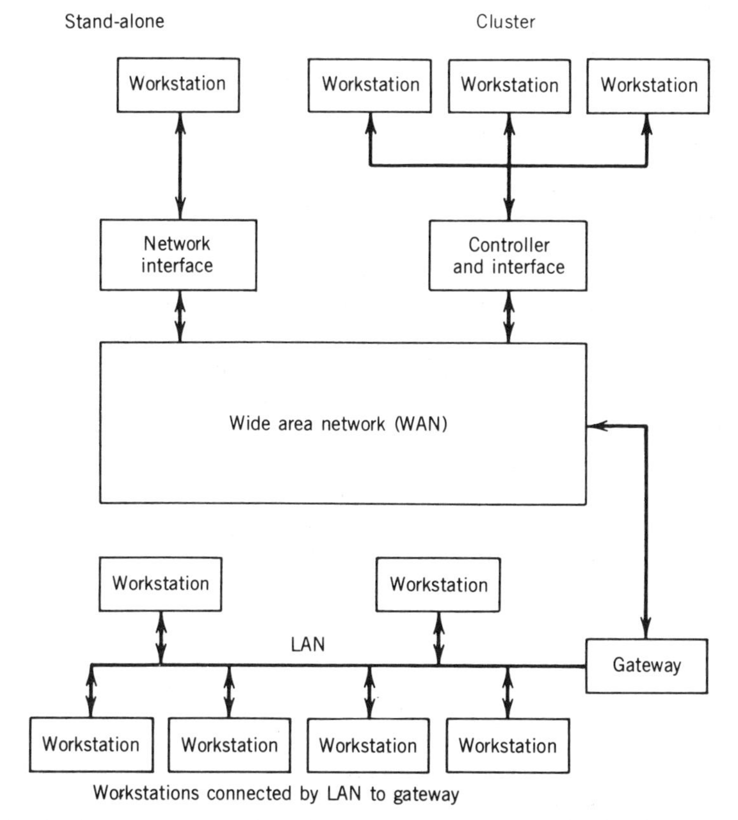

Figure 11.7 Graphic representation of three methods of connecting terminals or workstations to a wide area network (WAN).

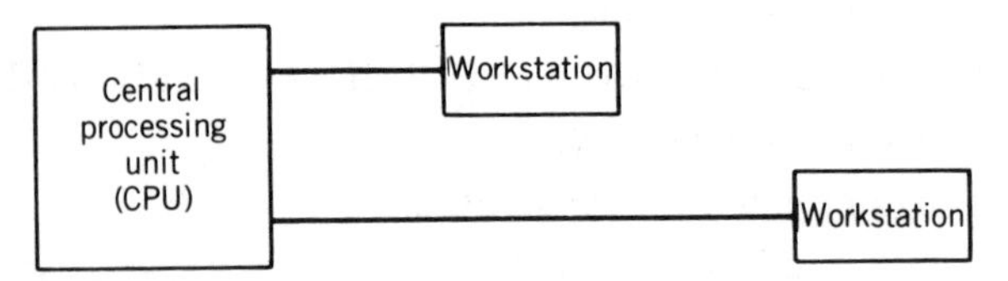

Figure 11.8A A simplified point-to-point network.

These are shown graphically in Figure 11.7.

4 SOME SIMPLIFIED NETWORK CONFIGURATIONS

Aside from LANs, which are covered in the next chapter, there are six methods of simple networking, as shown in Figure 11.8.

1. Point to point (Figure 11.8A).
2. Multipoint or multidrop (Figure 11.8B).
3. Star with host computer or CBX/EPABX at hub (Figure 11.8C).
4. Multistar/hierarchical (Figure 11.8D).
5. Ring network (Figure 11.8E).
6. Grid network (packet switched or circuit switched) (Figure 11.8F).

There are advantages and disadvantages of each. An advantage of a particular type of network in one situation may turn out to be a disadvantage in another. For instance, a multidrop (multipoint) network is simple and economical but should be used only when each access has a low traffic intensity. For the high traffic intensity and occupancy case, a point-to-point network is more attractive and its expense is a function of geographic dispersal.

The point-to-point network (Figure 11.8A) has the advantage of simplicity of design. As the number of accesses increases, a port contention device may be required. The disadvantage of point-to-point operation is the cost of the transmission media, particularly as the circuits are extended from feet to miles

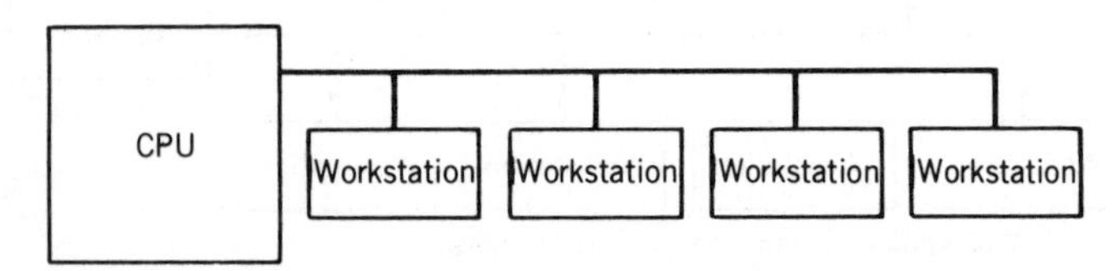

Figure 11.8B A simplified multipoint (multidrop) network.

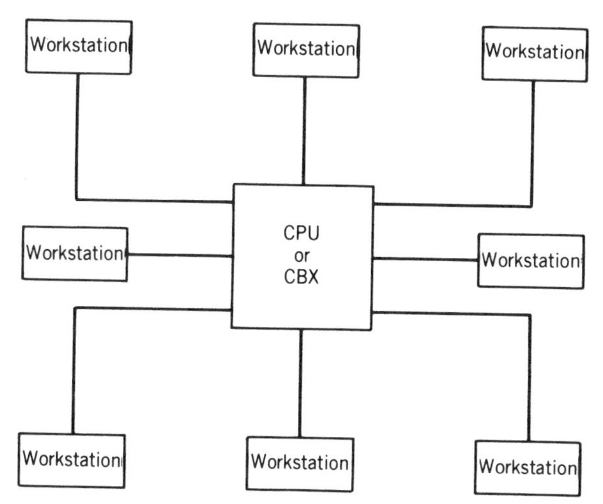

Figure 11.8C A simplified star network with CPU or CBX (computer-based PABX) at the hub of the star.

to hundreds or thousands of miles. For low-usage accesses dial-up connections (Figure 11.6) may reduce cost; local clustering is another alternative (Figure 11.7).

Multipoint or multidrop operation can solve the transmission media cost problem unless the connectivity route tends to meander a lot. That only one access at a time may use the medium also detracts from its utility. Depending on the operational regime and embedded firmware and software in each terminal, there are two common methods to access the host CPU: contention and polling. The first is simple random access where a user can get access of the medium for an indefinite period. Polling provides an access control mechanism where the host asks each access in turn if it has traffic and takes or passes traffic during the poll interval. These protocol techniques are covered in Section 7.4.

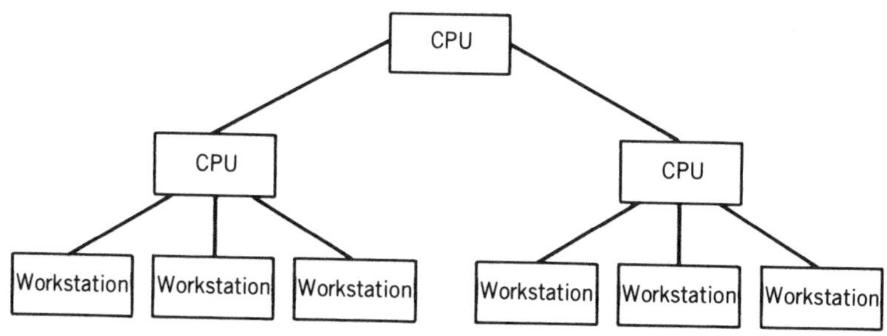

Figure 11.8D A simplified diagram of a multistar or hierarchical network.

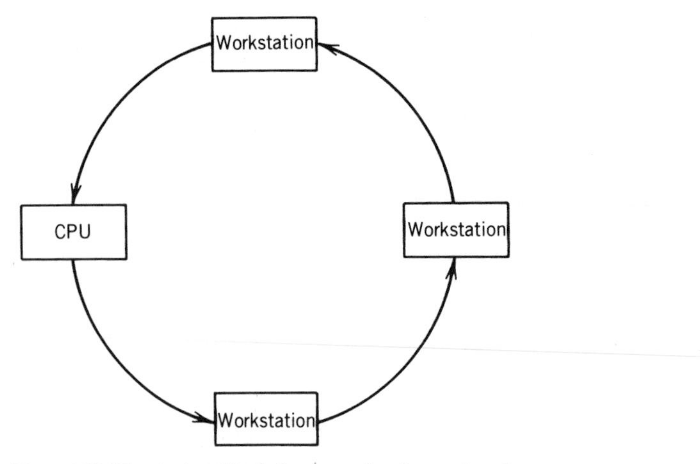

Figure 11.8E A simplified diagram of a ring network.

A star network (Figure 11.8C) has a processing and switching capability at its hub. This may be a CBX (computer-based PABX), which is designed to handle data, or a CPU with a message or circuit switching capability. In either case workstations can access the CPU or other workstations on a point-to-point basis through the hub. When a CBX is at the hub, one or more accesses off the CBX may be CPUs or other data devices (e.g., printers). Many CBX-based networks have data capabilities from 19.2 kbps to 64 kbps per line.

Figure 11.8D expands the star network concept shown in Figure 11.8C. Our present telephone network is a modified hierarchical network (Chapter 1). One

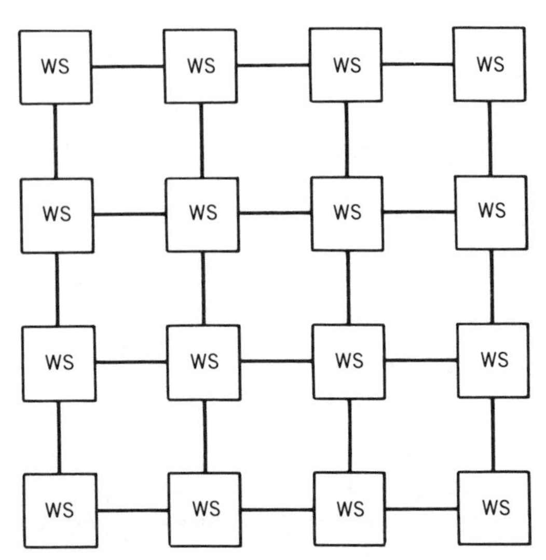

Figure 11.8F A grid network of workstations. [*Note*: A CPU can replace a workstation (WS) at any location or be colocated with a workstation.]

application for data communications places workstations at the lowest levels in the hierarchy. These workstations access nearby computers, which carry on the brunt of the processing and also provide a concentration capability for data transmission to a large, centrally located computer (at the top of the figure). Some of the same techniques used with the telephone network can be applied to a hierarchical data network. One is the application of high-usage (HU) and/or direct routes for high traffic intensity data relations.

Figure 11.8E illustrates a ring network. We note in the figure the unidirectional flavor of data flow. This is typical of the token ring LAN covered in Chapter 12. If a second ring is added, providing traffic flow in the other direction, survivability and availability are improved. If one ring should fail, the other can still operate. Such bidirectionality is typical of an FDDI fiber optic LAN (see Chapter 12).

In large networks survivability and availability can be enhanced still further with a grid network, shown in Figure 11.8F. This is because of the large number of alternative routing possibilities available with a grid network. It also lends itself well to packet communications, where each node has at least three inlets and outlets. Packets can travel on many distinct routes between any two nodes in the network.

5 THREE APPROACHES TO DATA SWITCHING

The three basic approaches to data switching are:

1. Circuit switching.
2. Message switching.
3. Packet switching.

Plain old telephone service, affectionately called POTS, uses circuit switching. A data switch can be set up in a similar manner, providing address information in the message header. The switch responds to the header by setting up a real (physical) or virtual circuit to the desired addressee. AutoDiN, a US Department of Defense digital network, offers circuit switching as one operational mode. The dial-up connection described previously uses the POTS circuit-switching capability of the telephone network to provide the required connectivity for data messages.

Message switching, often called store-and-forward switching, accepts messages from originators, stores the messages, and then forwards each message to the next node or destination when circuits become available. We wish to make two points when comparing circuit and message switching. Circuit switching provides end-to-end connectivity in near real time. Message switching does not. There usually is some delay as the message makes its way through the system to its destination. The second point deals with efficient use of expensive transmission links. A well-designed message-switching system keeps a uniform load throughout the working day and even into the night. Circuit-switching system, such as the telephone systems, are designed for busy hour loading and the system tends to loaf after hours.

Telex is a good example of message switching. Another example is the older "torn-tape" systems. Such systems were labor intensive, rather primitive in this day and age. With torn tape messages arriving at a switching node from an originator or relay node used paper tape to copy the messages in the ITA No. 2 code (Chapter 8). Headers were printed out laterally along the tape holes. Operators at a torn-tape center would read the header, tear off the tape, and insert it in the tape reader accessing the required outgoing circuit. If the reader was busy, the tape was placed in queue in accordance with the message precedence (priority). I've seen clothes pins and paper clips used to piece the tapes together in such queues.

Store-and-forward systems today use magnetic tape storage rather than paper tape. A processor arranges messages in the correct queue. Message processors "read" precedence and header information and automatically pass the data traffic to the necessary outgoing circuits.

Packet switching utilizes some of the advantages of message switching and circuit switching and mitigates some of the disadvantages of both. A data "packet" is a comparatively short block of message data of fixed length. Complete data messages are broken down into short packages, each with a header. These packets may be sent on diverse routes to their eventual destination, and each packet is governed by an ARQ error-correction protocol. Because packets often travel on diverse routes, they may not arrive at the far-end receiving node in sequential order. Thus the far-end node must have the capability to store incoming packets and rearrange them in sequential order. The destination node then reformats the message as it was sent by the originator and forwards it to the final destination user.

Packet switching can show considerably greater efficiency when compared to circuit and message switching. There is one caveat, however. For efficient packet-switching operation, a multiplicity of paths (more than three) must exist from the originating local node to the destination local switching node. Well-designed packet-switching systems can reduce delivery delays when compared to conventional message switching. Expensive transmission facilities tend to be more uniformly loaded than with other data switching methods. Adaptive routing becomes possible where a path between nodes has not been selected at the outset. Delivery delay is also reduced when compared to message switching because a packet switch starts forwarding packets before receipt of all the constituent packets making up a message. Path selection is a dynamic function of real-time conditions of the network. Packets advancing through the network can bypass trunks and nodes where congestion or failure exist.

A packet switch is a message processor. Packets are forwarded over optimum routes based on route condition, delay, and congestion. It provides error

TABLE 11.1 Summary of Data Switching Methods

Switching Method	Advantages	Disadvantages
Circuit switching	Mature technology Near real-time connectivity Excellent for inquiry and response Leased service attractive	High cost of switch Lower system utilization, particulary link utilization Privately owned service can only be justified with high traffic volume
Message switching	Efficient trunk utilization Cost-effective for low-volume leased service	Delivery delay may be a problem Not viable for inquiry and response Survivability problematical Requires large storage buffers
Packet switching	Efficiency Approaches near real-time connectivity Highly reliable, survivable Low traffic volume attractive for leased service	Multiple route and node network expensive Processing intensive Large traffic volume justifies private ownership

control and notifies the originator of packet receipt at the destination. Packet switching approaches the intent of real-time switching. Of course, there is no "real" connection from originator to destination. The connection is one form of a "virtual" connection.

Table 11.1 summarizes advantages and disadvantages of the three methods of data switching.

6 CIRCUIT OPTIMIZATION

Data links have sources and destinations with varying requirements regarding *quantity* of data and its urgency. Consider a so-called voice-grade line (analog) that is full period leased. It can support 2400 bps, 4800 bps, or possibly 9600 bps. It would be uneconomical to underutilize such an expensive facility. Suppose that urgency was not a consideration: the only requirement on a particular circuit was to send several thousands of bits per day. At 75 bps the *maximum* amount of bits that can be sent is

10 s	750 bits
1 min	4500 bits
10 min	45,000 bits
1 hr	270,000 bits
8 hr	2,160,000 bits

Depending on one's perspective, these are rather impressive quantities.

Another common data rate is 110 bps (see Chapter 8, start–stop ASCII with a 2-bit stop element). With this data rate the amount of data that can be sent is

10 s	1100 bits
1 min	6600 bits
10 min	66,000 bits
1 hr	396,000 bits
8 hr	3,168,000 bits

Here the reader should beware. A start–stop ASCII character, as stated, has 11 bits, as shown in Figure 11.9. Three bits per character are really overhead bits. These are the start space and two stop marks. This overhead is required for start–stop operation of the circuit and is of no use at all to the end data user. In fact, the ASCII parity bit can be placed in the same category. Thus we need 11 bits on line to transmit 7 bits of source information. Therefore the efficiency is 7/11. Now we turn to synchronous transmission where the data traffic is sent in frames, packets, or blocks. A typical format is shown in Figure 11.10. In this case the frame segment, called the text in the figure, contains the

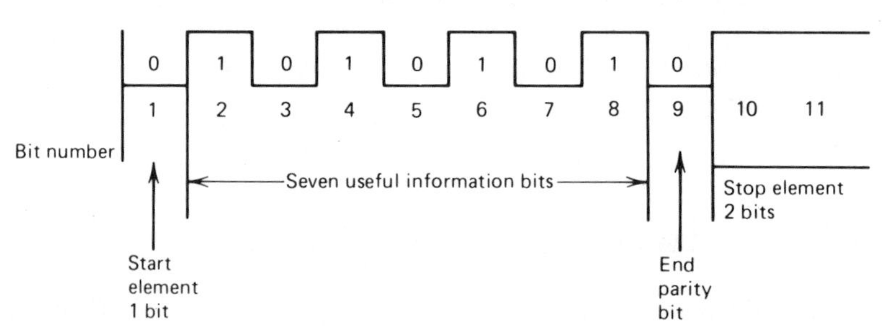

Figure 11.9 A start–stop bit sequence.

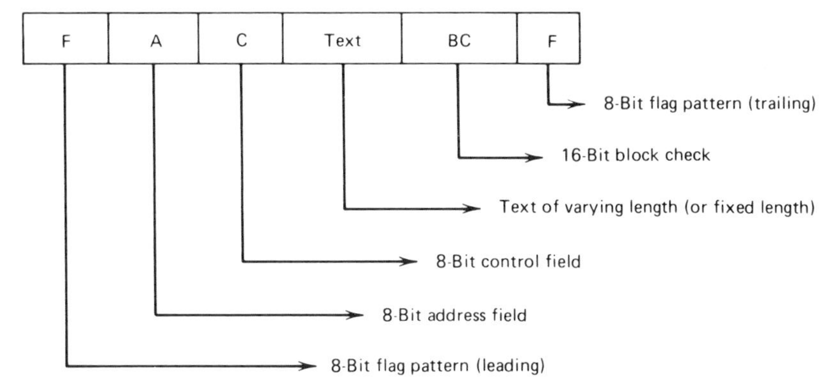

Figure 11.10 A typical data frame, block, or packet.

useful data for the destination user. We must amortize the text with the 48 overhead bits. Obviously a long frame amortizes the 48 overhead bits better than a short frame. Suppose that the text contained only 48 bits. We would be required to transmit 96 bits to achieve 48 useful bits. Here the efficiency is 48/98 or 50%. If the text contained 1028 bits, the efficiency improves to 1028/1076 or 95.5%. It would seem that long blocks should be favored over short blocks in the system design. *Beware*: On noisy circuits, where many block repeats are required, shorter blocks may turn out to be more efficient.

6.1 Throughput

The preceding discussion leads to the question of throughput. Throughput means different things to different people. We define throughput as the net *useful* bits put through per unit time. "Useful" is the key word. Useful to whom? The start and stop bits are useful to make the transmission system work. They are not useful to the data user who could care less how the system works. The same argument holds for the synchronous overhead bits (not that start and stop systems do not have additional overhead).

Turning back to the frame format in Figure 11.10, the text field contains the "useful" bits. As we will see later, even bits in this field are required for control overhead, which can be argued as *not useful* bits (noninformation characters). We also must try to partition the data communication system in some way to determine where the telecommunication responsibility leaves off and the data user responsibility begins. This point will be discussed in Section 7.

6.1.1 Effective Data Transfer: Transfer Rate of Information Bits. The American National Standards Institute (ANSI) recommends the use of the term "transfer rate of information bits" (TRIB) to quantify net data transfer rate.

$$\text{TRIB} = \frac{\text{Number of information bits accepted at destination}}{\text{Total time required to get those bits accepted}} \tag{11.1}$$

This formula has been used by Doll [2] with block transmission assumed. The formula is rewritten as

$$\text{TRIB} = [K(t)(M - C)]/N(t)[M/R + \Delta T] \tag{11.2}$$

where $K(t)$ is the information bits per character, M is the message block length in characters, R is the line transmission rate in characters per second, C is the average number of noninformation characters per block, $N(t)$ is the average number of transmissions required to get the block accepted at the destination (sink), and ΔT is the time between blocks in seconds.

If P is the probability of having to retransmit a block, then $N(t)$ can be expressed by

$$N(t) = 1/(1 - P) \tag{11.3}$$

Then

$$\text{TRIB} = K(t)[M - C][1 - P]/[(M/R) + \Delta T] \tag{11.4}$$

There is no direct reference to error rate in the formula; it is implied in the term $(1 - P)$. The reader will appreciate, after some reflection, that there is an optimum block length, given the channel error performance, that will optimize data transfer, at least as far as M is concerned, all other terms remaining constant.

We have discussed the term C in the previous subsection. These noninformation bits (overhead) can reduce the TRIB considerably. Delay is another factor. Among delays that can reduce TRIB are:

- Dial-up time (if the switched telephone network is used).
- Satellite channels and their inherent propagation delay.
- Type of ARQ, such as stop-and-wait, go-back-N, selective.
- Modem synchronization and handshaking delays.

TABLE 11.2 Some Available Transmission Options (In Order of Increasing Cost)

Bit Rate (bps)	Bandwidth (Hz)	Modulation (See Ch. 8)
75	120	VFCT (FSK)
150	240	VFCT (FSK)
300	480	VFCT (FSK)
600	960	VFCT (FSK)
1200	3100	FSK
2400	3100	QPSK
4800	3100	8-ary PSK
9600	3100	16-QAM
56,000	1/24 T1	baseband (ATT DDS)
64,000	approx. 64 kHz	baseband (ISDN rate)

A data link can be so overburdened with inefficiencies that, for example, a 2400-bps data link may only afford 100 bps of TRIB from source to destination.

6.2 Some Options to Fulfill Circuit Requirements

In this section we treat point-to-point service exclusively for off-premises connectivity. In general, cost is a function of the product of bit rate and distance. This assumes a tariff structure that is based on length of circuit. Table 11.2 gives a number of options that are available. If the data are perishable, such as on an inquiry/response system, the slow reaction of low bit rate systems may make such systems impractical. Typical inquiry/response systems are credit card verification, hotel and airline reservation systems, and electronic funds transfer (EFT). In some cases forms of data compression can speed up transaction time and thus take advantage of the less expensive lower bit rate techniques.

Another cost-sharing technique is to share a leased line such as the nominal 4-kHz channel (shown as 3100 Hz in Table 11.2) or a 56-kbps DDS implementation. One method divides the VF channel into frequency band segments using VFCT (voice-frequency carrier telegraph). This technique, using the CCITT standardized band division, is shown in Figures 11.11 and 11.12.

Another more effective technique is to use time-division multiplexing of the voice grade channel. The aggregate bit rate out of the multiplexer may be 2400 bps, 4800 bps, or 9600 bps, representing up to 32, 64, or 128 75-bps channels or 16, 32, or 64 150-bps channels, respectively. Many data multiplexers can accept a mix of input data rates based on the standard 75×2^n input data rate. However, overhead is often required to identify the far-end output ports as the proper destinations. If the input channels are start–stop, the start and stop signal elements are removed at the multiplexer and reinserted at the demultiplexer. The aggregate multiplexer output is synchronous. Of course,

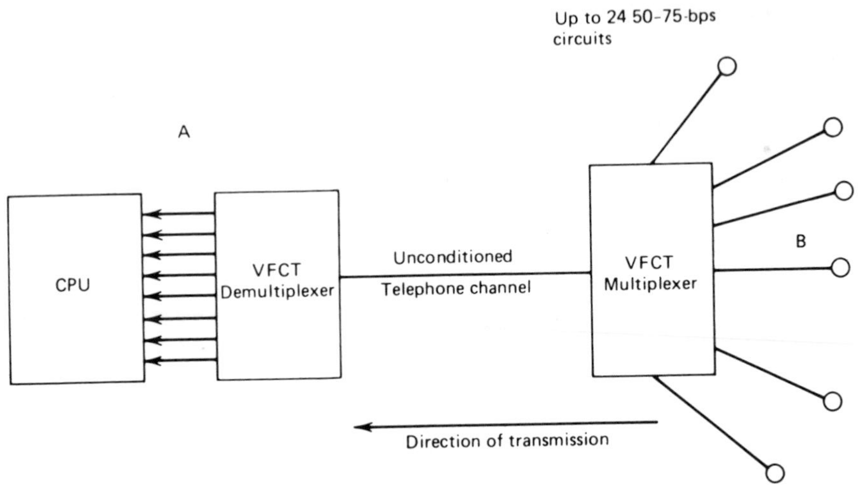

Figure 11.11 Frequency sharing of a voice grade circuit using VFCT techniques.

the requirements for equalization and/or conditioning apply as for any other higher bit rate (e.g., 4800 bps, 9600 bps) service.

Still another technique is the use of the statistical multiplexer or "stat-mux." In this case a greater number of input channels can be accommodated if those input channels have short bursty data traffic. Advantage is taken by the multiplexer of the idle periods between message bursts to accommodate the additional channels. As one can imagine, statistical multiplexers require a fair amount of message buffering and some additional overhead to indicate to the distant-end demultiplexer the origin (input channel port) of all data messages so that the data will be delivered to the proper output port corresponding to the near-end multiplexer input port.

DDS (digital data system) is an ATT offering on the digital (PCM) public switched network providing customers with synchronous data rates of 2400 bps, 4800 bps, 9600 bps, and 56 kbps with 99.9% channel availability [4] and 99.5% error-free seconds (EFS).

Another option shown in Table 11.2 is to lease an entire DS1 offering, which will provide a user with 24 full-duplex 56-kbps data channels or any mix of voice and data with an aggregate of 24 channels. In this case each 56-kbps data channel can be broken down into 2400-bps, 4800-bps, and 9600-bps time-division subchannels.

ISDN will offer 64-kbps clear channels (see Chapter 13). In North America with the large DS1 infrastructure, 64-kbps clear channel operation appears to be a far-off goal. These countries standardizing on the CEPT 30 + 2 PCM architecture will indeed have 64 kbps. Meanwhile North American ISDN will have to settle for 56-kbps clear channel ISDN. The interface between 56 kbps

Mean frequency (Hz)	420	540	660	780	900	1020	1140	1260	1380	1500	1620	1740	1860	1980	2100	2220	2340	2460	2580	2700	2820	2940	3060	3180	
Channel No.	001 101	002 102	003 103	004 104	005 105	006 106	007 107	008 108	009 109	010 110	011 111	012 112	013 113	014 114	015 115	016 116	017 117	018 118	019 119	020 120	021 121	022 122	023 123	024 124	In accordance with Recommendation R.31 } 50 bauds/ Recommendation R.35 } 120 Hz

Mean frequency (Hz)	480	720	960	1200	1440	1680	1920	2160	2400	2640	2880	3120	
Channel No.	201	202	203	204	205	206	207	208	209	210	211	212	Recommendation R.37 50 bauds } 240 Hz 100 bauds }

Mean frequency (Hz)	600	1080	1560	2040	2520	3000	
Channel No.	401	402	403	404	405	406	Recommendation R.38 A 200 bauds/480 Hz

Mean frequency (Hz)	540	900	1260	1620	1980	2340	2700	3060	
Channel No.	301	302	303	304	305	306	307	308	Recommendation R.38 B 200 bauds/360 Hz

Mean frequency (Hz)	420	540	660	780	900	1020	1140	1260	1560	2040	2340	2460	2640	2880	3120	
Channel No.	101	102	103	104	105	106	107	108	403	404	117	118	210	211	212	One example of the application of Recommendation R.36 2 channels-200 bauds/480 Hz 3 channels-100 bauds/240 Hz 10 channels-50 bauds/120 Hz

Figure 11.12 Five possible configurations of VFCT frequency/bit rate configurations. From CCITT Rec. R.70 bis. [3].

and 64 kbps for international operation is a consideration for the design engineer that should not be taken lightly.

7 DATA NETWORK OPERATION

7.1 Introduction

When more than one remote data terminal is required to input data to a CPU, the connecting data link(s) may be point to point and/or multipoint. Of course, if there were only two or three terminals sharing a single line, as in the case of multiport, operation could simply be on a "first come, first served" basis. This is known as *contention*, where terminals compete for access to a line. There is no discipline.

All data networks, big and small, require certain rules of operation or discipline. There is nothing wrong with contention when there is a small group of accesses. Rules or discipline is laid out in what is called a *protocol*. Included in these "rules" are interfaces. A good example is the source code used. We cannot expect users using the ASCII code to communicate with users with the EBCDIC code. Another rule is the convention of *sense* (see Chapter 8, Section 3), which establishes the electrical conditions of the 1 and 0, the mark, and space.

Protocols, as we know them today, extend across a whole gamut of rules and interfaces. We list several interface characteristics:

1. Synchronous or asynchronous transmission. If asynchronous, it will have a stop element characteristic. If synchronous, it will use a common sync pattern.
2. Transmission speed or data rate in bits per second and speed tolerance (e.g., 4800 bps $\pm 0.1\%$).
3. Modulation and waveform (e.g., FSK, NRZ)
4. Electrical interface, such as EIA RS-232D and CCITT Rec. V.24.
5. Error recovery scheme.

The error recovery scheme will probably utilize some form of ARQ. This subject was introduced in Chapter 8, Section 5.6. There are three common forms of ARQ. These are stop-and-wait ARQ, go-back-*n* ARQ and selective (continuous) ARQ. Stop-and-wait ARQ is the simplest. A transmitter sends a data block with the BCC (FCS)* appended (see Chapter 8, Section 5.4). The receiver stores the block, processes it for errors, and if it finds its calculated value for the BCC the same as the transmitted value, it sends an ACK (acknowledge) message to the transmitter. The transmitter then starts to send the second block of traffic (if any). With stop-and-wait ARQ the message

*FCS = frame check sequence.

accounting is simple and only one comparatively small data buffer is required. Of course, the stopping and waiting is inefficient. The lost time or dead time on the circuit can be expensive.

Go-back-*n* ARQ improves the efficiency somewhat. It assumes that all blocks are transmitted error free. The receiver continues to receive data blocks until one is detected in error. It continues to receive sequential blocks and at the same time initiates a message to the transmitter on its return circuit telling the transmitter to stop sending and "go-back-*n*" or go back *n* blocks and repeat *all* blocks from *n* forward.

Selective ARQ is yet more efficient. It is similar to go-back-*n* ARQ but asks for a repeat only of an identified block that is in error. Blocks or data frames are identified with "sequence numbers." There is usually a maximum number of blocks that can be sent without some form of acknowledgment from the receiver. Error recovery, including sequence numbering, is an important element in a data protocol.

7.2 Polling: An Operational Routine

Polling is an operational procedure mainly used on multipoint lines to invoke some sort of discipline. Ring or loop networks may also use polling. As we are aware, on a multipoint line at any one instant communication can be established with only one remote station. The order of use of the common line or who gets to use it and when is established in a polling procedure.

There are two types of polling that we cover here: roll-call polling and loop polling. With roll-call polling a control station, usually a CPU, queries each station on the multipoint line in a prearranged order of sequence. On receipt of a query signal directed to it, each station replies with a "no traffic" signal. If traffic is on hand to send at a particular station, the polling stops and the message is transmitted. Polling resumes after message receipt. It will be appreciated that data message traffic can pass either way, from a particular workstation to CPU or vice versa. The polling sequence can be varied as well. If we know that station B has twice the traffic as station A, then station B can be polled twice as often as station A. Likewise, there can be dynamic changes in the polling list, such as varying the list repetition of stations with the time of day.

Loop polling is a method of handing over the polling inquiry. Station A is polled, sends its traffic, and then hands over the polling inquiry to Station B, which goes through the same process, handing over the polling inquiry to Station C, so forth. In this case a strict, invariable sequence is followed. However, there is considerably less overhead involved in call and response. Contention and polling are also described in Chapter 12, Section 5.2.

There is also a procedure known as "selection," which is used with polling. The more common type is called "select hold," where the master or control station follows the roll call. Each station is double queried. First it is asked about its ability to receive, and if the answer is affirmative, the station is asked to pass traffic or traffic is passed to it. The second method is called "fast select," where traffic operations start without any prior checkout. The second method achieves more operating time on the circuit but may require more rigorous error control.

7.3 Introduction to Data Network Access and Control Procedures

The term "communications protocol" defines a set of procedures by which communication is accomplished within standard constraints. As shown in the following list, which outlines protocol topics, protocol deals with control functions.

1. Framing: frame makeup (format); block, message, or packet makeup.
2. Error control (note that this is an interface characteristic as well).
3. Sequence control: the numbering of messages (or blocks) to eliminate duplication, maintain proper sequence in the case of packet networks, and to maintain a proper record of identification of messages, especially in ARQ systems or for message servicing.
4. Transparency of the communication links, link control equipment, multiplexers, concentrators, modems, and so on. Transparency allows the use of any bit pattern the user wishes to transmit, even though these patterns resemble control characters or prohibited bit sequences such as long series of 1's or 0's.
5. Line control: determination, in the case of a half-duplex or multipoint line, of which station is going to transmit and which station is going to receive.
6. Idle patterns to maintain network synchronization.
7. Time-out control: which procedures to follow if message (or block or packet) flow ceases entirely.
8. Startup control: getting a network into operation initially or after some period of remaining idle for one reason or another.
9. Sign-off control: under normal conditions, the process of ending communication or transaction before starting the next transaction or message exchange.

At this juncture we distinguish data link control from user device control. The data link is defined as the configuration of equipment enabling end terminals in two different stations to communicate directly. The data link includes the paired DTEs, modems, or other signal converters and the interconnecting facilities. The user device may be a CPU, workstation, or other data peripheral. Pictorially we can illustrate the difference between data link control and user device control by the following diagram:

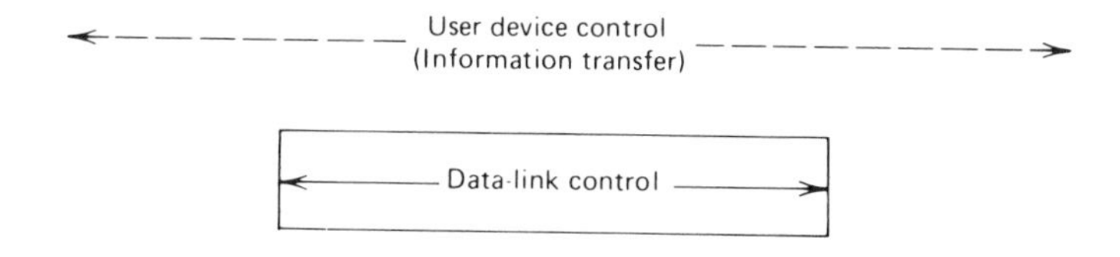

The current literature distinguishes *circuit connection* from *link connection*. Circuit connection is simply the establishment of an electrical path between two points (or multipoints) that want to communicate. We know from our previous discussions that the connection may be metallic (i.e., wire or cable), fiber optics, and/or radio in either the frequency or time domain (e.g., time slot in a frame). The mere establishment of an electrical connection does not mean that data communication can take place. Link establishment is a group of procedures that prepares the source to send data and the destination to receive that data.

In conventional telephony there are three distinct phases to a telephone call: (1) call setup, (2) information transfer, where the subscribers at each end of the connection carry on their conversation, and (3) call termination. A data link must essentially go through the same three procedures. The user control is analogous to the information transfer portion of the telephone call. Thus we introduce protocols.

7.4 Protocols

7.4.1 Basic Protocol Functions. Stallings [5] lists some basic protocol functions. Typical among these functions are:

- Segmentation and reassembly.
- Encapsulation.
- Connection control.
- Ordered delivery.
- Flow control.
- Error control.

In many respects these are really a restatement of the protocol topics listed in the previous section. A short description of each function is given below.

Segmentation and Reassembly. Segmentation refers to breaking up the data into blocks with some bounded size. Depending on the semantics or system, these blocks may be called frames or packets. Reassembly is the counterpart of segmentation, that is, putting the blocks or packets back into their original order. Another name used for a data block is *protocol data unit* (PDU).

Encapsulation. Encapsulation is the adding of control information on either side of the data *text* of a block. Typical control information is the *header*, which contains address information, sequence numbers, and error control.

Connection Control. There are three stages of connection control:

1. Connection establishment.
2. Data transfer.
3. Connection termination.

Some of the more sophisticated protocols also provide connection interrupt and recovery capabilities to cope with errors and other sorts of interruptions.

Ordered Delivery. PDUs are assigned sequence numbers to ensure an ordered delivery of the data at the destination. In a large network, especially if it operates in the packet mode, PDUs (packets) can arrive at the destination out of order. With a unique PDU numbering plan using a simple numbering sequence, it is a rather simple task for a long data file to be reassembled at the destination in its original order.

Flow Control. Flow control refers to the management of data flow from source to destination such that buffers do not overflow but maintain full capacity of all facility components involved in the data transfer. Flow control must operate at several peer layers of a protocol, as will be discussed later.

Error Control. Error control is a technique that permits recovery of lost or errored PDUs. There are three possible functions involved in error control:

1. Acknowledgment of each PDU or string of PDUs.
2. Sequence numbering of PDUs (e.g., missing numbers).
3. Error detection (see Chapter 8).

Acknowledgment may be carried out by returning to the source the source sequence number of a PDU. This ensures delivery of all PDUs to the destination. Error detection initiates retransmission of errored PDUs.

7.4.2 Open System Interconnection

7.4.2.1 Rationale. Interfacing data systems can be a complex matter. This is particularly true when dealing with CPUs and workstations from different vendors as well as variations in software and operating systems. To accom-

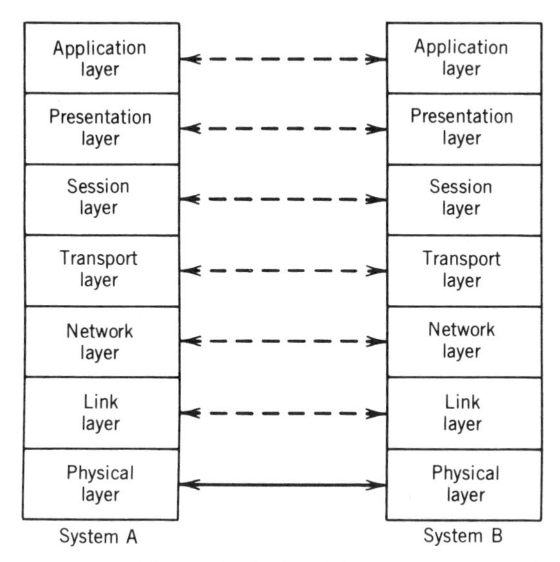

Figure 11.13 The OSI model.

modate the multitude of processing-related equipment and software, the International Standards Organization (ISO) developed the open systems interconnection (OSI) seven-layer model. This means that there are seven layers of interface starting at the communication input–output ports of a data device. These seven layers are shown in Figure 11.13.

The purpose of the model is to facilitate communication among data entities. It takes at least two to communicate. Thus we consider the model in twos, one entity at the left of the figure and one on the right. We use the term *peers*. Peers are corresponding entities on each side of Figure 11.13. A peer on one side (system A) communicates with its peer on the other side (system B) by means of a common protocol. For example, the transport layer system A communicates with its peer transport layer at system B. It is important to note that there is no *direct* communication between peer layers except at the physical layer (layer 1). That is, above the physical layer, each protocol entity sends data *down* to the next lower layer to get the data *across* to its peer entity on the other side. Even the physical layer need not be directly connected as in packet communications. Peer layers must share a common protocol to interface.

There are seven OSI layers, as shown in Figure 11.13. Any layer may be referred to an N layer. Within a particular system there are one or more active entities in each layer. An example of an entity is a process in a multiprocessing system. It could simply be a subroutine. Each entity communicates with entities above and below it across an interface. The interface is at a service access point (SAP). An $(N - 1)$ entity provides services to an N entity by use of primitives. A primitive [5] specifies the function to be performed and is used to pass data and control information.

CCITT Rec. X.210 [6] describes four types of primitive used to define the interaction between adjacent layers of the OSI architecture. A brief description of each of these primitives is given below.

Request. A primitive issued by a service user to invoke some procedure and to pass parameters needed to fully specify the service.

Indication. A primitive issued by a service provider either to invoke some procedure or to indicate that a procedure has been invoked by a service user at the peer service access point.

Response. A primitive issued by a service user to complete at a particular SAP some procedure invoked by an *indication* at that SAP.

Confirm. A primitive issued by a service provider to complete at a particular SAP some procedure previously invoked by a request at that SAP. CCITT Rec. X.210 [6] adds this note: Confirms and responses can be positive or negative depending on the circumstances.

The data that passes between entities is a bit grouping called a *data unit*. We discussed protocol data units (PDUs) earlier. Data units are passed downward from a peer entity to the next OSI layer, called the $(N - 1)$ layer. The lower layer calls the PDU a *service data unit* (SDU). The $(N - 1)$ layer adds control information, transforming the SDU into one or more PDUs. However, the identity of the SDU is preserved to the corresponding layer at the other end of the connection. This concept is shown in Figure 11.14.

When we discussed throughput in Section 6, it became apparent that throughput must be viewed from the eyes of the user. With OSI some form of encapsulation takes place at every layer above the physical layer. To a greater or lesser extent OSI is used on every and all data connectivities. The concept

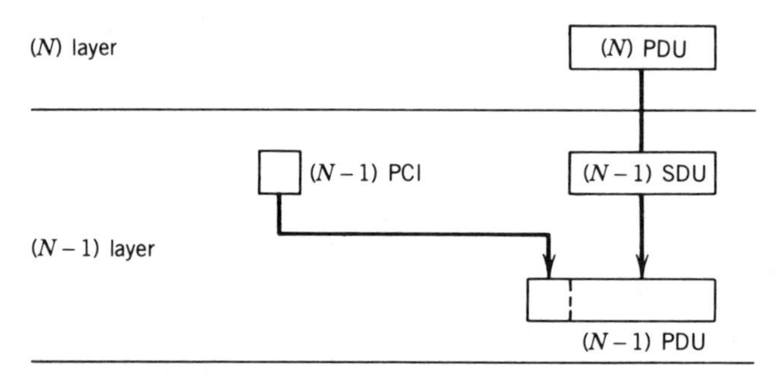

Figure 11.14 An illustration of mapping between data units in adjacent layers.

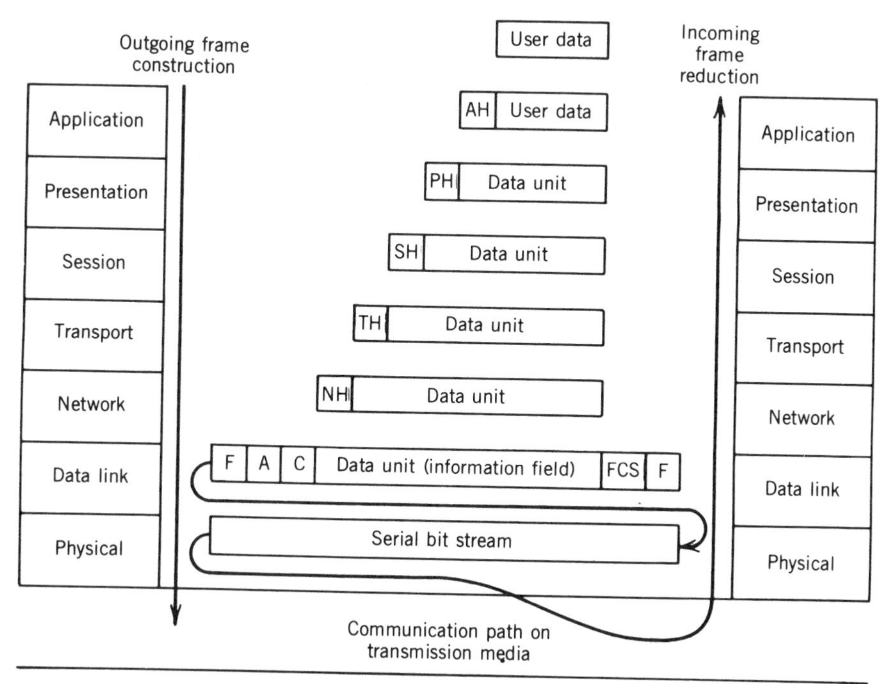

Figure 11.15 Buildup and breakdown of a data message following the OSI model. OSI encapsulates at every layer adding significant overhead.

of encapsulation, the adding of overhead, from layers 2 through 7 is shown in Figure 11.15.

7.4.2.2 Functions of OSI Layers

PHYSICAL LAYER. The physical layer is layer 1 and the lowest OSI layer. It provides the physical connectivity between two data end-users who wish to communicate. The services it provides to the data link layer are those required to connect, maintain, and disconnect the physical circuits that form the physical connectivity. The physical layer represents the traditional interface between data terminal equipment (DTE) and data communications equipment (DCE). See Chapter 8, Section 8.

The physical layer has four important characteristics:

1. Mechanical.
2. Electrical.
3. Functional
4. Procedural.

The mechanical aspects include the actual cabling and connectors necessary to connect the communications equipment to the media. Electrical characteristics cover voltage and impedance, balanced and unbalanced. Functional characteristics include connector pin assignments at the interface and the precise meaning and interpretation of the various interface signals and data set controls. Procedures cover sequencing rules that govern the control functions necessary to provide higher-layer services such as establishing a connectivity across a switched network.

Some applicable standards for the physical layer are:

- EIA RS-232D, RS-449, RS-422, and RS-423.
- CCITT Recs. V.10, V.11, V.24, V.28, X.20, X.21, and X.21 bits.
- ISO 2110, 2593, 4902, and 4903.
- US Fed. Stds. 1020A, 1030A, and 1031.
- US MIL-STD-188-114B.

DATA LINK LAYER. The data link layer provides services for reliable interchange of data across a data link established by the physical layer. Link layer protocols manage the establishment, maintenance, and release of data link connections. These protocols control the flow of data and supervise error recovery. A most important function of this layer is recovery from abnormal conditions. The data link layer services the network layer or logical link control (LLC; in the case of LANs) and inserts a data unit into the INFO portion of the data frame or block. A generic data frame generated by the link layer is shown in Figure 11.16.

Some of the more common data link layer protocols are:

- ISO HDLC, ISO 3309, 4375.
- CCITT LAP-B and LAP-D.
- IBM BSC, SDLC.
- DEC DDCMP.
- ANSI ADCCP (also US government).

NETWORK LAYER. The network layer moves data through the network. At relay and switching nodes along the traffic route, layering concatenates. In other words, the higher layers (above layer 3) are not required and are utilized only at user end points.

Flag	Address	Control	Information	FCS	Flag

Figure 11.16 Generalized data link layer frame.

The network layer carries out the functions of switching and routing, sequencing, logical channel control, flow control, and error recovery functions. We note the duplication of error recovery in the data link layer. However, in the network layer error recovery is network-wide, whereas on the data link layer error recovery is concerned only with the data link involved.

The network layer also provides and manages logical channel connections between points in a network such as virtual circuits across the public switched network (PSN). It will be appreciated that the network layer concerns itself with the network switching and routing function. On simpler data connectivities, where a large network is not involved, the network layer is not required and can be eliminated. Typical of such connectivities are point-to-point circuits, multipoint circuits, and LANS. A packet-switched network is a typical example where the network layer is required.

The best known standard for layer 3 is the CCITT Rec. X.25 layer 3 standard for packet operation. CCITT Rec. X.21 provides a standard for network layer functions for circuit-switched operation [7, 15].

LAYER 3.5: INTERNETWORK PROTOCOLS. Layer 3.5 is a sublayer of the OSI network layer and carries out the functions of interfacing two disparate networks. The sublayering is shown in Figure 11.17. This internet function is carried out by *gateways*.

There are two applicable protocols for internetworking: CCITT Rec. X.75 [8] and IP (internet protocol) [9]. This latter protocol was initially developed by the US Department of Defense Advanced Research Projects Agency (DARPA) and is now standardized by the Defense Department. A simpler standard has been developed by the ISO (ISO 8473).

The CCITT Rec. X.75 protocol is a subset of Rec. X.25 and assumes that all networks involved are based on the latter protocol. The IP provides datagram service; virtual-circuit service and fixed routing are typical of the

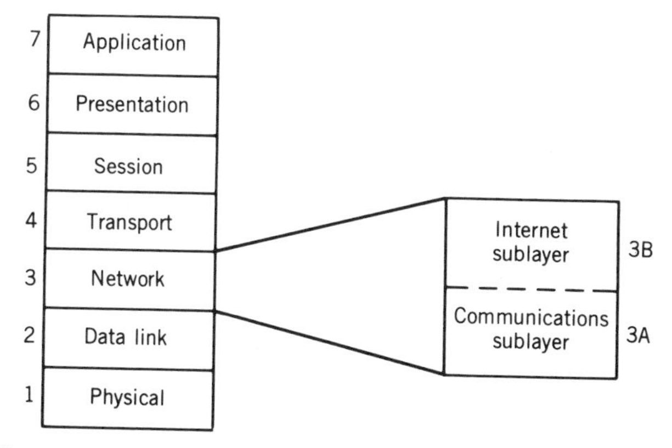

Figure 11.17 Sublayering of OSI layer 3 to achieve internetting.

Rec. X.75 protocol. The IP has broader application, and an IP gateway must know the access scheme of each network with which it interconnects.

TRANSPORT LAYER. The transport layer (layer 4) is the highest layer of the services associated with the provider of communication services. One can say that layers 1–4 are the responsibility of the communication system engineer. Layers 5, 6, and 7 are the responsibility of the data end-user. However, we believe that the telecommunication system engineer should have a working knowledge of all seven layers.

The transport layer has the ultimate responsibility for providing a reliable end-to-end data delivery service for higher-layer users. It is defined as an end system function, located in the equipment using network service or services. In this way its operations are independent of the characteristics of all the networks that are involved. Services that a transport layer provide are:

- *Connection Management*. This includes establishing and terminating connections between transport users. It identifies each connection and negotiates values of all needed parameters.
- *Data Transfer*. This involves the reliable delivery of transparent data between the users. All data is delivered in sequence with no duplication or missing parts.
- *Flow Control*. This is provided on a connection basis to ensure that data is not delivered at a rate faster than the user's resources can accommodate.

The TCP (transport control protocol) was the first working version of a transport protocol and was created by DARPA for DARPANET. All the features in TCP have been adopted in the ISO version. TCP is often lumped with the internet protocol and referred to as TCP/IP.

The ISO transport protocol messages are called transport protocol data units (TPDUs). There are connection management TPDUs and data transfer TPDUs. The applicable ISO references are ISO 8073 OSI (*Transport Protocol Specification*) and ISO 8072 OSI (*Transport Service Definition*).

SESSION LAYER. The purpose of the session layer is to provide the means for cooperating presentation entities to organize and synchronize their dialogue and to manage the data exchange. The session protocol implements the services that are required for users of the session layer. It provides the following services for users:

1. The establishment of session connection with negotiation of connection parameters between users.
2. The orderly release of connection when traffic exchanges are completed.

3. Dialogue control to manage the exchange of session user data.
4. A means to define activities between users in a way that is transparent to the session layer.
5. Mechanisms to establish synchronization points in the dialogue and, in case of error, resume from a specified point.
6. Interrupt a dialogue and resume it later at a specified point; possibly on a different session connection.

Session protocol messages are called session protocol data units (SPDUs). The session protocol uses the transport layer services to carry out its function. A session connection is assigned to a transport connection. A transport connection can be reused for another session connection if desired. Transport connections have a maximum TPDU size. The SPDU cannot exceed this size. More than one SPDU can be placed on a TPDU for transmission to the remote session layer.

Reference standards for the session layer are ISO 8327 [*Session Protocol Definition* (CCITT Rec. X.225)] and ISO 8326 [*Session Services Definition* (CCITT Rec. X.215)].

PRESENTATION LAYER. The presentation layer services are concerned with data transformation, data formatting, and data syntax. These functions are required to adapt the information handling characteristics of one application process to those of another application process.

The presentation layer services allow an application to interpret properly the data being transferred. For example, there are often three syntactic versions of the information to be exchanged between end-users A and B as follows:

- Syntax used by the originating application entity A.
- Syntax used by the receiving application entity B.
- Syntax used between presentation entities. This is called the transfer syntax [5].

Of course, it is possible that all three or any two of these may be identical. The presentation layer is responsible for translating the representation of information between the transfer syntax and each of the other two syntaxes as required.

The following standards apply to the presentation layer:

- ISO 8822 Connection-Oriented Presentation Service Definition.
- ISO 8823 Connection-Oriented Presentation Service Specification.

- ISO 8824 Specification of Abstract Syntax Notation One.
- ISO 8824 Specification of Basic Encoding Rules for Abstract Syntax Notation One.
- CCITT Rec. X.409 Message Handling Systems: Presentation Transfer Syntax and Notation.

APPLICATION LAYER. The application layer is the highest layer of the OSI architecture. It provides services to the application processes. It is important to note that the applications do not reside in the application layer. Rather, the layer serves as a window through with the application gains access to the communication services provided by the model.

This highest OSI layer provides to a particular application all services related to communication in such a format that easily interfaces with the user application and is expressed in concrete quantitative terms. These include identifying cooperating peer partners, determining the availability of resources, establishing the authority to communicate, and authenticating the communication. The application layer also establishes requirements for data syntax and is responsible for overall management of the transaction.

Of course, the application itself may be executed by a machine, such as a CPU in the form of a program, or by a human operator at a workstation.

The following standards apply to the application layer:

ISO 8449/3 Definition of Common Application Service Elements

ISO 8650 Specification of Protocols for Common Application Service Elements.

7.4.2.3 OSI Related to Specific Vendor Models. Table 11.3 relates the ISO model to several specific vendor architectures: IBM SNA, DECNET III, and Honeywell. It is interesting to note the terminology variance for essentially the same functions from HDLC and among the various vendors. The trend for the future is for all vendors to adopt eventually the ISO OSI model universally.

7.4.3 Overview of Several Data Link Layer Protocols

7.4.3.1 Basic Functions. Three basic functions are carried out by the link layer: (1) link establishment, (2) information transfer, and (3) termination. With these functions we assume that a physical connection, or its equivalent, has been established.

7.4.3.2 Generic Link Protocols. There are two or possibly three types of data link layer protocols: (1) character oriented, (2) bit oriented, and (3) hybrid protocols.

TABLE 11.3 Some Vendor Architectures Related to OSI

ISO	Honeywell DSA		Univac DCA		Burroughs BNA	IBM SNA	DECNET III	NCR DNA
A	Application		End user		Host services	User level	User	Users
P	Presentation	Message	Termination			Presentation services	Network applic. / Network mgmt.	Telecomm. access
S	Dialog		Common system		Port level	Data flow / Trans. control	Session control	
T	Transport	Comm. Mgmt.	System			Path control	End-to-end comm.	Comm. services
N	Routing		Route cont	Transport control	Router layer		Transport	Route mgmt.
L	Data Link		Data unit		Station control	Data link	Data link	Link control
P	Physical		Terminal Control		Physical interface	Control	Physical link	Circuit control

CHARACTER ORIENTED PROTOCOLS. Earlier link layer protocols were character oriented. Many of these are still in use today. If we examine the ASCII code set (Figure 8.3), we can quickly identify link control characters. Here are some examples from the figure.

SOH (start of header). This is a control character that identifies the beginning of the character sequence that constitutes the data message header.

STX (start of text). A control character that delimits that part of a data message that constitutes the text. The end of text delimiter is the ETX character.

EOT (end of transmission). This control character signifies the end of transmission, which may have contained one or more data messages.

SYN (synchronous idle). This control character is used to establish and maintain character synchronization between the DCE transmitter and the far-end DCE receiver.

The ACK and NACK control characters are used for ARQ operation.

Several character-oriented data link protocols are:

- IBM BSC (BISYNC or binary synchronous communication protocol) is probably one of the most widely used character-oriented protocols developed by a manufacturer. It is used for two-way half-duplex operation over point-to-point and multipoint circuits.
- ANSI X3.28 dates back to 1971. It is based on 10 ASCII control characters.
- ISO IS1745 is internationally accepted and is also based on the 10 ASCII/CCITT No. 5 control characters.

BIT-ORIENTED PROTOCOLS. A bit-oriented protocol uses positionally located control fields rather than code set combinations for supervisory control. Examples are the ANSI/US government ADCCP (advanced data communications control procedures), IBM's SDLC (synchronous data link control), and ISO's HDLC (high level data link control). This latter data link protocol is described in Section 7.4.3.4.

HYBRID PROTOCOLS. There are many possible variations of character-oriented and bit-oriented protocols. Typical is DDCMP (digital data communications message protocol) developed by Digital Equipment Corporation (DEC). It uses control characters and byte-length fields to supervise a link. It is sometimes referred to as a byte-count-oriented protocol.

Byte-count-oriented protocols such as DDCMP achieve transparency by keeping track of character count and transmit this information with each data block. The character (or byte) count is normally transmitted as a positionally located field usually following the SYN characters. The field length is in

SYN	SYN	Class	Count 14 bits	Flag 2 bits	Response 8 bits	Sequence 8 bits	Address 8 bits	BCC-1 16 bits	Text, up to 16,363 8-bit characters	BCC-2 16 bits

Figure 11.18 Basic DDCMP data message format (frame format). *Note*: BCC1 and BCC2 are referred to in DEC literature as CRC1 and CRC2. Both BCCs (FCSs) use CRC-16 error detection.

character increments and indicates the number of characters (bytes) that comprise the block. The receive end of the link then counts characters instead of searching for a control character to determine the location of the check characters and end of block [9, 10].

7.4.3.3 Digital Data Communications Message Protocol. As we mentioned, DDCMP is a data link layer protocol developed by DEC. It is a byte-oriented (hybrid) protocol that can be used on synchronous or asynchronous (start–stop) circuits. It can be used on half-duplex or full-duplex, point-to-point, or multipoint circuits. A DDCMP data message consists of two parts: a header containing control information and the text followed by a 16-bit BCC (FCS). Figure 11.18 illustrates the data message format of DDCMP.

The DDCMP protocol has three classes of messages: SOH, ENQ, and DLE, meaning "data," "control," and "maintenance," respectively. The class of messages is indicated by the sequence SOH, and this sequence appears in the field immediately following the synchronization sequences shown as SYN. After the "class," in Figure 11.18, is the "count," which gives the number of characters that follow for "data" or "maintenance" class messages. The count gives a counting of text characters (bytes). For control messages in lieu of the "count" per se, the 14 bits in that field take on another meaning. In this case the first 8 bits of the 14 bits designate the type of control message. The remaining 6 bits are filler bits, usually all zeros. However, in the case of NAK (negative acknowledgment) messages, these last six bits give the reason for the NAK. For instance:

000001	BCC header error
000010	BCC data error
001000	Receiver overrun
010001	Header format error

The first bit of the 2-bit flag informs the receiver that the message will be followed by sync characters. These are used, of course, to maintain synchronization and to avoid the filling of the receiver buffers until the next block or frame comes on line. The second bit is the select flag, which indicates that the last message in a sequence of messages is the final message from a particular transmitting station. Such a flag is useful on multipoint nets or half-duplex circuits where transmitters need to be turned on or off. The response field indicates the message number of the last message correctly received. This 8-bit field is used in the data class of messages and for ACK and NAK control messages. The sequence field contains information on message sequence numbers. It is useful on certain types of control message and in the data class of message. For message control, it asks the distant end whether it has received all messages correctly up through a certain sequence number. On data messages it contains the message sequence number of the message as assigned by the transmitting end. In multipoint operation the address field is used to identify the station to which the message is directed. For point-to-point operation, by placing a 1 in the first bit position, the receiving end ignores all other information in that field.

With DDCMP, data are transmitted in blocks or data messages that are sequentially numbered. If the receiving end sees no errors, it accepts the message as correct. It sends no ACK message on correct receipt and will continue to receive a string of up to 255 messages. However, if the receiving end finds a message in error, a NAK message is sent to the transmitting end indicating the sequence number of the last "good" message (or block) received. Errors may be found in the header CRC or "out of sequence" or in the data text CRC (BCC). The text of a DDCMP message may be of varying length with a maximum length of 16,363 8-bit characters (bytes) [10].

7.4.3.4 Common Bit-Oriented Protocols

INTRODUCTION. There are two widely used bit-oriented protocols: ISO HDLC and US ANSI/DOD ADCCP. There are virtually no differences between them [5]. LAP-B (CCITT X.25) and IBM's SDLC are subsets of HDLC.

STATIONS AND CONFIGURATIONS. Figure 11.19 shows two station configurations: balanced and unbalanced. Before proceeding with our discussion of bit-oriented protocol, we want to clarify some terminology and concepts. The model protocol for these discussions will be HDLC.

Primary Station. A logical primary station is an entity that has primary link control responsibility. It assumes responsibility for organization of data flow and for link level error recovery. Frames issued by the primary station are called *commands*.

Secondary Station. A logical secondary station operates under control of a primary station. It has no direct responsibility for control of the link but instead responds to primary station control. Frames issued by a secondary station are called *responses*.

Combined Station. A combined station combines the features of primary and secondary stations. It may issue both commands and responses.

Unbalanced Configuration (Figure 11.19a). An unbalanced configuration consists of a primary station and one or more secondary stations. It supports

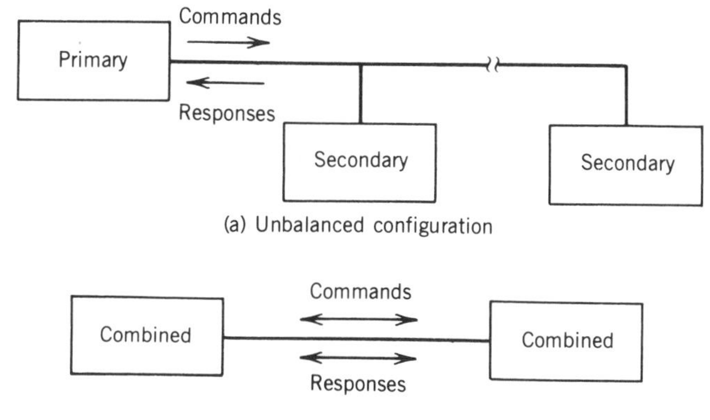

Figure 11.19 Link configuration applicable to bit oriented protocols. HDLC is the model.

both full-duplex and half-duplex operation and point-to-point and multipoint circuits.

Balanced Configuration (Figure 11.19b). A balanced configuration consists of two combined stations in which each station has equal and complementary responsibility of the data link. A balanced configuration operates only in the point-to-point mode and supports full-duplex and half-duplex operation.

MODES OF OPERATION. We describe three of the modes of operation used with HDLC.

With *normal response mode* (*NRM*) a primary station initiates data transfer to a secondary station. A secondary station transmits data only in response to a poll from the primary station. This mode of operation applies to an unbalanced configuration.

With *asynchronous response mode* (*ARM*) a secondary station may initiate transmission without receiving a poll from a primary station. It is useful on a circuit where there is only one active secondary station. The overhead of continuous polling is thus eliminated.

Asynchronous balanced mode (*ABM*) is a balanced mode that provides symmetric data transfer capability between combined stations. Each station operates as if it were a primary station, can initiate data transfer, and is responsible for error recovery. One application of this mode is hub polling, where a secondary station needs to initiate transmission.

THE HDLC FRAME. Figure 11.20 shows the HDLC frame format. Moving from left to right in the figure in the order of transmission, we have the flag field (F), which delimits the frame at both ends with the unique bit pattern 01111110. If

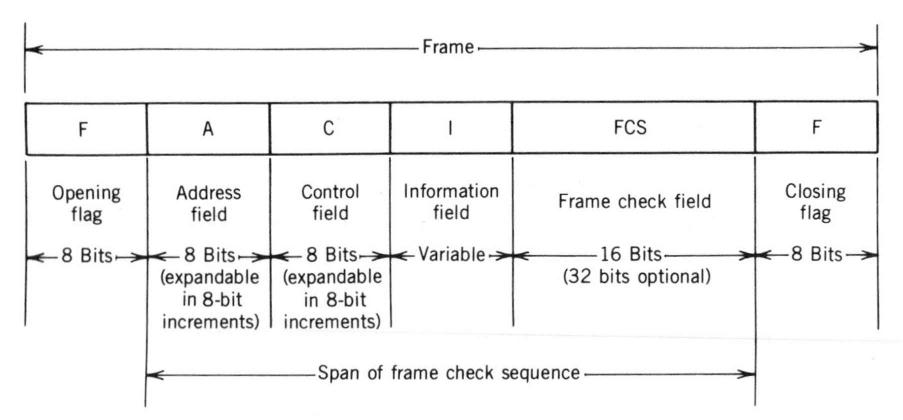

Figure 11.20 The HDLC frame format is typical of a bit-oriented protocol.

frames are sent sequentially, the closing flag of the first frame is the opening flag of the next frame.

We called the flag sequence unique. Receiving stations constantly search for the flag bit sequence to mark the beginning and end of a frame. There is no reason why this same sequence cannot appear somewhere in midframe. This would incorrectly tell a receiving station that the frame has ended and a new frame begun. Of course, this corrupts the whole frame and probably some subsequent frames. To avoid the problem, no sequence of six consecutive 1's bracketed by 0's is permitted. With the exception of the flags, transmitting stations insert a 0 after each occurrence of five consecutive 1's. This is called *zero insertion*.

After detection of an opening flag, a receiver monitors the bit stream for a pattern of five contiguous 1's. The sixth bit is examined. If that bit is a 0, it is deleted. If the sixth bit is a 1 and the seventh bit is a 0, the combination is accepted as a flag. If bits six and seven are both 1's, the transmitting station is assumed to be sending an abort condition [9].

The *address field* (A) immediately follows the opening flag of a frame and precedes the control field (C). Each station in the network normally has an individual address and a group address. A group address identifies a family of stations. It is used when data messages must be accepted from or destined to more than one user. Normally the address is 8 bits long, providing 256 bit combinations or addresses ($2^8 = 256$). In HDLC (and ADCCP) the address field can be extended in increments of 8 bits. When this is implemented, the least significant bit is used as an extension indicator. When that bit is 0, the following octet is an extension of the address field. The address field is terminated when the least significant bit of an octet is 1. Thus we can see that the address field can be extended indefinitely.

The *control field* (C) immediately follows the address field (A) and precedes the information field (I). The control field conveys commands, responses, and sequence numbers to control the data link. The basic control field is 8 bits long

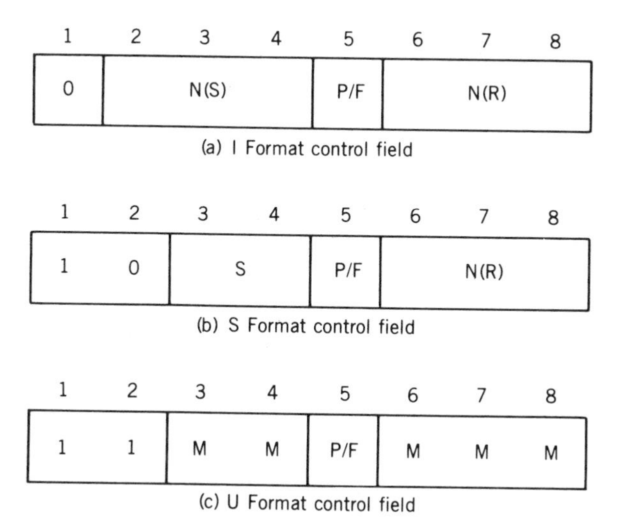

Figure 11.21 The three control field formats of HDLC.

and uses modulo 8 sequence numbering. There are three types of control field: (1) I frame (information frame), (2) S frame (supervisory frame), and (3) U frame (unnumbered frame). The three control field formats are shown in Figure 11.21.

Consider the basic 8-bit format as shown in Figure 11.21. The information flows from left to right. If the frame in Figure 11.20 has a 0 appear as the first bit in the control field, the frame is an I frame (see Figure 11.21a). If the bit is a 1, the frame is an S or a U frame, as shown in Figures 11.21b, c. If that first bit is a 1 followed by a 0, it is an S frame, and if the bit is a 1 followed by a 1, it is a U frame. These bits are called format identifiers.

Turning now to the information (I) frame (Figure 11.21a), its purpose is to carry user data. Bits 2, 3, and 4 of the control field in this case carry the *send* sequence count of transmitted messages (i.e., I frames).

We now digress to describe a *window* of frames. One generally thinks of a receiver on a point-to-point link with one buffer. This works well with stop-and-wait ARQ. Station X sends a message to station Y, which stores the message in a single buffer. On receipt of the entire bit string making up the message, the receiver processes the FCS for errors. If the frame is error free, the message or frame is acknowledged. We should note that stop-and-wait ARQ can be slow, tedious, and inefficient.

Suppose now that receiver Y has seven buffers (the number was picked arbitrarily). Thus Y can accept seven frames (or messages) and X is allowed to send seven frames without acknowledgment. To keep track of which frames have been acknowledged, each is labeled with a sequence number 0–7 (modulo 8). Station Y acknowledges a frame by sending the next sequence number expected. For instance, if Y sends a sequence number 3, this acknowledges frame number 2 and is awaiting frame number 3. Such a scheme can be used to acknowledge multiple frames. As an example, Y could receive frames 2, 3, and 4 and withhold all acknowledgments until frame 4 arrives. By sending sequence number 5, it acknowledges the receipt of frames 2, 3, and 4 all at once. Station X maintains a list of sequence numbers that it is allowed to send, and Y maintains a list of sequence numbers it is prepared to receive. These lists are thought of as a *window of frames*.

HDLC allows a maximum window size of 7, or 127 frames. In other words, a maximum number of 7, or 127 unacknowledged frames, can be sent or one less than the modulus 8 or 128. $N(S)$ is the sequence number of the next frame to be transmitted and $N(R)$ is the sequence number of the frame to be received.

Each frame carries a poll/final (P/F) bit. It is bit 5 in each of the three different types of control fields shown in Figure 11.21. The bit serves a function in both command and response frames. In a command frame it is referred to as a poll (P) bit; in a response frame as a final (F) bit. In both cases the bit is sent as a 1.

The P bit is used to solicit a response or sequence of responses from a secondary or balanced station. On a data link only one frame with a P bit set to 1 can be outstanding at any given time. Before a primary or balanced station can issue another frame with a P bit set to 1, it must receive a response frame from a secondary or balanced station with the F bit set to 1. In the NRM mode, the P bit is set to 1 in command frames to solicit response frames from the secondary station. In this mode of operation the secondary station may not transmit until it receives a command frame with the P bit set to 1.

Of course, the F bit is used to acknowledge an incoming P bit. A station may not send a final frame without prior receipt of a poll frame. As can be seen, P and F bits are exchanged on a one-for-one basis. Thus only one P bit can be outstanding at a time. As a result the $N(R)$ count of a frame containing a P or F bit set to 1 can be used to detect sequence errors. This capability is called *check pointing*. It can be used not only to detect sequence errors but to indicate the frame sequence number to begin retransmission when required.

Supervisory (S) frames, shown in Figure 11.21b, are used for flow and error control. Both go-back-*n* and selective ARQ can be accommodated. There are four types of supervisory or F frames:

1. Receive ready (RR): 1000 P/F $N(R)$.
2. Receive not ready (RNR): 1001 P/F $N(R)$.
3. Reject (Rej): 1010 P/F $N(R)$.
4. Selective reject (SRej): 1011 P/F $N(R)$.

The RR frame is used by a station to indicate that it is ready to receive information and acknowledge frames up to and including $N(R) - 1$. Also a primary station may use the RR frame as a command with the poll (P) bit set to 1.

The RNR frame tells a transmitting station that it is not ready to receive additional incoming I frames. It does acknowledge receipt of frames up to and including sequence number $N(R) - 1$. I frames with sequence number $N(R)$ and subsequent frames, if any, are not acknowledged. The Rej frame is used with go-back-*n* ARQ to request retransmission of I frames with frame sequence number $N(R)$, and $N(R) - 1$ frames and below are acknowledged.

Unnumbered frames are used for a variety of control functions. They do not carry sequence numbers, as the name indicates, and do not alter the flow or sequencing of I frames. Unnumbered frames can be grouped into the following four categories:

1. Mode-setting commands and responses.
2. Information transfer commands and responses.
3. Recovery commands and responses.
4. Miscellaneous commands and responses.

The information field follows the control field (Figure 11.20) and precedes the FCS field. The I field is present only in information (I) frames and some unnumbered (U) frames. The I field may contain any number of bits in any code, related to character structure or not. Its length is not specified in the standard (ISO 3309; Ref. 11). Specific system implementations, however, usually place an upper limit on I field size. Some implementations require that the I field contain an integral number of octets.

Frame check sequence (FCS). Each frame includes a frame check sequence (FCS). The FCS immediately follows the I field, or the C field if there is no I field, and precedes the closing flag (F). The FCS field detects errors due to transmission. The FCS field contains 16 bits, which are the result of a mathematical computation on the digital value of all bits excluding the inserted zeros (zero insertion) in the frame and including the address, control, and information fields.

It should be noted that previously we had called the FCS field the BCC or block check count. To most of us, FCS implies the use of CRC for error detection. BCC, to many of us, can have a wider implication. Namely it may mean only some form of parity check including CRC.

With most HDLC implementations, the FCS field is 16 bits long using the CCITT-recommended CRC (see Chapter 8, Section 5.4.1) (CCITT Rec. V.41, Ref. 13). In some situations that require stringent undetected error rate conditions and/or because of frame length, a 32-bit FCS may be used. This 32-bit CRC is similar to the one used with 802 series LANs; it is described in Chapter 12 [11].

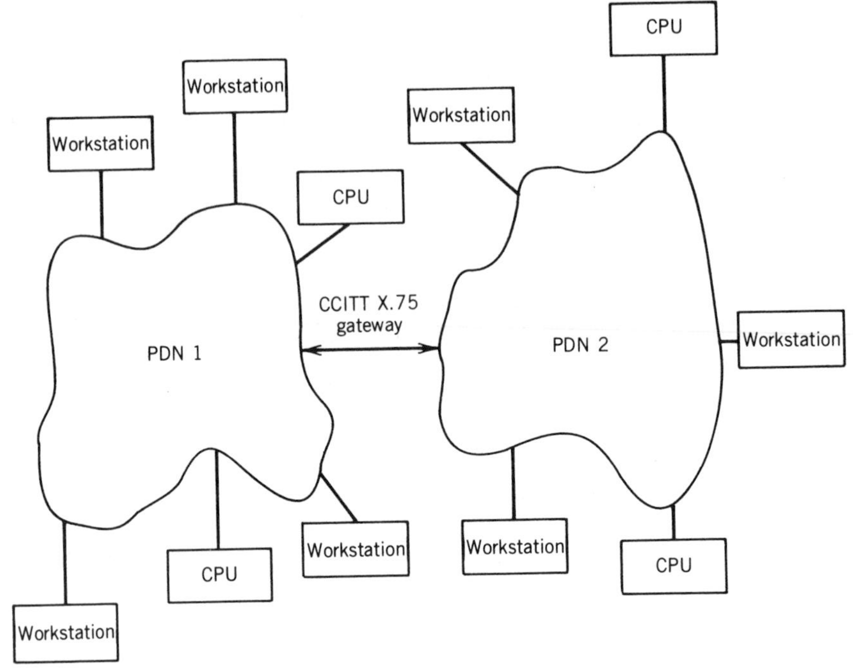

Figure 11.22 Rec. X.25 packet communications operates with the public switched data network (PDN). CPU = central processing unit or host computer.

7.5 X.25: A Packet-Switched Network Access Standard

7.5.1 Introduction to CCITT Rec. X.25. CCITT Rec. X.25 defines the procedures necessary for a packet mode data terminal to access the services provided by a packet-switched public data network (PDN). The original CCITT recommendation was approved in 1976 and subsequently has undergone a number of modifications. There are also several similar but not identical standards developed by the industry that have been modeled after Rec. X.25.

Data terminals defined by Rec. X.25 operate in a synchronous full-duplex mode with a data rate of 2400 bps, 4800 bps, 9600 bps, or 48,000 bps. The structure is generally compatible with the three lowest OSI layers.

Figure 11.22 shows how Rec. X.25 user accesses work with the PDN. At this juncture the reader is encouraged to review Section 5 before proceeding.

7.5.2 Rec. X.25 Architecture and Its Relationship to OSI. Recommendation X.25 spans the lowest three layers of the OSI model. Figure 11.23 illustrates the architecture and its relationship to OSI. It can be seen that Rec. X.25 is similar to and compatible with OSI up to the network layer. In this context there are differences at the network/transport layer boundary. CCITT

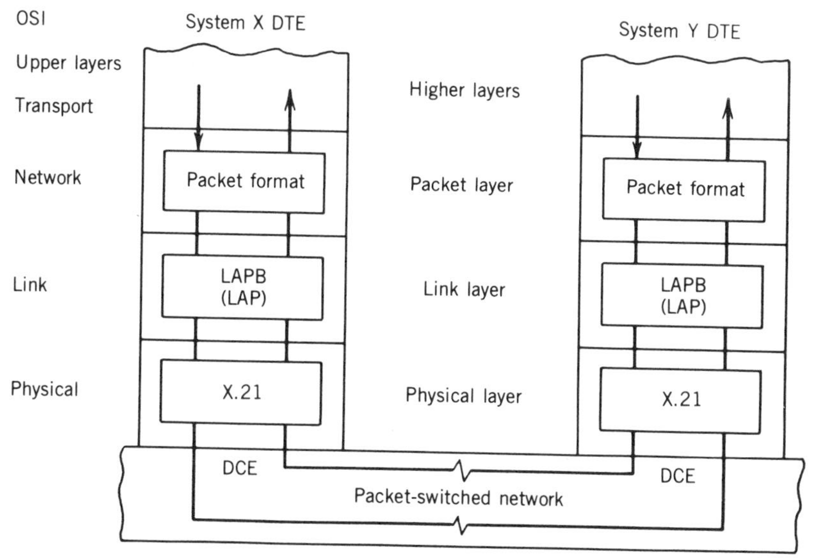

Figure 11.23 CCITT Rec. X.25 architecture and its relationship to OSI.

leans toward the view that the network and transport layer services are identical and that these are provided by the Rec. X.25 virtual circuits.

7.5.2.1 User Terminal Relationship to the PDN. CCITT Rec. X.25 calls the user terminal the DTE, and the DCE resides at the related PDN node. The entire recommendation deals with this DTE–DCE interface, not just the physical layer interface. For instance, a node (DCE) may connect to a related user (DTE) with one digital link, which is covered by SLP (single link procedure) or several links covered by MLP (multilink procedure). Multiple links from a node to a DTE are usually multiplexed on one transmission facility.

The user (DTE) to user (DTE) connectivity through the PDN based on OSI is shown in Figure 11.24. A three-node connection is illustrated in this example. Of course OSI layers 1–3 are Rec. X.25 specific. Note that the DTE protocol peers for these lower three layers are located in the PDN nodes and not in the distant DTE. The first DTE layer operating end to end is layer 4, the transport layer. Further, the Rec. X.25 protocol operates only at the interface between the DTE and its related PDN node and does not govern internodal network procedures.

7.5.3 The Three Layers of CCITT Rec. X.25

7.5.3.1 The Physical Layer. The physical layer is the lowest layer of Rec. X.25. Here the requirements are defined for the functional, mechanical, and

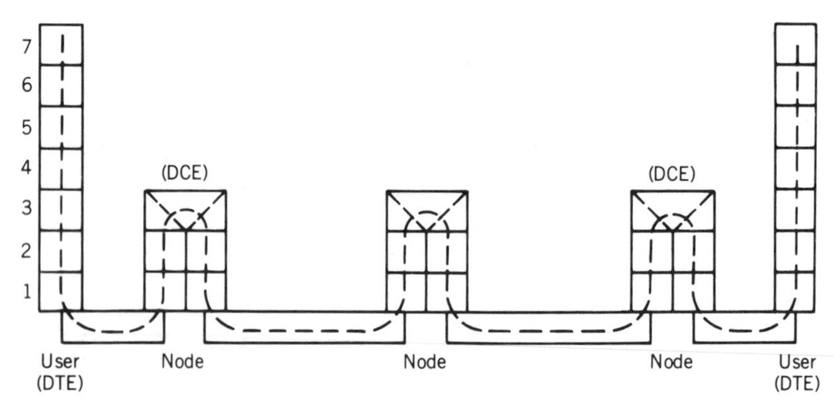

Figure 11.24 CCITT Rec. X.25 user (DTE) connects through the PDN to a distant user (DTE).

electrical interface DTE–DCE or equivalent communication facility. CCITT Rec. X.21 is the applicable standard for the interface, as called out in Rec. X.25. Rec. X.21 bis is allowed as a backward compatible interim. Rec. X.21 bis is similar to EIA RS-232D.

CCITT Rec. X.21 specifies a 15-pin DTE–DCE interface connector (refer ISO 4903). The electrical characteristics for this interface are the same as CCITT Recs. V.10 and V.11, depending on whether electrically balanced or unbalanced operation is desired.

7.5.3.2 CCITT Rec. X.25 Link Layer. The Rec. X.25 link layer specifies LAPB (link access protocol B). Earlier versions of Rec. X.25 permitted the use of LAP, which was based on the ISO ARM protocol. LAPB is preferred, but LAP is still permitted. LAPB is fully compatible with HDLC link layer access protocol, balanced asynchronous class (see Section 7.4.3.4). The information field in the LAPB (HDLC) frame carries the user data, in this case the layer 3 packet.

LAPB provides several options for link operation. These include modulo 8 or modulo 128 control fields. It also supports MLP or multilink procedures. MLP allows a group of links to be managed as a single transmission facility. It carries out the function of resequencing packets in the proper order at the desired destination. When MLP is implemented, an MLP control field of two octets in length is inserted as the first 16 bits of the information field. It contains a multilink sequence number and four control bits. See Figure 11.25.

7.5.3.3 Datagrams, Virtual Circuits, and Logical Connections. There are three approaches used with Rec. X.25 operation to manage the transfer and routing of packet streams: datagrams, virtual connections (VCs), and permanent virtual connections (PVCs). Datagram service uses optimal routing on a packet-by-packet basis, usually over diverse routes. In the virtual circuit approach, there are two operational modes: virtual connection and permanent

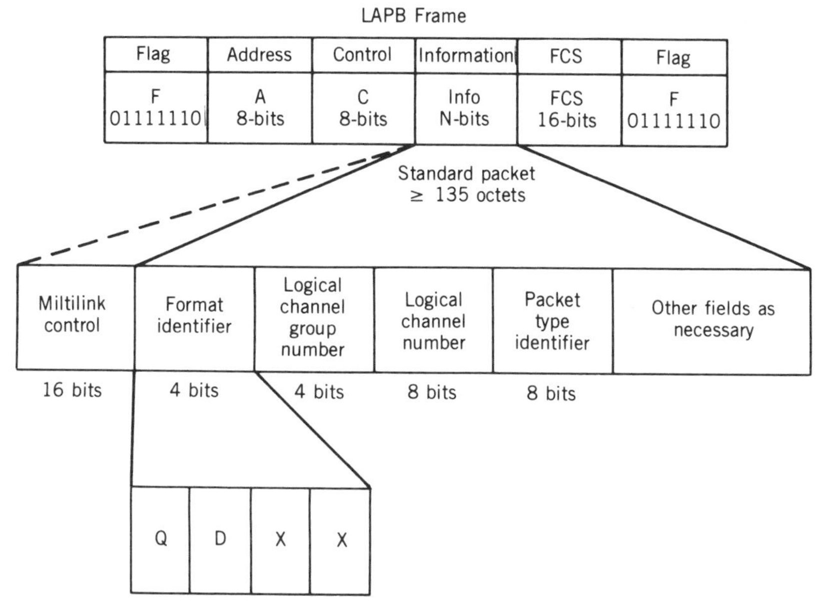

Figure 11.25 Basic Rec. X.25 frame structure. For the extended LAPB structure the control field will have 16 bits. There is also an extended modulo 128 structure for the Rec. X.25 packet. *Note*: For a basic data packet, XX = 01. For an extended (modulo 128) data packet, XX = 10. Q = qualifier bit; D = delivery confirmation bit.

virtual connection. These are analogous to a dial-up telephone connection and a leased line connection, respectively. With the virtual connection a logical connection is established before any packets are sent. The packet originator sends a call request to its serving node, which sets up a route in advance to the desired destination. All packets of a particular message traverse this route, and each packet of the message contains a virtual circuit identifier (logical channel number) and the packet data. At any one time each station can have more than one virtual circuit to any other station and can have virtual circuits to more than one station. With virtual circuits routing decisions are made in advance. With the datagram approach ad hoc decisions are made for each packet at each node. There is no call-setup phase with datagrams; there is with virtual connections. Virtual connections are advantageous for high community-of-interest connectivities, datagram for low community-of-interest relations.

Datagram service is more reliable because traffic can be alternately routed around network congestion points. Virtual circuits are fixed-routed for a particular call. Call-setup time at each node is eliminated on a packet basis with the virtual connection technique. Rec. X.25 also allows the possibility of setting up permanent virtual connections and is network assigned. This latter alternative is economically viable only for very high traffic relations; otherwise these permanently assigned logical channels will have long dormant periods.

7.5.4 Rec. X.25 Frame Structure: Layer 3, Packet Layer. The basic data link layer (LAPB) frame structure is shown in Figure 11.25. Its similarity to the HDLC structure (Figure 11.20) is apparent. For the Rec. X.25 case the packet is embedded in the information field, as mentioned. When applicable, the other part of the information field is the MLP, which is appended in front of the packet and is the first subfield in the information (I) field. Truly the MLP is not part of the actual packet and is governed by the layer-2 LAPB protocol.

7.5.4.1 Structure Common to All Packets. Table 11.4 shows 17 packet types involved in the Rec. X.25 DTE–DCE interface. Every packet transferred

TABLE 11.4 Packet Type Identifier

Packet type		Octet 3							
From DCE to DTE	From DTE to DCE	Bit							
		8	7	6	5	4	3	2	1
	Call setup and clearing								
Incoming call	Call request	0	0	0	0	1	0	1	1
Call connected	Call accepted	0	0	0	0	1	1	1	1
Clear indication	Clear request	0	0	0	1	0	0	1	1
DCE clear confirmation	DTE clear confirmation	0	0	0	1	0	1	1	1
	Data and interrupt								
DCE data	DTE data	X	X	X	X	X	X	X	0
DCE interrupt	DTE interrupt	0	0	1	0	0	0	1	1
DCE interrupt confirmation	DTE interrupt confirmation	0	0	1	0	0	1	1	1
	Flow control and reset								
DCE RR (modulo 8)	DTE RR (modulo 8)	X	X	X	0	0	0	0	1
DCE RR (modulo 128)[a]	DTE RR (modulo 128)[a]	0	0	0	0	0	0	0	1
DCE RNR (modulo 8)	DTE RNR (modulo 8)	X	X	X	0	0	1	0	1
DCE RNR (modulo 128)[a]	DTE RNR (modulo 128)[a]	0	0	0	0	0	1	0	1
	DTE REJ (modulo 8)[a]	X	X	X	0	1	0	0	1
	DTE REJ (modulo 128)[a]	0	0	0	0	1	0	0	1
Reset indication	Reset request	0	0	0	1	1	0	1	1
DCE reset confirmation	DTE reset confirmation	0	0	0	1	1	1	1	1
	Restart								
Restart indication	Restart request	1	1	1	1	1	0	1	1
DCE restart confirmation	DTE restart confirmation	1	1	1	1	1	1	1	1
	Diagnostic								
Diagnostic[a]		1	1	1	1	0	0	0	1
	Registration[a]								
	Registration request	1	1	1	1	0	0	1	1
Registration confirmation		1	1	1	1	0	1	1	1

[a]Not necessarily available on every network.

Note: A bit that is indicated as X may be set to either 0 or 1.

From Table 17 / X.25, Page 17, CCITT Rec. X.25, *Red Books*, Vol. VIII, Fascicle VIII.3.

across the DTE–DCE interface consists of at least three octets. These three octets contain a general format identifier, a logical channel identifier, and a packet type identifier. Other packet fields are appended as required. This is shown in Figure 11.25.

Now consider the general format identifier. The qualifier (Q) bit is not defined in CCITT Rec. X.25 but allows the user to distinguish between two types of data. The D bit, when set to 1, specifies end-to-end delivery confirmation. This confirmation is provided through the packet receive number ($P(R)$). The XX bits, when set to 01, indicate a basic packet (i.e., modulo 8), and when set to 10, an extended modulo 128 packet. The extension involves sequence number lengths.

The logical channel group and the logical channel number subfields identify logical channels with the capability of identifying up to 4096 channels (2^{12}). This permits a DTE to establish up to 4095 simultaneous virtual circuits through its DCE to other DTEs. As we have mentioned, this is usually done by multiplexing these circuits over a single transmission facility.

Permanent virtual circuits have permanently assigned logical channels, whereas those for virtual calls are assigned channels only for the duration of a call. The method of assignment is shown below.

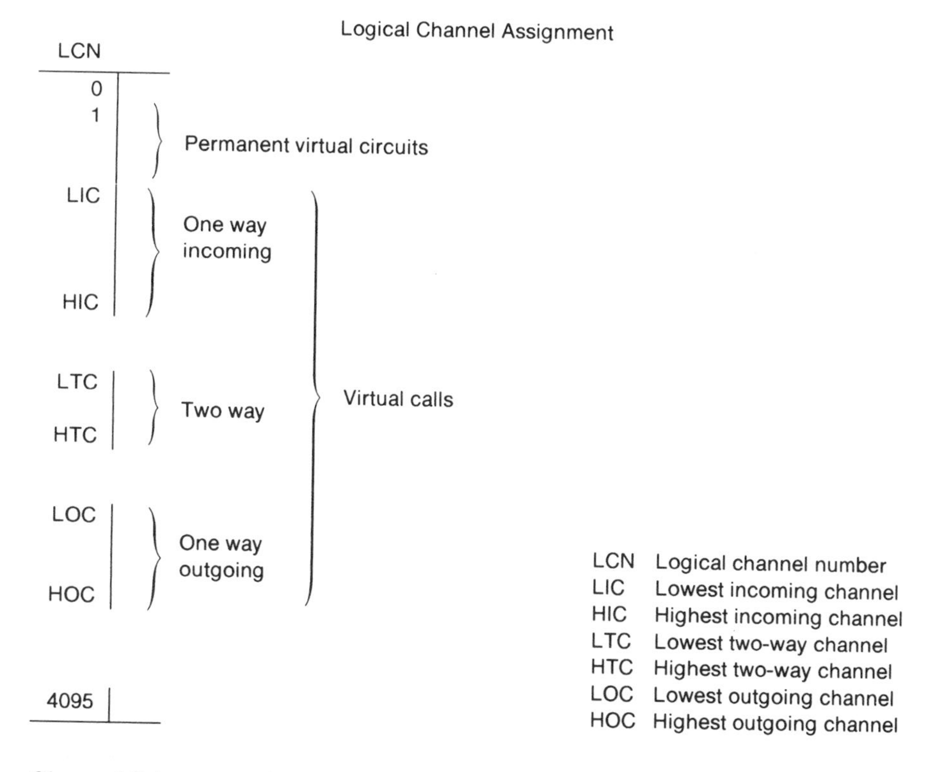

Channel 0 is reserved for restart and diagnostic functions. To avoid collision, the DCE starts assigning logical channel numbers at the low number end and

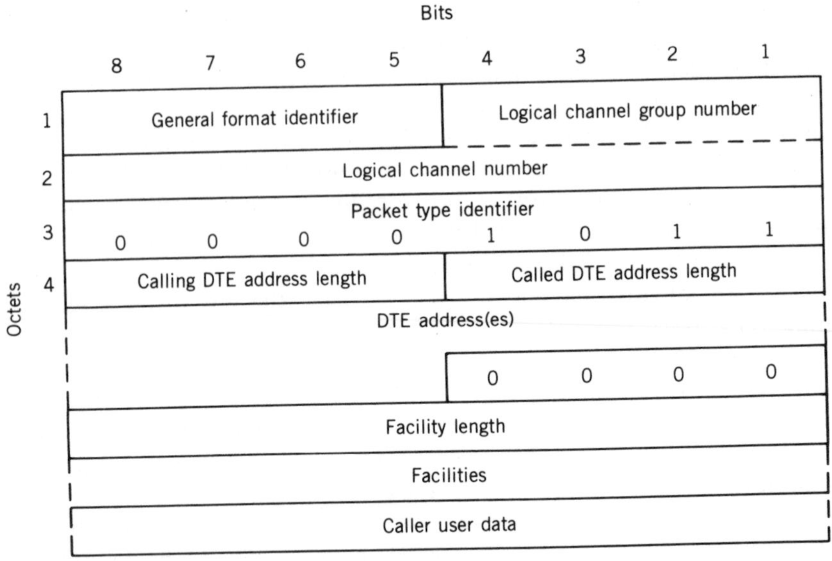

Figure 11.26 Call request and incoming call packet format. The general format identifier is coded 0X01 (modulo 8) or 0X10 (modulo 128). The figure is drawn assuming the total number of address digits present is odd. From Figure 2 / X.25, page 171, CCITT Rec. X.25, *Red Books*, Vol. VIII, Fascicle VIII.3.

the DTE from the highest number downward. There are one-way and both-way (two-way) circuits. The both-way circuits are reserved for overflow because of the chance of double seizure.

Octet 3 in Figure 11.25 is the packet type identifier subfield. The packet type and its corresponding coding is shown in Table 11.4. We can see in the table that packet types are identified in associated pairs carrying the same packet identifier (bit sequence). A packet from the calling terminal (DTE) to the network (DCE) is identified by one name. The associated packet delivered by the network to the called terminal (DTE) is referred to by another associated name.

7.5.4.2 Several Typical Packets

CALL REQUEST AND INCOMING CALL PACKET. This type of packet sets up the call for the virtual circuit. The format of the call request and incoming packet is shown in Figure 11.26. Octets 1–3 have been described in the previous subsection. Octet 4 consists of the address length field indicators for the called and calling DTE addresses. Each address length indicator is binary coded, and bit 1 or 5 is the low-order bit of the indicator. Octet 5 and the following octets consist of the called DTE address, when present, and then the calling DTE address, when present.

The facilities length field (one octet) indicates the length of the facilities field that follows. The facility field is present only when the DTE is using an optional user facility requiring some indication in the call request and incoming call packets. The field must contain an integral number of octets with a maximum length of 109 octets.

Optional user facilities are listed in CCITT Rec. X.2. There are 45 listed. Several examples are listed to give some idea of what is meant by *facilities* in CCITT Rec. X.25:

- Nonstandard default window.
- Flow control parameter negotiation.
- Throughput class negotiation.
- Incoming calls barred.
- Outgoing calls barred.
- Closed user group (CUG).
- Reverse charging acceptance.
- Fast select.

DTE AND DCE DATA PACKET. Of the 17 packets types listed in Table 11.4, only one truly carries user information, the DTE and DCE data packet. Figure 11.27 illustrates this packet format.

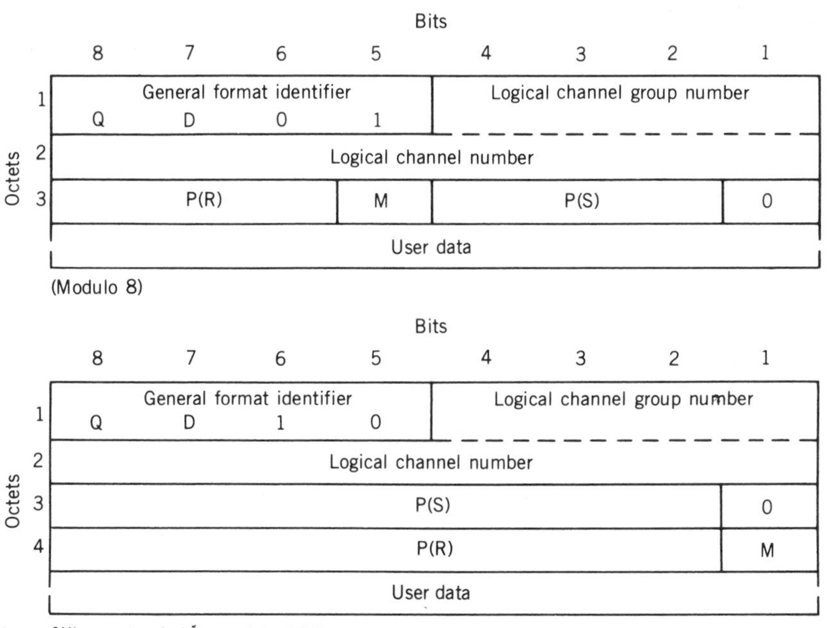

Figure 11.27 DTE and DCE data packet format. From CCITT Rec. X.25. D = delivery confirmation bit; M = more data bit; Q = qualifier bit. From Figure 6/X.25, Page 179, CCITT Rec. X.25, *Red Books*, Vol. VIII, Fascicle VIII.3.

Octets 1 and 2 have been described. Bits 6, 7, and 8 of octet 3 or bits 2–8 of octet 4, when extended, are used for indicating the packet received sequence number $P(R)$. It is binary coded and bit 6, or bit 2 when extended, is the low-order bit.

In Figure 11.27, M, which is bit 5 in octet 3 or bit 1 in octet 4, when extended, is used for more data (M bit). It is coded 0 for "no more data" and 1 for "more data to follow."

Bits 2, 3, and 4 of octet 3, or bits 2–8 of octet 3 when extended, are used for indicating the packet send sequence number $P(S)$. Bits following octet 3, or octet 4 when extended, contain the user data.

The standard maximum user data field length is 128 octets. CCITT Rec. X.25 (par. 4.3.2) states: "In addition, other maximum user data field lengths may be offered by Administrations from the following list: 16, 32, 64, 256, 512, 1024, 2048 and 4096 octets. . . . Negotiation of maximum user data field lengths on a per call basis may be made with the flow parameter negotiation facility."

7.5.5 Tracing the Life of a Virtual Call. A call is initiated by a DTE by the transfer to the network of a call request packet. It identifies the logical channel number selected by the originating DTE, the address of the called DTE (destination), and optional facility information and can contain up to 16 octets of user information. The facility and user fields are optional at the discretion of the source DTE. The receipt by the network of the call request packet initiates the call-setup sequence. This same call request packet is delivered to the destination DTE as an incoming call packet. The destination DTE in return sends a "call accepted" packet, and the source DTE receives a "call confirmation" packet. This completes the call-setup phase, and the data transfer can begin.

Data packets (Figure 11.27) carry the user data to be transferred. There may be one or a sequence of packets transferred during a virtual call. It is the M bit that tells the destination that the next packet is a logical continuation of the previous packet(s). Sequence numbers verify correct packet order and are the packet acknowledgment tools.

The last phase in the life of a virtual call is the call clearing (takedown). Either the DTE or the network can clear a virtual call. The "clear" applies only to the logical channel that was used for that call. Three different packet types are involved in the call-clearing phase. The clear request packet is issued by the DTE initiating the clear. The remote DTE receives it as a clear indication packet. Both the DTE and DCE then issue clear confirmation packets to acknowledge receipt of the clear packets.

Flow control is called out by the following packet types: receive ready (RR), receive not ready (RNR), and reject (Rej). Each of these packets is normally three octets long or four octets long using modulo 128 numbering. Sequence numbering also assists flow control.

8 NETWORK MANAGEMENT

8.1 Definition

Network management means somewhat different things to different people. One can argue that the term is synonymous with technical control. For this discussion, consider the terms the same.

Many conjure up a view of technical control, a military communications term, as banks of patch panels where all circuits of interest can be bridged or terminated for testing. They also can be rerouted in case of poor performance or failure. Network management also includes the traffic flow function and its control, although this function is automated in some of the higher-level protocols and in CCITT Signaling System No. 7.

8.2 Specific Technical Control Functions and Requirements

The investment made in technical control and network management facilities and capabilities relates directly to the importance and value of the network it serves. A well-engineered technical control with skilled technicians can save days of downtime of expensive assets. Many technical control functions are now being automated.

The following lists some specific functions of a tech control facility:

- Line and equipment failure detection.
- Capability to patch around failed lines (reroute) and failed equipment (i.e., patch in spare equipment).
- Capability to reconfigure assets to fulfill communication requirements as they occur.
- Diagnostics and quality measurements.
- Provision of emulation devices, in particular, protocol emulators.
- Loop-back capability with necessary terminations.
- Traffic recording for later analyses.

The tech control facility usually has an equipment and line status panel to determine equipment on line, line usage, spare lines, standby operational equipment, and failed equipment. A further refinement is a computer diagnostics readout capability to determine equipment failure to the card (circuit card assembly, or CCA) level. This may be a simple gathering together of BITE (built-in test equipment) data.

8.3 Main and Subsidiary Tech Control Facilities

The principal technical control facility is usually located at the network's operational hub. Many networks, however, have multiple hubs. In this case each hub should be provided with the appropriate technical control capability.

Subsidiary technical control facilities usually are more modest than principal facilities. These are located at small hubs and at remote facilities. The smaller facilities can be remotely controlled by a data link from the principal facility. Such a data link can be shared or separate. Subsidiary and remote tech control facilities should provide daily or even hourly status messages to the central technical control.

8.4 Basic Quality Measurement Techniques

8.4.1 Line and Equipment Access. The common means of manual access, as mentioned, is via patch panels. One set of jacks is for the equipment side and another for the line side. Still a third set of jacks may be used to access the digital side of the equipment (e.g., the DTE–DCE interface). What we mean here is that a jack splits the line connecting the equipment such as a modem and the line. On the equipment side we can "look back" at the equipment; on the line side we "look" outward to the line. This allows us to patch the equipment into another line and patch the line into other equipment. It also allows access to the line and equipment for testing and monitoring. Ordinarily, when in normal operation, a jack is *normaled-through*, meaning that with no plug in the jack the line is connected to the equipment.

There are often two jacks at each access: terminating and bridging. Terminating jacks are available for terminating the line in test equipment or for rerouting when a plug is inserted in the jack. A short connecting cable with a male plug on each end can be used for patching or rerouting. Test equipment can be plugged into the jack. In this case the test equipment terminates the line or equipment in its characteristic impedance. Bridging places a high impedance across the line, allowing measurements or monitoring without interrupting traffic.

Semiautomatic and automatic access can be made on either side (e.g., line or equipment) by means of crossbar switches. Semiautomatic, in this context, utilizes toggle switch actuation to access a particular line or equipment. A more elegant method is access actuation from the tech control PC-based workstation. Automatic access provides microprocessor-driven periodic measurements on each side of the line, which can give a record of performance data and utilization.

8.4.2 Analog Line Measurements. An audio signal generator with a level meter and appropriate variable attenuator is a most valuable tool. It can be used to verify line continuity to the distant end. When looped back at the distant end, it will give the two-way loss, which can be checked against specified loss.

Let's digress for a moment to explain *loopback* (Figure 11.28). Four-wire connectivity is assumed. Terminating jacks are used on the line side for each line, send and receive. A resistive attenuator is used to connect the receive side to the send side. The attenuator loss is fixed to the expected loss between send and receive, often 23 dB (e.g., +7 dB to −16 dB or 7 + 16). If the transmit

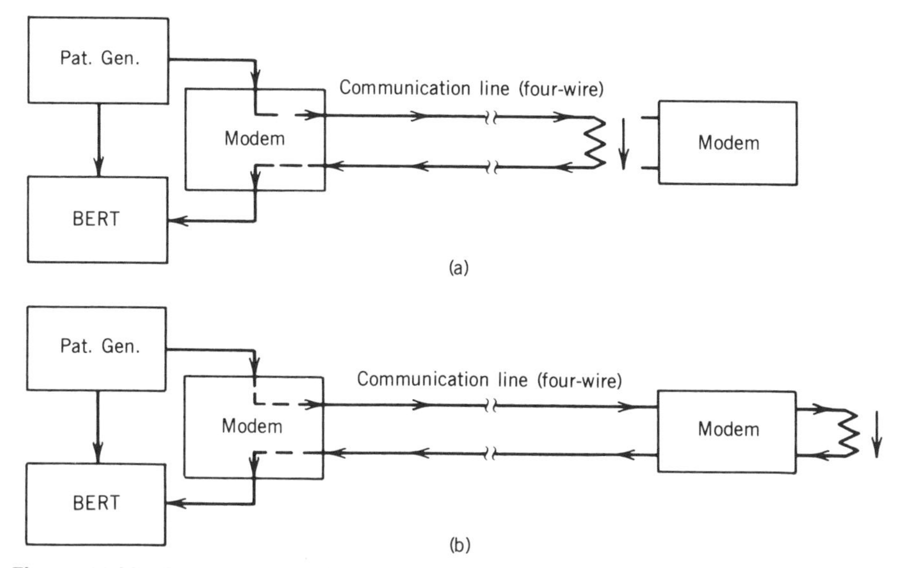

Figure 11.28 Concept of loopback. Example shown is BERT (bit error rate test). (a) Simple loopback: tests line and own modem only. (b) Full loopback: tests line, own modem, and distant modem. Pat. Gen. = pattern generator (P/O BERT set); BERT = bit error rate test set.

side sends a 1004-Hz tone, the same tone is returned on its receive leg. Its level can now be measured. The tone can be removed and the transmit side terminated in its characteristic impedance, and a noise meter can be connected to the receive side to measure circuit noise. This will be the sum of the noise contributions on the two legs, transmit and receive. By similar means a digital test pattern can be transmitted at the appropriate input to the modem (data send), and the BER can be read on the bit error rate test (BERT) set connected on the digital side of the modem (data receive). Figure 11.28 shows two ways of carrying out this vital BER test.

Often a tech control is provided with a precise audio signal source with outputs at 1004 Hz, 1000 Hz, and 800 Hz. A circuit under test is looped back, and the frequency is measured with an accurate counter. The result is the net frequency variation on the "go" and "return" circuits. To pinpoint which leg of the circuit is causing an out-of-tolerance frequency variation, measurement of frequency must be done at the distant end. The frequency should not vary by more than 1 Hz.

Tech control should also be equipped with a special level meter incorporating C-message and psophometric weighting networks, which can provide readings of line noise in dBmp, pWp, and dBrnC. It should also measure flat noise in dBm.

8.4.3 Digital Line Measurements. There are two connotations of *digital.* The first is digital data, usually with a serial NRZ waveform. This requires access between the DCE and DTE (Figure 11.28b). We want to access the digital data bit stream at baseband before it enters the modem or DCE or before it enters the DSU (digital service unit) if DDS is being used.

Figure 11.28 shows the most common digital data performance measurement technique. On the send side we connect a pattern generator at the specified bit rate. The pattern generator can transmit sequences in plain text, often what is called *foxes* or "the quick brown fox jumped over the lazy dog's back 1234567890987654321 times." It should also have the capability of transmitting a pseudorandom sequence pattern. At the distant end the circuit is looped back and a BER tester is connected where we can read the BER. The pattern generator is an integral part of the BERT. Also we would expect the BER to be worse in the loopback condition because errors are being inserted by the transmission media on both the "go" and "return" channels.

A distortion analyzer should also be provided to measure time distortion on a data bit stream. It should also measure peak distortion. An oscilloscope with sufficient rise time to accommodate the bit rates involved is another good test instrument for troubleshooting and eye pattern measurements.

Consider the second connotation of *digital*, namely, PCM such as DS0, DS1, and CEPT 30 + 2. If the technical control is responsible for DS0 or DS1 lines, then additional test equipment is required. Of course, digital patch bays or other semiautomatic access to digital lines and equipment should be available on the digital side of line and equipment. Audio patches should be available for the audio side of the equipment (PCM channel bank). For such installations involved with PCM, it is recommended that a PCM bank analyzer, such as the HP 3779B or the Wandel and Goltermann PCM-3, be acquired.

REVIEW QUESTIONS

1. List similarities and differences between speech telephony and data communications. As a minimum, consider switching, signaling, transmission, network topologies, and service quality.
2. Name at least ten applications of data communications.
3. Give at least seven of the basic considerations required for data network design (e.g., what we should know and quantify before we start the job).
4. Distinguish centralized from distributed processing, and describe how each affects data network design.
5. What impact has the PC had on data processing and data networks?
6. Define a *local area* regarding data communications. Compare it to the local serving area of telephony.

7. Give at least three ways a WAN can be implemented regarding end-user accesses.

8. In the design of a data network we need to know user requirements. List at least seven parameters or items required to quantify or qualify user requirements.

9. One economic expedient to provide data connectivity is to use the dial-up connection. Give at least three shortcomings of this approach. Why should we do it in the first place?

10. In the design of a data network, a first step is to develop a data traffic matrix. List at least five items that must be considered by the designer once traffic flow and intensities have been established.

11. A major element of any data network is the workstation. Give at least eight things we should know about workstations that will be incorporated into the proposed network.

12. Give three ways that workstations can be incorporated into a WAN.

13. The chapter gives six network topologies. Draw and identify all six. Discuss where each would be applied and why.

14. Compare point-to-point and multipoint operation. (Don't forget the economic side.)

15. Why would a hierarchical structure be attractive for a WAN?

16. List and describe the three types of data switching.

17. What is the principal requirement for efficient operation of a packet-switched data network?

18. Differentiate between a *real* and a *virtual* connection.

19. Considering economy and speed, what would be the most efficient means of transmission for data connectivity—numerous short data messages with a total accumulated 300,000 bits per day; several data messages in a day, each about 4000 bits long; or 20 Mbits a day consisting of many medium and long messages.

20. Argue the question: Should the "quality" of a data connectivity be measured in BER or throughput?

21. Define *throughput*. What are some factors that diminish throughput?

22. There are five essential fields in a generic data block, frame, or packet. Identify each and briefly describe their function.

23. TRIB is a function of what six items?

24. Name at least two methods to cost-share a leased VF channel for data transmission.

25. Why does North America have to settle for a 56-kbps channel for ISDN whereas European countries have a clear 64-kbps channel available?

26. When there is no line access discipline on a multipoint line, it is operated on a "first come, first served" basis. What is this called? What is the term for the simple discipline in which every access is granted its turn?

27. There are three types of ARQ described in the text. Name them and discuss their efficiency versus complexity (especially buffer size).

28. Differentiate between circuit connection and link connection.

29. Protocols deal with link control. Name at least seven topics that a protocol may embrace.

30. Name at least five protocol functions.

31. Discuss *encapsulation*. (What is it?) How does it affect throughput?

32. Relate *connection control* to the three phases of a telephone call.

33. Discuss sequence numbers in light of packet service and in light of ARQ.

34. What is a PDU? Define it in one sentence.

35. What is the major reason for flow control?

36. Name the seven OSI layers. How are peer layers connected?

37. Peer layers interface by means of a common ______?

38. What is an SAP?

39. What is the basic function of the data link layer?

40. Differentiate between a service data unit (SDU) and a protocol data unit (PDU).

41. What is a primitive? Name four primitives used in HDLC (one-word names).

42. What is the function of the network layer?

43. Where do internetwork protocols fit in the OSI picture? (Layer?)

44. Where (what layer) in OSI does the communication provider responsibility end?

45. Give at least four functions of the session layer.

46. What OSI layer is basically involved with syntax?

47. Name and describe the three basic or generic types of data link layer protocols. Briefly describe each.

48. In a DDCMP frame there are two BCCs (or FCSs). Why two? Considering question 47, what type of protocol is DDCMP?

49. HDLC deals with three types of stations (accesses). What are they? Briefly describe how each functions.

50. Considering the unique flag bit sequence that is so vital for proper operation of HDLC, how is transparency maintained?

51. What is a "window of frames" (HDLC)? Can you give maximum window sizes as specified in HDLC?

52. Discuss the HDLC control field and its impact on the three types of HDLC frames.

53. How many frames may be outstanding in HDLC where the P bit is set to 1?

54. How many bits are there, in most cases, in the HDLC FCS field?

55. Relate Rec. X.25 architecture to the OSI model. Describe how Rec. X.25 interacts with the PDN beyond the first entry point up to the destination node.

56. Define logical channels, virtual connections (two types), and datagrams. Compare these items with regard to network congestion and setup time. Trace a virtual call through its three basic stages.

57. LAPB relates to which OSI layer? Where does the Rec. X.25 packet fit in its frame structure?

58. What is the function of the MLP? Is it a layer 2 or layer 3 function in Rec. X.25?

59. There are three fields common to all Rec. X.25 packets. List these fields in proper sequence, and briefly describe the function of each.

60. In Rec. X.25 with extended modulo 128 operation, what field or fields in the packet are extended?

61. What is the purpose of the facility field in Rec. X.25?

62. Give at least five functions of a technical control facility.

63. What does the term "normaled-through" mean?

64. How is a BER test carried out?

65. Discuss how loopback can be used to measure quality and to troubleshoot a data circuit.

66. Differentiate between *terminating* and *bridging* jacks.

67. Describe at least two basic measurements that a tech control should be equipped to make on an analog circuit.

68. What are *foxes*?

REFERENCES

1. W. L. Harper and Robert C. Pollard, *Data Communications Desk Book: A Systems Analysis Approach*, Prentice-Hall, Englewood Cliffs, N.J., 1982.
2. Dixon R. Doll, *Data Communications: Facilities, Networks and Systems Design*, Wiley, New York, 1978.
3. Roger L. Freeman, *Telecommunication Transmission Handbook*, 2nd ed., Wiley, New York, 1981.
4. *Digital Data System Channel Interface Specifications*, Bell Sys. Tech. Ref. Pub 62310, ATT, Basking Ridge, N.J., 1983.
5. W. Stallings, *Handbook of Computer Communications Standards*, Vol. 1, Macmillan, New York, 1987.
6. *Open Systems Interconnection (OSI) Service Definition Conventions*, CCITT Rec. X.210, VIII Plenary Assembly, Malaga–Torremolinos, 1984.
7. *Data Communication Networks: Interfaces*, Vol. VIII.3, CCITT Recs. X.20–X.32, VIII Plenary Assembly, Malaga–Torremolinos, 1984.
8. *Data Communication Networks: Transmission, Signaling and Switching, Network Aspects, Maintenance and Administration Arrangements*, Vol. VIII.4, CCITT Recs. X.40–X.181, VIII Plenary Assembly, Malaga–Torremolinos, 1984.
9. James W. Conard, *Standards and Protocols for Communication Networks*, Carnegie Press, Madison, N.J., 1982.
10. John E. McNamara, *Technical Aspects of Data Communications*, 2nd ed., Digital Equipment Corporation, Bedford, Ma., 1982.
11. *High-Level Data Link Control Procedures*, ISO 3309, International Standards Organization, Geneva, 1979.
12. Roger L. Freeman, *Reference Manual for Telecommunications Engineering*, Wiley, New York, 1985.
13. *Code Independent Error Control Systems*, CCITT Rec. V.41, Vol. VIII.1, Data Communications over the Telephone Network, CCITT VIII Plenary Assembly, Malaga–Torremolinos, 1984.
14. *Interface between Data Terminal Equipment (DTE) and Data-Circuit-Terminating Equipment (DCE) for Terminals Operating in the Packet Mode and Connected to the Public Data Networks by Dedicated Circuits*, CCITT Rec. X.25, CCITT VIII Plenary Assembly, Malaga–Torremolinos, 1984.
15. *Interface Between Data Terminal Equipment (DTE) and Data Circuit-Terminating Equipment (DCE) for Synchronous Operation on Public Data Networks*, CCITT Rec. X.21, CCITT VIII Plenary Assembly, Malaga–Torremolinos, 1984.

16. *Use on Public Data Networks of Data Terminal Equipment (DTE) Which Is Designed for Interfacing Synchronous V-Series Modems*, CCITT Rec. X.21 bis, CCITT VIII Plenary Assembly, Malaga–Torremolinos, 1984.
17. *International Data Transmission Services and Optional User Facilities in Public Data Networks*, CCITT Rec. X.2, CCITT VIII Plenary Assembly, Malaga–Torremolinos, 1984.

12

LOCAL AREA NETWORKS

1 DEFINITION AND APPLICATIONS

Local area networks (LANs) interconnect workstations, computers, and/or other computer assets over a limited geographical area. Several LAN standards specify capability to serve up to a thousand or more devices on a single LAN. The geographical extension or "local area" may extend no more than several hundred feet (100 m) in certain cases and to over 6 mi (10 km) or more in other cases. The transmission media providing this connectivity may be wire pair, coaxial cable, or fiber optic cable. Several local area radio schemes have also been proposed. Other devices may be served by LANs as well, such as sensors (telemetry), digital telephones, facsimile, and video (television).

Data rates on current LANs vary from 1 Mbps to 100 Mbps. Future LANs on fiber hold the promise of hundreds of megabits per second. LAN data rates, the number of devices to be connected to a LAN, the spacing of those devices, and network extension depend on:

- The transmission medium used.
- Transmission technique (i.e., baseband or broadband).
- Network protocol.

Most LANs operate without error correction with bit error rates (BERs) specified in the range of 1×10^{-8} to 1×10^{-11}.

The most common application of a local area network is to interconnect terminals (workstations) with processing resources, where all connected devices reside in a single building or complex of buildings, and usually these resources have a common owner. Cost containment is a driving force toward implementation of LANs. A LAN permits effective cost sharing of high-value EDP equipment, such as mass storage, mainframe, or minicomputers and high-speed printers. A LAN also is an effective means to utilize distributed processing. Within certain limits it also permits future growth and evolution by adding processors, workstations, and other peripherals as demand requires.

Another advantage of LANs is that they permit coupling of multivendor equipments into an integral system. However, it should be pointed out that connectivity via OSI layer 1 does not guarantee compatibility of layer 2 and above of equipment of different vendors. Such compatibility can often be accomplished in conversion software.

With proper devices, LANs can be interconnected among themselves. Such devices are called bridges. They can also be connected to wide area networks (WANs; Chapter 11), such as ISDN (Chapter 13). The device that carries out this interface is called a gateway.

There are also two other related techniques for data device connectivity in the local area: HSLN (high-speed local network) and PBX or CBX for mixed voice and data application. The HSLN is used in what some define as back-end connectivity. In this context we mean the connection of computers (usually mainframe computers) and mass storage devices. Here cable runs are usually short and data rates higher than conventional LANs. The PBX or CBX (computer-based branch exchange) with tributaries or user lines performs the LAN function based on a star topology where the PBX serves double duty. It routes or switches both voice and data where data rates on each line are comparatively lower than bus- or ring-type LANs.

There are two generic transmission techniques utilized by LANs: baseband and broadband. Baseband transmission can be defined as the direct application of the baseband signal to the transmission medium. Broadband transmission, in this context, is where the baseband signal from the data device is translated in frequency to a particular band of frequencies in the RF or light spectrum. Broadband transmission requires a modem to carry out the translation. Baseband transmission may require some sort of signal conditioning device. With broadband LAN transmission we usually think of simultaneous multiple carriers that are separated in the frequency domain.

2 LAN TOPOLOGIES AND ARCHITECTURES

There are basically three types of LAN topology: star, bus, and ring. These are shown in Figure 12.1 along with the *tree network*, which is a subset of the conventional bus topology. A star configuration can be typified by a PBX or CBX or, alternatively, a processor or controller. Such a configuration, with its centralized control, requires that a user transmit a call request (e.g., off hook in telephony). The user must then indicate the destination to the CBX or controller, which then sets up the connection. In the case of a data circuit the destination information could be contained in the message header. This is really an extension of point-to-point operation. In certain applications a polling regime could be implemented. Figure 12.1 gives application-specific topology for LANs, whereas Figure 11.8 shows generic architecture.

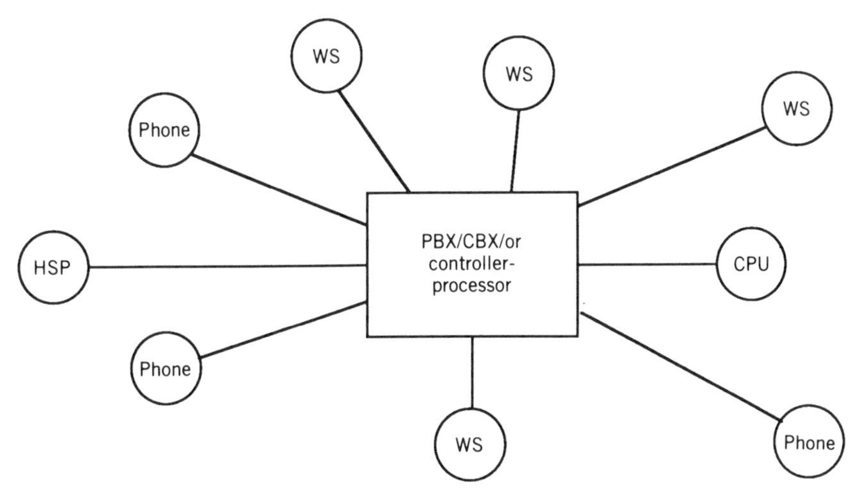

Figure 12.1A A star network.

The bus and tree topologies (Figures 12.1B, D) utilize a multiple access broadcast technique. Only one user at a time can access the network and all other users can copy the traffic. Data messages are broken down into packets, and each packet has a header containing source and destination (sink) addresses. Only those destinations included in the address actually copy the traffic.

A ring network (Figure 12.1C) is a closed loop. Adjacent accesses on the ring are connected on a point-to-point basis. Data circulates in one direction around the ring. A user waits until the network is free, then proceeds to send its traffic in packets similar to the bus technique. The destination senses the header and copies the data into a buffer, and the data continues to circulate

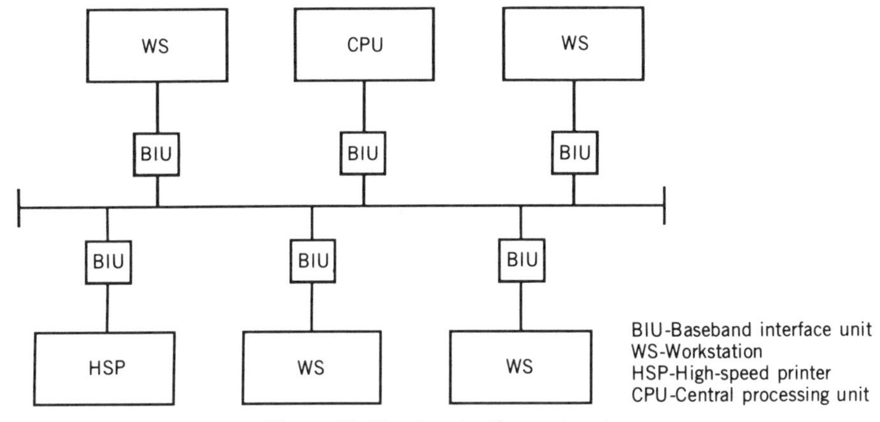

Figure 12.1B A typical bus network.

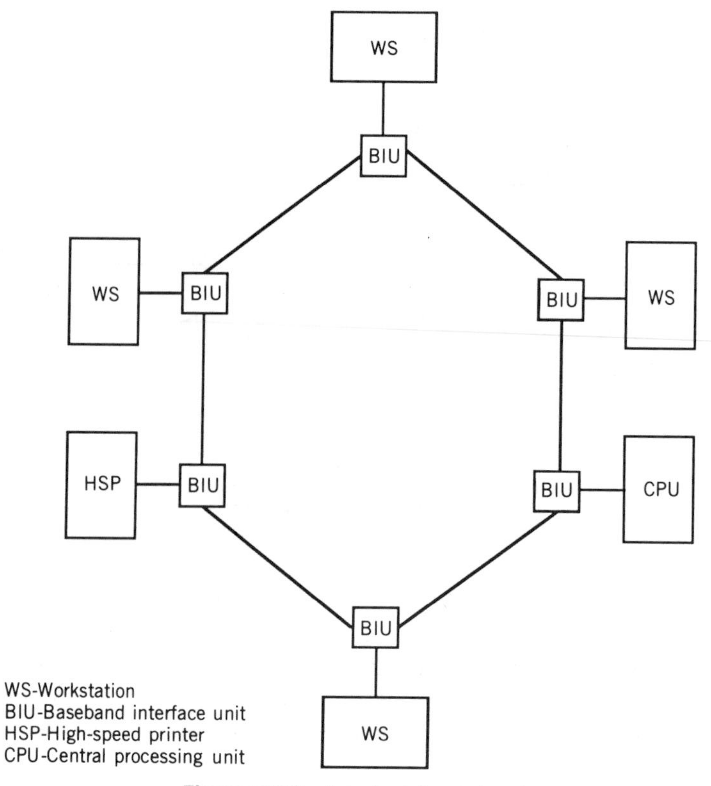

Figure 12.1C A typical ring network.

around the ring back to the source. The message return provides the source with a form of acknowledgment.

For bus and ring networks, where only one user at a time can access the network, some form of discipline must be imposed to ensure efficient and fair usage of the network. This discipline is embodied in a protocol. LAN protocols are described in Section 5.

3 TWO BROAD CATEGORIES OF LAN TRANSMISSION TECHNIQUES

There are two basic transmission techniques available for LANs and HSLNs. These are baseband and broadband, as mentioned earlier.

The baseband technique incorporates transmission of a digital waveform on a single medium, either wire pair or coaxial cable. Broadband systems utilize RF signals, which are FDM carriers, and the technology derives from cable television (CATV). In this case the medium is coaxial cable. Baseband lends itself to bus and ring topologies and broadband to bus and tree topologies.

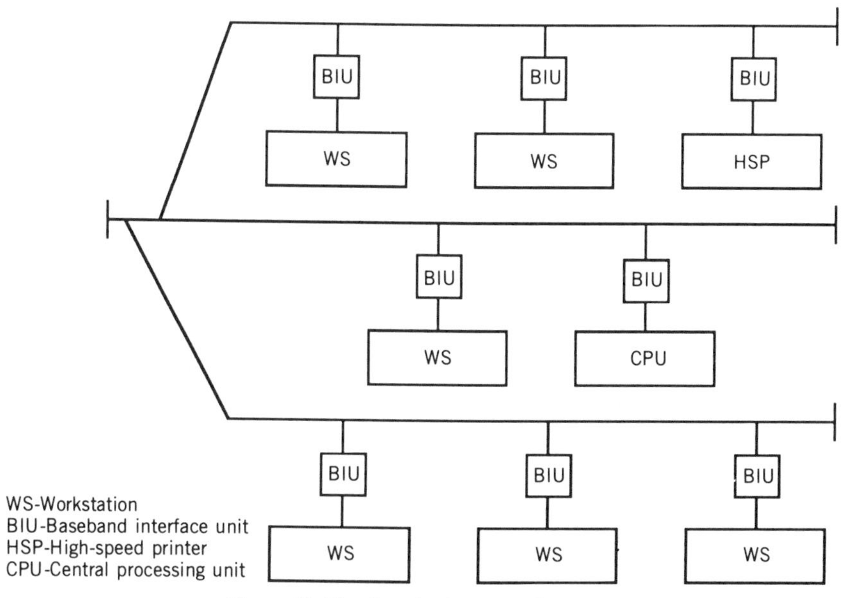

Figure 12.1D A typical tree configuration.

Broadband systems require a modem at each access; baseband systems do not. Baseband and broadband transmission techniques are compared in Table 12.1.

With baseband LANs, and actually with broadband LANs as well, we are dealing with multipoint operation. Two transmission problems arise as a result. The first deals with signal level and signal-to-noise (S/N) ratio and the second with standing waves. Each access on a common medium must have sufficient signal level and S/N such that copied signals have a BER in the range of 1×10^{-8} to 1×10^{-11}. If the medium is fairly long in extension and there are many accesses, the signal level must be high for an access to reach its most distant destination. The medium is lossy, particularly at the higher bit rates, and each access tap has an insertion loss. This leads to very high signal

TABLE 12.1 Comparison of Baseband and Broadband LAN Transmission

Item	Baseband	Broadband
Waveform	Digital: NRZ, RZ, Manchester	RF / FDM
Baseband	Bidirectional	Unidirectional
Topology	Bus or ring	Bus or tree
Access to medium	Tap	Modem or tap
Media	Wire pair, coaxial cable	Coaxial cable
LAN extension	Up to 2 km	Tens of km
Information type	Data only	Data, voice, facsimile, video
Utilization of bandwidth	Single signal occupies entire bandwidth	Multiple simultaneous signals in FDM structure

levels. These may be rich in harmonics and spurious, degrading error rate. On the other hand, with insufficient level, the S/N ratio degrades, which will also degrade error rate. A good level balance must be achieved for all users. Every multipoint connectivity must be examined. The number of multipoint connectivities can be expressed by $n(n - 1)$, where n is the number of accesses. If there are 100 accesses, there are 9900 possible connectivities to be analyzed to carry out signal balance. One way to simplify the job is to segment the network, placing a regenerative repeater at each segment boundary. This reduces the signal balance job to realizable proportions and ensures that a clean signal of proper level is available at each access tap.

For baseband LANs 50-Ω coaxial cable is favored over the more common 75-Ω cable. This lower impedance cable is less prone to signal reflections from access taps and provides better protection against low-frequency interference.

The effects of standing waves can be reduced by controlling the spacing between access taps. For example, the Ethernet technical summary [1] recommends spacings no less than 2.5 m for this 10-Mbps system. By following this placement rule, the technical summary says, the chance that objectionable standing waves will result is reduced to a very low (but not zero) probability. Again for Ethernet, up to 100 devices may be placed on a cable segment and the maximum segment length is 500 m. The segments are connected through regenerative repeaters, and the maximum end-to-end length is 2.5 km [1, 4].

One extremely important consideration for baseband transmission is that there is a single thread that can accommodate only one user at a time; otherwise there is a high probability of data message collision. Collision is where the traffic of two or more users interferes, corrupting the traffic of each.

3.1 Broadband Transmission Considerations

Broadband transmission permits multiple users to access the medium without collision. Broadband means that we take advantage of the medium's wide bandwidth. This wide bandwidth is broken down into smaller bandwidth segments in an analogous fashion to FDM. Each of these segments is assigned to a family of users. The statement regarding collision is correct if there are no more than two accesses per frequency segment connected on a point-to-point basis, where one access receives while the other transmits. If, in this case, we assume contention as the access protocol, then as the family increases in number, the chances of collision start to increase. With a little imagination, we can see that, with the proper switching scheme implemented, a user can join any family by simply switching to the proper frequency band of that family. All that is required is a change in modem frequency and possibly modulation waveform. There may also be certain protocol considerations as well.

Unlike their baseband counterparts, broadband systems can be designed to accommodate digital or analog voice, data from kilobit to multimegabit rates, video, and facsimile. Thus broadband systems are versatile. They are also much more expensive than their baseband counterparts and require a higher level of design engineering effort.

As mentioned, much of our present broadband technology derives from cable television technology. Total system bandwidths are of the order of 300–500 MHz. Each access requires a modem to modulate and demodulate the data or other user signal and to translate the modulated frequency to the assigned frequency slot on the cable.

This, then, is RF transmission and, by its very nature, must be one way or unidirectional. Thus a user can only access another user "downstream" from it. If we assume a single medium, usually a 75-Ω coaxial cable, then how does one access another user "upstream"? This is done using a similar approach to that of a two-way or interactive CATV (cable television) system, where two paths are provided on a single coaxial cable. This is done by splitting the cable spectrum into two frequency segments, one segment for one direction and the other for the opposite direction. At a cable terminating point, which some even call a *head end* from CATV terminology, a frequency translator converts and amplifies signals from one direction (frequency segment) into signals for transmission in the opposite direction (frequency segment). Another term for "head end" for broadband LANs is *central retransmission facility* (CRF). Figure 12.2 shows a typical frequency assignment plan for a broadband LAN using single cable operation.

There are two choices of topology for broadband LANs: bus and tree. The head end or CRF is located at some termination of the bus, and in the case of tree topology the CRF is located at its root, so to speak.

Another approach to achieve dual path operation is to use two cables; one provides the "go" path and the other the "return" path. For single-cable split-band operation, a rather large guard band is left in the center of the cable spectrum to ensure isolation between the two paths. Then we can see that with the provision of two-cable operation the usable bandwidth can be more than doubled. Modem operation is also simpler because, to access a particular net, only one frequency operation is required (i.e., the send and receive frequencies can be the same). With single-cable split-band operation send and receive frequencies must necessarily be different.

3.2 Fiber Optic LANs

Fiber optic LANs may be considered broadband in that a class of modem is required to place the digital signal on the fiber. The modem, of course, consists of a light source, detector, and the necessary driver and signal conditioning circuitry. With wavelength division multiplexing (WDM) we have a true broadband system. At this time single-wavelength operation prevails.

The type of fiber optic cable selected for a LAN is a cost trade-off. The extension of a LAN is generally short such that less expensive multimode fiber can be used, and the mature short wavelength technology permits other cost

Channel Designation	Frequency Range (MHz)	LAN use
T7	5.75-11.75	Data channels
T8	11.75-17.75	
T9	17.75-23.75	
T10	23.75-29.75	
T11	29.75-35.75	
T12	35.75-41.75	
T13	41.75-47.75	
T14	47.75-53.75	
—	53.75-54	Guardband
2	54-60	Off-air video channels
3	60-66	
4	66-72	
53 IRC	72-78	Off-air video (Ch. 5,6), IRC video channels, or FM band (88-108 MHz)
54 IRC	78-84	
55 IRC	84-90	
56 IRC	90-96	
57 IRC	96-102	
58 IRC	102-108	
59 IRC	108-114	
60 IRC	114-120	
A	120-126	CATV video channels
B	126-132	
C	132-138	
D	138-144	
E	144-150	
F	150-156	
G	156-162	
H	162-168	Data channels for midsplit (156.25 MHz shift)
I	168-174	Off-air video channels
7	174-180	
8	180-186	
9	186-192	
10	192-198	
11	198-204	Data channels for highsplit (192.25 MHz shift)
12	204-210	
13	210-216	
J	216-222	
K	222-228	
L	228-234	
M	234-240	
N	240-246	
O-up	246-up	CATV video or additional data

Figure 12.2 Frequency plan for a typical broadband LAN. Note that channel designation uses CATV channel designation nomenclature. From *IEEE Communications Magazine* **24** (6), 30 (June 1986) [2]. Courtesy of IEEE, New York.

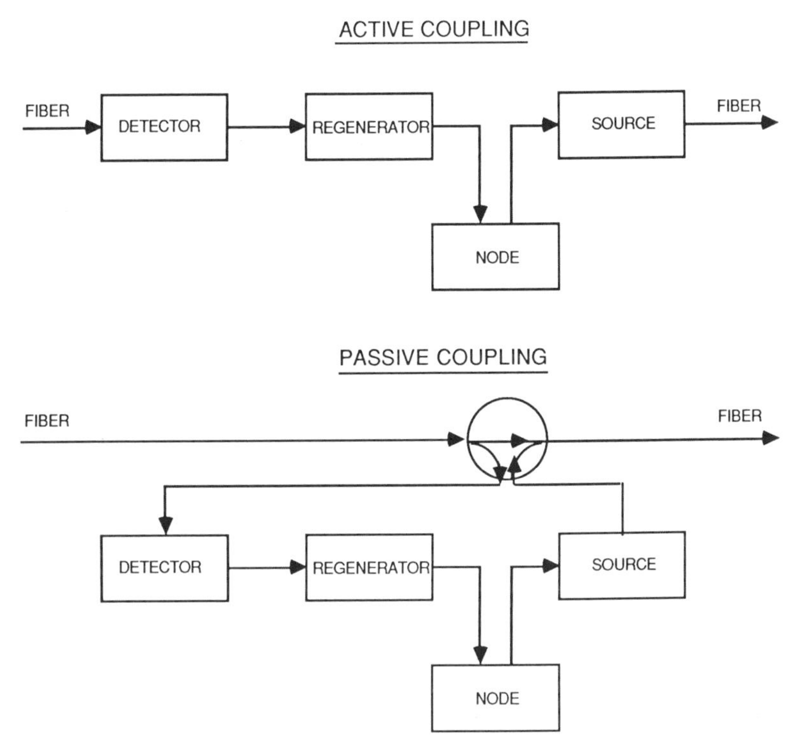

Figure 12.3 Coupling a node to optical fiber cable.

savings. Even plastic fiber may be considered. The losses of the fiber itself will generally be low due to the short distance operation even at 820-nm operation. A LAN that is 1 km long might display a fiber loss from 2 dB to 5 dB. The major contributor to loss is the taps if a fully passive network is to be implemented.

Fiber lends itself to all three basic LAN topologies: bus, ring, and star. There are two methods of coupling to the fiber: passive and active. The use of passive couplers leads to a more reliable system. With an active coupler the

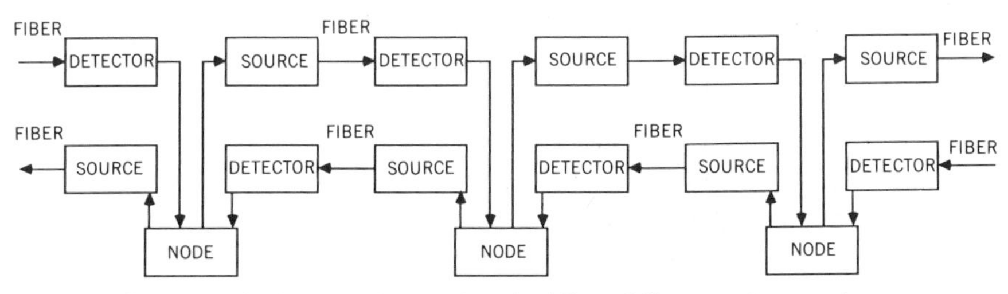

Figure 12.4A Bus topology with optical fiber, 2 fibers, active couplers.

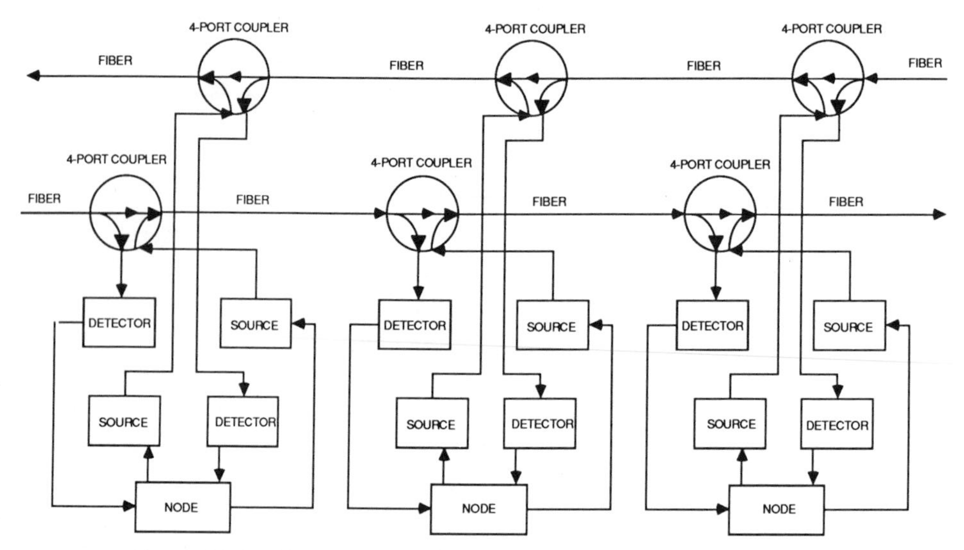

Figure 12.4B Bus topology with optical fiber, 2 fibers, passive couplers.

loss of one source or detector in an access causes the entire network to crash unless bypass switches are used. Figure 12.3 shows passive and active coupler implementations. Figure 12.4 illustrates typical bus topology using optical fiber. Figure 12.5 shows a ring topology, and Figure 12.6 shows a typical star configuration using optical fiber. In this case it is a segment of a multistar operating in a packet mode.

4 OVERVIEW OF SOME LAN PROTOCOLS

4.1 General

Many of the widely used LAN protocols have been developed in North America through the offices of the Institute of Electrical and Electronic Engineers (IEEE). The American National Standards Institute (ANSI) has subsequently accepted and incorporated these standards, and they now bear the ANSI imprimatur.

IEEE develops LAN standards in the IEEE 802 committee, which is currently organized into the following subcommittees:

802.1* High Level Interface.
802.2* Logical Link Control (LLC).
802.3* CSMA/CD Networks.

*Published standards.

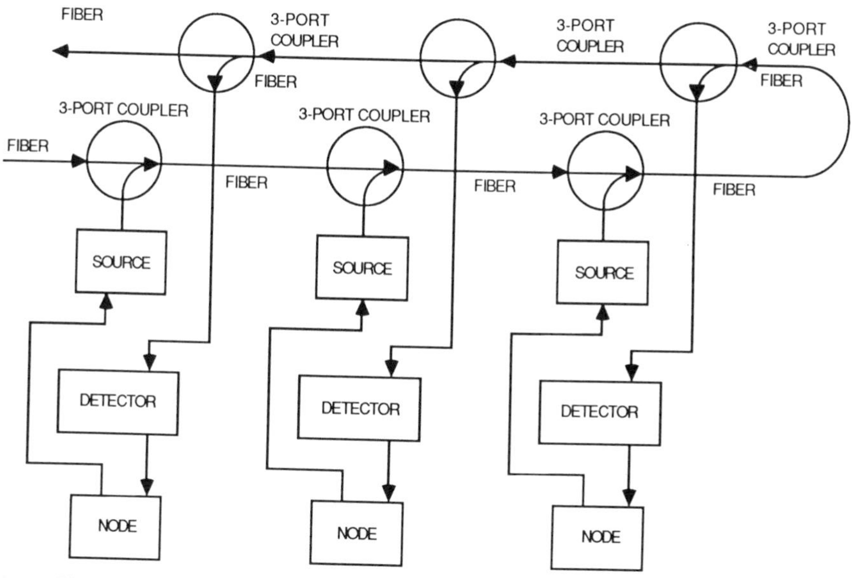

Figure 12.5 Ring topology with optical fibers using 3-port passive couplers.

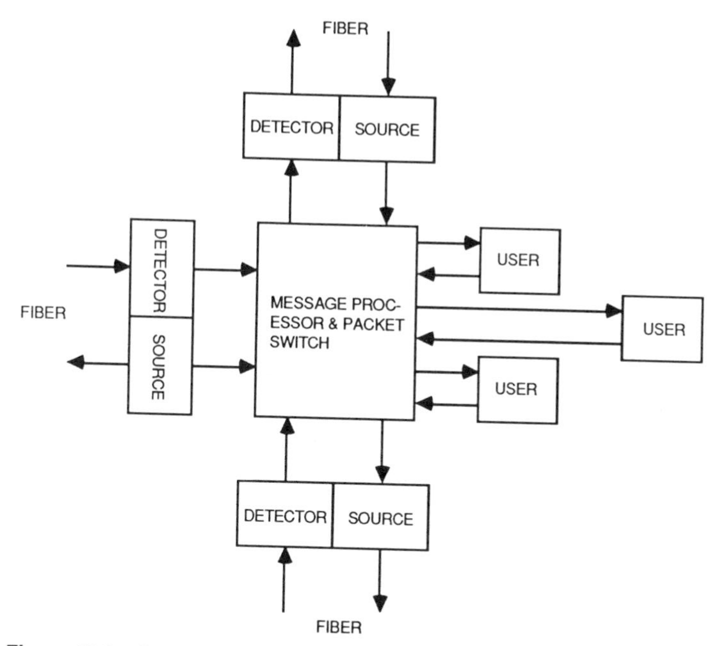

Figure 12.6 One section of a multiple star network with packet switching.

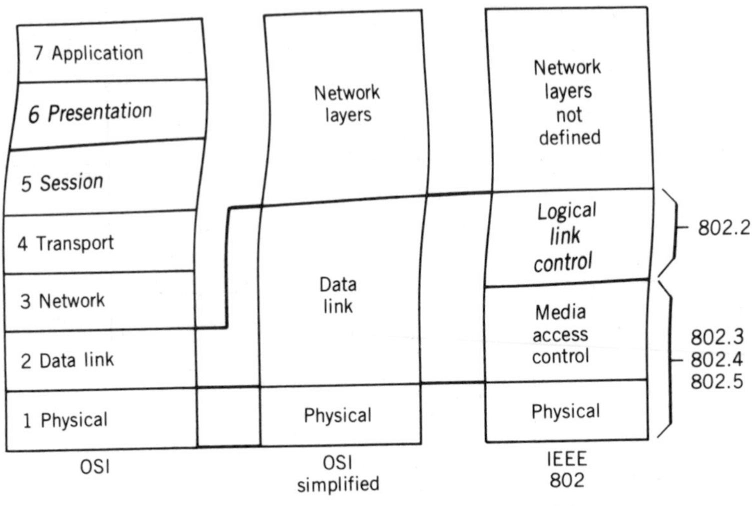

Figure 12.7 LAN architecture related to OSI.

802.4* Token Bus Networks.
802.5* Token Ring Networks.
802.6 Metropolitan Area Networks (MANs).
802.7 Broadband Technical Advisory Group.
802.8 Fiber Optic Technical Advisory Group.
802.9 Integrated Data and Voice Networks.

The fiber distributed data interface (FDDI) standards, which we discuss in Section 5.6, are being developed directly by ANSI.

4.2 How LAN Protocols Relate to OSI

LAN protocols utilize only OSI layers 1 and 2, the physical and data link layers, respectively. The data link layer is split into two sublayers: medium access control (MAC) and logical link control (LLC). These relationships to OSI are shown in Figure 12.7.

Stallings [3] presents an interesting and rational argument on the reasoning for limiting the layering to the first two OSI layers. There is no question that the functions of OSI layers 1 and 2 must be incorporated in a LAN architecture. We now ask, Why not layer 3? Layer 3, the network layer, is concerned with routing. There is no routing involved with LANs. There is a direct link involved between any two points. The other functions carried out by OSI layer 3—addressing, sequencing, and flow control—are carried out by layer 2 in LANs. The difference is that layer 2 performs these functions across a single link. OSI layer 3 carries out these functions across a sequence of links required

*Published standards.

to traverse a network. Of course, there is only one link required to traverse a LAN.

It would seem that layer 3 is required when viewed through an attached device. The reason is that the device sees itself attached to a network connecting multiple devices. One would think that ensuring delivery of a message to one or more accesses would be a layer 3 function. It was decided that, although the network provides services through layer 3, the characteristics of the network allow these functions to be performed in the first two layers.

As shown in Figure 12.7, the OSI data link layer is divided into two sublayers: logical link control and medium access control. These sublayers carry out four functions:

1. Provide one or more service access points (SAPs). A SAP is a logical interface between two adjacent layers.
2. Before transmission, assemble data into a frame with address and error-detection fields.
3. On reception, disassemble the frame and perform address recognition and error detection.
4. Manage communications over the link.

The first function and those related to it are performed by the LLC sublayer. The last three functions listed are handled by the MAC sublayer. In the next section we briefly describe how the LLC operates. It is common to all present IEEE 802 series standards.

4.3 Logical Link Control

The LLC provides services to the upper layers at a LAN station. The upper layers are user defined. The LLC provides three forms of services for its users:

1. Unacknowledged connectionless service.
2. Connection mode service.
3. Acknowledged connectionless service.

Some brief comments are required to clarify the functions and limitations of each service. With unacknowledged connectionless service a single service access initiates the transmission of a data unit to the LLC, the service provider. From the viewpoint of the LLC, previous and subsequent data units are unrelated to the present unit. There is no guarantee by the service provider of the delivery of the data unit to its intended user, nor is the sender informed if the delivery attempt fails. Further, there is no guarantee of ordered delivery.

This type of service supports point-to-point, multipoint, and broadcast modes of operation.

As we might imagine with connection mode service, a logical connection is established between two LLC users. During the data transfer phase of the connection the service provider at each end of the connection keeps track of the data units transmitted and received. The LLC guarantees that all data units will be delivered and that the delivery to the intended user will be ordered (e.g., in the sequence as presented to the source LLC for transmission). When there is a failure to deliver, it is reported to the sender.

Acknowledged connectionless service, although connectionless, provides immediate acknowledgment of receipt of each transmitted data unit. Data units are sent one at a time, and each transmitted unit must be acknowledged before the next data unit is sent.

IEEE (Ref. 14) defines the LLC as that part of a data station that supports the LLC functions of one or more logical links. The LLC generates command PDUs (protocol data units) and response PDUs for transmission and interprets received command PDUs and response PDUs. Specific responsibilities assigned to the LLC include:

1. Initiation of control signal interchange.
2. Interpretation of received command PDUs and generation of appropriate response PDUs.
3. Organization of data flow.
4. Actions regarding error-control and error-recovery functions in the LLC sublayer.

4.3.1 LLC Generic Primitives. Primitives can be viewed as commands or procedure calls with parameters [3]. In this context service primitives are of three generic types:

1. Request. The request primitive is passed from the *n* layer (or sublayer) to request that a service be initiated.
2. Indication. The indication primitive is passed from the *n* layer (or sublayer) to the *n* user to indicate an internal *n* layer (or sublayer) event that is significant to the *n* user. This event may be logically related to a remote service request, or may be caused by an event internal to the *n* layer (or sublayer).
3. Confirm. The confirm primitive is passed from the *n* layer (or sublayer) to the *n* user to convey the results of one or more associated previous service request(s). This primitive may indicate either failure to comply or some level of compliance. It does not necessarily indicate any activity at the remote peer interface.

4.3.2 LLC PDU Structure. User data are passed down to the LLC, which appends a header. The header contains control information, which is used to

DSAP Address	SSAP Address	Control	Information
8 bits	8 bits	y bits	8*M*bits

Figure 12.8 LLC PDU format. From IEEE 802.2 [14].

manage the protocol between the local LLC entity and the remote LLC. The combination of header and user data is called an LLC protocol data unit. This unit is passed to the MAC (see Figure 12.7), which appends another header and parity tail (FCS field).

There are two sets of addresses involved in the operation: MAC address, which identifies a station on the LAN, and LLC address, which identifies an LLC user. The LLC user is accessed through an LLC SAP (service access point). An LLC may serve one or more users, each with its own SAP.

The LLC PDU format is shown in Figure 12.8. It uses bit-oriented procedures. DSAP is the destination service access point, and SSAP is the source service access point. The control field has 16 bits ($y = 16$) for formats that include sequence numbering and 8 bits ($y = 8$) for formats that do not. The information field consists of M octets, where M can vary from 0 to some maximum value that is a function of the medium access methodology used.

The address fields are somewhat unique. Each is 8 bits long. Only seven of the eight bits are actual address bits. The least significant bit is the first bit delivered to and the first bit received from the MAC layer. In the case of the DSAP this first bit is used to identify the DSAP address as an individual or group (I/G) address. In the SSAP this bit identifies the LLC PDU as a command or a response (C/R), where 0 is command and 1 is response.

The control field consists of one or two octets that are used to designate command and response functions and contain sequence numbers when required.

5 LAN ACCESS PROTOCOLS

5.1 Introduction

In this context a protocol includes a means of permitting all users to access a LAN fairly and equitably. Access can be random or controlled. The random access schemes to be discussed include CSMA (carrier sense multiple access) and CSMA/CD, where CD stands for collision detection. The controlled access schemes that are described are token bus and token ring. It should be kept in mind that users accessing the network are unpredictable and the transmission capacity of the LAN should be allocated in a dynamic fashion in response to those needs.

5.2 Background: Contention and Polling

As discussed in Chapter 11, the simplest access method is contention, which is a random access technique. Contention, of course, would be commonly used on point-to-point service, typically a workstation on one end of a circuit and a CPU (computer) on the other, as shown in Figure 11.8A. Both ends are considered equal contenders for access and the technique operates as follows: when the line is free, either end can bid for control. If the other end has nothing to send, it accepts the bid and the bidder is given control of the line. Once the device has gained control of the line, it begins transmission. When the controlling device has completed its transmission, it releases the line. The line is once again open for bids from either side. If there is a *bid collision* (i.e., the two sides bid for the line simultaneously), one side backs down and relinquishes the line to the other side. Who relinquishes the line to the other side is determined by agreement in advance.

Now add one, two, three, or more users to the line, placing the line in a multipoint configuration, and the chances of a bid collision increase greatly. Such beforehand agreements on who backs down can get unwieldy, and line efficiency can drop significantly during back-down periods. Nevertheless, the CSMA and CSMA/CD LAN access techniques are direct outgrowths of this type of operation.

Polling is a form of controlled access. It is commonly used on multipoint configurations. Such a configuration is shown in Figure 11.8B. One station, usually the CPU, is assigned the responsibility of master station. The master station polls each remote station periodically, often sequentially, to determine if there is traffic to be transmitted. When a reply from the remote is in the affirmative, the traffic is then transmitted. Each station on the network has a unique address (e.g., bit sequence). This address is incorporated into the poll. Thus only one indicated remote station will respond to that address. In a similiar manner, traffic from the master station to a remote station incorporates the address of that remote station in the header of the message. In such a way that remote station and only that remote station will copy the traffic. This type of polling is called *roll-call polling* by some and *broadcast polling* by others. Because each remote station must respond to the poll (either negative or positive) with that additional overhead, roll-call polling is considered inefficient.

Loop polling is another form of polling. In this case the master station sends a poll request to the first remote station in the loop. If that station has traffic, it is sent to the master station. Once the traffic transmission terminates and all the traffic is sent and acknowledged, that remote station forwards the poll request to the next station in the loop. If the second station has traffic, it is transmitted, and if not, the poll request is sent to the third station, and so on until all remote stations have been polled. The process then starts all over again. The advantages of loop polling are that message transmission and polling operations overlap and no negative responses are transmitted.

As we will show later, token passing LAN access methods are a direct outgrowth of polling. All the access techniques used with bus, ring, and tree topologies are forms of time-division multiple access (TDMA). The CBX-(computer-based PABX) controlled star topology can also be considered a TDMA technique where a specific capacity is a dedicated connection.

5.3 CSMA and CSMA/CD Access Techniques

Carrier sense multiple access (CSMA) is a LAN access technique that some simplistically call "listen before transmit." This "listen before transmit" idea gives insight into the control mechanism. If user 2 is transmitting, user 1 and all others hear that the medium is occupied and refrain from using it. In actuality, when an access with traffic senses that the medium is busy, it backs off for a period of time and tries again. How does one control that period of time? There are three methods of control commonly used. These methods are called *persistence algorithms* and are outlined briefly below:

- Nonpersistent. The accessing station backs off a random period of time and then reattempts access.
- 1-persistent. The station continues to sense the medium until it is idle and then proceeds to send its traffic.
- p-persistent. The accessing station continues to sense the medium until it is idle, then transmits with some preassigned probability p. Otherwise it backs off a fixed amount of time, then transmits with a probability p or continues to back off with a probability of $(1 - p)$.

The algorithm selected depends on the desired efficiency of the medium usage and the complexity of the algorithm and resulting impact on firmware and software. With the nonpersistent algorithm collisions are effectively avoided because the two stations attempting to access the medium will back off, most probably with different time intervals. The result is wasted idle time following each transmission. The 1-persistent algorithm is more efficient by allowing one station to transmit immediately after another transmission. However, if more than two stations are competing for access, collision is virtually assured. The p-persistent algorithm lies between the other two and is a compromise attempting to minimize collisions and idle time.

It should also be noted that with CSMA, after a station transmits a message, it must wait for an acknowledgment from the destination. Here we must take into account the round-trip delay (2 × propagation time) and the fact that the acknowledging station must also contend for medium access. Another important point is that collisions can occur only when more than one user begins transmitting within the period of propagation time. Thus CSMA is an effective access protocol for packet transmission systems where the packet transmission time is much longer than the propagation time.

The inefficiency of CSMA arises from the fact that collisions are not detected until the transmissions from the two offenders have been completed. With CSMA/CD, which has collision detection, a collision can be recognized early in the transmission period and the transmissions can be aborted. As a result, channel time is saved and overall available channel utilization capability is increased.

CSMA/CD is sometimes called "listen while transmitting." It must be remembered that collisions can occur at any period during channel occupancy, and this includes the total propagation time from source to destination. Even at multimegabit data rates propagation time is not instantaneous; it remains constant for a particular medium, no matter what the bit rate is. With CSMA the entire channel is wasted. With CSMA/CD one offending station stops transmitting as soon as it detects the second offending station's signal. It can do this because all accesses listen *while* transmitting.

5.3.1 CSMA/CD Description. Carrier sense multiple access with collision detection is defined by IEEE Standard 802.3 [4]. It is based on the Ethernet approach initiated by Xerox Corporation, Digital Equipment Corporation, and Intel. Standard 802.3 closely resembles Ethernet with changes in the packet structure and an expanded set of physical layer options. Figure 12.9 relates CSMA/CD protocol layers to the conventional OSI reference model (see Chapter 11, Section 7.4.2) and identifies acronyms that we use in this description. The bit rates generally encompassed in CSMA/CD are between 1 Mbps and 20 Mbps. In general, the discussion covers only the 10-Mbps rate.

The medium is coaxial cable. A user connects to the cable by means of a medium access unit (MAU). This connects through an attachment unit interface (AUI) to the data terminal equipment (DTE). As shown in Figure 12.9, the DTE consists of the physical signaling sublayer (PLS), the medium access control (MAC) and the logical link control (LLC). The PLS is responsible for transferring bits between the MAC and the cable. It uses differential Manchester encoding for the data transfer. With such coding the datum 0 has a transition from high to low at midcell, while the datum 1 has the opposite transition. The line idle is a steady high condition with no transitions.

The MAC frame format is shown in Figure 12.10. The 7-octet preamble field is used for synchronization with received frame timing. One octet is assigned to the start frame delimiter (SFD). Two address fields follow the SFD: the destination address and the source address. Depending on the system design, each address field contains either 16 or 48 bits. However, at any given time the source and destination addresses are the same for all users on a particular LAN. There are two types of addresses: individual user and group address. In the group address category there are two subsets where all users of a particular LAN are addressed.

AUI - ATTACHMENT UNIT INTERFACE
MAU - MEDIUM ATTACHMENT UNIT
MDI - MEDIUM DEPENDENT INTERFACE
PMA - PHYSICAL MEDIUM ATTACHMENT

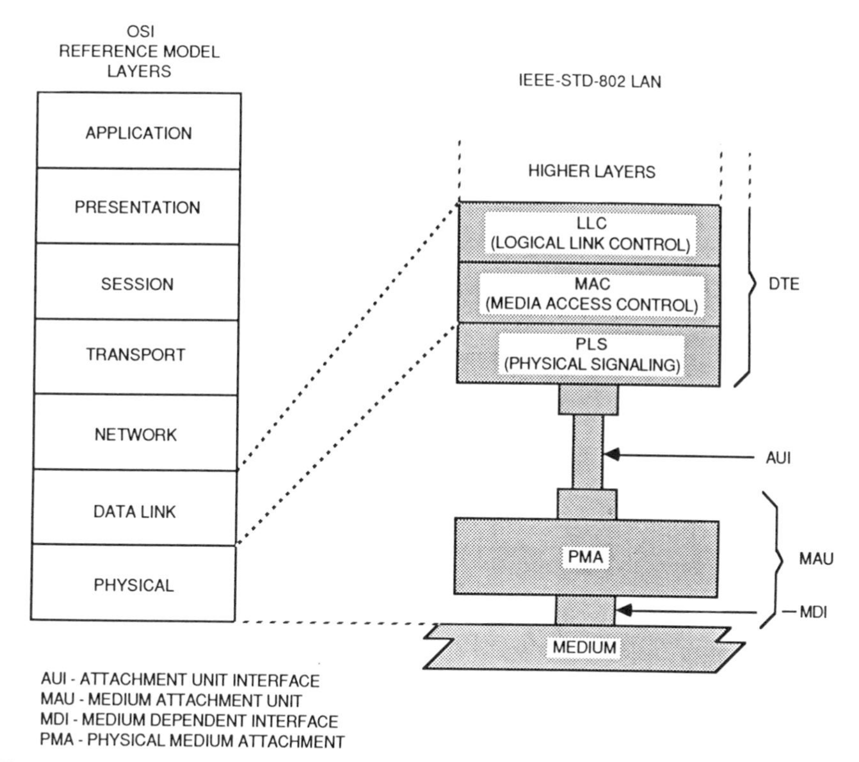

Figure 12.9 CSMA / CD LAN relationship to the OSI model. From IEEE / ANSI Std. 802.3 [3]. Courtesy of IEEE, New York.

The address fields are followed by a 2-octet length field whose value gives the number of LLC data octets in the data field. If the length field value is less than the minimum required for proper operation of the protocol, a sequence of octets is added, called a PAD field, at the end of the data field and just before the FCS (frame check sequence) field.

The next fields transmitted are the LLC and PAD fields. The LLC data field is fully transparent and contains *n* octets, where *n* is an integral number.

PREAMBLE (7 OCTETS)	SFD (1 OCTET)	DESTINATION ADDRESS (2 OR 6 OCTETS)	SOURCE ADDRESS (2 OR 6 OCTETS)	LENGTH (2 OCTETS)	LLC DATA	PAD	FCS (4 OCTETS)

LENGTH - GIVES NUMBER OF OCTETS IN DATA FIELD
LLC - LOGICAL LINK CONTROL (OSI LAYER 3 AND ABOVE)
FCS - FRAME CHECK SEQUENCE - A 32-BIT CRC
PAD - ADDS OCTETS TO ACHIEVE MINIMUM FRAME LENGTH WHERE NECESSARY
SFD - START FRAME DELIMITER

Figure 12.10 MAC frame format.

For a particular implementation there is a minimum frame size. The PAD field is adjusted to meet this minimum (i.e., it provides padding or extra bits to meet minimum frame length requirements). The maximum size of the data field is determined by the maximum frame size and address size parameters of the particular implementation.

The FCS field contains a 4-octet value for the cyclic redundancy check remainder value. The generating polynomial is

$$G(X) = X^{32} + X^{26} + X^{23} + X^{22} + X^{16} + X^{12} + X^{11} + X^{10} + X^{8} + X^{7} + X^{5} + X^{4} + X^{2} + X + 1. \quad (12.1)$$

A MAC frame is considered invalid if any one of the following conditions are met:

- Frame length is inconsistent with the field length.
- Not an integral number of octets in length.
- The bits of the received frame, exclusive of the FCS field, do not generate a CRC remainder value identical to the one received.

Invalid MAC frames are not passed to the LLC. Each octet of a MAC frame, with the exception of the FCS, is transmitted low-order bit first.

The minimum frame size is 512 bits for the 10-Mbps data rate [4]. This requires a data field of either 46 or 54 octets, depending on the size of the address field used. The minimum frame size is based on the *slot time*, which for the 10-Mbps data rate is 512 bit times. Slot time is the major parameter controlling the dynamics of collision handling and it is:

- An upper bound on the acquisition time of the medium.
- An upper bound on the length of a frame fragment generated by a collision.
- The scheduling quantum for retransmission.

To fulfill all three functions, the slot time must be larger than the sum of the physical round-trip propagation time and the MAC sublayer jam time. The propagation time for a 500-m segment of 50-Ω coaxial cable is 2165 ns, assuming that the velocity of propagation of this medium is $0.77 \times 300 \times 10^6$ m/s [4].

Collision is detected when voltage swings are higher than those generated by one user alone. These reinforced voltages are due to the presence of another signal on the medium. When a signal is detected during frame transmission, the transmission is not terminated immediately. Instead, the transmission continues using what is called a *jamming signal*. This collision enforcement or jam guarantees that the duration of the collision is sufficient to ensure detection by all active users on the network. The jam signal content is specified for particular implementations but should avoid being the 32-bit CRC value corresponding to the (partial) frame transmitted before initiation of the jamming signal.

When an access detects a collision, it continues to reattempt until a maximum number of attempts have been made and all have terminated due to collisions. When this occurs, the event is reported as an error. The scheduling of reattempts is determined by a controlled randomization process called *truncated binary exponential back-off*. This defines the delay time before reattempt. The delay is an integer multiple of the slot time. The number of slot times to delay before the nth retransmission attempt is chosen as a uniformly distributed integer r in the range of $0 \leq r < 2^k$, where $k = \min(n, 10)$. Algorithms used to generate the integer r should be designed to minimize correlation between the numbers generated by any two stations at any given time.

The following summarizes some of the parameters of CSMA/CD as specified in IEEE Std. 802.3 [4]:

Parameter	Value
Slot time	512 bit times
Interframe gap	9.2 μs
Attempt limit	16
Back-off limit	10
Jam size	32 bits
Maximum frame size	1518 octets
Minimum frame size	64 octets
Address size	48 bits

The baseband signals transmitted to the medium by the AUI use Manchester coding. Manchester coding is rich in transitions and is compatible with the coaxial cable transmission medium.

There are a number of different transmission techniques recommended by Standard 802.3 as shown below [12]:

10BASE5
10BASE2
10BROAD36
1BASE5.

The convention used for identifying these techniques uses the first number to indicate the data rate on the network in Megabits per second, BASE or BROAD to indicate baseband or broadband, respectively, and a final number that gives the actual link or segment length in meters.

5.4 Token Bus

A token bus LAN in its simplest version is a length of 75-Ω coaxial cable terminated at each end in its characteristic impedance. Users tap the cable with a coupler and connect to the coupler with a 37.5-Ω stub no longer than 350 mm. Three different transmission regimes are described in IEEE Std. 802.4 [6]: phase continuous FSK, phase coherent FSK, and multilevel duobinary

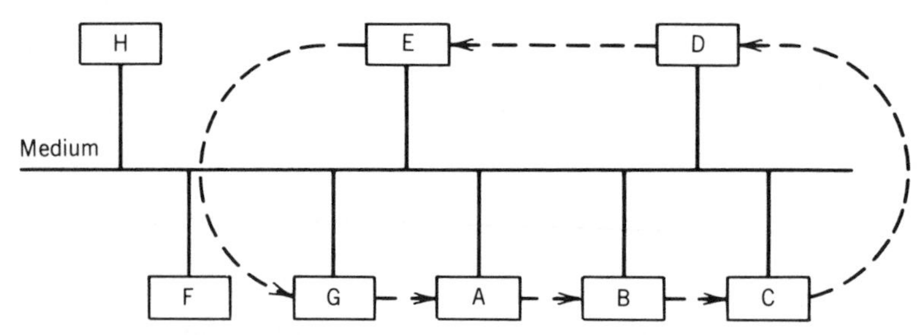

Figure 12.11 Token bus showing logical connectivity (dashed line).

AM/PSK. The coaxial cable can be extended by use of regenerative repeaters. In a similar manner tree topologies can be developed. The discussion below covers data rates of 1 Mbps, 5 Mbps, and 10 Mbps.

With the token bus access technique a short control packet known as the token regulates the right of access to the bus. A station holding the token has the exclusive right to use the network for a specified time period. During this period the station may poll other stations, receive responses, and pass data traffic. When the token holding station has completed operations or when its time period is up, it passes the token to the next station, its successor, in logical sequence.

Figure 12.11 illustrates a typical token bus, showing logical connectivity (dashed line) and that the access method is always sequential in a logical sense. The right to access the medium passes from user to user. It should be noted that physical connectivity has little impact on the order of the logical ring. In fact, stations can respond to a query from a token holder without being part of the *logical* ring. For example, stations H and F can receive data frames but cannot initiate a transmission because they cannot receive the token.

Token passing ensures equitable access to the network. Such a control also ensures against collision because only the user holding the token can access the medium.

The user access control equipment is shown functionally in Figure 12.12. "Station management," an important part of the control function, is shown on the right. The figure also shows the layer relationship with the OSI model. Also, the similarity with CSMA/CD is apparent. Key, again, is the medium access control.

Specific responsibilities of the MAC include ordered access to the medium, providing a means of admission and deletion of stations on the LAN and the handling of fault recovery. Among the faults handled are:

- Multiple tokens.
- Lost tokens.
- Token pass failure.

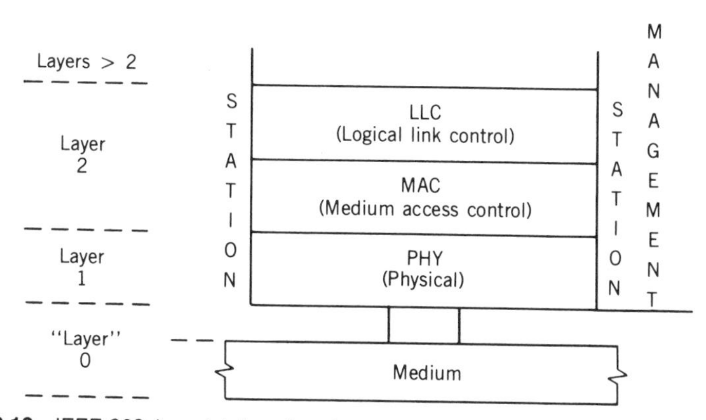

Figure 12.12 IEEE 802.4 model: functional relationship of user access control equipment.

- Nonresponsive stations (i.e., a station with an inoperative receiver).
- Duplicate station addresses.

It should be noted that no station takes on an exclusive monitoring and control function.

As with the CSMA/CD protocol, slot time is also an important parameter with the token-passing bus protocol. Slot time in this case is defined as the maximum time any station need wait for an immediate answer from another station. It is measured in octet times and is an integer. Slot time is twice the sum of the propagation delay, station delay, and safety margin. The *response window* equals the slot time. If a station waiting for a response hears a transmission start during the response window, that station does not transmit again until at least the received transmission terminates.

Holding the token gives the right to transmit. The token is passed from station to station in descending numerical order of station address. When a station hears a token frame addressed to itself, it "has the token" and may transmit data frames. After a station has completed transmitting data frames and has completed maintenance functions where necessary, the station passes the token to its successor by sending a *token MAC frame*. It then goes through a procedure to ensure that its successor has the token. For instance, after sending the token MAC frame, the station listens for evidence that the successor has heard the token frame and is active. If the sender hears a valid frame following the token, it can assume that its successor has the token and is transmitting. If the token sending station does not, it attempts to assess the state of the network.

If the token-sending station hears a noise burst or a frame with an incorrect FCS, it cannot be sure from the source address which station sent the frame. If a noise burst is heard, the token-passing station sets an internal indicator and continues to listen in the *check token pass state* up to four more slot times. If

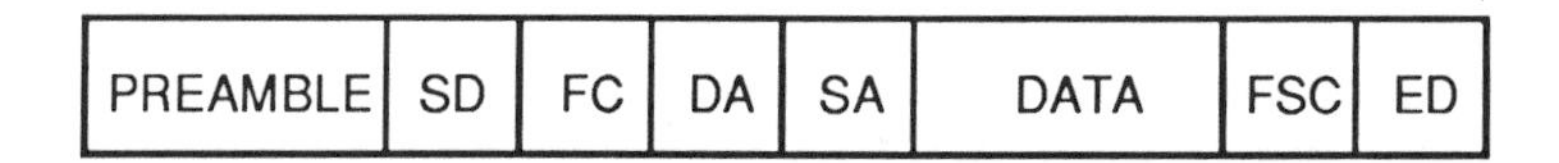

WHERE:		
	PREAMBLE -	PATTERN FOR SYNC AND SET RECEIVE LEVEL (1 OR MORE OCTETS)
	SD -	STARTING DELIMITER (1 OCTET)
	FC -	FRAME CONTROL (1 OCTET)
	DA -	DESTINATION ADDRESS(ES) (2 OR 6 OCTETS)
	SA -	SOURCE ADDRESS (2 OR 6 OCTETS)
	DATA -	INFORMATION (0 OR MORE OCTETS)
	FCS -	FRAME CHECK SEQUENCE (4 OCTETS)
	ED -	ENDING DELIMITER (1 OCTET)

Figure 12.13 Token bus frame format. From Ref. 6.

nothing is heard during the four-slot time delay, the station assumes that its successor has the token.

If the token holder does not hear a valid frame after sending the first time, it repeats the token pass procedure once more. If the successor does not transmit after the second attempt, it assumes that the successor has failed. The sender then sends a *who follows frame* with the successor's address in the data field of the frame. All active stations on the LAN then compare the value of the data field of a "who follows frame" with the address of their own predecessor, which is the station that normally would send them the token. The station whose predecessor is the successor of the sending station responds to the "who follows frame" by sending its address in a *set successor frame*. The station holding the token thus establishes a new successor, bridging the failed station out of the logical ring.

Figure 12.13 shows the MAC transmit frame format. The number of octets between the start delimiter (SD) and the end delimiter (ED) should be no greater than 8191. The abort sequence consists of an SD and an ED, each of which is 1 octet long [6].

The physical layer, as shown in Figure 12.12, connects to the medium via a short stub cable terminated in a T connector through which the main cable passes. The bit error rate at this interface is 1×10^{-8}, with a mean undetected bit error rate of 1×10^{-9}. The basic transmission medium is 75-Ω coaxial cable configured as an unbranched trunk. Extension of the topology is accomplished by means of active regenerative repeaters that are connected to span the branches.

Phase continuous FSK is one of the several transmission methods given in IEEE Std. 802.4 [6]. Successive symbols presented to the physical layer at the MAC interface are encoded, producing a three PHY-symbol code: H, L, and OFF. These symbols feed a two-tone FSK modulator where the H symbol is the higher frequency tone, the L symbol is the lower frequency tone, and the OFF symbol is no tone. The output signal is then ac-coupled to the coaxial cable.

IEEE Std. 802.4 standardizes the line data rate at 1 Mbps with a tolerance of $\pm 0.01\%$ for an originating station and $\pm 0.015\%$ for a repeater station. There are five possible symbols for the FSK implementation: 0, 1, nondata, pad-idle, and silence. Each of these MAC symbols is encoded into a pair of PHY-symbols from a different three symbols H, L, OFF code as follows:

1. Silence (OFF OFF).
2. Pad-idle. Pad-idle symbols are always octets. Each pair of pad-idle symbols is encoded as a sequence of LH, HL.
3. 0 (HL).
4. 1 (LH).
5. Nondata. The MAC layer transmits nondata symbols in pairs, which are encoded as the sequence LL HH.

The start frame delimiter subsequence is *nondata nondata 0 nondata nondata 0*, which is encoded as the sequence LL HH HL LL HH HL. The end-frame delimiter subsequence is *nondata nondata 1 nondata nondata 1*, which is encoded as the sequence LL HH LH HH LH.

The line signal corresponds to an FSK signal with its carrier frequency at 5.00 MHz, varying smoothly between two signaling frequencies of 3.75 MHz ± 80 kHz and 6.25 MHz ± 80 kHz. When transitioning between two signaling frequencies, the FSK modulator changes its frequency in a continuous and monotonic manner within 100 ns. The output signal level of the modulated carrier frequency is between $+54$ dB and $+60$ dB relative to 1 mV across 37.5 Ω. A S/N ratio of 20 dB at the receiver produces a BER of 1×10^{-8} or better. This is based on an in-band noise floor of $+4$ dB or less relative to 1 mV across 37.5 Ω.

Phase coherent FSK is another transmission method specified in IEEE Std. 802.4. The waveform in this case is somewhat different from that of continuous FSK. Again successive MAC symbols presented to the physical layer entity at its MAC interface are applied to an encoder that produces a three PHY symbol code: H, L, and OFF. The output is then applied to a two-tone FSK modulator that represents each H symbol as one full cycle of a tone whose period is exactly one-half of the MAC symbol period, each L as one-half cycle of a tone whose full-cycle period is exactly the MAC symbol period, and each OFF as no tone for the same half MAC symbol period. This modulated signal is then ac-coupled to the coaxial cable medium.

The standard data signaling rates for phase coherent FSK systems are 5 Mbps and 10 Mbps. The permitted tolerance for each signaling rate is $\pm 0.01\%$ for an originating station and $\pm 0.015\%$ for a repeater station. The coding is somewhat different from its continuous FSK counterpart.

For the 5-Mbps signaling rate the lower-tone FSK frequency is 5 MHz and the higher-tone FSK frequency is 10 MHz. For the 10-Mbps data rate the lower-tone FSK frequency is 10.0 MHz and the higher-tone FSK frequency is 20.0 MHz. The output level of the transmitted signal into a 75-Ω resistive load is +60 to +63 dB relative to 1 mV (dBmV). The S/N ratio at the receiver for a BER of 1×10^{-8} or better is 10 dB based on an in-band noise floor of −25 dBmV.

Turning now to a transmission loss plan for the LAN, IEEE Std. 802.4 does not specify insertion and tap loss values. However, some suggested typical values conforming to the specification are a tap insertion loss of 0.5 dB and a drop loss of 14 dB over the frequency range 0.1–30 MHz. If a LAN has 32 drops and operates with phase coherent FSK, the loss budget would be as follows:

Drop loss at first tap	14 dB
Insertion loss of 30 intervening taps	15 dB
Drop loss at 32nd tap	14 dB
Total lumped losses	43 dB

If we then assume a transmit level of +60 dB (relative to 1 mV) and a receive sensitivity threshold (BER = 1×10^{-8}) of −15 dB (relative to 1 mV), the cable loss should not exceed 32 dB over the band of interest (i.e., 60 − (−15) − 43 = 32). Because the band of interest extends to 30 MHz, the 32 dB should be measured at 30 MHz given the exponential loss characteristic with frequency of coaxial cable. It should be noted that these calculations hold for an unbranched cable topology.

Branched topologies may be implemented by impedance-matched nondirectional power splitters. These are three-port passive networks that divide the signal incident at one port into two equal parts that are transmitted to the other two ports. The insertion loss between any two ports of a typical splitter is 6.1 dB. When a LAN contains branches, a separate loss budget should be carried out for each possible end-to-end path to ensure that each user meets receive threshold requirements plus some margin.

Drop cables specified in IEEE Std. 802.4 for phase coherent FSK LANs have a 75-Ω impedance and typically should not exceed 30 m in length to maintain a drop cable loss of less than 1 dB. For phase continuous FSK the drop cable has a characteristic impedance of 37.5 Ω, and the drop cable has a much more limited length.

IEEE Std. 802.4 also specifies a broadband implementation for token pass bus using a CATV-like 75-Ω bidirectional broadband cable using midsplit, subsplit, or high-split techniques. It can incorporate 1-, 5-, and 10-Mbps data rates. The modulation waveform on the cable is multilevel AM/PSK, where the amplitude of each output pulse corresponds directly to the relative

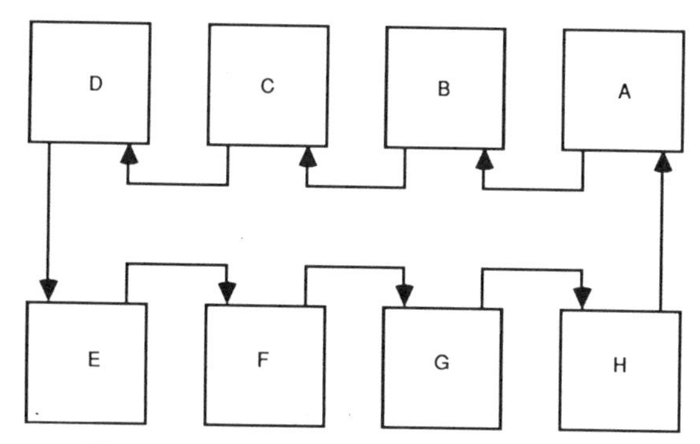

Figure 12.14 A token passing ring network.

numerical value of the associated PHY symbol. The conventional head end or CRF is required to provide each MAC entity with full-duplex send and receive capability on a single cable.

5.5 Token Ring

A typical token passing ring LAN is shown in Figure 12.14. The token ring operation, as specified in IEEE Std. 802.5 [7], has the capability of 1-Mbps or 4-Mbps data rate. A ring is formed by physically folding the medium back onto itself. Information is transmitted sequentially, bit by bit, from one active station to the next. Each station regenerates and repeats each bit and serves as a means of attaching one or more data terminals (CPUs, workstations) to the ring for the purpose of communicating with other devices on the network. As a traffic frame passes around the ring, all stations copy the traffic. Only those stations included in the address field pass that traffic on to the appropriate users that are attached to that station. The traffic frame continues onward to the originator, who then removes the traffic from the ring. The *pass-back* to the originator acts as a form of acknowledgment that the traffic had at least passed by the destination(s).

The sequential connection of stations removes the need to form a logical ring, as in token bus operation. A reservation scheme is used to accommodate priority traffic. Also, one station acts as ring monitor to ensure correct network operation. A monitor devolvement scheme to other stations is provided in case a monitor fails or drops off the ring (i.e., shuts down). A station on the ring can become inactive (i.e., close down), and a physical bypass is provided for this purpose.

A station gains the right to transmit frames onto the medium when it detects a token passing on the medium. Any station with traffic to transmit, on detection of the appropriate token, may capture the token by modifying it to a start-of-frame sequence and append the proper fields to transmit the first

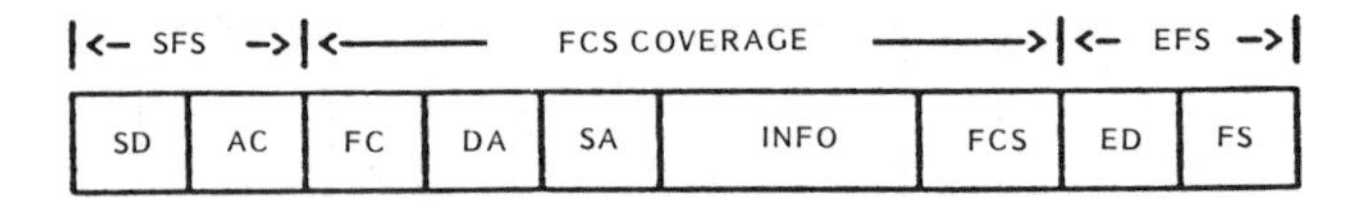

WHERE:

SFS - START-OF-FRAME SEQUENCE
SD - STARTING DELIMITER (1 OCTET)
AC - ACCESS CONTROL (1 OCTET)
FC - FRAME CONTROL (1 OCTET)
DA - DESTINATION ADDRESS (2 OR 6 OCTETS)
SA - SOURCE ADDRESS (2 OR 6 OCTETS)
INFO - INFORMATION (0 OR MORE OCTETS)*
FCS - FRAME CHECK SEQUENCE (4 OCTETS)
EFS - END-OF-FRAME SEQUENCE
ED - ENDING DELIMITER (1 OCTET)
FS - FRAME STATUS (1 OCTET)

* ALTHOUGH THERE IS NO MAXIMUM LENGTH FOR THE INFORMATION FIELD, THE TIME REQUIRED TO TRANSMIT A FRAME MAY BE NO GREATER THAN THE TOKEN HOLDING PERIOD ESTABLISHED FOR THE STATION.

NOTE: TOKEN CONSISTS OF SD, AC AND ED FIELDS

Figure 12.15 Frame format for token ring based on IEEE Std. 802.5 [7].

frame. At the completion of its information transfer and after appropriate checking for proper operation, the station initiates a new token, which provides other stations with the opportunity to gain access to the ring. Each station has a token holding timer that controls the maximum period of time a station may occupy the medium before passing the token on.

Figure 12.15 shows the frame format specified in IEEE Std. 802.5, and Figure 12.16 shows the token format. In these figures the left-most bit is transmitted first. The frame format, as shown in Figure 12.15, is used for transmitting both MAC and LLC messages to destination station(s). It may or may not contain an INFO field.

The starting delimiter (SD) consists of the symbol sequence JK0JK000, where J and K are nondata symbols and are described below. Both frames and tokens start with the SD sequence.

The access control (AC) is 1 octet long and contains 8 bits that are formatted PPPTMRRR. The first three bits, PPP, are the priority bits. These are used to indicate the priority of a token and therefore which stations are allowed to use the token. In a system designed for multiple priority, there are eight levels of priority available, where the lowest priority is PPP = 000, and the highest PPP = 111. The AC field contains the token bit T and the monitor bit M. If T = 0, then the frame is a token (Figure 12.16). The T bit is a 1 on

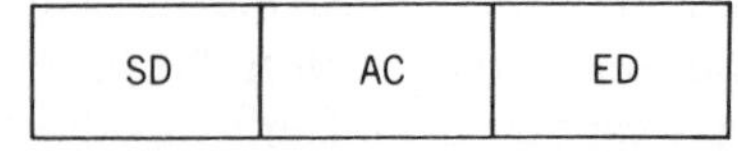

SD = Starting Delimiter (1 octet)
AC = Access Control (1 octet)
ED = Ending Delimiter (1 octet)

Figure 12.16 Token format based on IEEE Std. 802.5 [7].

all other frames. The M bit is set by the monitor as part of the procedures for recovering from malfunctions. The bit is transmitted as a 0 in all frames and tokens. The active monitor inspects and modifies this bit. All other stations repeat this bit as received. The three R bits are reservation bits. These bits allow stations with high-priority protocol data units (PDUs) to request that the next token issued be at the requested priority.

The next field in Figure 12.15 is the frame control (FC) field. It is one octet long. The first two bits in the field are the FF or frame-type bits, and the last six bits are control bits. The FF bits indicate the type of frame. If FF = 00, the frame is a MAC frame (i.e., it contains a MAC PDU). If FF = 01, it is an LLC frame (i.e., it contains an LLC PDU). The six control bits provide further control for either the MAC or LLC PDU.

Each frame (not token) contains a destination address (DA) and a source address (SA). Depending on the LAN system, each destination address may have 48-bit or 16-bit fields. In either case the first bit indicates whether as frame is addressed to an individual station (0) or to a group of stations (addresses) (1). The source address (SA) indicates the originator of the frame and may be 2 or 6 octets long, depending on the LAN system.

The INFO field carries the user data intended for the MAC, NMT (network management), or LLC. Although there is no maximum length specified for the information field, the time required to transmit a frame may be no greater than the token holding period that has been established. The format of the information field is indicated by the frame-type bits of the FC field.

The frame check sequence (FCS) is a 32-bit sequence based on the standard generator polynomial of degree 32 given in Section 5.3.1. It encompasses the FC, DA, SA, and INFO fields. Its transmission commences with the coefficient of the highest term.

The end delimiter (ED) is one octet in length and is transmitted as the sequence JK1JK1IE. The transmitting station transmits the delimiter as shown. Receiving stations consider the ending delimiter (ED) valid if the first six symbols JK1JK1 are received correctly. The I is the intermediate frame bit and is used to indicate whether a frame transmitted is a singular frame or whether it is a multiple frame transmission. The I bit is set at 0 for the singular frame case. The E bit is the error-detected bit. The E bit is transmitted as 0 by the station that originates the token, abort sequence, or frame. All stations on the ring check tokens and frames for errors such as FCS errors and nondata symbols. The E bit of tokens and frames that are repeated is set to 1 when a frame with an error is detected; otherwise the E bit is repeated as received.

The last field in the frame is the frame status (FS) field. It consists of one octet of the sequence ACrrACrr. The r bits are reserved for future standardization and are transmitted as 0's and their value is ignored by the receiver. The A bit is the address-recognized bit, and the C bit is the frame-copied bit. These two bits are transmitted as 0 by the frame originator. The A bit is changed to 1 if another station recognizes the destination address as its own or relevant group address. If it copies the frame into its buffer, it then sets the C bit to 1. When the frame reaches to the originator again, it may differentiate among three conditions:

1. Station nonexistent or nonactive on the ring.
2. Station exists but frame was not copied.
3. Frame copied.

Fill is used when a token holder is transmitting preceding or following frames, tokens, or abort sequences to avoid what would otherwise be an inactive or indeterminate transmitter state. Fill can be either 1's or 0's or any combination thereof and can be any number of bits in length within the constraints of the token holding timer.

IEEE Std. 802.5 describes a true baseband transmitting waveform using differential Manchester coding. It is characterized by the transmission of two line signal elements per symbol. An example of this coding is shown in Figure 12.17. The figure shows only the data symbols 1 and 0 where a signal element of one polarity is transmitted for one-half the duration of the symbol to be transmitted, followed by the contiguous transmission of a signal element of the opposite polarity for the remainder of the symbol duration. The following

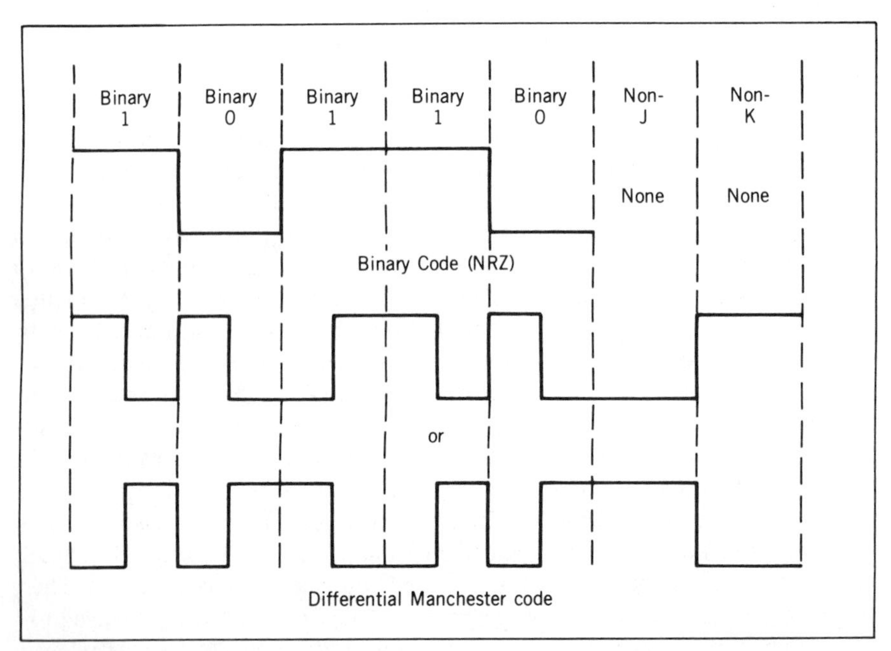

Figure 12.17 Differential Manchester coding format for symbols 1, 0, nondata J, and nondata K (IEEE Std. 802.5, Ref. 7).

advantages accrue from using this type of coding:

- The transmitted signal has no dc component and can be inductively or capacitively coupled.
- The forced midsymbol transition provides inherent timing information on the channel.

The nondata symbols J and K depart from the rule in that a signal element of the same polarity is transmitted for both signal elements of the symbol and therefore there is no midsymbol transition. A J symbol has the same polarity as the preceding symbol. The transmission of nondata symbols occurs in pairs (i.e., JK) to avoid accumulating a dc component.

All stations on the LAN ring are slaved to the active monitor station. They extract timing from the received data by means of a phase-locked loop. *Latency* is the time, expressed in number of bits transmitted, for a signal element to proceed around the entire ring. In order for the token to circulate continuously around the ring when all stations are in the repeat mode, the ring must have a latency of at least the number of bits in the token sequence, that is 24. Since the latency of the ring varies from one system to another and no a priori knowledge is available, a delay of at least 24 bits should be provided by the active monitor.

IEEE Std. 802.5 specifies that the connection of a station to the ring trunk cable be via shielded cable containing two balanced 150-Ω twisted pairs. The magnitude of the transmitted signal measured at the medium interface cable when terminated in 150 Ω resistive is specified as 3.0–4.0 V peak to peak. The amplitude of the positive and negative transmitted level should be balanced within 5%. The error rate required of the LAN is established by mutual agreement among users but in no case should it be worse than 1×10^{-8} [7].

5.6 Fiber Distributed Data Interface

Fiber distributed data interface (FDDI) is a LAN protocol that uses a fiber optic transmission medium. The data transmission rate is 100 Mbps. It uses a 4-out-of-5 code on the line such that the line modulation rate is 125 Mbaud. FDDI can support 500 stations linked by 100 km of cable.

FDDI operates in a token ring format and is an outgrowth of the IEEE 802.5 standard. Table 12.2 compares FDDI with IEEE Std. 802.5. Its similarity to the IEEE 802 series is obvious. Many of the acronyms are similar, such as LLC, MAC, and PHY.

FDDI II will be a follow-on of FDDI providing a circuit-switched mode of operation. It allocates time slots of FDDI to circuit-switched data in increments of 6.114-Mbps isochronous channels. Up to 16 isochronous channels may be assigned using a maximum of 98.304 Mbps of the total capacity. A 1-Mbps residual token channel capacity remains even when all the isochronous

TABLE 12.2 Token Ring Comparison

FDDI	IEEE 802.5
Half-duplex architecture and symbol- (or byte-) level manipulation	Full-duplex architecture and bit-level manipulation
Token sent immediately behind packet	Token sent only after source address has returned
Traffic regulated through timed token	Traffic regulated through priority and reservation bits on each packet
Uses 4-out-of-5 group coding, up to 10% dc component	Uses differential Manchester coding, no dc component
Ring is decentralized; individual clocks limit packet size	Centralized control with active monitor clock, allowing very long packets
Fiber optic medium	Wire-pair medium

channels are operational. Each of the 6.144-Mbps channels provides a full-duplex data highway, and each may be broken down still further into three 2.048-Mbps or four 1.536-Mbps data highways corresponding to CEPT and North American interfaces, respectively [8].

FDDI uses a counterrotating ring topology, as shown in Figure 12.18. Each class A type station is connected to both rings. A class B station has a simpler connectivity operating as a tail off of a concentrator. Concentrators may be used to attach stations to the ring. Each of these class B stations has a direct point-to-point connection to the concentrator. This allows any combination of class B stations to be switched out of the ring while retaining full connectivity for the remainder of the stations. Further, each class A station has a bypass switch that can be activated when a fault occurs or when it desires to drop out of the network.

FDDI frame and token formats are shown in Figure 12.19. Note the similarity to Figures 12.15 and 12.19. The FDDI frame format is used for transmitting both MAC and LLC messages to destination station(s). The frame may or may not have an information field. The MAC controls the maximum frame length, as required by the physical layer (PHY). For the purpose of counting frame length, all the fields in Figure 12.19, excluding the preamble (PA), are counted. The physical layer (PHY) of FDDI requires limiting the maximum frame length to 9000 symbols (4500 octets), including the 4 symbols of the preamble.

The following, which is taken from Ref. 9, is a brief description of the frame and token fields. Now refer to Figure 12.19. The preamble (PA) of a frame is transmitted by the frame originator as a minimum of 16 symbols of the idle pattern. Physical layers (PHY) of subsequent repeating stations may change the length of the idle pattern consistent with PHY clocking requirements. Thus repeating stations may see a variable length preamble that may be shorter or longer than the originally transmitted preamble. Stations are required to process frames of 12 or more symbols. A station is not required to process frames with a preamble of less than 12 symbols. However, if a

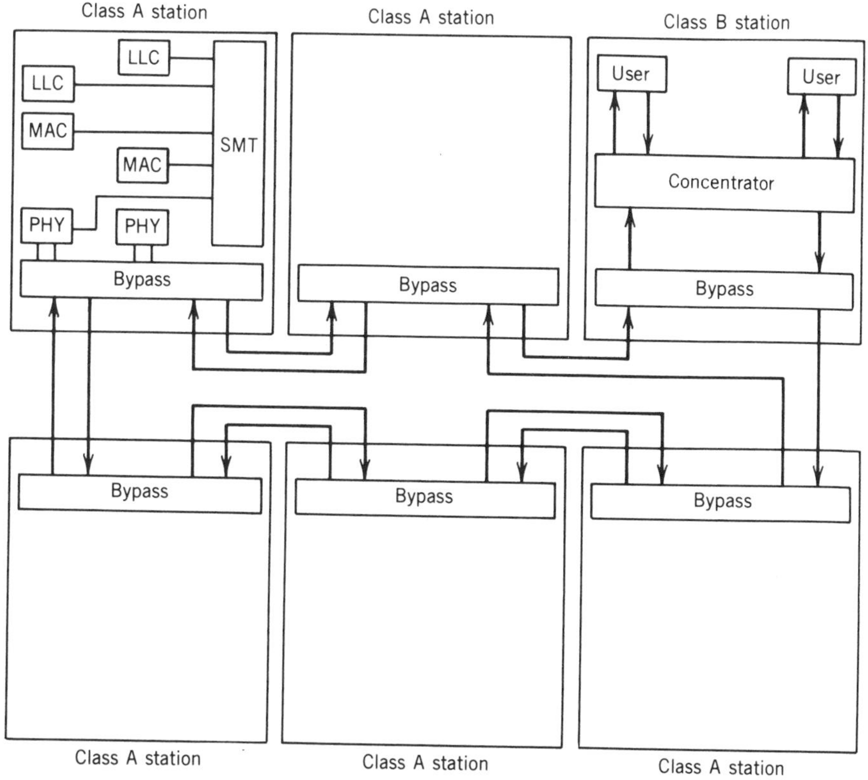

Figure 12.18 Typical FDDI (fiber) LAN layout showing counterrotating ring topology. SMT = station mangement; SMT carries out the supervisory function in an FDDI station.

particular station does not repeat such a frame, then that frame is stripped from the ring.

A starting delimiter (SD) contains a J and a K symbol (described later). No frame or token is considered valid unless it starts with this explicit sequence. After the SD there is the frame control (FC) field, which defines the type of frame and associated control functions. It contains a class bit (C), an address length bit (L), format bits (FF) and four control bits (ZZZZ).

Destination and source addresses (DA and SA) follow the FC. These addresses may be 16 bits or 48 bits in length. However, all stations must have a 16-bit address capability. A station with only a 16-bit address capability must be capable of functioning in a ring with stations concurrently operating with 48-bit addresses.

The information field (INFO) contains zero, one, or more data symbol pairs whose meaning is determined by the FC field and whose interpretation is made by the destination entity (e.g., MAC, LLC, or SMT). The length of the INFO field is variable. It does, however, have a maximum length restriction of 9000 symbols, including 4 symbols for the preamble.

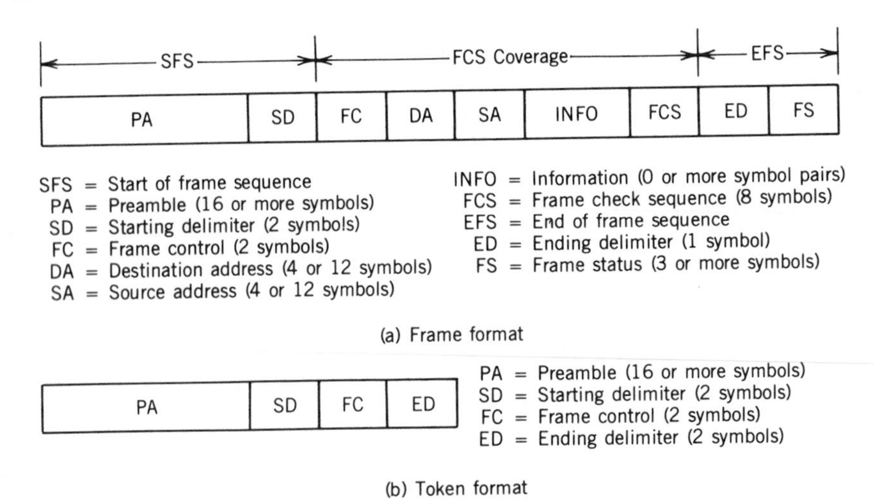

Figure 12.19 FDDI frame and token formats [9].

The frame check sequence (FCS) field is used to detect bits in error within the frame as well as erroneous addition or deletion of bits to the frame. The fields covered by the FCS include the FC, DA, SA, INFO, and FCS fields. The FCS generating polynomial is the ANSI 32-bit standard algorithm [eq. (12.1)].

The ending delimiter (ED) is the symbol T, which ends tokens and frames. Ending delimiters and optional control indicators form a balanced symbol sequence. These are transmitted in pairs so as to maintain octet boundaries. This is accomplished by adding a trailing symbol T as required. The ending delimiter of a token consists of two consecutive T symbols; the ending delimiter of a frame has a single T symbol.

The frame status (FS) field consists of an arbitrary length sequence of control indicator symbols (R and S). The FS field follows the ending delimiter of a frame. It ends if any symbol other than R and S is received. A trailing T symbol, if present, is repeated as part of the FS field. The first three control indicators of the FS field are mandatory. These indicators are indicating error detected (E), address recognized (A), and frame copied (C). The use of additional trailing control indicators in the FS field is optional and may be defined by the user.

The first R/S symbol that appears is the E indication, the second the A indication and the third the C indication. The final symbol is the terminating T symbol. Table 12.3 shows the FDDI 4B/5B coding scheme and hex equivalents and clarifies the special symbols given such as I, J, K, T, R, and S.

5.6.1 FDDI Timers and Timing Budget. The draft ANSI standard (ANSI X3T9, Ref. 9) states that each FDDI station will maintain three timers to regulate operation of the ring. The values of these timers are locally administered. They may vary from station to station on the ring, provided that the

TABLE 12.3 FFDI 4B / 5B Coding Scheme

Symbol	Item	Assignment	(5) Code	(4) Binary Equivalent
I	Line	Idle	11111	
J	Start delimiter	SD pair	11000	
K			10001	
		(HEX)		
0	Data	0	11110	0000
1	"	1	01001	0001
2	"	2	10100	0010
3	"	3	10101	0011
4	"	4	01010	0100
5	"	5	01011	0101
6	"	6	01110	0110
7	"	7	01111	0111
8	"	8	10010	1000
9	"	9	10011	1001
A	"	A	10110	1010
B	"	B	10111	1011
C	"	C	11010	1100
D	"	D	11011	1101
E	"	E	11100	1110
F	"	F	11101	1111
T	End delimiter terminates data stream		01101	
R	Control indicators	Denotes logical zero (reset)	01101	
S		Denotes logical one (set)	11001	

Note: Order of serial bit transmission is from left to right.
Extracted from Ref. 10.

applicable ring limits are not violated. These timers are the token holding timer (THT), the valid transmission timer (TVX), and the token rotation timer (TRT).

The following elements of timer calculation are taken from the FDDI standard [9]:

$$D_{\max} = 1.617 \text{ ms (default)} = \text{Maximum ring latency}$$

$D_{\max}$ is the maximum latency (i.e., circulation delay) for a start delimiter to travel around the ring expressed in time. It consists of the total ring cable delay plus the total station latency of all stations. It can accommodate a wide variety of topologies. For example: Considering only the path delay component for a 200-km path, the SD delay to travel this distance is 1.017 ms, assuming an approximate value of velocity of propagation of 5,085 ns/km. The 200-km path length limit allows for a total *cabled* ring length of 100 km, which accommodates the round-trip path length that exists between class B type connections as well as the total cable length for trunks formed by class A connections folded back onto themselves during times when configured as a chain. If we subtract from the total maximum ring latency (1.617 ms) the 1.017-ms path delay value, we are left with 0.600 ms. If we then assume a total number of physical connections to be 1000 and divide this value into 0.600 ms, we are left with a latency per station of 600 ns, or 15 symbols per physical connection.

Other values given in Ref. 9 are

$$M_{max} = 1000 \text{ (default)} = \text{Maximum number of MAC entities}$$

$$I_{max} = 25.0 \text{ ms} = \text{Maximum station physical insertion time}$$

$$A_{max} = 1.0 \text{ ms} = \text{Maximum signal acquisition time}$$

$$\text{Token time} = 0.00088 \text{ ms} = \text{Token length } (6 + 16 \text{ symbols})$$

$$L_{max} = 0.0035 \text{ ms} = \text{Maximum transmitter frame setup time}$$

$$F_{max} = 0.361 \text{ ms} = \text{Maximum frame time (maximum length of frame is 9000 symbols plus 16 preamble symbols)}$$

$$\textit{Claim FR} = 0.00256 \textit{ ms} = \textit{Claim frame length}$$

The claim FR is the time required to transmit a claim frame and its 16 symbol preamble. We assume a minimum length claim frame using long (48-bit) addresses. Continuing with the values:

$$S_{min} = 0.3645 \text{ ms} = \text{Minimum safety allowance}$$

which is the safety timing allowance for the recovery from random noise on the ring.

Let us return to the discussion of timers. The token holding timer (THT) controls how long a station may transmit asynchronous frames. The valid transmission timer (TVX) allows a station to recover from transient ring error situations. The time-out value of the TVX is determined such that:

$$\text{TVX} > D_{max} + \text{Token time} + F_{max} + S_{min} > 2.35 \text{ ms}$$

F_{max} is substituted for D_{max} in the above relation if it is greater than D_{max}.

The token rotation timer (TRT) is used to control the ring scheduling during normal operation and to detect and recover from serious ring error rate degradation.

5.6.2 FDDI Ring Operation. Access to the ring is controlled by passing the token around the ring (as in standard token ring practice). The token gives the downstream station (the receiving station relative to the station passing the token) the opportunity to transmit a frame or a sequence of frames. If a station wants to transmit, it strips the token from the ring before the frame control field of the token is repeated. After the captured token is completely received, the station begins transmitting its eligible queued frames. After completion of transmission, the station issues a new token for use by a downstream station.

Stations that have nothing to transmit at a particular time merely repeat the incoming symbol stream. While in the process of repeating the incoming symbol stream, the station determines whether the information is intended for that station. This is done by matching the destination address to its own address or a relevant group address. If a match occurs, subsequent received symbols up to the FCS are processed by the MAC or sent to the LLC.

Frame stripping is an important concept. Each transmitting station is responsible for striping frames from the ring that it originated. This is accomplished by stripping the remainder of each frame whose source address matches the station's own address from the ring and replacing it with idle symbols (symbol I in Table 12.3).

It will be noted that the process of stripping leaves remnants of frames consisting of the PA, SD, FC, DA, and SA fields, followed by idle symbols. This happens because the decision to strip a frame is based on recognition of the station's own address in the SA field. This cannot occur until after the initial part of the frame has already been repeated and passed on to the next downstream station. These remnants cause no ill effects. This is because of various specified criteria, including recognition of an ending delimiter (ED) must be met to indicate that a frame is valid. To the level of accuracy required for statistical purposes, these remnants can be distinguished from error or lost frames because they are always followed by an idle symbol (I) pattern. Such remnants are removed from the ring when they encounter the first transmitting station.

The FDDI draft specification distinguishes *asynchronous* from *synchronous* transmission. In this context asynchronous transmission is a class of data transmission service whereby all requests for service contend for a pool of dynamically allocated ring bandwidth and response time. Synchronous transmission is a class of data transmission service whereby each requester is preallocated a maximum bandwidth and guaranteed a response time not to exceed a specific delay.

5.6.3 FDDI Clocking. Each FDDI station has an independent transmit clock with a stability of 5×10^{-5}. On the receive side the PHY provides an elasticity buffer of 10 bits. The elasticity buffer is inserted between the receiver and the transmitter. The receiver employs a variable clock that tracks the clock of the incoming bit stream from the upstream transmitter. The elasticity buffer is reinitialized at each station during the preamble (PA) that precedes each frame or token. This has the effect of increasing or decreasing the length of the PA that is initially transmitted as 16 or more symbols. The 10-bit elasticity buffer allows transmission of frames up to 4500 octets in length without exceeding the limits of the buffer.

6 REPEATERS, BRIDGES, AND GATEWAYS

Repeaters connect common LAN segments. A repeater in this case is a regenerative repeater. It reconstitutes the electrical signal based on decision circuits and a clock. For instance, to extend a CSMA/CD 500-m segment, a repeater is installed at the segment end, bridging it to a second segment that also can be up to 500 m long. In accordance with Ref. 4, four repeaters may be placed in tandem, deriving five segments or a total extension of 2.5 km for an Ethernet–CSMA/CD LAN. Figure 12.20A illustrates a LAN repeater's relationship with OSI..

Figure 12.20B shows a bridge's relationship with OSI. The function of a bridge is to connect *dissimilar* LANs. As we have discussed previously, 802

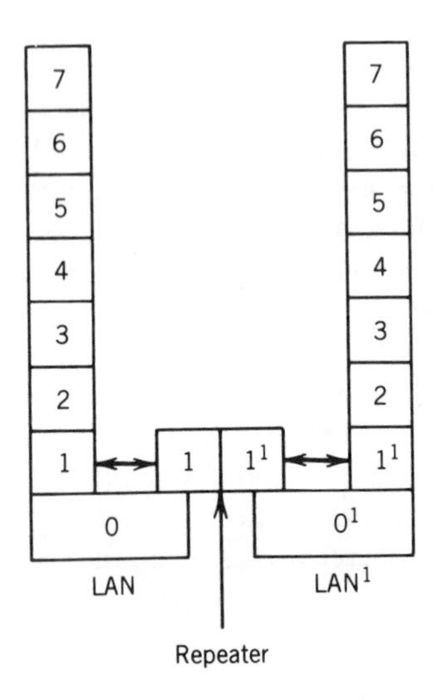

Figure 12.20A A LAN repeater extends a LAN. The LAN on the left-hand side is based on the same standard as the LAN on the right-hand side.

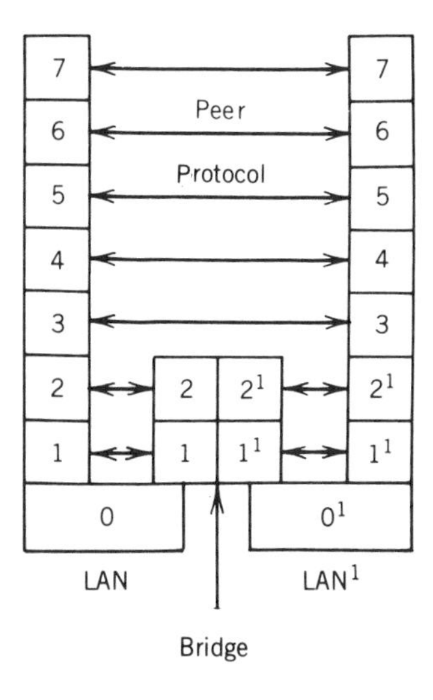

Figure 12.20B A bridge ties together two dissimilar LANs. The figure assumes that OSI levels 3–7 are compatible. The text refers to these as the "higher levels."

series LANs and FDDI require only the first two OSI layers. Thus a bridge has to carry out conversion of OSI layers 1 and 2 so that the protocols used at each peer level will interface and the two LANs will be capable of interworking. The figure assumes that what we have previously called "higher levels" are indeed compatible peer to peer.

A gateway provides an interface between a LAN and an external data network, typically a WAN. In this case all seven OSI levels may have to be

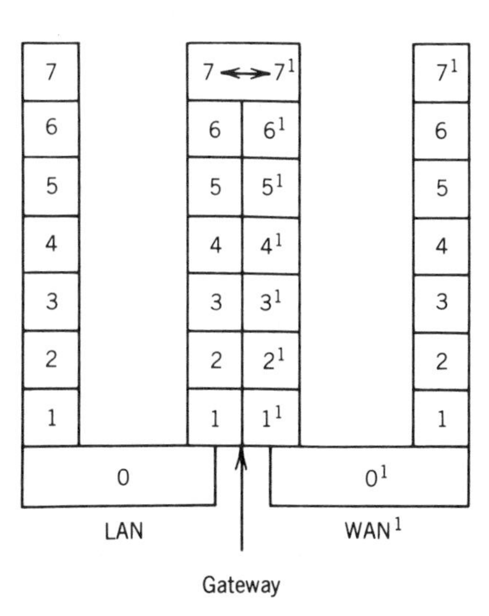

Figure 12.20C Gateways tie together data networks of different architecture. A common application is to interface a LAN with a WAN. The figure assumes that all seven OSI levels are incompatible.

made compatible at the gateway to permit peer-to-peer communication. This concept is shown in Figure 12.20C. The figure assumes incompatibility among all seven OSI layers.

7 PERFORMANCE ISSUES

In this section we discuss basic performance issues among the three IEEE 802 series LANs described in this chapter. In the case of bus topology we include CSMA/CD or Ethernet and token bus and for the ring topology, token ring. The underlying feature a LAN must possess is that, given a certain user group with specific traffic characteristics, the network must have adequate capacity for the expected load. In other words, the local area network should be able to keep up with the traffic load without undue delays.

There are two basic parameters that will gate this performance requirement: the data rate (R) on the medium and the average signal propagation delay (D) of the medium. Stallings [11] states that the most important parameter is the product of these two terms or $R \times D$. Other things held constant, a network's performance will be the same for equal value $R \times D$ products. Thus a 100-Mbps 1-km bus will have the same performance as a 10-Mbps 10-km bus assuming the same medium on each bus.

One thing that will help us to understand LAN performance is the number of bits that we can get onto the medium. Sometimes it is called the length of the transmission medium in bits. This can be stated as the number of bits that can be in transit between any two nodes when the leading edge of the first bit or symbol has just reached the receiving node. We then count the number of bits on the medium back to the transmitting node at any instant in time. If we assume a velocity of propagation of 2×10^8 m/s for the medium and a 10-Mbps Ethernet 500 m long, the leading edge takes 2.5×10^{-6} s to cover the 500 m. How many bits can be transmitted in that period of time where the duration of 1 bit is 1×10^{-7} s? The answer is $2.5 \times 10^{-6}/10^{-7} = 25$ bits. If we assume glass optical fiber has the same velocity of propagation and the total extension of the medium is 1 km at a data rate of 100 Mbps, then the length of this medium in bits is 500 bits. These calculations give us insight into collisions for CSMA and CSMA/CD systems. Given each sample case, a collision can occur (worst case) after a station has transmitted 25 bits for the former and 500 bits for the latter. Although these collision windows seem of short duration, as the number of accesses increases, the chances of collision begin to multiply.

This leads us to the next LAN performance measurement. The user should know where a LAN saturates and how it behaves under a heavy load of offered traffic and its effect on throughput. Overhead reduces throughput, which includes the overhead for controlling the protocol. Reference 11 makes three important points:

- CSMA/CD: Time wasted due to collisions; need for acknowledgment packets.
- Token ring: Time waiting for token if intervening stations have no traffic to send.
- Token bus: Time wasted on token transmissions and acknowledgment packets.

7.1 Comparisons of LAN Schemes

The following is a review of some advantages and disadvantages of CSMA/CD and token passing schemes.

CSMA/CD has properties that give advantages for lightly loaded networks where traffic requests are random in nature. CSMA/CD has the following advantages:

1. No waiting time once the bus is free. A user can transmit packets as soon as it has an opportunity.
2. No limitation on usage. A user can send another packet at once. Of course, this is subject to the operation of the CSMA/CD procedures.
3. User stations can be simplified because there is no need for a token monitor function at each station.
4. No protocol interchanges are required. There is no token passing. Because of this, overhead is reduced, simplifying user station design.
5. Performance for any user station is independent of the total number of stations on the bus. The only effect is when more than one station tries to use the bus at the same time.

Disadvantages of CSMA/CD are:

1. Inability to incorporate priority traffic such as voice, which has a real-time requirement. Immediate access to the bus cannot be guaranteed.
2. Performance of the bus degrades as the traffic load increases. As collisions increase, throughput degrades. Saturation tends to occur at about 50–60% of bus capacity [5].
3. The special coaxial cable required costs are currently six to ten times conventional coaxial cable [5].
4. It is doubtful that the system will be workable with 1024 users, the maximum specified [5].

Token passing has strengths where CSMA/CD has weaknesses and vice versa. Here are four advantages of token passing:

1. Performance can be calculated in the best and worst case because it is deterministic, and within these bounds access can be guaranteed.
2. Both token bus and token ring have means for handling traffic of different priorities.
3. Even under heavy loading, each user station will continue to have a fixed level of service, and throughput will stabilize.
4. The transmission medium is economical and readily available.

Disadvantages of token passing are:

1. There is a limitation on the traffic a user station can transmit before it must pass the token on and await its next turn.
2. There is a minimum time a station must wait after it passes the token on before it can transmit more traffic.
3. There is the additional overhead because token protocols require exchange of messages or, as a minimum, the token exchange.
4. User stations have the added complexity of the token monitor function, which adds cost.
5. Individual station performance is affected by the total number of active stations, regardless of their traffic transmission activity.

CSMA/CD is inefficient at higher traffic loads but is very efficient under light loading conditions. The converse is true for token passing, which is inefficient with light loading and efficient at the higher loading levels.

REVIEW QUESTIONS

1. Define a local area network. Contrast it with a wide area network.

2. Discuss how LANs can make multivendor processing equipment compatible.

3. What are the two basic underlying transmission techniques used for LANs? Compare these using a minimum of eight characteristics.

4. Name the three basic LAN topologies (i.e., network types). Name a fourth type that is a subset of one of the basic topologies.

5. What are the general ranges of bit error rates that can be expected from a well-designed LAN? How is the BER achieved (e.g., by ARQ, channel coding, S/N ratio)?

6. What are the two basic transmission problems that must be faced regarding the medium when designing a LAN?

7. If a LAN has 50 accesses, how many transmission connectivities must be analayzed?

8. Consider conventional broadband LANs. Discuss one- and two-cable operation. Include bandwidth utilization.

9. What is the function of a CRF or head end on a broadband LAN?

10. There are two methods of connecting user facilities to optical fiber LANs. What are they? Discuss the pros and cons of each.

11. What is a LAN access protocol?

12. Compare contention and polling. Describe how each leads to a particular LAN access protocol, such as token passing and CSMA/CD.

13. Why is loop polling more efficient than roll-call polling?

14. Relate the IEEE 802 LAN standards to the seven-layer OSI model. Describe the sublayering involved.

15. What is a SAP, and what is its function?

16. Describe the three basic communication services offered by the LLC. In the description show similarities and differences among the three.

17. Name at least three responsibilities of the LLC.

18. Describe the four fields of the LLC PDU and their functions.

19. How are collisions detected on Ethernet–CSMA/CD?

20. When a collision occurs with CSMA/CD, what happens? This should lead to a discussion of persistence algorithms and their relative efficiencies.

21. Given a 500-m length of coaxial cable with accesses at each extreme, in what time period can a collision occur? Assume a velocity of propagation of 2×10^8 m/s.

22. Describe differential Manchester coding and its application to baseband LAN transmission. Compare to conventional NRZ.

23. What is the function of the frame check sequence (FCS)?

24. In CSMA/CD operation, what is the function of the PAD field? Why is it necessary to have a minimum frame length?

25. Define *slot time*. Why is this an important parameter?

26. How are collisions avoided using a token passing scheme?

27. The MAC handles fault recovery with the token bus access scheme. Name at least four faults that can be handled by the MAC.

28. How does a user know that a traffic frame is directed to it?

29. Describe how a token holder handles a situation when its successor does not answer a token pass to it.

30. Differentiate between logical connectivity and physical connectivity. On which of the three LAN types can logical connectivity be used effectively?

31. In the token bus frame format we see the *data unit* field. Thinking of LLC, what would be inserted in that field?

32. Following the IEEE standards, is token bus truly a baseband system as we have defined it?

33. How does token ring differ from token bus? Name at least three differences.

34. Aside from the token, what are the three types of frames used on a token ring LAN?

35. Is token ring really a baseband system?

36. Define *latency* with respect to the token ring LAN.

37. What is the baud rate of FDDI? the bit rate?

38. Compare token ring operation based on IEEE Std. 802.5 and FDDI.

39. For both token ring operation and FDDI, name one simple way a transmitting station knows that its frame has been received by the intended destination.

40. What is *frame stripping*?

41. Regarding timing, what is the function of the elasticity buffer in FDDI?

42. Regarding LANs, give the purpose of a repeater, a bridge, and a gateway. Relate each to the OSI model.

43. Compare the operation of token passing schemes versus Ethernet–CSMA/CD schemes under light traffic loading and under heavy traffic loading conditions.

44. One measure of performance of a LAN is the $R \times D$ product. Discuss.

45. Define the term *primitive*. Name three generic primitives used on LANs.

REFERENCES

1. John F. Schoch, Yogen K. Delai, David D. Redell, and Ronald C. Crane, "Evolution of the Ethernet Local Computer Network," *Computer* (August 1982).
2. Richard N. Dunbar, "Design Considerations for Broadband Coaxial Cable Systems," *IEEE Commun. Mag.*, **24** (6) (June 1986).
3. W. Stallings, *Handbook of Computer-Communications Standards*, Vol. 2, *Local Area Networks*, Macmillan, New York, 1987.
4. *Carrier Sense Multiple Access with Collision Detection (CSMA/CD) Access Method and Physical Layer Specifications*, IEEE Std. 802.3, IEEE, New York, May 1986.
5. John McDonnel, "Internetworking and Advanced Protocols" (seminar), Network Technologies Group, Inc., Boulder, Col., 1985.
6. *Token-Passing Bus Access Method and Physical Layer Specifications*, IEEE Std. 802.4, IEEE, New York, March 1986.
7. *Token Ring Access Method and Physical Layer Specifications*, IEEE Std. 802.5, IEEE, New York, March 1985.
8. Floyd E. Ross, "FDDI: A Tutorial," *IEEE Commun. Mag.*, **24** (6) (May 1986).
9. "FDDI Token Ring Media Access Control," Draft of proposed Standard ANSI X3T9/84-100, American National Standards Institute, New York, February 1986.
10. V. Iyer and S. Joshi, "FDDI 100 Mbps Protocol Improves on 802.5 Specs 4 Mbps Limit," *EDN* (May 2, 1985).
11. William Stallings, "Local Area Network Performance," *IEEE Commun. Mag.*, **22** (2) (February 1984).
12. "LAN Protocol Analysis: Test Solutions for IEEE 802.3/Ethernet Local Area Networks" (seminar), Hewlett Packard, Colorado Telecommunications Division, Colorado Springs, Col., January 1987.
13. W. Stallings, ed., *Tutorial Local Area Network Technology*, 2nd ed., IEEE Computer Society Press, Washington, D.C., 1985.
14. *Logical Link Control*, IEEE Std. 802.2, IEEE Press, New York, 1985.

13

INTEGRATED SERVICES DIGITAL NETWORK

1 THE GOALS OF ISDN: BACKGROUND

The present analog telecommunications network is based on the 4-kHz voice channel. It has served well in providing speech telephony since the 1880s. In the nineteenth century the only other service was telegraph, which predated the telephone some 30 years. The two services evolved separately and distinctly. Before World War II there was some melding where telegraph and telex were carried as subcarriers on VF channels leased from telephone companies or administrations. This might be called the first move toward integrated services. However, it was probably done more for convenience and economy than for any forward thinking regarding integration.

Looking backward, telephony became ubiquitous, with a telephone in every office and in nearly every home. On the other hand, telegraphy evolved into telex but still took a backseat to telephony. Historically, facsimile was the next service that was integrated rapidly into the telephone network. Facsimile required a modem to make it compatible with analog telephony. In the office environment facsimile is often used in lieu of telex. Then in the 1950s computer-related data began to emerge, requiring some method of point-to-point relay. This relay facility was carried out by the ubiquitous telephone network. Again, a modem was required to integrate the service into the analog telephone network.

By this time the worldwide telephone network was in place and pervasive. Using that network turned out to be the most cost-effective method to communicate other information (i.e., other than speech telephony) from point X to point Y. Dial-up telephone connections provided one way of achieving switched service to transport that "other" information, whether point to point or multipoint, given the transmission limits of the VF channel traversing the analog network.

Digital telephony began to take hold after the development of the transistor in 1948. Solid-state circuitry, particularly in LSI, made pulse-code modulation (PCM) transmission and later PCM switching cost-effective. The first application of PCM was in the expansion of the trunk cable plant. In the United States it is estimated that by 1992 92% of the toll plant (long-distance plant) will be served by PCM. Certainly, by the year 2000 the US local telephone plant should follow suit.

PCM standards developed along two, or some could argue three, distinct routes. North America and Japan use a basic 24-channel system with in-band signaling and framing. Europe and much of the remainder of the world use a 30-channel format with 2 extra channels for separate channel signaling and framing. Japan has a distinct higher-level PCM hierarchy.

It would seem that a North American PCM design was more driven to serve speech telephony customers because more than 90% of the traffic on the network derives from such telephony customers. In-band signaling and framing corrupted the basic 64-kbps channel, and thus integrating other services, such as computer data, required a drop back to 56 kbps or less for the North American PCM. Typical of such a drop back is ATT's DDS (digital data system), which can offer only 56 kbps, not the more desirable 64 kbps.

ISDN has been developed to ease integration of all services, except full motion video, on one basic digital channel, namely 64 kbps. Whereas 4 kHz was the basic building block of analog telephony, 64 kbps is the basic building block of ISDN (integrated services digital network). The ISDN basic building block is designed to serve, among other services [1]:

- Digital voice.
- High-speed data, both circuit and packet switched.
- Telex/teletex.
- Telemetry.
- Facsimile.
- Slow-scan video.

The goal of ISDN is to provide an integrated facility to incorporate each of the services listed on a common 64-kbps channel and/or a combination of 64- and 16-kbps channels. This chapter provides an overview of the integration both for switching or signaling and transmission.

As you proceed through this chapter, you should be aware that we are dealing with user interfaces into an *existing* digital network (Chapter 10) in which CCITT signaling system no. 7 is operational (Chapter 14). In this chapter we are not dealing with the network itself, only how it affects the user, and how the ISDN user affects the network.

2 ISDN STRUCTURES

2.1 ISDN User Channels

Here we look from the user into the network. We consider two user classes: Residential and commercial. The following are the standard transmission structures for user access links:

- B-channel: 64 kbps.
- D-channel: 16 kbps.
- C-channel: 8 or 16 kbps.
- A-channel: 4 kHz analog VF channel.

The B-channel is the basic user channel and serves any one of the following traffic types:

- PCM-based digital voice channel.
- Computer digital data, either circuit or packet switched.
- A mix of multiplexed lower data rate traffic, such as vocoded (digital) low data rate voice and lower data rate computer data. However, in this category, the traffic must have the same destination.

The D-channel is a 16-kbps channel. It serves not only as the user signaling channel but also as a lower-speed data connectivity to the network. The A-channel serves as a transitional expedient to provide nominal 4-kHz analog connectivity to the network. The C-channel is associated with the A-channel to form a hybrid access arrangement.

There is an E-channel [2], which is a 64-kbps channel that is primarily used to carry signaling information for circuit switching by the ISDN. At the user–network interface, it is used only in the primary rate multiplexed channel structures as an alternative arrangement for multiple access configurations. The "primary rate" is discussed below.

The H-channels have the following bit rates:

- H0: 384 kbps.
- H1: 1536 kbps (H11) and 1920 kbps (H12).

The H-channel does *not* carry signaling information. Its purpose is to provide service for higher user data rates, such as digitized program channels, video (slow-scan) for teleconferencing, fast facsimile, and packet-switched data bit streams.

2.2 Basic and Primary User Interfaces

The *basic* interface structure is composed of two B-channels and a D-channel and is commonly referred to as "2B + D." The D-channel at this interface is 16 kbps. The B-channels may be used independently (i.e., two different simultaneous connections).

Appendix I to CCITT Rec. I.412 states that the basic access may also be B + D or D.

The *primary* rate B-channel interface structures are composed of *n* B-channels and one D-channel, where the D-channel in this case is 64 kbps. There are two primary B-channel data rates:

- 1.544 Mbps = 23B + D.
- 2.048 Mbps = 30B + D.

For the user–network access arrangement containing multiple interfaces, it is possible for the D-channel in one structure to carry signaling for B-channels in another primary rate structure without an activated D-channel. When a D-channel is not activated, the designated time slot may or may not be used to provide an additional B-channel, depending on the situation, such as 24B with 1.544 Mbps.

An alternative primary rate interface B-channel structure is composed of B-channels and an E-channel:

- 1.544 Mbps = 23B + E.
- 2.048 Mbps = 30B + E.

There are a number of H-channel interface structures covered in CCITT Rec. I.412. For the H0 structure, a D-channel may or may not be present and if present, is always 64 kbps. At the 1.544-Mbps rate interface the H0 channel structures are 4H0 or 3H0 + D. The use of the additional capacity has not as yet been determined (CCITT *Red Books*, 1984). The H11 and H12 structures use D-channels from other structures to carry signaling, if required.

3 USER ACCESS AND INTERFACE

3.1 General

ISDN, when fully implemented, may be described as ubiquitous and universal. These terms, of course, can be argued. Sixty-four kbps is the principal user data rate. We can argue that a much higher rate should be considered (i.e., such as B-ISDN). Consider the trend in LANs. For many users a 1-Mbps LAN is not a high enough bit rate, and the trends seem to be 10 Mbps and 100 Mbps, such as with FDDI.

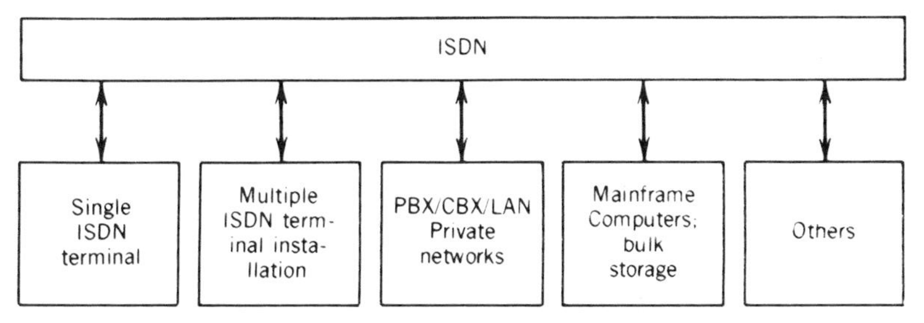

Figure 13.1 ISDN generic users.

Figure 13.1 shows generic ISDN user connectivity to the network. We can select either basic or primary service (e.g., 2B + D, 23B + D, or 30B + D) to connect to the ISDN network.

The objectives of any digital interface design, and specifically of ISDN access and interface, are:

1. Electrical and mechanical specification.
2. Channel structure and access capabilities.
3. User–network protocols.
4. Maintenance and operation.
5. Performance.
6. Services.

Figure 13.2 shows the ISDN reference model. It delineates interface points for the user. In the figure NT1, or network termination 1, provides the physical layer interface; it is essentially equivalent to OSI layer 1. The functions of the

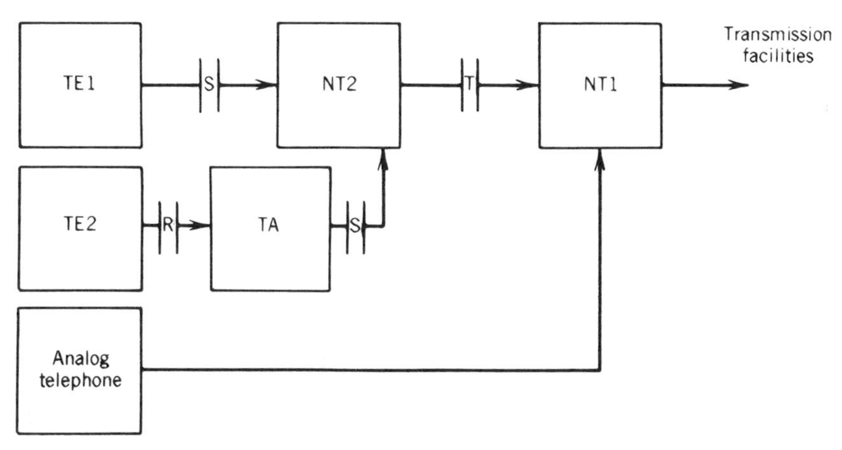

Figure 13.2 ISDN reference model.

physical layer (layer 1) include:

- Transmission facility termination.
- Layer 1 maintenance functions and performance monitoring.
- Timing.
- Power transfer.
- Layer 1 multiplexing.
- Interface termination, including multidrop termination employing layer 1 contention resolution.

Network termination 2 (NT2) can be broadly associated with OSI layers 1, 2, and 3. Among the examples of equipment that provide NT2 functions are user terminal controllers, LANs, and PABXs. Among the NT2 functions are [3, 4]:

- Layers 1, 2, and 3 protocols processing.
- Multiplexing (layers 2 and 3).
- Switching.
- Concentration.
- Interface termination and other layer 1 functions.
- Maintenance functions.

A distinction must be drawn here between North American and European practice. As we are aware, in Europe the telecommunication administrations are, in general, national monopolies that are government controlled. In North America (i.e., United States and Canada) they are private enterprises, often very competitive. Thus in Europe the NT1 and NT2 functions are combined and called NT12, and the equipment involved is the property of the telecommunication administration. Of course, in North America they are separate; one belongs to the user and the other to the telephone company.

TE1 in Figure 13.2 is the terminal equipment and has an interface that must comply with the ISDN user–network interface specifications. Terminal equipment (TE) covers functions broadly belonging to OSI layer 1 and higher OSI layers. Among this equipment are digital telephones, computer workstations (DTE), and other devices in the user end-equipment category.

TE2 refers to equipment that does *not* meet ISDN terminal–network interface specifications and that requires interface modifications to adapt the equipment to ISDN. A terminal equipment adapter (TA) provides the necessary conversion functions to permit TE2-type terminal equipment to interface with ISDN.

Reference points T, S, and R are used to identify the interface available at those points. T and S are identical electrically, mechanically, and from the point of view of protocol. Point R relates to the TA interface or, in essence, it is the interface of that nonstandard (i.e., non-ISDN) device.

We will return to user–network interfaces once we set the stage for ISDN protocols looking into the network from the user.

4 ISDN PROTOCOLS AND PROTOCOL ISSUES

When implemented, ISDN will provide both circuit and packet switching. For the circuit-switching case the B-channel is transparent to the network permitting the user to utilize any protocol desired so long as there is end-to-end agreement on the protocol employed.

It is the D-channel that carries out the network control function for its related B-channels. Essentially, this is separate channel signaling embodied in CCITT signaling system No. 7 (SS No. 7, see Chapter 14). Thus the D-channel is used for call establishment (setup) and termination (take-down) and access to network facilities.

The B-channel, in the case of circuit switching, is serviced by NT1 or NT2, using OSI layer 1 functions only. The D-channel carries out OSI layer 1, 2, and 3 functions such that the B-channel protocol established by a family of end-users will generally make layer 3 null in the B-channel where the networking function is carried out by the associated D-channel.

With packet switching two possibilities emerge. The first basically relies on the B-channel to carry out OSI layer 1, 2, and 3 functions at separate packet-switching facilities (PSFs). The D-channel is used to set up the connection to the local switching exchange at each end of the connection. This type of packet-switched offering provides 64-kbps service. The second method utilizes the D-channel exclusively for lower data rate packet-switched service where the local interface can act as a CCITT Rec. X.25 data communication equipment (DCE).

Figure 13.3 is a simplified conceptual diagram of ISDN circuit switching. It shows the B channel riding on the digital network and the D channel, which is used for signaling. The D channel is a separate channel and may traverse several STPs (see Chapter 14) and may be quasi-associated or disassociated from the B channel. Figure 13.4 is a detailed diagram of the same ISDN circuit switching concept. The reader should note the following in the Figure. (1) Only users at each end have a peer-to-peer relationship available for all seven OSI layers on the B-channel. As the call is routed through the system, there is only layer 1 (physical layer) interaction at each switching node along the call route. (2) The D-channel requires the first three OSI layers for call setup to the local switching center at each end of the circuit. (3) The D-channel signaling data are turned over to SS No. 7 at the near- and far-end local switching centers. (4) SS No. 7 also utilizes the first three OSI layers for circuit establishment, which requires the transfer of control information. In SS No. 7 terminology this is called the *message transfer part*. There is a fourth layer called the *user part*. There is a telephone user part and a data user part, depending on whether the associated B-channel is in telephone or data service for the user.

Figure 13.5 shows packet service for lower data rates where only the D-channel is involved.

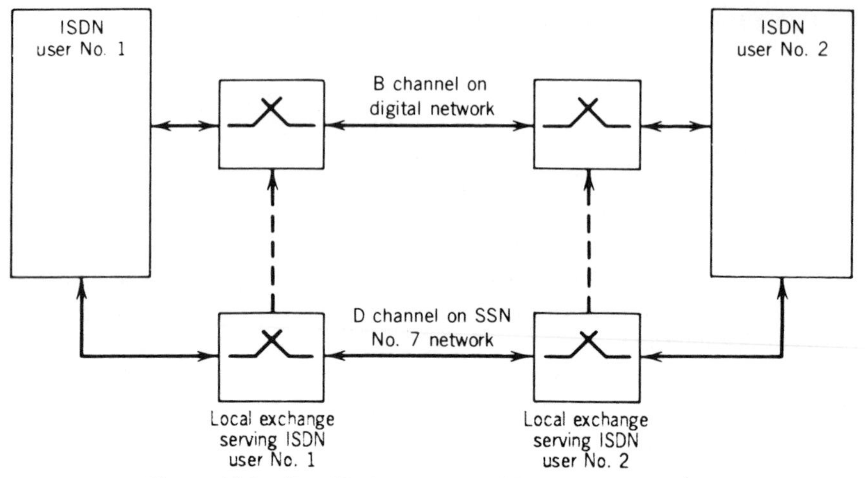

Figure 13.3 Simplified concept of ISDN circuit switching.

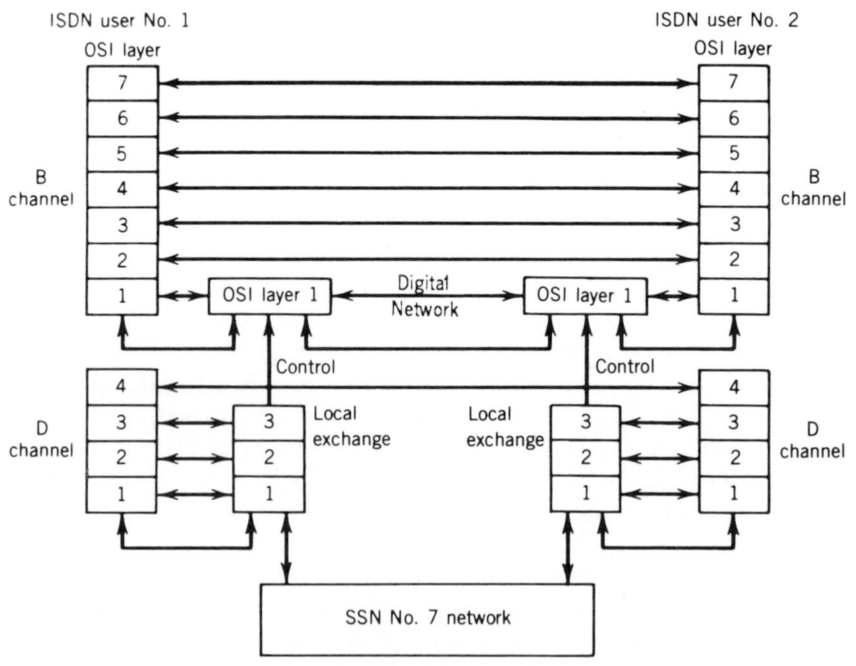

Figure 13.4 Detailed diagram of ISDN circuit switching concept. Traffic connectivity on channel B; its related signaling on the D channel.

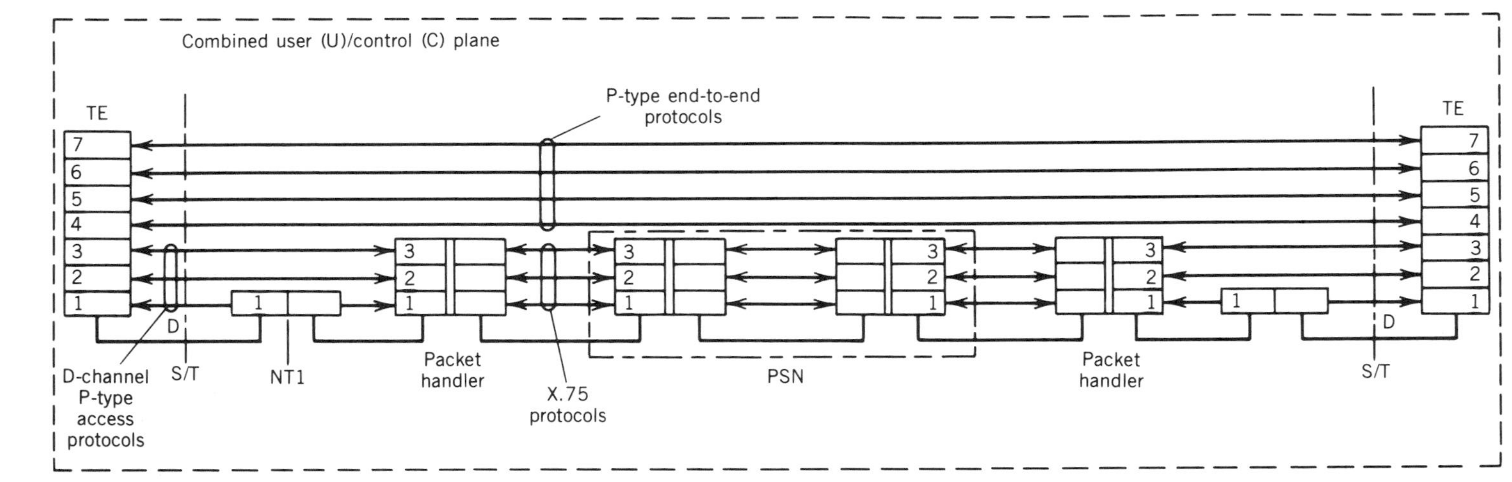

Figure 13.5 Packet service for lower data rates using the D-channel. From Ref. 6. Figure 8A / I.320, page 86, CCITT Rec. I.320, *Red Books*, Vol. III, Fascicle III.5. Courtesy of ITU – CCITT, Geneva.

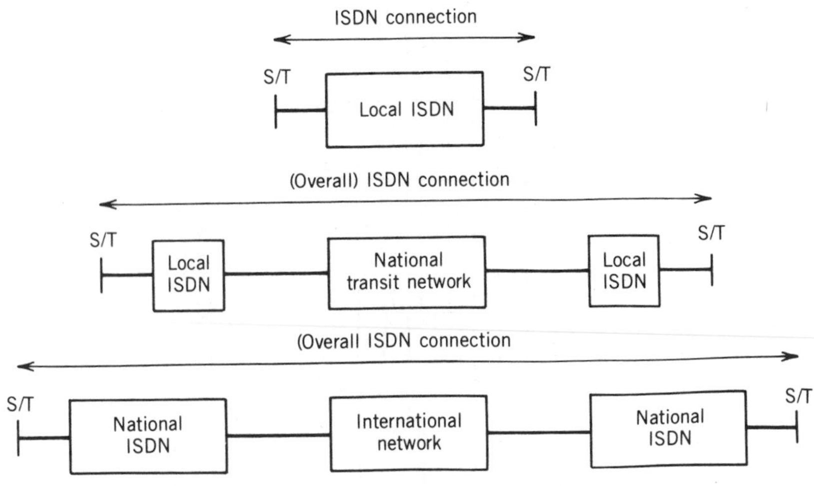

Figure 13.6 ISDN network connectivity.

5 ISDN NETWORKS

In the context of networking, ISDN can be seen as a group of access attributes connecting a user at either end to its local serving exchange (see Figure 13.6). Conceptually this is shown in the middle diagram of Figure 13.6. We envision the national transit network as the existing national digital network now in place. This network, whether based on the North American DS1 through DS4 architecture or the CEPT hierarchy, provides or will provide two necessary attributes:

1. 64-kbps channelization.
2. Separate channel signaling based on SS No. 7 or its North American version.

Connections from the user at the local connecting exchange interface include:

- Basic service 2B + D = 192 kbps (includes overhead).
- Primary service 23B + D/30B + D = 1.544/2.048 Mbps.

Figure 13.7 shows an ISDN connection involving several networks. ISDN networks are thus the present telephone company/administration networks that have been upgraded to meet attributes 1 and 2 in software, processing and electrically.

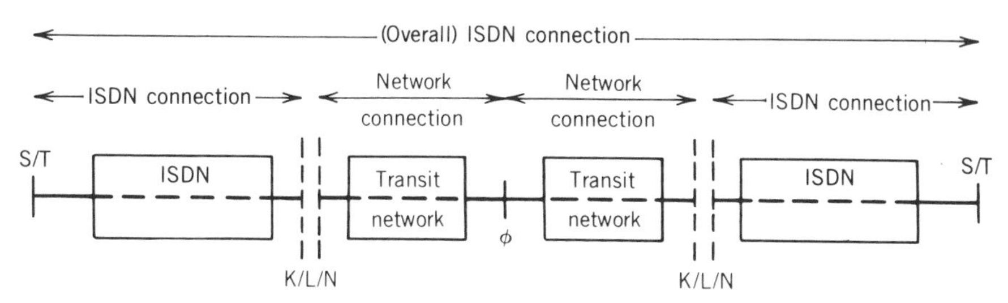

Figure 13.7 ISDN connection involving several networks. CCITT Rec. I.310 states that reference points K/L/N need further study. Courtesy of ITU-CCITT, Geneva. *Note*: Reference points are defined in Figure 13.2 and Rec. 1.310 and 1.411. Reference point ∅ may or may not be an ISDN defined reference point.

6 ISDN PROTOCOL STRUCTURES

6.1 ISDN and OSI

Figure 13.8 revisits the seven OSI layers discussed in Chapter 11. ISDN concerns itself with the first three layers only, as shown in Figure 13.9 [6].

Layers 4–7 are peer-to-peer connections and are the responsibility of end-users to ensure compatible interfaces.

The D-channel and its signaling and control function is the exception to the above statement. The D-channel interfaces with CCITT SS No. 7 at its serving exchange. The D-channels handle three types of information: Signaling (s), interactive data (p), and telemetry messages (t).

The layering of the D-channel protocol has followed the intent of the OSI reference model. The handling of p and t data can be adapted to the OSI model; the s data, by its very nature, cannot. Figure 13.10 shows the correspondence between D-channel signaling protocols and SS No. 7 levels and the OSI seven-layer model.

SS No. 7 interexchange protocol has been specified into a four-layer structure. Layers 1, 2, and 3 encompass the message transfer part (MTP) and layer 4 the user part (UP). This is discussed further in Chapter 14.

6.2 Layer 1 Interface: Basic Rate

The S/T interface of Figures 13.6 and 13.7 or layer 1 physical interface requires a balanced metallic transmission medium for each direction of transmission capable of supporting 192 kbps (Ref. 8, CCITT Rec. I.430). This is the NT interface shown in Figure 13.2 as NT1.

Layer 1 provides the following services to layer 2 for ISDN operation:

- The transmission capability by means of appropriately encoded bit streams for both B- and D-channels and also any timing and synchronization functions that may be required.

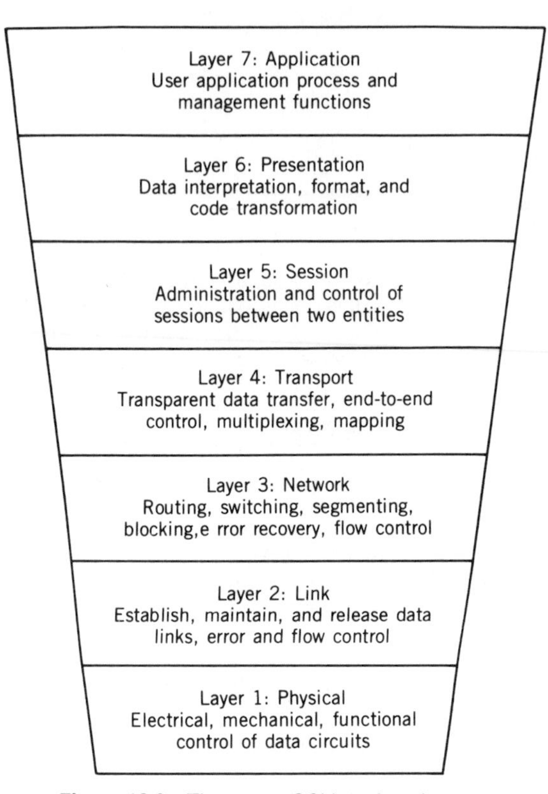

Figure 13.8 The seven OSI interface layers.

- The signaling capability and the necessary procedure to enable customer terminals and/or network terminating equipment to be deactivated when required and reactivated when required.
- The signaling capability and necessary procedures to allow terminals to gain access to the common resource of the D-channel in an orderly fashion while meeting the performance requirements of the D-channel signaling system.
- The signaling capability and procedures and necessary functions at layer 1 to enable the maintenance functions to be performed.
- An indication to the higher layers of the status of layer 1.

6.2.1 Primitives between Layer 1 and Other Entities. Primitives represent in an abstract way the logical exchange of information and control between layer 1 and other entities. They do not specify or constrain the implementation of entities or interfaces.

The primitives to be passed across the boundary between layers 1 and 2 or to the management entity are defined and summarized in Table 13.1. The

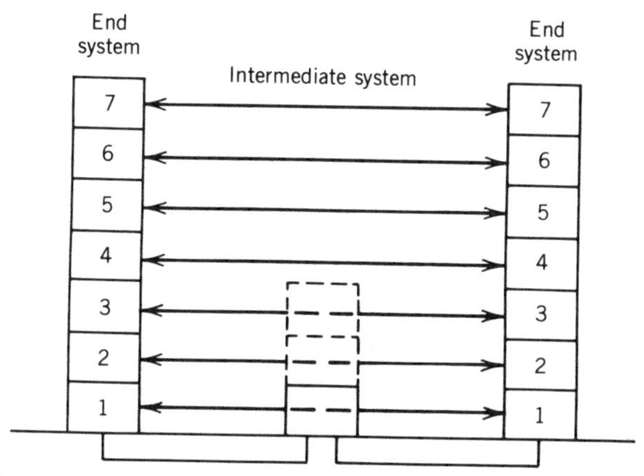

Figure 13.9 A generic communication context. *Note*: The end system protocol blocks may reside in subscriber's TE or network exchanges or other equipment related to an ISDN. From CCITT Rec. I.320, Figure 4/I.320, page 82, *Red Books*, Vol. III, Fascicle III.5.

parameter values associated with these primitives are also summarized in the table. CCITT Rec. X.211 describes the syntax and use of these primitives.

6.2.2 Interface Functions. The S or T interface functions consist of three time division multiplexed bit streams: two 64-kbps B-channels and one 16-kbps D-channel for an aggregate bit rate of 192 kbps. Of this 192 kbps the 2B + D configuration accounts for only 144 kbps. The remaining 48 kbps are overhead bits whose function will be described briefly below.

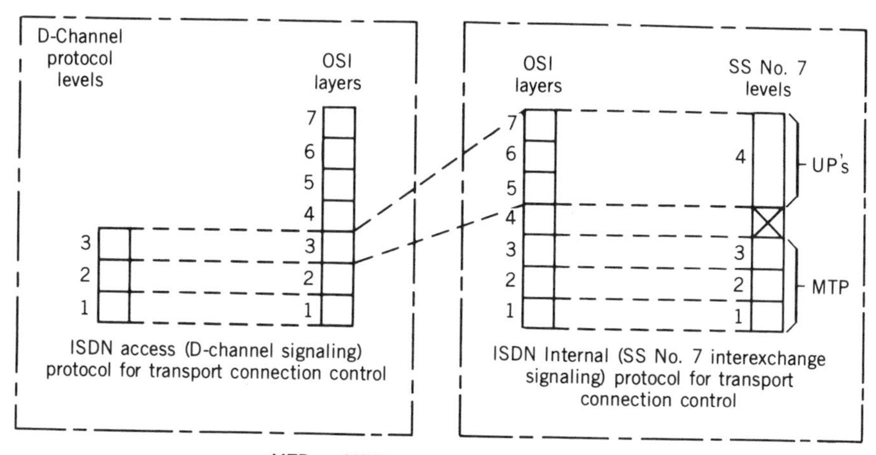

Figure 13.10 Correspondence between SS No. 7 and OSI model. From Ref. 7. Copyright © IEEE NY 1982.

TABLE 13.1 Primitives Associated with Layer 1

	Specific Name			Parameter		
Generic Name	Request	Indication	Response	Priority Indicator	Message Unit	Message Unit Contents
L1 ↔ L2						
PH-DATA	X[a]	X	–	X[b]	X	Layer 2 peer-to-peer message
PH-ACTIVATE	X	X	–	–	–	
PH-DEACTIVATE	X	X	–	–	–	
M ↔ L1						
MPH-ERROR	–	X*	X°	–	X	* Type of error or recovery from a previously reported error. ° Abandon search for framing.

[a]PH-DATA-REQUEST implies underlying negotiation between layer 1 and layer 2 for the acceptance of the data.
[b]Prior indication applies only to the request type.
From CCITT Rec. I.430, Table 1 / I.430, page 143, *Red Books* Vol. III, Fascicle III.5.

The functions covered at this interface include bit timing at 192 kbps to enable the TE and NT to recover information from the aggregate bit stream. Octet timing provides 8-kHz octet timing for the NT and TE. There is also the frame alignment function, which provides information to enable NT and TE to recover the time-division multiplexed channels (i.e., 2B + D multiplexed). Other functions include D-channel access control, power feeding, deactivation, and activation.

Interchange circuits are required, of which there is one in each direction of transmission (i.e., to and from NT); they are used to transfer digital signals across the interface. All the functions described above, except for power feeding, are carried out by means of a digitally multiplexed signal, described in the next section.

6.2.3 Frame Structure. In both directions of transmission the bits are grouped into frames of 48 bits each (Ref. 8, CCITT Rec. I.430). The frame structure is identical for all configurations, whether point to point or point to multipoint. However, the frame structures are different for each direction of transmission. These structures are shown in Figure 13.11, with explanatory notes given in Table 13.2.

6.2.4 Line Code. For both directions of transmission pseudoternary coding is used with 100% pulse width, as shown in Figure 13.12. Coding is performed such that a binary 1 is represented by no line signal, whereas a binary 0 is represented by a positive or negative pulse. The first binary signal

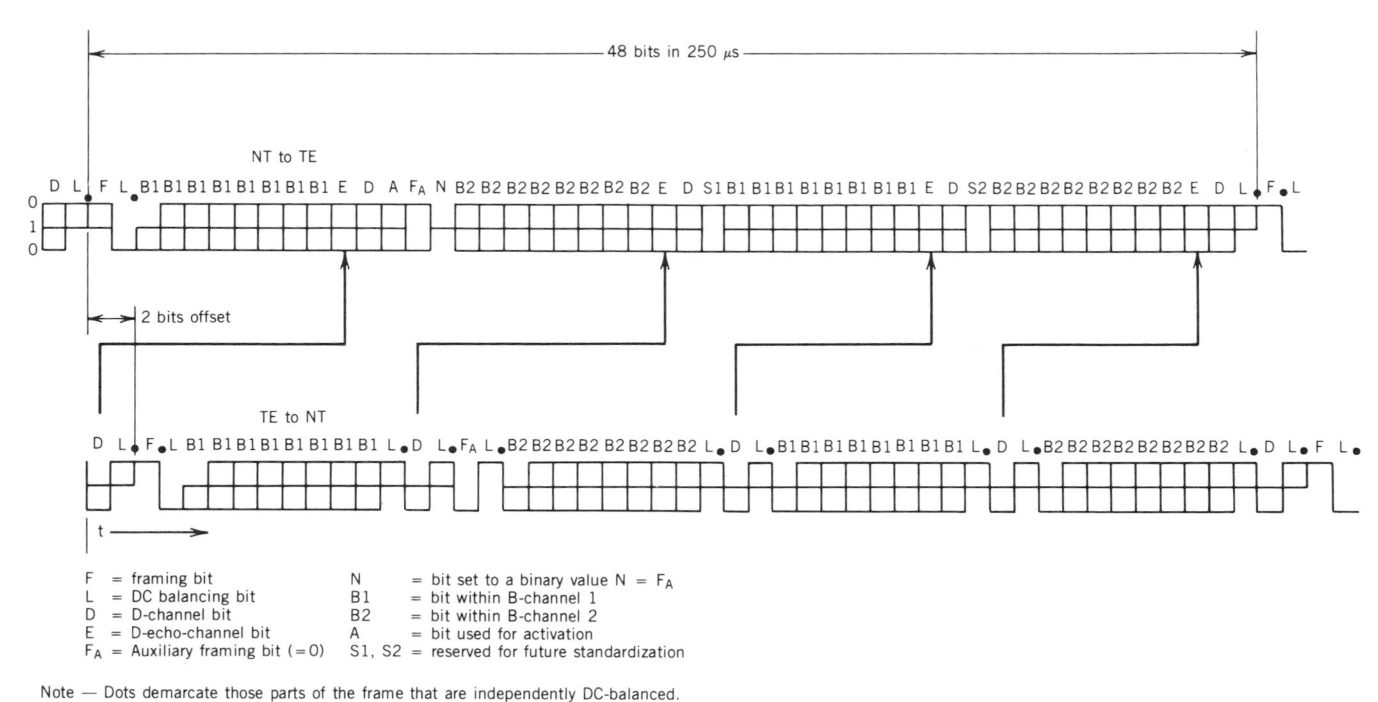

Figure 13.11 Frame structure at reference points S and T. From CCITT Rec. I.340, Figure 3/I.340, page 143, *Red Books*, Vol. III, Fascicle III.5.

TABLE 13.2 Notes on Bit Positions and Groups in Figure 13.11

Bit Position	Group
Terminal to network: Each frame consists of the following group of bits; each individual group is dc-balanced by its last bit (L-bit).	
1 and 2	Framing signal with balance bit
3–11	B1-channel with balance bit (first octet)
12 and 13	D-channel bit with balance bit
14 and 15	Auxiliary framing with balance bit
16–24	B2-channel with balance bit (first octet)
25 and 26	D-channel bit with balance bit
27–35	B1-channel with balance bit (second octet)
36 and 37	D-channel bit with balance bit
38–46	B2-channel with balance bit (second octet)
47 and 48	D-channel with balance bit
Network to terminal: Frames transmitted by the network (NT) contain an echo channel (E-bits) used to retransmit the D-bits received from the terminals. The D-echo channel is used for D-channel access control. The last bit of the frame (L-bit) is used for balancing each complete frame. The bits are grouped as follows:	
1 and 2	Framing signal with balance bit
3–10	B1-channel (first octet)
11	E, D-echo-channel bit
12	D-channel bit
13	Bit A used for activation
14	F_A auxiliary framing bit
15	N bit[a]
16–23	B2-channel (first octet)
24	E, D-echo-channel bit
25	D-channel bit
26	S1, reserved for future standardization[b]
27–34	B1-channel (second octet)
35	E, D-echo-channel bit
36	D-channel bit
37	S2, reserved for future standardization[b]
38–45	B2-channel (second octet)
46	E, D-echo-channel bit
47	D-channel bit
48	Frame balance bit

From Ref. 8. Courtesy of ITU–CCITT, Geneva.

[a]As defined in Section 6.3 of Ref. 8.

[b]S1 and S2 are set to binary 0.

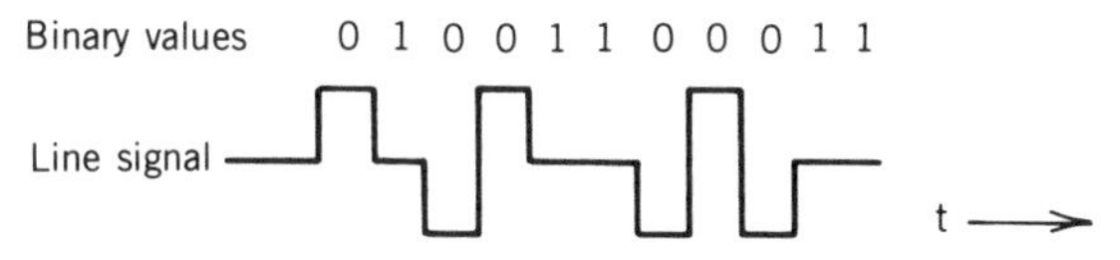

Figure 13.12 ISDN pseudoternary line code.

following the framing balance bit is the same polarity as the framing balance bit. Subsequent binary 0's alternate in polarity. A balance bit is 0 if the number of binary 0's following the previous balance bit is odd. A balance bit is a binary 1 if the number of binary 0's following the previous balance bit is even.

6.2.5 Timing Considerations. The NT derives its timing from the network clock. A TE synchronizes its bit, octet, and frame timing from the received bit stream from the NT and uses this derived timing to synchronize its transmitted signal (CCITT Rec. I.430 Ref. 8).

6.3 Layer 1 Interface: Primary Rate

This interface is described in CCITT Rec. I.431, and the following material is abridged from that recommendation. It is applicable to user–network interfaces at 1.544-Mbps and 2.048-Mbps primary data rates. It is valid only for a point-to-point configuration. This means that only one transmitter (source) and one receiver (sink) are connected to the interface.

Here again the B-channel has bidirectional transmission at 64 kbps. For the binary rate the D- or E-channel also is a full-duplex 64-kbps channel. The NT derives timing and framing alignment from the incoming bit stream from its associated digital network node. The bit stream contains the appropriate frame alignment signals.

The frame in this case is 193 bits long and is shown in Figure 13.13 You will recognize the 193-bit frame with its familiar F bit (framing bit) as the DS1 frame for the primary 1.544-Mbps rate (CCITT Rec. G.704). Each time slot consists of eight consecutive bits, numbered 1 to 8.

Table 13.3 shows the multiframe structure, which is 24 frames long. The e(1) through e(6) bits are used for error checking, as described in CCITT Rec.

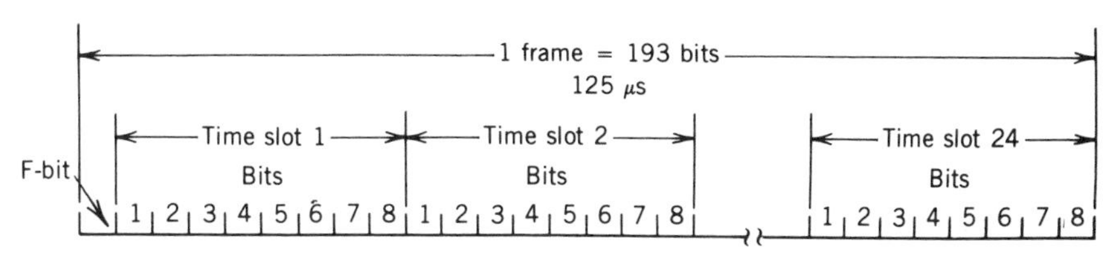

Figure 13.13 Frame structure of 1.544-Mbps interface. From Ref. 9. Courtesy of ITU–CCITT, Geneva.

TABLE 13.3 Multiframe Structure

Multiframe Frame Number	Multiframe Bit Number	F Bits		
		FAS[a]	Assignments: m Bit use[b]	Assignments: e Bit use[c]
1	0	–	m	–
2	193	–	–	e_1
3	386	–	m	–
4	579	0	–	–
5	772	–	m	–
6	965	–	–	e_2
7	1158	–	m	–
8	1351	0	–	–
9	1544	–	m	–
10	1737	–	–	e_3
11	1930	–	m	–
12	2123	1	–	–
13	2316	–	m	–
14	2509	–	–	e_4
15	2702	–	m	–
16	2895	0	–	–
17	3011	–	m	–
18	3281	–	–	e_5
19	3474	–	m	–
20	3667	1	–	–
21	3860	–	m	–
22	4053	–	–	e_6
23	4246	–	m	–
24	4439	1	–	–

[a]Frame alignment signal (. . . 001011 . . .).
[b]The use of m bits is for further study (for example, for maintenance and operational information).
[c]The use of the bits e_1 and e_6 is for further study.
From CCITT Rec. I.431, Table 1 / I.431, page 180, *Red Books*, Vol. III, Fascicle III.5

G.704. Along with the m bits, the use of these bits is to be defined in the future by CCITT. One possible use of the m bits is for maintenance and operation.

Time slot 24 is assigned to the D- or E-channel when either of these channels is present [9].

As in the basic service, timing is derived from the network clock at the NT, which distributes the timing to TE users on the data bit stream to the TE.

The interface for the 2.048-Mbps primary rate conforms to CCITT Rec. G.704. There are 32 time slots per frame, numbered 0 to 31, and the number of bits per frame is 256 (8×32). The 8-bit frame alignment signal shown in Table 13.4 occupies position 1 to 8 in time slot 0 of every other frame in the CEPT30 + 2 format.

The frame alignment signal is the 7-bit sequence 0011011 shown in Table 13.4. In order to avoid simulation of the frame alignment signal by bits 2 to 8

TABLE 13.4 Allocation of Bit Numbers 1 to 8 of the Frame

Alternate frames \ Bit Number	1	2	3	4	5	6	7	8
Frame containing the frame alignment signal	S_i[a]	0	0	1	1	0	1	1
Frame not containing the frame alignment signal	S_i[a]	1[b]	A[c]	S_n[c]	S_n	S_n	S_n	S_n

[a]The use of S_i will be defined later.
[b]The bit is fixed at 1 to assist in avoiding simulations of the frame alignment signal.
[c]The use of the c_n bits will be defined later.
From CCITT Rec. I.431, Table 2 / I.431, Page 182, *Red Books*, Vol. III, Fascicle III.5.

of channel time slot 0 in frames not containing the frame alignment signal, bit 2 in those channel times slots is fixed at 1.

Time slot 16 is assigned to the D- or E-channel when either of these channels is present. Again the NT derives its clock from the network clock. The TE synchronizes its timing from the bit stream received from the NT and then synchronizes its transmit signal accordingly [9].

6.4 Overview of the Layer 2 Interface: Link Access Procedure for the D-Channel

The link access procedure (LAP) for the D-channel (LAPD) is used to convey information between layer 3 entities across the ISDN user–network interface using the D-channel. Besides CCITT Recs. I.440 and 441, the following recommendations and standards are referenced:

- CCITT Recs. X.200 and X.210 (OSI).
- CCITT Rec. X.25 (LAPB).
- ISO 3309 and ISO 4335 (HDLC standards for frame structure and elements and procedures).

A service access point (SAP) is a point at which the data link layer provides services to its next higher OSI layer or layer 3. Associated with each data link layer (OSI layer 2) is one or more data link connection end points (see Figure 13.14). A data link connection end point is identified by a data link connection end point identifier as seen from layer 3 and by a data link connection identifier (DLCI) as seen from the data link layer.

Cooperation between data link layer entities is governed by a specific protocol to the applicable layer. In order for information to be exchanged between two or more layer 3 entities, an association must be established between layer 3 entities in the data link layer using a data link layer protocol. This association is provided by the data link layer between two or more SAPs,

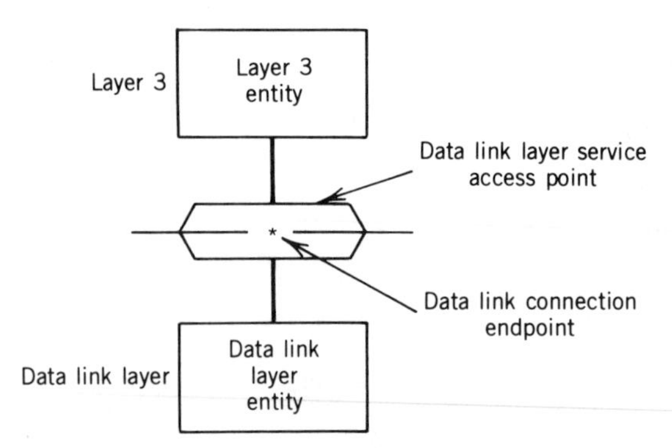

Figure 13.14 Entities, service access points (SAPs), and end points. From Ref. 10.

as shown in Figure 13.15 [10]. Data link message units are conveyed between data link layer entities by means of a physical connection.

Layer 3 uses *service primitives* to request service from the data link layer. A similar interaction takes place between layer 2 and layer 1.

Between the data link layer and its adjacent layers there are four types of service primitives:

1. Request.
2. Indication.
3. Response.
4. Confirm.

Their functions are shown diagrammatically in Figure 13.16.

The REQUEST primitive is used where a higher layer is requesting service from the next lower layer. The INDICATION primitive is used by a layer providing service to notify the next higher layer of activities related to the REQUEST primitive. The RESPONSE primitive is used by a layer to ac-

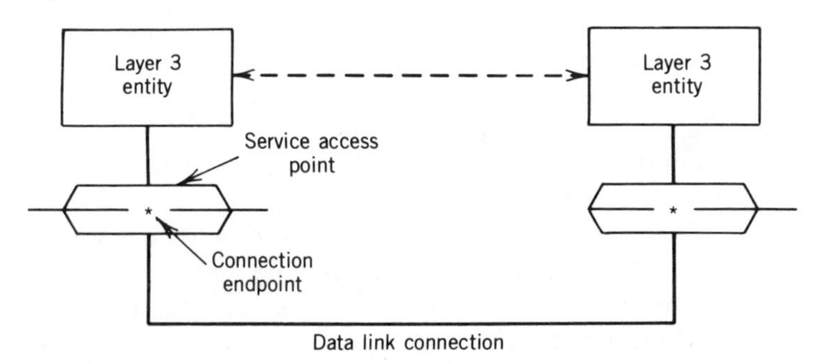

Figure 13.15 Data link connections between two or more SAPs. From Ref. 10.

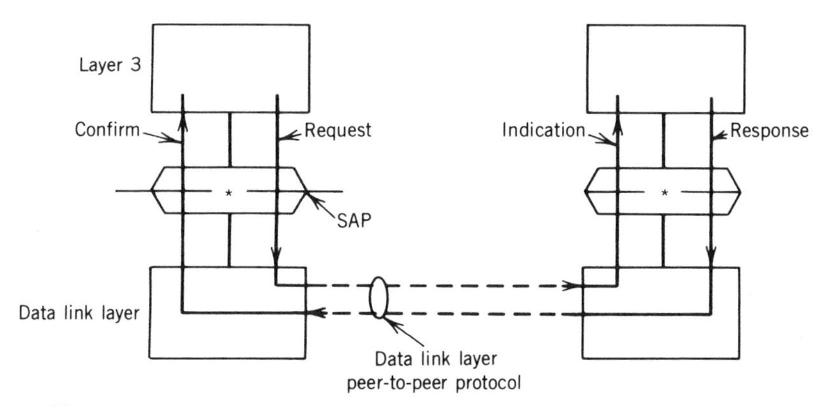

Figure 13.16 Functions of service primitives. *Note*: The same principle applies for data link layer – physical layer interactions. From Ref. 10.

knowledge receipt from a lower layer of the INDICATION primitive. The CONFIRM primitive is used by the layer providing the requested service to confirm that the requested activity has been completed (see Chapter 12, Section 4.3.1).

Figure 13.17 shows the data link layer reference model. All data link layer messages are transmitted in frames delimited by flags, where a flag is a unique

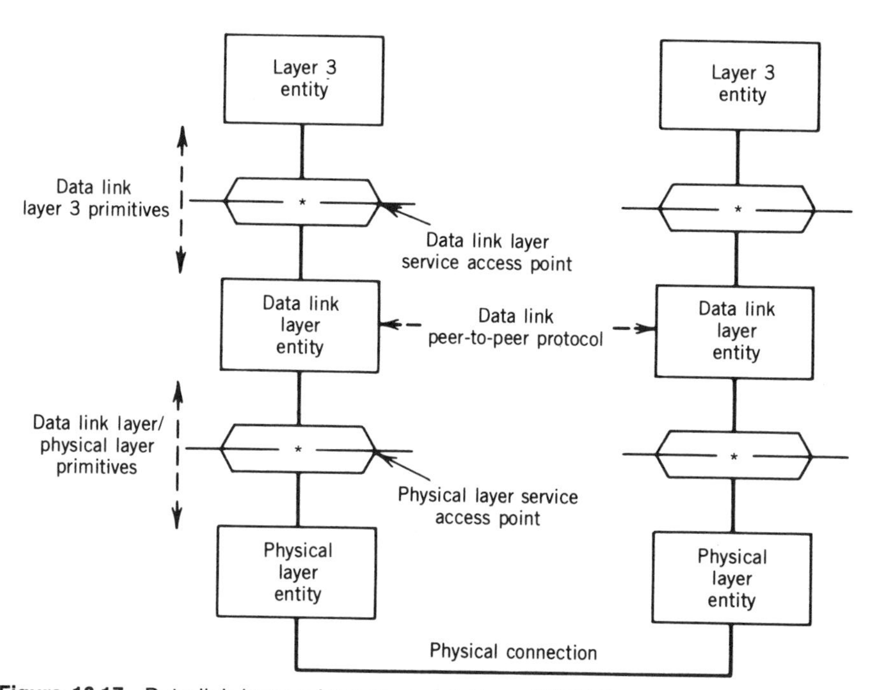

Figure 13.17 Data link layer reference model. From CCITT Rec. I.440, Figure 5/I.440, Page 189, *Red Books*, Vol. III, Fascicle III.5.

bit pattern. The frame structure is defined in CCITT Rec. I.441 and is briefly described in this section.

The LAPD includes functions for:

1. The provision of one or more data link connections on a D-channel. Discrimination between the data link connections is by means of a data link connection identifier (DLCI) contained in each frame.
2. Frame delimiting, alignment, and transparency, allowing recognition of a sequence of bits transmitted over a D-channel as a frame.
3. Sequence control, which maintains the sequential order of frames across a data link connection.
4. Detection of transmission, format, and operational errors on a data link.
5. Recovery from detected transmission, format, and operational errors. Notification to the management entity of unrecovered errors.
6. Flow control.

There is unacknowledged and acknowledged operation. With unacknowledged operation information is transmitted in unnumbered information (UI) frames. At the data link layer the UI frames are unacknowledged. Transmission and format errors may be detected, but no recovery mechanism is defined. Flow control mechanisms are also not defined. With acknowledged operation layer 3 information is transmitted in frames that are acknowledged at the data link layer. Error recovery procedures based on retransmission of unacknowledged frames are specified. For errors that cannot be corrected by the data link layer, a report to the management entity is made. Flow control procedures are also defined.

Unacknowledged operation is applicable for point-to-point and broadcast information transfer. However, acknowledged operation is applicable only for point-to-point information transfer.

There are two forms of acknowledged information that are defined:

- Single-frame operation.
- Multiframe operation.

For single-frame operation layer 3 information is sent in sequenced information 0 (SI0) and sequenced information 1 (SI1) frames. No new frame is sent until an acknowledgment has been received for a previously sent frame. This means that only one unacknowledged frame may be outstanding at a time. With multiple-frame operation layer 3 information is sent in numbered information (I) frames. A number of I frames may be outstanding at the same time. Multiple-frame operation is initiated by a multiple-frame establishment procedure using set asynchronous balanced mode/set asynchronous balanced mode extended (SABM/SABME) command [10].

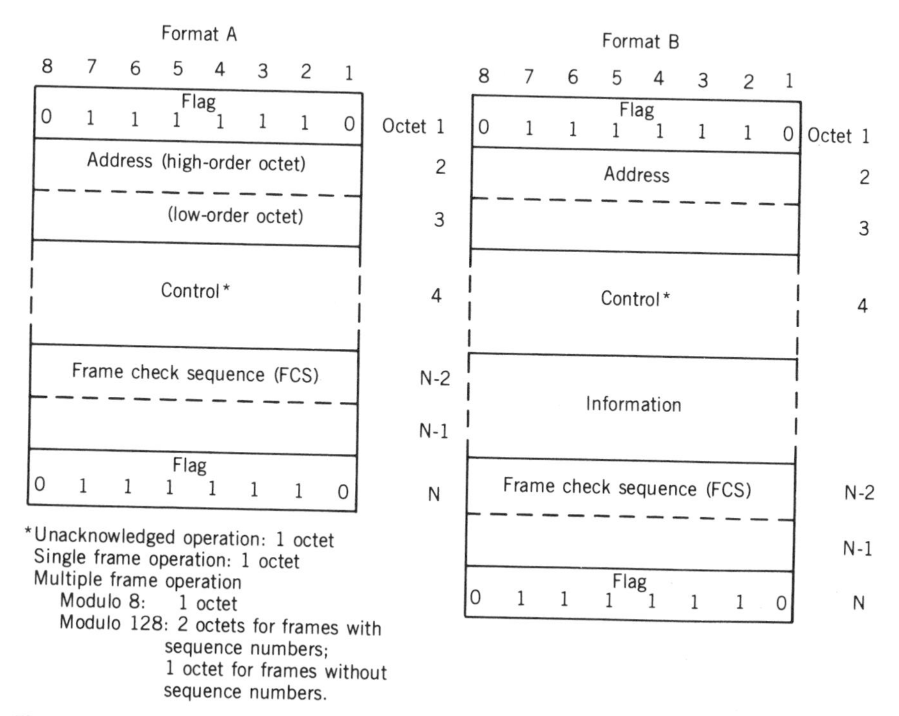

Figure 13.18 Frame formats for layer 2 frames. From CCITT Rec. I.441, Figure 1/I.441, Page 202, *Red Books*, Vol. III, Fascicle III.5.

6.4.1 Layer 2 Frame Structure for Peer-to-Peer Communication. There are two frame formats used on layer 2 frames:

1. Format A, for frames where there is no information field.
2. Format B, for frames containing an information field.

These two frame formats are shown in Figure 13.18.

The following discussion briefly describes the frame content (sequences and fields).

Flag Sequence. All frames start and end with a flag sequence consisting of one 0 bit followed by six contiguous 1 bits and one 0 bit. These flags are called the opening and closing flag.

Address Field. As shown in Figure 13.18, the address field consists of two octets and identifies the intended receiver of a command frame and the transmitter of a response frame. A single octet address field is used for LAPB operation.

Control Field. The control field consists of one or two octets. It identifies the type of frame, either command or response. It contains sequence numbers where applicable. These types of control field formats are specified:

1. Numbered information transfer (I format).
2. Supervisory functions (S format).
3. Unnumbered information transfers and control functions (U format).

Information Field. The information field of a frame, when present, follows the control field and precedes the frame check sequence (FCS). The information field contains an integer number of octets:

- For a SAP supporting signaling, the default value is 128 octets (CCITT Rec. I.441).
- For SAPs supporting packet information, the default value is 260 octets (CCITT Rec. I.441).

Frame Check Sequence Field. The FCS field is a 16-bit sequence and is the 1's implement of the modulo 2 sum of:

1. The remainder of X raised to the k power:

   $$X^{15} + X^{14} \ldots X^{1} + 1$$

   divided by the generating polynomial

   $$X^{16} + X^{12} + X^{5} + 1$$

   where k is the number of bits in the frame existing between but not including the final bit of the opening flag and the first bit of the FCS, excluding bits inserted for transparency, and
2. The remainder by modulo 2 division by the generating polynomial

   $$X^{16} + X^{12} + X^{5} + 1$$

   of the product of X^{16} by the content of the frame defined in (1).

Transparency, mentioned in (1) ensures that a flag or abort sequence is not imitated within a frame. On the transmit side the data link layer examines the frame content between the opening and closing flag sequences and inserts a 0 bit after all sequences of five contiguous 1 bits (including the last five bits of the FCS). On the receive side the data link layer examines the frame contents between the opening and closing flag sequences and discards any 0 bit that directly follows five contiguous 1 bits.

The following comments clarify the semantics and usage of primitives. Primitives consist of commands and their respective responses associated with the services requested of a lower layer. The general syntax of a primitive is:

XX-generic name-type: parameters

where *XX* designates the layer providing the service. For the data link layer, *XX* is DL, PH for the physical layer, or MDL for the management entity to the data link layer interface. Table 13.5 gives the primitives associated with the data link layer.

6.5 Overview of Layer 3

The layer 3 protocol provides the means to establish, maintain, and terminate network connections across an ISDN between communicating application entities. A detailed description of layer 3 protocol may be found in CCITT Recs. I.451 and Q.931. Reference should also be made to CCITT Rec. I.311. This discussion is based on the CCITT VIII Plenary Assembly (October 1984). It states that Recs. Q.931, I.451, and I.320 are not presently completely consistent in their structure of protocols. Further study is required to enhance these recommendations in order to resolve the inconsistencies.

Layer 3 utilizes functions and services provided by its data link layer, as described in Section 6.4 under LAPD functions. These necessary layer 2 support functions are listed and briefly described below:

- Establishment of the data link connection.
- Error-protected transmission of data.
- Notification of unrecoverable data link errors.
- Release of data link connections.
- Notification of data link layer failures.
- Recovery from certain error conditions.
- Indication of data link layer status.

Layer 3 performs two basic categories of functions and services in the establishment of network connections. The first category directly controls the connection establishment. The second category includes those functions relating to the transport of messages in addition to the functions provided by the data link layer. Among these additional functions are the provision of rerouting of signaling messages on an alternative D-channel (where provided) in the event of D-channel failure. Other possible functions include multiplexing and message segmenting and blocking.

The D-channel layer 3 protocol is designed to carry out establishment and control of circuit-switched and packet-switched connections. Also, services

TABLE 13.5 Primitives Associated with the Data Link Layer

	Type				Parameters		
Generic Name	Request	Indication	Response	Confirm	Priority Indicator	Message Unit	Message Unit Contents
L3 ↔ L2 (Layer 3 / data link layer boundary)							
DL-Establish	X	X	[a]	[a]	–	[a]	Choice of single / multiple frame operation
DL-Release	X	X	[a]	[a]	–	[a]	Choice of single / multiple frame operation
DL-Data	X	X	–	–	–	X	Network layer peer-to-peer message
DL-Unit data	X	X	–	–	–	X	
M ↔ L2 (Management entity / data link layer boundary)							
MDL-Assign	X	X	–	[a]	–	X	TEI value
MDL-Remove	X	–	–	[a]	–	X	TEI value
MDL-Error	–	X	X	–	–	X	Reason for error message
MDL-Unit data	X	X	–	–	–	X	Management function peer-to-peer message
L2 ↔ L1 (data link layer / physical layer boundary)							
PH-Data	X	X	–	–	X	X	Data link layer peer-to-peer message
PH-Activate	X[b]	X	–	–	–	–	
PH-Deactivate	X[b]	X	–	–	–	–	

[a]For further study.
[b]Use requires further study.

From CCITT Rec. I.441, Table 5/I.441, page 215, *Red Books*, Vol. III, Fascicle III.5.

involving the use of connections of different types, according to user specifications, may be effected through "multimedia" call control procedures.

Functions performed by layer 3 include:

1. The processing of primitives for communicating with the data link layer.
2. Generation and interpretation of layer 3 messages for peer level communications.
3. Administration of timers and logical entities (e.g., call references) used in the call control procedures.
4. Administration of access resources, including B-channels and packet-layer logical channels (e.g., CCITT Rec. X.25).
5. Checking to ensure that services provided are consistent with user requirements, such as compatibility, address, and service indicators.

The following functions may also be performed by layer 3:

1. *Routing and relaying*. Network connections exist either between users and ISDN exchanges or between users. Network connections may involve intermediate systems that provide relays to other interconnecting subnetworks. Routing functions determine an appropriate route between layer 3 addressees.
2. *Network connection*. This function includes mechanisms for providing network connections making use of data link connections provided by the data link layer.
3. *Conveying user information*. This function may be carried out with or without the establishment of a circuit-switched connection.
4. *Network connection multiplexing*. Layer 3 provides multiplexing of call control information for multiple calls onto a single data link connection.
5. *Segmenting and blocking*. Layer 3 may segment and/or block layer 3 information for facilitating information transfer.
6. *Error detection*. Error detection functions are used to detect procedural errors in the layer 3 protocol. Error detection in layer 3 uses, among other information, error notification from the data link layer.
7. *Error recovery*. This includes mechanisms for recovering from detected errors.
8. *Sequencing*. This includes mechanisms for providing sequenced delivery of layer 3 information over a given network connection when requested. Under normal conditions layer 3 ensures the delivery of information in the sequence it is submitted by the user [12].
9. *Flow control*. Refer to CCITT Rec. I.451 [13].
10. *Reset*. CCITT Rec. I.450 states that this function is for further study.

6.5.1 Layer 3 Specification. The layer 3 specification is contained in CCITT Rec. I.451 [13]. Throughout this specification reference is made to B-channels as far as circuit-switched calls are concerned. The application of call control procedures defined in the recommendation to other channel types is not excluded. Recommendation I.451 states that further study on extending the application to other channel types is needed.

Table 13.6 lists messages for circuit-mode connections. Each of these messages is defined in the recommendation, and the content elements of each are given in 30 accompanying tables. One typical table is presented in Table 13.7. The following are several explanatory notes for Table 13.7. The letters "M" and "O" mean *mandatory* and *optional*, respectively. Letters "n" and

TABLE 13.6 Messages for Circuit-Mode Connections

Call establishment messages
ALERTing
CALL PROCeeding
CONNect
CONNect ACKnowledge
SETUP
SETUP ACKnowledge
Call information phase messages
RESume
RESume ACKnowledge
RESume REJect
SUSPend
SUSPend ACKnowledge
SUSPend REJect
USER INFOrmation
Call disestablishment message
DETach
DETach ACKnowledge
DISConnect
RELease
RELease COMplete
Miscellaneous messages
CANCel
CANCel ACKnowledge
CANCel REJect
CONgestion CONtrol
FACility
FAcility ACKnowledge
FACility REJect
INFOrmation
REGister
REGister ACKnowledge
REGister REJect
STATUS

From CCITT Rec. I.451, Table 1/I.451, page 258, *Red Books*, Vol. III, Fascicle III.5.

TABLE 13.7 Setup Message Contents

Message type: SETUP Direction: both Information Element	Direction	Type	Length
Protocol discriminator	both	M	1
Call reference	both	M	1 – ?
Message type	both	M	1
Bearer capability[b]	both	M	4 – ?
Channel identification	u → n / n → u	0 / M	3 – ?
CCITT-standardized facilities	both	0	3 – ?
Network-specific facilities	both	0	3 – ?
Terminal capabilities	u → n	(c)	3
Display	n → u	0	3 – ?
Keypad	u → n	0	3 – ?
Signal	n → u	0	3 – ?
Switchhook	u → n	0	3
Origination address	both	0	4 – ?
Destination address	both	0	4 – ?
Redirecting address	n → u	0	4 – ?
Transit network selection	u → n	0	3 – ?
Low-layer compatibility[b]	both	0	4 – ?
High-layer compatibility[b]	both	0	3 – ?
User – user information	both	0	3[a]

[a]The maximum length of user – user information element is network dependent and is 34 or 130 octets.
[b]The bearer service and compatibility information elements may be used to describe a CCITT telecommunication service, if appropriate.
[c]"M" for stimulus terminals; not included in functional equipment.
From CCITT Rec. I.451, Table 25/I.451, *Red Books*, Vol. III, Fascicle III.5.

"u" refer to *network* and *user*, respectively, and give the direction of traffic such as n → u (network to user) and u → n (user to network). The question mark in the table refers to the length in octets and means that the maximum length is undefined.

7 ISDN PACKET MODE REVIEW

Although nB + D operation for voice and data connectivity by ISDN is fairly well detailed in the 1984 CCITT I recommendations, the packet data mode is not so well defined. CCITT Rec. I.462 [14] is the relevant recommendation. Reference is made therein to CCITT Recs. X.25 and X.213.

Two scenarios are given in CCITT Rec. I.462:

1. Minimum integration scenario.
2. Maximum integration scenario.

These scenarios, described briefly in what follows, are the basis on which the support of CCITT Rec. X.25 DTEs by the ISDN should be standardized according to CCITT Rec. I.462. These scenarios are also the basis on which the support of packet mode TEs by an ISDN has been standardized, since a Rec. X.25 DTE and its terminal adaptor (TA) are always equivalent to a packet mode TE1. Therefore every reference in this context to the combination of a Rec. X.25 DTE and its TA should also be considered as being the equivalent of a packet mode TE1. However, some TE1s may have more capability than that available from a TE2 plus a TA. Similarly, Rec. I.462 covers the support of NT2s operating in the packet mode.

7.1 Minimum Integration Scenario

Figure 13.19 is a functional block diagram of the minimum integration scenario as described in CCITT Rec. I.462. This scenario refers to a transparent handling of packet calls through an ISDN. In this case only access through the B-channel is possible. In this context the only support an ISDN gives to packet messages is a physical 64-kbps circuit, nonswitched or switched transparent network connection between the appropriate public switched packet data network (PSPDN) port and the Rec. X.25 user terminal adaptor on the customer premises.

Figure 13.19 has an upper and lower portion. The upper portion provides switched access to the PSPDN. Here the Rec. X.25 DTE is connected to an ISDN port at the PSPDN IP (ISDN port). The IP is also able to set up the physical B-channel through the ISDN. The lower portion of the figure illustrates the semipermanent access. In this case the Rec. X.25 DTE is connected to the corresponding ISDN port at the PSPDN (IP′). The TA performs only the necessary physical channel rate adaption between the user rate at reference point R and the 64-kbps B-channel rate.

For the switched access alternative (upper portion of the figure), originating calls are set up over the B-channel toward the PSPDN port using the ISDN signaling procedure before initiating Rec. X.25 level 2 and level 3 functions. This can be done by exploiting either hot line (i.e., direct call) or complete selection methods. Moreover, the TA performs user rate adaption at 64 kbps. Depending on the data rate adaption technique employed, a complementary function may be needed at the ISDN access port (IP) of the PSPDN, such as HDLC interframe flag stuffing.

For calls originated by the PSPDN, the same considerations apply. As shown in Figure 13.19, the ISDN port of the PSPDN includes both rate adaption, if required, and path setup functions.

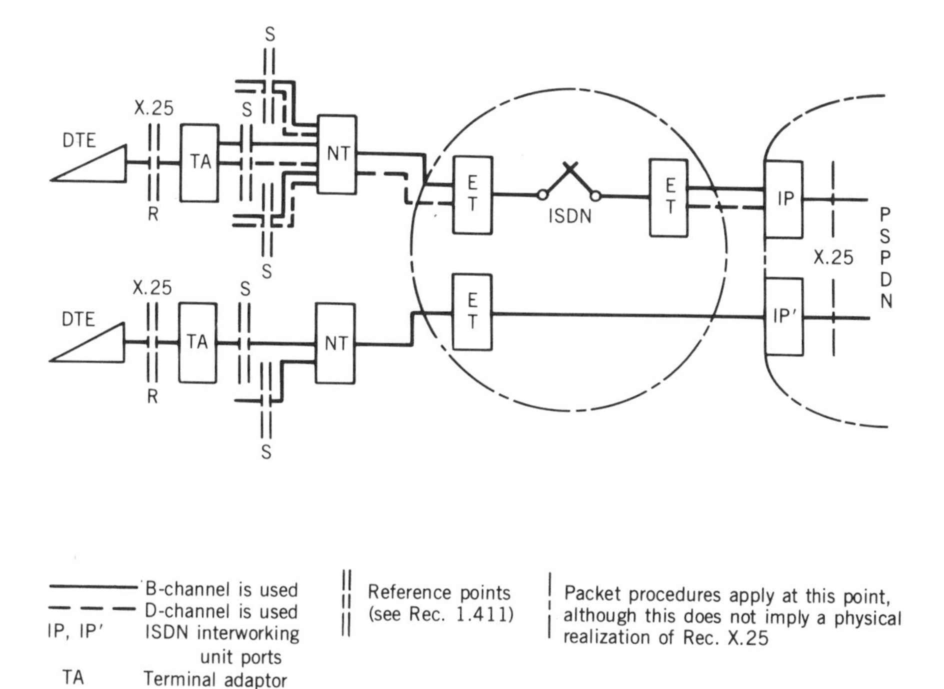

Figure 13.19 Minimum integration scenario. *Note*: This figure is only an example of many possible configurations and is included as an aid to the text describing the various interface functions. From CCITT Rec. I.462, Figure 1/I.462, page 417, *Red Books*, Vol. III, Fascicle III.5.

When required, DTE identification may be provided to the PSPDN by using the call establishment protocols of Rec. I.451 (see Section 6.5). In addition, DCE identification may be provided to the DTE, when required, using these same protocols.

7.2 Maximum Integration Scenario

This scenario occurs when a packet-handling (PH) function is provided with the ISDN (Figure 13.20). It relates to the case of Rec. X.25 procedures conveyed through the B-channel. Here the packet call or message is routed within an ISDN to some packet-handling function where complete processing of Rec. X.25 messages or calls can be carried out [14].

Multiple terminals can be supported at the customer's premises, where any one of the terminals can use the B-channel on a per call or per message basis at different times. Simultaneous multiple terminal operation of packet TEs in the B-channel is made possible by using level 3 multiplexing (See Section 6.5).

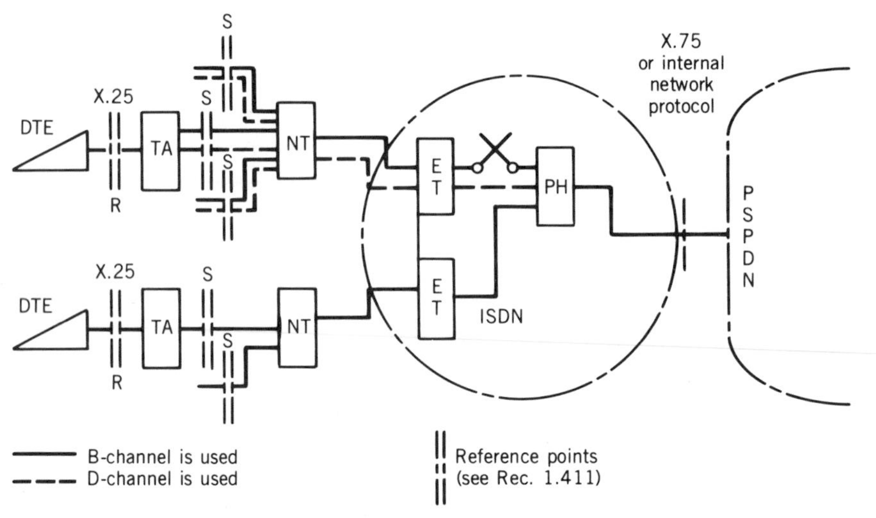

Figure 13.20 Maximum integration scenario. *Note*: This figure is only an example of many possible configurations and is included as an aid to the text describing the various interface functions. When a PSPDN coexists with an ISDN, as an interim solution, the node of the PSPDN connected to the ISDN may contain the PH function in some implementations. In this case the PSPDN is logically considered as belonging to the ISDN. In any event, international interworking will be done according to Rec. X.75. (Access Via B-channel) From CCITT Rec. I.462, Figure 2/I.462, Page 418, *Red Books*, Vol. III, Fascicle III.5.

CCITT Rec. I.462 states that the use of level 1 and level 2 multiplexing is for further study.

The PH function may be accessed in various ways depending on the related ISDN implementation alternative. In any case, a B-channel connection is set up toward a PH port supporting the necessary processing for B-channel packet service, standard Rec. X.25 functions for level 2 and level 3, path setup functions for level 1, and possible rate adaption.

Figure 13.21 illustrates the case where Rec. X.25 packet level procedures are conveyed via the D-channel. In this case a number of Rec. X.25 DTEs can operate simultaneously through the D-channel by using address discrimination at layer 2. The accessed PH port is still able to support the standard Rec. X.25 packet level 3 procedures.

CCITT Rec. I.462 [14] states that it is also important to note that the procedures for accessing a PSDTS (packet-switched data transmission service) through an ISDN user interface over a B- or D-channel are independent of where the service provider chooses to locate the packet handling functions, such as:

- In a remote exchange or packet-switching module in an ISDN.
- In the local exchange.
- In the NT2.

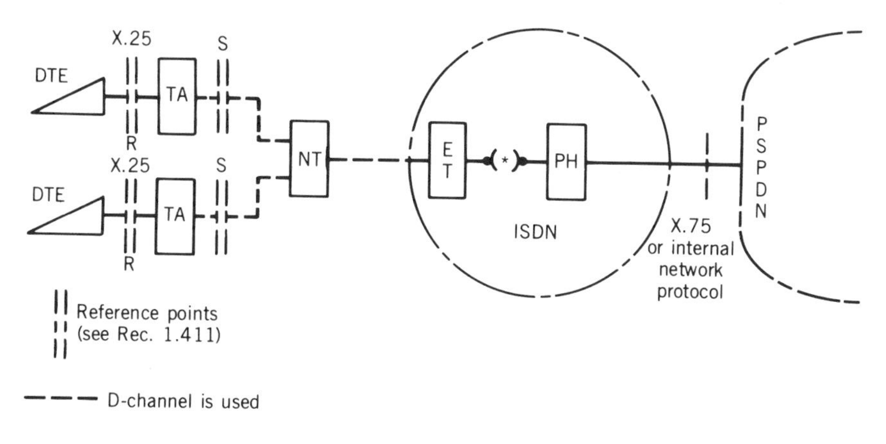

Figure 13.21 Maximum integration scenario, access via the D-channel. *Note*: This figure is only an example of many possible configurations and is included as an aid to the text describing the various interface functions. When a PSPDN coexists with an ISDN, as an interim solution, the node of the PSPDN connected to the ISDN may contain the PH function in some implementations. In this case the PSPDN is logically considered as belonging to the ISDN. In any event, international interworking will be done according to Rec. X.75. It is possible that an ISDN packet service provider may support a single X.25 DTE on the D-channel using a TA that makes use of the single-octet D-channel addressing exception of Rec. I.441. In this case the support of multiple X.25 DTEs is precluded. From CCITT Rec. I.462, Figure 3/I.462, page 419, *Red Book*, Vol. III, Fascicle III.5.

It should be noted, however, that the procedures for packet access through the B-channel or D-channel are different.

8 COMMENTS ON ISDN FEASIBILITY AND IMPLEMENTATION

8.1 Reticence to Accept New Services

At this juncture the demand for ISDN would seem limited to business service. At present there is no ubiquitous public data network or public switched data network. A fully digital INTERLATA toll network goal is about the year 2000 in the contiguous United States. There is no firm plan to implement SS No. 7, which is required to be in place for implementation of ISDN as we have described it. However, in the United States, at least one common carrier (toll) has announced SS 7 implementation across its entire network.

Once there is a data network infrastructure covering North America and the world, would the customer demand support it? For instance, it would seem that the 2B + D basic rate is aimed at small business and residential customers where the present demand is the least if not nil. Primary service (i.e., 23B + D or 30B + D) offers an attractive alternative to services such as ATT DDS (nonswitched). However, many large businesses and government entities are setting up private networks exactly tailored to their particular needs. Some of these networks are bypass and others, such as FTS-2000, are not. Once these private networks are in place, it will be a very hard sell to replace these facilities with ISDN. Of course, there may be some possibility that ISDN would augment such private networks.

Other potential business customers may await B-ISDN (broadband ISDN) with its 150-Mbps capability [15, 16]. B-ISDN standardization has a far way to go.

8.2 Subscriber Plant Shortcomings

A number of US Bell Operating Companies (BOCs) such as Nynex report that some existing copper pair subscriber loops cannot support narrowband ISDN data rates. It has been heard on the street that perhaps 50% of the subscriber pairs cannot support 192 kbps. These narrowband rates were designed to use existing wire-pair systems in the local area. Leapfrogging to fiber may be the only answer. Such as alternative would be a very large and expensive undertaking, considering that over 50% of the total public switched network investment is in local plant subscriber distribution [16].

8.3 Present ISDN Standardization

There are essentially two separate and distinct PCM systems in the world today, each with a large existing infrastructure. These were discussed and compared in Chapter 9. For ubiquitous international ISDN service, this lack of one singular basic standard leads to a serious shortcoming. One of the most important salient differences is that the European CEPT 30 + 2 system provides clear 64-kbps channels, whereas the North American system provides a corrupted 64-kbps channel. It is corrupted by in-band supervisory signaling by robbing bit 8 of each sixth frame and the single framing bit added to each frame. This, then, has led to two ISDN basic standards to fit each system, or:

2B + D, where B is 56 kbps (United States)

23B + D, where B is 56 kbps (United States)

2B + D, where B is 64 kbps (Europe)

30B + D, where B is 64 kbps (Europe)

One should also note that ISDN is really based on the separate channel signaling concept. CEPT 30 + 2 uses separate channel signaling; DSX (United States) does not. Thus we have 23B + D rather than 24B + D.

Another important aspect is ISDN standards interpretation. There are numerous local ISDN experimental and developmental systems, each of which has its own ISDN flavor. This is particularly true in the United States, which has no governing monopoly such as the ATT Bell System before divestiture [15].

It would seem that ISDN, particularly in North America, is going to be hard to sell to achieve any sort of universality. The only way the picture may change is if the federal government subsidizes the service implementation.

REVIEW QUESTIONS

1. Name at least three services other than speech telephony that already have been integrated into the PSN (predating ISDN).
2. Identify at least two shortcomings of PCM as implemented today regarding its suitability for ISDN. (Clue: Think standards.)
3. Name at least five communication services that ISDN will support.
4. Distinguish *primary rate* from *basic rate*.
5. Define the B- and D-channels, including 2B + D, 23B + D, and 30B + D.
6. What is the aggregate bit rate of 2B + D?
7. Distinguish among NT1, NT2, and NT12. Show European versus North American differences and supporting rationale.
8. Name the OSI layers involved in an ISDN voice connection for the B-channel and for the D-channel.
9. What is a TA and what purpose does it serve?
10. For ISDN to be a reality, what signaling system has to be implemented in the national (digital) network?
11. How does an ISDN user (e.g., a TE) derive its timing?
12. Define *primitive* in the context of ISDN protocols.
13. Describe pseudoternary coding as used as a line signal for ISDN.
14. What is the function of the balancing bit in an ISDN frame?
15. How many time slots does the 23B + D frame structure have? What is the period of this frame?
16. Why must the simulation of a frame alignment signal be prevented?
17. Define an SAP.

18. What is the basic method of error detection for ISDN frames? Name at least one other way that errors are detected in ISDN protocols.

19. Define the purpose of a flag in a frame address field, in a control field, in an information field, and in an FCS.

20. Name and describe at least three functions carried out by layer 3 in ISDN.

21. Describe two ways a user can operate in the packet mode (we refer here to channels only).

22. Distinguish the two ways the PH function can be carried out.

23. Give at least two reasons why it may be hard to sell ISDN in North America in view of attaining widespread use nearly equal to that of today's telephone.

24. Discuss the so-called nonstandardization of ISDN from two distinct points of view.

25. Why is the US 64-kbps channel a corrupted channel?

REFERENCES

1. William Stallings, ed., *Tutorial: Integrated Services Digital Networks (ISDN)*, IEEE Computer Society Press, Washington, D.C., 1985.
2. *ISDN User–Network Interfaces: Interface Structures and Access Capabilities*, CCITT Rec. I.412, CCITT VIII Plenary Assembly, Malaga-Torremolinos, Spain, October 1984.
3. *Service Aspects of ISDNs*, CCITT Rec. I.210, CCITT VIII Plenary Assembly, Malaga-Torremolinos, Spain, October 1984.
4. *General Aspects and Principles Relating to Recommendations on ISDN User–Network Interfaces*, CCITT Rec. I.410, CCITT VIII Plenary Assembly, Malaga-Torremolinos, Spain, October 1984.
5. William Stallings, "Integrated Services Digital Network," in his *Tutorial: Integrated Services Digital Networks (ISDNs)*, IEEE Computer Society Press, Washington, D.C., 1985.
6. *ISDN Protocol Reference Model*, CCITT Rec. I.320, CCITT VIII Plenary Assembly, Malaga-Torremolinos, Spain, October 1984.
7. M. Decina, "Progress towards User Arrangements in Integrated Services Digital Networks," *IEEE Trans. on Commun.* (September 1982). Reprinted in Ref. 1.
8. *Basic User–Network Interface: Layer 1 Specification*, CCITT Rec. I.430, CCITT VIII Plenary Assembly, Malaga-Torremolinos, Spain, October 1984.
9. *Primary Rate User–Network Interface: Layer 1 Specification*, CCITT Rec. I.431, CCITT VIII Plenary Assembly, Malaga-Torremolinos, Spain, October 1984.
10. *ISDN User–Network Interface Data Link Layer: General Aspects*, CCITT Rec. I.440, CCITT VIII Plenary Assembly, Malaga-Torremolinos, Spain, October 1984.

11. *ISDN User–Network Interface Data Link Layer Specification*, CCITT Rec. I.441, CCITT VIII Plenary Assembly, Malaga-Torremolinos, Spain, October 1984.
12. *ISDN User–Network Interface Layer 3: General Aspects*, CCITT Rec. I.450, CCITT VIII Plenary Assembly, Malaga-Torremolinos, Spain, October 1984.
13. *ISDN User–Network Interface Layer 3 Specification*, CCITT Rec. 1.451 CCITT VIII Plenary Assembly, Malaga-Torremolinos, Spain, October 1984.
14. *Support of X.21 and X.21 bis Based Data Terminal Equipment (DTEs) by an Integrated Services Digital Network (ISDN)*, CCITT Rec. I.462, CCITT VIII Plenary Assembly, Malaga-Torremolinos, Spain, October 1984.
15. "ISDN in the US: An Assessment," *Telecommunications* (December 1987).
16. Tom Valovic, "Fourteen Things You Should Know about ISDN," *Telecommunications* (December 1987).
17. IEEE Journal on Selected Areas in Communications, Special Issue on Integrated Services Digital Network: Recommendations and Field Trials, I," IEEE Computer Society, New York, May 1986.
18. IEEE Journal on Selected Areas in Communications, Special Issue on "Integrated Services Digital Network: Technology and Implementations, II," IEEE Computer Society, New York, November 1986.

14

CCITT SIGNALING SYSTEM NO. 7

1 INTRODUCTION

CCITT Signaling System No. 7 (SS No. 7) was initially developed to meet the advanced signaling requirements of the all-digital network based on the 64-kbps channel. SS No. 7 is a common-channel signaling system much like CCS and CCIS discussed in Chapter 4. These forerunners of SS No. 7 operated at 2400 bps and were implemented on standard analog VF channels. They do not have sufficient capacity, nor are they compatible with the present evolving digital network and, in particular, with ISDN. The reader will note, however, that there are many similarities in topology and message structure between CCIS/CCITT No. 6 and SS No. 7 [1].

Simply put, CCITT SS No. 7 is described as an international standardized general-purpose common-channel signaling system:

- Optimized for operation with digital networks where switches use stored-program control (SPC), such as the DMS-100 described in Chapter 10.
- That can meet present and future requirements of information transfer for interprocessor transactions with digital communications networks for call control, remote control, network data base access and management, and maintenance signaling.
- That provides a reliable means of information transfer in correct sequence without loss or duplication.

(CCITT 1980 *Yellow Book*, Ref. 2).

CCITT SS No. 7, in the years since 1980, has become known as the signaling system for ISDN. This it is. Without the infrastructure of SS No. 7 embedded in the digital network, there will be no ISDN with ubiquitous access. One important point is to be made. CCITT SS No. 7, in itself, is the choice for signaling in the digital PSN (public switched network) without ISDN. It can stand on its own in this capacity.

SS No. 7 is a data communications system designed for only one purpose: signaling. It is *not* a general-purpose system. We then must look at SS No. 7 as (1) a specialized data network and (2) a signaling system.

2 SS NO. 7 RELATIONSHIP TO OSI

CCITT SS No. 7 relates to OSI (Chapter 11, Section 7.4.2) up to a certain point. One group believes that SS No. 7 should be fully compatible with the seven layers of OSI. However, the CCITT working groups responsible for the SS No. 7 concept and design were concerned with delay, whether for the data or telephone user of the digital PSN or ISDN. Recall from Chapter 4 that postdial delay is one of the principal measures of performance of a signaling system. To minimize delay, the seven layers of OSI were truncated at layer 4. In fact, CCITT Rec. Q.709 specifies no more than 2.2 seconds of post-dial delay for 95% of calls. To accomplish this, a limit is placed on the number of relay points, called STPs, that can be traversed by a signaling message and by the inherent design of SS No. 7 as a four layer system. Figure 14.1 relates SS No. 7 protocol layers to OSI.

We should note that SS No. 7 layer 3 signaling network functions include signaling message handling functions and network management functions. Figure 14.2 shows the general structure of SS No. 7 signaling system.

Schlanger [3] makes the following pertinent observations:

- "Signaling is typically performed to create a communications subnetwork for a 'network end user.' As such, some argue that the entire reference

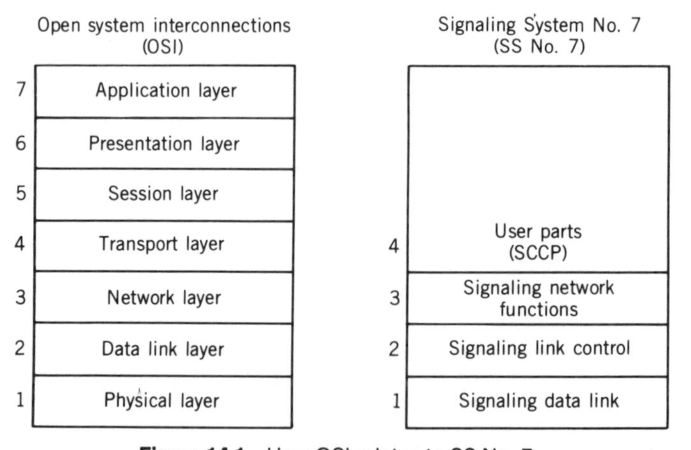

Figure 14.1 How OSI relates to SS No. 7.

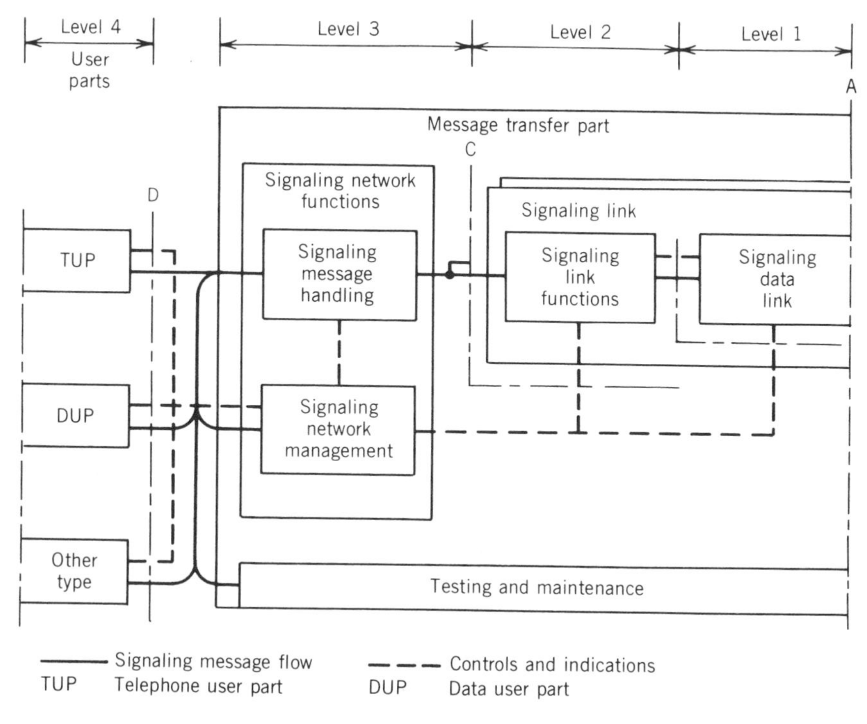

Figure 14.2 General structure of CCITT SS No. 7 functions. From CCITT Rec. Q.701 Figure 2 / Q.701, page 7, *Red books*, 1984, Vol. VI, Fascicle VI.7.

model of SS No. 7, as a protocol within the communications subnetwork, should only exist at OSI layer 3 (the network layer) and below."

- "The applications processes within a communications network invoke protocol functionality to communicate with one another in much the same way as 'end users.' Thus, the same seven-layer reference model is felt to apply in this application context."
- "The signaling system protocol is felt to encompass operations, administration and maintenance (OA & M) activities related to telecommunications. Because craftspeople can be involved in such activities (truly end users) as well as OA & M application processes, the distinction between network layer entities and end users becomes 'fuzzy.'"

There seem to be various efforts to force-fit SS No. 7 into OSI from layer 4 and above. These efforts have resulted in the sublayering of layer 4 into an SCCP (signaling connection control part) and the user parts.

Section 3 briefly describes the basic functions of the four SS No. 7 layers, which are covered in more detail in Sections 4 through 6.

3 SIGNALING SYSTEM STRUCTURE

Figure 14.2, which illustrates the basic structure of SS No. 7 shows two parts of the system: the message transfer part (MTP) and the user parts. There are three user parts: Telephone User (TUP), Data User (DUP), and other. "Other," in this context, refers to the ISDN user part (ISUP). Figures 14.1 and 14.2 show layers 1, 2, and 3, which make up the MTP. The next several paragraphs describe the functions of each of these layers from a system viewpoint.

Layer 1 defines the physical, electrical, and functional characteristics of the signaling data link and the means to access it. In the digital network environment the 64-kbps digital path is the normal basic connectivity. The signaling link may be accessed by means of a switching function that provides the capability of automatic reconfiguration of signaling links.

Layer 2 carries out the signaling link function. It defines the functions and procedures for the transfer of signaling messages over one individual signaling data link. A signaling message is transferred over the signaling link in variable length signal units. A signal unit consists of transfer control information in addition to the information content of the signaling message. The signaling link functions include:

- Delimination of a signal unit by means of flags.
- Flag imitation prevention by bit stuffing.
- Error detection by means of check bits included in each signal unit.
- Error control by retransmission and signal unit sequence control by means of explicit sequence numbers in each signal unit and explicit continuous acknowledgments.
- Signaling link failure detection by means of signal unit error monitoring and signaling link recovery by means of special procedures.

Layer 3, signaling network functions, in principle, defines such transport functions and procedures that are common to and independent of individual signaling links. There are two categories of functions in Layer 3:

1. Signaling message handling functions. During message transfer, these functions direct the message to the proper signaling link or user part.
2. Signaling network management functions. This controls real-time routing, control, and network reconfiguration, if required.

Layer 4 is the user part. Each user part defines the functions and procedures peculiar to the particular user, whether telephone, data, or ISDN user part.

The *signal message* is defined by CCITT Rec. Q.701 as an assembly of information, defined at layer 3 or 4, pertaining to a call, management transaction, etc., which is then transferred as an entity by the message transfer function. Each message contains "service information," including a service indicator identifying the source user part and possibly whether the message relates to international or national application of the user part.

The *signaling information* of the message contains user information, such as data or call control signals, management and maintenance information, and type and format of message. It also includes a "label." The label enables the message to be routed by layer 3 through the signaling network to its destination and directs the message to the desired user part or circuit.

On the signaling link such signaling information is contained in the *message signal units* (MSUs), which also include transfer control functions related to layer 2 functions on the link.

There are a number of terms used in SS No. 7 literature that should be understood before we proceed further.

Signaling Points. Nodes in the network that utilize common-channel signaling.

Signaling Relation (similar to traffic relation). Any two signaling points for which the possibility of communication between their corresponding user parts exist are said to have a signaling relation.

Signaling Links. Signaling links convey signaling messages between two signaling points.

Originating and Destination Points. The originating and destination points are the locations of the source user part function and location of the receiving user part function, respectively.

Signaling Transfer Point (STP). An STP is a point where a message received on one signaling link is transferred to another link.

Message Label. Each message contains a label. In the standard label, the portion that is used for routing is called the *routing label*. The routing label includes:

- Destination and originating points of the message.
- A code used for load sharing, which may be the least significant part of a label component that identifies a user transaction at layer 4.

The standard label assumes that each signaling point in a signaling network is assigned an identification code according to a code plan established for the purpose of labeling.

Message Routing. Message routing is the process of selecting the signaling link to be used for each signaling message. Message routing is based on analysis of the routing label of the message in combination with predetermined routing data at a particular signaling point.

Message Distribution. Message distribution is the process that determines to which user part a message is to be delivered. The choice is made by analysis of the service indicator.

Message Discrimination. Message discrimination is the process that determines, on receipt of a message at a signaling point, whether or not the point is the destination point of that message. This decision is based on analysis of the destination code of the routing label in the message. If the signaling point is the destination, the message is delivered to the message destination function. If not, the message is delivered to the routing function for further transfer on a signaling link.

3.1 Signaling Network Management

Signaling network management is made up of three technical areas:

- Signaling traffic management.
- Signaling link management.
- Signaling route management.

The signaling traffic management functions are

- Control message routing, including, where necessary, message route modification to maintain connectivity.
- Ensuring message flow in conjunction with modifications of message routing.
- Flow control actions to minimize effects of overload.

The signal link management functions controls the locally connected link sets, including link restoration. This function supplies link availability data of local links to the signaling traffic management function.

The signaling route management function is used only when the signaling link operates in the quasi-associated mode. The quasi-associated mode refers to signaling links that use a route other than the route of normal message traffic. The route management task is to transfer information about changes in the availability of signaling routes in the signaling network to enable remote signaling points to take appropriate signaling traffic management actions.

4 THE SIGNALING DATA LINK (LAYER 1)

The signaling data link is a full-duplex channel (i.e., it has two data transmission channels operating together in opposite directions at the same data rate). These digital channels terminate in digital switches or their terminating equipment, providing an interface to signaling terminals. CCITT SS No. 7 can operate over both terrestrial and satellite transmission links.

The standard bit rate is 64 kbps. The 64 kbps digital signaling channels entering a digital exchange via a multiplex structure are switchable as semipermanent channels in the exchange. For a signaling data link derived from a 2048-kbps digital path, time slot 16 is used. No bit inversion is performed.

When a signaling data link is derived from an 8448-kbps digital path, time slots 67 to 70 are used in descending order of priority. The signaling bit rate is again 64 kbps, and no bit inversion is performed.

The 1984 edition of the CCITT recommendations states that the 1.544-Mbps digital path is left for further study (CCITT Rec. Q.702, Ref. 4).

5 THE SIGNALING LINK (LAYER 2)

The signaling link is OSI layer 2 (Figure 14.3). Seven functions can be identified for this layer:

1. Signal unit delimitation.
2. Signal unit alignment.
3. Error detection.
4. Error correction.
5. Initial alignment.
6. Signaling link error monitoring.
7. Flow control.

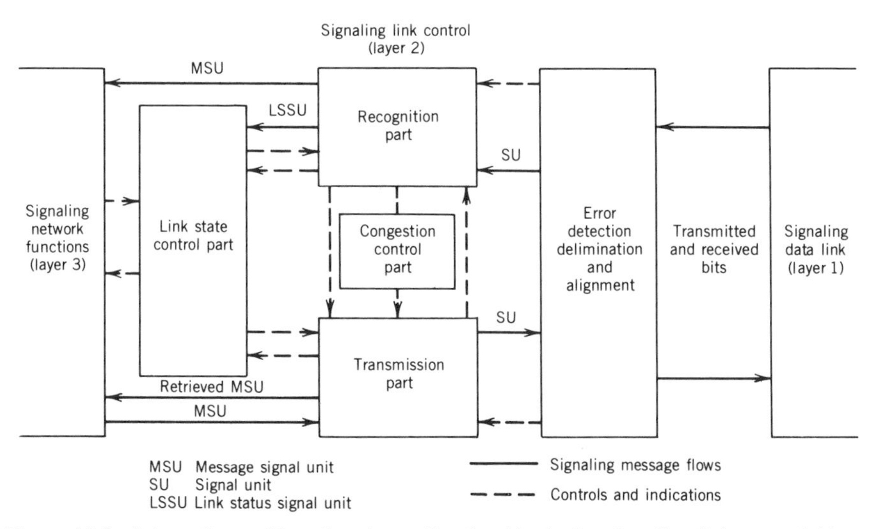

Figure 14.3 Interactions of functional specification blocks for signaling link control. Note: The MSUs LSSU, and SUs do not include all error-control information. From CCITT Rec. Q.703. Figure 1 / Q.703, page 23, *Red Books* 1984, Vol. VI, Fascicle VI.7.

The seven functions are coordinated by the link state control, as shown in Figure 14.3. Each function is described in what follows.

Signal unit delimitation. Signaling messages delivered by higher layers are transferred over the signaling link in variable length *signaling units*. These signaling units include transfer control information for proper operation of the signaling link in addition to the actual signaling information. The beginning and end of each signaling unit are indicated by a unique 8-bit pattern called a flag (see Chapter 11). Measures are taken to ensure that the pattern cannot be imitated elsewhere in a unit. These flags are delimiters.

Loss of alignment occurs when a bit pattern is disallowed by the delimitation procedure (more than six consecutive 1's) or when a certain maximum length of a signal unit is exceeded. Loss of alignment will cause a change in the mode of operation of the *signal unit error monitor*.

Error detection. Error detection is carried out by a 16-bit frame check sequence (FCS) using a cyclic redundancy check (CRC) technique. These techniques are described in Chapter 11. The CRC used in SS No. 7 is based on the CCITT standard that uses the generating polynomial $X^{16} + X^{12} + X^5 + 1$.

Error correction. Error correction with CCITT SS No. 7 uses ARQ (see Chapter 11, Section 7.1). SS No. 7 utilizes some application-specific terminology as described below. CCITT Rec. Q.703 specifies two forms of error correction: the basic method and preventive cyclic retransmission method. The *basic method* is used on links where the one-way propagation delay is less than 15 ms. The *preventive cyclic retransmission method* applies where the one-way propagation time is equal to or greater than 15 ms, such as on satellite, intercontinental, and transcontinental links.

The *basic method* was described in Chapter 11 as "go-back-*n*" ARQ. CCITT defines this method as a noncompelled positive/negative acknowledgment retransmission error-correction system where unacknowledged signal units are retransmitted only *once*. The preventive cyclic retransmission method is designed for implementation on degraded circuits. When a predetermined number of retained, unacknowledged signal units exist, the transmission of new signal units is interrupted and the retained units are retransmitted cyclically until the number of unacknowledged units is reduced.

Initial alignment procedure is used for first-time initialization and alignment after a link failure. The procedure is based on the compelled exchange of status information between the two signaling points concerned and the provision of a prove-in period.

Signaling link error monitoring provides two functions. One operates while the link is in service and provides one criterion for taking the link out of service. The other is used during the initial alignment procedure. These functions are called *significant error rate monitor* and *alignment error rate monitor*, respectively. The first uses signal unit error count as the criterion that is incremented and decremented using the "leaky bucket" principle, and the secondary is strictly a linear count of signal unit errors. *Link state control functions* provide directives to the other signaling link functions, as shown in Figure 14.3.

Flow control is initiated when congestion is detected at the receiving end of a signaling link. When this occurs, the receiving end notifies the distant transmit end of the condition by means of an appropriate status signal unit, and it withholds acknowledgments of all incoming message signal units. When the congestion abates, acknowledgments resume.

5.1 Basic Signal Unit Format

Signaling and other information originating from a user part is transferred over the signaling link by means of signal units. There are three types of signal units used in SS No. 7:

1. Message signal unit (MSU).
2. Link status signal unit (LSSU).
3. Fill-in signal unit (FISU).

These units are differentiated by means of the *length indicator*. MSUs are retransmitted in case of error; LSSUs and FISUs are not. The MSU carries signaling information; the LSSU provides link status information, and the FISU is used during the link idle state—they fill in.

The signaling information field (SIF) is variable in length and carries the signaling information generated by the user part. All other fields are of fixed length. Figure 14.4 shows the basic formats of the three types of signal units. As shown in the figure, the message transfer control information encompasses eight fixed-length fields in the signal unit that contains information required for error control and message alignment. These eight fields are described below. In the figure we start from right to left, which is the direction of transmission.

The opening *flag* indicates the start of a signal unit. The opening flag of one signal unit is normally the closing flag of the previous signal unit. The flag bit pattern is 01111110. The *forward sequence number* (FSN) is the sequence number of the signal unit in which it is carried. The *backward sequence number* (BSN) is the sequence number of a signal unit being acknowledged. The value of the FSN is obtained by incrementing (modulo 128) the last assigned value by 1. The FSN value uniquely identifies a message signal unit until its delivery is accepted without errors and in correct sequence by the receiving terminal. The FSN of a signal unit other than an MSU assumes the value of the FSN of the last transmitted MSU. The maximum capacity of sequence numbers is 127 message units before reset (modulo 128).

Positive acknowledgment is accomplished when a receiving terminal acknowledges the acceptance of one or more MSUs by assigning an FSN value of the latest accepted MSU to the BSN of the next signal unit sent in the opposite direction. The BSNs of subsequent signal units retain this value until a further MSU is acknowledged, which will cause a change in the BSN sent.

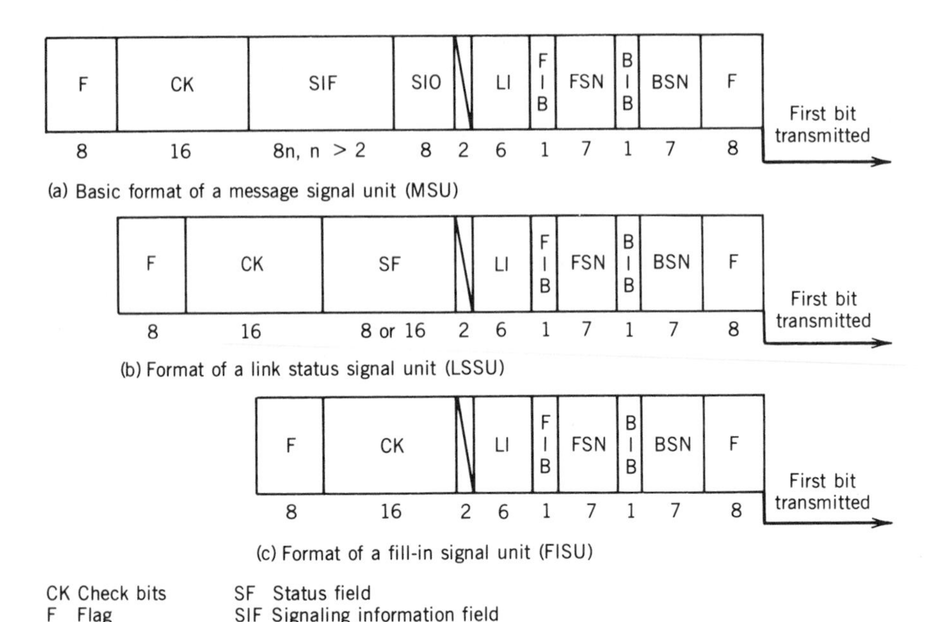

Figure 14.4 Signal unit formats. From CCITT Rec. Q.703. Figure 3 / Q.703, page 27, *Red Books*, Vol. VI, Fascicle VI.7.

The acknowledgment to an accepted MSU also represents an acknowledgment to all, if any, previously accepted, though not yet acknowledged, MSUs.

Negative acknowledgment is accomplished by inverting the backward indicator bit (BIB) value of the signal unit transmitted. The BIB value is maintained in subsequently sent signal units until a new negative acknowledgment is to be sent. The BSN assumes the value of the FSN of the last accepted signal unit.

As we can now discern, the forward indicator bit (FIB) and the backward indicator bit together with the FSN and BSN are used in the basic error-control method to perform signal unit sequence control and acknowledgment functions.

The *length indicator* (LI) is used to indicate the number of octets following the length indicator octet and preceding the check bits and is a binary number in the range of 0–63. The length indicator differentiates between three types of signal units as follows:

Length indicator = 0	Fill-in signal unit
Length indicator = 1 or 2	Link status signal unit
Length indicator ≥ 2	Message signal unit

The *service information octet* (SIO) is divided into a *service indicator* and a *subservice field*. The service indicator is used to associate signaling information for a particular user part and is present only in MSUs. Each is 4 bits long. For example, a service indicator with a value 0100 relates to the telephone user part, and 0101 to the ISDN user part. The subservice field portion of the SIO contains two network indicator bits and two spare bits. The network indicator discriminates between international and national signaling messages. It can also be used to discriminate between two national signaling networks, each having a different routing label structure. This is accomplished when the network indicator is set to 10 or 11.

The *signaling information field* (SIF) consists of an integral number of octets greater than or equal to 2 and less than or equal to 62. In national signaling networks it may consist of up to 272 octets. Of these 272 octets information blocks of up to 256 octets in length may be accommodated, accompanied by a label and other possible housekeeping information that may, for example, be used by layer 4 to link such information blocks together.

The *link status signal unit* (LSSU) provides link status information between signaling points. The status field can be made up of one or two octets. CCITT Rec. Q.703 shows application of the one-octet field in which the first three bits (from right to left) are used (bits A, B, and C) and the remaining five bits are spare. These first 3-bit values follow:

Bits			Status	
C	B	A	Indication	Meaning
0	0	0	0	Out of alignment
0	0	1	N	Normal alignment
0	1	0	E	Emergency alignment
0	1	1	OS	Out of service
1	0	0	PO	Processor outage
1	0	1	B	Busy

From CCITT Rec. Q.703 (Ref. 5).

6 SIGNALING NETWORK FUNCTIONS AND MESSAGES (LAYER 3)

6.1 Introduction

This section describes the functions and procedures relating to the transfer of messages between signaling points that are signaling network nodes. These nodes are connected by signaling links (layers 1 and 2), described in Sections 4 and 5. Another important function of layer 3 is to inform the appropriate entities of a fault and, as a consequence, carry out a rerouting of messages through the network. The signaling network functions are broken down into two basic categories:

- Signaling message handling.
- Signaling network management.

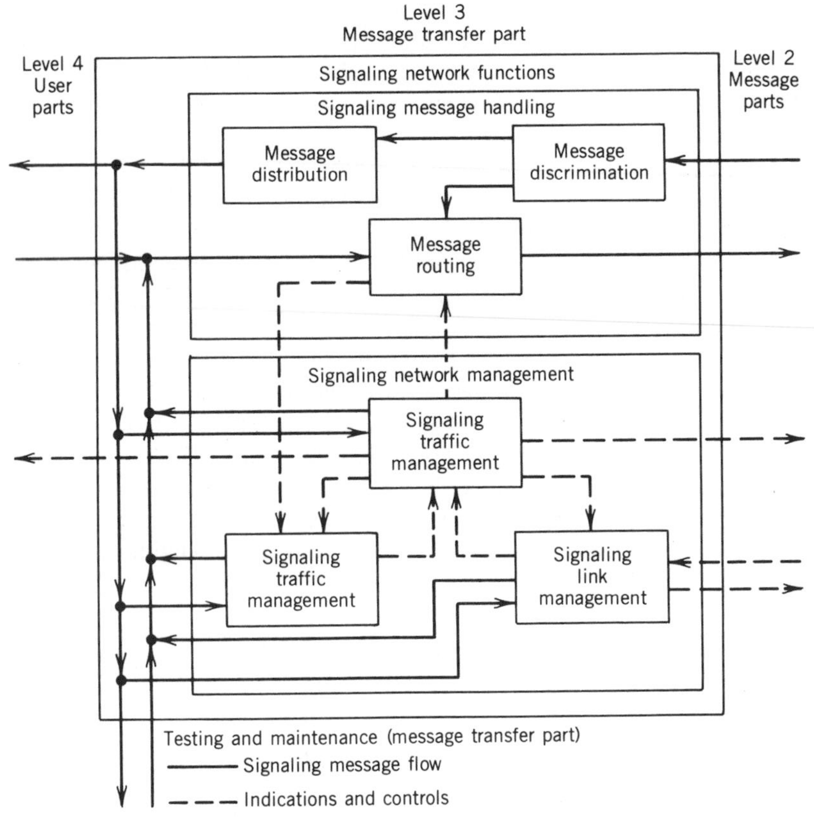

Figure 14.5 Signaling network functions. From CCITT Rec. Q.704, Figure 1 / Q.704, Page 74, *Red Books*, Vol. VI Fascicle VI.7.

The functional interrelations between these functions are shown in Figure 14.5. The discussion that follows is based on CCITT Rec. Q.704 [6].

6.2 Overview of Signaling Message Handling

The signaling message handling functions ensure that signaling messages originated by a particular user part at an originating signaling point are delivered to the indicated user part at the destination signaling point as indicated by the originator (user part).

These functions are carried out based on label information contained in the message, which explicitly identifies the originator and destination identified by the sending user part. This is called the *routing label* and is described in

Section 6.3. As shown in Figure 14.5, the signaling message handling functions are divided into:

- The *signaling routing function*, which is used at each signaling point to determine the routing of the message to its destination.
- The *message discrimination function*, which is used at each signaling point to determine whether or not a message is destined for itself or for onward routing. In the latter case the message is transferred to the message routing function.
- The *message distribution function*, which is used at a signaling point, when a signaling message is destined for itself, to direct it to the appropriate user part.

6.3 Description of Message Handling

The interrelationships of the three basic signaling message handling functions are shown in Figure 14.6, namely routing, discrimination, and distribution. In the figure we see two directions for signaling message traffic: from the user part (layer 4) at the left, which enters the routing function to determine the appropriate route, or from the network (at the right), where there are two possibilities for a message at a given signaling point. These are local delivery to a user part and forward transfer when the message is destined from some point onward in the network. This decision is carried out by the discrimination function. If the message is for local delivery, it is passed to the distribution function. If it is destined for another node, it is passed to the routing function. The distribution function determines delivery to the appropriate user part.

The processing of the three functions is based on that part of the label called the *routing label*, on the service indicator (SI), and in national networks on the national indicator.

6.3.1 Routing Label. The standard routing label consists of 32 bits (4 octets) and appears at the beginning of the signaling information field (see Figure 14.4). The structure of the label is shown in Figure 14.7. The label is

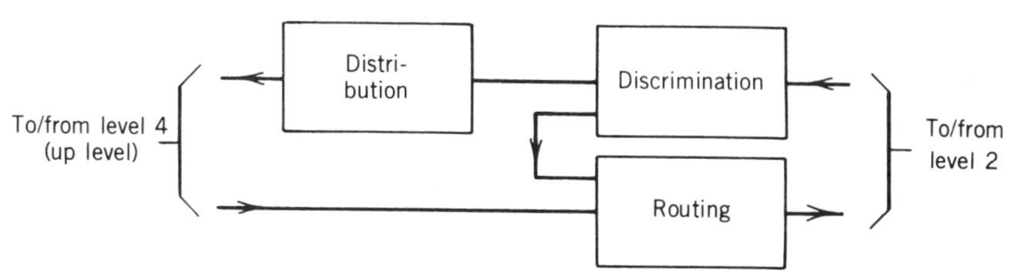

Figure 14.6 Message routing, discrimination, and distribution.

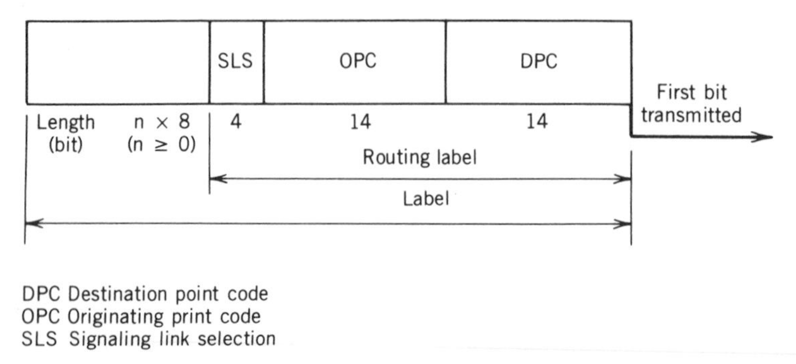

Figure 14.7 Routing label structure. From CCITT Rec. Q.704, Figure 3 / Q.704, Page 76, *Red Books*, Vol. VI, Fascicle VI.7.

analogous to the dialed digits of a telephone in our present telephone network. As shown in Figure 14.7, there is a *destination point code* (DPC), which indicates the destination of the message, and the *originating point code* (OPC), which identifies the originator. The coding of each is pure binary, and within each field the least significant bit occupies the first position and is transmitted first. A unique numbering scheme is used for the signaling points in the international network irrespective of the user parts connected to each signaling point.

The *signaling link selection* (SLS) field is used, where appropriate, in the performance of *load sharing*. Each signaling point will have routing information that allows it to determine the signaling link over which a message has to be sent on the basis of the destination point code and the signaling link selection field and in some cases on the basis of the national indicator. Typically the DPC is associated with more than one signaling link, which may be used to carry the message. The selection of the particular signaling link is made by means of the SLS field, thus effecting load sharing.

There are two basic cases for load sharing:

- Load sharing between links belonging to the same link set.
- Load sharing between links not belonging to the same link set.

Figure 14.8 illustrates the concept of load sharing between links of the same link set in the associated mode of operation. Figure 14.9 conceptually illustrates the second case. The load sharing rule used for a particular signaling

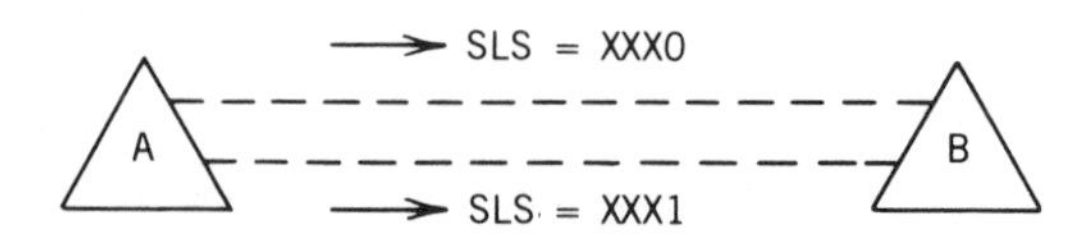

Figure 14.8 Example of load sharing within a link set.

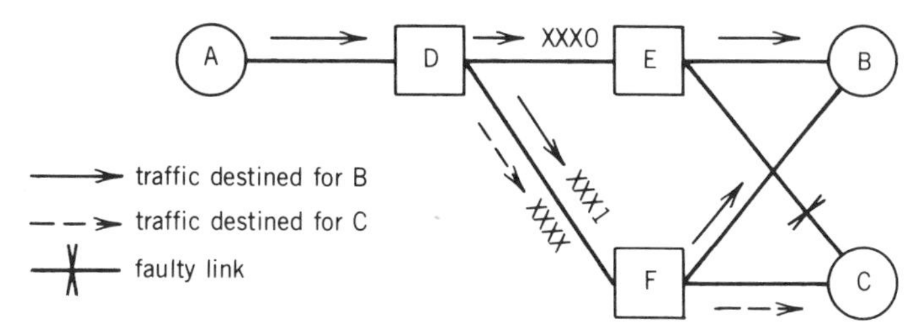

Figure 14.9 Example of load sharing between link sets.

relation may or may not apply to all signaling relations that use one of the signaling links. For instance, in Figure 14.9 traffic destined to B is shared between links DE and DF with a given signaling link selection field assignment, whereas that destined to C is sent only on link DF due to the failure of link EC.

As a result of the message routing function under normal conditions, all messages having the same routing label (e.g., call-setup messages related to a given circuit) are routed via the same signaling links and signal transfer points (STPs).

6.4 Signaling Network Management

This basic function deals with maintaining signaling service in view of three possible fault or degraded operation conditions:

- Loss of signaling link.
- Loss of signaling point.
- Degraded operation due to congestion.

As mentioned previously, there are three subsidiary functions used to carry out signaling network management: traffic management, link management, and route management. The signaling traffic management function is used to direct signaling traffic from a link or route to one or more different links or routes or to deload temporarily a link or route in the case of congestion at a signaling point. The following procedures are involved with this function:

- Changeover.
- Change back.
- Forced rerouting.
- Controlled rerouting.
- Signaling traffic flow control.

The signaling link management function is used to restore failed links, activate idle links (not yet aligned), and deactivate aligned signal links. This function involves the following procedures:

- Signaling link activation.
- Link set activation.
- Automatic allocation of signaling terminals and signaling data links.

The signaling route management function distributes information about signaling network status in order to block or unblock signaling routes. Six procedures for route management are described in CCITT Rec. Q.704 [6].

7 SIGNALING NETWORK STRUCTURE

The simplest method of routing signaling links is by the associated network where signaling links are routed with their associated traffic links. This approach is practical for the most elementary signaling network, consisting of originating and destination signaling points connected by a single signaling link or a signaling link set where several links share the signaling traffic node.

For technical and economic reasons a simple associated network may not be suitable and a *quasi-associated network* may be implemented. In this case the information between originating and destination signaling points may be transferred via a number of signaling transfer points. Such a network may be represented by a mesh network, and other networks are either a subset of the mesh network or structured using this network or its subsets as components. The network components are signaling links and signaling points and may be combined in many different ways to form a signaling network.

CCITT Rec. Q.705 [7] details the worldwide signaling network that is structured into two functionally independent levels. These are the national and international levels. The structure makes a clear division of responsibility for signaling network management and allows numbering plans for signaling points of the international and different national networks to be independent of one another.

A signaling point may be assigned to one of three categories:

1. National signaling point (NSP) only, encompassed in the national signaling point numbering plan.
2. International signaling point (ISP) only, identified by a signaling point code as part of the international signaling point numbering plan.
3. A node that serves both functions and belongs to both the national and international signaling networks, identified by a specific signaling point code in each of the signaling networks.

In the international signaling network the number of signaling transfer points between an originating and a destination signaling point should not exceed 2

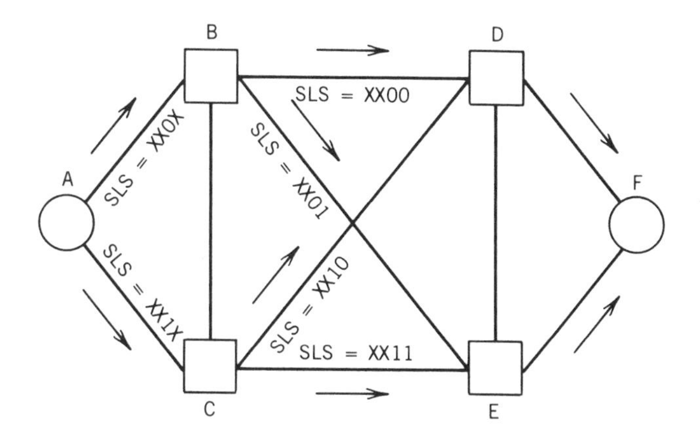

Normal message routes from A to F
— A → B → D → F (SLS = XX00)
— A → C → D → F (SLS = XX10)
— A → B → E → F (SLS = XX01)
— A → C → E → F (SLS = XX11)

SLS Signaling link selection code in the routing label.
Assumption: There is only one link between adjacent signaling points.

Figure 14.10 An example of routing in the absence of failures. Note mesh connection of STPs and the coding of the SLS. From CCITT Rec. Q.705. Figure A.3 / Q.705, page 205, *Red Books*, Vol. VI, Fascicle VI.7.

under normal conditions. In failure situations this number could be 3 or 4 for a short period of time. Such a constraint is intended to limit the complexity of administration of the international signaling network. A 14-bit code is used for the identification of signaling points. The international allocation scheme (numbering), which identifies the signaling points, is described in CCITT Rec. Q.708 [8].

Figure 14.10 shows a typical mesh connection of STPs with routing in the absence of failures.

8 SIGNALING PERFORMANCE: MESSAGE TRANSFER PART

8.1 Basic Performance Parameters

CCITT Rec. Q.706 [9] breaks down SS No. 7 performance into three parameter groups:

1. Message delay (Section 10) and availability.
2. Signaling traffic load.
3. Error rate.

Availability. The unavailability of a signaling route set should not exceed 10 min per year.

Undetected Errors. Not more than 1 in 10^{10} of all signal unit errors will go undetected in the message transfer part.

Lost Messages. Not more than 1 in 10^{7} messages will be lost due to failure of the message transfer part.

Messages Out of Sequence. Not more than 1 in 10^{10} messages will be delivered out of sequence to the user part due to failure in the message transfer part. This includes message duplication.

8.2 Traffic Characteristics

Labeling Potential. There are 16,384 identifiable signaling points.

Loading Potential. Loading potential is restricted by the following factors:

- Queuing delay.
- Security requirements (redundancy with changeover).
- Capacity of sequence numbering (127 unacknowledged signal units).
- Signaling channels using bit rates under 64 kbps.

8.3 Transmission Parameters

The message transfer part operates satisfactorily with the following error performance:

- Long-term error rate on the signaling data links of less than 1×10^{-6}.
- Medium-term error rate of less than 1×10^{-4}.

9 NUMBERING PLAN FOR INTERNATIONAL SIGNALING POINT CODES

This numbering plan, described in CCITT Rec. Q.708 [8], has no direct relationship with telephone, data, or ISDN numbering. A 14-bit binary code is used for identification of signaling points, as mentioned before. An international signaling point code (ISPC) is assigned to each signaling point in the international signaling network. The breakdown of these 14 bits into fields is shown in Figure 14.11. The assignment of the signaling network codes is administered by the CCITT.

Turning to Figure 14.11, an ISPC is represented in decimal form in each subfield as Z–UUU–V where Z, UUU, and V correspond to bits NML, K–D,

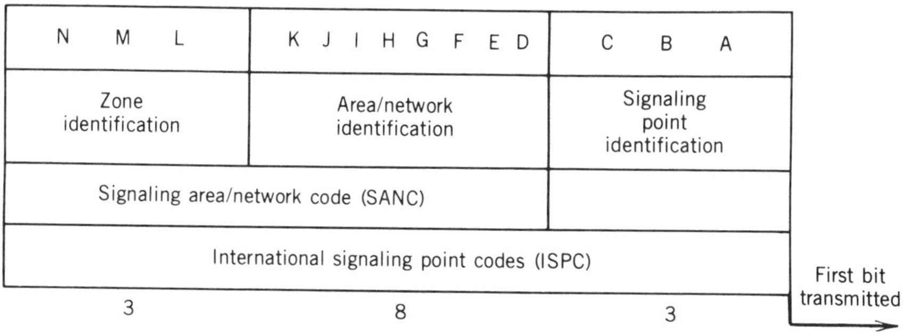

N M L	K J I H G F E D	C B A
Zone identification	Area/network identification	Signaling point identification
Signaling area/network code (SANC)		
International signaling point codes (ISPC)		
3	8	3

First bit transmitted →

Figure 14.11 Format of international signaling point code. From CCITT Rec. Q.708. Figure 1 / Q.708, page 238, *Red Books*, Vol. VI, Fascicle VI.7.

and CBA, respectively. Z is the zone identification code. There are six zones (numbered 2–7). UUU is the area or network identification. Examples of Z–UUU are: United States 3–020, Canada 3–004, United Kingdom 2–068, West Germany 2–124, Spain 2–028, India 4–004, South Africa 6–110, Argentina 7–044, and Australia 5–010.

10 HYPOTHETICAL SIGNALING REFERENCE CONNECTIONS

The hypothetical signaling reference connection (HSRC) is composed of signaling points and STPs that are connected in series by signaling data links to produce a signaling connection. The number of signaling points and STPs depends on the size of the network. There are two important parameters we derive from the HSRC: (1) the number of signaling points and STPs in a signaling connection, and (2) the signaling message transfer delay, which will directly affect postdial delay.

There is an important relationship between the HSRC and the maximum number of links incorporated in the telephone routing plan of CCITT Recs. Q.13 and E.171. By limiting the number of signaling points required in a connection, we can reduce the signaling delay by considering that the signaling point delay forms the largest component of the total signaling delay. The number of STPs in an HSRC is a function of the number of signaling points and the signaling network topology used to connect these signaling points. For an international connection CCITT Rec. Q.709 recommends that no more than two STPs be used in a signaling relation. The maximum number of STPs from on average-size country to an average-size country for 95% of connections is 7 or less. The maximum overall signaling delay is shown in Table 14.1 Somewhat more than half of the signaling delays are attributed to the two national components, one on each end of an international connection.

TABLE 14.1 Maximum Overall Signaling Delays

		Delay (ms)	
		Message Type	
Country Size	Percent of Connections	Simple (e.g., Answer)	Processing Intensive (e.g., IAM)
Large country to large country	Mean	1170	1800
	95	1450	2220
Large country to average-size country	Mean	1170	1800
	95	1450	2220
Average-size country to average-size country	Mean	1170	1800
	95	1470	2240

From CCITT Rec. Q.709 (Ref. 10) Table 5 / Q 709, page 238, *Red Books*, Vol. VI, Fascicle VI.7.

11 SIGNALING CONNECTION CONTROL PART

11.1 Introduction

The signaling message control part (SCCP) modifies the connectionless sequenced transport service provided by the MTP (message transfer part) for those user parts requiring enriched connectionless or connection-oriented service to transfer signaling and other related information between nodes. The combined service (MTP + SCCP) is called the *network service part*.

The MTP addresses are limited to a 14-bit signaling point code. For signaling applications this is adequate. All signaling messages have a format that allows extraction of additional information required for message distribution (e.g., user part discrimination). However, this can lead to additional complexity when nonsignaling applications are supported by the SS No. 7 signaling network. Each of these nonsignaling applications could be designed as user parts and could provide their own routing and distribution functions. However, this could result in a number of applications that implement the same function and would not be in the spirit of the OSI reference model [11].

This has brought about the addition of another layer, the SCCP. To the MTP the SCCP appears to be another user part. The SCCP layer carries out the OSI function of the network layer.

Five classes of network transport service are provided by the SCCP:

0 Basic sequenced connectionless
1 Sequence connectionless (like the MTP)
2 Basic connection oriented
3 Flow control connection oriented
4 Error recovery and flow control, connection oriented.

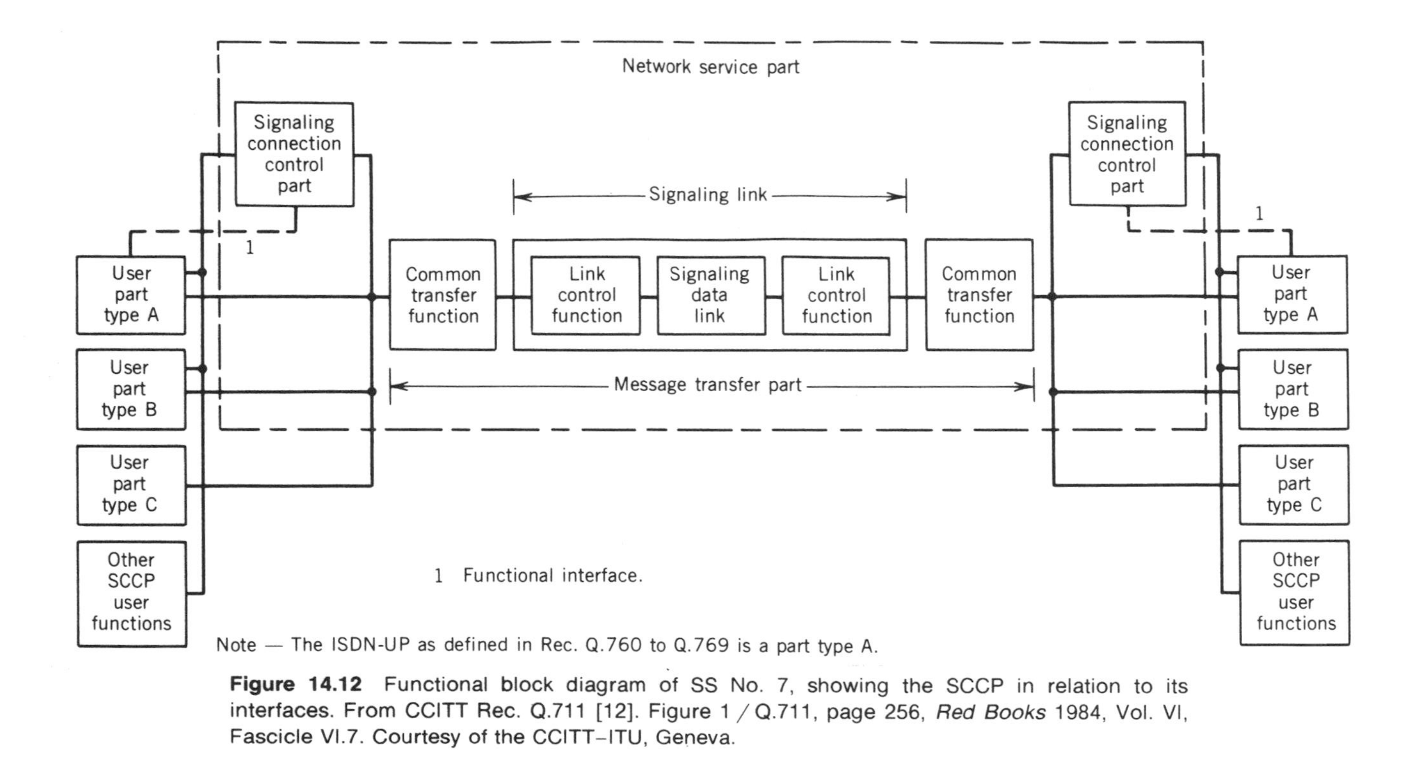

Figure 14.12 Functional block diagram of SS No. 7, showing the SCCP in relation to its interfaces. From CCITT Rec. Q.711 [12]. Figure 1 / Q.711, page 256, *Red Books* 1984, Vol. VI, Fascicle VI.7. Courtesy of the CCITT–ITU, Geneva.

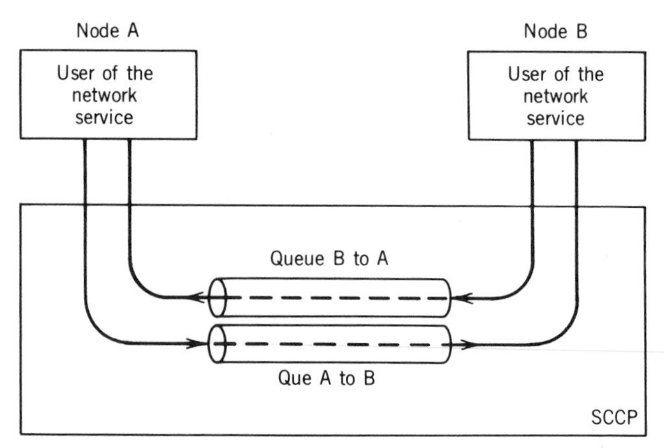

Figure 14.13 Model for internodal communication with the SCCP. From CCITT Rec. Q.711 [12]. Figure 5 / Q7.11, page 259, *Red Books* 1984, Vol. VI, Fascicle VI.7.

The SCCP accepts signaling point codes in a similar fashion as the MTP. It accepts global titles (such as dialed telephone numbers) or subsystem numbers (such as addresses). With access to routing information, the SCCP can associate any of these addresses with a signaling point code that the MTP has acted on. At the destination signaling point a signaling message is handled by the destination SCCP, which then delivers the message to the end-user.

Figure 14.12 is a functional block diagram of SS No. 7, showing the function of the SCCP in relation to the MTP and the user parts.

11.2 Peer-to-Peer Communication

A protocol facilitates the exchange of information between two peers of the SCCP. The protocol provides the means for:

- Setup of logical signaling connections.
- Release of logical signaling connections.
- Transfer of data with or without logical signaling connections.

A signaling connection is modeled in the abstract by a pair of queues. The protocol elements are objects on the queue added by the origination service user. Each queue represents a flow control function. Figure 14.13 illustrates the SCCP model for connection-oriented service.

11.3 Primitives and Parameters

Primitives consist of commands and their respective responses associated with the services requested, in this case for the SCCP. Such primitives and parame-

TABLE 14.2 Network Service Primitives for Connection-Oriented Services

Primitives		
Generic Name	Specific Name	Parameters
N-CONNECT, CONNECT	Request Indication Response Confirmation	Called address Calling address Responding address Receipt confirmation selection Expedited data selection Quality of service parameter set User data Connection identification[a]
N-DATA	Request Indication	Confirmation request User data Connection identification
N + EXPEDITED DATA	Request Indication	User data Connection identification[a]
N-DATA ACKNOWLEDGE (for further study)	Request Indication	Connection identification[a]
N-DISCONNECT	Request Indication	Originator Reason User data Responding address Connection identification[a]
N-RESET	Request Indication Response Confirmation	Originator Reason Connection identification[a]
N-NOTICE (for further study)	Indication	Reason Connection identification

[a] In Rec. X.213, Section 5.3, this parameter is implicit.

From CCITT Rec. Q.711, Table 1 / Q.711, page 261, *Red Books*, Vol. VI, Fascicle VI.7.

ters are an attempt to conform to OSI, as expressed in the CCITT X.200 series, especially CCITT Rec. X.213. Table 14.2 gives an overview of the primitives to the upper layers and the corresponding parameters for the temporary connection-oriented network service. Figure 14.14 shows an overview state transition diagram for the sequence of primitives at a connection end point.

11.4 Connection-Oriented Functions: Temporary Signaling Connections

11.4.1 Connection Establishment. A connection setup uses service primitives described in Section 11.3. The following are the principal functions used in the connection establishment phase by the SCCP to set up a signaling

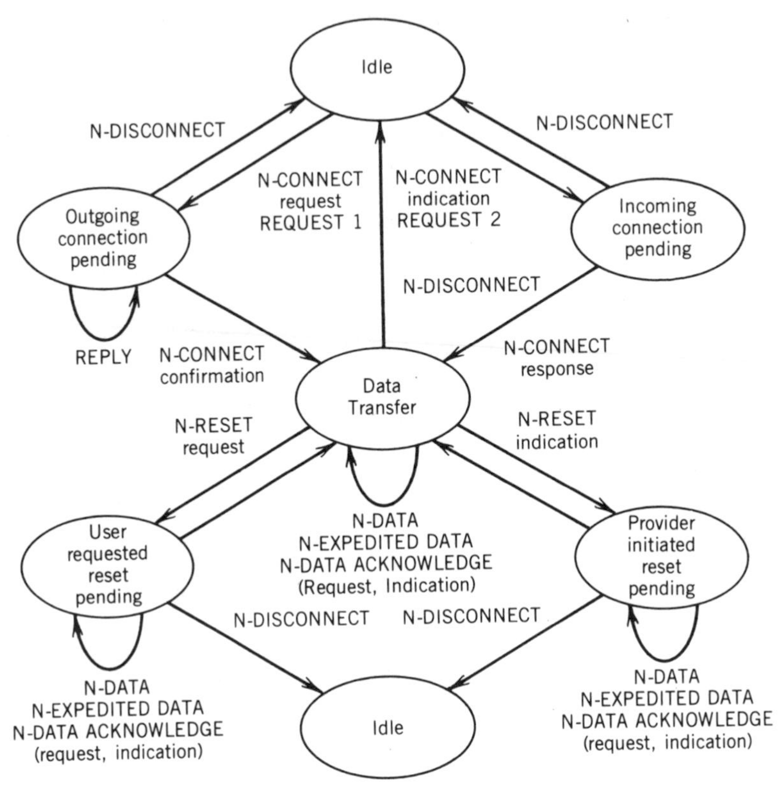

Figure 14.14 State transition diagram for the sequence of primitives at a connection end point (basic transitions). From CCITT Rec. Q.711, Figure 6 / Q.711, page 262, *Red Books*, Vol. VI, Fascicle VI.7.

connection:

- Set up signaling connection.
- Establish the optimum size of the network protocol data units (NPDUs)
- Map network address onto signaling relations.
- Select operational functions during data transfer phase, such as layer service selection.
- Provide means to distinguish network connections.
- Transport user data (within the request).

11.4.2 Data Transfer Phase. The data transfer phase functions provide the means for a two-way simultaneous transport of messages between two end points of a signaling connection. The principal data transport phase functions are listed below. These are used or not used in accordance with the result of the selection function performed in the connection establishment phase. The items with an asterisk require further study (as reported in CCITT Rec. Q.711 [12]).

- Concatenation and separation.*
- Segmenting and reassembly.
- Flow control.
- Error detection and error correction.*
- Connection identification.
- NSDU (network service data unit) delimiting (M-bit).
- Expedited data.
- Missequence detection.
- Sequence recovery.*
- Reset.
- Receipt confirmation.*

11.4.3 Connection Release Functions. Release functions disconnect the signaling connection regardless of the current phase of the connection. The release may be performed by an upper-layer stimulus or by maintenance of the SCCP itself. The release can start at each end of the connection (symmetric procedure). Of course, the principal function of this phase is disconnection (from CCITT Rec. Q.711, Ref. 12).

11.5 Structure of the SCCP

The basic SCCP structure is shown in Figure 14.15. It consists of four functional blocks as follows (from CCITT Rec. Q.714):

1. SCCP connection-oriented control. This controls the establishment and release of signaling connections for data transfer on signaling connections.
2. SCCP connectionless control. This provides the connectionless transfer of data units.
3. SCCP management. This functional block provides the capability, in addition to the signal route management and lower control functions of the MTP, to handle the congestion or failure of either the SCCP user or signaling route to the SCCP user.
4. SCCP routing. On receipt of the message from the MTP or from the functions listed above, SCCP routing either forwards the message to the MTP for transfer or passes the message to the functions listed above. A message whose called party address is a local user is passed to functions 1, 2, or 3, whereas one destined for a remote user is forwarded to the MTP for transfer to the distant SCCP.

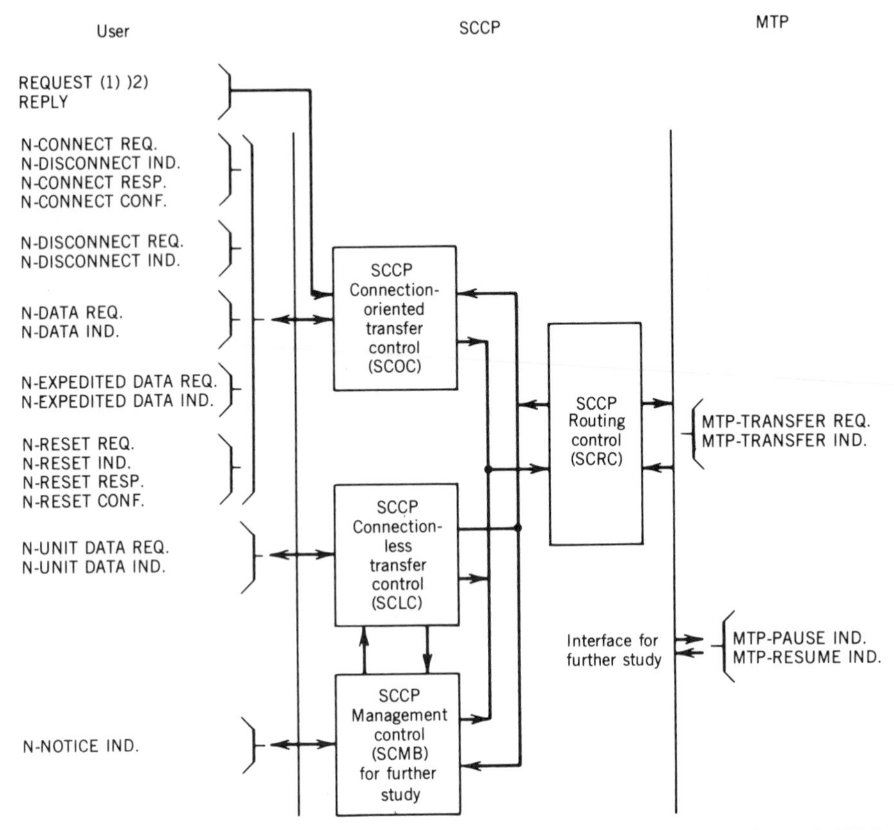

Figure 14.15 SCCP overview block diagram. From CCITT Rec. Q.714 [13]. Figure 1 / 714, page 307, *Red Books*, Vol. VI, Fascicle VI.7.

12 USER PARTS

12.1 Introduction

CCITT SS No. 7 user parts, along with the routing label, carry out the basic signaling functions. We should look again at Figure 14.4. There are two fields in this figure we need to discuss: the SIO (service information octet) and the SIF (signaling information field). In the paragraphs that follow we briefly cover the telephone user part (TUP) and the ISDN user part (ISUP). As shown in Figure 14.16, the user part, OSI layer 4, is contained in the signaling information field to the left of the routing label. CCITT Rec. Q.723 deals with the sequence of three sectors (fields and subfields of the standard basic message signal unit shown in Figure 14.4).

Turning now to Figure 14.16, we have, from right to left the service information octet, the routing label, and user information subfields (after the routing label in the SIF). The SIO is an octet in length, made up of two

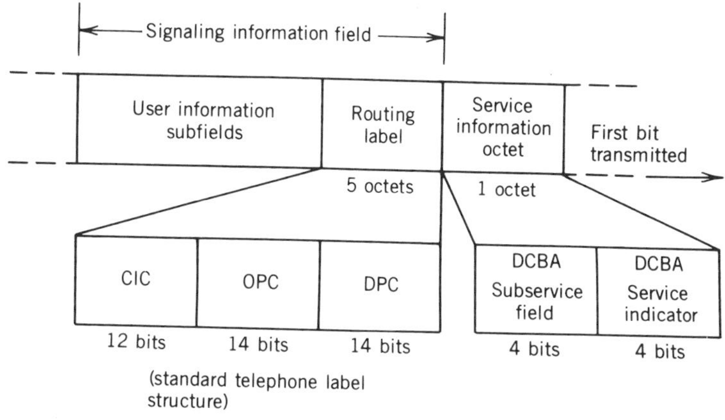

Figure 14.16 Signaling information field preceded by service information octet. Sequence is right to left, least significant bit transmitted first. DPC = destination point code; OPC = originating point code; CIC = circuit identification code.

subfields: service indicator (4 bits) and subservice field (4 bits). The service indicator is 4 bits long with 16 distinct bit combinations with the following meanings (read from right to left) (Ref. CCITT Rec. Q.704, Ref. 6):

Bits DCBA	Meaning
0000	Signaling network management message
0001	Signaling network testing and maintenance
0010	Spare
0011	SCCP
0100	Telephone user part
0101	ISDN user part
0110	Data user part (call- and circuit-related message)
0111	Data user part (facility registration and cancellation)
Remainder (8 sequences)	Spare

It is evident that the SIO directs the signaling message to the proper layer 4 entity, whether SCCP or user part. This is called message distribution.

The subservice indicator contains the network bits C and D and two spare bits, A and B. The network indicator is used by signaling message handling functions determining the relevant version of the user part. If the network indicator is set at 00 or 01, the two spare bits, coded 00, are available for possible future needs. If these two bits are coded 10 or 11, the two spare bits are for national use, such as message priority as an optional flow procedure. The network indicator provides discrimination between international and national usage (bits D and C).

The routing label forms part of every signaling message:

- To select the proper signaling route.
- To identify the particular transaction by the user part (the call) to which the message pertains.

The label format is shown in Figure 14.16. The DPC is the destination point code (14 bits), which indicates the signaling point for which the message is intended. The OPC (originating point code) indicates the source signaling point. The CIC (circuit identification code) indicates the one circuit (speech circuit in the TUP case) among those directly interconnecting the destination and originating points.

For the OPC and the DPC unambiguous identification of signaling points is carried out by means of an allocated code. Separate code plans are used for international and national networks. The CIC, as shown in the figure, is applicable only to the TUP. CCITT Rec. Q.704 shows a signaling link selection (SLS) field following (to the left) the OPC. The SLS is 4 bits long and is used for load sharing. For the data user part there is the bearer identification code. For the ISDN user part the routing label is not well defined (CCITT, 1984). CCITT Rec. Q.763 [14] places the CIC after the routing label.

12.2 Telephone User Part (TUP)

The core of signaling information is contained in the SIF (Figure 14.16). The TUP label was described briefly in Section 3.1. Several signal message formats and codes are described in what follows. These follow the label.

One typical message of the TUP is the initial address message (IAM); its format is shown in Figure 14.17. A brief description is given of each subfield, providing further insight into how SS No. 7 operates.

Common to all signaling messages are the subfields H0 and H1. These are the heading codes, each consisting of 4 bits, giving 16 code possibilities in pure binary coding. H0 identifies the specific message group to follow. "Message group" means the type of message. Some samples of message groups are:

Message Group Type	H0 Code
Forward address messages	0001
Forward setup messages	0010
Backward setup messages	0100
Unsuccessful backward setup messages	0101
Call supervision messages	0110
Node-to-node messages	1001

H1 contains a signal code or identifies the format of more complex messages. For instance, there are four types of address message identified by H0 = 0001,

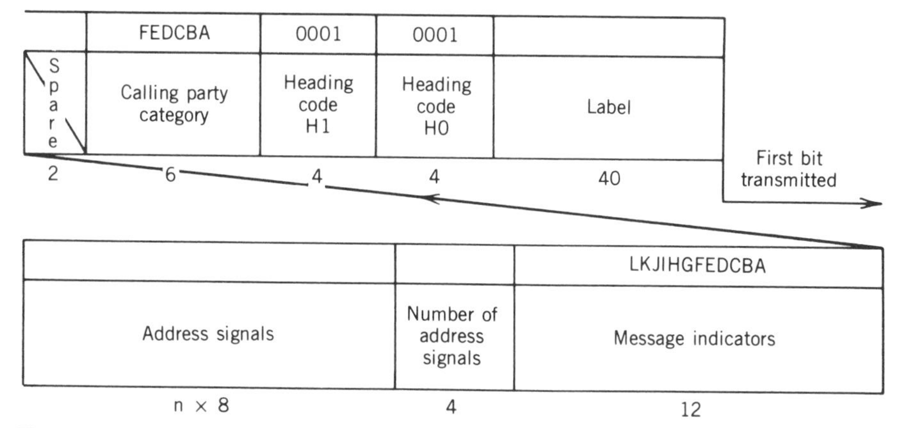

Figure 14.17 Initial address message format. From CCITT Rec. Q.723 [15]. Figure 3/Q.723, page 23, *Red Books*, Vol. VI, Fascicle VI.8.

and H1 identifies the type of message, such as:

Address Message Type	H0	H1
Initial address message	0001	0001
IAM with additional information	0001	0010
Subsequent address message	0001	0011
Subsequent address message with signal unit	0001	0100

Moving from right to left in Figure 14.17, after H1 we have the calling party subfield consisting of 6 bits. It identifies the language of the operator (Spanish, English, Russian, etc.). An English-speaking operator is coded 000010. It also differentiates the calling subscriber from one with priority, a data call, or a test call. A data call is coded 001100, and a test call 001101. Fifty of the 64 possible code groups are spare.

Continuing to the left in the figure, 2 bits are spare for international allocation. Then there is the message indicator, where the first 2 bits, B and A, give the nature of the address. This is information given in the forward direction indicating whether the associated address or line identity is an international, national (significant), or subscriber number. A subscriber number is coded 00, an international number 11, and a national (significant) number 10.

Bits D and C are the circuit indicator. 00 in this location indicates that there is no satellite circuit in the connection. Remember that the number of space satellite relays in a speech telephone connection is limited to one relay link through a satellite because of propagation delay.

Bits F and E are significant for common-channel signaling systems such as CCIS, CCS No. 6, and SS No. 7. The associated voice channel operates on a separate circuit. Does this selected circuit for the call have continuity? The bit sequence FE is coded:

Bits F and E	Meaning
00	Continuity check not required
01	Continuity check required on this circuit
10	Continuity check performed on previous circuit
11	Spare

Bit G gives echo suppressor information. When coded 0, it indicates that the outgoing half-echo suppressor is not included, and when coded 1, that the outgoing half-echo suppressor is included. Bit I is the redirected call indicator. Bit J is the all-digital path required indicator. Bit K tells whether any path may be used or whether only SS No. 7 controlled paths may be used. Bit L is spare.

The next subfield has 4 bits and gives the number of address signals contained in the initial address message. The last subfield contains address signals where each digit is coded by a 4-bit group as follows:

Code	Digit	Code	Digit
0000	0	1000	8
0001	1	1001	9
0010	2	1010	Spare
0011	3	1011	Code 11
0100	4	1100	Code 12
0101	5	1101	Spare
0110	6	1110	Spare
0111	7	1111	ST

The most significant address signal is sent first. Subsequent address signals are sent in successive 4-bit fields. As shown in Figure 14.17, the subfield contains *n* octets. A filler code of 0000 is sent to fill out the last octet, if needed. Recall that the ST signal is the "end of pulsing signal" and is often used on semi-automatic circuits.

Besides the initial address message, there is the subsequent address message used when all address digits are not contained in the IAM. The subsequent address message is an abbreviated version of the IAM. There is a third type of address message, the initial address message with additional information. This is an extended IAM providing such additional information as network capability, user facility data, additional routing information, called and calling address, and closed under group (CUG). There is also the forward setup message, which is sent after the address messages and contains further information for call setup.

CCITT SS No. 7 is rich with backward information messages. In this group are backward setup request; successful backward setup information message group, which includes charging information; unsuccessful backward setup information message group, which contains information on unsuccessful call setup; call supervision message group; circuit supervision message group; and the node-to-node message group (CCITT Recs. Q.722 and Q.723, Refs. 15 and 16).

Label capacity for the telephone user part is given in CCITT Rec. Q.725 [17] as 16,384 signaling points and up to 4096 speech circuits for each signaling point.

12.3 ISDN User Part

12.3.1 Introduction. CCITT SS No. 7 is an integral segment of ISDN. Without SS No. 7 implemented in the network, ISDN does not work. The ISUP encompasses the signaling functions required to provide switched services and user facilities for voice and nonvoice applications in the ISDN. We must not forget that ISDN handles voice as well as data and other services. The ISUP is also suited for application to dedicated telephone and circuit-switched data networks and in analog and mixed analog and digital networks. In what follows we provide a brief description of several key elements of the ISUP.

12.3.2 Services Supported. The basic service offered by the ISUP is the control of circuit-switched network connections between subscriber line exchange terminations. The standard connection types are 64-kbps transparent and nontransparent connections. The nontransparent mode is used for voice communications where the connection may use bit-manipulating devices such as echo suppressors. Allowance has been made to accommodate additional types, such as subrate channels. The 64-kbps transparent connection may be used to carry any one of the standard user classes defined in CCITT Rec. X.1. In addition to the basic service, the ISUP also supports the following:

- User-accessed calling party address identification.
- User-accessed called party address identification.
- Redirection of calls.
- Connection when free waiting allowed.
- Completion of calls to busy subscribers.
- Malicious call identification.
- Closed user groups (CUGs).

12.3.3 End-to-End Signaling. End-to-end signaling is used typically between ISUPs located in call originating and call terminating local exchanges to request or respond to requests for additional call-related information or to

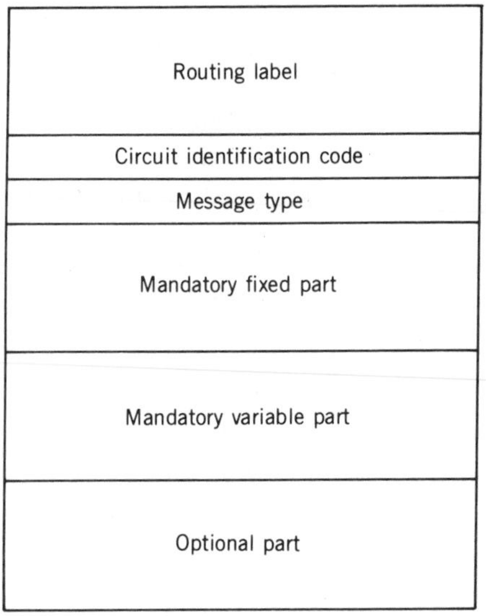

Routing label
Circuit identification code
Message type
Mandatory fixed part
Mandatory variable part
Optional part

Figure 14.18 ISDN user part message parts. From CCITT Rec. Q.763 [14], Figure 1/Q.763, page 154, *Red Books*, Vol. VI, Fascicle VI.8.

transfer user-to-user information transparently through the network. The means for connection-oriented or connectionless transport of end-to-end signaling information is provided by the SCCP and the message transfer part of SS No. 7 as described in previous sections. An alternative method for transporting end-to-end signaling information is called *pass-along*. The service is provided within the ISDN user part and is independent of the SCCP. With this method, signaling information is sent along the signaling path of a previously established physical connection.

12.3.4 Formats and Codes. ISDN user part messages are carried on the signaling link by means of signaling units described in previous sections. The service indicator (SI) is coded 0101 for these types of messages. The signaling information field (see Figure 14.16) of each message signal unit containing an ISUP message consists of an integral number of octets and includes the following parts (see Figure 14.18):

- Routing label.
- Circuit identification code.
- Mandatory fixed part.
- Mandatory variable part.
- Optional part, which may contain fixed-length and variable-length parameter fields.

The routing label was covered in section 6.3.1. The *message type code* consists of one octet field. The message type code uniquely defines the function and format of each ISDN user part message (see Table 14.3).

TABLE 14.3 ISUP Message Type Parameters

Message Type	Code
Address complete	00000110
Answer	00001001
Blocking	00010011
Blocking acknowledgment	00010101
Call modification completed	00011101
Call modification request	00011100
Charging	Note
Circuit group blocking	00011000
Circuit group blocking acknowledgment	00011010
Circuit group unblocking	00011001
Circuit group unblocking acknowledgment	00011011
Closed user group selection and validation request	00100101
Closed user group selection and validation response	00100110
Continuity	00000101
Continuity check request	00010001
Delayed release	00100111
Facility accepted	00100000
Facility deactivated	00100010
Facility information	00100011
Facility reject	00100001
Facility request	00011111
Forward transfer	00001000
Information	00000100
Information request	00000011
Initial address	00000001
Pass along	00101000
Pause	00001101
Reject connect modify	00011110
Release	00001011
Release complete	00010000
Released	00001111
Reset circuit	00010010
Reset circuit group	00010111
Reset circuit group acknowledgment	00101001
Resume	00001110
Subsequent address	00000010
Unblocking	00010100
Unblocking acknowledgment	00010110
Unsuccessful backward setup information	00001010
User-to-user information	

Note: for further study.
From CCITT Rec. Q.763 [14], Table 3 / Q.763, page 159, Red Books, Vol. III Fascicle VI.8.

12.3.4.1 Formatting Principles. Each message consists of a number of PARAMETERs. Each PARAMETER has a NAME coded as a single octet. The length of a parameter may be fixed or variable and a LENGTH INDICATOR of one octet may be included. A general format is shown in Figure 14.19. A listing of parameter names is given in Table 14.4, which gives the reader some appreciation of the richness of SS No. 7 ISUP signaling possibilities.

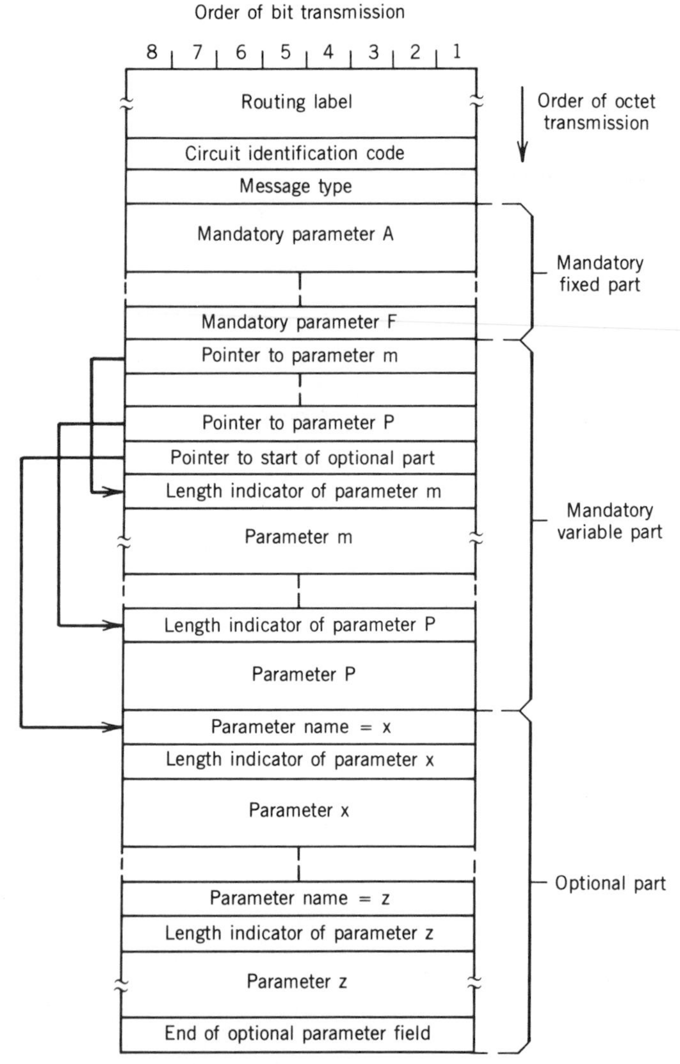

Figure 14.19 ISUP general signaling message format. From CCITT Rec. Q.763 [14]. Figure 3 / Q.763, page 157, *Red Books*, Vol. VI, Fascicle VI.8.

13 SIGNALING PROCEDURES

13.1 Introduction

To give the reader some grasp of typical SS No. 7 operation, we briefly describe a sample signaling procedure following the three phases of a "call": call setup, data or conversation phase, and call takedown. This description is taken from CCITT Rec. Q.764—successful call setup–forward address signaling–en bloc operation [18].

TABLE 14.4 Parameter Name Codes

Parameter Name	Code
Address presentation restriction indicators	00011111
Backward call indicators	00010001
Call modification indicators	00010111
Call reference	00000001
Called party address	00000100
Calling party address	00001010
Calling party's category	00001001
Cause indicator	00010100
Charge information[b]	
Circuit group supervision message type indicator	00010101
Closed user group check response indicators	00011100
Closed user group interlock code	00011010
Compatibility information[b]	
Connection request	00001101
Continuity indicators	00010000
End of optional parameters	00000000
Facility indicator	00011000
Facility information indicators	00011001
Forward call indicators	00000111
Index	00011011
Information indicators	00001111
Information request indicators	00001110
Nature of connection indicators	00000110
Optional forward call indicators	00001000
Original address	00001011
Range and status	00010110
Redirection address	00001100
Redirection indicator	00010011
Signaling point code[a]	00011110
Subsequent address	00000101
Transmission medium requirements indicators	00000010
User service information	00011101
User-to-user information[b]	

[a]For national use only.
[b]For further study.

From CCITT Rec. Q.763 [14], Table 4 / Q.763, page 160, *Red Books,* Vol. VI, fascicle VI.8.

13.2 Actions at Originating Exchange

1. Circuit selection. When the originating exchange has received the complete selection information from the calling party and has determined that the call is to be routed to another exchange, selection of a suitable free interexchange circuit takes place and an initial address message is sent to the appropriate destination. The necessary routing information is either stored at the originating exchange or at a remote database to which a request may be made. Examination of the setup message from the calling party will determine the requirements for the connection (e.g., 64 kbps and SS No. 7 for data or

voice and data connections) and the ability to request "call modification." This information, which includes transmission medium requirement and forward call indicators, is included in the initial address message to permit correct routing at intermediate exchanges. The seizing function at the receiving exchange is implicit in the reception of the IAM.

2. Address information sending sequence. The sending sequence of address information on international calls will be the country code (not sent to an incoming international exchange) followed by the national (significant) number. On national connections the address information may be the local number or the national (significant) number, as required by the administration (telephone company, common carrier) concerned. Calls to international operator positions are indicated by code 11 or code 12. The end of pulsing (ST) signal is used on "en bloc" operations.

3. Initial address message. The IAM, in principle, contains all the information that is required to route the call to the destination exchange and connect the call to the desired unit. All IAMs include a protocol control indicator (PCI). The originating exchange will set the parameters in the PCI to indicate:

- The type of end-to-end signaling to be accommodated.
- The availability of SS No. 7 signaling.
- The use of the ISUP.
- Whether further information is available, such as calling line identity.

The originating exchange may also include in the IAM:

- A local reference and point code (of the originating exchange) to enable the destination exchange to establish an end-to-end connection.
- The calling party address, if this is to be passed forward without being requested.
- Other information related to the user or facilities and network utilities.

4. Transfer of information by end-to-end protocol. As an alternative to the inclusion of call-setup user facility information in the IAM, any such information not to be examined at intermediate exchanges may be passed end to end from the originating exchange to the destination exchange.

5. Completion of transmission path. Through-connection of the transmission path is completed at the originating exchange immediately after sending the IAM, except in cases where conditions on the outgoing circuit prevent it.

13.3 Actions Required at an Intermediate Exchange

1. Circuit selection. An intermediate exchange, on receipt of an IAM, analyzes the destination address and other routing information to determine

the routing of the call. The intermediate exchange then seizes a free interexchange circuit (signaling trunk) and sends the IAM to the succeeding exchange. When no echo suppressor or nature-of-circuit indicator is received from a preceding circuit using a signaling circuit with fewer facilities, the indicators will be considered as received "no" unless positive knowledge is available.

2. Initial address message. An intermediate exchange examines the protocol control indicator and, if the call is still permitted, modifies the parameters in the PCI according to the capabilities that can be provided (e.g., the TUP has been used instead of the ISUP). The PCI should be modified to indicate to the succeeding exchanges that different user parts have been utilized.

3. Completion of transmission path. Through-connection of the transmission path is completed at an intermediate exchange immediately after an IAM has been sent, except in those cases where conditions on an outgoing circuit prevent it.

4. Congestion. In case of congestion at an intermediate exchange, the exchange will send an unsuccessful backward setup information message to the preceding exchange indicating congestion and initiating release of the call at that exchange.

13.4 Actions Required at the Destination Exchange

1. Selection of called party. On receipt of an IAM, the destination exchange analyses the destination address to determine to which party the call should be connected. It also checks the party's line condition and performs various checks, using, for example, the service indicator received from the calling terminal to verify whether the connection is permitted. These checks include correspondence of compatibility checks, such as checks associated with user facilities. At this point certain call-setup information may need to be obtained by an end-to-end protocol. Examination of the protocol control indicator shows whether an end-to-end interchange is feasible and which end-to-end technique may be used (e.g., pass-along or SCCP). When the connection is allowed, the destination exchange alerts the called party using the setup message in accordance with the applicable interface protocol.

2. Connection not allowed. If the call cannot be connected due to, for example, the called party being busy, a call supervision message indicating the reason is sent to the preceding exchange.

13.5 Call Release (Normal)

The call release (takedown) procedures are based on a three-message approach whereby the release message is transmitted through the network as quickly as possible. The same procedures are used in the network irrespective of whether they are initiated by the calling subscriber, the called subscriber, or the network. The normal release procedure can be prevented by the network if this is required on a particular call.

13.5.1 Release Initiated by Calling Subscriber

1. Actions at originating exchange. On receipt of a request to release the call from a calling subscriber, the originating exchange immediately starts the release of the switched path and at the same time sends a release message to the succeeding exchange. When the path has been fully disconnected, a released message is sent to the succeeding exchange and a timer is started to ensure that a release complete message is received from the succeeding exchange within time $T(1)$ (4–15 s).

2. Actions at intermediate exchange. On receipt of the release message from the preceding exchange, an intermediate exchange will:

- Start a timer $T(12)$ to ensure that a released message is received from the preceding exchange.
- Immediately start the release of the switched path and at the same time send a release message to the succeeding exchange.

When the path has been fully disconnected, a released message is sent to the succeeding exchange and a timer is started to ensure that a release complete message is received from the succeeding exchange within time $T(1)$. When the path has been fully disconnected and a released message has been received from the preceding exchange, a release complete message is returned to the preceding exchange.

3. Actions at the destination exchange. On receipt of a release message from the preceding exchange, the destination exchange:

- Starts a timer $T(12)$ to ensure that a released message is received from the preceding exchange.
- Starts the release of the switched path. When the path has been fully disconnected and a release message has been received from the preceding exchange, a release complete message is returned to the preceding exchange.

4. Charging. Charging is stopped on receipt of the release message (or released message if a release message has not been received) at the charging exchange(s). It should be noted that charging normally begins when the exchange(s) controlling charges received the ANSWER message from the network. The signals "answer charge" and "answer no charge" are used only as a result of the first off-hook signal from the called party (from CCITT Rec. Q.764, Ref. 18).

14 INTERWORKING

Interworking refers to how one signaling system interoperates with another signaling system. CCITT Rec. Q.764 states that when SS No. 7 interworks with other internationally specified signaling systems, the following rules on switch-through apply (see Table 14.5).

TABLE 14.5 Interworking Rules for Switch-Through

No. 7 → No. 7	When no continuity check is to be made on the outgoing circuit, through connection should occur after sending the initial address message. When continuity check is to be made on the outgoing circuit, through connection should happen after residual check tone has propagated through the return path of the speech circuit (see Rec. Q.724, Section 7.3).
No. 6 → No. 7, No. 5 → No. 7,	When no continuity check is to be made on the outgoing circuit, through connection can happen after sending the initial address message.
R1 → No. 7, No. 7 → No. 6,	When a continuity check is to be made on the out going circuit, through connection can happen after residual check tone has propagated through the return path of the speech circuit (see Rec. Q.724, Section 7.3).
R2 → No. 7	Through connection should occur after sending address complete.
No. 7 → No. 5, No. 7 → R1	Through connection can occur after sending ST (end of pulsing) signal and removal of a possible check loop.
No. 7 → R2	Through connection should occur after receipt of address-complete message.

From CCITT Rec. Q.764 [18], page 193, *Red Books,* Vol. VI, Fascicle VI.8.

REVIEW QUESTIONS

1. What is the principal rationale for developing and implementing CCITT Signaling System No. 7?
2. Describe the SS No. 7 relationship with OSI. Why does SS No. 7 truncate at OSI layer 4?
3. Give the two primary "parts" of SS No. 7. Briefly describe the function of each generic part.
4. With the SS No. 7 implementation, OSI layer 4 is subdivided into two sublayers. Name them.

5. Name the three user parts of SS No. 7.
6. What is the normal basic connectivity of No. 7? (Hint: Bit rate.)
7. Layer 2 of SS No. 7 functions as the signaling link. Name five of the seven functions of layer 2. In one sentence describe what layer 2 does.
8. What are the two basic categories of functions in layer 3 of SS No. 7? What is the basic purpose of layer 3? (Hint: Think OSI.)
9. What is a signaling relation?
10. How are signaling points identified?
11. Interrelate message routing, distribution, and discrimination.
12. What does a routing label do?
13. What is the purpose of signaling unit delimitation?
14. What is the *basic* method of error correction in SS No. 7?
15. Besides the CRC function, what is the second method used to detect errors in a signaling message?
16. Flow control notifies the distant transmit end of a congestion condition by means of a status signal unit (SSU). How does flow control operate? How does it slow traffic flow to reduce congestion?
17. There are three types of signal units used in SS No. 7. What are they? Define the basic function of each.
18. Differentiate forward and backward sequence numbers. When does reset occur for FSN? In essence, what do FSN and BSN accomplish?
19. What is the function of the service indicator octet?
20. The routing label is analogous to what in our present telephone system? Name the three basic pieces of signaling information that the routing label provides.
21. There are three functions carried out by the signaling network management. What are they?
22. Differentiate between an *associated* and *quasi-associated* signaling network.
23. What is the function of an STP? Differentiate between an STP and a signaling point.
24. Why do we wish to limit the number of STPs in a signaling connection on a particular relation?
25. What are the three measures of performance of SS No. 7?

26. How is the capacity of SS No. 7 measured? With what potential?

27. What is the biggest contributor to signaling delay in the SS No. 7 network?

28. What was the basic rationale for implementing the SCCP in SS No. 7? It equates to which OSI layer?

29. What are *primitives*?

30. Regarding user parts, what is the function of the SIO? of the network indicator?

31. What does the heading code (H0, H1) tell us?

32. What is the purpose of circuit continuity? Explain.

33. Address signals (such as telephone number digits) are sent digit by digit embedded in the last subfield in the SIF. How are they represented? (Hint: Per digit.)

34. Give two examples of backward information in SS No. 7.

35. What facilitates end-to-end signaling typically for ISUP (i.e., what "part" of SS No. 7)?

36. Give at least three functions of the protocol control indicator, as used, for example, in the IAM.

37. Give the actions taken at an intermediate exchange during call setup using SS No. 7.

38. Give the actions taken by the destination exchange during call setup using SS No. 7.

39. Describe the actions taken for call release (circuit takedown).

40. What stops the charging on a call during the call release stage?

REFERENCES

1. W. Stallings, *Tutorial: Integrated Services Digital Network (ISDN)*, IEEE Computer Society, Washington, D.C., 1985.
2. *Specifications of Signaling System No. 7*, Fasc. VI.6, CCITT Recommendations (*Yellow Books*), VII Plenary Assembly, Geneva, 1980.
3. G. G. Schlanger, "An Overview of Signaling System No. 7," *IEEE J. Selected Areas in Comm.*, **7** (3) (May 1986).
4. *Signaling Data Link*, CCITT Rec. Q.702, Vol. VI, Fasc. VI.7, "Specifications of Signaling System No. 7," CCITT VIII Plenary Assembly, Malaga-Torremolinos, 1984.

5. *Signaling Link*, CCITT Rec. Q.703, CCITT VIII Plenary Assembly, Malaga-Torremolinos, 1984.
6. *Signaling Network Functions and Messages*, CCITT Rec. Q.704, CCITT VIII Plenary Assembly, Malaga-Torremolinos, 1984.
7. *Signaling Network Structure*, CCITT Rec. Q.705, CCITT VIII Plenary Assembly, Malaga-Torremolinos, 1984.
8. *Numbering of International Signaling Point Codes*, CCITT Rec. Q.708, CCITT VIII Plenary Assembly, Malaga-Torremolinos, 1984.
9. *Message Transfer Part Signaling Performance*, CCITT Rec. Q.706 CCITT VIII Plenary Assembly, Malaga-Torremolinos, 1984.
10. *Hypothetical Signaling Reference Connection*, CCITT Rec. Q.709, CCITT VIII Plenary Assembly, Malaga-Torremolinos, 1984.
11. Walter C. Roehr, Jr., "Signaling System Number 7," in *Tutorial: Integrated Services Digital Network (ISDN)*, W. Stallings, ed., IEEE Computer Society, Washington, D.C., 1985.
12. *Functional Description of the Signaling Connection Control Part (SCCP) of Signaling System No.* 7, CCITT Rec. Q.711, Vol. VI, Fasc. VI.7, "Specifications of Signaling System No. 7," CCITT VIII Plenary Assembly, Malaga-Torremolinos, 1984.
13. *Signaling Connection Control Part Procedures*, CCITT Rec. Q.714, CCITT VIII Plenary Assembly, Malaga-Torremolinos, 1984.
14. *Formats and Codes (ISDN User Part)*, CCITT Rec. Q.763, Vol. VI, Fasc. VI.8, "Specifications of Signaling Systems No. 7," CCITT VIII Plenary Assembly, Malaga-Torremolinos, 1984.
15. *Formats and Codes (Telephone User Part)*, CCITT Rec. Q.723, CCITT VIII Plenary Assembly, Malaga-Torremolinos, 1984.
16. *General Function of Telephone Messages and Signals*, CCITT Rec. Q.722, CCITT VIII Plenary Assembly, Malaga-Torremolinos, 1984.
17. *Signaling Performance in the Telephone Application*, CCITT Rec. Q.725, CCITT VIII Plenary Assembly, Malaga-Torremolinos, 1984.
18. *Signaling Procedures (ISDN User Part)*, CCITT Rec. Q.764, CCITT VIII Plenary Assembly, Malaga-Torremolinos, 1984.

15

TELECOMMUNICATION PLANNING

1 GENERAL

The "planning function" is vital to a telecommunication operating company or administration. In the last 20 years or so it has been found equally as vital in the industrial and government environments. However, the two areas must be distinguished and treated separately.

An operating company is often a monopoly faced with a demand for service yet with limited resources to satisfy the demand. Ideally, because it is a monopoly, it is overseen by a watchdog commission such as the Federal Communications Commission (FCC) in the United States. Industry–government entities require optimal communication facilities (with communications taking on an ever widening meaning). Such facilities must again be under the constraint of limited resources. The industry–government telecommunication manager has the prime responsibility of making communications serve the organization with minimum cost and maximum results. The early portion of this chapter deals with the operating telephone company or administration and its planning function. The latter part of the chapter discusses the industry–government problem.

2 GROSS PLANNING ON A NATIONAL SCALE

2.1 Economics

Telecommunication planning is largely economic. For this discussion we consider this "economics" in three levels: (1) macroeconomics, (2) midrange economy, and (3) microeconomics. Macroeconomics (our definition) deals with the amount of wealth a country spends in the telecommunication sector. Of course, the stage of development of a particular country and whether the overall economy is controlled, free-running, or somewhere in between will be strongly contributing factors. It is very important for evolving nations to achieve a balance among economic sectors for efficient development. Telecommunications should be given its proper place with the other sectors, such as transport, industry, agriculture, and social programs. We define "midrange economy" as the method used by an operating company (or administration) to raise capital. We must distinguish capital from tariff revenue. Oversimplifying, we could say that for a privately owned telephone company in a free economy, capital investment (e.g., new plant) will be financed by internally generated funds* and from the sale of securities. The dividends and interest to be paid on those securities will derive from the tariff revenue.

There is the other extreme, where the telecommunication operating administration is owned by the government. In this case (oversimplifying again) capital for new plant may be taken from tax coffers, probably a legislated budget for the year. Then tariff revenues (all or a portion thereof) are returned to the national treasury. Many administrations or "telephone" companies are in a gray area. Spain is an example. Its telephone company (CTNE) is government controlled but issues stock and is operated as any other industry with an annual report to the stockholders.

We define "microeconomics" as plant extension. It deals with costs and return on investment for the expansion or upgrading of a telephone plant. It is project oriented or area oriented. An example may be the installation and cutover of several new exchanges with revamped serving areas to meet demand in a certain metropolitan area several years hence. Such projects and improvements or plant extension follow economic and technical plans. Economic plans permit the extending of capital and will show payback to management. Technical plans ensure compatibility and coordinated upgrading and modernization. The microeconomics issue deals with the problem of getting the best for the least expenditure. Therefore the telecommunication planners, even at this low stage, are just as involved with finances as they are with engineering. In the case of commercial common carrier systems (telephone administrations) the telecommunication planners will probably resort to expressing the cost of projects in terms of present value (worth) of annual charges, usually expressed as PVAC (PWAC), depending on which side of the ocean they are.

3 BASIC PLANNING

We would expect upper management of a telephone company to make certain policy decisions that will affect the planning group. For instance, one policy decision might state that 99% of telephones may be reached by direct subscriber dialing within 5 years, or that the entire toll network will be digital by

*In most cases these funds are retained earnings.

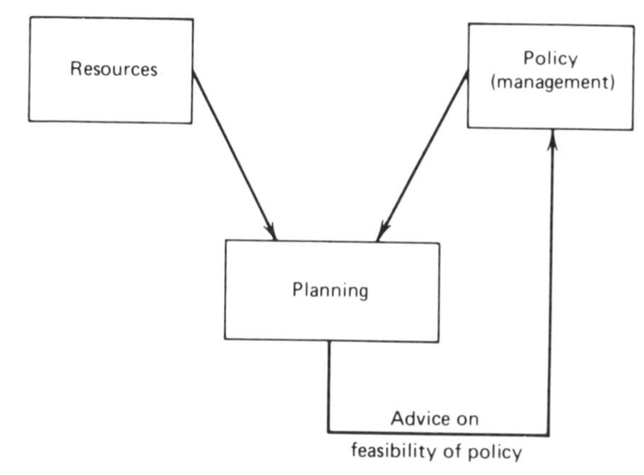

Figure 15.1 How "planning" interworks with management and budgetary limitations inside a telecommunication company (administration).

the year 2000 based on CEPT 30 + 2 PCM and its related hierarchy. The financial impact of such plans must be verified by the planning group for feasibility before they are made policy. The successful planning group interworks with management and other departments in the preparation of plans and on plant extension projects. This interrelationship is shown in Figure 15.1.

The basis of all planning is economic. The greater part of technical planning consists of selecting from various possible schemes to achieve a level of development and extension, those schemes that are the most economical. However, economy is not the only consideration. Planning includes overall organization, work force expansion, and training; controlling the flow of work; estimating the impact on revenue of tariff changes; advising on the improvement of service quality (Chapter 1, Section 13); and assisting and advising other departments on the estimation of overall profitability, cash-flow requirements, and future growth. In some cases the financial resources may not be available to carry out certain planning objectives, thus requiring their deferral or cancellation. In other cases the money may be on hand and the planning group may be asked how best it may be spent to encourage development and/or improve quality of service. Engineering planning, therefore, is interactive and part of a complex cycle of operations whose primary task is to keep the business running in sound financial condition.

Engineering planning starts with (1) subscriber and traffic forecasts and (2) technology forecasts. With the present network as a basis, fundamental plans are developed. Actual plant extension and modernization projects result from annual capital expenditure programs and short-term plans. Inputs to the formation of these programs and action plans result from guidelines and objectives set down in fundamental plans. These plans state general policy and milestones, and they also reflect the pressures requiring short-term plant relief. Before these plans and projects are put into action, a financial review is carried

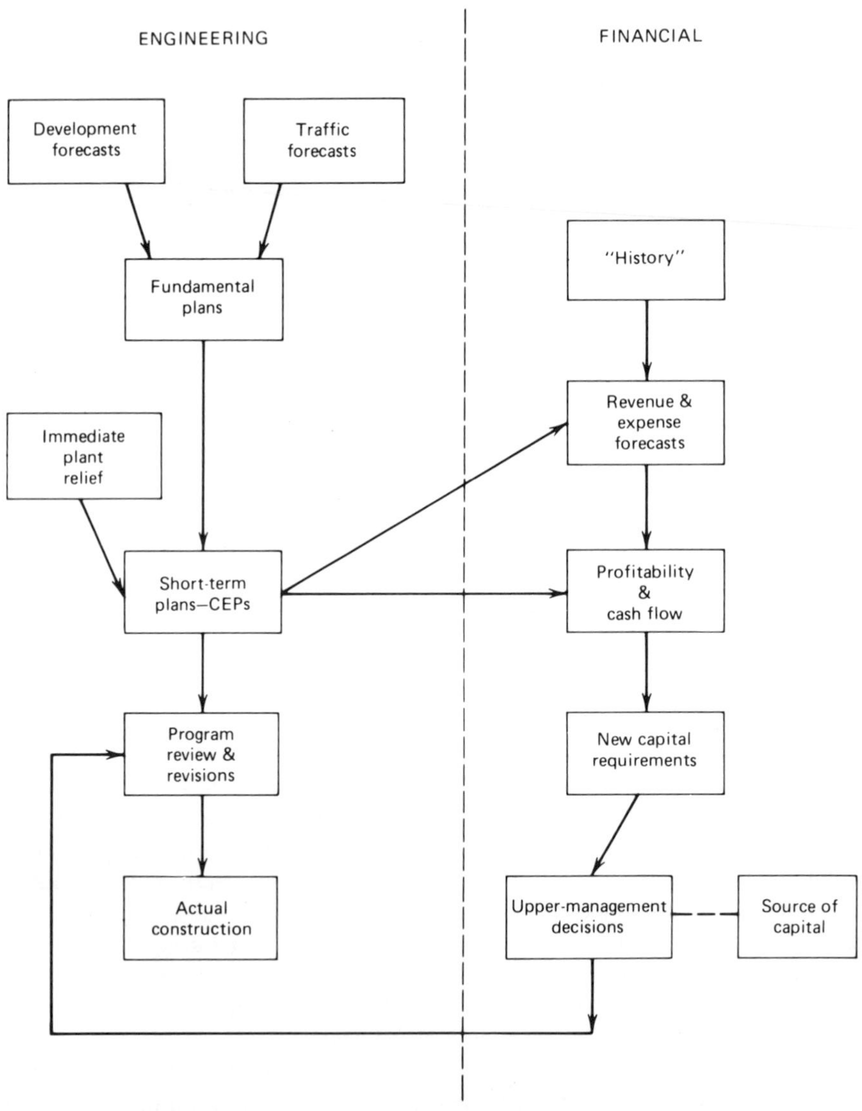

Figure 15.2 Telecommunication planning in a telecommunication operating company (administration): a plan from inception to execution with actual construction (CEP = capital expenditure program).

out to determine the impact on revenue and to allocate resources. These basic planning functions are illustrated in the flowchart in Figure 15.2.

4 TYPES OF FUNDAMENTAL PLANNING

We like to distinguish between two types of fundamental plan, namely, fundamental development plans and fundamental technical plans. Fundamental plans are plans that serve as a foundation for detailed or highly specific plans. Fundamental development plans state the means for satisfying expected demand for new lines and services or service improvements. These plans specify *quantity* of plant (as distinguished from quality). Fundamental technical plans set technical standards and detail technical guidelines. They state the techniques to be followed to ensure the required flexibility of a network and compatibility among its parts and to guarantee that service performance meets the desired standard. The important attribute of technical plans is that they specify *quality* of plant. The planning group faces two problems simultaneously: (1) satisfying demand for service (new subscribers) and (2) upgrading and improving the network, concomitant with the offering of new services, such as ISDN.

5 THE STARTING POINTS

Long-range forecasts provide the initial guidelines to satisfy future demand. Objectives of quality of service (Chapter 1, Section 13) establish the point of departure for network improvement [7].

5.1 Forecasts: An Introduction

Forecasting is carried out by two quite different methods. The first is done on a local area basis and is often referred to as "block-by-block" forecasting. It is a continuous process requiring periodic area surveys to determine actual growth on the local scale. Here the forecaster is watchful for new home construction, new industry, zoning laws, population movements, and the general economic health of the area. The second method is based on historical information. One simple way of using historical information is simply to extrapolate linearly future telephone lines and stations from past growth. Such forecasts usually are fairly valid for up to about 3 years in the future. Another way of using historical information is on the basis of "analogy." With analogy, telephone growth forecast is based on other quantities whose growth is more predictable, such as the number of automobiles per capita, the business index, and the use of electrical power. Another quite acceptable method, when working on a national basis, is to select another country with similar characteristics but that is 5 or 10 years ahead in telephone development. Own country trend in growth will probably fairly well match that country. Another favorite index is the gross domestic product (GDP), which is used in the CCITT *Economic Studies at the National Level in the Field of Telecommunications* [4]. Once cyclic and seasonal factors have been removed from past statistics, an underlying trend can often be recognized, and growth can be expressed mathematically by one of the five equations listed as follows:

Linear	$Y = A + Bt$
Parabolic	$Y = A + BC + Ct^2$
Exponential	$Y = Ae^{Bt}$
Gompertz function	$\log Y = A - Br^t$
Logistic function	$Y = \dfrac{1}{A + Br^t}$

where A, B, and C are constants, the value of r should be taken as something less than unity. t is time, and Y is the value to be forecast (e.g., subscriber density). Linear growth is the extrapolation we mentioned previously. Line growths are rarely linear and approximate a constant percentage per year (e.g., exponential).

Application of the parabolic function to express growth trends is attractive because it is a compromise between the linear and the exponential functions. It adjusts better to past data because it has three constants: A, B, and C. Further, the parabolic curve can be fitted to approximate exponential growth initially and less rapid growth deeper in the forecast period. However, in the final analysis the parabolic curve will not prove any more accurate than the exponential if the same historical data are used to construct both functions. The Gompertz and logistic functions are applied when saturation is expected. When there is some doubt regarding the accuracy of the theoretical functions that we have reviewed, it is often helpful to resort to finding the best regression line (i.e., the line that gives the least mean square error when calculated in terms of the growth equation selected). Then the telecommunication planner selects the growth equation with the greatest correlation coefficient. Of course, we are trying to correlate historical data and future growth with current and future telephone stations and lines, which is our final interest.

When there are waiting lists for telephone service, accurate forecasting is made even more difficult. The existence of waiting lists is the rule, not the exception, even in many advanced industrialized nations. Nevertheless, where there is a waiting list, immediate pressure is taken off the back of the forecaster for accurate short-term forecasts. This is because upper management will set a policy regarding short-term relief of the waiting list. It might state that it is the policy of the administration to install 100,000 new lines each year. Waiting lists do not reflect true demand. Short-term demand (2 or 3 years) equals waiting lists, plus hidden demand, minus abandonments. Hidden demand may reach 40% of the waiting list, and abandonments 20%.

Traffic forecasting (see Chapter 2, Sections 1, 11, and 12) is yet another important aspect of basic planning, although the actual traffic forecasts are usually made by specialists in traffic departments. In the case of local exchanges, future calling rate and holding times per line remain (see Table 2.7) fairly constant over the years. Distinguish, though, among business, residence, and rural lines. With care, future traffic can be forecast quite accurately from past history in the local area on a per line basis. Again, as discussed in Chapters 1 and 2, the importance of keeping good traffic data is patently evident. Leaving aside cyclic calling (holidays) and emergencies (flood, blackouts, hurricanes, etc.), two occurrences that can vary the traffic forecasts are (1) service improvements and (2) tariff changes to stimulate usage. One example of the former that was quite striking was the traffic jump when earth station service was instituted in Chile and Argentina. International traffic jumped four to six times prior values. On the other hand, tariff changes do not usually affect the busy hour. Such changes are usually to stimulate usage of the telephone system during periods of low usage, such as on nights, weekends, and holidays. Also, local traffic forecasts sometimes err if care is not taken with cutovers of new exchanges and the impact on traffic forecast accuracy. There is a particular liability on the side of error when there is a mix of new service and transferred service from existing exchanges needing relief.

Long-distance or toll traffic forecasting also extrapolates past history (traffic data) to determine future calling rate and erlangs of traffic. Long-distance service is also sensitive to tariffs. One rule of thumb valid for a large number of countries outside of North America [2] is that 4–6% of the local calling rate will give some measure of the toll calling rate. Or we could also say that the average residential subscriber pays about the same for local as for long-distance service (half of the typical bill for each). Of course, this figure will vary from country to country, depending on toll tariff structure and particularly on the size of the local area and tariff area increments. Toll service usage shows constant growth in more developed nations and, in particular, North America. The calling rate for toll calls in the United States runs from 10% to 15% or more of the local rate on a typical residential bill. If calling rate were counted on bulk service situations, such as the wide-area telephone service (WATS) in the United States, the equivalent toll figures would be even larger. In all these latter cases more than half the telephone bill is for toll service. (*Note*: Compare this with accumulated capital investment share shown in Chapter 2, Section 1, where we show over 65% in local area.)

5.2 Quality of Service

Standards and objectives of quality of service (described in Chapter 1, Section 13) serve as the principal starting point guideline for the formation of fundamental technical plans. Consider the several service quality factors listed here:

1. Transmission quality (level, crosstalk, echo, etc.).
2. Dial-tone delay, and postdial delay.
3. Grade of service (lost calls).
4. Fault incidence and service deficiency.
5. Adaptation of the system to the subscriber
6. Billing errors (method of billing and its administration).

When a quality-of-service requirement is set, one of two possible approaches can be taken:

1. Design for a maximum permitted impairment in the most unfavorable case.
2. Design for a certain range of impairments occurring as a result of chance combination of elements, such as a majority of subscribers' opinions being favorable [1, 7].

The latter is often referred to as "statistical design." Both approaches have weaknesses. The first may require unnecessarily high performance to satisfy those rare unfavorable cases. Some subscribers may be very unhappy with the second approach, which will be considerably more influenced by *variability* of plant performance. Such variability is particularly felt in the areas of signaling and transmission and should be taken into account in the planning of those technical areas. There are three possible causes of variability: (1) manufacturing standardization and quality control, (2) multiple usage of elements (e.g., links switched in tandem), and (3) day-to-day variability (e.g., temperature, loading, and other variations). The third item is also strongly influenced by maintenance and its intervals, standards, and quality. Technical planning in North America essentially follows the lines of statistical design. In Europe, those countries following European practice, and with CCITT, maximum impairment design is used, although often modified where, for instance, a certain standard will be met in perhaps 95% of the situations.

6 FUNDAMENTAL TECHNICAL PLANS

There are at least six fundamental technical plans that should be prepared and periodically updated by a telephone company or administration:

1. Numbering (Chapter 3, Section 16).
2. Routing (Chapters 1, 2, and 6).
3. Transmission (Chapters 5 and 9).

4. Switching (Chapters 3 and 10).
5. Charging (Chapter 3) (sometimes coupled with switching): rates and tariffs.
6. Signaling (Chapters 4 and 14).

6.1 Numbering

"Numbering" provides for the assignment of telephone numbers over a period of growth for the plant. That period should be 40 years or more, with a review of the numbering plan every 10 years to check its validity and conformity to updated forecasts. The plan should meet three major constraints:

- The numbering should be easily understood by the subscriber.
- It should be compatible with existing and planned switching equipment.
- It must be fully interworkable with international numbering schemes.

In developing a numbering plan, the planner will be limited by:

- Existing numbering practices.
- Switching equipment installed and in use.
- Pertinent CCITT recommendations.
- Services offered, such as PABX; in-dialing; abbreviated dialing; and special services, such as fire, police, "information," direct dial, and others that tend to block numbers.
- Economics and economic trade-offs.

6.2 Routing

Chapters 1, 2, and 6 discussed routing techniques and network design. A routing plan is closely related to the switching plan. For most planning purposes a network is designed on 20-year development figures without reference to existing equipment. One rule of thumb [1] states that system will generally grow to 4 times its size in 20 years. To comply with a 20-year objective of network design to the meeting of certain routing philosophies, intermediate network designs may be advisable for 10-year and even 5-year periods.

A routing plan should include as a minimum:

1. Description of hierarchy.
2. Definition of full direct, high-usage, and overflow routes and the criteria for choosing among them.
3. Specifications of grades of service on trunk and local routes.

4. Principal route layouts geographically and rules for survivability (route diversity) to lessen consequences of breakdowns or disaster.
5. Guidelines for the selection of transmission media for major and minor routes.

6.3 Switching and Charging Plans

A switching plan is closely related to a routing plan. One essential element of a switching plan is to define the number of links for various connections and the dependencies of one class of exchange on another. The plan must also specify the facilities required at each class of exchange. Likewise, it sets out rules for combination exchanges. One example of a combination is the direct connection of subscribers to higher-level exchanges, where a local exchange may also serve as a toll exchange. Alternative routing capabilities should also be specified, that is, how many alternative route attempts will be made at a node before returning the ATB (all trunks busy) signal.

At least three headings are required in a switching plan: (1) national long-distance switches, (2) local urban switches, and (3) rural switches. International gateways should also be specified. The depreciation period of a switch is 20–25 years, although some switches see more than a 40-year lifetime. A switching plan is often written for 20 years. The plan should also spell out the services offered and that will be offered (as future objectives) in addition to basic telephone service. Typical in this group is CENTREX. A future objective may be ISDN packet switching.

The switching plan should consider aspects of centralization and distributed switching, particularly in the local area. Digital switching, compared to its analog counterpart, permits greater switch capacity per unit building area and leads to many economic advantages, particularly in administration and maintenance. One limiting factor for a local switch is the size of the mainframe terminating subscriber loops. In major metropolitan areas switches can accommodate 100,000–300,000 subscribers. The one major disadvantage to this centralized approach is fire, explosion, flooding, or sabotage. The impact on customers is massive and disastrous. Restoral time is long and expensive.

Distributed switching is more survivable and the mainframe of more amenable size. Fewer customers are affected during a disaster. Restoral is easier, faster, and cheaper and lends itself more to interim mobile emergency facilities.

Distributed switching can mean more and smaller exchanges. It can also take on the connotation of the use of satellites, outside plant modules, and concentrators. Economic trade-off studies can be used to determine the most cost-effective approach among centralized, distributed, and hybrid combinations. However, it is difficult to put a financial value of the impact on customers of the loss of telephone service and the resulting legal actions because of a disaster in a large exchange.

Another item for the switching plan is the provision of transportable (van-mounted) switches to provide limited restoral during a disaster. Each van should be a module so that any reasonable size switch can be built up. The van ensemble should also include signaling interface equipment and a limited grouping of transmission equipment.

Charging plans are often incorporated in switching plans. The importance of a coherent, well-organized charging plan cannot be overemphasized. It has repercussions on switching, signaling, and numbering plans. Numbering usually determines tariff rate; signaling supervision determines call duration, and the charging function is carried out by the switch.

Charging involves one of two types of billing: detail billing or bulk billing. Detail billing involves the planner in:

1. Traffic service positions (TSPs) for credit card calling, information, coin-box operation (pay phone), charge reporting, and third-party charges.
2. Provision of facilities for automatic calling number identification (ANI).
3. Deciding charging equipment locations. Automatic message accounting can reside in local exchanges (LAMA) when toll traffic from such an exchange is sufficient to justify a separate installation. Centralized automatic message accounting (CAMA) is favored where a CAMA facility serves several exchanges. Rural areas may present a special case because of the cost of automatic message accounting equipment.
4. Making suitable arrangements for charge reporting.

Bulk billing involves the planner in:

1. Selection of a technical solution of sending metering pulses over trunks without interfering with revenue-bearing traffic.
2. Location of charge meters (e.g., subscriber premises, local exchange, or centralized).
3. Coin-box (pay phone) selection and provision of necessary exchange facilities for coin checking on long-distance calls.
4. Credit card and third-party billing.
5. Interaction of charging mechanics and tariffs, such as coexistence of bulk billing and detailed billing systems [1].

6.4 The Transmission Plan

The most essential requirement of a transmission plan for a telephone company or administration is that it enable all subscribers to talk to each other satisfactorily. It should reflect or improve on CCITT requirements for international calling.

The transmission plan deals with various aspects of telephone transmission impairments. It will set standards and objectives for volume, crosstalk, noise, bandwidth, and amplitude and phase distortion. We find that the most important transmission factor for a speech telephone system is volume (i.e., the receive level at the subscriber instrument). The internationally accepted unit of perceived volume is the *corrected reference equivalent* (see Chapter 2, Section 2.3). Loss in the telephone network reduces subscriber receive level. Inherently, telephone systems require loss to avoid singing and to reduce echo on long circuits. Economy dictates the type of subscriber loop now in use. They are lossy. A transmission plan assigns losses across the network. As we discussed in Chapter 6, one key to reducing loss in the overall network is the balance return loss achievable in two-wire to four-wire conversions (i.e., hybrids, term sets). The concept of virtual switching point used by CCITT is vital to understanding a transmission plan. Basic signal levels at four-wire switches, minimum stability values, and tolerances should also be included in the transmission plan. Existing and future design standards of subscriber loops should also be incorporated in the plan. The future changeover to fiber is yet another issue.

Modern telephone networks are designed primarily for speech transmission. But more and more these networks are being given the task of carrying other information, such as telegraph, data, and facsimile (see Chapter 8). The transmission plan must also cover these aspects. Loading of FDM carrier systems, envelope delay distortion (group delay), impulse noise, signal-to-noise ratio, the characteristics of conditioned circuits, and so on must be clearly expressed in the plan. Absolute delay has become important and limits the use of earth-to-satellite transmission systems for certain types of signaling systems (e.g., compelled signaling). It also inhibits the use of stop-and-wait ARQ on data circuits.

The transmission plan must also include user and network performance objectives of an all-digital network. A complex problem arises for the planners and their staff as to the extension of digitization. In general, by its very nature, ISDN assumes digital extension to the user. A practical interim step is to extend to the local exchange (see Chapter 10, Section 5.2). When and how to extend digitization to the subscriber, considering the state of the copper outside plant, becomes a major issue. The use of fiber optic cable is the probable solution and a very costly one.

An all-digital network requires a number of new requirement parameters, objectives, and guidelines to be incorporated in the transmission plan. Among these are:

- Error rate end to end expressed as a time distribution or in error-free seconds (EFS).
- Allocation of the error rate (i.e., allocation among contributors, such as switching nodes and local plant).

- Network synchronization plan.
- Slip rate criteria.
- A new loss and stability plan.
- New corrected reference equivalent maximums expressed as a distribution over total number of possible connections. This should be a notably improved value from that of the analog network it is replacing.
- Accommodation of ISDN and the implementation of CCITT SS No. 7.
- Quantization noise.

An interim transmission plan for a hybrid network should seriously be considered. By a hybrid network we mean one that is partially digital, partially analog. This interim plan must state values of quantization distortion (noise) expressed as a time distribution and will necessarily limit the number of A/D and D/A conversions (analog-to-PCM and PCM-to-analog).

The implementation of broadband ISDN (BISDN) will surface in still another aspect of transmission planning. BISDN will require, among other items, some standard digital video parameters.

The transmission plan must treat the network end to end. In the past we were essentially concerned with maximum subscriber loop loss. With the advent of ISDN we must now concern ourselves with the ability of the subscriber loop to handle ISDN 2B + D (192 kbps), 23B + E (1.544 Mbps), or 30B + E (2.048 Mbps). BISDN will require fiber connectivity to the subscriber. This leads to establishing more detailed technical requirements for subscriber plant than we have had in the past. Among these requirements are bit rates, line formats, and BER [8].

Transmission plans must be iterated with the switching and signaling plans to produce a coherent and cost-effective telecommunication system.

6.5 Signaling

Signaling is a particularly ticklish matter for the planner. A well-thought-out and properly implemented fundamental signaling plan can pay off in spades to the administration. Signaling is dependent on switching, and vice versa. Remember that the planners have inherited a switching plant that is from 0 to 30 years or more of age. With the variation of plant, they have inherited a number, perhaps five or more, of signaling systems. These require various kinds of appliqué units (black boxes) for compatibility. Thus the fundamental signaling plan should fix at the outset standard signaling systems, for both the long-distance (toll) and local areas. A policy should be expressed in the plan for milestones to convert to the standard. Likewise, a standard approach should be adopted for signal conversions. One example may be in the local area where we would find exchanges arranged for loop signaling. When a pulse-code modulation (PCM) trunk system is installed, a choice is available to substitute loop signaling by E and M signaling (see Chapter 4), or of converting E and M to loop signaling with appliqué units and remaining with a standard incoming relay set at the switch.

We remember from Chapter 4 that a telephone call involves two types of signaling, line signaling for call supervision and interregister signaling. Both require special study for the plan. That plan, among other items, must consider special facilities for operators, antispoofing devices, and malicious call interception. The transmission aspects of signaling are also a major consideration of the plan, such as levels, signal-to-noise ratio, talk-down, and the use of out-of-band signaling.

A signaling plan should have a 10–15 year-duration. With the phasing in of digital transmission and switching, the shorter period may be more practical. As a minimum, a signaling plan should include:

- A standard interregister signaling system for adoption across the entire network and a milestone plan for total phase-in.
- A standard line-signaling system and a plan for total phase-in.
- A standard local-area signaling criterion and its interface with the toll network (both line and interregister).

It is most important that the signaling plan be coordinated with the charging and tariff plans, the numbering plans, the transmission plans, the switching plans, and the routing plans.

At this juncture it would seem that CCITT SS No. 7 (Chapter 14) will be universally implemented, at least for circuit switching. It must be in place for ISDN to work. How much national flavor will be added to SS No. 7 remains an open question. Of course, to properly implement SS No. 7, certain specific transmission criteria will have to be complied with, such as BER from the local exchange on one end of a connection to the local exchange on the other to reduce the number of block repetitions of SS No. 7 data blocks. Switches must be modified to accommodate this new signaling system.

7 NOTES ON OUTSIDE PLANT PLANNING AND DESIGN

7.1 Importance of the Outside Plant

Although about 40% of the total telephone company (administration) investment is in outside plant, it is often relegated to the position of passing mention or just neglected entirely in the literature. It is often forgotten that all telephone conversations originate and terminate in subscriber instruments that are connected to local exchanges by wire or cable. Thus it is important that the outside plant be just as well designed as the switching and transmission subsystems if service is to be satisfactory. Likewise, as we saw in Chapter 2, the close coordination of outside plant planning with exchange placement cannot be overemphasized. For this discussion *outside plant* is defined as all telephone plant between exchange buildings (e.g., local trunks) and between exchange buildings and subscriber premises. It also includes the subscriber subset. Outside plant is primarily concerned with cable, such as service connections or drop wires (to subscriber equipment), subscriber cables, interexchange trunk (junction) cables, and long-distance cables. It also includes terminal cabinets, pole lines, conduit systems, duct work, and rural distribution systems.

7.2 Subscriber Plant

Figure 15.3 illustrates the principal elements of a subscriber distribution network. In essence it is the outside plant cable system connecting the subscriber to the local exchange. This alone represents 12–16% of the total telephone plant investment (see Chapter 2, Section 1). Working our way out from the exchange, there is the main cable that connects to a branch splice or cabinet (for cross-connects). Feeder cables connect outward from the main cable from the cross-connection cabinets to the distribution cables. Drop wires connect from the various distribution points and the distribution cable. It is the drop wire that services the subscriber.

The selection of cable route, the cable sizing and assignment, requires a refined engineering effort to minimize cost and optimize service. Outside plant engineers use the term "cable fill" when referring to the number of pairs in use in a specific cable. The percentage of fill is the number of pairs in use divided

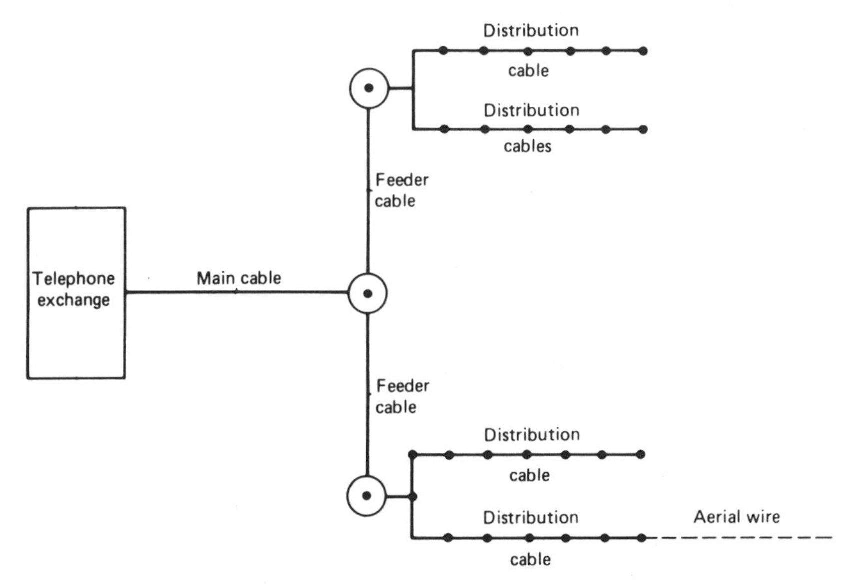

Figure 15.3 Principal elements of a subscriber distribution network: ● distribution point; ⊙ cross-connect box or branch splice.

by the total number of pairs available in the cable. The period of fill is the time to fill the cable to capacity. For small growth rates the period of fill is 12–15 years. Many feeder cables have a period of fill of 7–8 years, and for main cables, which have an annual growth rate exceeding 300 pairs per year, an optimum period of fill may be 5 years. Total fill of a cable beyond 85% is found to be uneconomical. The remaining 15% are saved for spares or are pairs found to be unserviceable.

Subscriber loop design from the transmission viewpoint was discussed in Chapter 2, where we found that the principal limiting factor to exchange area size was economic, depending more on the maximum permissible length of subscriber loops without conditioning to achieve longest possible extension. A typical local exchange serving area could be defined as one with a population of 10,000 inhabitants at the end of the forecast period. Assume that the area meets transmission and signaling requirements in all cases. Thus the subscriber plant design would involve the selection (of):

- Distribution design: the choices regarding the use of pairs or star-quaded conductors, rigid and flexible layouts, flexibility joints or cabinets, multiple tees and bridged taps, dedicated conductors, and the amount of dedication.
- Construction design: the selection of construction criteria (i.e., aerial, ducted or buried, choice of materials for sheaths, conductors, insulation, and block or drop wires).

7.2.1 Distribution Design. There are two basic approaches in the design of subscriber loop network, *rigid* and *flexible*. Rigid networks are those where all pairs are continuous from the exchange MDF (main distribution frame) out to the distribution point (see Figure 15.3). No provision is made for intermediate distribution points. The flexible network uses intermediate cross-connection points that interconnect feeder and distribution cables. As the term implies, such a network is more flexible but more expensive. However, in the long run the flexible network may turn out to be more economical because feeder cables can be operated at a higher fill than distribution cable. On a rigid network the overall fill is usually lower. Telephone companies usually use some of both design approaches or what might be called a "compromise." The rigid approach is used to connect subscribers comparatively close to the exchange out to a cable route distance of 600–3000 ft (200–1000 m). Beyond these distances the outer portions of the network use cabinets and the flexible approach. The use of multiple tees and bridge taps is deprecated. The multiple tee has application where extensive multiparty service exists, but such service offerings have become fewer in most parts of the world. Bridge taps offer another form of flexibility but place additional transmission impairment in loops when used of the order of 0.8 dB/km. The needed flexibility can be better provided by using cross-connects with the flexible approach.

In many areas telephone subscribers are highly mobile, moving in, moving out, changing requirements, and so forth. This has resulted in the expenditure of large amounts by telephone companies for rearrangements. To reduce this cost, administrations have attempted various schemes to "dedicate" plant, that is, to dedicate MDF space, cable pairs, and station equipment to each premise even before its occupancy. Full dedication has not proved economical, especially for optimum use of main and feeder cable pairs. Dedication does prove out in many cases in flexible areas where plant is dedicated in distribution pairs with the use of cross-connect cabinets at the junction of distribution and feeder cables. This allows expensive main and feeder cable to fill more to exhaust before new cable must be put into service. In other words, dedicated pairs that are unused are not sitting idle in this portion of the network.

Flexible systems require careful planning and reasonably accurate 20-year forecasts to be successful. Forecast demand should be broken down into units of demand consisting of 10–15 acres (4–6 ha), except where population density is very low. These "units" represent cabinet districts at the end of the forecast period. Ideally, a cabinet installed at the beginning of a forecast period would serve throughout the entire period. Cabinet unit districts and their size and shape should be chosen for minimum rearrangement of feeder and distribution cables as new cabinets are added during the period to satisfy growth requirements.

Once cabinet districts have been established, the outside plant planners must then concern themselves with routing main and feeder cables to the cabinet locations and the sizing of each cable. Estimates of the approximate year of installation of each cable is part of the long-range plan. The short-range plans define specific expansion years. Main cables normally have a 5-year planning interval, and feeder cables a 10-year interval. Distribution cables are added to care for year-to-year growth and sized to provide for the estimated useful life of the cable in its district under the working conditions found.

The type and size of distribution points vary with the application or concentration of subscribers. Table 15.1 gives the basic types of distribution point now available and their application. Large buildings with a fairly high telephone population will use some form of distribution frame similar to the main distribution frame used at the exchange. For any type of distribution point, service or drop wires should not exceed about 150 ft (50 m) in length. If that length is exceeded, cable extension should be considered. Distribution points external to buildings should accommodate a maximum of 25 subscribers.

7.2.2 Notes on Outside Plant Construction. There is underground and buried construction. Underground construction refers to cables in conduit or duct works. Conduit or duct works offer the cable added protection. Also, on routes over which a succession of cables are required, say, at 5- to 10-year intervals, rather than excavating streets with possible resulting cable damage, advanced provision of spare ducts is advisable. Such spare ducts in place

TABLE 15.1 Distribution Points and Their Application

Terminal Types	Application	Distribution-Cable Type	Service-Wire Type
Main distribution frame	Large office buildings, commercial enterprises	Duct	Internal cable
Cross-connection cabinet	Apartment buildings, medium-size office buildings	Duct	Internal cable
Box with unsealed or sealed chamber on pole or building	Residential or light business	All types	Drop or block wire
Box with unsealed chamber flush with ground surface	Medium residential	Buried	Buried
Pedestal	Medium residential	Buried	Buried
Ready-access messenger support	Light residential	Aerial	Aerial
Encapsulated in cable splice	All residential	Buried	Buried

Source: Ref. 1.

permit the placing of new cables without movement or damage to cable already in place. Conduit systems, rather than direct buried cable, are almost always indicated in metropolitan areas where the cable lays are near other services, such as power, sewage, water, and steam.

Buried construction, that is, the direct laying of cable underground, is cheaper than conduit cable. Buried cable is used where there is less chance of damage or where there is little possibility that future additional cables will be laid. Buried cable is recommended, therefore, in areas of lower population density, such as residential neighborhoods where there is no pole line. Cable depth is about 18–20 in. (60 cm) below the surface in residential areas, whether in ducts or directly buried. Main route cable with a high concentration of service is placed at a greater depth, 3–5 ft (100–130 cm).

Modern cable laying is done by either "trenching" or "plowing." Trenching is where a trench is dug, and the cable is placed in the trench, which is subsequently backfilled. "Plowing" refers to the insertion of a cable into the ground through a hollow plow blade or by pulling it into the ground just behind the plow blade. Plowing is more economical than trenching for depths down to 25–30 in. (ca. 75 cm). For depths greater than this, larger tractors are required, and the resulting cost is often greater than that for plowing. Of course, plowing is difficult when there are many underground obstacles, particularly rock ledges. Aerial construction may be indicated as more economical where there is an existing pole line and where such aerial plant can be maintained for at least 20 years. These conditions are now found almost exclusively in rural areas [1].

7.3 Fiber Optics in the Subscriber Distribution Plant

Extending fiber to the subscriber will take place. The question is when. Today the subscriber distribution plant consists of cabled copper pairs and some aluminum pairs. The investment is very large, amounting to 40–50% of the value of the total existing telecommunication plant. Some of these cables have been installed and in use for over 50 years. Many pairs or the entire cable in the older plant give marginal operation for speech transmission and will not be able to support even the modest bit rates of 2B + D.

Motivation to use fiber will be subscriber demand for more bit rate capacity. As we pointed out, much of the existing plant will not be able to handle ISDN bit rates. We see ISDN, then, as a principal stimulating factor to convert from copper pairs to fiber cable. Of course, BISDN will require fiber, and in this case it is our belief that coaxial cable will no longer be a contender.

Fiber may come in from another direction—CATV (cable TV). There is a growing trend to use optical fiber rather than coaxial cable for CATV systems.

In the context of extending fiber to the subscriber, let us stress that we mean digital connectivity, not analog. The question of handling BORSCHT at the subscriber equipment was discussed in Chapter 10.

8 INSTITUTIONAL, CORPORATE, AND INDUSTRIAL COMMUNICATIONS

8.1 Overview

Telecommunications planning and management today in the institutional, corporate, and industrial sectors are a far cry from what they were 30 or 20 or even 10 years ago. Once there was no planning, and management consisted only of seeing to the provision of a PBX large enough to satisfy office needs, to make periodic changes in extensions, and to pay telephone and telex bills. It is a far different story today. Now the corporate communication manager is involved or will be involved with over a dozen activities such as:

1. Private automatic branch exchange (which is digital and computer based) and its dimensioning, multiple service offerings, cost, interface with the PSN, and own networking capabilities. Another concern is the data transport and switching capabilities of the PABX.
2. Qualifying and quantifying local and long-distance telephone service requirements. The planner must optimize that service with the services offered by the local telephone company (or administration), such as CENTREX, DDS, WATS, foreign exchange (FX) lines, local access trunks, local tie lines, and special billing arrangements. Offerings of specialized common carriers must also be considered for the most cost-effective service.

3. Telex and other special telegraph services. (We note that telex is rapidly falling out of favor because of the better service available from point-to-point facsimile over the ubiquitous telephone).
4. Internal data services without intervention of the telephone company (administration). Here we mean the internal data network on company or institution premises. The user will deal with LANs, data loops, buses, and WAN interfaces (see Chapters 8, 11, and 12).
5. Data services, with distant workstations, CPUs, etc. The user will make decisions on centralized or distributed processing in conjunction with corporate EDP management. A compromise between the two will have to be struck. Then consideration would be given to workstation selection, network design and protocols, data administration, choice of transmission media, possibilities of concentration and/or TDM, to achieve the lowest net EDP communication cost with the most efficient service (see Chapter 11).
6. Facsimile service.
7. Paging communications and its incorporation with the PABX for access and control.
8. Office security systems, CCTV, handie-talkie, alarms, and so on.
9. Cryptographic systems and computer security.
10. Transport dispatching and other mobile communication applications.
11. Private corporate, industrial, and institutional networks, such as own VHF–UHF and microwave broadband systems, multichannel systems, and fiber optic cable systems. The user will need to compare savings with cost of use of common carrier and its security, the quality and grade of service, and so on.
12. Advanced building preparation for communications optimization. See Figure 15.4.
13. Savings on long-distance service with specialized common carriers, some of which supply switching as well. These could be dedicated corporate circuits or semidedicated or equivalent common-carrier circuits.
14. Specialized common carriers not requiring local telephone company access (bypass), VSAT networks to cite one example (see Chapter 7, Section 7.8).

The list is not exhaustive, especially considering how entire office procedures, including accounting, are being melded into an overall "communication" system. The list was made considering a minimally regulated society, especially in the telecommunications sector. For instance, in many countries today even the idea of a specialized common carrier is unthought-of. All forms of communication in such places are government monopolies. Changes are being forced on these more controlled societies because of the competitiveness

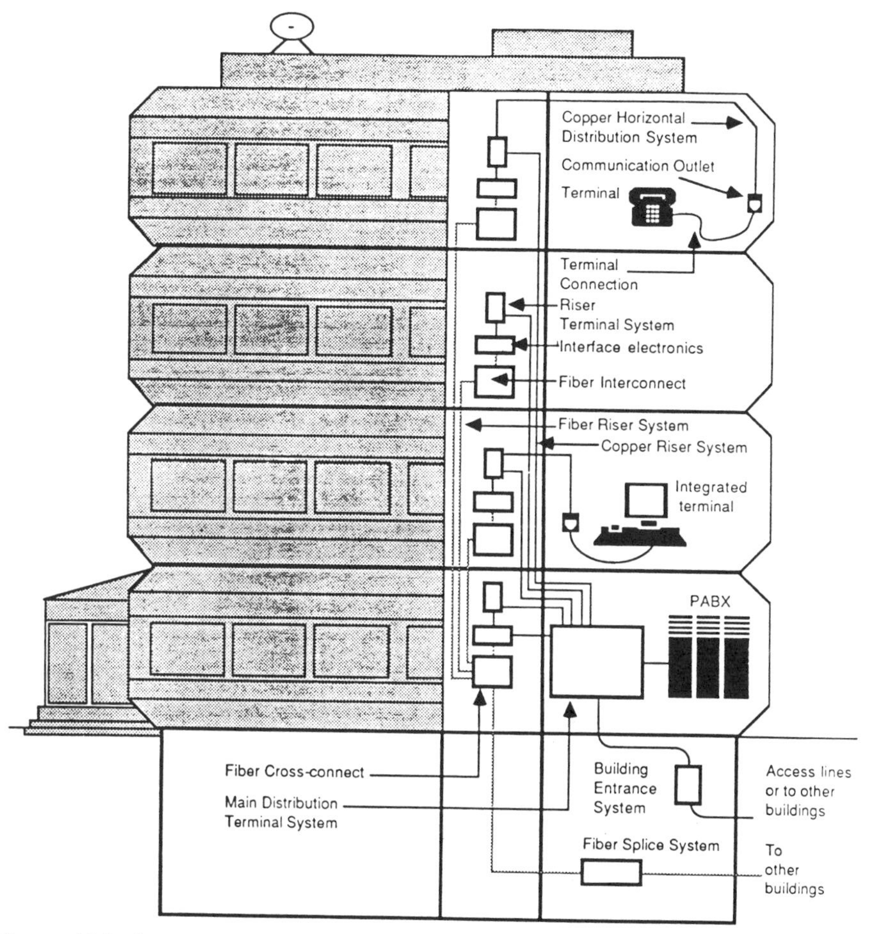

Figure 15.4 Integrated building distribution network: major network components (integrated copper and fiber-based systems). From Ref. 11. Courtesy of Northern Telecom Inc.

of the more "open" societies, possibly due in no small part on more efficient communications found typically in North America. These changes are now becoming evident, such as in Great Britain.

8.2 Typical PABX Considerations

A major responsibility of the corporate, industrial, and government telecommunication planner-manager is the selection and later expansion and maintenance of a digital PABX system or systems. Its design should be such to allow DS1 or CEPT 30 + 2 FX foreign exchange interface. The "hooks" should be present to incorporate ISDN readily.

8.2.1 Trade-Offs and Issues

1. Digital architecture. Some less expensive PABXs are based on a quasidigital technique using time-division multiplexing and pulse amplitude modulation (PAM). We would opt for the more costly PCM-based PABX using the standard 64-kbps PCM voice channel and the DS1 or CEPT 30 + 2 architecture. This will allow for ISDN, when implemented, and for other digital network interfaces with the outside world.

2. Blocking. How many lines may be in use simultaneously before blocking occurs? Consider the office busy hour. Multiply the number of user stations by 0.5 and divide by 2 (i.e., 2 stations are involved with an internal call). This calculation produces a conservative traffic intensity value in erlangs. Multiply the erlang value by 36 to derive traffic intensity in cent call-seconds (ccs). Generally, a manufacturer will specify a blocking probability and probability of dial tone delay less than 3 s. Blocking probabilities are usually expressed for line to line, line to trunk, and trunk to line. All assume that a called telephone is idle. The blocking probability specified should include a statement on the traffic intensity formula used: Erlang B, Erlang C, Poisson, or Engset (Chapter 1, Section 5). Nonblocking PABX configurations are also available from many manufacturers, at added cost. The planner must decide if the added cost is justified compared to a typical blocking probability of 0.01 for busy season busy hour. In most cases it is not.

3. Voice and data support. A PABX may support voice and data over separate lines, competitively over a single line from a user station, or simultaneously over a single line. If most workstation positions will not use data or use it in place of voice when it is used, competitive or duplicative path systems may be more cost-effective.

4. End instrument considerations. Where voice and data can use the same line together or alternately, a single unit should provide connection of a terminal and a telephone end instrument to the line. There is a cost and value trade-off for separate end instruments or a combined unit for voice and data. The cost impact of data growth should be projected over time to aid in the decision process. Somewhat more expensive choices made now may save large sums in the future. Consider that a system that is cost-effective with predominantly voice traffic may become much less so as data terminations are added in the future.

5. Modular architecture. Many digital PABXs have a modular architecture, which permits the addition of line groups with little or no expansion of the central switching equipment. This is done by adding peripheral equipment, multiplexers, and/or remote concentrators. Such an architecture allows users to grow from a relatively small to a very large system while keeping incremental cost per unit nearly uniform. Modularity may extend to the main switching facility itself. It also can be expanded on a modular basis. Some ATT and Northern Telecom PABXs have this feature.

6. Internal interfaces. The following list some of the internal interfaces that must be considered:

- Electrical interfaces supported for data connectivity, such as RS-232D, RS-422, 423, and CCITT V.10, V.11, V.21, V.24, and V.35.
- Availability of 64-kbps clear channels.
- Data rates supported (e.g., 75×2^n, 16 kbps, etc.)
- Synchronous or asynchronous, clocking.

7. Security considerations. We mean here the protection of sensitive computer traffic routed through the PABX. Voice crosstalk is usually not an issue in modern systems.

8. Battery backup. The planner should consider the impact of the failure of prime power on internal and external telephone service. A backup power supply, such as a no-break storage battery based supply with an ac charger, may be a valuable enhancement. It should be noted that human stress levels go up when the lights go out. The ability to dial out for an ambulance during a power failure to save a heart attack victim will more than pay for the enhancement.

8.2.2 Features. The modern PABX is computer based (SPC) and can offer a large group of features, some at extra charge. Among these features are:

1. Station-to-station calling, 3-, 4-, or 5-digit dialing.
2. Station-to-trunk calling.
   a. Direct outward dialing (DOD). Foreign exchange and WATS trunks can also be provided. Second dial tone is used.
   b. Through dialing. Restricted lines may require placing calls through an attendant or an attendant may give direct dial connection on request.
   c. WATS and CCSA (common control switching arrangement) using single or sometimes two-digit access.
3. Trunk-to-station calling.
   a. Incoming trunk calls handled by an attendant.
   b. Direct inward dialing (DID).
4. Tie trunks. These are one-way or two-way circuits for interconnecting two PABX systems.
5. Tandem switching. This is another tie trunk arrangement that permits tie trunk–to–tie trunk connections and tie trunk–to–local exchange (central office) or special service trunk connections. Such calls may be completed with or without attendant intervention, depending on strapping of the PABX.
6. Special power failure arrangements; battery.

7. Station hunting. With this option a call is routed to an idle PABX station in a prearranged group when the called station is busy.
8. Camp-on busy. When the attendant extends a call to a busy station, the trunk will automatically camp on the busy line and the party to which it is connected will hear a distinct tone indicating the camp-on condition; the calling trunk party will hear a ringback tone. Connection between the trunk calling party and the station is made when the called station goes off hook.
9. Station restriction. This is a class-of-service item denying access of certain stations (extensions) to some special features, such as direct access to WATS or FX lines, or other features such as camp-on or call transfer.
10. Toll restriction. Several variations are available. For instance, access may be permitted by all stations to local area dial tone and some stations may be prohibited from making automatically dialed toll calls.
11. Key system features.
12. Off-premise extensions. Some PABX loops can be extended from the nominal 900-Ω or 1000-Ω limit to a 2000-Ω limit.
13. Night answering service.
14. Line lockout. Some PABX systems employ 100% line lockout to prevent control equipment from being held when it is no longer required. Busy tone or reorder tone is applied to the line to indicate lockout. (In European parlance, lockout is referred to as "time out").
15. Timed recall. Timed recall puts an attendant on line when the called party does not answer on an incoming trunk for some predetermined time, often 30 s.
16. Consultation hold. The PABX system should allow the called party to put a calling party on hold while the called party drops off the line to carry out another task, such as a consultation or a document search.
17. Call transfer, automatic by called party or by attendant.
18. Add-on conference.
19. Intercept (vacant level intercept and vacant number intercept).
20. Music on hold.
21. Touch-tone dialing.
22. Paging service.
23. Call forwarding (follow-me) and call forward on no answer.
24. Busy line call back and busy trunk call back.
25. Priority override.
26. Automatic identification outward dialing (AIOD). This feature records calling number and called number and time for billing. This is usually used by supervisors to check on spurious usage of company telephones by employees.

27. Dial call hold. A PABX should be able to place a second call after another call on the same line is put on hold.

8.3 Modern PABX Systems

8.3.1 Integrated Telephone and Data Services. The telecommunication manager has an option of incorporating a data switching capability as well as voice into a PABX. Obviously adding a data capability can be a cost increment, often of the order of $100 to $300 additional cost per line. Under certain conditions a data switching capability may be desirable. A PABX can replace one or more LANs or augment LANs presently installed. How widespread the use of the PC is may well drive the decision.

Is there a requirement to provide PC connectivity? for all PCs? some PCs? in isolated areas of common interest? What are the PC workstation to mainframe computer requirements? PC connectivity to distant offices? Will one or more LANs suffice without resorting to routing via a PABX? Is electronic mail a requirement? All these are questions the telecommunication managers or planners should be asking themselves.

The PABX is an interesting alternative to a LAN. Yes, data rates are usually much lower with the PABX. On the other side of the coin, connectivity is point to point, so one does not have to brave the world of contention or wait for a turn with a token. The token turn allows use only for a short fixed period of time. Not so with PABX connectivity; the connection can be forever.

One large New England computer company utilizes a private telecommunication network based on four DMS-100 Northern Telecom switches (Chapter 10) and many dozens of Meridian SL-1 data/voice PABXs. Almost every employee has or will have access to a workstation (PC) and all are interconnected. All of the company's field offices around the world can access the system via either voice or data (electronic mail, database access, resource sharing). LANs are widely used and can route through the central communication system by means of gateways. The design, implementation, and operation of such a system is a very large and costly undertaking. However, as the importance of effective information transfer grows, such systems will become more prevalent.

8.4 Brief Overview of a Typical Voice / Data PABX

The Northern Telecom Meridian SL-1 has evolved into a very effective voice/data switch for the office and factory environment. It can accommodate 30 to 7000 users. The switch is capable of accessing the public switched network (PSN) via nominal 4-kHz analog facilities and North American DS1 or CEPT 30 + 2 digital facilities [11].

The SL-1 can provide a 40-MHz packet transport system that allocates bandwidth on demand. It also can provide 2.56-Mbps twisted wire-pair distribution to the desktop. The SL-1 accommodates both standard analog telephones and single-channel (64-kbps) digital telephones with simultaneous data transmission capability with data rates up to 19.2 kbps. It also is a hub for a LAN with 2.56-Mbps accesses for PC workstations. Voice messaging is another feature of the SL-1 [11]. There is considerable functional similarity between the SL-1 family of PABXs and the DMS-100 described in Chapter 10.

The SL-1 provides the following grade of service values [11]:

Intraoffice calls	0.04 (called line idle)
Incoming calls	0.01 (called line idle)
Outgoing calls	0.02 (ATB)
Dial tone delay	$<$ 3 s for 98.5% of calls.

The grade of service values shown above can be met with 160 terminations per 600 ccs (16.66 erlangs) or about 0.1 erlang per line or 0.2 erlang per termination pair. These values should be compared with the traffic intensity values of Table 2.9.

The SL-1 can be supplied in nonblocking configurations.

8.4.1 Technical Characteristics of the SL-1

8.4.1.1 Signaling Parameters

Digitone Signaling

Digit	Frequency $\pm 1.5\%$ (Hz)
1	697 + 1209
2	697 + 1336
3	697 + 1477
4	770 + 1209
5	770 + 1336
6	770 + 1477
7	852 + 1209
8	852 + 1336
9	852 + 1477
0	941 + 1336
*	941 + 1209
#	941 + 1477

Minimum interdigital interval	40 ms
Minimum pulse duration	40 ms
Maximum speed	12.5 digits / s

Tone	Frequency (Hz)
Dial tone	350–440
Audible ringback	440–480
High tone	480
Low tone	480–620
Miscellaneous tone	440

From Ref. 11.

8.4.1.2 Transmission Parameters (North American DSX PCM). The following tables are from Ref. 11.

Insertion Loss at 1020 Hz

Connection Type	Nominal Insertion Loss	Loss Dispersion
Line to line	5 dB	±1.0 dB
Line to trunk	1 dB	±0.7 dB
Trunk to trunk	1 dB	±0.7 dB

Frequency Response (Amplitude Distortion)

	Frequency Response (dB) at Frequency				
	200 Hz	300 Hz	3000 Hz	3200 Hz	3400 Hz
Connection	Min Max	Min Max	Min Max	Min Max	Min Max
Line to line or line to trunk or trunk to trunk	0.0 − 5.0	−5.0 + 1.0	−0.5 + 1.0	−0.5 + 1.5	−0.0 + 3.0

Values are stated relative to loss at 1000 Hz. The symbol + denotes more loss; the symbol − denotes less loss than that measured at 1000 Hz.

Overload Levels

Type of Circuit	Nominal Overload Level (dBm)	
	Receive (A / D)	Transmit (D / A)
Line	+7	+2
Trunk	+3	+6

Receive and transmit relate to switch.

Tracking (Linearity)

Input Signal (dB) Below Overload	Tracking Error (dB)	
	Maximum	Average
3 to 40	±0.5	±0.25
40 to 53	±1.0	±0.5

Signal at 1020 Hz. Maximum specification 99% of all connections.

Transhybrid Loss (Return Loss)

Two-Wire Port[a]	Transhybrid Loss (dB)[b]	
	200–3400 Hz	500–2500 Hz
Line	> 17	> 19
Trunk	> 18	> 21

[a]Two-wire port termination: 600 Ω, except for EIA-compatible trunk packs that need to be terminated with 350 Ω in series with 100 Ω, 0.21 μF.

[b]Measurement of transhybrid loss (THL) is made from equal-level (transmit and receive) four-wire port toward the two-wire port. The THL is the equivalent specification for design return loss for non-mu-law PBXs.

Input Impedance

Connection from Four-Wire Trunk to Port	Reference Impedance[a]	Frequency Range	Minimum Return Loss
Line	600 Ω	200–500 Hz	20 dB
		500–3400 Hz	26 dB
Trunk	600 Ω	200–500 Hz	20 dB
		500–1000 Hz	26 dB
		1000–3400 Hz	30 dB

[a]A reference impedance of 600 Ω resistive or of 600 Ω in series with 2.16 μF capacitance is acceptable for trunk ports.

Idle Channel Noise

Connection Type	C-Message Weighted (dBrnC)	3 kHz Flat (dBm)
Line to line	< 20	< 38
Line to trunk	< 20 at line	< 38 at line
	< 23 at trunk	< 42 at trunk
Trunk to trunk	< 20	< 38

Longitudinal Balance

Frequency (Hz)	Minimum Balance (dB)	Average Balance (dB)
200	58	63
500	58	63
1000	58	63
3000	53	58

Measured according to IEEE Standard 455-1983. Requirement applies to trunks only.

Impulse Noise

Connection	Number of Counts Above 55 dBrnC
All	0

For test purposes a 5-min counting interval is used.

Intermodulation Distortion

Connection Type	Distortion Limits (dB Below Received Level)[a]		Test Signal Input Level (dBm)[b]
	R2	R3	
Line-to-line	40	43	−9
Line-to-Trunk	45	53	−9 at line
			−13 at trunk
Trunk-to-Trunk	45	53	−13

[a]Four-tone method is used.
[b]Test signal input level is the composite power level of all four tones.

Envelope Delay Distortion

Bandwidth (Hz)	Envelope Delay Distortion (μs)	
	Line to Line	Line to Trunk and Trunk to Trunk
800–2700	750	735
1000–2600	380	190
1150–2300	300	150

Quantization Distortion

Input Level Below Overload	Signal / Distortion Ratio (dB)
3–33	33
33–43	27
43–48	22

Input signal is 1 kHz sine wave; output measured with C-message weighting.

Crosstalk

Connection Type	Minimum Crosstalk Attenuation (dB)
Line to line	75
Line to trunk	75
Trunk to trunk	75

Input frequency range 200–3200 Hz, 0 dBm level.

Extracted from Ref. 11. Courtesy of Northern Telecom, Inc.

8.4.2 The Meridian SL-1 as a Data Switch. The architectural design of the SL-1 is such that the PABX can accommodate many of the requirements for switching data as well as voice. It offers 300 Mbps of circuit-switched data bandwidth and 40 Mbps of packet-switched bandwidth to create a high-speed media-independent local area network. The architecture of the SL-1 allows all communications to be handled in a single integrated system. For example, voice communications and slower-speed data are processed at 64 kbps, while higher-speed files and data transfer between PCs and PC servers are accomplished at much higher data rates. The SL-1 can also provide local area networking with dynamic allocation of bandwidth and distribution speeds up to 2.56 Mbps. Figure 15.5 shows the basic architecture of the SL-1 PABX, and Figure 15.6 illustrates the data services supported by the SL-1 [11].

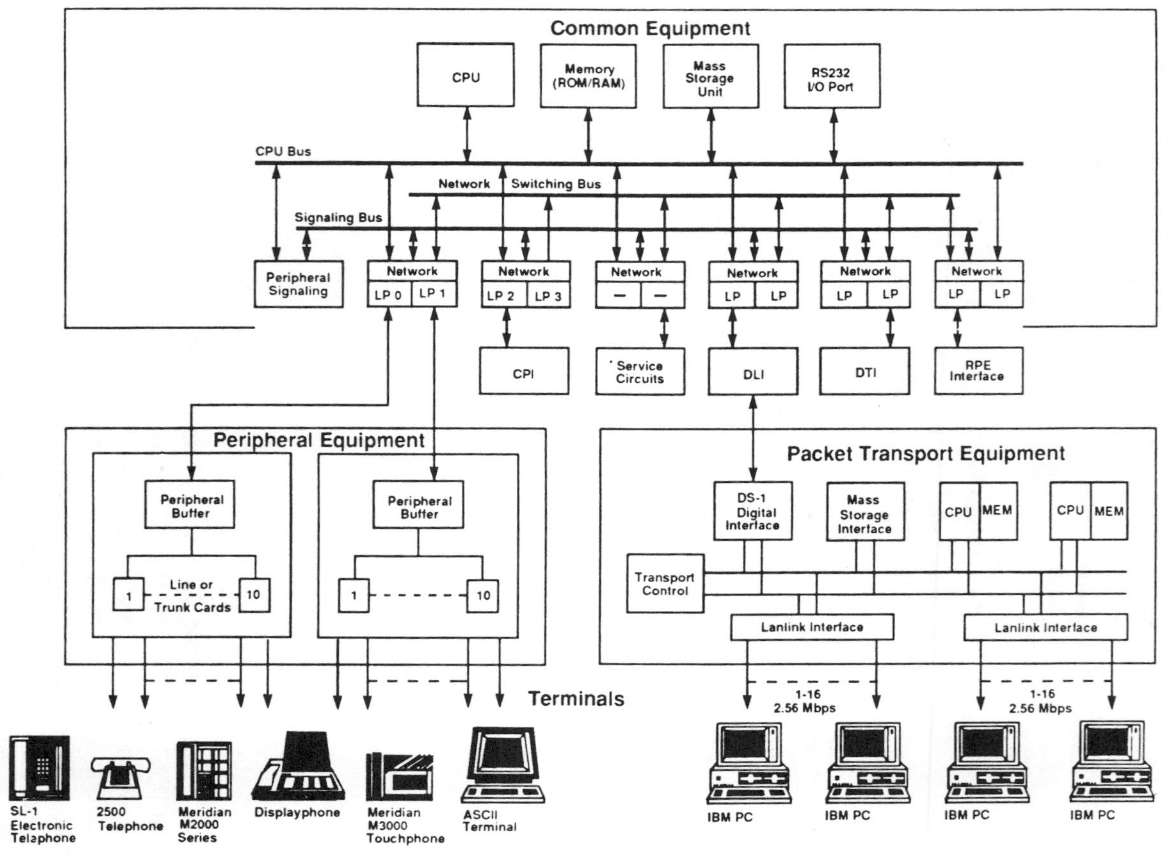

Figure 15.5 Basic architecture of the SL-1 PABX.

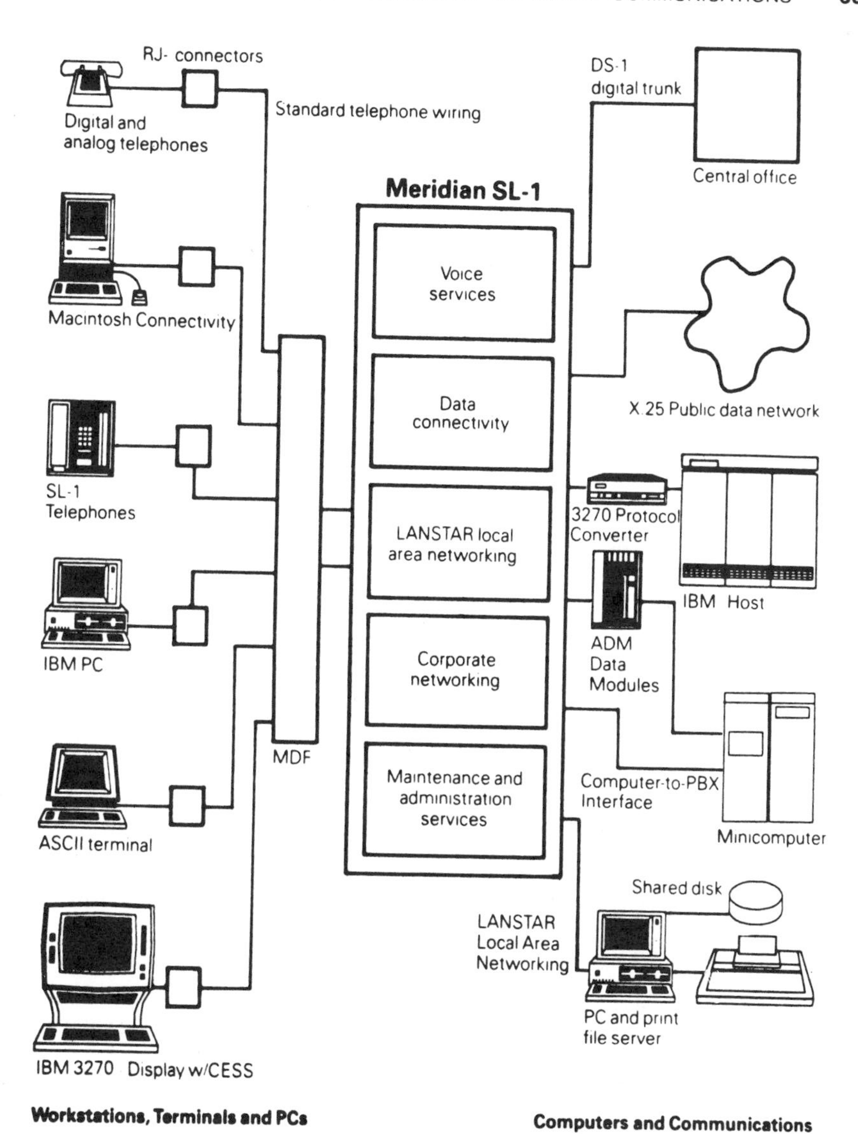

Figure 15.6 Data services supported by the SL-1.

The SL-1 offers many of the features required of an office data switch. Several of these features are described in the following paragraphs.

1. Digital telephones. The digital telephone has an internal single-channel codec (64 kbps). The 64 kbps voice and asynchronous data are multiplexed at the telephone set and transmitted over a single twisted pair to the SL-1. It supports asynchronous data up to 19.2 kbps with an RS-232D interface. The distance limit of a data terminal or digital telephone is about 3000 ft.

2. Displayphone 220. This offering requires two twisted pairs. It provides simultaneous voice and data operation. Data rates are 300–1200 bps. It emulates the DEC VT100 and VT220 and the IBM PC and compatibles. Access to the SL-1 is through RS-422 data ports.

3. Add-on data module (ADM). This module permits integrated (simultaneous) voice and data operation when colocated with an electronic telephone subset. It requires two twisted pairs. The ADM can also be used as a stand-alone device to host computers, printers, and modems. It can support asynchronous operation up to 19.2 kbps (RS-232D interface) and synchronous operation up to 56 kbps (V.35, DDS interfaces). The distance limit from the PABX is up to 4000 ft.

4. Personal computer interface (PCI) card. This card is installed in a PC port providing RS-422 interface with the SL-1 and accepts asynchronous data rates up to 19.2 kbps. A PC is typically limited to 9600 bps. The PCI requires two-pair wiring with a maximum distance from the SL-1 of about 4000 ft. One additional pair can be used to connect a 500/2500 type telephone set. Employing the PCI card, a PC user may access an IBM host computer by making a data call through the appropriate 3270 protocol converter. The PCI also allows access to modem pooling, Rec. X.25 PAD (packet assembler disassembler), and Rec. X.75 gateway. Figure 15.7 illustrates these applications.

5. LANSTAR local area networking. LANSTAR is a Northern Telecom trade name. A LAN with a star architecture is described in Chapter 12. Reference 11 states that LANSTAR provides a variety of connection options to host computers, as shown in Figure 15.8. Host ports may be synchronous or asynchronous, using one of the following OSI layer 1 standards: RS-232D (V.24), RS-422, or V.35. Port contention is an inherent feature of the SL-1, allowing a lower number of host computer ports to be shared by a larger number of terminal users. Host computer ports may be configured in hunt groups under a single directory number, and terminal users may use "autodial," "hotline," and "speed call" features. When all host ports are in use, the user is placed in queue for the next available port. Host computers may be located up to 4000 ft away from the SL-1. Depending on the interface module used, host accesses are with asynchronous data rates up to 19.2 kbps and 56 kbps synchronous.

6. Multichannel data system (MCDS). The MCDS is a single-rack add-on module (ADM) system that provides efficient connection of multiple computer ports. The MCDS consists of multiple ADM equivalent circuit cards that are rack mounted with a common power supply. The MCDS provides an asynchronous answer-only interface between multiport computers and SL-1 PABX data line cards. The host port cannot originate a data connection using the MCDS.

The MCDS operates without operator intervention. Each port can automatically adjust itself to the calling terminal's data rate and operates indepen-

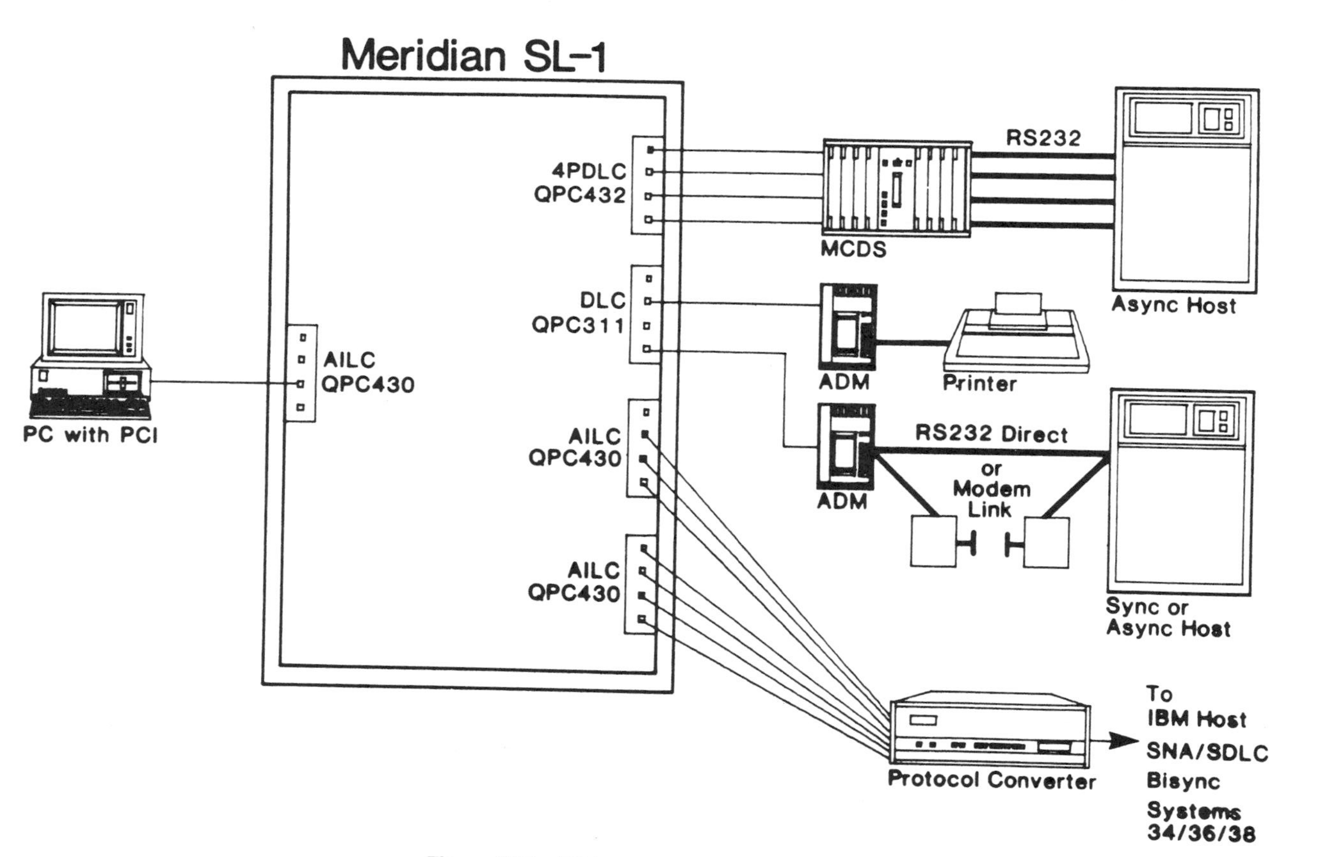

Figure 15.7 PC interface card applications.

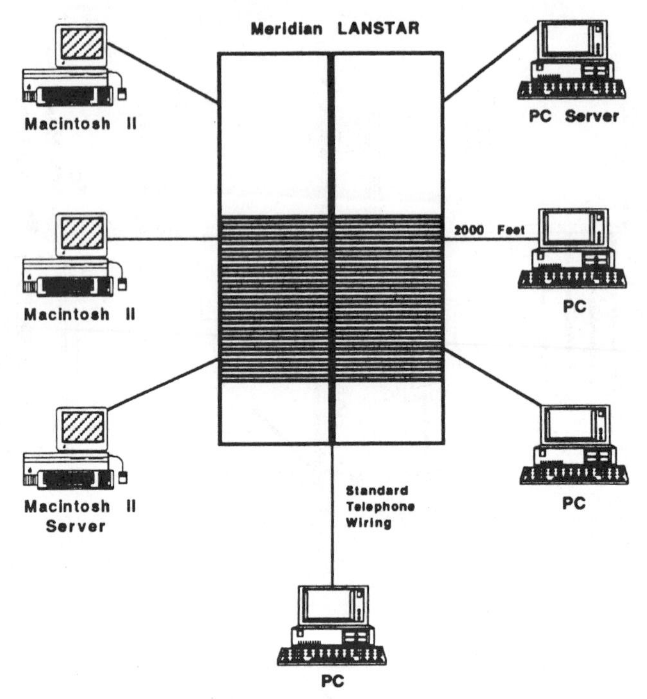

Figure 15.8 SL-1 LANSTAR architecture.

dently of other ports. The MCDS, when fully equipped, has a capacity of 64 ports, and each port can interface to different host computers operating at different data rates and different data formats. If desired, all ports can serve a single host computer.

Each MCDS port is connected using one twisted pair, and the MCDS can be located up to 4000 ft away from the associated SL-1 PABX. The MCDS is linked to the MDF of the SL-1 using standard 25-pair cable. RS-232 cables are used to connect to the host ports. Figure 15.9 shows the architecture of the MCDS and its associated SL-1 host computer and data peripherals.

7. Computer-to-PABX interface (CPI). The CPI provides a multiplex interface and switched access among 24 terminals and a host computer via the SL-1 integrated services network. This is shown in Figure 15.10 [11].

REVIEW QUESTIONS

1. In a free economy, we are constantly faced with the problem of service versus ______.

2. The text considers the "economics" of telecommunications on three levels. Name the levels and discuss each briefly.

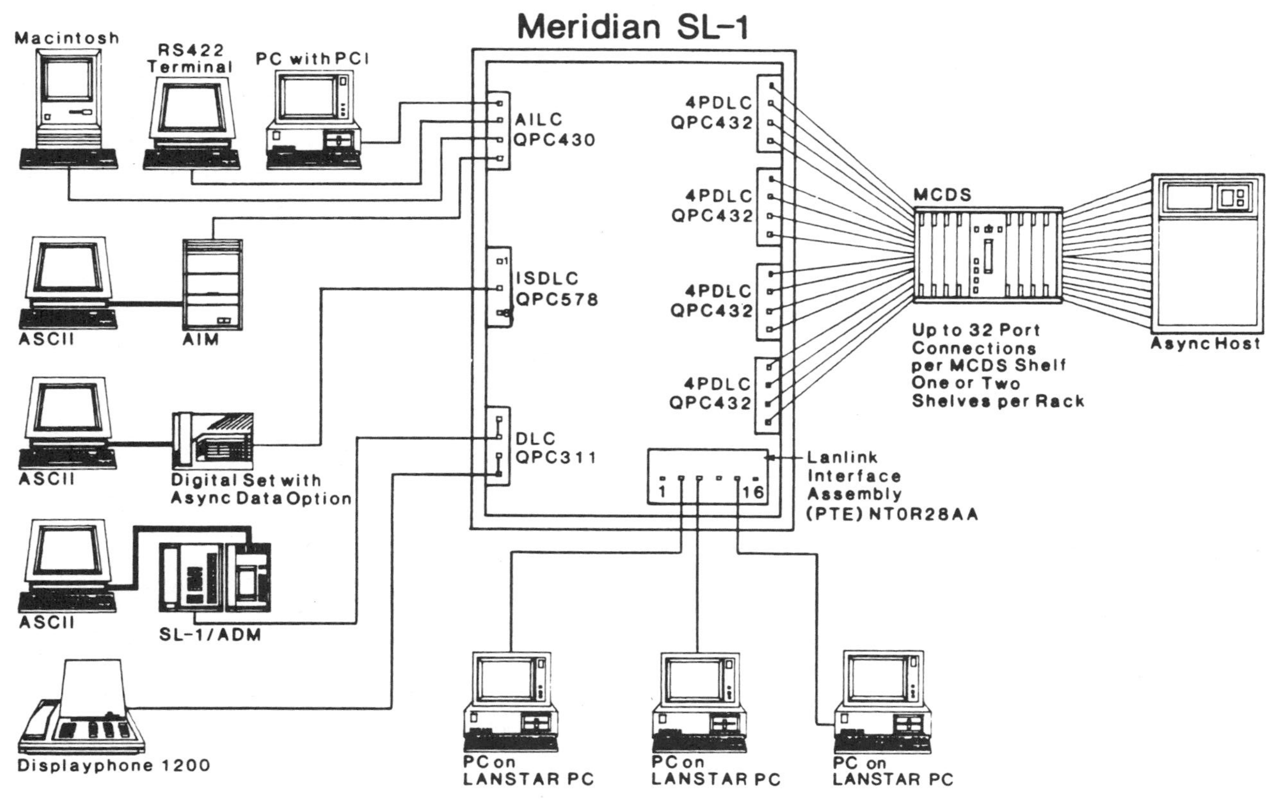

Figure 15.9 Multichannel data system connectivity.

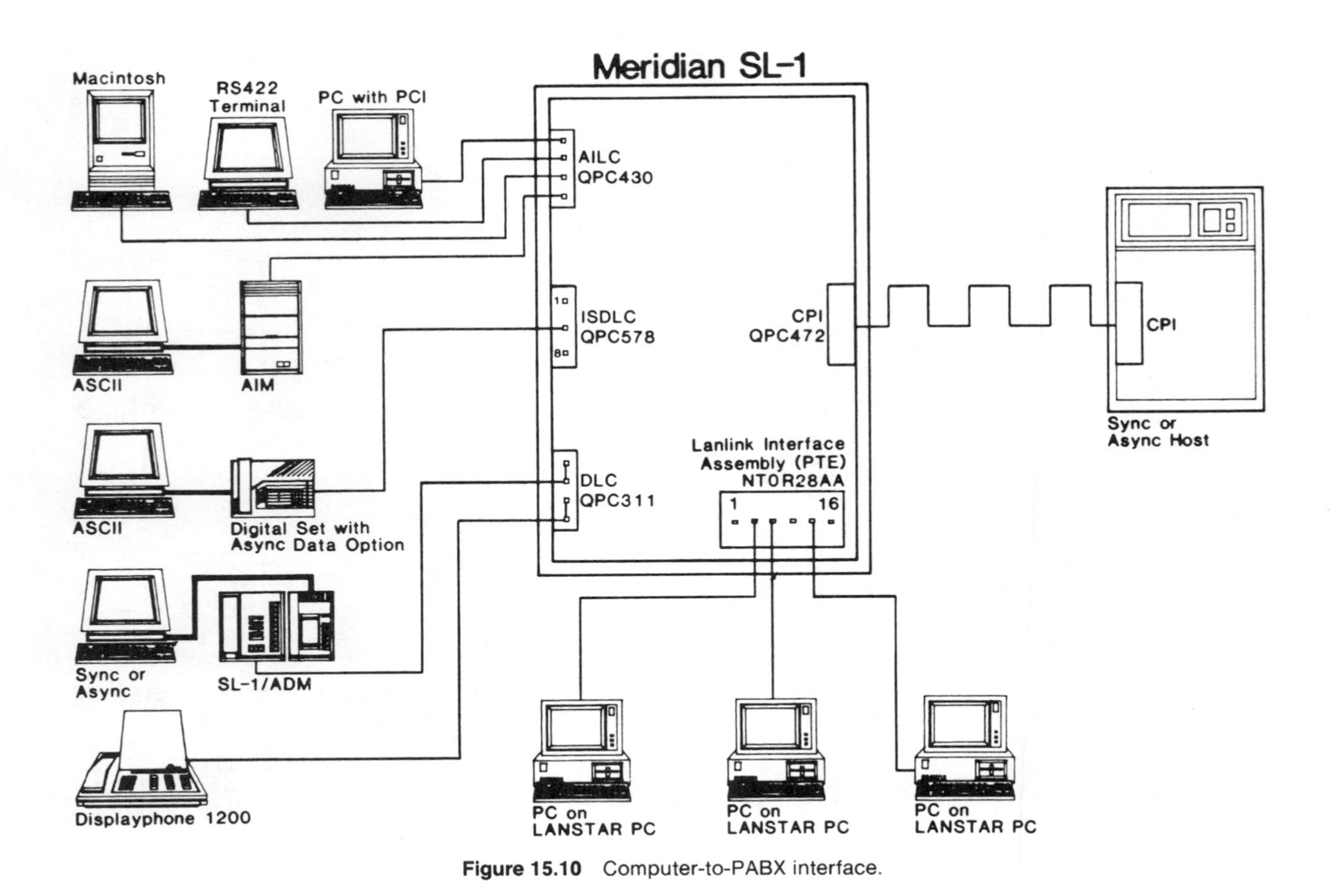

Figure 15.10 Computer-to-PABX interface.

3. Compare telecommunications economy in the open, competitive society with telecommunications operating as a government monopoly. First discuss the primary source of finance for expanding and modernizing plant.

4. What do basic (telecommunication) plans really deal with?

5. In the area of basic planning, there are two starting points. What are they?

6. What is the starting point with fundamental planning?

7. Discuss linear projections and their validity period (approximate). What is the analogy method and what things is it based on?

8. What are some of the parallel indexes we can use for gross national planning?

9. When dealing with traffic forecasts, what three classes of traffic must be distinguishied?

10. List at least four quality-of-service factors.

11. List the six fundamental technical plans of a telephone company or administration.

12. List at least four factors that will limit or otherwise affect the numbering plan.

13. As a minimum, what should a routing plan include? List at least four factors.

14. The switching and charging plans are usually incorporated into one plan. List three reasons why this should be.

15. Argue centralized versus distributed switching. What factors bode well for centralized and for distributed switching?

16. We have placed a switch in a van with its appropriate emergency power. The switch is to be used for restoral during an emergency. What are the two principal interface issues that the system engineer faces when designing this emergency restoral facility?

17. What are the two basic types of billing found around the world? Argue the pros and cons of each from both the subscriber and the administration viewpoint.

18. What does a transmission plan really deal with ______?

19. By their very nature, telephone networks are lossy. Why are they that way?

20. When dealing with absolute delay, what types of service does it affect? (There are several.)

21. Name at least seven technical parameters to be incorporated in a transmission plan for the *digital* network.

22. How does a signaling plan affect a charging and tariff plan?

23. When discussing "outside plant," what is the meaning of *cable fill* and *exhaust*? Differentiate between the two terms.

24. Why would we consider a subscriber cable to have reached exhaust when only 85% of the pairs are in use?

25. When describing the subscriber loop network, there are two approaches to its design. What are they? Discuss their pros and cons.

26. Why is it preferable to use conduit systems rather than direct burial in metropolitan areas?

27. Fiber, rather than copper, in the subscriber plant is ideal. What is holding us back from making the conversion right away? Can you give at least one driving factor that would help accelerate the change?

28. What are some of the crucial changes in the office environment that have caused such radical changes in office telecommunications?

29. Why should the digital regime of an office PABX conform to telephone company or administration interface when the office telephones are analog and present interconnect with the PSN is analog?

30. Give at least ten telecommunication-related activities in which the telecommunication manager will be involved.

31. Give one simple, conservative methodology to quantify traffic intensity in the office or factory environment.

32. Name at least seven factors that should be considered when selecting a PABX.

33. Why should a PABX use (storage) batteries in a similar manner as a telephone company or administration switch?

34. How can a PABX offer so many options in this day and age when some 20 years ago few if any of these options were available?

35. What are some of the options open to the telecommunications manager for connecting PCs?

36. Argue the pros and cons for using a PABX in lieu of a conventional LAN.

37. What are some typical blocking probabilities one might expect to encounter in a well-designed PABX?

38. Give at least eight transmission parameters of a state-of-the-art PABX.

REFERENCES

1. *Telecommunications Planning*, ITT Laboratories (Madrid), January 1974.
2. *National Telephone Networks for the Automatic Service*, CCITT, ITU, Geneva, 1964.
3. R. Chapius, "Common Carrier Telecommunications in the World Economy," *Telecommunications Journal* (October 1972).
4. *Economic Studies at the National Level in the Field of Telecommunications*, ITU, Geneva, 1968 (amended 1972).
5. *Local Telephone Networks*, CCITT, ITU, Geneva, 1968.
6. *US Federal Communications Commission Rules and Regulations*, Parts 31 and 32, US Government Printing Office, Washington, D.C., 1983.
7. "Telecommunications Quality," *IEEE Communications* (entire issue) (October 1988).
8. *Notes on the BOC Intra-LATA Networks: 1986*, Bell Communications Research TR-NPL-000275, Holmdel, N.J., 1986.
9. International Telephone and Telegraph Company, *Reference Data for Radio Engineers*, 6th ed., Howard W. Sams, Indianapolis, 1986.
10. *Current Planning and the Construction Budget*, Construction Department, American Telephone and Telegraph Co., New York, 1982.
11. *Meridian SL-1 Consultant Handbook*, Northern Telecom Inc., Richardson, Texas, April 1988.
12. Robert F. Gellerman, *Subscriber Financing of Telecommunication Investments: IntelCom 1979*, Horizon House, Dedham, Ma. 1979.

INDEX